Springer Texts in Sta

Series Editors:
G. Casella
S. Fienberg
I. Olkin

MW00835333

For further volumes:
http://www.springer.com/series/417

Dennis D. Boos • L.A. Stefanski

Essential Statistical Inference

Theory and Methods

Springer

Dennis D. Boos
Department of Statistics
North Carolina State University
Raleigh, North Carolina
USA

L.A. Stefanski
Department of Statistics
North Carolina State University
Raleigh, North Carolina
USA

ISSN 1431-875X
ISBN 978-1-4614-4817-4 ISBN 978-1-4614-4818-1 (eBook)
DOI 10.1007/978-1-4614-4818-1
Springer New York Heidelberg Dordrecht London

Library of Congress Control Number: 2012949516

Printed on acid-free paper

Springer is part of Springer Science+Business Media (www.springer.com)

To our advisors, Bob Serfling (DDB) and Ray Carroll and David Ruppert (LAS), who set our careers on track and generously and expertly guided us through our early professional years
and
To our spouses, Kathy (DDB) and Donna (LAS), who graciously and patiently tolerated us both then and now

Preface

The origins of this book date back to the 1990s when we co-taught a semester-long advanced inference course. Our goal then and now is to give an accessible version of classical likelihood inference plus modern topics like M-estimation, the jackknife, and the bootstrap. The last chapter on classical permutation and rank methods is not "modern" but certainly is timeless. Our worldview is that models are important for statistical framing of scientific questions, but the M-estimation and resampling approaches facilitate robust inference in the face of possible misspecification. The Bayesian chapter is a newer addition intended to give a solid introduction to an increasingly important general approach. Most of the book, however, is clearly frequentist.

A typical semester course consists of Chaps. 1–6 plus selections from Chaps. 7–12. We have sprinkled R code throughout the text and also in problems.

We expect students to have taken a first-year graduate level mathematical statistics course from a text like *Statistical Inference* by Casella and Berger. But measure theory is not a requirement and only shows up briefly when discussing almost sure convergence in Chap. 5 and in the Chap. 2 appendix on exponential families.

Although intended for second-year graduate students, many of the chapters can serve as references for researchers. In particular, Chap. 9 on Monte Carlo studies, Chap. 10 on the jackknife, and Chap. 12 on permutation methods contain results and summaries that are not easily accessible elsewhere.

We thank many generations of students for careful reading and constructive suggestions.

Raleigh, NC, USA

Dennis D. Boos
Leonard A. Stefanski

Contents

Part I
Introductory Material

Chapter 1
Roles of Modeling in Statistical Inference

1.1 Introduction

A common format for a course in advanced statistical inference is to accept as given a parametric family of statistical or probability models and then to proceed to develop what is an almost exclusively model-dependent theory of inference. The usual focus is on the development of efficient estimators of model parameters and the development of efficient (powerful) tests of hypotheses about those unknown parameters.

The model and the likelihood implied by it, along with maximum likelihood for parameter estimation and likelihood ratio statistics for hypothesis testing, provide an integrated and coherent approach to inference that is applicable to a wide variety of problems.

For example, a simple parametric family of models for a sample of independent observations $Y_1, \ldots, Y_n$ is that each observation Y_i has density $f(y; \boldsymbol{\theta})$ belonging to the normal (Gaussian) location-scale family

$$\{f(y; \boldsymbol{\theta}) = \frac{1}{\sqrt{2\pi}\sigma} \exp\left\{-\frac{(y-\mu)^2}{2\sigma^2}\right\}, -\infty < \mu < \infty, 0 < \sigma < \infty\}. \quad (1.1)$$

More concise language is to say that $Y_1, \ldots, Y_n$ are independent and identically distributed (iid) from the $\mathrm{N}(\mu, \sigma^2)$ family. The likelihood of $\boldsymbol{Y} = (Y_1, \ldots, Y_n)^T$ is $L(\boldsymbol{\theta} \mid \boldsymbol{Y}) = \prod_{i=1}^n f(Y_i; \boldsymbol{\theta})$ leading to maximum likelihood estimators $\widehat{\mu} = \overline{Y}$ and $\widehat{\sigma} = \{n^{-1} \sum_{i=1}^n (Y_i - \overline{Y})^2\}^{1/2}$. The likelihood ratio statistic for $H_0 : \mu = \mu_0$ is equivalent to the square of the usual one-sample t statistic.

The model/likelihood package is straightforward and so easy to use that it is often forgotten that it is a wholly self-contained and self-justifying theory. Once the probabilistic model is fully specified, it is, in a wide variety of cases (although not all), a foregone conclusion that maximum likelihood estimators are most efficient and that likelihood ratio tests are most powerful, at least asymptotically. Thus it should be remembered that the mathematical optimality of likelihood-based

D.D. Boos and L.A. Stefanski, *Essential Statistical Inference: Theory and Methods*, Springer Texts in Statistics, DOI 10.1007/978-1-4614-4818-1_1,

procedures (maximum likelihood estimation, likelihood ratio testing) is almost always dependent on the correctness of the assumed model.

In applications it is generally recognized that all models are approximations, and that proposed models are tentative at best, and often subject to reevaluation and modification throughout the course of the data analysis. Given the tentative nature of modeling in applications, it would seem logical to regard as equally suspect the attendant statistical theory that itself is often totally dependent on the assumed model through the likelihood.

A healthy skepticism of an assumed model and the inferences dictated by it is only possible with a thorough understanding of the various parts of a model and their roles in the analysis of the data, especially with regard to inferences of a probabilistic (as opposed to descriptive) nature, e.g., statements of precision (standard errors) and significance (p-values) or their Bayesian counterparts. Understanding the various parts of a model and their roles in data analysis is also the first step toward the development and understanding of methods of inference other than those based on a full likelihood.

Although these other approaches include such methods as conditional likelihood, partial likelihood, and quasi likelihood—methods whose very names suggest their less-than-full-likelihood flavor—they also include methods that are combinations of likelihood-dictated procedures and non-model-based (also called design-based or what are essentially method-of-moments) methods of inference. The key feature of these latter methods is that a model may be used for certain aspects of the data analysis, e.g., the definition and determination of estimates of those parameters of primary importance or the construction of a test statistic, whereas the (often secondary) problem of assessing precision of an estimate, or assigning a significance level to a test statistic, may be handled with non-model-based methods including method-of-moments, and resampling methods such as the jackknife and bootstrap.

It is important to understand the advantages and disadvantages of deviating from a full-likelihood specification, or from a complete exploitation of the assumed likelihood, for only then is it possible to understand the full implication of model assumptions on conclusions, and just as importantly, to tailor analyses to specific problems where a hybrid analysis may be more desirable than a totally likelihood-based analysis. The starting point for developing such an understanding is the recognition of the various parts of a statistical model and their roles in statistical inference.

1.2 The Parts of a Model

We start by noting the impossibility of the task of dissecting statistical models into individual distinct components in general. Nevertheless it is often the case that upon taking a very broad perspective of a model, it is possible to roughly classify certain parts of a model into two or more components depending on their importance or degree of relevance to the data analysis (Cox and Snell 1981, pp. 17–18).

In order to understand the classification it is useful to take a step back and consider the scientific investigation producing the data to be analyzed. It is seldom the case that an experiment, survey, observational study, etc., is undertaken for the stated purpose of fitting a statistical model to the data. Remember that a statistician's job is to analyze data, not models. There is almost always a more fundamental question motivating the research and data collection, e.g., what is the nature of the dependence of a response variate on a predictor variate (regression), or what is the nature of the relationship between two or more variates when there is no distinction as to response or predictor (correlation).

Bearing in mind that the global objective of data analysis is to address the questions motivating the research, it is seen that statistical modeling is just a means for accomplishing this task. It is not surprising then that the primary objective of a statistical model is to address, as directly as possible, the research questions of greatest interest. Thus for example if an experiment is conducted to learn how a response variate Y depends on a predictor X, the most important part of the statistical model is the specified form of the regression function $E(Y \mid X = x)$. If, as is often adequate in applied work, this specification is linear in the predictor variable,

$$E(Y \mid X = x) = \beta_0 + \beta_1 x, \tag{1.2}$$

then the motivating research question is for the most part satisfactorily answered once sensible and defensible estimates of β_0 and β_1 are obtained. There may in fact be no need to imbed the regression model (1.2) in a probabilistic model. Least squares estimation, for example, is a sensible and defensible method of obtaining regression estimates quite apart from its connection to normal theory likelihood methods.

However, in many scientific investigations the very notion of "defensible" evokes the notion of reproducibility, which is fundamental to the experimental method. Very few scientific investigations are of a nature where exact reproducibility is possible, and thus some measure of the precision of the results is required. The formulation of an appropriate statistical model for the data provides an objective means of obtaining such measures of precision.

The most common statistical model for data $\{X_i, Y_i\}$, encompassing the linear regression specification (1.2) is the normal theory, simple linear regression model,

$$Y_i = \beta_0 + \beta_1 X_i + e_i \qquad e_1, \ldots, e_n \text{ are iid } \mathrm{N}(0, \sigma^2). \tag{1.3}$$

The distributional assumptions on the equation errors $e_1, \ldots, e_n$ serve foremost to provide a means of assessing the precision of the estimates of β_0 and β_1, or in assigning significance to hypotheses about these regression parameters. These assumptions may also serve to "fine tune" the method of estimation in some cases. For example, the homogeneous error structure in (1.3, p. 5) indicates unweighted, or ordinary least squares estimation, whereas a heterogeneous error structure would suggest consideration of some form of weighted least squares.

Insofar that it is more important to get a reasonable representation and estimate of the regression function, the linear regression specification in (1.3, p. 5) is more important than the distributional specification, as the latter usually has, and ideally should have, only a secondary effect on the conclusions of the data analysis.

It is generally the case that specification of regression structure is "easier" than specification of variation structure or distributional form. A desirable feature of data analysis is to arrive at conclusions that are most robust (insensitive) to those assumptions that are most likely to be misspecified (or likely to be most misspecified).

The simple linear regression model (1.3) provides a nice example of an essentially two-part model, the primary part is the regression specification; the secondary part is the distributional specification. It should come as no surprise that in more complicated models the distinction is not so clear-cut and that depending on emphasis, a particular part of a model may be either primary or secondary in the sense described above. Regression models with non-homogeneous error structures provide a good illustration of this.

In the analysis of a regression problem, it is often noted that the variation of the response variate around its mean is not constant, but rather increases with increasing mean. Poisson data exhibit this feature — recall that for the Poisson distribution the mean and variance are equal. For certain applications it is important to model this aspect of the data quite apart from any consideration of improving estimates of the regression function by using weighted least squares. A popular and useful model for capturing variation of this type assumes that variance is proportional to a power of the mean.

Thus we entertain a model with a three-tiered structure. At the first, and most important level is the regression structure, say for simplicity the linear regression in (1.2, p. 5). Added to this is the assumption that variation is proportional to a power of the mean,

$$\mathrm{Var}(Y \mid X = x) = \tau^2(\beta_0 + \beta_1 x)^{2\theta}, \tag{1.4}$$

where $\tau > 0$ and $\theta \geq 0$ are additional unknown parameters. The model is completed with the assumption that observations are independent and normally distributed. Thus a concise specification of the full model is

$$Y_i = \beta_0 + \beta_1 X_i + \tau(\beta_0 + \beta_1 X_i)^{\theta} e_i, \qquad e_1, \ldots, e_n \text{ are iid N}(0,\ 1). \tag{1.5}$$

Note that this model implies $\beta_0 + \beta_1 x > 0$ for all possible x because only positive numbers can be raised to fractional powers. A model like (1.5) where the variance of the errors is not constant is called *heteroscedastic* (in contrast to a model with constant or *homoscedastic* errors).

Although this is a very reasonable model for a large number of regression problems, seldom is it the case that the model parameters β_0, β_1, τ, and θ would be estimated from the likelihood

$$L(\beta_0, \beta_1, \tau, \theta) = \prod_{i=1}^{n} \frac{1}{\sqrt{2\pi}\, \tau(\beta_0 + \beta_1 X_i)^{\theta}} \exp\left[-\frac{\{Y_i - (\beta_0 + \beta_1 X_i)\}^2}{2\tau^2(\beta_0 + \beta_1 X_i)^{2\theta}}\right]. \tag{1.6}$$

The problem with using (1.6) for estimation is that misspecification of either the distribution of the errors, or the variance function has an adverse effect on estimation of the regression function $E(Y \mid X = x) = \beta_0 + \beta_1 x$, the quantity most likely to be of greatest interest. This problem is apparent from the fact that the parameters β_0 and β_1 appear in both the mean and variance function.

The model (1.5, p. 6) is an example where in general it is advised not to use the likelihood for estimation (i.e., maximum likelihood estimation is undesirable); but, for example, the full model may be used for the problem of setting prediction intervals, or in the related problem of setting calibration intervals (interval estimation of X_0 corresponding to a "new" value of the response, say Y_0).

In the preceding paragraphs we have made the distinction between those parts of a model that are directly related to, and allow us to address, the important research question(s) (primary part), and those parts of a model that play a secondary role in the data analysis often by way of providing an objective assessment of precision or significance of the important conclusions (secondary and tertiary parts). There is a good chance that most students have encountered a logically different model dichotomy in previous statistics courses, especially in the study of regression models. In the simple linear regression model (1.3, p. 5) it is common to refer to $\beta_0 + \beta_1 X_i$ as the systematic component of the model, and e_i as the random component.

With (1.3, p. 5), as with many regression models, the systematic component is often equivalent to what we have identified as the primary part, and the random component is our secondary part. This breakdown is not always the case as can be seen with the heteroscedastic regression model (1.5) where it could be possible for the variance function (1.4) to be of primary interest. From a very practical point of view, we should be most concerned with those aspects of a model that are most related to, and have the greatest effect on, the main conclusions drawn from the data. Thus for many problems it is more instructive to think in terms of (and use the terminology of) primary and secondary parts as we have defined them, if for no other reason than to ensure that the focus of the data analysis is on those research questions of greatest interest.

1.3 The Roles of a Model in Data Analysis

In the preceding discussion of modeling we have already mentioned the roles of a model in connection with the various parts of a model. This section summarizes the main points made above with respect to modeling and data analysis .

- The model provides a means of addressing research questions of interest, often via the definition of parameters directly relevant to the research issues.
- The model usually implies a likelihood, and estimators can often be defined by maximizing the likelihood. Thus in situations where no other method of estimation is immediately apparent, one can always resort to modeling followed by maximum likelihood estimation.

- With estimates in hand, whether obtained by maximum likelihood or some other method (e.g., method of moments), a model, again often via the likelihood, provides a means of assessing precision of the estimates, i.e., the calculation of standard errors of the estimates. In most cases large sample approximations are used in the calculation of standard errors.
- Related to the calculation of standard errors (at least asymptotically) is the determination of statistical significance, usually in the form of calculating a p-value of a hypothesis test. Often the p-value is the frequentist probability under a null hypothesis model of observing a test statistic that is at least as extreme as the value of the statistic calculated from the data. In such cases it is clear that determination of the p-value depends on the model.

In addition to its role in providing a vehicle for data analysis, a model can also play a direct role in the scientific investigation by providing the researcher with a simplified theoretical version of the process or phenomenon under investigation. In effect the model serves as a manageable surrogate for the real phenomenon in the critical assessment of existing theories or in the development of more comprehensive theories. In this latter case it is usually true that some or even all of the parts of the model are indicated by relevant scientific theory.

1.4 A Consulting Example

A graduate student in civil engineering came to the Statistics Department with a modeling problem concerning flood predictions on rivers in North Carolina. His basic question was how to best model the relationship between watershed area and maximum flood waters that can be expected over a 100 year period at the "gauging station" associated with each watershed. Actually, he had a possible model in mind, but he did not know how to fit the model with the data he had. First a little background.

At the time there were about 140 gauging stations on rivers in North Carolina where the water level was continuously recorded. The area of the watershed that feeds into the river before the gauging station was known, as well as the cross-sectional area of the river at the station. Thus the flow (in gallons per second) past each station could be determined. The hydrology literature suggested a model of the form $Q = kA^{\eta-1}$, linearized to

$$\log Q = \log k + (\eta - 1) \log A, \tag{1.7}$$

where Q is the 100-year maximum flow and A is the watershed area, and k and η are unknown parameters. (Note that we use "log" for natural logarithm.)

The student had collected one data point for each of the 140 stations equal to the maximum flow observed *during the time the station had been keeping records*. However, these times ranged from 6 to 83 years, and most were between 20 and 60 years. The problem is that the observed maximum from a station having only

10 years of data tends to be smaller than the maximum from a station having 80 years of data, which would tend to be smaller than the maximum from a station having 100 years of data. This is problematic because the model is defined in terms of 100-year maximum flows.

Three problems were identified to think about:

1. How should one estimate/predict the 100-year maximum flow Q for each station given the observed data?
2. What assumptions about random errors should be added to the mean model in (1.7)?
3. What stations should be used in fitting the regression model?

It might be worthwhile for the reader to pause for a moment and reflect on what is the model and what are its primary and secondary parts. However, as the story continues, we will see added complexities, and models within models.

Let us attempt to work on the first problem above, estimation of Q. After further discussion, we discovered that at some stations we would be able to get yearly maximum flows for the recording period. For example, at a station observed for 36 years, we could obtain $X_1, \ldots, X_{36}$, where each X_i is a yearly maximum flow rate. If we had a full 100 years of data, then we would just order the data $X_{(1)} \le X_{(2)} \le \ldots \le X_{(100)}$, and then set $Q = X_{(100)}$. The problem is that we do not have $X_{(100)}$ for any stations. However, we do know something about its distribution function

$$P(X_{(100)} \le t) = P(X_1 \le t, \ldots, X_{100} \le t) = \prod_{i=1}^{100} P(X_i \le t) = [F(t)]^{100}, \quad (1.8)$$

where $F(t) = P(X_i \le t)$, the distribution function of X_i, and we have made the assumption that all the yearly maximums are independent and identically distributed (iid). The median of this distribution is t_0 such that $[F(t_0)]^{100} = 1/2$, or $F(t_0) = (1/2)^{.01} = .993$. So the median of the distribution of $X_{(100)}$ is the .993 quantile of $F(t)$. We might consider using an estimate of this value for Q, the maximum over 100 years.

We have reduced problem one above to estimating the .993 quantile of the yearly maximum distribution at a station based on $X_1, \ldots, X_n$, where n may be considerably smaller than 100. A little reflection shows that a nonparametric approach is not feasible because of the small sample size. Thus, we need to assume that the yearly maxima follow some parametric distribution. A natural distribution for maxima is the location-scale extreme value distribution

$$P(X_i \le t) = \exp\left\{-\exp\left(-\frac{t-\mu}{\sigma}\right)\right\}, \qquad -\infty < t, \mu < \infty,\ 0 < \sigma < \infty. \quad (1.9)$$

If we set this distribution function equal to .993 and solve for t, we obtain

$$Q_{.993}(\mu, \sigma) = \sigma\,[-\log\{-\log(.993)\}] + \mu.$$

Finally, plugging in estimates $(\widehat{\mu}, \widehat{\sigma})$ such as maximum likelihood estimates yield $Q_{.993}(\widehat{\mu}, \widehat{\sigma})$, our solution to problem one above.

Thus we have added an additional stage of modeling in order to get the data needed to fit the main model (1.7, p. 8). The remaining two problems on the type of errors to add to (1.7, p. 8), possibly correlated due to nearness in space, and which stations to use (a statistical design question), are interesting, but we leave them for homework or class discussion.

1.5 Notation, Definitions, Asymptotics, and Simulations

In this section we give a few definitions and asymptotic results in order to discuss an example in the next section. More formal development of asymptotic results is found in Chapter 5. The first subsection introduces some general language used for models and classes of distributions.

1.5.1 Families of Distributions

A *parametric model* or *parametric family* is a set of distributions indexed by a finite dimensional parameter $\boldsymbol{\theta}^T = (\theta_1, \ldots, \theta_b)$. Typically we use b for the dimension but usually change to p for regression situations (to be in concert with standard usage). We might describe a family of distributions as a set of distribution functions $\{F(y; \boldsymbol{\theta}), \boldsymbol{\theta} \in \boldsymbol{\Theta}\}$ or a set density functions $\{f(y; \boldsymbol{\theta}), \boldsymbol{\theta} \in \boldsymbol{\Theta}\}$. The normal family defined in (1.1, p. 3) is one example. Another example is the exponential scale model defined in terms of the kernel $F(y) = \{1 - \exp(-y)\}I(0 \leq y < \infty)$ as

$$\{F(y; \sigma) = F(y/\sigma),\ \ 0 < \sigma < \infty\}.$$

It is called a *scale family* and σ is a scale parameter because the distribution function of the rescaled σY, for a random variable Y with distribution function $F(y)$, is $P(\sigma Y \leq y) = P(Y \leq y/\sigma) = F(Y/\sigma)$. The normal family in (1.1, p. 3) is a *location-scale family* because it has the basic form $F((y - \mu)/\sigma)$. A family of the form $F(y - \mu)$ is called a *location family* with kernel $F()$.

A *semiparametric family* is a family of distributions indexed by a finite dimensional parameter $\boldsymbol{\theta}$ and some infinite dimensional parameter such as an unknown function belonging to a large class of functions. For example, a semiparametric location family is

$$\{f(y; \mu, f_0) = f_0(y - \mu), -\infty < \mu < \infty, f_0 \in C\},$$

where C is the class of continuous unimodal densities, symmetric about 0, μ is the finite-dimensional parameter and $f_0()$ is the infinite-dimensional parameter. A semiparametric linear regression model example is $Y_i = \boldsymbol{x}_i^T \beta + e_i$, where the density of e belong to C. In contrast, a *nonparametric* regression model example is $Y_i = g(\boldsymbol{x}_i) + e_i$, where again the density of e is in C, but the function g is assumed to be in the class of continuous functions on $[a,b]$ with m continuous derivatives.

More recent nonparametric research has focused mostly on nonparametric regression models, whereas classical nonparametrics involving rank and permutation methods, also called distribution-free methods, often uses semiparametric models. For example, in the two-independent-samples case with distribution functions F and G, the hypotheses might be totally nonparametric $H_0 : F(y) = G(y)$ versus $H_a : F(y) \neq G(y)$, where there are no restrictions on F or G. But often, one assumes a semiparametric *shift model* $G(y) = F(y - \Delta)$ and tests $H_0 : \Delta = 0$ versus $H_0 : \Delta \neq 0$ with F is unknown.

1.5.2 Moments and Cumulants

The jth population moment of a random variable is denoted by μ'_j; for the random variable Y,

$$\mu'_j = E(Y^j), \quad j = 1, 2, \ldots .$$

Of course, for $j = 1$ we often follow the standard convention that $\mathrm{E}(Y) = \mu$ instead of μ'_1. The jth population central moment of a random variable is denoted by μ_j,

$$\mu_j = E[\{Y - E(Y)\}^j], \quad j = 1, 2, \ldots .$$

Note that $\mu_1 \equiv 0$. Unless stated otherwise, a reference to μ'_j or μ_j implies the assumption that the moment exists and is finite, in other words, that

$$\mathrm{E}(|Y^j|) < \infty \quad \text{or} \quad E\{|Y - E(Y)|^j\} < \infty.$$

The jth sample moment and sample central moment are denoted by m'_j and m_j,

$$m'_j = \frac{1}{n}\sum_{i=1}^{n} Y_i^j \quad \text{and} \quad m_j = \frac{1}{n}\sum_{i=1}^{n}(Y_i - \overline{Y})^j .$$

The jth population cumulant of a random variable is denoted by κ_j, $j = 1, 2, \ldots$, assuming the defining moments exist. In terms of population moments the first four cumulants are

$$\begin{aligned}
\kappa_1 &= \mu'_1 = \mu \quad \text{(mean)} \\
\kappa_2 &= \mu'_2 - \mu^2 = \mu_2 \quad \text{(variance)} \\
\kappa_3 &= \mu'_3 - 3\mu\mu'_2 - 2\mu^3 = \mu_3 \\
\kappa_4 &= \mu'_4 - 4\mu\mu'_3 - 3(\mu'_2)^2 + 12\mu^2\mu'_2 - 6\mu^4 = \mu_4 - 3\mu_2^2. \qquad (1.10)
\end{aligned}$$

The only nonzero cumulants of a normal random variable are $\kappa_1 = E(Y)$ and $\kappa_2 = \text{var}(Y)$. The cumulants are derived from the cumulant generating function $k(t) = \log\{m(t)\}$, where $m(t)$ is the usual moment generating function, $m(t) = \text{E}\{\exp(tY)\}$. That is, $\kappa_j = k^{(j)}(0), \; j = 1, 2, \ldots$, where $k^{(j)}(t)$ denotes the jth derivative of $k(t)$.

When needed for clarity, an additional subscript is added to moments and cumulants to identify the random variable, e.g., $\mu'_{\text{Y,j}}$ is the jth moment of Y, $\mu_{\text{X,j}}$ is the jth central moment of X, and $\kappa_{\text{W,j}}$ is the jth cumulant of W.

The ratio of moments, Skew $= \mu_3/(\mu_2)^{3/2}$, is called the skewness coefficient and sometimes designated by $\sqrt{\beta_1}$ or γ_3 or α_3. Symmetric densities such as the normal have Skew $= 0$ when μ_3 exists. For non-symmetric densities such as the extreme value (Skew $= 1.14$) and the exponential (Skew $= 2$), Skew measures the lack of symmetry. Similarly, the kurtosis coefficient, Kurt $= \mu_4/\mu_2^2$, measures tail mass (equivalently peakedness) relative to the normal distribution for which Kurt $= 3$. Other names for Kurt are β_2 and α_4; the coefficient of excess is $\gamma_4 = \text{Kurt} - 3$. See Ruppert (1987) for a readable exposition on kurtosis.

A third ratio of moments is the coefficient of variation, $(\mu_2)^{1/2}/\mu$, the standard deviation divided by the mean. Scientists often use the coefficient of variation because it intuitively measures variation relative to the mean, i.e., it is a relative standard deviation.

1.5.3 *Quantiles and Percentiles*

For a random variable X with strictly increasing continuous distribution function $F(x) = P(X \leq x)$, the pth quantile (also called the 100*pth percentile) is the x value η_p that satisfies $F(\eta_p) = p$, or simply $\eta_p = F^{-1}(p)$. For example, the standard exponential distribution function is $F(x) = 1 - \exp(-x)$, $x \geq 0$. Thus, solving $F(\eta_p) = 1 - \exp(-\eta_p) = p$ leads to $\eta_p = -\log(1-p)$. At $p = 0.2$, $\eta_{0.2} = -\log(1 - 0.2) = 0.223$.

Discrete random variables have discontinuous F, and even continuous F can have intervals over which F is constant. In these cases we need a suitable definition of F^{-1} to define quantiles. For any distribution function F, a definition of F^{-1} that covers all possible cases is

$$F^{-1}(t) = \inf\{y : F(y) \geq t\}, \tag{1.11}$$

where inf $=$ infimum is the greatest lower bound of the set. This inverse $F^{-1}(t)$ is left-continuous and is the same as the usual inverse of a function when F is continuous and strictly increasing. Figure 1.1 illustrates definition (1.11) using a modification to the standard exponential distribution function. Basically, $F(x) = 1 - \exp(-x)$ has been modified between 0.4 and 1.5 to have a jump at $x = 0.4$ and a flat spot between $x = -\log(1 - .8) = 1.09$ and $x = 1.5$. To get an inverse value at p, we move horizontally to the right at height p until we reach $F(x)$ and then project down to the x axis. At $p = 0.2$, $\eta_{0.2}$ is the usual exponential

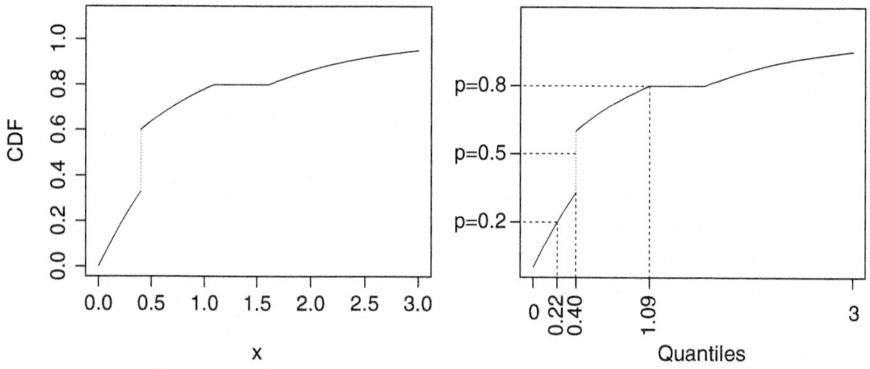

Fig. 1.1 Modified standard exponential distribution function on left with quantiles at $p = (0.2, 0.5, 0.8)$ added on right

quantile $F^{-1}(0.2) = -\log(1 - 0.2) = 0.223$ because $F(x) = 1 - \exp(-x)$ on the first portion of the curve. But at $p = 0.5$ (actually for any $p \in [0.33, 0.60]$), the horizontal dashed line intersects the vertical line at $x = 0.4$ and thus $F^{-1}(p) = 0.4$. Perhaps the most non-intuitive case is the flat spot at $p = 0.8$. Any x in the interval where $F(x) = 0.8$ might have been used as the quantile, but the definition above chooses the leftmost x of the flat region to be $F^{-1}(0.8) = 1.09$.

Moving to a sample, $X_1, \ldots, X_n$, there are numerous definitions of a sample pth quantile. The simplest approach is to use $F_n^{-1}(p)$, where $F_n(x) = n^{-1}\sum_{i=1}^{n} I(X_i \leq x)$ is the sample distribution function. The sample distribution function F_n is the nonparametric maximum likelihood estimator of F (see Section 2.2.6, p. 45) and has many desirable properties. However, the sample distribution function has jumps at the distinct sample values and is flat in-between. Thus, $F_n^{-1}(p)$ is the $[np]$th ordered sample value where $[\cdot]$ is the greatest integer function, whereas most other definitions are weighted averages of two adjacent ordered sample values. The R function `quantile` allows nine options based on the discussion in Hyndman and Fan (1996).

1.5.4 Asymptotic Normality and Some Basic Asymptotic Results

Suppose that $\widehat{\boldsymbol{\theta}}$ is a b-dimensional estimator such that for some sequence of nonrandom p-vectors, $\{\boldsymbol{\theta}_n\}_{n=1}^{\infty}$, the centered and scaled statistic $n^{1/2}\left(\widehat{\boldsymbol{\theta}} - \boldsymbol{\theta}_n\right)$ converges in distribution to a $N_b(\mathbf{0}, \boldsymbol{\Omega})$ distribution, that is, a b-variate normal distribution with mean $\mathbf{0}$ and covariance matrix $\boldsymbol{\Omega}$. "Convergence in distribution" to a b-variate normal distribution means that

$$P(n^{1/2}(\widehat{\boldsymbol{\theta}} - \boldsymbol{\theta}_n) \leq \boldsymbol{x}) \longrightarrow P(\boldsymbol{X} \leq \boldsymbol{x})$$

at each b-vector $\boldsymbol{x}$ where $\boldsymbol{X}$ is a normal random b-vector with mean $\mathbf{0}$ and covariance $\boldsymbol{\Omega}$ and the inequality $\boldsymbol{X} \le \boldsymbol{x}$ is component-wise. More concise notation is $n^{1/2}(\widehat{\boldsymbol{\theta}} - \boldsymbol{\theta}_n) \xrightarrow{d} \boldsymbol{X}$. The asymptotic variance is $\text{Avar}(\widehat{\boldsymbol{\theta}}) = \boldsymbol{\Omega}/n$, and the asymptotic mean is $\text{Amean}(\widehat{\boldsymbol{\theta}}) = \boldsymbol{\theta}_n$. Note that we use "variance" generically here to mean either a scalar variance or a covariance matrix. We say that the asymptotic distribution of $\widehat{\boldsymbol{\theta}}$ is $\text{N}_b(\boldsymbol{\theta}_n, \boldsymbol{\Omega}/n)$ or that $\widehat{\boldsymbol{\theta}}$ is asymptotically normal with asymptotic mean $\boldsymbol{\theta}_n$ and asymptotic variance $\boldsymbol{\Omega}/n$. We often write this as $\widehat{\boldsymbol{\theta}} \sim \text{AN}_b(\boldsymbol{\theta}_n, \boldsymbol{\Omega}/n)$.

For two estimators of a scalar parameter θ, the asymptotic relative efficiency of $\widehat{\theta}_1$ to $\widehat{\theta}_2$ is the limit of the ratio of the asymptotic variance of $\widehat{\theta}_2$ to the asymptotic variance of $\widehat{\theta}_1$,

$$\text{ARE}(\widehat{\theta}_1, \widehat{\theta}_2) = \frac{\text{Avar}(\widehat{\theta}_2)}{\text{Avar}(\widehat{\theta}_1)}.$$

Another important concept is convergence in probability. A random sequence Y_n converges in probability to c if $P(|Y_n - c| > \epsilon) \to 0$ as $n \to \infty$ for each $\epsilon > 0$. Concise notation for this convergence is $Y_n \xrightarrow{p} c$. Convergence in probability for a vector-valued random sequence is equivalent to convergence in probability of each component. An important result for convergence in probability is the continuity theorem.

Theorem 1.1 (Continuity). *Suppose that* $\boldsymbol{Y}_n \xrightarrow{p} \boldsymbol{c}$ *and* $\boldsymbol{g}$ *is continuous at* $\boldsymbol{c}$. *Then* $\boldsymbol{g}(\boldsymbol{Y}_n) \xrightarrow{p} \boldsymbol{g}(\boldsymbol{c})$.

The next two theorems are extremely useful for obtaining asymptotic distributions.

Theorem 1.2 (Slutsky's Theorem). *If* $X_n \xrightarrow{d} X$, $Y_n \xrightarrow{p} a$, *and* $Z_n \xrightarrow{p} b$, *where* a *and* b *are constants, then* $Y_n X_n + Z_n \xrightarrow{d} aX + b$.

In this following theorem, the vector-valued derivative $g'(\boldsymbol{\theta}) = \partial g(\boldsymbol{\theta})/\partial \boldsymbol{\theta}$ is a row vector (see the Appendix on derivative notation, p. 531).

Theorem 1.3 (Delta Method). *Suppose that* $\widehat{\boldsymbol{\theta}} \sim AN_b(\boldsymbol{\theta}, \boldsymbol{\Omega}/n)$ *and that* g *is a real-valued function on* $\mathbf{R}^b$ *possessing a derivative* $g'(\boldsymbol{\theta})$ *in a neighborhood of* $\boldsymbol{\theta}$ *and continuous at* $\boldsymbol{\theta}$ *with at least one component nonzero at* $\boldsymbol{\theta}$. *Define* $\widehat{\gamma} = g(\widehat{\boldsymbol{\theta}})$ *and* $\gamma = g(\boldsymbol{\theta})$. *Then*

$$\widehat{\gamma} \sim AN\{\gamma, g'(\boldsymbol{\theta})\boldsymbol{\Omega} g'(\boldsymbol{\theta})^T/n\}.$$

Intuition for this result is readily seen by approximating $g(\widehat{\boldsymbol{\theta}})$ by a Taylor series,

$$\begin{aligned}\widehat{\gamma} &= g(\widehat{\boldsymbol{\theta}}) \\ &\approx g(\boldsymbol{\theta}) + g'(\boldsymbol{\theta})(\widehat{\boldsymbol{\theta}} - \boldsymbol{\theta}), \end{aligned} \tag{1.12}$$

from which it follows (assuming that the remainder in (1.12) can be ignored) that

$$n^{1/2}(\widehat{\gamma} - \gamma) \approx g'(\boldsymbol{\theta}) n^{1/2}(\widehat{\boldsymbol{\theta}} - \boldsymbol{\theta}) \xrightarrow{d} \text{N}\{0, g'(\boldsymbol{\theta})\boldsymbol{\Omega}\, g'(\boldsymbol{\theta})^T\}.$$

1.5.5 Simulation Methods

Computer simulation is perhaps the most important tool for the 21st century statistician. We use it throughout the book, both in examples and problems, often accompanied with relevant R code.

We synonymously use the terms *simulation*, *Monte Carlo simulation*, and *Monte Carlo methods* to refer to the use of randomly generated data for estimating quantities of interest, often expectations. For example, consider estimating the mean of a distribution with density $f(y)$ and distribution function $F(y)$,

$$E(Y) = \int y f(y) dy.$$

A simple approach to approximate this integral is to generate independent $Y_1, \ldots, Y_N$ from $f(y)$ and estimate it with $N^{-1} \sum_{i=1}^{N} Y_i$. The estimated variance of this estimator is simply s_{N-1}^2/N, where $s_{N-1}^2 = (N-1)^{-1} \sum_{i=1}^{N} (Y_i - \overline{Y})^2$ is the sample variance.

The R computing language is designed to facilitate the generation of random variables. For many distributions there are functions available. For example, to generate a vector y of $N = 1000$ independent normal($\mu = 3, \sigma^2 = 10$) random variables, the code is simply

```
y <- rnorm(1000,mean=3,sd=sqrt(10)).
```

Knowing the order of the arguments allows one the shortened version

```
y <- rnorm(1000,3,sqrt(10)).
```

A slightly harder example is the extreme value location-scale distribution with distribution function $F(y; \mu, \sigma) = \exp[-\exp\{-(y-\mu)/\sigma\}]$ used in (1.9, p. 9). Here $E(Y) = \sigma c + \mu$ where c is the mean for the case $\mu = 0$ and $\sigma = 1$ (this is true for any location-scale distribution). Now to generate from the extreme value distribution, we use the fact that $F^{-1}(U)$ for a uniform(0,1) random variable U has exactly the distribution function F for any F. This result is an important companion result to the *probability integral transformation* result: $F(Y)$ has a uniform(0,1) distribution for continuous random variables Y with distribution function F. However, the inverse distribution function result is stronger because it is true for any distribution function, continuous or discrete. See Problem 1.9 (p. 21) for a proof.

For the extreme value distribution function with ($\mu = 0, \sigma = 1$), we set $\exp\{-\exp(-y)\} = t$ and solve for y yielding

$$F^{-1}(t) = -\log\{-\log(t)\}.$$

Here is the R code and result:

```
set.seed(157)        # sets starting point for random numbers
N <- 1000000         # sample size
U <- runif(N)        # generates vector of N uniform(0,1) rv's
Y <- -log(-log(U))   # vector of N extreme value rv's
mean(Y)              # mean
[1] 0.5750475
sd(Y)/sqrt(N)        # standard error of mean
[1] 0.001282290
```

In this case we know that $c = 0.577$, Euler's number. The estimate 0.575 is within two standard errors, .0026, of 0.577.

A more complex problem is estimating parameters of the sampling distribution of an estimator. We use the estimator from Section 1.4 as an example,

$$Q_{.993}(\widehat{\mu}, \widehat{\sigma}) = \widehat{\sigma}[-\log\{-\log(.993)\}] + \widehat{\mu}.$$

Recall that $Q_{.993}(\widehat{\mu}, \widehat{\sigma})$ estimates the median of the maximum of a sample of size n from an extreme value distribution. We know the true value is $Q_{.993}(\mu, \sigma)$. A question of interest might be the bias of this estimator. Thus we seek to estimate $E\{Q_{.993}(\widehat{\mu}, \widehat{\sigma})\} - Q_{.993}(\mu, \sigma)$. In R we need to generate a number of samples of size n and compute $Q_{.993}(\widehat{\mu}, \widehat{\sigma})$ for each sample. Here we generate 100 samples for the case $\mu = 0, \sigma = 1$.

```
dextval <- function(x,mu,sigma){    # extreme value density
  y <- exp(-(x-mu)/sigma)
  y*exp(-y)/sigma
}
l.extval <- function(theta,y){      # neg. of loglike
  -sum(log(dextval(y,theta[1],theta[2])))
}
ext.val.mle <- function(y){         # finds mle
# starting values from method of moments
 sig.start<-sqrt(6)*sd(y)/pi
 theta.start<-mean(y)-sig.start*.577
 nlm(l.extval,c(theta.start,sig.start),y=y)$estimate
}
set.seed(367)
n<-20                               # sample size
nrep<-100                           # number of reps
# each row of data is a sample of size n
data<-matrix(-log(-log(runif(n*nrep))),nrow=nrep,ncol=n)
out<-t(apply(data,1,ext.val.mle))   # nrep by 2 matrix, mle
parm<- -log(-log(.993))             # target parameter
est<-out[,2]*(-log(-log(.993)))+out[,1]  # vector of est
mean(est-parm)                 # estimated bias
[1] -0.2913428
sd(est-parm)/sqrt(nrep)        # standard error of est. bias
[1] 0.1011618
```

1.6 Example: A Simple Mean/Variance Model

In this introductory chapter we have explained the importance of model assumptions in statistical inference. In particular, we have defined the primary part of the model as that part directly related to the scientific question of interest, whereas the secondary part is essentially all the remaining components of the model. These secondary components play a large role in the type of methods used and inferences drawn.

To illustrate these ideas, we now describe the maximum likelihood estimator and method-of-moment estimator in a simple normal model where the variance depends on the mean and the coefficient of variation is assumed known. This example is adapted from Carroll and Ruppert (1988, pp. 21–22).

1.6.1 The Assumed Model and Associated Estimators

The assumed model for constructing a likelihood and defining inferential procedures is

$$Y_1, \ldots, Y_n \text{ iid N}(\mu,\, c^2\mu^2), \quad c^2 \text{ known.} \tag{1.13}$$

Note that c is the coefficient of variation. The method-of-moments (MOM) estimator of μ does not depend on the variance assumption, but its variance does:

$$\widehat{\mu}_{\text{MOM}} = \overline{Y}, \quad \text{Var}(\widehat{\mu}_{\text{MOM}}) = \frac{c^2\mu^2}{n}. \tag{1.14}$$

Using model (1.13), the likelihood is

$$\begin{aligned} L(\mu \mid \{Y_i\}_1^n) &= \prod_{i=1}^{n} \frac{1}{c\mu\sqrt{2\pi}} \exp\left\{-\frac{(Y_i-\mu)^2}{2c^2\mu^2}\right\} \\ &= c^{-n}\mu^{-n}(2\pi)^{-n/2} \exp\left\{-\sum_{i=1}^{n} \frac{(Y_i-\mu)^2}{2c^2\mu^2}\right\}, \end{aligned} \tag{1.15}$$

and the maximum likelihood estimator (MLE) and asymptotic variance are given by

$$\widehat{\mu}_{\text{MLE}} = \frac{\left(\overline{Y}^2 + 4c^2 m_2'\right)^{1/2} - \overline{Y}}{2c^2}, \quad \text{Avar}(\widehat{\mu}_{\text{MLE}}) = \frac{c^2\mu^2}{n(1+2c^2)}, \tag{1.16}$$

where recall $m_2' = n^{-1}\sum_{i=1}^{n} Y_i^2$.

1.6.2 Model-Based Comparison of MOM to MLE

Under model (1.13), both $\widehat{\mu}_{\text{MOM}}$ and $\widehat{\mu}_{\text{MLE}}$ are consistent for μ. The asymptotic (large sample) relative efficiency of $\widehat{\mu}_{\text{MLE}}$ with respect to $\widehat{\mu}_{\text{MOM}}$ is

$$\text{ARE}(\widehat{\mu}_{\text{MLE}}, \widehat{\mu}_{\text{MOM}}) = \frac{\text{Avar}(\widehat{\mu}_{\text{MOM}})}{\text{Avar}(\widehat{\mu}_{\text{MLE}})} = 1 + 2c^2. \tag{1.17}$$

It is no surprise that the MLE is more efficient than the MOM. For example, if $c = 1$, then maximum likelihood is three times as efficient as method-of-moments.

1.6.3 A Non-Model-Based Comparison of MOM to MLE

We now study $\widehat{\mu}_{\text{MOM}} = \overline{Y}$ and $\widehat{\mu}_{\text{MLE}}$ in (1.16) under an enlarged model that drops the normality assumption and the mean/variance relationship in (1.13, p. 17):

$$Y_1, \ldots, Y_n \text{ iid } E(Y_1) = \mu, \; E|Y_1|^4 < \infty, \tag{1.18}$$

Results are stated in terms of μ and the central moments.

The method-of-moments estimator ($\overline{Y}$) is of course still unbiased for estimating μ and its variance is μ_2/n. The maximum likelihood estimator based on assuming normality converges in probability to

$$\frac{\left(\mu^2 + 4c^2\mu^2 + 4c^2\mu_2\right)^{1/2} - \mu}{2c^2} \tag{1.19}$$

and has asymptotic variance

$$\text{Avar}(\widehat{\mu}_{\text{MLE}}) = \frac{1}{nv_1}\left(\frac{\mu_2 v_2^2}{4c^4} + \frac{\mu_3 v_2}{c^2} + \mu_4 - \mu_2^2\right), \tag{1.20}$$

where

$$v_1 = \mu^2(1 + 4c^2) + 4c^2\mu_2, \quad v_2 = \mu(1 + 4c^2) - \sqrt{v_1}.$$

From expression (1.19), it is clear that the MLE is consistent if and only if the variance is exactly equal to $c^2\mu^2$, i.e., $\mu_2 = c^2\mu^2$; however, normality is not required. When $\mu_2 = c^2\mu^2$, then v_1 and v_2 are $v_1 = \mu^2(1 + 2c^2)^2$ and $v_2 = 2c^2\mu$, and (1.20) is

$$\text{Avar}(\widehat{\mu}_{\text{MLE}}) = \frac{c^2\mu^2}{n(1 + 2c^2)^2}\left\{1 + 2c^2 + 2c(\text{Skew}) + c^2(\text{Kurt} - 3)\right\}, \tag{1.21}$$

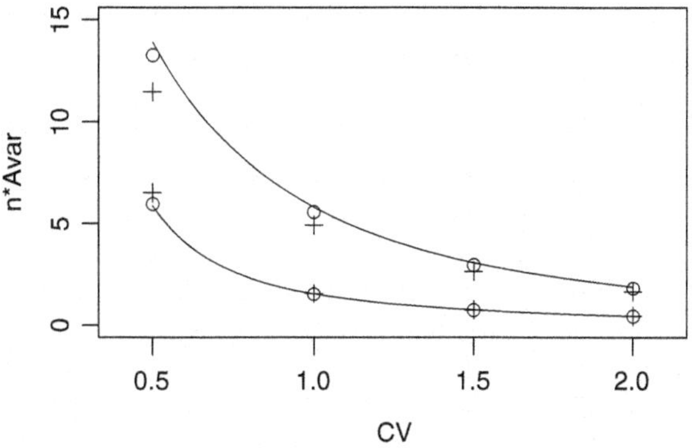

Fig. 1.2 Asymptotic variances times n (lines) and estimated variances times n (+ for $n = 20$ and o for $n = 100$) of $\widehat{\mu}_{\text{MLE}}$ in (1.16, p. 17) for normal ($\mu = 1$, variance = 4) (lower line) and exponential data with $\mu = 1$ (upper line). Plotted estimates are based on 10,000 Monte Carlo replications with standard error approximately 1.4% of the estimates

where recall that Skew $= \mu_3/\mu_2^{3/2}$ and Kurt $= \mu_4/\mu_2^2$ are the standardized third and fourth moments, respectively. Now the asymptotic relative efficiency of $\widehat{\mu}_{\text{MLE}}$ with respect to $\widehat{\mu}_{\text{MOM}}$ is

$$\text{ARE}(\widehat{\mu}_{\text{MLE}}, \widehat{\mu}_{\text{MOM}}) = \frac{(1+2c^2)^2}{1+2c^2+2c(\text{Skew})+c^2(\text{Kurt}-3)}. \tag{1.22}$$

It is apparent from (1.22) that for certain nonzero values of Skew and Kurt, the asymptotic relative efficiency of maximum likelihood to method of moments can be less than one, i.e., there are distributions for which method of moments is more efficient. This would not be important if such values of Skew and Kurt were uncommon in practice; however, Carroll and Ruppert (1988, pp. 21–22) argue that data sets for which method of moments is more efficient than, or nearly as efficient as, maximum likelihood occur commonly in practice. For example, exponential distributions have $c = 1$, Skew $= 2$, and Kurt $= 9$, in which case $\text{ARE}(\widehat{\mu}_{\text{MLE}}, \widehat{\mu}_{\text{MOM}}) = 9/13$.

1.6.4 Checking Example Details

The asymptotic variance expression (1.20, p. 18) follows by applying Theorem 1.3 (p. 14) to $\widehat{\mu}_{\text{MLE}}$ in (1.16, p. 17), noting that $\widehat{\mu}_{\text{MLE}}$ may be viewed as a function $g(\widehat{\boldsymbol{\theta}})$, where $\widehat{\boldsymbol{\theta}} = (\overline{Y}, m_2')^T$ and $\boldsymbol{\Omega}$ is the covariance matrix of (Y_1, Y_1^2). Alternatively, $\widehat{\mu}_{\text{MLE}}$ may be viewed as a function of $\widehat{\boldsymbol{\theta}} = (\overline{Y}, m_2)^T$, where $\boldsymbol{\Omega}_{11} = \mu_2$, $\boldsymbol{\Omega}_{12} = \mu_3$, and $\boldsymbol{\Omega}_{22} = \mu_4 - \mu_2^2$ are taken from Example 5.36 (p. 256).

Either derivation of (1.20, p. 18) is tedious and subject to mistakes. One way to check the calculations is to simulate samples and estimate the actual variance. To illustrate, Figure 1.2 plots the asymptotic variance from (1.20, p. 18) multiplied by n for normal data with mean $\mu = 1$ and variance=4 (lower line) and for exponential data with $\mu = 2$ and thus variance=4 (upper line). Multiplying the asymptotic variance by n allows us to keep results for various values of n on the same plot. The plotted points are from Monte Carlo variance estimates for $n = 20$ (+) and $n = 100$ (o).

For each of 10,000 generated samples from each distribution and sample size, the estimator $\widehat{\mu}_{\text{MLE}}$ was calculated and the sample variance computed and multiplied by n. For the normal data, the asymptotic variances are close to the estimated variances even at $n = 20$. For exponential data, larger sample sizes are required, but clearly by $n = 100$ the estimated variances are close to the asymptotic variance curves. Note that $\widehat{\mu}_{\text{MLE}}$ is consistent in these cases when $c = 2$ for the normal data and $c = 1$ for the exponential data. However, the curves and estimates are valid even when $\widehat{\mu}_{\text{MLE}}$ is inconsistent. From Section 9.3.2 (p. 370), we find that the approximate relative standard errors of the plotted Monte Carlo estimates are $(2/10000)^{1/2} = .014$, or 1.4% of the estimates.

This comparison of asymptotic variances and finite sample variances is a paradigm we want to emphasize as one of the key approaches of modern statistical research. The asymptotic expressions are useful for evaluating new estimators, comparing estimators, and basic understanding of the problem at hand. Monte Carlo estimates are useful to evaluate and compare estimators in finite samples. Finally, comparing asymptotic expressions with Monte Carlo estimates helps confirm that both the asymptotic expressions are correct and that the computations in the Monte Carlo study are correctly coded. However, it should be kept in mind that computational results alone do not constitute proof.

1.7 Problems

1.1. In generalized linear models there is typically a *linear predictor* $\eta_i = x_i^T \boldsymbol{\beta}$, a link function $g()$ such that $\eta_i = g(\mu_i)$, where μ_i is the mean of Y_i (and thus $\mu_i = g^{-1}(x_i^T \boldsymbol{\beta})$). Further, assuming independence of the Y_i and a particular exponential family form for the density of Y_i leads to a likelihood and to maximum likelihood estimators. Explain what part or parts of the above description make up the primary part of the model.

1.2. A pharmaceutical company is interested in how the variability of drug tablet weights depends on manufacturing factors A and B. The proposed model is

$$Y_{ijk} = \mu_{ij} + \theta_{ij}\epsilon_{ijk},$$

where the ϵ_{ijk} are iid $N(0, \sigma^2)$ and $\theta_{ij} = \alpha_i \beta_j \tau_{ij}$ with $\sum_i \log \alpha_i = 0$, $\sum_j \log \beta_j = 0$, $\sum_i \log \tau_{ij} = 0$, and $\sum_j \log \tau_{ij} = 0$. What are the primary and secondary parts of the model?

1.3. How does the model specification in (1.5, p. 6) affect the

a. prediction of a new Y at a given x value?
b. the variance of that prediction?

1.4. Let Y be a random variable with moment generating function $m_Y(t)$, defined for all t in a neighborhood of $t = 0$. The cumulant generating function of Y is $k(t) = \log(m(t))$ and defines the cumulants of Y in the same fashion as the moment generating function defines the moments, i.e., $\kappa_j = k^{(j)}(0)$, $j = 1, 2, \ldots$, where $k^{(j)}(t)$ denotes the jth derivative of $k(t)$.

a. Derive the first four cumulants in terms of the first four (noncentral) moments.
b. Express κ_2, κ_3, and κ_4 in terms of the central moments.
c. Express the skewness coefficient Skew $= \mu_3/(\mu_2)^{3/2}$ and the kurtosis coefficient Kurt $= \mu_4/\mu_2^2$ in terms of the cumulants.

1.5. Find the first four cumulants of:

a. a Poisson(λ) random variable;
b. a N(μ, σ^2) random variable.

Then find the skewness and kurtosis coefficients for these two random variables.

1.6. Prove that if Y is a random variable with all of its cumulants of order greater than two equal to zero, i.e., $0 = \kappa_3 = \kappa_4 = \cdots$, then Y has a normal distribution.

1.7. Suppose that Z has cumulant generating function $k_Z(t)$ with $E(Z) = 0$ and $\text{Var}(Z) = 1$. Let $Y = \sigma Z + \mu$. Find the cumulant generating function of Y, $k_Y(t)$, in terms of $k_Z(\cdot)$. Use this to prove that all cumulants except κ_1 are location invariant, i.e., do not depend on μ.

1.8. Show that $\text{Skew}(a + bY) = \text{Skew}(Y)$ and $\text{Kurt}(a + bY) = \text{Kurt}(Y)$. For an iid random sample $Y_1, \ldots, Y_n$, show that $\text{Skew}(\overline{Y}) = \text{Skew}(Y_1)/n^{1/2}$ and $\text{Kurt}(\overline{Y}) = 3 + \{\text{Kurt}(Y_1) - 3\}/n$. Give a heuristic explanation of why skewness is more important than kurtosis when the normal distribution is used to approximate the distribution of a sample mean.

1.9. Using the definition (1.11, p. 12), Serfling (1980, p. 3) gives the result that $F(y) \geq t$ if and only if $y \geq F^{-1}(t)$ for any distribution function. Use that result to prove: for any distribution function F, the distribution function of $F^{-1}(U)$, where U is a uniform(0,1) random variable, is $F(y)$. That is, show that $P(F^{-1}(U) \leq y) = F(y)$ for all y.

1.10. Recall that the location-scale extreme value distribution has distribution function $F(t) = F_0((t - \mu)/\sigma)$, where $F_0(x) = \exp\{-\exp(-x)\}$. Give an expression for the pth quantile, $F^{-1}(p)$.

1.11. Related to the bias estimation example of Section 1.4 for a sample of size $n = 20$ from an extreme value distribution with $\mu = 0, \sigma = 1$, estimate the bias of the method of moments estimator $Q_{.993}(\widehat{\mu}_{mm}, \widehat{\sigma}_{mm})$, where $\widehat{\mu}_{mm} = \overline{Y} - 0.577\widehat{\sigma}_{mm}$, and $\widehat{\sigma}_{mm} = \sqrt{6}s_{n-1}/\pi$. In addition, estimate the mean squared error, in R code `mean((est-parm)^2)`.

1.12. Suppose that $Y_1, \ldots, Y_n$ are iid Poisson(λ) and that $\widehat{\lambda} = \overline{Y}$. Define $\sigma^2_{\widehat{\lambda}} = \text{var}(\widehat{\lambda})$. Consider the following two estimators of $\sigma^2_{\widehat{\lambda}}$,

$$\widehat{\sigma}^2_{\widehat{\lambda}} = \frac{\widehat{\lambda}}{n} \text{ and } \widetilde{\sigma}^2_{\widehat{\lambda}} = \frac{s^2}{n},$$

where $s^2 = (n-1)^{-1} \sum_{i=1}^{n} (Y_i - \overline{Y})^2$ and note that in general

$$\text{var}(s^2) = \{\text{Var}(Y_1)\}^2 \left[\frac{2}{n-1} + \frac{(\text{Kurt} - 3)}{n} \right]. \tag{1.23}$$

We want to compare the asymptotic efficiency of the two estimators of $\sigma^2_{\widehat{\lambda}}$ above.

Note that $\sigma^2_{\widehat{\lambda}} \to 0$, and that both $\widehat{\sigma}^2_{\widehat{\lambda}}$ and $\widetilde{\sigma}^2_{\widehat{\lambda}}$ converge in probability to 0. When studying the consistency and efficiency of estimators of asymptotic variances that are converging to 0 at rate n^{-1}, it is necessary to normalize everything by multiplying by n. Thus you are to compare the consistency and efficiency of $n\widehat{\sigma}^2_{\widehat{\lambda}}$ and $n\widetilde{\sigma}^2_{\widehat{\lambda}}$ as estimators of $\lim_{n\to\infty} n\sigma^2_{\widehat{\lambda}}$.

a. When the Poisson model holds, are both estimators consistent?
b. When the Poisson model holds, give the asymptotic relative efficiency of the two estimators. (Actually here you can compare exact variances since the estimators are so simple.)
c. When the Poisson model does not hold but assuming second moments exist, are both estimators consistent?

1.13. Consider the model (1.13, p. 17). Define $\widehat{\mu}_{\text{WLS}}$ as that value of μ minimizing

$$\rho(\mu) = \sum_{i=1}^{n} \left\{ \frac{(Y_i - \mu)^2}{\sigma^2 \mu^2} \right\}. \tag{1.24}$$

Do all calculations assuming the full model (1.13, p. 17).

a. Show that $\widehat{\mu}_{\text{WLS}}$ converges in probability, but that its limit is not μ, i.e., $\widehat{\mu}_{\text{WLS}}$ is not a consistent estimator of μ.
b. Find $E\{\rho'(\mu)\}$ and show that it is not equal to zero.
c. Find the estimator obtained by minimizing

$$\rho_*(\mu) = \sum_{i=1}^{n} \left\{ \frac{(Y_i - \mu)^2}{\sigma^2 \mu^2} + 2\log(\mu) \right\} \tag{1.25}$$

and denote it by $\widehat{\mu}_{*\text{WLS}}$.

d. Is $\widehat{\mu}_{*\mathrm{WLS}}$ a consistent estimator of μ? If so find its asymptotic variance and compare to that of $\widehat{\mu}_{\mathrm{MOM}}$ and $\widehat{\mu}_{\mathrm{MLE}}$.
e. Find $E\{\rho'_*(\mu)\}$. Is it equal to zero?

1.14. Use Theorem 1.1 (p. 14) to justify the limiting expression (1.19, p. 18) for $\widehat{\mu}_{\mathrm{MLE}}$. Then compute the value of the limiting expression for the four situations and two distributions in Figure 1.2 (p. 19).

1.15. Use the delta method to derive the asymptotic variance expression (1.16, p. 17) given by

$$\frac{\sigma^2\mu^2}{n(1+2\sigma^2)}$$

when the true model is $\mathrm{N}(\mu, \sigma^2\mu^2)$.

1.16. Use the delta method to derive the asymptotic variance expression in (1.20, p. 18).

1.17. To check on the asymptotic variance expression in (1.20, p. 18), generate 1000 samples of size $n = 20$ from the exponential density $(1/2)\exp(-y/2)$ that has mean $\mu = 2$ and variance=4 so that in terms of the model $\sigma^2 = 1$. For each sample, calculate $\widehat{\mu}_{\mathrm{MLE}}$ in (1.16, p. 17). Then compute (1.20, p. 18) for this exponential density and compare it to $n = 20$ times the sample variance of the 1000 values of $\widehat{\mu}_{\mathrm{MLE}}$. Repeat for $n = 50$.

Part II
Likelihood-Based Methods

Chapter 2
Likelihood Construction and Estimation

2.1 Introduction

After a statistical model for the observed data has been formulated, the likelihood of the data is the natural starting point for inference in many statistical problems. This function typically leads to essentially automatic methods of inference, including point and interval estimation, and hypothesis testing. In fact, likelihood methods are generally asymptotically optimal provided the assumed statistical model is correct. Thus, the likelihood is the foundation of classical model-based statistical inference. In this chapter we describe, and often derive, basic likelihood methods and results. We start with construction of the likelihood in numerous examples and then move to various inferential techniques. The development in this chapter follows the classical frequentist approach to inference. However, the Bayesian approach to inference is equally dependent on the likelihood, and the material in this chapter is essential to understanding the Bayesian methods in Chapter 4.

If $\boldsymbol{Y} = (Y_1, \ldots, Y_n)^T$ has joint density or probability mass function $f(\boldsymbol{y}; \boldsymbol{\theta})$, with $\boldsymbol{\theta}^T = (\theta_1, \ldots, \theta_b)$, then the likelihood function is just the joint density (or probability mass function) evaluated at the observed data points, $L(\boldsymbol{\theta} \mid \boldsymbol{Y}) = f(\boldsymbol{Y}; \boldsymbol{\theta})$. Note that capital letters, e.g., $\boldsymbol{Y}$, denote random variables, observed or not, and lower-case letters, e.g., $\boldsymbol{y}$, denote arguments of mathematical functions. The parameter vector $\boldsymbol{\theta}$ is listed first in L because L is primarily viewed as a function of $\boldsymbol{\theta}$ for given data vector $\boldsymbol{Y} = (Y_1, \ldots, Y_n)^T$. The notation here suggests that the Y_i are scalars, but they could just as well be vectors. Henceforth, we stop using the term "probability mass function" and just use "density" unless it is important to emphasize that the data are discrete random variables.

For independent Y_i, the likelihood factors into

$$L(\boldsymbol{\theta} \mid \boldsymbol{Y}) = \prod_{i=1}^{n} f_i(Y_i; \boldsymbol{\theta}),$$

D.D. Boos and L.A. Stefanski, *Essential Statistical Inference: Theory and Methods*, Springer Texts in Statistics, DOI 10.1007/978-1-4614-4818-1_2,

where we denote the density of Y_i by f_i to indicate that each Y_i can have a different density, as would be the case with regression models. For an independent and identically distributed (iid) sample $Y_1, \ldots, Y_n$ with the density of Y_1 given by $f(y; \boldsymbol{\theta})$, the likelihood has the familiar form $L(\boldsymbol{\theta} \mid \boldsymbol{Y}) = \prod_{i=1}^{n} f(Y_i; \boldsymbol{\theta})$.

We discuss a variety of likelihoods in this chapter that are more complex than the simple product likelihoods associated with iid data. In such models it is important to remember that the likelihood is defined in terms of the density of the observed data to be analyzed. For emphasis, we state the following

Key Concept: In all situations the likelihood is the joint density of the observed data to be analyzed.

Thus, for example, in models for censored or missing data, the likelihood is not the density of the so-called "complete" data that includes the censored or missing values. Rather it is the density of only those components of the data that are observed and used in the statistical analysis. Note also that we usually write $L(\boldsymbol{\theta} \mid \boldsymbol{Y})$ as a random function depending on the data $\boldsymbol{Y}$ rather than as $L(\boldsymbol{\theta} \mid \boldsymbol{y})$, although the latter may be appropriate in a context where only the mathematical properties of $L(\boldsymbol{\theta} \mid \boldsymbol{y})$ are being described.

It is apparent by its definition, that likelihood construction is an exercise in the calculus of random variables and distributions; i.e., the essential problem is that of determining the joint probability density of the data to be analyzed. Such problems constitute a core topic covered in introductory mathematical statistics courses and texts; for example, Casella and Berger (2002).

Of the relevant techniques taught in such courses, we prefer using the so-called *distribution function method* for deriving the density of random variables and vectors. This technique is more general and more useful than the *Jacobian method* commonly emphasized in mathematical statistics courses. For example, suppose that the random variable Y has distribution function $F_Y(y; \boldsymbol{\theta}) = P(Y \le y)$, but we desire the density of the random variable $X = g(Y)$. (Note that *distribution function* is often called *cumulative distribution function*, we prefer the short name.) Then the distribution of X is simply

$$F_X(x; \boldsymbol{\theta}) = P\{X \le x\} = P\{g(Y) \le x\},$$

where the latter is a probability calculated from the distribution of Y. Often g is a strictly increasing function, in which case g^{-1} exists and

$$F_X(x; \boldsymbol{\theta}) = P\{g(Y) \le x\} = P\{Y \le g^{-1}(x)\} = F_Y\{g^{-1}(x); \boldsymbol{\theta}\}. \tag{2.1}$$

Finally, in the case that Y has a differentiable distribution function with density $f_Y(y; \boldsymbol{\theta}) = (d/dy)F_Y(y; \boldsymbol{\theta})$ and g^{-1} is differentiable, the density of X is found by differentiating both sides of (2.1) with respect to x resulting in

$$f_X(x; \boldsymbol{\theta}) = f_Y\{g^{-1}(x); \boldsymbol{\theta}\} \frac{dg^{-1}(x)}{dx}.$$

The expression for $f_X(x;\boldsymbol{\theta})$ above is identical to that obtained by the Jacobian method. Note that the absolute value of the derivative of $g^{-1}(x)$ that appears in the general Jacobian method formula is not necessary in our case due to the assumption that g is increasing.

We further illustrate the utility of this method with a transformation model likelihood. Consider an iid sample $Y_1,\ldots,Y_n$, and a parametric transformation $h(y,\alpha)$ strictly increasing in y for each α. The model assumptions are such that the transformed data $h(Y_1,\alpha),\ldots,h(Y_n,\alpha)$ are iid with common density $f(y;\boldsymbol{\theta})$ and distribution function $F(y;\boldsymbol{\theta})$. Both α and $\boldsymbol{\theta}$ are parameters in this model. The problem is to find the likelihood of the observed data $Y_1,\ldots,Y_n$. In light of the "Key Concept" above, we need the density of Y_i to construct the likelihood. Under the stated monotonicity assumption on h, $\{t \le y\}$ is equivalent to $\{h(t,\alpha) \le h(y,\alpha)\}$. Thus the distribution function of Y_i is,

$$P\{Y_i \le y\} = P\{h(Y_i,\alpha) \le h(y,\alpha)\} = F\{h(y,\alpha);\boldsymbol{\theta}\}$$

Taking the derivative with respect to y, the density of Y is

$$f_Y(y;\boldsymbol{\theta},\alpha) = f\{h(y,\alpha);\boldsymbol{\theta}\}\frac{\partial h(y,\alpha)}{\partial y},$$

resulting in the likelihood

$$L(\boldsymbol{\theta},\alpha;\, \boldsymbol{Y}) = \prod_{i=1}^{n} f\{h(Y_i,\alpha);\boldsymbol{\theta}\}\left\{\left.\frac{\partial h(y,\alpha)}{\partial y}\right|_{y=Y_i}\right\}.$$

A common mistake in this kind of problem is to start working with $h(Y_i;\alpha)$ instead of with the observed data Y_i.

2.1.1 Notes and Notation

It is generally accepted, e.g., Hald (1998), that Fisher (1912) was the first to use the likelihood to obtain estimators by finding the value of $\boldsymbol{\theta}$ that maximizes $L(\boldsymbol{\theta} \mid \boldsymbol{Y})$. In a series of articles in the 1920's, R. A. Fisher showed that the maximum likelihood estimator (MLE), denoted herein by $\widehat{\boldsymbol{\theta}}_{\mathrm{MLE}}$, is generally optimal or at least optimal in large samples (Fisher 1922). We do not dwell on questions of optimality, but focus our attention on constructing likelihoods in a variety of situations in the following section.

On a practical note, the MLE ($\widehat{\boldsymbol{\theta}}_{\mathrm{MLE}}$) of the $b \times 1$ parameter vector $\boldsymbol{\theta}$ is usually calculated by optimizing the *log likelihood function*, $\log(L(\boldsymbol{\theta} \mid \boldsymbol{Y}))$, henceforth denoted by $\ell(\boldsymbol{\theta})$. For suitably differentiable likelihoods, optimization is usually accomplished by differentiating $\ell(\boldsymbol{\theta})$, with respect to $\boldsymbol{\theta}$ to obtain the *likelihood score function* $\boldsymbol{S}(\boldsymbol{\theta}) = \{\ell'(\boldsymbol{\theta})\}^T$ where $\ell'(\boldsymbol{\theta}) = \partial\ell(\boldsymbol{\theta})/\partial\boldsymbol{\theta}$, and then solving the *likelihood equations*, $\boldsymbol{S}(\boldsymbol{\theta}) = \boldsymbol{0}_{b\times 1}$.

Finally note that we use the convention that the derivative of the scalar function $\ell(\boldsymbol{\theta})$ with respect to the $b \times 1$ argument $\boldsymbol{\theta}$ is the $1 \times b$ row vector $\ell'(\boldsymbol{\theta}) = \partial\ell(\boldsymbol{\theta})/\partial\boldsymbol{\theta} = (\partial\ell/\partial\theta_1, \ldots, \partial\ell/\partial\theta_b)$. However, we prefer to define the likelihood equations in terms of the $b \times 1$ column-vector score function $\boldsymbol{S}(\boldsymbol{\theta})$. The Appendix (p. 531) contains a summary of the vector calculus notation and results used in the remainder of this book.

2.2 Likelihood Construction

2.2.1 Discrete IID Random Variables

The use of the likelihood function in parameter estimation is easiest to understand in the case of discrete random variables. Suppose that each of the n iid random variables in the sample $Y_1, \ldots, Y_n$ have probability mass function

$$f(y;\boldsymbol{\theta}) = P_{\boldsymbol{\theta}}(Y_1 = y), \qquad y = y_1, y_2, \ldots.$$

For example, $f(y;\boldsymbol{\theta})$ might be the Poisson probability mass function with $\theta = \lambda$, and

$$f(y;\lambda) = \frac{\lambda^y e^{-\lambda}}{y!}, \qquad y = 0, 1, \ldots.$$

Because a probability mass function evaluated at y is just the probability that the event $\{Y = y\}$ occurs, the likelihood is

$$L(\boldsymbol{\theta} \mid \boldsymbol{Y}) = \prod_{i=1}^{n} f(Y_i;\boldsymbol{\theta}) = \prod_{i=1}^{n} P_{\boldsymbol{\theta}}(Y_i^* = Y_i \mid Y_i) \tag{2.2}$$

where $Y_1^*, \ldots, Y_n^*$ are iid random variables having the same distribution as, but mutually independent of $Y_1, \ldots, Y_n$. In other words the likelihood is *the probability of getting the sample actually obtained (or to be obtained)*. We use $Y_1^*, \ldots, Y_n^*$ so that we can use the language of probability rather than the notation of probability mass functions.

Why should one choose $\widehat{\boldsymbol{\theta}}$ to maximize $L(\boldsymbol{\theta} \mid \boldsymbol{Y})$? In effect, $\widehat{\boldsymbol{\theta}}_{\text{MLE}}$ is the parameter value that makes the observed data most probable, or most likely, under the assumed family of probability mass functions. Thus $f(y;\widehat{\boldsymbol{\theta}}_{\text{MLE}})$ is the probability mass function that is most consistent with the observed data.

Example 2.1 (Fetal lamb movements). Leroux and Puterman (1992, p. 546) give data on counts of movements in five-second intervals of one fetal lamb ($n = 240$ intervals):

No. of movements:	0	1	2	3	4	5	6	7
Counts :	182	41	12	2	2	0	0	1

Note that this table is just a concise way to exhibit the 240 ordered observed values $0, \ldots, 0, 1, \ldots, 1, 2, \ldots, 2, 3, 3, 4, 4, 7$. Under the assumption that movements follow a Poisson process in time, interval counts are independent Poisson random variables, and the likelihood is

$$L(\boldsymbol{\theta} \mid \boldsymbol{Y}) = \prod_{i=1}^{n} f(Y_i; \lambda) = \prod_{i=1}^{n} \lambda^{Y_i} e^{-\lambda} / Y_i! = \lambda^{n\overline{Y}} e^{-n\lambda} \left(\prod_{i=1}^{n} Y_i! \right)^{-1},$$

where $\boldsymbol{\theta}$ consists of just the scalar λ. Equating the derivative of the log likelihood with respect to λ to zero and solving results in $\widehat{\lambda}_{\text{MLE}} = \overline{Y} = 86/240 = .358$. Using this estimate, the expected counts at y are $nf(y; \widehat{\lambda}_{\text{MLE}})$:

No. of movements	0	1	2	3	4	5	6	7
Observed Counts	182	41	12	2	2	0	0	1
Expected-Poisson	167.7	60.1	10.8	1.3	0.1	0	0	0

Note that the Poisson model under predicts at $y = 0$ and over predicts at $y = 1$. A more formal evaluation of this observation is provided by a chi-squared goodness-of-fit test of the null hypothesis that a Poisson model is correct:

$$\chi^2(\text{GOF}) = \sum_{i=1}^{4} \frac{(O_i - E_i)^2}{E_i} = 16.6, \tag{2.3}$$

based on grouping cells with movements > 2 into a fourth group. The p-value is .00025 using the χ^2 distribution with $4 - 1 - 1 = 2$ degrees of freedom, suggesting the Poisson model is inappropriate. The p-value constructed in this manner is not entirely correct because it does not account for the grouping of cells 3–7. The problem of grouping is described in Example 2.5 (p. 34) where it is shown that a more appropriate p-value is .0001.

To account for the inflated number of zeroes, consider the mixture of a point mass at 0 and the Poisson: $P(Y_1 = y) = pI(y = 0) + (1 - p)f(y; \lambda)$, where $I(A)$ is the indicator function of the event A, so that $I(y = 0) = 1$ for $y = 0$ and $I(y = 0) = 0$ for $y \neq 0$. Alternatively,

$$P(Y_1 = y) = \begin{cases} p + (1 - p)f(0; \lambda), & y = 0, \\ (1 - p)f(y; \lambda), & y = 1, 2, \ldots. \end{cases}$$

This is called a zero-inflated Poisson model, or ZIP for short. Letting n_0 denote the number of zeroes in the sample, the likelihood is

$$L(\lambda, p \mid \boldsymbol{Y}) = [p + (1 - p)f(0; \lambda)]^{n_0} (1 - p)^{n - n_0} \prod_{I(Y_i > 0)} f(Y_i; \lambda).$$

Using the `nlm` function in R to minimize $-\ell(\boldsymbol{\theta})$ with respect to $\boldsymbol{\theta} = (\lambda, p)^T$ results in $\widehat{p}_{\text{MLE}} = .577$ and $\widehat{\lambda}_{\text{MLE}} = .847$. The ZIP model expected counts are

No. of movements	0	1	2	3	4	5	6	7
Observed Counts	182	41	12	2	2	0	0	1
Expected-ZIP	182	36.9	15.6	4.4	0.9	.2	0	0

and the chi-squared goodness of fit statistic is 1.35 with approximate p-value = .24. Thus there is little evidence of lack of fit with the ZIP model. ♦

2.2.2 *Multinomial Likelihoods*

The multinomial distribution is a generalization of the binomial distribution that arises often in practice. Recall that the binomial distribution results from counting the number of successes in n independent trials where each trial has only two possible outcomes, success or failure. The probability of a success is usually denoted p. For a multinomial distribution , there are $k \geq 2$ possible outcomes. One way to conceptualize multinomial data is to consider independently tossing n balls into k different urns, where p_i is the probability of the ball landing in the ith urn on each toss, $i = 1, \ldots, k$. Doing so results in N_i balls in the ith urn, $i = 1, \ldots, k$, $\sum_{i=1}^k N_i = n$. We say that $(N_1, \ldots, N_k)$ are distributed as multinomial$(n; p_1, p_2, \ldots, p_k)$. The likelihood is

$$L(\boldsymbol{p} \mid N_1, \ldots, N_k) = \frac{n!}{N_1!N_2!\cdots N_k!} p_1^{N_1} p_2^{N_2} \cdots p_k^{N_k}, \tag{2.4}$$

where $\sum_{i=1}^k N_i = n$, $\sum_{i=1}^k p_i = 1$, and $0 < p_i < 1$, $i = 1, \ldots, k$. Note that for $k = 2$ we can identify p of the binomial with p_1 and $1 - p$ with p_2. In fact, because $p_k = 1 - \sum_{i=1}^{k-1} p_i$, there are only $k - 1$ freely varying parameters in the multinomial density and likelihood. Although the symmetry in (2.4, p. 32) is aesthetically appealing, it is almost always better in applications and theory to rewrite (2.4, p. 32) with $1 - \sum_{i=1}^{k-1} p_i$ substituted for p_k. For example, when finding maximum likelihood estimators, that substitution avoids the need for constrained optimization.

Note that any subset of $(N_1, \ldots, N_k)$ also has a multinomial distribution . In particular, N_i has a binomial distribution with parameters p_i and n. Thus $\text{E}(N_i) = np_i$ and $\text{Var}(N_i) = np_i(1 - p_i)$. The covariance between N_i and N_j for $i \neq j$ is $-np_i p_j$. This can be shown by first noting that when $n = 1$,

$$\text{E}(N_i - p_i)(N_j - p_j) = P(N_i = 1, N_j = 1) - p_i p_j = -p_i p_j$$

because when $n = 1$, both $N_i = 1$ and $N_j = 1$ cannot occur. The result for general n follows by further noting that a multinomial$(n; p_1, p_2, \ldots, p_k)$ vector is equal in distribution to a sum of n independent multinomial$(n = 1; p_1, p_2, \ldots, p_k)$ vectors.

The maximum likelihood estimator of p_i is the sample proportion N_i/n. More interesting multinomial likelihoods arise when the p_i are modeled as a function of a lesser number of parameters $\theta_1, \ldots, \theta_m$, $m < k - 1$, as shown in the following examples.

Example 2.2 (Hardy-Weinberg equilibrium). Suppose that the genotypes of a certain gene with two alleles a and A are labeled AA, Aa, and aa. If a sample of n individuals results in N_{AA} individuals with genotype AA, N_{Aa} with genotype Aa, and N_{aa} individuals with genotype aa, then (N_{AA}, N_{Aa}, N_{aa}) are multinomial$(n; p_{AA}, p_{Aa}, p_{aa})$. Thus, $k{=}3$ and $p_1{=}p_{AA}$, $p_2{=}p_{Aa}$, and $p_3{=}p_{aa}$. Now, Hardy-Weinberg Equilibrium refers to a random mating scheme whereby these genotype probabilities can be written in terms of a single allele probability p_A (the population proportion of A alleles): $p_{AA} = p_A^2$, $p_{Aa} = 2p_A(1 - p_A)$, $p_{aa} = (1 - p_A)^2$. Substituting these modeled probabilities into (2.4) gives the likelihood for p_A:

$$L(p_A \mid N_{AA}, N_{Aa}, N_{aa})$$

$$= \frac{n!}{N_{AA}!N_{Aa}!N_{aa}!}(p_A^2)^{N_{AA}} \left[2p_A(1 - p_A)\right]^{N_{Aa}} \left[(1 - p_A)^2\right]^{N_{aa}}.$$

Maximizing this likelihood leads to $\widehat{p}_A = (2N_{AA} + N_{Aa})/(2n)$, the sample frequency of A alleles. Numerous other multinomial models arising in population genetics are presented in Weir (1996). ♦

Example 2.3 (Capture-recapture removal design). To estimate fish survival during a specified length of time (e.g., one month), a common approach in capture-recapture methodology is to use a removal design. Suppose that in a closed population (e.g., a lake), n fish are caught, tagged, and released back into the lake. One month later N_1 of the tagged fish are removed from the lake (untagged fish may also be removed at the same time). Assuming that the probability that a tagged fish survives for a month is s, and the probability that it is removed at the end of month one is p (given it survived until that point), then the probability of one of the original tagged fish being removed at the end of month one is sp. Continuing, at the end of the second month N_2 tagged fish are removed, and the probability of an original tagged fish being removed is then $s^2(1 - p)p$, that is, the probability of surviving two months (s^2) times the probability of not being removed at the end of month one $(1 - p)$ times the probability of removal at the end of two months. Suppose that we continue this process one more time and let $N_4 = n - N_1 - N_2 - N_3$ be the fish that were not removed. Then (N_1, N_2, N_3, N_4) are multinomial$(n, p_1 = sp, p_2 = s^2(1 - p)p, p_3 = s^3(1 - p)^2 p, p_4 = 1 - sp - s^2(1 - p)p - s^3(1 - p)^2 p)$. Multinomial models from the capture-recapture literature are studied in Brownie et al. (1985). ♦

Example 2.4 (Two-way contingency tables). Consider a two-way table with r rows and c columns and cell counts N_{ij}, $i = 1, \ldots, r$, $j = 1, \ldots, c$. The usual situation is that $(N_{11}, \ldots, N_{rc})$ are multinomial$(n; p_{11}, \ldots, p_{rc})$. The hypothesis of independence of rows and columns is $p_{ij} = p_{i+}p_{+j}$, $i = 1, \ldots, r$, $j = 1, \ldots, c$, where $p_{i+} = \sum_{j=1}^{c} p_{ij}$ and $p_{+j} = \sum_{i=1}^{r} p_{ij}$ are the marginal probabilities. If the table is *square*, $r = c$, then another hypothesis of interest is *symmetry*, $p_{ij} = p_{ji}$ for $i \neq j$. These and many other models for contingency tables appear in Agresti (2002). ♦

Multinomial models can be used to describe samples from any discrete distribution with a finite support. For example, suppose that $Y_1, \ldots, Y_n$ are each from a binomial distribution with probability of success θ and 4 trials. So the possible values of Y_i are $0, 1, 2, 3, 4$. If we let N_j be the number of the $Y_1, \ldots, Y_n$ equal to j, $j = 0, \ldots, 4$, then $(N_0, N_1, N_2, N_3, N_4)$ has a multinomial distribution with $p_j = \binom{4}{j}\theta^j(1-\theta)^{4-j}$, $j = 0, \ldots, 4$. This idea is important enough to set apart:

Key Concept: For an iid sample $Y_1, \ldots, Y_n$ from any discrete distribution with finite support, the data can be viewed as multinomial data where the counts are the number of occurrences of each of the finite number of possible values, and the likelihood is obtained from (2.4, p. 32). There are not many common examples of finite discrete distributions, but often such structure is imparted by grouping the large values in a data set as illustrated in the following example related to Example 2.1 (p. 31).

Example 2.5 (Grouped likelihood: discrete data). The goodness-of-fit analysis in Example 2.1 (p. 31) to test whether the Poisson is a reasonable model for the fetal lamb data uses expected counts based on the full maximum likelihood estimate $\widehat{\lambda}_{\text{MLE}} = .358$. Calculating expected counts in this way is expedient, but not supported by asymptotic theory. Chernoff and Lehmann (1954) showed that the limiting null-hypothesis distribution of the chi-square statistic with counts based on the full MLE converges to the sum of a chi-squared random variable with $k - p - 1$ degrees of freedom and an additional weighted sum of p chi-squared random variables with one degree of freedom, where p is the number of parameters estimated. Thus using χ^2_{k-p-1} critical values for such a chi-square statistic is liberal in the sense that the probability of rejection under the null hypothesis exceeds the nominal significance level. A conservative approach is to use χ^2_{k-1} critical values. For the fetal lamb data the conservative analysis uses χ^2_3 critical values, and the conservative p-value corresponding to the $\chi^2(\text{GOF}) = 16.6$ in Example 2.1 (p. 31) is .00087, c.f., to the liberal p-value $= .00025$.

An approach that leads to the correct asymptotic chi-squared distribution with $k - p - 1$ degrees of freedom for the chi-square statistic is to use expected counts based on the maximum likelihood estimator from the grouped data. The grouped-data multinomial likelihood for the fetal lamb data is

$$L_{\text{G}}(\lambda) = \frac{240!}{182!41!12!5!}\left[\frac{\lambda^0 e^{-\lambda}}{0!}\right]^{182}\left[\frac{\lambda^1 e^{-\lambda}}{1!}\right]^{41}\left[\frac{\lambda^2 e^{-\lambda}}{2!}\right]^{12}\left[\sum_{i=3}^{\infty}\frac{\lambda^i e^{-\lambda}}{i!}\right]^{5},$$

and is maximized at $\widehat{\lambda}_{\text{GMLE}} = .335$. Using $\widehat{\lambda}_{\text{GMLE}}$ to calculate expected counts in the chi-squared statistic yields $\chi^2_{\text{G}}(\text{GOF}) = 18.4$ with p-value $= .00010$ from the χ^2_2 distribution. Note that in a strict sense using the grouped-data likelihood contradicts the principle that the likelihood is always the density of the observed data. However, the grouped-data likelihood is correct for the *grouped* observed data.

There are other variations on the grouped-data likelihood approach that result in goodness-of-fit statistics having the standard chi-squared distribution. For example, one can minimize the chi-squared statistic

$$\chi^2(\lambda) = \sum_{i=1}^{4}\frac{\{O_i - E_i(\lambda)\}^2}{E_i(\lambda)}$$

with respect to λ, obtaining $\widehat{\lambda} = .381$ with minimum value 16.0. Recall that $E_i(\lambda)$ is just the probability of falling in the ith category, i.e., $0, 1, 2$, or > 2 under the Poisson(λ) model. The estimator $\widehat{\lambda}$ is asymptotically equivalent to $\widehat{\lambda}_{\text{GMLE}}$. Also, comparing the minimum chi-squared statistic value 16.0 to a χ^2_2 distribution (resulting in p-value = .00033) is justified asymptotically. In fact, there is a large class of statistics that can be minimized to obtain estimators that are asymptotically equivalent to $\widehat{\lambda}_{\text{GMLE}}$ (Cressie and Read 1984; Read and Cressie 1988).

The asymptotically equivalent estimators of λ for a fixed grouping structure are less efficient than the full maximum likelihood estimator $\overline{Y}$. However, grouping often produces estimators that are more *robust* (insensitive) to outliers than the full MLE (Simpson 1987, Lindsay 1994).

The multiple analyses of the fetal lamb data illustrate the limited ability of large sample distribution approximations (even when justified by asymptotic theory) to distinguish between small p-values. An alternative approach described in Chapter 11 that is often more accurate in finite samples is based on the *parametric bootstrap*. Briefly, a parametric bootstrap p-value is obtained by randomly sampling B times from a Poisson distribution with $\lambda = \widehat{\lambda} = .358$, computing $\chi^2(\text{GOF})$ for each sample, and calculating the fraction of the $\chi^2(\text{GOF})$ test statistics that exceed the observed $\chi^2(\text{GOF}) = 16.6$.

Grouping to obtain multinomial data with parameter-dependent cell probabilities is a generally-applicable strategy. The resulting grouped-data MLE is generally less efficient than the full-data MLE. However, there are certain other advantages as noted above, especially those related to robustness to extremes. Another is to construct goodness-of-fit statistics for parametric models of continuous data which we illustrate in Section 3.5, (p. 149). ♦

2.2.3 Continuous IID Random Variables

We now consider likelihoods for iid sample data $Y_1, \dots, Y_n$ for which Y_1 is a continuous random variable with density $f(y; \boldsymbol{\theta})$. The following example illustrates a typical application.

Example 2.6 (Hurricane data). For 36 hurricanes that had moved far inland on the East Coast of the U.S. in 1900–1969, Larsen and Marx (2001, p. 320) give maximum 24-hour precipitation levels during the time they were over mountains:

```
> hurr.rain
 [1]  31.00  2.82  3.98  4.02  9.50  4.50 11.40 10.71
 [9]   6.31  4.95  5.64  5.51 13.40  9.72  6.47 10.16
[17]   4.21 11.60  4.75  6.85  6.25  3.42 11.80  0.80
[25]   3.69  3.10 22.22  7.43  5.00  4.58  4.46  8.00
[33]   3.73  3.50  6.20  0.67
```

A histogram suggests that the gamma density,

$$f(y; \alpha, \beta) = \frac{1}{\Gamma(\alpha)\beta^{\alpha}} y^{\alpha-1} e^{-y/\beta},$$

is a reasonable candidate model for these data. The gamma likelihood is

$$L(\boldsymbol{\theta} \mid \boldsymbol{Y}) = \prod_{i=1}^{n} \frac{1}{\Gamma(\alpha)\beta^{\alpha}} Y_i^{\alpha-1} e^{-Y_i/\beta} = \{\Gamma(\alpha)\}^{-n} \beta^{-n\alpha} \left\{\prod Y_i\right\}^{\alpha-1} e^{-\sum Y_i/\beta},$$

with log likelihood

$$\ell(\boldsymbol{\theta}) = -n \log \Gamma(\alpha) - n\alpha \log \beta + (\alpha - 1) \sum \log Y_i - \frac{\sum Y_i}{\beta},$$

where $\boldsymbol{\theta} = (\alpha, \beta)^T$. The log likelihood surface is shown in Figure 2.1. To maximize the log likelihood, we use the *R* minimization function `nlm`:

```
dgam<-function(x,alpha,beta=1.){dgamma(x/beta,alpha)
                                        /beta}
glike<-function(theta,x){
         -sum(log(dgam(x,theta[1],theta[2])))}
nlm(glike,c(1.59,4.458),x=hurr.rain)
$minimum
[1] 102.3594
$estimate
[1] 2.187214 3.331862
```

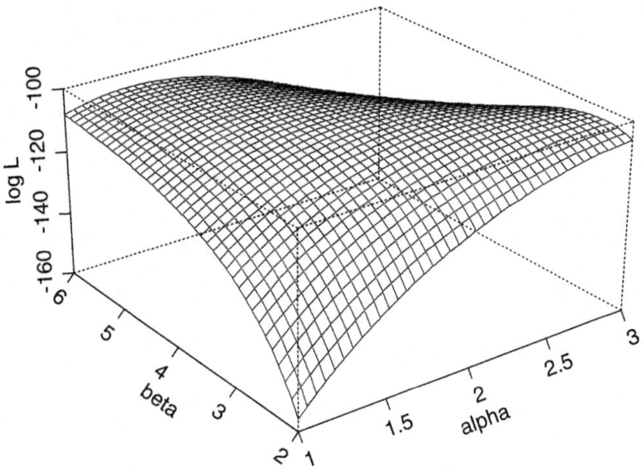

Fig. 2.1 Gamma log likelihood for the hurricane data

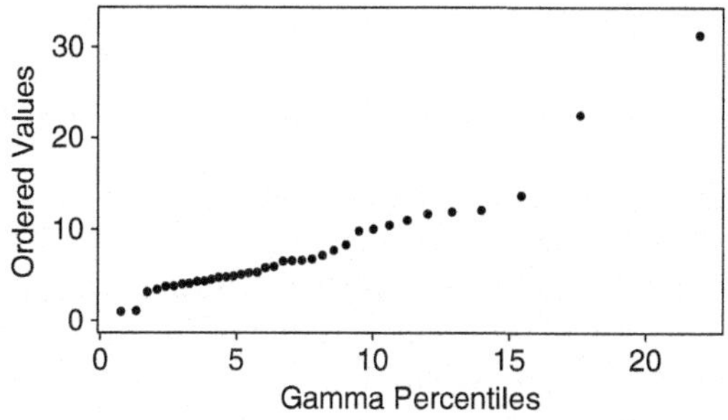

Fig. 2.2 Gamma QQ plot for the hurricane data

The starting values 1.59 and 4.58 are the method of moment estimators that solve $\overline{X} = 7.29 = \alpha\beta$ and $s^2 = 33.41 = \alpha\beta^2$. To check the fit we plot sample quantiles versus the quantiles of the fitted model (QQ plot):

```
qgam<-function(q,alpha,beta=1.){qgamma(q,alpha)*beta}
plot(qgam(ppoints(hurr.rain),2.187,3.332),
     sort(hurr.rain))
```

A QQ plot should be approximately linear if the model fits the data well. The plot in Figure 2.2 shows that the largest values in the sample are somewhat larger than to be expected from a gamma distribution. Thus the visual evidence suggests that a gamma distribution may not be an adequate model for these data. ♦

2.2.3a The Connection Between Discrete and Continuous Likelihoods

The continuous-data likelihood $L(\boldsymbol{\theta} \mid \boldsymbol{Y}) = \prod_{i=1}^{n} f(Y_i; \boldsymbol{\theta})$ looks the same as that for discrete data, but it is not a probability as it is for discrete data; see (2.2, p. 30). A continuous distribution assigns zero probability to any single point. Thus for continuous data the probability of any particular iid sample is zero, and the probabilistic intuition underlying maximum likelihood that is readily apparent with discrete data (2.2, p. 30) is not as evident for continuous data. However, the intuition underlying maximum likelihood is revealed by considering certain normalized limits of probabilities.

We start by recalling the definition of the derivative $g'(x)$ of a function $g(x)$,

$$g'(x) = \lim_{h \to 0} \frac{g(x+h) - g(x)}{h},$$

provided the limit exists (meaning that the limits as $h \to 0^+$ and $h \to 0^-$ exist and are equal). When the derivative exists a second, equivalent definition of the derivative of g at x is

$$g'(x) = \lim_{h \to 0+} \frac{g(x+h) - g(x-h)}{2h}. \tag{2.5}$$

To see the equivalence just add and subtract $g(x)$ and take limits. The two-sided definition of $g'(x)$ is sometimes more useful in applied work than the usual one-sided definition, e.g., when calculating derivatives numerically. We use it here to show that the probabilistic intuition for maximum likelihood with discrete data holds for continuous data as well.

Consider the two-sided derivative definition applied to a differentiable distribution function. If F is the distribution function of a continuous random variable Y having density $f(y)$, then wherever f is continuous,

$$f(y) = \lim_{h \to 0+} \frac{F(y+h) - F(y-h)}{2h} = \lim_{h \to 0+} \frac{P(Y \in (y-h, y+h])}{2h}. \tag{2.6}$$

In fact it is sometimes appropriate to use the limit in the right-hand side above as the definition of the density. This is because densities are not unique and certain versions of the density do not result in sensible likelihoods. However, the version of the density obtained from (2.6) generally does lead to a sensible likelihood. For example, consider the logistic distribution $F(y) = \{1 + \exp(-y)\}^{-1}$. The usual form of the density associated with F is $f(y) = \exp(-y)\{1 + \exp(-y)\}^{-2}$. However, $f_*(y) = f(y)I(y \neq 2) + 0.3I(y = 2)$ is nonnegative and for all y

$$\int_{-\infty}^{y} f_*(t)\, dt = F(y),$$

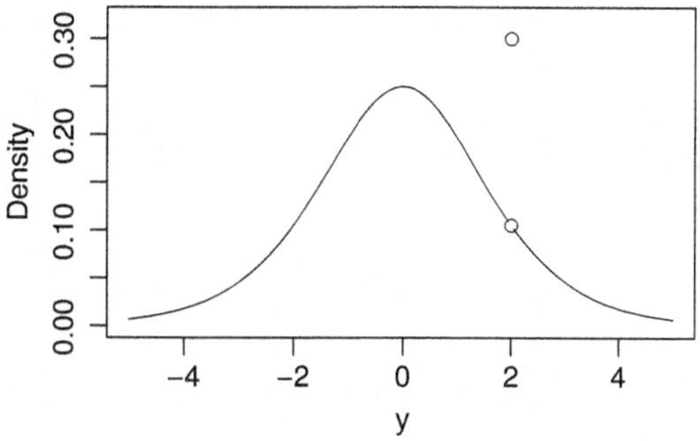

Fig. 2.3 Modified logistic density

so that $f_*(y)$ is also a valid probability density associated with F. A graph of $f_*(y)$ appears in Figure 2.3. Note that in a sample of size $n = 1$ from the location model $f(y;\mu) = f_*(y-\mu)$, the f_*-likelihood is maximized at $Y_1 - \mu = 2$. Thus $\widehat{\mu}_{*\text{MLE}} = Y_1 - 2$ instead of the usual $\widehat{\mu}_{\text{MLE}} = Y_1$. This simple example illustrates that all versions of a density are not equally useful for maximum likelihood estimation. The key point here is that the cumulative distribution function is unique, and using it to define the density via (2.6) generally results in the "natural" or "regular" version of the density function. This is true for the logistic distribution and also for any other distribution that is continuously differentiable. Using (2.6) to define a density when the continuous distribution function is not continuously differentiable is not as straightforward.

There is a version of (2.6) for bivariate data as well. If X and Y are both continuous and have joint distribution function $F_{X,Y}$ and density function $f_{X,Y}$, then wherever $f_{X,Y}$ is continuous

$$\begin{aligned} f_{X,Y}(x,y) &= \lim_{h\to 0+}\left(\frac{1}{2h}\right)^2\Big\{F_{X,Y}(x+h,y+h) - F_{X,Y}(x-h,y+h) \\ &\qquad - F_{X,Y}(x+h,y-h) + F_{X,Y}(x-h,y-h)\Big\} \\ &= \lim_{h\to 0+}\frac{P(X\in(x-h,x+h],\,Y\in(y-h,y+h])}{(2h)^2}. \end{aligned}$$

Also, if X is discrete and Y is continuous, then because the events $\{X \in (x-h, x+h]\}$ and $\{X = x\}$ are equivalent for h sufficiently small, we only need to normalize for the continuous component, and the joint mixed density/mass function is obtained via

$$f_{X,Y}(x,y) = \lim_{h\to 0+}\frac{P(X = (x-h,x+h],\,Y\in(y-h,y+h])}{2h}.$$

Having introduced the symmetric definition of derivative (2.5, p. 38) as a useful way to think about densities of continuous random variables, we are now ready to talk about likelihoods. Assume that $f(y;\boldsymbol{\theta})$ is the derivative of the distribution function $P_{\boldsymbol{\theta}}(Y_1 \le y) = F(y;\boldsymbol{\theta})$. Then we have

$$
\begin{aligned}
L(\boldsymbol{\theta} \mid \boldsymbol{Y}) &= \prod_{i=1}^{n} f(Y_i;\boldsymbol{\theta}) \\
&= \prod_{i=1}^{n} \lim_{h\to 0+} \left\{ \frac{F(Y_i + h;\boldsymbol{\theta}) - F(Y_i - h;\boldsymbol{\theta})}{2h} \right\} \\
&= \lim_{h\to 0+} \prod_{i=1}^{n} \left\{ \frac{F(Y_i + h;\boldsymbol{\theta}) - F(Y_i - h;\boldsymbol{\theta})}{2h} \right\} \\
&= \lim_{h\to 0+} \left(\frac{1}{2h}\right)^{n} \prod_{i=1}^{n} P_{\boldsymbol{\theta}}(Y_i^* \in (Y_i - h, Y_i + h] \mid Y_i),
\end{aligned}
$$

where again $Y_1^*, \ldots, Y_n^*$ are iid with the same distribution and independent of $Y_1, \ldots, Y_n$. Thus, for small h the likelihood is approximately equal to a constant $(2h)^{-n}$ multiple of the probability of a $2h$-neighborhood of the observed data. In other words, the likelihood is proportional to the probability of obtaining a new sample that is close to the sample actually obtained. So, similar to the case of discrete data, maximizing the $2h$-likelihood to obtain $\widehat{\boldsymbol{\theta}}_h$ is the same as choosing the density, $f(y;\widehat{\boldsymbol{\theta}}_h)$, that assigns the greatest probability to a neighborhood of the observed data $\boldsymbol{Y}$. Of course this makes sense only if $\widehat{\boldsymbol{\theta}}_h$ converges to $\widehat{\boldsymbol{\theta}}_{\text{MLE}}$ as $h \to 0$ as is generally the case.

In the discrete case, note that the likelihood is obtained via a similar limiting process, but without the constant factor $(2h)^{-1}$,

$$
\begin{aligned}
\lim_{h\to 0+} \prod_{i=1}^{n} \{F(Y_i + h;\boldsymbol{\theta}) - F(Y_i - h;\boldsymbol{\theta})\} &= \prod_{i=1}^{n} \{F(Y_i^+;\boldsymbol{\theta}) - F(Y_i^-;\boldsymbol{\theta})\} \\
&= \prod_{i=1}^{n} f(Y_i;\boldsymbol{\theta}) \\
&= L(\boldsymbol{\theta} \mid \boldsymbol{Y}).
\end{aligned}
$$

Here it helps to remember that the distribution function of a discrete random variable has jumps of size $f(y;\boldsymbol{\theta})$ at the possible values y. Note also that at the jump points y, $F(y^+;\boldsymbol{\theta})$ and $F(y^-;\boldsymbol{\theta})$ denote the limits of $F(t;\boldsymbol{\theta})$ as t converges to y from the right and left, respectively.

The $2h$-method of calculating likelihoods allows us to construct likelihoods for more complicated problems where a mathematically rigorous definition of likelihood would require measure theoretic concepts. Our general working definition of the likelihood for independent data $Y_1, \ldots, Y_n$, where Y_i has distribution function $F_i(y; \boldsymbol{\theta})$, is then

$$L(\boldsymbol{\theta} \mid \boldsymbol{Y}) = \lim_{h \to 0+} \left(\frac{1}{2h}\right)^m \prod_{i=1}^{n} \{F_i(Y_i + h; \boldsymbol{\theta}) - F_i(Y_i - h; \boldsymbol{\theta})\}, \qquad (2.7)$$

where $1 \le m \le n$ depends on the number of continuous components in the data. We now illustrate this definition with a class of models having both discrete and continuous components.

2.2.4 *Mixtures of Discrete and Continuous Components*

Data Y such as daily rainfall, or snowfall, or the weight of fish caught on a commercial fishing expedition, often have a number of zeroes (no rain, no snow, no luck), and the amounts greater than zero are best modeled by a continuous distribution. Such data are often modeled by a distribution

$$F_Y(y; p, \boldsymbol{\theta}) = P(Y \le y) = pI(0 \le y) + (1 - p)F_T(y; \boldsymbol{\theta}), \qquad (2.8)$$

that is a mixture of a point mass at zero (note that $I(0 \le y)$ is the distribution function of a degenerate random variable at zero), and a continuous positive random variable T having distribution function $F_T(y; \theta)$, e.g., a Weibull random variable. Alternatively, (2.8) can be expressed as

$$F_Y(y; p, \boldsymbol{\theta}) = \begin{cases} 0, & y < 0, \\ p, & y = 0, \\ p + (1 - p)F_T(y; \boldsymbol{\theta}), & y > 0. \end{cases}$$

We now obtain the likelihood of an iid sample $Y_1, \ldots, Y_n$ from the distribution (2.8) using the definition (2.7, p. 41). Letting n_0 be the number of zeroes in the data, and with $m = n - n_0$ the number of non-zero Y_i, we have

$$\begin{aligned} L(\boldsymbol{\theta} \mid \boldsymbol{Y}) &= \lim_{h \to 0+} \left(\frac{1}{2h}\right)^m \prod_{i=1}^{n} \{F_Y(Y_i + h; p, \boldsymbol{\theta}) - F_Y(Y_i - h; p, \boldsymbol{\theta})\} \\ &= \lim_{h \to 0+} \{F_Y(h; p, \boldsymbol{\theta}) - F_Y(-h; p, \boldsymbol{\theta})\}^{n_0} \\ &\quad \times \lim_{h \to 0+} \prod_{Y_i > 0} \left\{ \frac{F_Y(Y_i + h; p, \boldsymbol{\theta}) - F_Y(Y_i - h; p, \boldsymbol{\theta})}{2h} \right\} \\ &= \lim_{h \to 0+} \{p + (1 - p)F_T(h; \boldsymbol{\theta})\}^{n_0} \end{aligned}$$

$$\times \lim_{h \to 0+} \prod_{Y_i > 0} \left\{ \frac{(1-p)F_T(Y_i + h; \boldsymbol{\theta}) - (1-p)F_T(Y_i - h; \boldsymbol{\theta})}{2h} \right\}$$

$$= p^{n_0}(1-p)^{n-n_0} \prod_{Y_i > 0} f_T(Y_i; \boldsymbol{\theta}).$$

In the second step the zero Y_i are replaced by 0 and factored out of the full product leaving only the product over the positive Y_i. The third step uses the fact that for h sufficiently small, $Y_i - h > 0$ for the nonzero Y_i and the discrete part of the distribution function subtracts out of F_Y, that is $F_Y(Y_i + h; p, \boldsymbol{\theta}) - F_Y(Y_i - h; p, \boldsymbol{\theta}) = (1-p)F_T(Y_i + h; \boldsymbol{\theta}) - (1-p)F_T(Y_i - h; \boldsymbol{\theta})$ for $Y_i > 0$. The likelihood is intuitive consisting of a Bernoulli component for the n_0 zeroes and a continuous component for the $n - n_0$ nonzero values. Consideration of the log likelihood makes it apparent that $\widehat{p}_{\text{MLE}} = n_0/n$ and that $\widehat{\boldsymbol{\theta}}_{\text{MLE}}$ is obtained in the usual way for a sample of size $n - n_0$ having density $f_T(y; \boldsymbol{\theta})$.

Example 2.7 (Snowfall in Seattle). Siegel (1985) gives total January snowfall in inches for Seattle from 1906 to 1960:

```
> snow
 [1]  4.1  2.1 23.3  3.1  0.0  0.0  0.0  0.0  0.0
[10]  0.0  0.3 12.2  4.1  6.3  2.6  0.0  4.8 10.9
[19]  1.0  3.8  6.6 10.1  0.0 11.5  0.0  1.0  2.3
[28]  1.1  1.3 18.4  0.0  0.0  0.0 14.6  0.7  0.0
[37]  0.3 11.3  0.0  0.0  0.0  7.0 16.9  5.0  2.5
[46]  0.0  0.0  0.0  6.5 17.4  0.8  0.0 31.0  0.0
[55]  1.0
```

Because there are 21 months of January with no snow in the 55 years, $\widehat{p}_{\text{MLE}} = 21/55$. A Weibull distribution was fit to the nonzero snowfall amounts. Because −log(Weibull) has a location-scale extreme value distribution, we display in the left panel of Figure 2.4 an extreme value QQ plot of the −log transformed data. The right panel of Figure 2.4 displays the estimated Weibull distribution function with the empirical distribution function of the original data jittered to avoid ties. Neither plot indicates a problem with the fit.

♦

2.2.5 *Proportional Likelihoods*

Our working definition (2.7, p. 41) for the likelihood is a bit vague because the factor $(2h)^{-m}$ depends on the number of discrete and continuous components. This vagueness should not be overly disconcerting, though, because likelihoods are equivalent for point estimation as long as they are proportional and the constant of proportionality does not depend upon unknown parameters. In fact, some

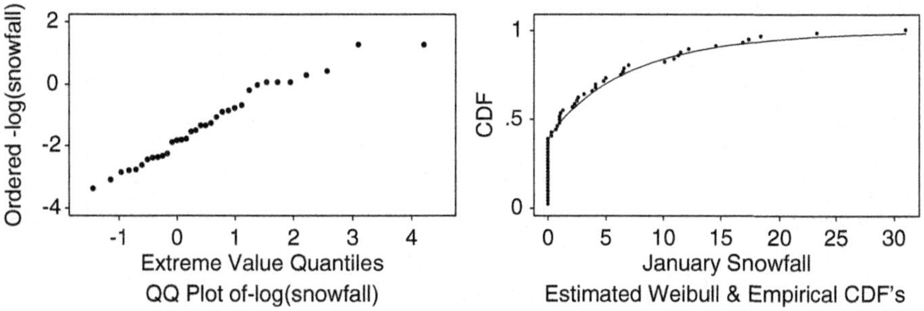

Fig. 2.4 Assessing fit of Weibull to nonzero snowfall data

statisticians make the much stronger statement that all inference about $\boldsymbol{\theta}$ should be the same as long as likelihoods are proportional. The strong Likelihood Principle is considered again at the end of this subsection. We proceed by looking at a series of examples.

Suppose that $Y_1, \ldots, Y_n$ are iid from a continuous distribution with density $f_Y(y; \boldsymbol{\theta})$. Consider the transformed data $X_i = g(Y_i)$, $i = 1, \ldots, n$, where g is a known, increasing, continuously differentiable function. Because g is one-to-one, we can reconstruct Y_i from X_i and vice versa. Thus the two data sets $\{Y_1, \ldots, Y_n\}$ and $\{X_1, \ldots, X_n\}$ are equivalent in the sense that each contains exactly the same information. Because of this equivalence, it is intuitive that likelihood inference based on one data set should be identical to inference based on the other. We now show that this is the case.

The density of X_i is $f_X(x; \boldsymbol{\theta}) = f_Y(h(x); \boldsymbol{\theta})h'(x)$, where $h = g^{-1}$. Thus the likelihood of the X sample is

$$
\begin{aligned}
L(\boldsymbol{\theta} \mid \boldsymbol{X}) &= \prod_{i=1}^{n} f_Y(h(X_i); \boldsymbol{\theta})h'(X_i) \\
&= \prod_{i=1}^{n} f_Y(Y_i; \boldsymbol{\theta})h'(g(Y_i)) \\
&= \prod_{i=1}^{n} f_Y(Y_i; \boldsymbol{\theta})\frac{1}{g'(Y_i)} \qquad \left(\text{since } \frac{dh(x)}{dx} = \frac{dg^{-1}(x)}{dx} = \frac{1}{g'(g^{-1}(x))}\right) \\
&= L(\boldsymbol{\theta} \mid \boldsymbol{Y}) \left\{ \prod_{i=1}^{n} \frac{1}{g'(Y_i)} \right\}. \qquad (2.9)
\end{aligned}
$$

Note that the two likelihoods are proportional as functions of $\boldsymbol{\theta}$ for all Y_i. This implies that maximum likelihood estimates and likelihood ratio tests are identical whether derived from $L(\boldsymbol{\theta} \mid \boldsymbol{Y})$ or $L(\boldsymbol{\theta} \mid \boldsymbol{X})$.

A similar line of reasoning ensues when there are sufficient statistics that are not one-to-one functions of the data. For example, suppose that we have a sample $Y_1, \ldots, Y_n$ of iid Bernoulli random variables with parameter p. Noting that the density of Y_i is $p^y(1-p)^{1-y}$, the likelihood is $L(p \mid \boldsymbol{Y}) = p^S(1-p)^{n-S}$, where $S = \sum Y_i$. The binomial likelihood of the sufficient statistic S is $L(p \mid S) = \binom{n}{S} p^S(1-p)^{n-S}$. Thus $L(p \mid \boldsymbol{Y})$ and $L(p \mid S)$ differ by the factor $\binom{n}{S}$ that is independent of p for all S, and similar to the transformation example in (2.9), this difference has no consequence for inference on p.

Note that in each of the previous two examples the sampling plan and data are the same, yet we could find multiple different but indisputably equivalent likelihoods in each case (equivalent via their proportionality). However, when different sampling plans give rise to proportional likelihoods, their equivalence is debatable, as illustrated in the next example.

Example 2.8 (Likelihood Principle example). This example is taken from Berger and Wolpert (1984, p. 19–20), who cite Lindley and Phillips (1976). Consider data obtained from two different sampling plans. The first is the usual iid Bernoulli sampling with $n = 12$ and sufficient statistic-likelihood

$$\binom{12}{S} p^S(1-p)^{12-S},$$

where $S = \sum Y_i$. The second sampling plan is *negative binomial*, wherein Bernoulli random variables Y_i are observed until 3 zeroes are obtained. For negative binomial sampling $S = \sum Y_i$ is a sufficient statistic with likelihood,

$$\binom{S+2}{S} p^S(1-p)^3.$$

The ratio of the binomial likelihood to the negative-binomial likelihood is

$$\frac{\binom{12}{S}}{\binom{S+2}{S}}(1-p)^{9-S}.$$

Unlike (2.9), this ratio depends on the parameter p except in the particular case $S = 9$. So for any value of $S \neq 9$, the likelihoods are not proportional and there is no reason to expect the two likelihoods to result in the same inferences.

However, suppose that $S=9$ in both cases, resulting in likelihoods $220p^9(1-p)^3$ and $55p^9(1-p)^3$ that are clearly proportional. Should all inferences be the same for these two different experiments? Certainly $\widehat{p}_{\text{MLE}} = 9/12$ for both, but for testing $H_0 : p = 1/2$ versus $H_a : p > 1/2$, the p-value for the first experiment is .0730 and .0337 for the second (confidence intervals also differ). Some statisticians, primarily Bayesians, maintain that proportional likelihoods contain the same information and thus should necessarily result in the same inferences. They cite the discrepancy in p-values and confidence intervals as evidence that frequentist methods are not logical approaches to inference because they can lead to different conclusions even when the likelihoods are proportional (and thus the inferences are not based solely on the likelihood). For example, the difference in p-values above is due to the fact that their calculation depends on the null probability mass functions and the sample spaces for S ($\{0, 1, \ldots, 12\}$ for binomial sampling and $\{0, 1, \ldots\}$ for negative binomial sampling). The key point is that the p-value calculations do not depend solely on the observed likelihood.

The belief that proportional likelihoods should lead to equivalent inferences is canonized in the *Likelihood Principle* here reproduced verbatim from Berger and Wolpert (1984, p. 19) (note that they use x where we have used Y or S):

> THE LIKELIHOOD PRINCIPLE. All the information about $\boldsymbol{\theta}$ from an experiment is contained in the likelihood function for the actual observation x. Two likelihood functions for $\boldsymbol{\theta}$ (from the same or different experiments) contain the same information about $\boldsymbol{\theta}$ if they are proportional to each other.

Consideration of the philosophy behind The Likelihood Principle or of the controversies surrounding it is not addressed in this book. We merely note that proportional likelihoods always lead to the same $\widehat{\boldsymbol{\theta}}_{\text{MLE}}$ and contain the same information *within the same experiment*, as illustrated by the transformed data and sufficient statistic examples. ♦

2.2.6 The Empirical Distribution Function as an MLE

In this next example, we essentially ignore the factor $(2h)^{-m}$ in deriving the maximum likelihood estimator. Suppose that $Y_1, \ldots, Y_n$ are iid with distribution function $F(y)$. Here $F(y)$ is the unknown parameter. We merely require $F(y)$ to have the properties of a distribution function: it is nonnegative, nondecreasing, right-continuous, and satisfies $0 \le F(y) \le 1$ for all $y \in (-\infty, \infty)$ with $\lim_{y \to -\infty} F(y) = 0$ and $\lim_{y \to \infty} F(y) = 1$. Thus the parameter space is the set of all distribution functions.

Ignoring the factor $(2h)^{-m}$, an approximate likelihood for F is

$$L_h(F \mid \boldsymbol{Y}) = \prod_{i=1}^{n} \{F(Y_i + h) - F(Y_i - h)\},$$

where h is assumed to be a small positive constant. The following argument assumes that there are no ties in the sample; the results apply more generally, but requires a modified proof. In the absence of ties we can assume that h is small enough to ensure that $[Y_i - h, Y_i + h]$ does not contain Y_j for any $j \neq i$. Let $p_{i,h} = F(Y_i + h) - F(Y_i - h)$. Then $L_h(F \mid \boldsymbol{Y}) = \prod_{i=1}^{n} p_{i,h}$. Note that $L_h(F \mid \boldsymbol{Y})$ is maximized only if each $p_{i,h} > 0$, $i = 1, \ldots, n$. Since increasing $p_{i,h}$ increases $L_h(F \mid \boldsymbol{Y})$, we want the $p_{i,h} > 0$ to be as large as possible while still satisfying $\sum_{i=1}^{n} p_{i,h} \leq 1$. Of course this implies that $\sum_{i=1}^{n} p_{i,h} = 1$. Thus our goal is to maximize $\prod_{i=1}^{n} p_{i,h}$ over all $p_{1,h}, \ldots, p_{n,h}$ subject to the constraints $p_{i,h} > 0$ and $\sum_{i=1}^{n} p_{i,h} = 1$. Using the method of Lagrange multipliers, we solve the optimization problem by finding the stationary points of

$$g(p_{1,h}, \ldots, p_{n,h}, \lambda) = \sum_{i=1}^{n} \log(p_{i,h}) + \lambda \left(\sum_{i=1}^{n} p_{i,h} - 1 \right),$$

by solving

$$\frac{\partial g}{\partial p_{i,h}} = \frac{1}{p_{i,h}} + \lambda = 0, \qquad i = 1, \ldots, n,$$

$$\frac{\partial g}{\partial \lambda} = \sum_{i=1}^{n} p_{i,h} - 1 = 0.$$

The first n equations Imply that $p_{i,h} = -1/\lambda$, which upon substitution into the last equation yields $\lambda = -n$. Thus any distribution function $\widehat{F}_h(y)$ satisfying $\widehat{F}_h(Y_i + h) - \widehat{F}_h(Y_i - h) = 1/n$, $i = 1, \ldots, n$, maximizes $L_h(F|\boldsymbol{Y})$. Problem 2.10 shows that all such $\widehat{F}_h(y)$ converge in distribution to $\widehat{F}_{\text{EMP}}(y) = n^{-1} \sum_{i=1}^{n} I(Y_i \leq y)$ as $h \to 0$. Thus we take as the MLE of $F(y)$, the empirical distribution function, i.e,

$$\widehat{F}_{\text{MLE}}(y) = F_{\text{EMP}}(y) = \frac{1}{n} \sum_{i=1}^{n} I(Y_i \leq t).$$

2.2.7 *Likelihoods from Censored Data*

2.2.7a Type I Censoring

Suppose that a random variable X is normally distributed with mean μ and variance σ^2, but whenever $X \leq 0$, all we observe is that it is less than or equal to 0. We say that X has been *censored* at 0. If the sample value is set to 0 in the censored cases, then we can define the observed variate Y by

$$Y = \begin{cases} 0, & X \leq 0, \\ X, & X > 0. \end{cases}$$

At $y = 0$ the distribution function of Y is

$$F_Y(0) = P(Y = 0) = P(\sigma Z + \mu \le 0) = P(Z \le -\mu/\sigma) = \Phi(-\mu/\sigma),$$

where Z is a standard normal random variable with distribution function Φ. For $y > 0$, $F_Y(y) = P(X \le y) = \Phi((y-\mu)/\sigma)$, and $F_Y(y) = 0$ for $y < 0$. Thus, suppose we have a sample $Y_1, \ldots, Y_n$, and, as before, let n_0 be the number of sample values that are 0. Then with $m = n - n_0$

$$\begin{aligned} L_h(\boldsymbol{\theta} \mid \boldsymbol{Y}) &= \left(\frac{1}{2h}\right)^m \prod_{i=1}^{n} \{F_Y(Y_i + h; \boldsymbol{\theta}) - F_Y(Y_i - h; \boldsymbol{\theta})\} \\ &= \left\{\Phi\left(\frac{h-\mu}{\sigma}\right) - 0\right\}^{n_0} \\ &\quad \times \prod_{Y_i > 0} \left\{\frac{\Phi((Y_i + h - \mu)/\sigma) - \Phi((Y_i - h - \mu)/\sigma)}{2h}\right\} \end{aligned}$$

$$\stackrel{h \to 0+}{\longrightarrow} \left\{\Phi\left(\frac{-\mu}{\sigma}\right)\right\}^{n_0} \prod_{Y_i > 0} \left\{\frac{1}{\sigma}\phi\left(\frac{Y_i - \mu}{\sigma}\right)\right\} = L(\boldsymbol{\theta} \mid \boldsymbol{Y}).$$

Tobin (1958) introduced this model in a regression setting, and later it was named the *Tobit model* after Tobin and its similarity to probit models. Of course, censoring at zero is not unique; any threshold value L_0 might be appropriate depending on the situation. In general, this type of censoring at a fixed point is called *Type I censoring*, and we have just discussed *left* censoring. *Type II censoring* refers to situations where the first r ordered values of a sample are observed or more generally some specified subset of the ordered values are observed (see Problem 2.12, p. 110).

It is worth mentioning that censoring is different from *truncation*. For example, we might sample only households that have above a certain income, say L_0. That is, our sample $Y_1, \ldots, Y_n$ of incomes is only for incomes above L_0, and in a sense we are unaware of data below L_0. If all incomes have distribution function $F(x; \boldsymbol{\theta})$, then for $y > L_0$

$$P(Y_1 \le y \mid Y_1 > L_0) = \frac{P(Y_1 \le y, Y_1 > L_0)}{P(Y_1 > L_0)} = \frac{F(y; \boldsymbol{\theta}) - F(L_0; \boldsymbol{\theta})}{1 - F(L_0; \boldsymbol{\theta})}.$$

Taking derivatives, the likelihood of $Y_1, \ldots, Y_n$ is then

$$L(\boldsymbol{\theta} \mid \boldsymbol{Y}) = \prod_{i=1}^{n} \left\{\frac{f(y; \boldsymbol{\theta})}{1 - F(L_0; \boldsymbol{\theta})}\right\}.$$

This is just an iid-data likelihood, where the densities are adjusted to take into account that $Y_i > L_0$.

Now back to Type I censoring, we might have censoring on the left at L_0 and censoring on the right at R_0, but observe all values of X between L_0 and R_0. Suppose that X has density $f(x;\boldsymbol{\theta})$ and distribution function $F(x;\boldsymbol{\theta})$, and that

$$Y_i = \begin{cases} L_0, & X_i \le L_0, \\ X_i, & L_0 < X_i < R_0, \\ R_0, & X_i \ge R_0. \end{cases}$$

If we let n_L and n_R be the number of X_i values $\le L_0$ and $\ge R_0$, respectively, then the likelihood of the observed data $Y_1, \ldots, Y_n$ is

$$L(\boldsymbol{\theta} \mid \boldsymbol{Y}) = \{F(L_0;\boldsymbol{\theta})\}^{n_L} \left\{ \prod_{L_0 < Y_i < R_0} f(Y_i;\boldsymbol{\theta}) \right\} \{1 - F(R_0;\boldsymbol{\theta})\}^{n_R}.$$

We could also let each X_i be subject to its own censoring values L_i and R_i. The notation just gets a bit more onerous. However, right censoring is the most common; so let us restrict to that case and define $Y_i = \min(X_i, R_i)$. In addition, define indicator variables $\delta_i = I(X_i \le R_i)$ that are equal to one if we observe X_i and are zero if X_i is censored. Then the likelihood can be written in the fairly simple form

$$L(\boldsymbol{\theta} \mid \boldsymbol{Y}) = \prod_{i=1}^{n} f(Y_i;\boldsymbol{\theta})^{\delta_i} \left[1 - F(R_i;\boldsymbol{\theta})\right]^{1-\delta_i}.$$

Example 2.9 (Equipment failure times). Lawless (1982, Table 1.1.2) gives data from Bartholomew (1957) on pieces of equipment that are started at different times and later regularly checked for failure. By a fixed date when the study ended, three of the items had not failed and therefore were censored. The data in days to failure are as follows:

Y:	2	72	51	60	33	27	14	24	4	21
δ:	1	0	1	0	1	1	1	1	1	0

Note that values 72, 60, and 21 are censoring times, not failure times, and indicated by $\delta = 0$. For simplicity, suppose that the failure times follow an exponential distribution, $F(x;\sigma) = 1 - \exp(-x/\sigma), x \ge 0$. Then

$$L(\sigma \mid \boldsymbol{Y}) = \prod_{i=1}^{n} \left[\frac{1}{\sigma}\exp(-Y_i/\sigma)\right]^{\delta_i} \left[\exp(-Y_i/\sigma)\right]^{1-\delta_i} = \left(\frac{1}{\sigma}\right)^{n-n_R} \exp(-n\overline{Y}/\sigma),$$

where n_R is the number of observations censored on the right. Taking logarithms, we have

$$\ell(\sigma) = \log L(\sigma \mid \boldsymbol{Y}) = -(n - n_R)\log\sigma - \frac{n\overline{Y}}{\sigma},$$

and differentiating with respect to σ and equating to zero leads to

$$S(\sigma) = \frac{-(n - n_R)}{\sigma} + \frac{n\overline{Y}}{\sigma^2} = 0,$$

and

$$\widehat{\sigma}_{\text{MLE}} = \left(\frac{n}{n - n_R}\right)\overline{Y} = \left(\frac{10}{10 - 3}\right) 30.8 = 44.0. \qquad \blacklozenge$$

2.2.7b Random Censoring

In the situations described above, the censoring times L and R were considered fixed time points. In medical studies, however, patients often enter the studies at different times that are modeled as random variables. Along with a fixed end date of the study, this leads to random right censoring times $R_1, \ldots, R_n$. The notation stays the same, $Y_i = \min(X_i, R_i)$ and $\delta_i = I(X_i \le R_i)$, but we assume that the censoring times are independent of $X_1, \ldots, X_n$ and iid with distribution function $G(t)$ and density $g(t)$. Now consider the contribution to the likelihood due to $(Y_i, \delta_i = 1)$ and recall that X_i has density $f(x; \boldsymbol{\theta})$:

$$\begin{aligned}
\frac{P(Y_i \in (y-h, y+h], \delta_i = 1)}{2h} &= \frac{P(X_i \in (y-h, y+h], X_i \le R_i)}{2h} \\
&= \left(\frac{1}{2h}\right) \int_{-\infty}^{\infty} \int_{-\infty}^{\infty} \Big[I(y - h < t \le y + h, t \le r) \\
&\qquad \times f(t; \boldsymbol{\theta}) g(r) \Big] \, dt \, dr \\
&= \left(\frac{1}{2h}\right) \int_{y-h}^{y+h} \left[\int_{-\infty}^{\infty} I(t \le r) g(r) dr \right] f(t; \boldsymbol{\theta}) dt \\
&= \left(\frac{1}{2h}\right) \int_{y-h}^{y+h} [1 - G(t)] \, f(t; \boldsymbol{\theta}) dt \\
&\longrightarrow [1 - G(y)] \, f(y; \boldsymbol{\theta})
\end{aligned}$$

The last line is by the Fundamental Theorem of Calculus. Next consider the contribution due to $(Y_i, \delta_i = 0)$:

$$\begin{aligned}
\frac{P(Y_i \in (y-h, y+h], \delta = 0)}{2h} &= \frac{P(R_i \in (y-h, y+h], X_i > R_i)}{2h} \\
&= \left(\frac{1}{2h}\right) \int_{y-h}^{y+h} \left[\int_{-\infty}^{\infty} I(t > r) f(t; \boldsymbol{\theta}) dt \right] g(r) dr
\end{aligned}$$

$$= \left(\frac{1}{2h}\right) \int_{y-h}^{y+h} [1 - F(r; \boldsymbol{\theta})]\, g(r) dr$$

$$\longrightarrow [1 - F(y; \boldsymbol{\theta})]\, g(y)$$

Putting together these two types of contributions to the likelihood, we have

$$L(\boldsymbol{\theta} \mid \boldsymbol{Y}, \boldsymbol{\delta}) = \left\{ \prod_{i=1}^{n} f(Y_i; \boldsymbol{\theta})^{\delta_i} [1 - F(Y_i; \boldsymbol{\theta})]^{1-\delta_i} \right\} \prod_{i=1}^{n} \left\{ g(Y_i)^{1-\delta_i} [1 - G(Y_i)]^{\delta_i} \right\}$$

$$= \prod_{i=1}^{n} \left\{ f(Y_i; \boldsymbol{\theta})^{\delta_i} [1 - F(Y_i; \boldsymbol{\theta})]^{1-\delta_i} g(Y_i)^{1-\delta_i} [1 - G(Y_i)]^{\delta_i} \right\}.$$

Note that the first part of the likelihood is the same as we derived for the fixed censoring times situation. Note also that the unknown censoring distribution G is not needed to estimate $\boldsymbol{\theta}$.

2.3 Likelihoods for Regression Models

2.3.1 Linear Model

Consider the familiar normal linear model,

$$Y_i = \boldsymbol{x}_i^T \boldsymbol{\beta} + e_i \qquad i = 1, \ldots, n, \tag{2.10}$$

where $e_1, \ldots, e_n$ are iid $N(0, \sigma^2)$, and the $\boldsymbol{x}_1, \ldots, \boldsymbol{x}_n$ are known nonrandom p-vectors, the first component of which is usually the constant "1" corresponding to an intercept. In this section we use p for the dimension of $\boldsymbol{\beta}$, and thus $\boldsymbol{\theta} = (\boldsymbol{\beta}^T, \sigma)^T$ has dimension $b = p + 1$. The likelihood is

$$L(\beta, \sigma \mid \{Y_i, \boldsymbol{x}_i\}_{i=1}^n) = \prod_{i=1}^{n} \frac{1}{\sqrt{2\pi}\sigma} \exp\left\{ -\frac{\left(Y_i - \boldsymbol{x}_i^T \boldsymbol{\beta}\right)^2}{2\sigma^2} \right\}$$

$$= \left(\frac{1}{\sqrt{2\pi}\sigma}\right)^n \exp\left\{ -\sum_{i=1}^{n} \frac{\left(Y_i - \boldsymbol{x}_i^T \boldsymbol{\beta}\right)^2}{2\sigma^2} \right\}. \tag{2.11}$$

Using the distribution-function method, the likelihood is found by first calculating the distribution function of Y_i,

$$P(Y_i \le y) = P(x_i^T \boldsymbol{\beta} + e_i \le y) = P(e_i \le y - x_i^T \boldsymbol{\beta}) = \Phi\left(\frac{y - \boldsymbol{x}_i^T \boldsymbol{\beta}}{\sigma}\right),$$

where Φ is the standard normal distribution function, then differentiating with respect to y to get the density of Y_i, replacing y by Y_i, and taking the product over i.

Taking logarithms of (2.11) shows that the maximum likelihood estimator of $\boldsymbol{\beta}$ is the same as the least squares estimator, $\widehat{\boldsymbol{\beta}}_{\text{MLE}} = \widehat{\boldsymbol{\beta}}_{\text{LS}} = (\boldsymbol{X}^T\boldsymbol{X})^{-1}\boldsymbol{X}^T\boldsymbol{Y}$, where $\boldsymbol{X} = (\boldsymbol{x}_1, \ldots, \boldsymbol{x}_n)^T$, and for simplicity $\boldsymbol{X}^T\boldsymbol{X}$ is assumed to be nonsingular. Setting the derivative of the log likelihood with respect to σ equal to zero and solving produces $\widehat{\sigma}^2_{\text{MLE}} = n^{-1}\sum_{i=1}^n \widehat{e}_i^2$, where $\widehat{e}_i = Y_i - \boldsymbol{x}_i^T\widehat{\boldsymbol{\beta}}_{\text{MLE}}$, $i = 1, \ldots, n$. However, common practice is to use the unbiased estimator $(n-p)^{-1}\sum_{i=1}^n \widehat{e}_i^2$ in place of $\widehat{\sigma}^2_{\text{MLE}}$.

For sampling designs wherein $(\boldsymbol{x}_i, Y_i)$ are sampled from a population of $(\boldsymbol{x}, Y)$ pairs, the predictors are also random variables. In such cases the linear model specification (2.10) is understood to specify the conditional distribution of Y_i given $\boldsymbol{X}_i = \boldsymbol{x}_i$. Thus if marginally, $\boldsymbol{X}_1, \ldots, \boldsymbol{X}_n$ are iid with parametric density $f_X(\boldsymbol{x}; \boldsymbol{\tau})$, then the full likelihood is

$$L(\boldsymbol{\beta}, \sigma, \boldsymbol{\tau} \mid \{Y_i, \boldsymbol{X}_i\}_{i=1}^n) = \left(\frac{1}{\sqrt{2\pi}\sigma}\right)^n \exp\left\{-\sum_{i=1}^n \frac{(Y_i - \boldsymbol{X}_i^T\boldsymbol{\beta})^2}{2\sigma^2}\right\} \prod_{i=1}^n f_X(\boldsymbol{X}_i, \boldsymbol{\tau}). \quad (2.12)$$

Note that the log likelihood decomposes into two parts $\ell(\boldsymbol{\beta}, \sigma, \boldsymbol{\tau}) = \ell_1(\boldsymbol{\beta}, \sigma) + \ell_2(\boldsymbol{\tau})$ where $\ell_1(\boldsymbol{\beta}, \sigma)$ is identical in structure to the logarithm of (2.11). It follows that if $\boldsymbol{\tau}$ does not depend functionally on $\boldsymbol{\beta}$ or σ as is usually assumed, then likelihood-based inference for $\boldsymbol{\beta}$ and σ is identical for (2.11) and (2.12) (e.g., the partial derivatives with respect to $\boldsymbol{\beta}$ and σ of $\ell(\boldsymbol{\beta}, \sigma, \boldsymbol{\tau})$ and $\ell_1(\boldsymbol{\beta}, \sigma)$ are identical, neither one depending $\boldsymbol{\tau}$). When the density of the $\boldsymbol{X}_i$ does not depend on the parameters $\boldsymbol{\beta}$ and σ, the sample values $\boldsymbol{X}_1, \ldots, \boldsymbol{X}_n$ are said to be *ancillary* and play no role in likelihood methods. The estimators and information matrix are the same as before, and the asymptotic results are the same. For exact finite sample results, it is traditional to condition on $\boldsymbol{X}_1, \ldots, \boldsymbol{X}_n$, in which case results are the same as in the fixed constants case. See Sampson (1974) for a comparison of exact conditional and unconditional inference in the normal linear model.

For modeling certain data, e.g., when Y_i are maxima, it makes sense to assume the additive model (2.10, p. 50) but with nonnormal errors. Assuming that the errors have the scale-family density $\sigma^{-1} f_e(t/\sigma)$, then the likelihood is

$$L(\beta, \sigma \mid \{Y_i, \boldsymbol{x}_i\}_{i=1}^n) = \prod_{i=1}^n \frac{1}{\sigma} f_e\left(\frac{Y_i - \boldsymbol{x}_i^T\boldsymbol{\beta}}{\sigma}\right). \quad (2.13)$$

For example, the extreme value density

$$f_e(t) = \exp(-t)\exp\{-\exp(-t)\},$$

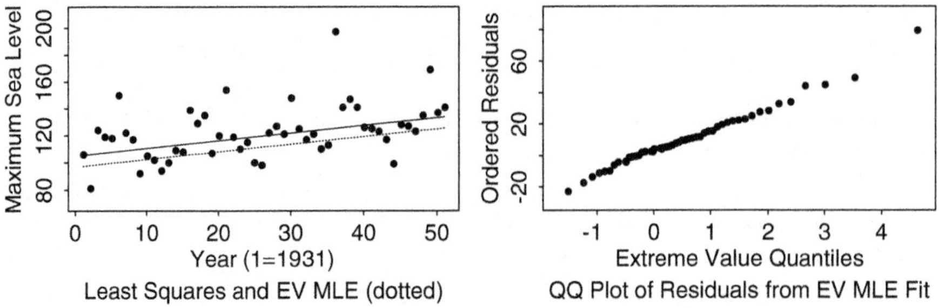

Fig. 2.5 Simple linear regression modeling of Venice maximum sea levels 1931–81

is a natural candidate model when the Y_i are maxima. With the extreme value error model it is no longer true that least squares and maximum likelihood estimators are equal. Both are consistent for the non-intercept components of $\boldsymbol{\beta}$, but the asymptotic relative efficiency of the least squares estimator compared to the extreme value maximum likelihood estimator is only .61 *assuming the extreme value model is correct*. Thus, there is good reason to go to the extra trouble of using the extreme value MLE when the Y_i are maxima and the extreme value error model is justified.

Example 2.10 (Venice sea levels). The annual maximum sea levels in Venice for 1931–1981 are given in Pirazzoli (1982) and analyzed by Embrechts et al. (1985) using extreme value likelihood methods. The data are as follows:

```
> venice
 [1] 103  78 121 116 115 147 119 114  89 102  99
[12]  91  97 106 105 136 126 132 104 117 151 116
[23] 107 112  97  95 119 124 118 145 122 114 118
[34] 107 110 194 138 144 138 123 122 120 114  96
[45] 125 124 120 132 166 134 138
```

The left panel of Figure 2.5 plots the data and the least squares line, $y = 104.9 + .567\text{year}$, and the extreme value maximum likelihood line, $y = 96.8 + .563\text{year}$, where the year variable is from 1 to 51 for years 1931 to 1981. The right panel of Figure 2.5 gives an extreme value QQ plot of the residuals and suggests that the extreme value model is reasonable. The slope estimates are very close, but the standard errors of the slopes are .177 (LS) and .136 (MLE), consistent with the greater efficiency of the latter when the extreme value model is appropriate.

The difference in intercepts is explained by the fact that the extreme value error distribution has mean $\approx .577\sigma$. Thus we expect that $\widehat{\beta}_{1,\text{LS}} \approx \widehat{\beta}_{1,\text{MLE}} + .577\widehat{\sigma}$. The MLE of σ from the extreme value fit is 14.5, and thus $\widehat{\beta}_{1,\text{MLE}} + .577\widehat{\sigma} = 96.8 + .577(14.5) = 105.17 \approx 104.9$ as expected. ♦

2.3.2 Additive Errors Nonlinear Model

The standard nonlinear regression model is very similar to (2.10, p. 50) but $Y_i = \boldsymbol{x}_i^T\boldsymbol{\beta} + e_i$ is replaced by $Y_i = g(\boldsymbol{x}_i, \boldsymbol{\beta}) + e_i$, where g is a known function. Common examples are the *exponential growth model* $g(\boldsymbol{x}_i, \boldsymbol{\beta}) = \beta_0 \exp(\beta_1 x_i)$ and the *logistic growth model* $g(x_i, \boldsymbol{\beta}) = \beta_0 (1 + \beta_1 \exp\{-\beta_2 x_i\})^{-1}$. For normal errors, the likelihood is the same as (2.11, p. 50) with $\boldsymbol{x}_i^T\boldsymbol{\beta}$ replaced by $g(\boldsymbol{x}_i, \boldsymbol{\beta})$ and maximum likelihood again leads to least squares estimators, but the estimator of $\boldsymbol{\beta}$ no longer has a closed form and must be calculated numerically. The maximum likelihood estimator of σ has the same form as in the linear case, i.e., the average squared residual.

2.3.3 Generalized Linear Model

Generalized linear models introduced by Nelder and Wedderburn (1972) are another important class of nonlinear models that generalize the normal linear model. Suppose that Y_i has log density

$$\log f(y_i; \theta_i, \phi) = \frac{y_i\theta_i - b(\theta_i)}{a_i(\phi)} + c(y_i, \phi). \tag{2.14}$$

The density of Y_i is almost an exponential family density except for the *dispersion* term $a_i(\phi)$, where a_i is a known function and ϕ is possibly an unknown parameter. In exponential family language, θ_i is called the *natural* or *canonical* parameter.

For those unfamiliar with exponential families, it is worth taking a moment to show how standard densities fit into the framework of (2.14). Consider first the Bernoulli density

$$f(y; p) = p^y(1-p)^{1-y}, \quad y = 0, 1.$$

Taking natural logarithms, we have

$$\begin{aligned}\log f(y; p) &= y \log p + (1-y)\log(1-p)\\ &= y \log\left(\frac{p}{1-p}\right) + \log(1-p).\end{aligned}$$

Thus $a_i(\phi) = 1, c(y_i, \phi) = 0, \theta = \log\{p/(1-p)\}$ and then $p{=}1/\{1{+}\exp(-\theta)\}$, so that

$$b(\theta) = -\log(1-p) = -\log\left\{1 - \frac{1}{1{+}\exp(-\theta)}\right\} = \log\{1 + \exp(\theta)\}.$$

Next consider the normal(μ, σ^2) density:

$$\begin{aligned}\log f(y; \mu, \sigma) &= -\log\left(\sqrt{2\pi}\sigma\right) - \frac{(y-\mu)^2}{2\sigma^2} \\ &= \frac{y\mu - \mu^2/2}{\sigma^2} - \log\left(\sqrt{2\pi}\sigma\right) - \frac{y^2}{2\sigma^2}.\end{aligned}$$

Thus $\theta = \mu$, $b(\theta) = \mu^2/2 = \theta^2/2$, $a_i(\phi) = \sigma^2$, and $c(y_i, \phi) = \log(\sigma\sqrt{2\pi}) - y^2/(2\sigma^2)$. Note that in the general exponential family framework of Appendix B (p. 97), the generalized linear model representation of the normal density is closer to the representation of a normal density with known variance σ^2, Example 2.29 (p. 98), than the two-parameter version, Example 2.31 (p. 100).

Returning to the general form (2.14), we first use the fact that the derivative of (2.14) must have expectation zero, to obtain $\mu_i = E(Y_i) = b'(\theta_i)$. Next, using the information identity often studied in a first course in mathematical statistics,

$$E\left\{\frac{\partial}{\partial\theta_i}\log f(Y_i; \theta_i, \phi)\right\}^2 = E\left\{-\frac{\partial^2}{\partial\theta_i^2}\log f(Y_i; \theta_i, \phi)\right\}$$

leads to $\text{Var}(Y_i) = b''(\theta_i)a_i(\phi)$. Note that since the variance must be positive, $b(\theta_i)$ is a strictly convex function and $b'(\theta_i)$ is monotone increasing with a unique inverse b'^{-1}.

For generalized linear models there is a *link* function g between the mean of Y_i and the *linear predictor* $\eta_i = \boldsymbol{x}_i^T\boldsymbol{\beta}$ such that $g(\mu_i) = \boldsymbol{x}_i^T\boldsymbol{\beta}$. Thus, the formal definition of the density in terms of the linear predictor has the awkward term $\theta_i = b'^{-1}\left\{g^{-1}(\boldsymbol{x}_i^T\boldsymbol{\beta})\right\}$. Fortunately, the most important cases have $g(\mu_i) = \theta_i = \boldsymbol{x}_i^T\boldsymbol{\beta}$. These g are called *natural* or *canonical* link functions and are found by setting $g(\mu_i) = b'^{-1}(\mu_i)$. Examples are $g(\mu_i) = \mu_i$ for normal data, $g(\mu_i) = \log\{\mu_i/(1-\mu_i)\}$ for Bernoulli data, and $g(\mu_i) = \log(\mu_i)$ for Poisson data.

For canonical-link models, the log likelihood of $(Y_1, \boldsymbol{x}_1), \ldots, (Y_n, \boldsymbol{x}_n)$, where the Y_i are independent with log density (2.14) and the $\boldsymbol{x}_i$ are known nonrandom vectors, is

$$\log L(\boldsymbol{\beta}, \phi \mid \{Y_i, \boldsymbol{x}_i\}_{i=1}^n) = \sum_{i=1}^n \left\{\frac{Y_i\boldsymbol{x}_i^T\boldsymbol{\beta} - b(\boldsymbol{x}_i^T\boldsymbol{\beta})}{a_i(\phi)} + c(Y_i, \phi)\right\}. \tag{2.15}$$

We now consider three important models. For normal data, we have already shown $b(\theta_i) = \theta_i^2/2$. Choosing the canonical link $g(\mu_i) = \mu_i$ (since $b'(\theta_i) = \theta_i$) and $a_i(\phi) = \sigma^2$ leads directly to (2.10, p. 50) with N(0,σ^2) errors, the normal linear model.

For Bernoulli data, $Y_i = 0$ or 1 with $P(Y_i = 1) = p_i = \mu_i$, and $a_i(\phi) = 1$ and $b(\theta_i) = \log\left(1 + e^{\theta_i}\right)$. Thus $b'(\theta_i) = (1 + e^{-\theta_i})^{-1}$ and the canonical link is $g(p_i) = \log\{p_i/(1-p_i)\}$. Because p_i is modeled as $(1+e^{-\boldsymbol{x}_i^T\boldsymbol{\beta}})^{-1}$ and $(1+e^{-t})^{-1}$ is the distribution function of a logistic random variable, this model is called *logistic regression*. In dose-response settings and other applications, it is common to use the normal distribution function Φ in place of the logistic, in which case $g(p_i) = \Phi^{-1}(p_i)$ and the model is called *probit regression* or *probit analysis*. (However, the probit link is not a canonical link and is not further discussed here.)

Often the data are such that each unique $\boldsymbol{x}_i$ is associated with m_i binary responses, in which case the data are naturally represented as $(\boldsymbol{x}_i, Y_{i1}, \ldots, Y_{im_i})$, for $i = 1, \ldots, n$. When $Y_{i1}, \ldots, Y_{im_i}$ are iid Bernoulli(p_i), then sufficiency can be invoked to further reduce the stored data to the pairs $\left(\boldsymbol{x}_i, \overline{Y}_{i\cdot}\right)$, $i = 1, \ldots, n$. In this case the $\overline{Y}_{i\cdot}$ follow model (2.14, p. 53) with the same definitions as for Bernoulli data except that $\boldsymbol{V} = \text{diag}\{m_1 p_1(1-p_1), \ldots, m_n p_n(1-p_n)\}$, $a_i(\phi) = 1/m_i$, and $c(y_i, \phi)$ differs accordingly.

Often the Bernoulli response variables $Y_{i1}, \ldots, Y_{im_i}$ are obtained by some type of cluster sampling, thus violating the independence assumption and usually resulting in positive pairwise correlation (e.g., $\overline{Y}_{i\cdot}$ is the proportion of family members contracting the flu in a given time period). In this case the $m_i\overline{Y}_{i\cdot}$ are not binomially distributed but are *over-dispersed* or have *extra-binomial variation*. The generalized linear model handles this case by letting $a_i(\phi) = \phi/m_i$ so that $\text{Var}(\overline{Y}_{i\cdot}) = \phi p_i(1-p_i)/m_i$ and $\boldsymbol{V} = \text{diag}\{m_1 p_1(1-p_1), \ldots, m_n p_n(1-p_n)\}/\phi$.

For Poisson data, $b(\theta_i) = e^{\theta_i}$ and $a_i(\phi) = 1$, and thus $b'(\theta_i) = e^{\theta_i}$. Inverting b' leads to the canonical link $g(\mu_i) = \log(\mu_i)$ and modeling the mean μ_i by $e^{\boldsymbol{x}_i^T\boldsymbol{\beta}}$. The variance of Y_i is μ_i. Often data initially thought to be Poisson are over-dispersed, and it is important to add the over-dispersion parameter ϕ so that $\text{Var}(Y_i) = \phi\mu_i$.

2.3.4 Generalized Linear Mixed Model (GLMM)

In recent years there has been a large effort focused on extending generalized linear models (GLMs) to include random effect terms. These extended models have a GLM structure conditional on a vector of random effects $\boldsymbol{U}$. That is, the log conditional density of the dependent variable Y_i has the form (in the canonical link case)

$$\log f_{Y_i|\boldsymbol{u}}(y_i|\boldsymbol{U}, \boldsymbol{x}_i, z_i, \boldsymbol{\beta}, \tau) = \frac{y_i\eta_i - b(\eta_i)}{\tau} + c(y_i, \tau), \tag{2.16}$$

where $\eta_i = \boldsymbol{x}_i^T\boldsymbol{\beta} + z_i^T\boldsymbol{U}$ is the linear predictor now enhanced to include the random effects $\boldsymbol{U}$ via the terms $z_i^T\boldsymbol{U}$ where z_i are known nonrandom predictors. To complete the specification, assume that the random effects have density $f_{\boldsymbol{u}}(\boldsymbol{u}; \boldsymbol{\nu})$, where $\boldsymbol{\nu}$ is a vector of parameters. Such a model is called a generalized linear mixed model (GLMM). In the normal distribution case, this leads to the usual linear mixed model $\boldsymbol{Y} = \boldsymbol{X\beta} + \boldsymbol{ZU} + \boldsymbol{e}$, here written in the familiar vector form where $\boldsymbol{X} = (\boldsymbol{x}_1^T, \ldots, \boldsymbol{x}_n^T)$ and $\boldsymbol{Z} = (z_1^T, \ldots, z_n^T)$. If the Y_i are conditionally independent, then the likelihood of the observations $Y_1, \ldots, Y_n$ is

$$L(\boldsymbol{\beta}, \tau, \boldsymbol{\nu} \mid \{y_i, x_i, z_i\}_{i=1}^n) = \int \prod_{i=1}^n f_{Y_i|\boldsymbol{u}}(y_i|\boldsymbol{u}, \boldsymbol{x}_i, z_i, \boldsymbol{\beta}, \tau) f_{\boldsymbol{u}}(\boldsymbol{u}; \nu) d\boldsymbol{u}. \tag{2.17}$$

Integration over $\boldsymbol{u}$ is required because $\boldsymbol{U}$ is not observed, and the likelihood is the density of the observed data. In the normal case, the integration is tractable, but in general this likelihood is not easy to calculate or maximize. Breslow and Clayton (1993) and McCullagh (1997) give some of the important approaches for finding maximum likelihood estimators, but this remains an active area of research.

2.3.5 Accelerated Failure Model

Accelerated Failure Models comprise an important class of regression models for censored data. For the case of random right censoring, the model is

$$\log T_i = \boldsymbol{x}_i^T \boldsymbol{\beta} + \sigma e_i,$$

where we observe $Y_i = \min(\log T_i, \log R_i)$, and R_i is a censoring time that is assumed independent of T_i. Typical models for the errors e_i are standard normal, or logistic, or

$$f_e(z) = e^z e^{-e^z}, \quad -\infty < z < \infty,$$

the density of the logarithm of a standard exponential random variable. The name "Accelerated Failure Models" comes from the fact that for the failure time $T_i = e^{\boldsymbol{x}_i^T \boldsymbol{\beta}} e^{\sigma e_i}$, the role of $\boldsymbol{x}_i$ can be to accelerate (shorten) the time to failure. Using notation as before, the likelihood for error density f_e and distribution function F_e is

$$L(\boldsymbol{\beta}, \sigma \mid \{Y_i, \delta_i, \boldsymbol{x}_i\}_{i=1}^n) = \prod_{i=1}^n \left[\frac{1}{\sigma} f_e(r_i)\right]^{\delta_i} [1 - F_e(r_i)]^{1-\delta_i}.$$

where $\delta_i = I(Y_i = \log T_i)$ and $r_i = (Y_i - \boldsymbol{x}_i^T \beta)/\sigma$.

Example 2.11 (Censored survival times). Lawless (1982, p. 318) gives data from Glasser (1965) on survival times of patients with primary lung tumors and two covariates, age and lung performance status (ps). The data are as follows: log survival time=(1.94, 2.23, 1.94, 1.98, 2.23, 1.59, 2.13, 1.80, 2.32, 1.92, 2.15*, 2.05*, 2.48*, 2.42*, 2.56*, 2.56*) (* for censored); age=(42, 67, 62, 52, 57, 58, 55, 63, 44, 62, 51, 64, 54, 64, 54, 57); performance status=(4, 6, 4, 6, 5, 6, 6, 7, 5, 7, 7, 10, 8, 3, 9, 9). A scatter plot of log survival time versus age and ps is not very suggestive of a significant linear relationship. Moreover, a least squares fit ignoring the censoring suggests that neither covariate is important. However, maximization of the accelerated failure likelihood (in SAS PROC LIFEREG) with f_e the standard normal yields

$$\log T = 1.614 - .006(\text{age}) + .102(\text{ps}),$$

with standard error .06 for the performance status coefficient estimate .102. Thus the Wald statistic (studied in detail in Chapter 3) is T_W =10.2/6=1.7, which is suggestive of a positive effect due to performance status, but not conclusive, especially in light of the liberal tendencies of T_W statistics as discussed in Section 3.2.4 (p. 132). ◆

2.4 Marginal and Conditional Likelihoods

The presence of nuisance parameters is a common problem in likelihood analysis of parametric models, especially when the number of nuisance parameters is large. In these cases $\boldsymbol{\theta}^T = (\boldsymbol{\theta}_1^T, \boldsymbol{\theta}_2^T)$ where $\boldsymbol{\theta}_1$ contains the parameters of interest and typically has small dimension, and $\boldsymbol{\theta}_2$ contains the nuisance parameters. When the dimension of $\boldsymbol{\theta}_2$ is large, maximum likelihood estimators of the parameters of interest ($\boldsymbol{\theta}_1$) are often biased in small samples and can be inconsistent in large samples. We present two alternatives to full maximum likelihood-based inference in such cases. Each method depends on finding a one-to-one transformation of the data $\boldsymbol{Y}$ to $(\boldsymbol{V}, \boldsymbol{W})$ such that either:

$$f_Y(\boldsymbol{y}; \boldsymbol{\theta}_1, \boldsymbol{\theta}_2) = f_{W,V}(\boldsymbol{w}, \boldsymbol{v}; \boldsymbol{\theta}_1, \boldsymbol{\theta}_2) = f_{W|V}(\boldsymbol{w}|\boldsymbol{v}; \boldsymbol{\theta}_1, \boldsymbol{\theta}_2) f_V(\boldsymbol{v}; \boldsymbol{\theta}_1); \quad (2.18)$$

or

$$f_Y(\boldsymbol{y}; \boldsymbol{\theta}_1, \boldsymbol{\theta}_2) = f_{W,V}(\boldsymbol{w}, \boldsymbol{v}; \boldsymbol{\theta}_1, \boldsymbol{\theta}_2) = f_{W|V}(\boldsymbol{w}|\boldsymbol{v}; \boldsymbol{\theta}_1) f_V(\boldsymbol{v}; \boldsymbol{\theta}_1, \boldsymbol{\theta}_2). \quad (2.19)$$

The key common feature of the likelihood factorizations in (2.18) and (2.19) is that one component of each contains only the parameter of interest. In (2.18) the marginal density of $\boldsymbol{V}$, $f_V(\boldsymbol{V}; \boldsymbol{\theta}_1)$, depends on $\boldsymbol{\theta}_1$ only, and a likelihood derived from it is called a *marginal likelihood* and results in marginal likelihood inference. In (2.19) the conditional density of $\boldsymbol{W}$ given $\boldsymbol{V}$, $f_{W|V}(\boldsymbol{w}|\boldsymbol{v}; \boldsymbol{\theta}_1)$, depends on $\boldsymbol{\theta}_1$ only, and a likelihood derived from it is called a *conditional likelihood* and results in conditional likelihood inference. Both types of alternative likelihoods can be useful in practice.

We illustrate the usefulness of marginal and conditional likelihoods in the context of the famous Neyman-Scott example of an inconsistent maximum likelihood estimator of the error variance in one-way analysis of variance (Neyman and Scott 1948).

2.4.1 Neyman-Scott Problem

Let $Y_{ij}, i = 1, \ldots, n, j = 1, 2$ be independent normal random variables with possibly different means μ_i but the same variance σ^2. Thus $\boldsymbol{\theta} = (\sigma^2, \mu_1, \ldots, \mu_n)^T$

has dimension $n+1$. The maximum likelihood estimator of μ_i is $\widehat{\mu}_{i,\mathrm{MLE}} = (Y_{i1}+Y_{i2})/2$, and

$$\begin{aligned}\widehat{\sigma}^2_{\mathrm{MLE}} &= \frac{1}{2n}\sum_{i=1}^{n}\sum_{j=1}^{2}\left(Y_{ij}-\widehat{\mu}_{i,\mathrm{MLE}}\right)^2 \\ &= \frac{1}{2n}\sum_{i=1}^{n}\left\{\left(Y_{i1}-\frac{Y_{i1}+Y_{i2}}{2}\right)^2+\left(Y_{i2}-\frac{Y_{i1}+Y_{i2}}{2}\right)^2\right\} \\ &= \frac{1}{n}\sum_{i=1}^{n}\frac{1}{4}\left(Y_{i1}-Y_{i2}\right)^2. \end{aligned} \tag{2.20}$$

Using (2.20) we can see that $\mathrm{E}(\widehat{\sigma}^2_{\mathrm{MLE}}) = \sigma^2/2$ and that $\widehat{\sigma}^2_{\mathrm{MLE}} \xrightarrow{p} \sigma^2/2$ as $n \to \infty$ by the Weak Law of Large Numbers.

The failure of maximum likelihood in this example is due to the fact that the number (n) of nuisance parameters ($\mu_1, \ldots, \mu_n$) grows proportionally with sample size. One way to handle the problem here is to define $V_i = (Y_{i1} - Y_{i2})/\sqrt{2}$ and $W_i = (Y_{i1}+Y_{i2})/\sqrt{2}, i = 1, \ldots, n$. This is a one-to-one transformation of the data, and $V_1, \ldots, V_n$ are iid normal$(0, \sigma^2)$ and independent of $(W_1, \ldots, W_n)$ which are iid normal$(\sqrt{2}\mu_i, \sigma^2)$. Because of the independence, the marginal distribution of $\boldsymbol{V}$ is the same as the conditional distribution of $\boldsymbol{V}$ given $\boldsymbol{W}$. Thus this transformation fits into either (2.18) or (2.19), and the marginal (or conditional) likelihood of $V_1, \ldots, V_n$,

$$L(\sigma|\boldsymbol{V}) = \prod_{i=1}^{n}\frac{1}{\sigma}\phi\left(\frac{V_i}{\sigma}\right) = (2\pi)^{-n/2}\sigma^{-n}\exp\left(-\frac{1}{2\sigma^2}\sum_{i=1}^{n}V_i^2\right),$$

leads directly to $\widehat{\sigma}^2_{\mathrm{MMLE}} = 2\widehat{\sigma}^2_{\mathrm{MLE}}$, the usual unbiased ANOVA estimator. The use of $\boldsymbol{V}$ here is an example of the general restricted maximum likelihood (REML) method (see, e.g., Harville 1977, and Problem 2.22, p. 112).

2.4.2 *Marginal Likelihoods*

A marginal likelihood approach is simple provided that you can find a statistic $\boldsymbol{V}$ whose distribution is free of the nuisance parameter $\boldsymbol{\theta}_2$. The notation in (2.18) is helpful to get a broad perspective, but it is not necessary to actually find $\boldsymbol{W}$ in order to use the method. For example, Pace and Salvan (1997, p. 136–137) give the distribution of the sample correlation coefficient r from a sample of bivariate normal pairs. This marginal distribution of r depends only on the underlying correlation coefficient ρ and leads to an approximate marginal maximum likelihood estimator $\widehat{\rho} = r\left\{1-(1-r^2)/2n\right\}$. Note that here we do not need to find a $\boldsymbol{W}$ to complement $\boldsymbol{V}$. In general, the main reason to find the complementary $\boldsymbol{W}$ might be to gain intuition about the possible information loss for using only $\boldsymbol{V}$.

There does not seem to be a general approach for finding appropriate marginal likelihoods. However, the situation is different for conditional likelihoods in which it is often possible to exploit the existence of sufficient statistics. We now give some examples of constructing conditional likelihoods.

2.4.3 Neyman-Scott Problem via Explicit Conditional Likelihood

The solution to the Neyman-Scott problem in Section 2.4.1, (p. 57) used properties of the transformed variables V_i and W_i, but did not explain how the appropriate transformations were chosen. The key in this example and many others is to identify sufficient statistics for the nuisance parameters under the assumption that the parameter of interest is known. It follows that the distribution of the data given these sufficient statistics does not depend on the nuisance parameters whether the parameter of interest is known or not. For the Neyman-Scott problem we know that $T_i = Y_{i1} + Y_{i2}$ is sufficient for μ_i, $i = 1, \ldots, n$ in the $\mathrm{N}(\mu_i, \sigma^2)$ family assuming that σ^2 is known. It follows that the conditional distribution of the data vector $\boldsymbol{Y}$ given the vector of sufficient statistics $\boldsymbol{T} = (T_1, \ldots, T_n)$ does not depend on the nuisance parameter vector $\boldsymbol{\mu} = (\mu_1, \ldots, \mu_n)$. Thus we seek the conditional distribution of $\boldsymbol{Y}$ given $\boldsymbol{T}$.

Because of independence of (Y_{i1}, Y_{i2}, T_i) over i and the fact that $Y_{i,1} + Y_{i2} = T_i$, the required conditional distribution is readily determined from the conditional distributions of Y_{i1} given T_i, $i = 1, \ldots, n$. By normality, Y_{i1} given T_i is normal with mean $T_i/2$ and variance $\sigma^2/2$. Using these results it is easy to determine that the conditional maximum likelihood estimator of σ^2 is

$$\widehat{\sigma}^2_{\text{CMLE}} = \frac{2}{n} \sum_{i=1}^{n} (Y_{i1} - T_i/2)^2,$$

which is identical to the marginal maximum likelihood estimator of σ^2 derived in Section 2.4.1, (p. 57).

In the Neyman-Scott problem the conditioning statistics T_i did not depend on any unknown parameters. There are problems where the appropriate conditioning statistics do depend on parameters, and we now give such an example taken from the class of generalized linear measurement error models. For other examples, see Stefanski and Carroll (1987), Carroll et al. (2006) and Tsiatis and Davidian (2001).

2.4.4 Logistic Regression Measurement Error Model

Suppose that conditioned on the predictor variable X, the binary response variable Y and the measured predictor W follow the simple logistic regression measurement error model

$$P(Y = 1|X) = F(\alpha + \beta X),$$
$$W = X + U,$$

where F is the logistic distribution function $F(t) = \{1 + \exp(-t)\}^{-1}$, and the measurement error $U \sim \mathrm{N}(0, \sigma_u^2)$ independently of X and Y with σ_u^2 known. Observed data consist of independent pairs (Y_i, W_i), $i = 1, \ldots, n$. The parameters of interest are α and β. In the *functional* version of this model the unobserved true predictors $X_1, \ldots, X_n$ are regarded as unknown nuisance parameters; in the *structural* version $X_1, \ldots, X_n$ are regarded as iid random variables from a population with unknown nuisance distribution F_X.

In either version of the model the "statistic" $T_i = W_i + (Y_i - 1/2)\sigma_u^2\beta$ is sufficient for the unobserved X_i in the model in which X_i is regarded as a parameter and β is known. We put "statistic" in quotes to emphasize the fact that it is not a statistic in the strict sense (because it depends on an unknown parameter). However, it is a statistic in the model in which X_i is regarded as a parameter and β is known and this in turn means that the conditional distribution of (Y_i, W_i) given T_i (and X_i when it modeled as a random variable) does not depend on X_i (and this is the case whether β is known or not). Thus as in the Neyman-Scott problem, conditioning eliminates dependence on the unknown nuisance parameters $X_1, \ldots, X_n$ (or F_X in the case of the structural model).

The conditional distribution of (Y_i, W_i) given T_i is determined by that of Y_i given T_i which can be shown to be

$$P(Y_i = y|T_i = t) = yF(\alpha + \beta t) + (1 - y)\{1 - F(\alpha + \beta t)\}.$$

The conditional likelihood estimating equations are formed from

$$\frac{\partial}{\partial(\alpha, \beta)^T} \log\{P(Y_i = y|T_i = t)\} = \{y - F(\alpha + \beta t)\}\begin{pmatrix}1\\ t\end{pmatrix}$$

resulting in the conditional likelihood estimating equations for (α, β) given by

$$\sum_{i=1}^{n}[Y_i - F\{\alpha + \beta T_i(\beta)\}]\begin{pmatrix}1\\ T_i(\beta)\end{pmatrix} = \begin{pmatrix}0\\ 0\end{pmatrix},$$

where we have written $T_i(\beta)$ instead of T_i as a reminder that the conditioning statistic depends on the parameter β. These estimating equations are unbiased and the general theory of M-estimation (Chapter 7) can be applied to obtain the asymptotic distribution of the estimators found by solving the estimating equations; see Stefanski and Carroll (1987), and Carroll et al. (2006), for further details.

2.4.5 General Form for Exponential Families

It is no coincidence that the previous two examples involve exponential family distributions. The structure of exponential families is such that it is often possible to exploit their properties to eliminate nuisance parameters. Our final example uses Theorem 2.6 (p. 104) of Appendix B, that states that if Y has density of form (2.60, p. 99) with $\boldsymbol{\eta} = (\boldsymbol{\theta}_1^T, \boldsymbol{\theta}_2^T)^T$,

$$f(\boldsymbol{y}; \boldsymbol{\theta}_1, \boldsymbol{\theta}_2) = h(\boldsymbol{y}) \exp\left\{\sum \theta_{1i} W_i + \sum \theta_{2j} V_j - A(\boldsymbol{\theta}_1, \boldsymbol{\theta}_2)\right\},$$

then the conditional distribution of $\boldsymbol{W}$ given $\boldsymbol{V}$ has an exponential family distribution that depends only on $\boldsymbol{\theta}_1$. In principle, then, exponential families often provide an automatic procedure for finding $\boldsymbol{W}$ and $\boldsymbol{V}$. The details are not always simple, but the results are often well worth the effort. Our final important example is conditional logistic regression.

2.4.6 Conditional Logistic Regression

For binary independent Y_i, the standard logistic regression model is

$$P(Y_i = 1) = p_i(\boldsymbol{x}_i, \boldsymbol{\beta}) = \frac{1}{1 + \exp(-\boldsymbol{x}_i^T \boldsymbol{\beta})} = \frac{\exp(\boldsymbol{x}_i^T \boldsymbol{\beta})}{1 + \exp(\boldsymbol{x}_i^T \boldsymbol{\beta})}.$$

The likelihood is

$$\begin{aligned} L(\boldsymbol{\beta} \mid \boldsymbol{Y}, \boldsymbol{X}) &= \prod_{i=1}^{n} p_i(\boldsymbol{x}_i, \boldsymbol{\beta})^{Y_i} \{1 - p_i(\boldsymbol{x}_i, \boldsymbol{\beta})\}^{1-Y_i} \\ &= \prod_{i=1}^{n} \left\{\frac{\exp(\boldsymbol{x}_i^T \boldsymbol{\beta})}{1 + \exp(\boldsymbol{x}_i^T \boldsymbol{\beta})}\right\}^{Y_i} \left\{1 - \frac{\exp(\boldsymbol{x}_i^T \boldsymbol{\beta})}{1 + \exp(\boldsymbol{x}_i^T \boldsymbol{\beta})}\right\}^{1-Y_i} \\ &= \frac{\exp\left\{\sum_{i=1}^n Y_i (\boldsymbol{x}_i^T \boldsymbol{\beta})\right\}}{\prod_{i=1}^n \left\{1 + \exp(\boldsymbol{x}_i^T \boldsymbol{\beta})\right\}} \\ &= c(\boldsymbol{X}, \boldsymbol{\beta}) \exp\left(\sum_{j=1}^{p} \beta_j \sum_{i=1}^{n} x_{ij} Y_i\right), \end{aligned} \tag{2.21}$$

where $T_j = \sum_{i=1}^n x_{ij} Y_i, \ j = 1, \dots, p$ are clearly sufficient statistics for this exponential family . Now suppose that $\theta_1 = \beta_k$ is the parameter of interest, treating the others as nuisance parameters. Then, in our notation $W_1 = T_k = \sum_{i=1}^n x_{ik} Y_i$ and $V = (T_1, \dots, T_{k-1}, T_{k+1}, \dots, T_p)^T$, and the conditional density of interest

$$P(T_k = t_k | T_1 = t_1, \dots, T_{k-1} = t_{k-1}, T_{k+1} = t_{k+1}, \dots, T_p = t_p)$$

$$= \frac{c(t_1, \dots, t_p) \exp(\beta_k t_k)}{\sum_u c(t_1, \dots, t_{k-1}, u, t_{k+1}, \dots, t_p) \exp(\beta_k u)} \tag{2.22}$$

depends only on β_k. These results were first exposited in Cox (1970), and later made computationally feasible by Hirji et al. (1987). This use of conditioning is called *exact conditional logistic regression* and is especially useful for small data sets where asymptotic approximations are not valid.

2.5 The Maximum Likelihood Estimator and the Information Matrix

Having learned to construct the likelihood in a variety of situations, it is now time to use the likelihood to make inferences about model parameters. The theory and methodology of parameter estimation, hypothesis testing, and confidence intervals based on the likelihood are both elegant and practical. In many situations, this methodology is the natural starting point for statistical inference. A central component of this theory is the information matrix $\boldsymbol{I}(\boldsymbol{\theta})$. Thus, after a brief introduction to the definitions of $\widehat{\boldsymbol{\theta}}_{\text{MLE}}$, $\boldsymbol{I}(\boldsymbol{\theta})$, and related likelihood quantities, we focus on computation of $\boldsymbol{I}(\boldsymbol{\theta})$ and then move on to methods for finding $\widehat{\boldsymbol{\theta}}_{\text{MLE}}$ (Section 2.6).

Because they encompass a majority of the likelihoods encountered in practice, we often restrict attention to likelihoods that are continuously differentiable. That is, the row vector of partial derivatives $\ell'(\boldsymbol{\theta}) = \partial \log L(\boldsymbol{\theta} \mid \boldsymbol{Y})/\partial \boldsymbol{\theta}$ exists and is continuous for all $\boldsymbol{\theta}$ in Θ. The partial derivative of the log likelihood plays an important role in likelihood inference via the *likelihood score function*, $\boldsymbol{S}(\boldsymbol{\theta}) = \{\ell'(\boldsymbol{\theta})\}^T$, also denoted by $\boldsymbol{S}(\boldsymbol{Y}, \boldsymbol{\theta})$ when the dependence on the data needs to be emphasized. Thus we define for future reference

$$\boldsymbol{S}(\boldsymbol{\theta}) = \boldsymbol{S}(\boldsymbol{Y}, \boldsymbol{\theta}) = \{\ell'(\boldsymbol{\theta})\}^T = \left\{ \frac{\partial \log L(\boldsymbol{\theta} \mid \boldsymbol{Y})}{\partial \boldsymbol{\theta}} \right\}^T. \tag{2.23}$$

Note that $\boldsymbol{S}(\boldsymbol{\theta})$ and $\boldsymbol{\theta}$ have the same dimension.

Generally, the maximum likelihood estimator $\widehat{\boldsymbol{\theta}}_{\text{MLE}}$ is the value (or values) of $\boldsymbol{\theta}$ where the maximum (over the parameter space Θ) of $L(\boldsymbol{\theta} \mid \boldsymbol{Y})$ is attained, i.e.,

$$L(\widehat{\boldsymbol{\theta}}_{\text{MLE}} \mid \boldsymbol{Y}) \geq L(\boldsymbol{\theta} \mid \boldsymbol{Y}), \qquad \text{for all } \boldsymbol{\theta} \in \Theta. \tag{2.24}$$

Under the assumption that the log likelihood is continuously differentiable, then any $\widehat{\boldsymbol{\theta}}_{\text{MLE}}$ satisfying (2.24) also satisfies the *likelihood equations*,

$$\boldsymbol{S}(\boldsymbol{\theta}) = \boldsymbol{0}. \tag{2.25}$$

The likelihood equations are often used to calculate $\widehat{\boldsymbol{\theta}}_{\text{MLE}}$ using an equation-solving method such as Newton-Raphson iteration. In fact, there are likelihoods where a finite $\widehat{\boldsymbol{\theta}}_{\text{MLE}}$ satisfying (2.24) does not exist, but a local maximum identified by solving (2.25) works well (see Example 2.28, p. 97). Moreover, all the regular asymptotic properties of likelihood-based inference methods follow from the fact that $\widehat{\boldsymbol{\theta}}_{\text{MLE}}$ solves (2.25), i.e., $\boldsymbol{S}(\widehat{\boldsymbol{\theta}}_{\text{MLE}}) = \mathbf{0}$.

Ideally $\widehat{\boldsymbol{\theta}}_{\text{MLE}}$ satisfying (2.24) or solving (2.25) is unique, or at least there is a principled strategy for choosing a single solution from among the possibly multiple values. These issues are discussed in Section 2.6 (p. 80) and Appendix 2.7 (p. 90). For many models used in practice, $\widehat{\boldsymbol{\theta}}_{\text{MLE}}$ is uniquely defined by (2.24) and solves (2.25), and we generally assume these properties, and deal with the exceptional cases as they arise. However, we present a simple example illustrating just such an exceptional case.

Example 2.12 (Exponential threshold model). Suppose that $Y_1, \ldots, Y_n$ are iid from the exponential distribution with threshold parameter μ,

$$f(y;\mu) = \begin{cases} e^{-(y-\mu)} & \mu < y < \infty, \\ 0 & \text{otherwise,} \end{cases} \tag{2.26}$$

for $-\infty < \mu < \infty$. The likelihood is $L(\mu \mid \boldsymbol{Y}) = e^{-n\overline{Y}} e^{n\mu} \prod_{i=1}^{n} I(\mu < Y_i)$, where $I(A)$ is the indicator function of the event A. Thus, the likelihood is zero for any value of $\mu \geq Y_{(1)}$ where $Y_{(1)} = \min\{Y_1, \ldots, Y_n\}$. Figure 2.6 displays a plot of the likelihood for the artificial data set $\{2.47, 2.35, 2.23, 3.53, 2.36\}$.

It is evident from Figure 2.6 that the supremum of $L(\mu \mid \boldsymbol{Y})$ is obtained as μ increases to the point of discontinuity $\widehat{\mu}_{\text{MLE}} = Y_{(1)} = 2.23$, but that $L(2.23 \mid \boldsymbol{Y}) = 0$, so that the MLE $\widehat{\mu}_{\text{MLE}}$ does not satisfy (2.24). Obviously neither does it solve (2.25) as the derivative of $L(\mu \mid \boldsymbol{Y})$ does not exist at $\mu = 2.23$. Changing the definition of the density in (2.26) by replacing $\mu < y < \infty$ with $\mu \leq y < \infty$ results in the MLE satisfying (2.24) but the likelihood is still not differentiable. This example is continued in Problem 2.31 where it is shown that the maximizer of the corresponding $2h$-likelihood, $\widehat{\mu}_{2h\text{-MLE}}$, satisfies the $2h$-likelihood version of (2.24) regardless of whether $\mu < y < \infty$ or $\mu \leq y < \infty$ is used in the definition of the model, and furthermore that for small h, $\widehat{\mu}_{2h\text{-MLE}} = Y_{(1)} - h$ and thus converges to $\widehat{\mu}_{\text{MLE}} = Y_{(1)}$ as $h \to 0$. ♦

Often, the parameter of interest is a function of $\boldsymbol{\theta}$, say $\boldsymbol{\tau} = \boldsymbol{g}(\boldsymbol{\theta})$. An important property of maximum likelihood estimation is that $\widehat{\boldsymbol{\tau}}_{\text{MLE}} = \boldsymbol{g}(\widehat{\boldsymbol{\theta}}_{\text{MLE}})$ is the maximum likelihood estimator of $\boldsymbol{\tau} = \boldsymbol{g}(\boldsymbol{\theta})$. This property follows immediately as spelled out in Zehna (1966) and Casella and Berger (2002, p. 320).

Most of the discussion in the rest of this chapter involves the asymptotic distribution of $\widehat{\boldsymbol{\theta}}_{\text{MLE}}$ under "regular" conditions discussed more thoroughly in Chapter 6. These regularity conditions preclude situations like Example 2.12 above in which the support (all y values for which the density is positive) depends on

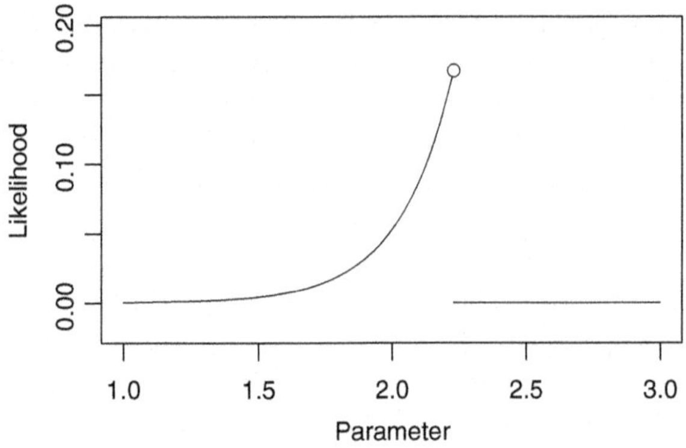

Fig. 2.6 Likelihood of a sample of size $n = 5$ from (2.26)

a parameter value. Under these regularity conditions, the maximum likelihood estimator for an iid sample with density $f(y;\boldsymbol{\theta})$ satisfies

$$\sqrt{n}\left(\widehat{\boldsymbol{\theta}}_{\text{MLE}} - \boldsymbol{\theta}\right) \xrightarrow{d} \text{N}_b\left\{0, \boldsymbol{I}(\boldsymbol{\theta})^{-1}\right\}, \qquad \text{as } n \to \infty, \tag{2.27}$$

where N_b denotes a b-variate multivariate normal distribution, and $\boldsymbol{I}(\boldsymbol{\theta})$ is the Fisher information matrix

$$I_{ij}(\boldsymbol{\theta}) = \text{E}\left[\left\{\frac{\partial}{\partial\theta_i}\log f(Y_1;\boldsymbol{\theta})\right\}\left\{\frac{\partial}{\partial\theta_j}\log f(Y_1;\boldsymbol{\theta})\right\}\right], \tag{2.28}$$

or using notation for derivatives with respect to a vector,

$$\boldsymbol{I}(\boldsymbol{\theta}) = \text{E}\left[\left\{\frac{\partial}{\partial\boldsymbol{\theta}^T}\log f(Y_1;\boldsymbol{\theta})\right\}\left\{\frac{\partial}{\partial\boldsymbol{\theta}}\log f(Y_1;\boldsymbol{\theta})\right\}\right], \tag{2.29}$$

where we use the convention that $\partial \log f(Y_1;\boldsymbol{\theta})/\partial\boldsymbol{\theta}^T = \{\partial \log f(Y_1;\boldsymbol{\theta})/\partial\boldsymbol{\theta}\}^T$. The latter vector appears frequently in likelihood inference and thus it is convenient to have a notation for it. We use $\boldsymbol{s}(Y_i, \boldsymbol{\theta})$ where

$$\boldsymbol{s}(y, \boldsymbol{\theta}) = \left\{\frac{\partial}{\partial\boldsymbol{\theta}}\log f(y;\boldsymbol{\theta})\right\}^T. \tag{2.30}$$

With this notation (2.29) can be written more succinctly as

$$\boldsymbol{I}(\boldsymbol{\theta}) = \text{E}\left[\boldsymbol{s}(Y_1, \boldsymbol{\theta})\boldsymbol{s}(Y_1, \boldsymbol{\theta})^T\right], \tag{2.31}$$

Often students are most familiar with $\boldsymbol{I}(\boldsymbol{\theta})$ for the $b = 1$ case of a single real parameter and the Cramér-Rao lower bound, which states that in finite samples any unbiased estimator must have variance greater than or equal to $\{nI(\theta)\}^{-1}$.

The approximate normality result in (2.27) is related but different, stating that the lower bound is achieved in an asymptotic distribution sense by $\widehat{\boldsymbol{\theta}}_{\mathrm{MLE}}$. In fact, an asymptotic optimality result for maximum likelihood estimation is that $\boldsymbol{\Sigma} - \boldsymbol{I}(\boldsymbol{\theta})^{-1}$ is nonnegative definite, where $\boldsymbol{\Sigma}$ is the asymptotic covariance matrix of any other consistent estimator for $\boldsymbol{\theta}$.

According to (2.27), $\boldsymbol{I}(\boldsymbol{\theta})^{-1}$ is the variance matrix of the limiting distribution of $n^{1/2}(\widehat{\boldsymbol{\theta}}_{\mathrm{MLE}} - \boldsymbol{\theta})$. It is possible that in finite samples (of any size) $\boldsymbol{I}(\boldsymbol{\theta})^{-1}$ differs substantially from $\mathrm{Var}(n^{1/2}\widehat{\boldsymbol{\theta}}_{\mathrm{MLE}})$, as some elements of the exact variance could be infinite or undefined. However, even in such cases the $\mathrm{N}_b\{0, \boldsymbol{I}(\boldsymbol{\theta})^{-1}\}$ distribution often provides a good approximation to the distribution of $n^{1/2}(\widehat{\boldsymbol{\theta}}_{\mathrm{MLE}} - \boldsymbol{\theta})$ in finite samples, and thus $\boldsymbol{I}(\boldsymbol{\theta})^{-1}$ (and its variants defined below) is useful for assessing the variability of estimators. However, it is important to distinguish between the finite-sample expectations $E(\widehat{\boldsymbol{\theta}}_{\mathrm{MLE}})$ and variances $\mathrm{Var}(\widehat{\boldsymbol{\theta}}_{\mathrm{MLE}})$ and the means and variances of the normal distribution used to approximate the distribution of $\widehat{\boldsymbol{\theta}}_{\mathrm{MLE}}$. It is customary to do so by using the modifier "asymptotic." Thus we say that the asymptotic mean of $n^{1/2}(\widehat{\boldsymbol{\theta}}_{\mathrm{MLE}} - \boldsymbol{\theta})$ is zero and the asymptotic variance of $n^{1/2}(\widehat{\boldsymbol{\theta}}_{\mathrm{MLE}} - \boldsymbol{\theta})$ is $\boldsymbol{I}(\boldsymbol{\theta})^{-1}$; or equivalently that the asymptotic mean and variance of $\widehat{\boldsymbol{\theta}}_{\mathrm{MLE}}$ are $\mathbf{0}$ and $\{n\boldsymbol{I}(\boldsymbol{\theta})\}^{-1}$ respectively. Finally, we note that the latter should not be confused with $\lim_{n\to\infty} E(\widehat{\boldsymbol{\theta}}_{\mathrm{MLE}})$ and $\lim_{n\to\infty} \mathrm{Var}(\widehat{\boldsymbol{\theta}}_{\mathrm{MLE}})$ even when these are well defined.

The typical regularity conditions ensuring that (2.27) holds (Chapter 6) also imply that

$$\mathrm{E}\left\{\frac{\partial}{\partial\theta_i}\log f(Y_1;\boldsymbol{\theta})\right\} = 0,$$

so that the diagonal elements of $\boldsymbol{I}(\boldsymbol{\theta})$ are equal to $\mathrm{Var}\{\partial \log f(Y_1;\boldsymbol{\theta})/\partial\theta_i\}$, and the off-diagonal elements equal

$$\mathrm{Cov}\left\{\frac{\partial}{\partial\theta_i}\log f(Y_1;\boldsymbol{\theta}),\ \frac{\partial}{\partial\theta_j}\log f(Y_1;\boldsymbol{\theta})\right\}.$$

If, in addition to the aforementioned regularity conditions, $\ell(\boldsymbol{\theta})$ is twice differentiable, then it transpires that for all i and j,

$$\mathrm{E}\left\{-\frac{\partial^2}{\partial\theta_i\partial\theta_j}\log f(Y_1;\boldsymbol{\theta})\right\} = \mathrm{E}\left[\left\{\frac{\partial}{\partial\theta_i}\log f(Y_1;\boldsymbol{\theta})\right\}\left\{\frac{\partial}{\partial\theta_j}\log f(Y_1;\boldsymbol{\theta})\right\}\right],$$

so that in addition to the defining relationships in (2.28) and (2.29) we have the equivalent definitions

$$I_{ij}(\boldsymbol{\theta}) = \mathrm{E}\left\{-\frac{\partial^2}{\partial\theta_i\partial\theta_j}\log f(Y_1;\boldsymbol{\theta})\right\}, \tag{2.32}$$

or using vectors

$$
\begin{aligned}
\boldsymbol{I}(\boldsymbol{\theta}) &= \mathrm{E}\left\{-\frac{\partial}{\partial \boldsymbol{\theta}} \boldsymbol{s}(Y_i, \boldsymbol{\theta})\right\} \\
&= \mathrm{E}\left\{-\frac{\partial^2}{\partial \boldsymbol{\theta}\, \partial \boldsymbol{\theta}^T} \log f(Y_1; \boldsymbol{\theta})\right\}. \qquad (2.33)
\end{aligned}
$$

The equivalence of (2.29) and (2.33) in the context of (2.27) depends critically on the assumed model being correct, i.e., $Y_1, \ldots, Y_n$ are iid $f(y; \boldsymbol{\theta})$. When the assumed model is not correct (as is often the case), the matrices defined by the expectations in the right-hand sides of (2.29) and (2.33) are not equal and (2.27) is no longer true. The maximum likelihood estimator is still usually asymptotically normal, but generally with a different asymptotic variance and sometimes with a different asymptotic mean. Asymptotic results for such cases are described in Chapter 7.

For practical inference about $\boldsymbol{\theta}$, the main consideration is to be able to calculate $\boldsymbol{I}(\boldsymbol{\theta})$ for a family of densities and to estimate it from the sample. Generally one uses estimates of $\{n\boldsymbol{I}(\boldsymbol{\theta})\}^{-1}$ to get standard errors for $\widehat{\boldsymbol{\theta}}_{\mathrm{MLE}}$ because the convergence in distribution result (2.27) implies that $\widehat{\boldsymbol{\theta}}_{\mathrm{MLE}}$ is approximately $\mathrm{N}_b(\boldsymbol{\theta}, \{n\boldsymbol{I}(\boldsymbol{\theta})\}^{-1})$.

The most obvious estimate of the information matrix is $\boldsymbol{I}(\widehat{\boldsymbol{\theta}}_{\mathrm{MLE}})$ because it is the maximum likelihood estimator of $\boldsymbol{I}(\boldsymbol{\theta})$. However, $\boldsymbol{I}(\boldsymbol{\theta})$ is not always known or easily computed, but it can usually be estimated. Note that for an iid sample $Y_1, \ldots, Y_n$, with log likelihood

$$
\log L(\boldsymbol{\theta} \mid \boldsymbol{Y}) = \sum_{i=1}^{n} \log f(Y_i; \boldsymbol{\theta}),
$$

taking two derivatives and dividing by n, results in

$$
\begin{aligned}
\overline{\boldsymbol{I}}(\boldsymbol{Y}, \boldsymbol{\theta}) &= \frac{1}{n}\sum_{i=1}^{n}\left\{-\frac{\partial}{\partial \boldsymbol{\theta}} \boldsymbol{s}(Y_i, \boldsymbol{\theta})\right\} \\
&= \frac{1}{n}\sum_{i=1}^{n}\left\{-\frac{\partial^2}{\partial \boldsymbol{\theta}\, \partial \boldsymbol{\theta}^T} \log f(Y_i; \boldsymbol{\theta})\right\}, \qquad (2.34)
\end{aligned}
$$

where the over bar "—" indicates averaging.

It follows from (2.32) or (2.33) that $E\{\overline{\boldsymbol{I}}(\boldsymbol{Y}, \boldsymbol{\theta})\} = \boldsymbol{I}(\boldsymbol{\theta})$ when $Y_i \sim f(y; \boldsymbol{\theta})$, $i = 1, \ldots, n$. Thus $\overline{\boldsymbol{I}}(\boldsymbol{Y}, \widehat{\boldsymbol{\theta}}_{\mathrm{MLE}})$ is a natural estimator of $\boldsymbol{I}(\boldsymbol{\theta})$. The matrix $n\overline{\boldsymbol{I}}(\boldsymbol{Y}, \widehat{\boldsymbol{\theta}}_{\mathrm{MLE}})$ is generally called either the *sample information matrix* or the *observed information matrix*. The estimator $\overline{\boldsymbol{I}}(\boldsymbol{Y}, \widehat{\boldsymbol{\theta}}_{\mathrm{MLE}})$ is convenient because it is a byproduct of the common iterative numerical methods used to calculate $\widehat{\boldsymbol{\theta}}_{\mathrm{MLE}}$. Also, Efron and Hinkley (1978) suggest some philosophical reasons for preferring $\{n\overline{\boldsymbol{I}}(\boldsymbol{Y}, \widehat{\boldsymbol{\theta}}_{\mathrm{MLE}})\}^{-1}$ to $\{n\boldsymbol{I}(\widehat{\boldsymbol{\theta}}_{\mathrm{MLE}})\}^{-1}$ as estimators of the asymptotic covariance matrix of $\widehat{\boldsymbol{\theta}}_{\mathrm{MLE}}$.

There is one additional empirical version of (2.31) that has applications when model robustness is an issue (Chapter 7),

$$\overline{I}^*(Y,\theta) = \frac{1}{n}\sum_{i=1}^{n} s(Y_i,\theta)s(Y_i,\theta)^T$$
$$= \frac{1}{n}\sum_{i=1}^{n}\left\{\frac{\partial}{\partial\theta^T}\log f(Y_i;\theta)\right\}\left\{\frac{\partial}{\partial\theta}\log f(Y_i;\theta)\right\}. \tag{2.35}$$

It follows from (2.31) that $E\{\overline{I}^*(Y,\theta)\} = I(\theta)$. However, although this estimator plays an important role in Chapter 7, it is generally inefficient as an estimator of $I(\theta)$ and thus seldom used unless robustness concerns outweigh efficiency concerns.

When $Y_1,\ldots,Y_n$ are independent but not identically distributed (inid), as, for example, in many regression models, the likelihood score function is $S(\theta) = \sum_{i=1}^{n} s_i(Y_i,\theta)$ where, similar to the definition in (2.30),

$$s_i(y,\theta) = \left\{\frac{\partial}{\partial\theta}\log f_i(y;\theta)\right\}^T. \tag{2.36}$$

In this case $\overline{I}(Y,\theta)$ is defined as

$$\overline{I}(Y,\theta) = \frac{1}{n}\sum_{i=1}^{n}\left\{-\frac{\partial}{\partial\theta}s_i(Y_i,\theta)\right\}$$
$$= \frac{1}{n}\sum_{i=1}^{n}\left\{-\frac{\partial^2}{\partial\theta\,\partial\theta^T}\log f_i(Y_i;\theta)\right\}. \tag{2.37}$$

The only difference between (2.34) and (2.37) is that $f(Y_i;\theta)$ in the former is replaced by $f_i(Y_i;\theta)$ in the latter. Then we define the average expected information matrix (also called the average Fisher information matrix) for the whole sample to be

$$\overline{I}(\theta) = \mathrm{E}\{\overline{I}(Y,\theta)\} = \frac{1}{n}\sum_{i=1}^{n}\mathrm{E}\left\{-\frac{\partial^2}{\partial\theta\,\partial\theta^T}\log f_i(Y_i;\theta)\right\}. \tag{2.38}$$

In the iid case $\overline{I}(\theta) = I(\theta)$. Of course, we can also use the analog of (2.29, p. 64), to make this computation, resulting in

$$\overline{I}(\theta) = \frac{1}{n}\sum_{i=1}^{n}\mathrm{E}\left[s_i(Y_i,\theta)s_i(Y_i,\theta)^T\right]$$
$$= \frac{1}{n}\sum_{i=1}^{n}\mathrm{E}\left[\left\{\frac{\partial}{\partial\theta^T}\log f_i(Y_i;\theta)\right\}\left\{\frac{\partial}{\partial\theta}\log f_i(Y_i;\theta)\right\}\right]. \tag{2.39}$$

The empirical analogue of (2.39),

$$\overline{\boldsymbol{I}}^*(\boldsymbol{Y},\boldsymbol{\theta}) = \frac{1}{n}\sum_{i=1}^{n} \boldsymbol{s}_i(Y_i,\boldsymbol{\theta})\boldsymbol{s}_i(Y_i,\boldsymbol{\theta})^T$$
$$= \frac{1}{n}\sum_{i=1}^{n} \left\{\frac{\partial}{\partial\boldsymbol{\theta}^T}\log f_i(Y_i;\boldsymbol{\theta})\right\}\left\{\frac{\partial}{\partial\boldsymbol{\theta}}\log f_i(Y_i;\boldsymbol{\theta})\right\}. \qquad (2.40)$$

plays an important role in Chapter 7, but like its iid counterpart, (2.35), is generally an inefficient estimator of $\boldsymbol{I}(\boldsymbol{\theta})$, and used only when model robustness is an issue.

Finally, there are situations where the average information is not as useful as the total information in the sample; for example, when defining test statistics in non-iid situations. For such cases we define

$$\boldsymbol{I}_{\mathrm{T}}(\boldsymbol{Y},\boldsymbol{\theta}) = -\frac{\partial^2}{\partial\boldsymbol{\theta}\,\partial\boldsymbol{\theta}^T}\log L(\boldsymbol{\theta}\mid\boldsymbol{Y}) \qquad (2.41)$$

and

$$\boldsymbol{I}_{\mathrm{T}}(\boldsymbol{\theta}) = \mathrm{E}\left\{-\frac{\partial^2}{\partial\boldsymbol{\theta}\,\partial\boldsymbol{\theta}^T}\log L(\boldsymbol{\theta}\mid\boldsymbol{Y})\right\} \qquad (2.42)$$

for its expectation. In multi-sample situations, where there may be a series of independent samples with different sample sizes, $\boldsymbol{I}_{\mathrm{T}}(\boldsymbol{\theta})$ avoids dealing with notation problems caused by differing sample sizes. In simpler situations, such as a single sample of independent observations of size n, $\boldsymbol{I}_{\mathrm{T}}(\boldsymbol{\theta})$ reduces to $n\overline{\boldsymbol{I}}(\boldsymbol{\theta})$. The advantage of the seemingly redundant definition $\overline{\boldsymbol{I}}(\boldsymbol{\theta})$ is that for asymptotic analysis, $\overline{\boldsymbol{I}}(\boldsymbol{\theta})$ typically converges to a finite limit as $n \to \infty$, whereas $\boldsymbol{I}_{\mathrm{T}}(\boldsymbol{\theta})$ diverges and thus asymptotic results are more easily stated in terms of $\overline{\boldsymbol{I}}(\boldsymbol{\theta})$ and its limit.

To summarize the various information quantities, we started with $\boldsymbol{I}(\boldsymbol{\theta})$ for the information (also called expected or Fisher information) in one data point with density $f(y;\boldsymbol{\theta})$. The average expected information in a sample of independent data points is denoted $\overline{\boldsymbol{I}}(\boldsymbol{\theta})$ and can be computed by either (2.37, p. 67) or (2.38, p. 67). For n iid data points, $\overline{\boldsymbol{I}}(\boldsymbol{\theta}) = \boldsymbol{I}(\boldsymbol{\theta})$ and $\boldsymbol{I}_{\mathrm{T}}(\boldsymbol{\theta}) = n\overline{\boldsymbol{I}}(\boldsymbol{\theta})$. The various empirical or sample versions of these quantities are averages or sums over the sample; we add $\boldsymbol{Y}$ to the definitions $\overline{\boldsymbol{I}}(\boldsymbol{Y},\boldsymbol{\theta})$ and $\boldsymbol{I}_{\mathrm{T}}(\boldsymbol{Y},\boldsymbol{\theta})$ to emphasize that they depend on the sample data. Taking expectations with respect to the sample values eliminates the dependence on $\boldsymbol{Y}$ and yields $\overline{\boldsymbol{I}}(\boldsymbol{\theta})$ and $\boldsymbol{I}_{\mathrm{T}}(\boldsymbol{\theta})$, respectively.

In the likely event that the sundry versions of and notations for information matrices seem like information overload, Table 2.1 provides a concise summary of the key likelihood and information quantities for three classes of data types: independent

Table 2.1 Summary of important likelihood quantities and their notations for iid, inid, and general data sets. Recall that $s(y, \theta) = \partial \log f(y;\theta)/\partial\theta^T$, and $s_i(y, \theta) = \partial \log f_i(y;\theta)/\partial\theta^T$ as defined in (2.30, p. 64) and (2.36, p. 67).

	Data Type		
	iid	inid	General
$L(\theta \mid Y)$	$\prod_{i=1}^n f(Y_i;\theta)$	$\prod_{i=1}^n f_i(Y_i;\theta)$	$f(Y;\theta)$
$\ell(\theta) = \log L(\theta \mid Y)$	$\sum_{i=1}^n \log f(Y_i;\theta)$	$\sum_{i=1}^n \log f_i(Y_i;\theta)$	$\log f(Y;\theta)$
$S(\theta) = \frac{\partial}{\partial\theta^T}\ell(\theta)$	$\sum_{i=1}^n s(Y_i,\theta)$	$\sum_{i=1}^n s_i(Y_i,\theta)$	$\frac{\partial}{\partial\theta^T}\log f(Y;\theta)$
$I_T(Y,\theta) = -\frac{\partial}{\partial\theta}S(\theta)$	$-\sum_{i=1}^n \frac{\partial}{\partial\theta}s(Y_i,\theta)$	$-\sum_{i=1}^n \frac{\partial}{\partial\theta}s_i(Y_i,\theta)$	$-\frac{\partial}{\partial\theta}S(\theta)$
$I_T(\theta) = E\{I_T(Y,\theta)\}$	$nI(\theta)$	$n\overline{I}(\theta)$	$I_T(\theta)$
$\overline{I}(Y,\theta) = \frac{1}{n}I_T(Y,\theta)$	$\overline{I}(Y,\theta)$	$\overline{I}(Y,\theta)$	–
$\overline{I}(\theta) = E\{\overline{I}(Y,\theta)\}$	$I(\theta)$	$\overline{I}(\theta)$	–
$\overline{I}^*(Y,\theta)$	$\frac{1}{n}\sum_{i=1}^n s(Y_i,\theta)s(Y_i,\theta)^T$	$\frac{1}{n}\sum_{i=1}^n s_i(Y_i,\theta)s_i(Y_i,\theta)^T$	–
$\overline{I}^*(\theta) = E\{\overline{I}^*(Y,\theta)\}$	$I(\theta)$	$\overline{I}(\theta)$	–

and identically distributed (iid); independent and not identically distributed (inid); and the general case that allows for dependence or nonidentical distributions.

2.5.1 Examples of Information Matrix Calculations

We now look at a variety of fairly simple examples of calculating $I(\theta)$ analytically (as opposed to numerically). For twice differentiable likelihoods the information matrix can be determined using either (2.29, p. 64) or (2.33, p. 66). In specific cases one formula is often simpler to use than the other, but there is no general guidance on this matter. Of course, calculating $I(\theta)$ using both definitions provides a check for errors.

Example 2.13 (Binomial distribution). The binomial probability mass function is

$$f(y;p) = \binom{n}{y} p^y (1-p)^{n-y}, \quad y = 0, 1, \ldots n.$$

Differentiating the log density results in

$$\begin{aligned} s(y,p) = \frac{\partial}{\partial p} \log f(y;p) &= \frac{\partial}{\partial p}(\text{constant} + y \log p + (n-y)\log(1-p)) \\ &= \frac{y}{p} - \frac{(n-y)}{1-p} = \frac{y-np}{p(1-p)}, \end{aligned}$$

and using (2.31) leads to

$$I(p) = I_{\mathrm{T}}(p) = \operatorname{Var}\{s(Y,p)\} = \frac{np(1-p)}{\{p(1-p)\}^2} = \frac{n}{p(1-p)}.$$

Alternatively, as

$$-\frac{\partial}{\partial p} s(y,p) = \frac{np(1-p) + (y-np)(1-2p)}{\{p(1-p)\}^2},$$

using (2.33, p. 66) results in

$$I(p) = I_{\mathrm{T}}(p) = \mathrm{E}\left\{-\frac{\partial}{\partial p} s(y,p)\right\} = \frac{np(1-p)}{\{p(1-p)\}^2} = \frac{n}{p(1-p)}.$$

Note that $I(p) = I_{\mathrm{T}}(p)$ in this case, but our preferred notation is $I_{\mathrm{T}}(p)$ because often the binomial variable Y is the sum (either explicitly or implicitly) of n independent Bernoulli(p) variables, i.e., $Y = X_1 + \cdots + X_n$. The information of a single Bernoulli variate X_i is $I(p) = 1/\{p(1-p)\}$, and $I_{\mathrm{T}}(p) = n/\{p(1-p)\}$ for the entire Bernoulli sample. Thus the total information is the same whether the data are viewed as a Bernoulli n-sample or as a single binomial observation.

Finally note that for binomial data Y, solving $s(Y,p) = 0$ is equivalent to solving

$$\frac{Y-np}{p(1-p)} = 0$$

and so $\widehat{p}_{\mathrm{MLE}} = Y/n$. Thus $\operatorname{Var}(\widehat{p}_{\mathrm{MLE}}) = p(1-p)/n$, corresponding exactly to $I_{\mathrm{T}}(p)^{-1}$. ♦

Example 2.14 (Multinomial distribution). For the multinomial model with parameters $(n; p_1, \ldots, p_k)$ where $p_k = 1 - \sum_{i=1}^{k-1} p_i$ (see p. 32), computations similar to that for the binomial give the $k-1$ by $k-1$ information matrix for $\boldsymbol{p} = (p_1, \ldots, p_{k-1})^T$,

$$\boldsymbol{I}_{\mathrm{T}}(\boldsymbol{p}) = n \begin{pmatrix} \frac{1}{p_1} + \frac{1}{p_k} & \frac{1}{p_k} & \frac{1}{p_k} & \cdots & \frac{1}{p_k} \\ \frac{1}{p_k} & \frac{1}{p_2} + \frac{1}{p_k} & \frac{1}{p_k} & \cdots & \frac{1}{p_k} \\ \cdot & \cdot & \cdot & \cdot & \cdot \\ \frac{1}{p_k} & \frac{1}{p_k} & \cdots & \frac{1}{p_k} & \frac{1}{p_{k-1}} + \frac{1}{p_k} \end{pmatrix}.$$

In matrix notation, $\boldsymbol{I}_{\mathrm{T}}(\boldsymbol{p}) = n\{\mathrm{diag}(1/p_1, \ldots, 1/p_{k-1}) + \mathbf{1}\mathbf{1}^T/p_k\}$, for which it can be shown that $\boldsymbol{I}_{\mathrm{T}}(\boldsymbol{p})^{-1} = \{\mathrm{diag}(\boldsymbol{p}) - \boldsymbol{p}\boldsymbol{p}^T\}/n$, or

$$\boldsymbol{I}_{\mathrm{T}}(\boldsymbol{p})^{-1} = \frac{1}{n} \begin{pmatrix} p_1(1-p_1) & -p_1 p_2 & -p_1 p_3 & \cdots & -p_1 p_{k-1} \\ -p_2 p_1 & p_2(1-p_2) & -p_2 p_3 & \cdots & -p_2 p_{k-1} \\ \cdot & \cdot & \cdot & \cdot & \cdot \\ -p_{k-1} p_1 & -p_{k-1} p_2 & -p_{k-1} p_3 & \cdots & p_{k-1}(1-p_{k-1}) \end{pmatrix}.$$

This is the covariance matrix of the first $k - 1$ sample proportions. For certain purposes, adding a row and column for the kth sample proportion to this matrix to obtain the covariance matrix of all k sample proportions is convenient. However, the resulting covariance matrix is singular. ♦

Example 2.15 (Normal distribution). For the normal family both parameterizations $\boldsymbol{\theta}^T = (\mu, \sigma)$ and $\boldsymbol{\theta}^T = (\mu, \sigma^2)$ are used in practice, each leading to a different information matrix . Here we use $\boldsymbol{\theta}^T = (\mu, \sigma)$. The logarithm of the $\mathrm{N}(\mu, \sigma^2)$ density is

$$\log f(y; \mu, \sigma) = \text{constant} - \log \sigma - \frac{1}{2\sigma^2}(y - \mu)^2,$$

and

$$\frac{\partial}{\partial \mu} \log f(y; \mu, \sigma) = \frac{y - \mu}{\sigma^2},$$

so that the (1,1) element of $\boldsymbol{I}(\boldsymbol{\theta})$ is

$$\mathrm{Var}\left(\frac{Y - \mu}{\sigma^2}\right) = \frac{1}{\sigma^2}.$$

Similarly

$$\frac{\partial}{\partial \sigma} \log f(y; \mu, \sigma) = -\frac{1}{\sigma} + \frac{(y - \mu)^2}{\sigma^3},$$

and the (2,2) element of $\boldsymbol{I}(\boldsymbol{\theta})$ is

$$\mathrm{Var}\left(-\frac{1}{\sigma} + \frac{(Y - \mu)^2}{\sigma^3}\right) = \frac{2\sigma^4}{\sigma^6} = \frac{2}{\sigma^2}.$$

The off-diagonal elements of $\boldsymbol{I}(\boldsymbol{\theta})$ are 0 because the covariance between $Y - \mu$ and $(Y - \mu)^2$ is zero. Thus

$$\boldsymbol{I}(\mu, \sigma) = \frac{1}{\sigma^2} \begin{pmatrix} 1 & 0 \\ 0 & 2 \end{pmatrix}. \tag{2.43}$$

For an iid N(μ, σ^2) sample the MLEs of μ and σ are $\overline{Y}$ and $s_n = \left\{n^{-1} \sum_{i=1}^{n} (Y_i - \overline{Y})^2\right\}^{1/2}$. In large samples these MLEs are approximately normally distributed with asymptotic covariance matrix

$$\frac{\boldsymbol{I}(\mu, \sigma)^{-1}}{n} = \frac{\sigma^2}{n} \begin{pmatrix} 1 & 0 \\ 0 & 1/2 \end{pmatrix}.$$

Plugging in s_n for σ gives large sample standard errors $s_n/\sqrt{n}$ for $\widehat{\mu} = \overline{Y}$ and $s_n/\sqrt{2n}$ for $\widehat{\sigma} = s_n$. ◆

Example 2.16 (Extreme value distribution). The extreme value density is $f(y; \mu, \sigma) = f_0\left\{(y - \mu)/\sigma\right\}/\sigma$, where

$$f_0(x) = e^{-x} e^{-e^{-x}}, \quad -\infty < x < \infty.$$

If Y has density $f(y; \mu, \sigma)$, then $X = (Y - \mu)/\sigma$ has density $f_0(x)$. The first moment of X is

$$\begin{aligned} \text{E}(X) &= \int_{-\infty}^{\infty} x e^{-x} e^{-e^{-x}} \, dx \\ &= \int_{0}^{\infty} -(\log u) e^{-u} \, du = \gamma, \end{aligned}$$

where $\gamma = .577215\ldots$ is Euler's constant and the latter integral is obtained via the transformation of variables $u = \exp(-x)$. Similarly

$$\text{E}(X^2) = \int_0^{\infty} (\log u)^2 e^{-u} \, du = \frac{\pi^2}{6} + \gamma^2.$$

Thus the variance of X is $\pi^2/6 = 1.6449$ (to four decimals). Since $Y = \sigma X + \mu$, we have $\text{E}(Y) = \sigma\gamma + \mu$ and $\text{Var}(Y) = 1.6449\sigma^2$. Note that μ is not the mean of Y and σ^2 is not the variance of Y. The identifications $\text{E}(Y) = \mu$ and $\text{Var}(Y) = \sigma^2$ only occur in location-scale families like the normal when $\int x f_0(x) \, dx = 0$ and $\int x^2 f_0(x) \, dx = 1$. Alternatively we can define a standardized form for the extreme value density, $f_*(x) = f_0\left\{(\pi/\sqrt{6})(x + \gamma)\right\} \pi/\sqrt{6}$, with the property that μ and σ^2 of $f_*\left\{(x - \mu)/\sigma\right\}/\sigma$ are the mean and variance.

The logarithm of the density of Y is

$$\log f(y; \mu, \sigma) = -\log \sigma - \left(\frac{y - \mu}{\sigma}\right) - \exp\left\{-\left(\frac{y - \mu}{\sigma}\right)\right\},$$

so that

$$\frac{\partial}{\partial \mu} \log f(y;\mu,\sigma) = \frac{1}{\sigma} - \frac{1}{\sigma} \exp\left\{-\left(\frac{y-\mu}{\sigma}\right)\right\},$$

and

$$\frac{\partial^2}{\partial \mu^2} \log f(y;\mu,\sigma) = -\frac{1}{\sigma^2} \exp\left\{-\left(\frac{y-\mu}{\sigma}\right)\right\}.$$

Using the second derivative definition of the information matrix (2.33, p. 66), the (1,1) element is

$$\begin{aligned} \mathrm{E}\left[\frac{1}{\sigma^2}\exp\left\{-\left(\frac{Y-\mu}{\sigma}\right)\right\}\right] &= \mathrm{E}\left\{\frac{1}{\sigma^2}\exp(-X)\right\} \\ &= \frac{1}{\sigma^2}\int_{-\infty}^{\infty} e^{-x}e^{-x}e^{-e^{-x}}\,dx \\ &= \frac{1}{\sigma^2}\int_{0}^{\infty} ue^{-u}\,du = \frac{1}{\sigma^2}\Gamma(2) = \frac{1}{\sigma^2}. \end{aligned}$$

Similarly

$$\begin{aligned} \frac{\partial^2}{\partial\sigma\,\partial\mu}\log f(y;\mu,\sigma) &= -\frac{1}{\sigma^2} + \frac{1}{\sigma^2}\exp\left\{-\left(\frac{y-\mu}{\sigma}\right)\right\} \\ &\quad - \frac{1}{\sigma^2}\left(\frac{y-\mu}{\sigma}\right)\exp\left\{-\left(\frac{y-\mu}{\sigma}\right)\right\}, \end{aligned}$$

and the (1,2) element of the information matrix is

$$\begin{aligned} \frac{1}{\sigma^2} - \frac{1}{\sigma^2}\mathrm{E}\left(e^{-X}\right) + \frac{1}{\sigma^2}\mathrm{E}\left(Xe^{-X}\right) &= \frac{1}{\sigma^2} - \frac{1}{\sigma^2} - \frac{1}{\sigma^2}\int_0^{\infty}(\log u)ue^{-u}\,du \\ &= -\frac{1}{\sigma^2}(1-\gamma) \approx -\frac{.423}{\sigma^2}. \end{aligned}$$

Finally,

$$\begin{aligned} \frac{\partial^2}{\partial\sigma^2}\log f(y;\mu,\sigma) &= \frac{1}{\sigma^2} - \frac{2}{\sigma^2}\left(\frac{y-\mu}{\sigma}\right) + \frac{2}{\sigma^2}\left(\frac{y-\mu}{\sigma}\right)\exp\left\{-\left(\frac{y-\mu}{\sigma}\right)\right\} \\ &\quad - \frac{1}{\sigma^2}\left(\frac{y-\mu}{\sigma}\right)^2\exp\left\{-\left(\frac{y-\mu}{\sigma}\right)\right\}. \end{aligned}$$

The (2,2) element of the information matrix is

$$\frac{1}{\sigma^2}\left\{-1+2\gamma+2(1-\gamma)+(1-\gamma)^2+\frac{\pi^2}{6}-1\right\} = \frac{1}{\sigma^2}\left\{(1-\gamma)^2+\frac{\pi^2}{6}\right\} \approx \frac{1.824}{\sigma^2}.$$

Thus the information matrix for the extreme value distribution is (to three decimal places)

$$I(\mu,\sigma) = \frac{1}{\sigma^2}\begin{pmatrix} 1 & -.423 \\ -.423 & 1.824 \end{pmatrix},$$

and the MLEs for an iid sample of size n are approximately normal with mean $(\mu,\sigma)^T$ and variance

$$\frac{I(\mu,\sigma)^{-1}}{n} = \frac{\sigma^2}{n}\begin{pmatrix} 1.109 & .257 \\ .257 & .608 \end{pmatrix}. \tag{2.44}$$

♦

Example 2.17 (Normal error regression models). Consider the normal regression model likelihood (2.11, p. 50). The components are independent but not identically distributed. Thus the relevant definition of the information matrix is (2.38, p. 67), which for the normal linear model results in

$$\overline{I}(\beta,\sigma) = \frac{1}{\sigma^2}\begin{pmatrix} X^T X/n & 0 \\ 0 & 2 \end{pmatrix}. \tag{2.45}$$

Recall that this is the average expected information matrix for the entire sample and is comparable to $I(\theta)$ in iid samples. Inverting $I_T(\beta,\sigma) = n\overline{I}(\beta,\sigma)$, shows that the asymptotic variance of $\widehat{\beta}$ is the exact finite-sample variance, $(X^T X)^{-1}\sigma^2$, and that the asymptotic variance of $\widehat{\sigma}_{\text{MLE}}$ is $\sigma^2/2n$, the same as for the maximum likelihood estimator of σ from an iid $N(\mu,\sigma^2)$ sample. The latter two large sample variances are equal because neither one incorporates a finite-sample degrees-of-freedom adjustment.

The information matrix for the nonlinear model $Y_i = g(x_i,\beta) + e_i$ of (2.3.2, p. 53) is

$$\overline{I}(\beta,\sigma) = \frac{1}{\sigma^2}\begin{pmatrix} G^T G/n & 0 \\ 0 & 2 \end{pmatrix}, \tag{2.46}$$

where G is the matrix with elements $[G]_{ij} = \partial g(x_i,\beta)/\partial\beta_j$. ♦

Example 2.18 (Generalized Linear Models). Consider the likelihood (2.15, p. 54). Taking derivatives with respect to β, the score equation for estimating β is

$$S(\beta,\phi) = \sum_{i=1}^{n} \frac{\{Y_i - b'(x_i^T\beta)\}\, x_i}{a_i(\phi)} = \sum_{i=1}^{n} \frac{(Y_i - \mu_i)\, x_i}{a_i(\phi)} = 0. \tag{2.47}$$

Taking another derivative with respect to β, we have the average information matrix for β:

$$\overline{I}(Y,\beta) = X^T V X/n, \tag{2.48}$$

where $X^T V X = \sum_{i=1}^{n} \left[b''(x_i^T \beta)/a_i(\phi)\right] x_i x_i^T$ and

$$V = \text{diag}\left\{\text{Var}(Y_1)/a_1(\phi)^2, \ldots, \text{Var}(Y_n)/a_n(\phi)^2\right\}.$$

Note that $\overline{I}(Y, \beta) = X^T V X/n$ does not depend on the Y_i's and thus $\overline{I}(Y, \beta) = \overline{I}(\beta)$; that is, the average Fisher information for β is the same as the average observed information when using the canonical link. We have slightly abused notation here because $\overline{I}(\beta)$ often depends on ϕ as well. The full average expected information matrix has the form

$$\overline{I}(\beta, \phi) = \begin{pmatrix} X^T V X/n & 0 \\ 0 & I_n(\phi) \end{pmatrix}.$$

The off-diagonal value 0 follows from taking a derivative of (2.47) with respect to ϕ and then an expectation. The information submatrix $\overline{I}(\phi)$ relating to ϕ is rarely mentioned in the generalized linear model literature.

For logistic regression, $\overline{I}(\beta) = X^T V X/n$ where the diagonal matrix is $V = \text{diag}\left[p_1(1 - p_1), \ldots, p_n(1 - p_n)\right]$.

For Poisson data, $b(\theta_i) = e^{\theta_i}$ and $a_i(\phi) = 1$, and thus $b'(\theta_i) = e^{\theta_i}$. Inverting b' leads to the canonical link $g(\mu_i) = \log(\mu_i)$ and modeling the mean μ_i by $e^{x_i^T \beta}$. The variance of Y_i is μ_i and $\overline{I}(\beta) = X^T V X/n$ where $V = \text{diag}\left[\mu_1, \ldots, \mu_n\right]$. ♦

2.5.2 *Variance Cost for Adding Parameters to a Model*

For the extreme value location-scale model the asymptotic $\text{corr}(\widehat{\mu}_{\text{MLE}}, \widehat{\sigma}_{\text{MLE}}) \approx 0.31$, whereas for the normal location-scale model the asymptotic correlation is $\text{corr}(\widehat{\mu}_{\text{MLE}}, \widehat{\sigma}_{\text{MLE}}) = 0$ (in fact $\text{corr}(\widehat{\mu}_{\text{MLE}}, \widehat{\sigma}_{\text{MLE}}) = 0$ in finite samples). One consequence of the nonzero correlation in the extreme value model is that for estimating μ there is a cost in terms of increased variability for jointly estimating σ. The calculations in Example 2.16 (p. 72) show that when σ is known the asymptotic variance of $\widehat{\sigma}_{\text{MLE}}$ is σ^2/n; whereas if σ is unknown and estimated jointly with μ, then the asymptotic variance of $\widehat{\sigma}_{\text{MLE}}$ is $1.109\sigma^2/n$. In this model the asymptotic variance inflation in estimating μ due to estimating σ is not large. However, it is indicative of a pervasive phenomenon with important practical implications:

Key Concept: Whenever parameters are added to a model, the diagonal elements of the inverse information matrix are always greater than or equal to the corresponding elements of the simpler model.

This follows from properties of the inverse of a partitioned matrix. For example, suppose that the information matrix I be partitioned as

$$I = \begin{pmatrix} I_{11} & I_{12} \\ I_{21} & I_{22} \end{pmatrix},$$

where I_{11} is a scalar corresponding to the partition $\boldsymbol{\theta} = (\theta_1, \boldsymbol{\theta}_2^T)^T$ where θ_1 is scalar. If $\boldsymbol{\theta}_2$ is known, then the information matrix (a scalar in this case) for θ_1 is I_{11} and the asymptotic variance of the MLE of θ_1 is $1/(nI_{11})$. When $\boldsymbol{\theta}_2$ is unknown and jointly estimated the asymptotic variance of the MLE of θ_1 is $1/(nI^{(11)})$ where $I^{(11)} = (I_{11} - \boldsymbol{I}_{12}\boldsymbol{I}_{22}^{-1}\boldsymbol{I}_{21})^{-1}$, the (1,1) element of $\boldsymbol{I}^{-1}$. Clearly $I_{11}^{-1} \leq (I_{11} - \boldsymbol{I}_{12}\boldsymbol{I}_{22}^{-1}\boldsymbol{I}_{21})^{-1}$ if and only if $I_{11} \geq I_{11} - \boldsymbol{I}_{12}\boldsymbol{I}_{22}^{-1}\boldsymbol{I}_{21}$, or equivalently if and only if $\boldsymbol{I}_{12}\boldsymbol{I}_{22}^{-1}\boldsymbol{I}_{21} \geq 0$. The latter inequality always holds because $\boldsymbol{I}_{22}$ is a covariance matrix and thus nonnegative definite. Also the inequality is strict whenever $\boldsymbol{I}_{12} \neq \boldsymbol{0}$ and $\boldsymbol{I}_{22}$ is positive definite. It follows that the asymptotic variance of the maximum likelihood estimator of θ_1 when the other parameters are known is less than or equal to the asymptotic variance of the estimator of θ_1 when at least one of the other parameters is jointly estimated with θ_1. The inequality is generally strict unless $\boldsymbol{I}_{12} = \boldsymbol{0}$ (e.g., as in the normal location-scale model).

We illustrate the variance inflation phenomenon first with a general location-scale model and then with some three-parameter models.

2.5.2a Location-Scale Models

Suppose that

$$f(y; \mu, \sigma) = \frac{1}{\sigma} f_0\left(\frac{y-\mu}{\sigma}\right),$$

where $f_0(x)$ is a density on $(-\infty, \infty)$. Straightforward calculations yield

$$\boldsymbol{I}(\mu,\sigma) = \frac{1}{\sigma^2}\begin{pmatrix} \displaystyle\int_{-\infty}^{\infty}\left\{\frac{f_0'(x)}{f_0(x)}\right\}^2 f_0(x)dx & \displaystyle\int_{-\infty}^{\infty} x\left\{\frac{f_0'(x)}{f_0(x)}\right\}^2 f_0(x)dx \\ \displaystyle\int_{-\infty}^{\infty} x\left\{\frac{f_0'(x)}{f_0(x)}\right\}^2 f_0(x)dx & \displaystyle\int_{-\infty}^{\infty}\left\{1 + x\frac{f_0'(x)}{f_0(x)}\right\}^2 f_0(x)dx \end{pmatrix}.$$

Recall that μ and σ^2 are the mean and variance only when $\int_{-\infty}^{\infty} x f_0(x)dx = 0$ and $\int_{-\infty}^{\infty} x^2 f_0(x)dx = 1$. When the base density is symmetric about zero, $f_0(x) = f_0(-x)$, then the off-diagonal term $\boldsymbol{I}_{12} = 0$ as it is the integral of an odd function about zero, and $\widehat{\mu}_{\rm MLE}$ and $\widehat{\sigma}_{\rm MLE}$ are uncorrelated. Thus when $f_0(x)$ is symmetric about 0, there is no asymptotic variance inflation cost due to estimating σ in addition to μ.

The normal distribution is symmetric with information matrix given by (2.43, p. 72) and therefore there is no asymptotic variance inflation for estimating μ with unknown σ. In contrast, there is an asymptotic variance inflation in the asymmetric extreme value distribution, as noted in the example at the beginning of this section.

2.5.2b Three-Parameter Models

The next three examples are three-parameter models and illustrate clearly the asymptotic costs of estimating a third parameter compared to just estimating two of the parameters.

Example 2.19 (Transformed generalized gamma). The random variable X has a generalized gamma distribution (Stacy 1962) when $-\log X$ has distribution function $F(t;\mu,\sigma,\lambda) = F_\lambda\left((t-\mu)/\sigma\right)$, where

$$F_\lambda(x) = 1 - \frac{1}{\Gamma(\lambda)}\int_0^{\exp(-x)} u^{\lambda-1}e^{-u}du.$$

For $\lambda = 1$, $F(t;\mu,\sigma,1)$ is the extreme value location-scale distribution. For the model with μ, σ and λ unknown with the true value of $\lambda = 1$, the information matrix is

$$\boldsymbol{I}(\mu,\sigma,\lambda=1) = \frac{1}{\sigma^2}\begin{pmatrix} 1 & -.423 & -1 \\ -.423 & 1.824 & -.577 \\ -1 & -.577 & 1.645 \end{pmatrix},$$

with inverse

$$\boldsymbol{I}(\mu,\sigma,\lambda=1)^{-1} = \sigma^2\begin{pmatrix} 43.8 & 20.9 & 34.0 \\ 20.9 & 10.6 & 16.4 \\ 34.0 & 16.4 & 27.0 \end{pmatrix}.$$

Comparison to the extreme value location-scale model inverse information matrix in (2.44, p. 74) reveals a huge asymptotic variance inflation due to estimating λ. For example, the ratio of asymptotic variances for $\widehat{\mu}_{\text{MLE}}$ when $\lambda = 1$ is unknown and estimated and when $\lambda = 1$ is correctly assumed known, is $43.8/1.109 = 39.5$. ◆

Example 2.20 (Burr II distribution). With location-scale parameters, the Burr II distribution function is

$$F_\lambda(t;\mu,\sigma,\lambda) = \left[1 + \exp\left\{-\frac{(t-\mu)}{\sigma}\right\}\right]^{-\lambda}.$$

At $\lambda = 1$, this is the logistic location-scale model distribution. For the model with μ, σ and λ unknown with the true value of $\lambda = 1$, the information matrix is (to three decimal places)

$$\boldsymbol{I}(\mu,\sigma,\lambda=1) = \frac{1}{\sigma^2}\begin{pmatrix} .333 & 0 & .5 \\ 0 & 1.430 & -.5 \\ .5 & -.5 & 1.0 \end{pmatrix}, \tag{2.49}$$

with inverse

$$I(\mu,\sigma,\lambda=1)^{-1} = \sigma^2 \begin{pmatrix} 32.9 & -7.0 & -20.0 \\ -7.0 & 2.3 & 4.7 \\ -20.0 & 4.7 & 13.3 \end{pmatrix}.$$

Because $I_{1,2} = I_{2,1} = 0$ in (2.49) we know that the asymptotic variance of $\widehat{\mu}_{\text{MLE}}$ for the case where $\lambda = 1$ is correctly assumed is $1/(nI_{1,1}) = \sigma^2/(.333n) = 3\sigma^2/n$. Thus the variance inflation when using $\widehat{\mu}_{\text{MLE}}$ with $\lambda = 1$ unknown and jointly estimated is $32.9/3 = 11.0$. ♦

Example 2.21 (Box-Cox model). In a famous paper, Box and Cox (1964) studied the use of power transformations in data analysis. The Box-Cox family of transformations is given by, for $Y > 0$,

$$Y^{(\lambda)} = \begin{cases} \dfrac{Y^\lambda - 1}{\lambda}, & \lambda \neq 0, \\ \log Y, & \lambda = 0, \end{cases}$$

and is defined so that $Y^{(\lambda)}$ is continuous as $\lambda \to 0$. The usual assumption is that $Y^{(\lambda)}$ has a $\text{N}(\mu,\sigma^2)$ distribution. For example, with $\lambda = 1/3$, then $(Y^{1/3} - 1)/(1/3) = 3(Y^{1/3} - 1)$ has a $\text{N}(\mu,\sigma^2)$ distribution. The normality assumption is necessarily violated in theory for any $\lambda \neq 0$ because the logarithmic transformation ($\lambda = 0$) is the only member of the family that maps the positive real axis onto the whole real line. Nevertheless normality is often a useful working assumption. The quality of the approximations to normality that can be achieved when $\lambda \neq 0$ has been studied by Hernandez and Johnson (1980).

For this example we consider the case when the true $\lambda = 0$ ($Y \sim$ lognormal) so that the normality assumption is exactly valid. For this model when all three parameters (μ,σ,λ) are estimated, the information matrix of (μ,σ,λ) is

$$I(\mu,\sigma,\lambda=0) = \frac{1}{\sigma^2} \begin{pmatrix} 1 & 0 & -\tau_1 \\ 0 & 2 & -2\sigma\mu \\ -\tau_1 & -2\sigma\mu & \tau_2 \end{pmatrix},$$

where $\tau_1 = (\sigma^2 + \mu^2)/2$ and $\tau_2 = \left(7\sigma^4 + 10\sigma^2\mu^2 + \mu^4\right)/4$. The inverse is

$$I(\mu,\sigma,\lambda=0)^{-1} = \begin{pmatrix} \eta_1 & \dfrac{\mu\sigma\eta_2}{3} & \dfrac{\eta_1}{3} \\ \dfrac{\mu\sigma\eta_2}{3} & \left(\dfrac{\sigma^2}{2} + \dfrac{2\mu^2}{3}\right) & \dfrac{2\mu}{3\sigma} \\ \dfrac{\eta_1}{3} & \dfrac{2\mu}{3\sigma} & \dfrac{2}{3\sigma^2} \end{pmatrix}.$$

where $\eta_1 = (7\sigma^2 + 2\mu^2 + \mu^4/\sigma^2)/6$ and $\eta_2 = (1 + \mu^2/\sigma^2)$. Of course when $\lambda = 0$ is correctly assumed, the variance of $\widehat{\mu}_{\mathrm{MLE}}$ is σ^2/n. Thus the asymptotic variance inflation factor when using $\widehat{\mu}_{\mathrm{MLE}}$ with $\lambda = 0$ but unknown is $\eta_1/\sigma^2 = (7 + 2\mu^2/\sigma^2 + \mu^4/\sigma^4)/6$, which can be arbitrarily large depending on (μ, σ).

Bickel and Doksum (1981) touched off a heated debate by arguing that standard errors for $\widehat{\mu}_{\mathrm{MLE}}$ should be derived from $\boldsymbol{I}(\mu, \sigma, \lambda)^{-1}$, whereas the accepted practice at the time was to ignore the fact that λ was jointly estimated. Hinkley and Runger (1984) countered that the accepted practice is correct because the analysis should be conditional on the scale (as determined by λ) chosen, even if that scale is chosen based on the data. ♦

2.5.3 *The Information Matrix for Transformed and Modeled Parameters*

Suppose that the density $f(y;\boldsymbol{\theta})$ has b dimensional parameter $\boldsymbol{\theta}$, $b \times b$ information matrix $\boldsymbol{I}(\boldsymbol{\theta})$, and we are interested in either a one-to-one transformation $\boldsymbol{\theta} = \boldsymbol{g}(\boldsymbol{\beta})$ where $\boldsymbol{\beta}$ also has dimension b, or a reduced-parameter model where $\boldsymbol{\theta} = \boldsymbol{g}(\boldsymbol{\beta})$ and $\boldsymbol{\beta}$ has dimension $s < b$. We show that for either case,

$$\boldsymbol{I}(\boldsymbol{\beta}) = \left\{\frac{\partial \boldsymbol{g}(\boldsymbol{\beta})}{\partial \boldsymbol{\beta}}\right\}^T I(\boldsymbol{g}(\boldsymbol{\beta}))\left\{\frac{\partial \boldsymbol{g}(\boldsymbol{\beta})}{\partial \boldsymbol{\beta}}\right\} \tag{2.50}$$

by using the vector calculus chain rule for the derivative of the real-valued function $\ell(\boldsymbol{\beta}) = \log f(y;\boldsymbol{\theta}) = \log f(y;\boldsymbol{g}(\boldsymbol{\beta}))$. Doing so results in

$$\frac{\partial \ell(\boldsymbol{\beta})}{\partial \boldsymbol{\beta}} = \frac{\partial \log f(y;\boldsymbol{g}(\boldsymbol{\beta}))}{\partial \boldsymbol{g}}\frac{\partial \boldsymbol{g}(\boldsymbol{\beta})}{\partial \boldsymbol{\beta}},$$

where

$$\frac{\partial \log f(y;\boldsymbol{g}(\boldsymbol{\beta}))}{\partial \boldsymbol{g}} = \frac{\partial}{\partial \boldsymbol{\theta}}\log f(y;\boldsymbol{\theta})\bigg|_{\boldsymbol{\theta}=\boldsymbol{g}(\boldsymbol{\beta})},$$

and

$$\frac{\partial \boldsymbol{g}(\boldsymbol{\beta})}{\partial \boldsymbol{\beta}} = \begin{pmatrix} \dfrac{\partial g_1(\boldsymbol{\beta})}{\partial \beta_1} & \dfrac{\partial g_1(\boldsymbol{\beta})}{\partial \beta_2} & \cdots & \dfrac{\partial g_1(\boldsymbol{\beta})}{\partial \beta_s} \\[2ex] \dfrac{\partial g_2(\boldsymbol{\beta})}{\partial \beta_1} & \dfrac{\partial g_2(\boldsymbol{\beta})}{\partial \beta_2} & \cdots & \dfrac{\partial g_2(\boldsymbol{\beta})}{\partial \beta_s} \\ \vdots & \vdots & & \vdots \\ \dfrac{\partial g_b(\boldsymbol{\beta})}{\partial \beta_1} & \dfrac{\partial g_b(\boldsymbol{\beta})}{\partial \beta_2} & \cdots & \dfrac{\partial g_b(\boldsymbol{\beta})}{\partial \beta_s} \end{pmatrix}.$$

Then using (2.29, p. 64) leads to (2.50).

Example 2.22 (Normal: (μ, σ) transformed to (μ, σ^2)). The information matrix for the normal distribution with $\boldsymbol{\theta} = (\mu, \sigma)^T$ was given earlier to be $\boldsymbol{I}(\mu, \sigma) = \text{diag}(1/\sigma^2, 2/\sigma^2)$. If we prefer the parameter to be $\boldsymbol{\beta} = (\mu, \sigma^2)^T$, then

$$\boldsymbol{\theta} = \begin{pmatrix} g_1(\mu, \sigma^2) \\ g_2(\mu, \sigma^2) \end{pmatrix} = \begin{pmatrix} \mu \\ \sqrt{\sigma^2} \end{pmatrix},$$

$\partial \boldsymbol{g}(\boldsymbol{\beta})/\partial \boldsymbol{\beta} = \text{diag}\{1, 1/(2\sigma)\}$ and

$$\boldsymbol{I}(\boldsymbol{\beta}) = \begin{pmatrix} 1 & 0 \\ 0 & 1/(2\sigma) \end{pmatrix} \begin{pmatrix} 1/\sigma^2 & 0 \\ 0 & 2/\sigma^2 \end{pmatrix} \begin{pmatrix} 1 & 0 \\ 0 & 1/(2\sigma) \end{pmatrix} = \frac{1}{\sigma^2} \begin{pmatrix} 1 & 0 \\ 0 & 1/(2\sigma^2) \end{pmatrix}$$

♦

Example 2.23 (Normal model: constant coefficient of variation). In Chapter 1 (1.13, p. 17) we discussed the $N(\mu, \sigma_0^2 \mu^2)$ model with σ_0^2 known. To get the information $I(\mu)$ using (2.50, p. 79), let $\boldsymbol{\theta} = (\mu, \sigma)^T$ with $\boldsymbol{I}(\mu, \sigma) = \text{diag}(1/\sigma^2, 2/\sigma^2)$ as in the last example, and set $\theta_1 = \mu = \beta = g_1(\beta)$ and $\theta_2 = \sigma = \sigma_0 \beta = g_2(\beta)$. Then

$$I(\mu) = I(\boldsymbol{\beta}) = \begin{pmatrix} 1 & \sigma_0 \end{pmatrix} \begin{pmatrix} 1/(\sigma_0^2 \mu^2) & 0 \\ 0 & 2/(\sigma_0^2 \mu^2) \end{pmatrix} \begin{pmatrix} 1 \\ \sigma_0 \end{pmatrix} = \frac{1 + 2\sigma_0^2}{\sigma_0^2 \mu^2}.$$

Note that $\{nI(\mu)\}^{-1}$ corresponds to the asymptotic variance in (1.16, p. 17). ♦

Example 2.24 (Hardy-Weinberg model). In the Hardy-Weinberg example (Example 2.2, p. 33), $\boldsymbol{\theta} = (p_{AA}, p_{Aa})^T$ is the full multinomial parameterization, and the model in terms of $\beta = p_A$ is $p_{AA} = g_1(p_A) = p_A^2$, $p_{Aa} = g_2(p_A) = 2p_A(1 - p_A)$. We leave the computation of $I_T(p_A) = 2n/\{p_A(1 - p_A)\}$ for Problem b, p. 118. ♦

2.6 Methods for Maximizing the Likelihood or Solving the Likelihood Equations

Maximum likelihood estimation requires maximization of the log likelihood $\ell(\boldsymbol{\theta}) = \log L(\boldsymbol{\theta} | \boldsymbol{Y})$, which in most cases means taking derivatives and solving the likelihood equations,

$$\boldsymbol{S}(\boldsymbol{\theta}) = \frac{\partial}{\partial \boldsymbol{\theta}^T} \ell(\boldsymbol{\theta}) = 0. \tag{2.51}$$

In simple cases, we can solve (2.51) analytically, and Section 2.6.1 discusses briefly the use of the profile likelihood to simplify solving (2.51) either analytically or numerically. When an analytic solution does not exist, there are two main approaches to solving (2.51) for $\widehat{\boldsymbol{\theta}}_{\text{MLE}}$: standard optimization methods like Newton methods; or the EM Algorithm, a clever statistically-based iteration procedure introduced by Dempster et al. (1977). We briefly discuss Newton methods in Section 2.6.2 and then focus on the EM Algorithm in Section 2.6.3. In addition there are a variety of computer search methods that we do not discuss, but a review may be found in Fouskakis and Draper (2002).

2.6.1 *Analytical Methods via Profile Likelihoods*

In certain problems it is possible to maximize the log likelihood for part of $\boldsymbol{\theta} = (\boldsymbol{\theta}_1^T, \boldsymbol{\theta}_2^T)^T$, say $\widetilde{\boldsymbol{\theta}}_2(\boldsymbol{\theta}_1)$, without actually knowing the value of $\boldsymbol{\theta}_1$. The notation $\widetilde{\boldsymbol{\theta}}_2(\boldsymbol{\theta}_1)$ means that for any value $\boldsymbol{\theta}_1$, $\widetilde{\boldsymbol{\theta}}_2(\boldsymbol{\theta}_1)$ maximizes the likelihood with respect to $\boldsymbol{\theta}_2$. The *profile likelihood* is merely the usual likelihood with $\widetilde{\boldsymbol{\theta}}_2(\boldsymbol{\theta}_1)$ inserted for $\boldsymbol{\theta}_2$ — thus, the profile likelihood is a function of only $\boldsymbol{\theta}_1$. The advantage here is that one need only maximize the profile likelihood $L(\boldsymbol{\theta}_1, \widetilde{\boldsymbol{\theta}}_2(\boldsymbol{\theta}_1))$ with respect the lower-dimensional parameter $\boldsymbol{\theta}_1$. Doing so results in $\widehat{\boldsymbol{\theta}}_1$ and then $\widehat{\boldsymbol{\theta}}_2 = \widetilde{\boldsymbol{\theta}}_2(\widehat{\boldsymbol{\theta}}_1)$.

As an example, recall the gamma log likelihood

$$\ell(\alpha, \beta) = -n \log \Gamma(\alpha) - n\alpha \log \beta + (\alpha - 1) \sum \log Y_i - \frac{\sum Y_i}{\beta}$$

shown in Figure 2.1 (p. 37) for the hurricane data. Taking a partial derivative with respect to β gives the score equation

$$S_2(\alpha, \beta) = -\frac{n\alpha}{\beta} + \frac{\sum_{i=1}^n Y_i}{\beta^2} = 0.$$

Solving this equation leads to $\widetilde{\beta}(\alpha) = \overline{Y}/\alpha$. Now substituting $\widetilde{\beta}(\alpha) = \overline{Y}/\alpha$ for β in the log likelihood leads to the profile log likelihood

$$\ell\left(\alpha, \widetilde{\beta}(\alpha)\right) = -n \log \Gamma(\alpha) - n\alpha(\log \overline{Y} - \log \alpha) + (\alpha - 1) \sum_{i=1}^{n} \log Y_i - n\alpha.$$

Figure 2.7 shows the profile log likelihood for the hurricane data. From this plot it is easy to locate (visually and numerically) $\widehat{\alpha}_{\text{MLE}} = 2.19$ and substitute to get $\widehat{\beta}_{\text{MLE}} = 7.29/2.19 = 3.33$, the same as found before by using the R function `nlm`.

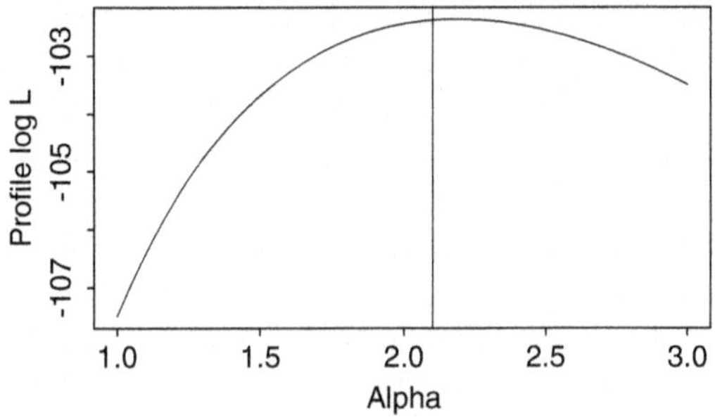

Fig. 2.7 Gamma profile log likelihood for the hurricane data

Profiling is most advantageous when the first maximization can be done analytically, leaving only second maximization for numerical methods. However, because each maximization is lower dimensional, it is sometimes useful even when both optimizations are done numerically.

2.6.2 Newton Methods

Taylor expansion of (2.51) about a "current value" $\boldsymbol{\theta}^{(\nu)}$ leads to

$$\mathbf{0} = \boldsymbol{S}(\boldsymbol{\theta}) \approx \boldsymbol{S}(\boldsymbol{\theta}^{(\nu)}) + \left\{ \frac{\partial}{\partial \boldsymbol{\theta}} \boldsymbol{S}(\boldsymbol{\theta}) \bigg|_{\boldsymbol{\theta}=\boldsymbol{\theta}^{(\nu)}} \right\} \{\boldsymbol{\theta} - \boldsymbol{\theta}^{(\nu)}\}.$$

Letting

$$\boldsymbol{I}_{\mathrm{T}}(\boldsymbol{Y}, \boldsymbol{\theta}^{(\nu)}) = \left\{ \frac{-\partial}{\partial \boldsymbol{\theta}} \boldsymbol{S}(\boldsymbol{\theta}) \bigg|_{\boldsymbol{\theta}=\boldsymbol{\theta}^{(\nu)}} \right\}$$

and solving the approximate equation for $\boldsymbol{\theta}$, call it $\boldsymbol{\theta}^{(\nu+1)}$, yields

$$\boldsymbol{\theta}^{(\nu+1)} = \boldsymbol{\theta}^{(\nu)} + \left\{ \boldsymbol{I}_{\mathrm{T}}(\boldsymbol{Y}, \boldsymbol{\theta}^{(\nu)}) \right\}^{-1} \boldsymbol{S}(\boldsymbol{\theta}^{(\nu)}). \tag{2.52}$$

Iteration of (2.52) often results in a convergent sequence providing the starting value is good, e.g., $\boldsymbol{\theta}^{(0)}$ might be a method of moments estimator. The iteration (2.52) proceeds until the difference $\|\boldsymbol{\theta}^{(\nu+1)} - \boldsymbol{\theta}^{(\nu)}\|$, or relative difference $\|\boldsymbol{\theta}^{(\nu+1)} - \boldsymbol{\theta}^{(\nu)}\| / \|\boldsymbol{\theta}^{(\nu)}\|$, is small. Stopping rules based on the magnitudes of $\|\boldsymbol{S}(\boldsymbol{\theta}^{(\nu)})\|$ or $\|\ell(\boldsymbol{\theta}^{(\nu+1)}) - \ell(\boldsymbol{\theta}^{(\nu)})\| / \|\ell(\boldsymbol{\theta}^{(\nu)})\|$ are also used (see Monahan 2001, Sec. 8.4).

If we replace $\boldsymbol{I}_{\mathrm{T}}(\boldsymbol{Y}, \boldsymbol{\theta}^{(\nu)})$ in (2.52) by its expectation $\boldsymbol{I}_{\mathrm{T}}(\boldsymbol{\theta}^{(\nu)})$, then the method is called Fisher scoring. This replacement is useful in cases like probit regression where $\boldsymbol{I}_{\mathrm{T}}(\boldsymbol{\theta})$ is simpler than $\boldsymbol{I}_{\mathrm{T}}(\boldsymbol{Y}, \boldsymbol{\theta})$.

Under certain conditions, Newton methods have local quadratic convergence, which means that for some $c > 0$

$$\|\boldsymbol{\theta}^{(\nu+1)} - \widehat{\boldsymbol{\theta}}_{\mathrm{MLE}}\| \leq c \|\boldsymbol{\theta}^{(\nu)} - \widehat{\boldsymbol{\theta}}_{\mathrm{MLE}}\|^2.$$

Thus, at least near the solution, convergence is fast for a Newton method. One downside is that $\boldsymbol{I}_{\mathrm{T}}(\boldsymbol{Y}, \boldsymbol{\theta}^{(\nu)})$ or $\boldsymbol{I}_{\mathrm{T}}(\boldsymbol{\theta}^{(\nu)})$ might be costly to calculate at each step in which case secant methods can be used in place of derivatives, but the quadratic convergence is then lost. It is also possible that a Newton step in (2.52) jumps too far. Thus, backtracking and other modifications are suggested. Chapter 8 of Monahan (2001) is a good place to get an overview of these approaches.

It is also possible to just use the first iterate in (2.52) as an estimate, called a "one-step estimator,"

$$\widehat{\boldsymbol{\theta}}_1 = \boldsymbol{\theta}^{(1)} = \boldsymbol{\theta}^{(0)} + \left\{\boldsymbol{I}_{\mathrm{T}}(\boldsymbol{Y}, \boldsymbol{\theta}^{(0)})\right\}^{-1} \boldsymbol{S}(\boldsymbol{\theta}^{(0)}).$$

This estimator has the same asymptotic properties as the full $\widehat{\boldsymbol{\theta}}_{\mathrm{MLE}}$ as long as $\boldsymbol{\theta}^{(0)}$ is consistent.

Example 2.25 (Newton iteration for logistic regression). Consider the usual logistic regression model with Y_i distributed as binomial (m_i, p_i) and $\log\{p_i/(1 - p_i)\} = \boldsymbol{x}_i^T \boldsymbol{\beta}$. Then

$$\boldsymbol{S}(\boldsymbol{\beta}) = \frac{\partial}{\partial \boldsymbol{\beta}^T} \ell(\boldsymbol{\beta}) = \sum_{i=1}^{n} \boldsymbol{x}_i (Y_i - m_i p_i) = \boldsymbol{0},$$

$$\boldsymbol{I}_{\mathrm{T}}(\boldsymbol{Y}, \boldsymbol{\beta}) = -\frac{\partial^2}{\partial \boldsymbol{\beta} \partial \boldsymbol{\beta}^T} \ell(\boldsymbol{\beta}) = \sum_{i=1}^{n} m_i p_i (1 - p_i) \boldsymbol{x}_i \boldsymbol{x}_i^T$$

and

$$\beta^{(\nu+1)} = \beta^{(\nu)} + \left\{\sum_{i=1}^{n} m_i p_i^{(\nu)} (1 - p_i^{(\nu)}) \boldsymbol{x}_i \boldsymbol{x}_i^T\right\}^{-1} \sum_{i=1}^{n} \boldsymbol{x}_i (Y_i - m_i p_i^{(\nu)}),$$

where $p_i^{(\nu)} = \left\{1 + \exp(-\boldsymbol{x}_i^T \beta^{(\nu)})\right\}^{-1}$. See Agresti (2002, p. 147) for interpretation of this Newton iteration as Iteratively Reweighted Least Squares. ♦

2.6.3 *EM Algorithm*

In a ground-breaking paper, Dempster et al. (1977) put together in coherent form an approach to solving the likelihood equations that a number of people had discovered earlier in particular applications. Their paper popularized the "EM Algorithm" ("E" for "Expectation" and "M" for "Maximization") and led to many new statistical applications as well as modifications and extensions of the algorithm, and even a number of books on the subject, e.g., Little and Rubin (1987), and McLachlan and Krishnan (1997).

The basic idea of the EM Algorithm is to view the observed data $\boldsymbol{Y}$ as incomplete, that somehow there is missing data $\boldsymbol{Z}$ that would make the problem simpler if we had it. In some cases $\boldsymbol{Z}$ could truly be missing data, but in others it is just additional data that we wish we had. The first step is to write down the joint likelihood of the "complete" data $(\boldsymbol{Y}, \boldsymbol{Z})$, call it $L_C(\boldsymbol{\theta} \mid \boldsymbol{Y}, \boldsymbol{Z})$. Since we do not have $\boldsymbol{Z}$, we actually cannot compute $L_C(\boldsymbol{\theta}|\boldsymbol{Y}, \boldsymbol{Z})$. We always need to maximize an objective function that depends only on $\boldsymbol{\theta}$ and the observed data $\boldsymbol{Y}$. However, remember that conditional expectations are functions of the conditioning variate. Thus, the "E" step of the EM Algorithm is to compute the conditional expectation of $\log L_C(\boldsymbol{\theta} \mid \boldsymbol{Y}, \boldsymbol{Z})$ given $\boldsymbol{Y}$ assuming the true parameter value is $\boldsymbol{\theta}^{(\nu)}$. Define

$$\begin{aligned} Q(\boldsymbol{\theta}, \boldsymbol{\theta}^{(\nu)}, \boldsymbol{Y}) &= \mathrm{E}_{\boldsymbol{\theta}^{(\nu)}} \{\log L_C(\boldsymbol{\theta} \mid \boldsymbol{Y}, \boldsymbol{Z}) \mid \boldsymbol{Y}\} \\ &= \int \log L_C(\boldsymbol{\theta} \mid \boldsymbol{Y}, z) f_{Z|Y}(z \mid \boldsymbol{Y}, \boldsymbol{\theta}^{(\nu)}) dz, \end{aligned} \tag{2.53}$$

where we have written the expectation in the second line as if $\boldsymbol{Z}|\boldsymbol{Y}$ has a continuous density, but this is for notational convenience only. It may look odd to have both $\boldsymbol{\theta}^{(\nu)}$ and $\boldsymbol{\theta}$ in the expectation, but this is a key aspect of the procedure; in reality $\boldsymbol{\theta}^{(\nu)}$ is the current value in an iteration. The "M" step of the EM Algorithm is to maximize $Q(\boldsymbol{\theta}, \boldsymbol{\theta}^{(\nu)}, \boldsymbol{Y})$ with respect to $\boldsymbol{\theta}$ with $\boldsymbol{\theta}^{(\nu)}$ fixed. This produces a new value $\boldsymbol{\theta}^{(\nu+1)}$ that is then reinserted in the "E" step and so on. We summarize the EM Algorithm as follows:

- E Step: Calculate $Q(\boldsymbol{\theta}, \boldsymbol{\theta}^{(\nu)}, \boldsymbol{Y})$.
- M Step: Calculate $\boldsymbol{\theta}^{(\nu+1)}$ that maximizes $Q(\boldsymbol{\theta}, \boldsymbol{\theta}^{(\nu)}, \boldsymbol{Y})$ with respect to $\boldsymbol{\theta}$.

Under some conditions it is possible to show that the likelihood is increased at each step of the iteration and that convergence is guaranteed. Before getting to such details, it helps to see several examples.

2.6.3a EM Algorithm for Two-Component Mixtures

Suppose that $Y_1, \ldots, Y_n$ are iid from the mixture density

$$f(y; \boldsymbol{\theta}) = p f_1(y; \mu_1, \sigma_1) + (1-p) f_2(y; \mu_2, \sigma_2), \tag{2.54}$$

where f_1 and f_2 are normal densities with means μ_1 and μ_2, respectively, and variances σ_1^2 and σ_2^2. Thus the parameter vector is $\boldsymbol{\theta}^T = (\mu_1, \mu_2, \sigma_1, \sigma_2, p)$. The usual log likelihood,

$$\ell(\boldsymbol{\theta}) = \sum_{i=1}^{n} \log \{p f_1(Y_i; \mu_1, \sigma_1) + (1-p) f_2(Y_i; \mu_2, \sigma_2)\}, \tag{2.55}$$

is simple to write down, but not so simple to maximize. (Actually, it is well known that this log likelihood has maxima on the boundaries of the parameter space, and is not well-behaved. However, solutions to the likelihood equations with good properties do exist, and we want to find these, and the EM Algorithm accomplishes this.) To use the EM Algorithm, we define independent Bernoulli random variables Z_i, where $P(Z_i = 1) = p$, and represent Y_i as

$$Y_i = Z_i X_{1i} + (1 - Z_i) X_{2i},$$

where $X_{11}, \ldots, X_{1n}$ are iid from $f_1(y; \mu_1, \sigma_1)$, $X_{21}, \ldots, X_{2n}$ are iid from $f_2(y; \mu_2, \sigma_2)$, and all the random variables are independent of $Z_1, \ldots, Z_n$. It is a simple exercise to show that the Y_i defined in this fashion have the mixture density (2.54). We also note that this representation is also useful for computer generation of random variables from the mixture density.

The first step of the EM Algorithm is to write down the joint likelihood of the complete data $(\boldsymbol{Y}, \boldsymbol{Z})$. We find in Problem 2.45 (p. 120) that the joint density of (Y_i, Z_i) is given by

$$\{p f_1(y; \mu_1, \sigma_1)\}^z \{(1 - p) f_2(y; \mu_2, \sigma_2)\}^{1-z}.$$

Thus the complete data log likelihood is

$$\log L_C(\boldsymbol{\theta} \mid \boldsymbol{Y}, \boldsymbol{Z}) = \sum_{i=1}^{n} \{Z_i \log f_1(Y_i; \mu_1, \sigma_1) + (1 - Z_i) \log f_2(Y_i; \mu_2, \sigma_2)$$

$$+ Z_i \log p + (1 - Z_i) \log(1 - p)\}.$$

The conditional expectation of the E step is given by

$$Q(\boldsymbol{\theta}, \boldsymbol{\theta}^{(v)}, \boldsymbol{Y}) = \mathrm{E}_{\boldsymbol{\theta}^{(v)}} \{\log L_C(\boldsymbol{\theta} \mid \boldsymbol{Y}, \boldsymbol{Z}) \mid \boldsymbol{Y}\} = \sum_{i=1}^{n} \Big\{ w_i^{(v)} \log f_1(Y_i; \mu_1, \sigma_1)$$

$$+ (1 - w_i^{(v)}) \log f_2(Y_i; \mu_2, \sigma_2)$$

$$+ w_i^{(v)} \log p + (1 - w_i^{(v)}) \log(1 - p) \Big\},$$

where

$$w_i^{(v)} = \mathrm{E}_{\boldsymbol{\theta}^{(v)}}(Z_i \mid Y_i) = \frac{p^{(v)} f_1(Y_i; \mu_1^{(v)}, \sigma_1^{(v)})}{p^{(v)} f_1(Y_i; \mu_1^{(v)}, \sigma_1^{(v)}) + (1 - p^{(v)}) f_2(Y_i; \mu_2^{(v)}, \sigma_2^{(v)})}.$$

This conditional expectation for the Z_i is found in the second part of Problem 2.45.

Substituting the form of the normal densities (but ignoring the constant 2π) leads to

$$Q(\boldsymbol{\theta}, \boldsymbol{\theta}^{(\nu)}, \boldsymbol{Y}) = \sum_{i=1}^{n} \left[w_i^{(\nu)} \left\{ -\log \sigma_1 - \frac{(Y_i - \mu_1)^2}{2\sigma_1^2} \right\} + (1 - w_i^{(\nu)}) \left\{ -\log \sigma_2 - \frac{(Y_i - \mu_2)^2}{2\sigma_2^2} \right\} + w_i^{(\nu)} \log p + \left\{ 1 - w_i^{(\nu)} \right\} \log(1 - p) \right].$$

Notice that in terms of minimizing $Q(\boldsymbol{\theta}, \boldsymbol{\theta}^{(\nu)}, \boldsymbol{Y})$ in $\boldsymbol{\theta}$, the weights $w_i^{(\nu)}$ are just constants. Thus, taking a derivative with respect to μ_1 is similar to working with the log likelihood of a single normal distribution,

$$\frac{\partial}{\partial \mu_1} Q(\boldsymbol{\theta}, \boldsymbol{\theta}^{(\nu)}, \boldsymbol{Y}) = \sum_{i=1}^{n} w_i^{(\nu)} \frac{(Y_i - \mu_1)}{\sigma_1^2} = 0.$$

Solving this yields $\mu_1^{(\nu+1)} = \sum w_i^{(\nu)} Y_i / \sum w_i^{(\nu)}$. Similarly, we obtain get $\mu_2^{(\nu+1)} = \sum (1 - w_i^{(\nu)}) Y_i / \sum (1 - w_i^{(\nu)})$. The updating formulas for σ_1 and σ_2 are also simple but left for Problem 2.46 (p. 120). Taking the derivative of $Q(\boldsymbol{\theta}, \boldsymbol{\theta}^{(\nu)}, \boldsymbol{Y})$ with respect to p yields

$$\frac{\partial}{\partial p} Q(\boldsymbol{\theta}, \boldsymbol{\theta}^{(\nu)}, \boldsymbol{Y}) = \sum_{i=1}^{n} \left\{ \frac{w_i^{(\nu)}}{p} - \frac{(1 - w_i^{(\nu)})}{1 - p} \right\} = 0.$$

Solving this last equation gives $p^{(\nu+1)} = (1/n) \sum w_i^{(\nu)}$.

In this case the EM Algorithm has turned local maximization of the difficult log likelihood (2.55, p. 84) into an iteration process with very simple updates. The normal distribution played only a small role until the M step where it leads to simple update formulas. Let us go back and make the problem a bit more general in order to gain some insight into how the EM Algorithm works. Let $\boldsymbol{\theta} = (\boldsymbol{\gamma}, p)^T$ and let $f_1(y; \boldsymbol{\gamma})$ and $f_2(y; \boldsymbol{\gamma})$ be arbitrary densities so that the log likelihood is

$$\ell(\boldsymbol{\theta}) = \sum_{i=1}^{n} \log f(Y_i; \boldsymbol{\theta}) = \sum_{i=1}^{n} \log \{ p f_1(Y_i; \boldsymbol{\gamma}) + (1 - p) f_2(Y_i; \boldsymbol{\gamma}) \}.$$

Taking derivatives with respect to $\boldsymbol{\gamma}$ leads to the vector likelihood equation

$$\frac{\partial}{\partial \boldsymbol{\gamma}^T}\ell(\boldsymbol{\theta}) = \sum_{i=1}^{n} \left\{ p \frac{\partial}{\partial \boldsymbol{\gamma}^T} f_1(Y_i; \boldsymbol{\gamma}) + (1-p) \frac{\partial}{\partial \boldsymbol{\gamma}^T} f_2(Y_i; \boldsymbol{\gamma}) \right\} \Big/ f(Y_i; \boldsymbol{\theta}) = 0. \tag{2.56}$$

The Q function of the EM Algorithm is given by

$$Q(\boldsymbol{\theta}, \boldsymbol{\theta}^{(\nu)}, \boldsymbol{Y}) = \sum_{i=1}^{n} \Big\{ w_i^{(\nu)} \log f_1(Y_i; \boldsymbol{\gamma}) + (1 - w_i^{(\nu)}) \log f_2(Y_i; \boldsymbol{\gamma})$$
$$+ w_i^{(\nu)} \log p + (1 - w_i^{(\nu)}) \log(1-p) \Big\}.$$

Taking derivatives with respect to $\boldsymbol{\gamma}$ leads to

$$\frac{\partial}{\partial \boldsymbol{\gamma}} Q(\boldsymbol{\theta}, \boldsymbol{\theta}^{(\nu)}, \boldsymbol{Y}) = \sum_{i=1}^{n} \left\{ \frac{w_i^{(\nu)}}{f_1(Y_i; \boldsymbol{\gamma})} \frac{\partial}{\partial \boldsymbol{\gamma}^T} f_1(Y_i; \boldsymbol{\gamma}) + \frac{(1 - w_i^{(\nu)})}{f_2(Y_i; \boldsymbol{\gamma})} \frac{\partial}{\partial \boldsymbol{\gamma}^T} f_2(Y_i; \boldsymbol{\gamma}) \right\} = 0.$$

Finally, replacing $w_i^{(\nu)}$ by its definition $p^{(\nu)} f_1(Y_i; \boldsymbol{\theta}^{(\nu)})/f(Y_i; \boldsymbol{\theta}^{(\nu)})$ and $\boldsymbol{\gamma}$ by $\boldsymbol{\gamma}^{(\nu+1)}$ when this last vector equation is solved, the equation becomes

$$\sum_{i=1}^{n} \left\{ p^{(\nu)} r_1 \frac{\partial}{\partial \boldsymbol{\gamma}^T} f_1(Y_i; \boldsymbol{\gamma}^{(\nu+1)}) + (1 - p^{(\nu)}) r_2 \frac{\partial}{\partial \boldsymbol{\gamma}^T} f_2(Y_i; \boldsymbol{\gamma}^{(\nu+1)}) \right\} \Big/ f(Y_i; \boldsymbol{\theta}^{(\nu)}) = 0, \tag{2.57}$$

where $r_j = f_j(Y_i; \boldsymbol{\gamma}^{(\nu)})/f_j(Y_i; \boldsymbol{\gamma}^{(\nu+1)})$, $j = 1, 2$. As the EM Algorithm converges, r_1 and r_2 are each converging to 1, and this last equation looks very similar to (2.56). This similarity is not a proof of convergence, but it suggests that the EM Algorithm gives the same solution as solving (2.57).

2.6.3b EM Algorithm for Right Censored Data

Recall that in the right censoring context we observe $(Y_1, \delta_1), \ldots, (Y_n, \delta_n)$, where $Y_i = \min(X_i, R_i)$, X_i has the density $f(x; \boldsymbol{\theta})$ that we want to estimate, R_i is a censoring time, and $\delta_i = I(X_i \le R_i)$ indicates whether Y_i is a censored observation or is X_i. To make things simpler, we assume that the censoring times R_i are not random. The likelihood was given before as

$$L(\boldsymbol{\theta} \mid \boldsymbol{Y}) = \prod_{i=1}^{n} f(Y_i; \boldsymbol{\theta})^{\delta_i} \{1 - F(R_i; \boldsymbol{\theta})\}^{1-\delta_i}.$$

In some cases this likelihood is hard to maximize, and the EM Algorithm gives an alternate approach. For notational simplicity, let us relabel the data so that $Y_1, \ldots, Y_{n_u}$ are the uncensored observations and $Y_{n_u+1}, \ldots, Y_n$ are the censored ones. It then makes sense to take $\boldsymbol{Z}$ as the unknown $X_{n_u+1}, \ldots, X_n$. Assuming that the censoring times are known, the "complete" data $(\boldsymbol{Y}, \boldsymbol{\delta}, \boldsymbol{Z})$ are equivalent to just the full X sample. (Note that we use the notation $(\boldsymbol{Y}, \boldsymbol{\delta}, \boldsymbol{Z})$ to mean $\{Y_i, \delta_i, Z_i\}_{i=1}^n$.) Thus the complete data log likelihood can be written as

$$\log L_C(\boldsymbol{\theta} \mid \boldsymbol{X}) = \sum_{i=1}^{n} \log f(X_i; \boldsymbol{\theta}) = \sum_{i=1}^{n_u} \log f(X_i; \boldsymbol{\theta}) + \sum_{i=n_u+1}^{n} \log f(X_i; \boldsymbol{\theta}).$$

The conditional expectation given the observed data $(\boldsymbol{Y}, \boldsymbol{\delta})$ is

$$\begin{aligned} Q(\boldsymbol{\theta}, \boldsymbol{\theta}^{(\nu)}, \boldsymbol{Y}, \boldsymbol{\delta}) &= \mathrm{E}_{\boldsymbol{\theta}^{(\nu)}} \{\log L_C(\boldsymbol{\theta} \mid \boldsymbol{X}) \mid \boldsymbol{Y}, \boldsymbol{\delta}\} \\ &= \sum_{i=1}^{n_u} \log f(Y_i; \boldsymbol{\theta}) + \sum_{i=n_u+1}^{n} \mathrm{E}_{\boldsymbol{\theta}^{(\nu)}} \{\log f(X_i; \boldsymbol{\theta}) \mid Y_i, \delta_i\}. \end{aligned}$$

Since $X_i = Y_i$ in the first sum, its value is left unchanged by the conditional expectation, but we have replaced X_i by Y_i. To proceed further we assume that the data are exponentially distributed so that $f(x; \boldsymbol{\theta}) = \sigma^{-1} \exp(-x/\sigma)$ for $x \geq 0$. We already know from Example 2.9 (p. 48) that the maximum likelihood estimator is $\sum_{i=1}^{n} Y_i / n_u$, but it is instructive to see how the EM Algorithm arrives at the same estimator. (Some of this material was adapted from Example 1.3 of McLachlan and Krishnan 1997.)

Substituting the exponential form $\log f(x; \sigma) = -\log \sigma - x/\sigma$, we have

$$Q(\sigma, \sigma^{(\nu)}, \boldsymbol{Y}, \boldsymbol{\delta}) = -n \log \sigma - \frac{1}{\sigma} \sum_{i=1}^{n_u} Y_i - \frac{1}{\sigma} \sum_{i=n_u+1}^{n} \mathrm{E}_{\sigma^{(\nu)}} (X_i \mid X_i > R_i).$$

The conditional density of X_i given $X_i > R_i$ is $\sigma^{-1} \exp(-x/\sigma) \exp(R_i/\sigma) I(x > R_i)$, and the conditional expectation in the last expression is simply $\sigma^{(\nu)} + R_i$ (Problem 2.52, p. 122, asks for a derivation of this result). Substituting for the conditional expectation, we have

$$Q(\sigma, \sigma^{(\nu)}, \boldsymbol{Y}, \boldsymbol{\delta}) = -n \log \sigma - \frac{1}{\sigma} \sum_{i=1}^{n_u} Y_i - \frac{1}{\sigma} \sum_{i=n_u+1}^{n} \left(\sigma^{(\nu)} + R_i\right).$$

Noting that for the censored observations $Y_i = R_i$, the above equation simplifies to

$$Q(\sigma, \sigma^{(\nu)}, \boldsymbol{Y}, \boldsymbol{\delta}) = -n \log \sigma - \frac{1}{\sigma} \sum_{i=1}^{n} Y_i - \frac{1}{\sigma} (n - n_u) \sigma^{(\nu)}.$$

Finally, taking the derivative of this last expression with respect to σ, we obtain

$$-\frac{n}{\sigma} + \frac{1}{\sigma^2}\left\{\sum_{i=1}^{n} Y_i + (n - n_u)\sigma^{(\nu)}\right\} = 0,$$

leading to the update formula

$$\sigma^{(\nu+1)} = \frac{1}{n}\left\{\sum_{i=1}^{n} Y_i + (n - n_u)\sigma^{(\nu)}\right\}.$$

As $\nu \to \infty$, we expect both $\sigma^{(\nu+1)}$ and $\sigma^{(\nu)}$ to converge to $\widehat{\sigma}_{\text{MLE}}$. Substituting $\widehat{\sigma}_{\text{MLE}}$ for $\sigma^{(\nu+1)}$ and $\sigma^{(\nu)}$ in the update formula, shows that $\widehat{\sigma}_{\text{MLE}}$ satisfies

$$\widehat{\sigma}_{\text{MLE}} = \frac{1}{n}\left\{\sum_{i=1}^{n} Y_i + (n - n_u)\widehat{\sigma}_{\text{MLE}}\right\},$$

with solution $\widehat{\sigma}_{\text{MLE}} = \sum_{i=1}^{n} Y_i / n_u$, as found previously. Note that if we subtract the last display from the update formula, we obtain

$$\sigma^{(\nu+1)} - \widehat{\sigma}_{\text{MLE}} = \left(1 - \frac{n_u}{n}\right)\left(\sigma^{(\nu)} - \widehat{\sigma}_{\text{MLE}}\right),$$

and taking absolute values,

$$\left|\sigma^{(\nu+1)} - \widehat{\sigma}_{\text{MLE}}\right| = (1 - n_u/n)\left|\sigma^{(\nu)} - \widehat{\sigma}_{\text{MLE}}\right|.$$

Thus the convergence here is linear with constant $(1 - n_u/n)$. McLachlan and Krishnan (1997, Sec. 3.9) argue that a linear rate of convergence is typical for the EM Algorithm.

2.6.3c Convergence Results for the EM Algorithm

The EM Algorithm is often useful when the complete data likelihood has the form of an exponential family. In fact Dempster et al. (1977) gave their first presentation in terms of exponential families. The examples is Sections 2.6.3a (p. 84) and 2.6.3b (p. 87) are both of this type. Often, as in these examples, the M step is straightforward and basically inherited from the exponential family, but the E step is often challenging and not necessarily aided by the exponential family structure.

Theorem 1 of Dempster et al. (1977) essentially says that

$$\ell(\boldsymbol{\theta}^{(\nu+1)}) \geq \ell(\boldsymbol{\theta}^{(\nu)}) \tag{2.58}$$

where equality holds if and only if both

$$Q(\boldsymbol{\theta}^{(\nu+1)}, \boldsymbol{\theta}^{(\nu)}) = Q(\boldsymbol{\theta}^{(\nu)}, \boldsymbol{\theta}^{(\nu)}) \quad \text{and} \quad f_{Z|Y}(\boldsymbol{Z}|\boldsymbol{Y}, \boldsymbol{\theta}^{(\nu+1)}) = f_{Z|Y}(\boldsymbol{Z}|\boldsymbol{Y}, \boldsymbol{\theta}^{(\nu)}).$$

Thus, the EM Algorithm is guaranteed to increase the log likelihood at each step. If $\ell(\boldsymbol{\theta})$ is bounded from above, then (2.58) guarantees that the sequence $\ell(\boldsymbol{\theta}^{(\nu)})$ converges to some l^*. However, convergence of $\ell(\boldsymbol{\theta}^{(\nu)})$ does not guarantee that the sequence of $\boldsymbol{\theta}^{(\nu)}$ converges to $\widehat{\boldsymbol{\theta}}_{\text{MLE}}$. Wu (1983) points out that the results of Dempster et al. (1977) on this convergence are incorrect, and gives a number of technical conditions to obtain this convergence. We refer the reader to Wu (1983) and McLachlan and Krishnan (1997, Ch. 3) for these results.

The EM Algorithm is a very useful method for finding maximum likelihood estimators. Update formulas are often simple when the complete data likelihood has an exponential family form. Because of (2.58) it tends to be more stable than Newton methods, although it can be considerably slower in certain situations. The EM Algorithm does not require derivatives of the log likelihood, but on the other hand, $\boldsymbol{I}_{\text{T}}(\boldsymbol{Y}, \widehat{\boldsymbol{\theta}})$ is not available at the end to use for standard errors.

2.7 Appendix A – Uniqueness of Maximum Likelihood Estimators

In this appendix we give some fairly technical results on uniqueness of $\widehat{\boldsymbol{\theta}}_{\text{MLE}}$. Parts of this subsection are at a somewhat higher mathematical level than the rest of Chapter 2.

For given data $\boldsymbol{Y}$ and likelihood $L(\boldsymbol{\theta}|\boldsymbol{Y})$, there is no guarantee that the likelihood has a unique maximizing value. The best situation is when the likelihood equations $\boldsymbol{S}(\boldsymbol{\theta}) = \partial\ell(\boldsymbol{\theta})/\partial\boldsymbol{\theta}^T = \boldsymbol{0}$ have a *unique* root $\widehat{\boldsymbol{\theta}}$, where $\ell(\boldsymbol{\theta})$ is the log likelihood function. A good source concerning this uniqueness is Makelainen et al. (1981). Here we give several results from that paper and then some examples.

2.7.1 *Definitions*

Let Θ be an open set of R^b. A sequence $\boldsymbol{\theta}^{(1)}, \boldsymbol{\theta}^{(2)}, \ldots$, in Θ is said to *converge to the boundary*, $\partial\Theta$, of Θ if for every compact set $K \subset \Theta$, there exists an integer $k_0 \geq 1$ such that $\boldsymbol{\theta}^{(k)} \notin K$ for every $k \geq k_0$. When $\Theta = R^b$, this condition is equivalent to $\lim_{k\to\infty} \| \boldsymbol{\theta}^{(k)} \| = \infty$. A real-valued function f defined on Θ is said to be *constant on the boundary*, $\partial\Theta$, if $\lim_{k\to\infty} f(\boldsymbol{\theta}^{(k)}) = c$ for every sequence $\boldsymbol{\theta}^{(1)}, \boldsymbol{\theta}^{(2)}, \ldots$, in Θ converging to $\partial\Theta$, where c is an extended real number (possibly ∞ or $-\infty$). We use the notation $\lim_{\theta\to\partial\Theta} f(\boldsymbol{\theta}) = c$ to denote that f is constant on the boundary, i.e., equal to c on the boundary.

2.7.2 Main Results

The following theorem tells us that a unique maximum likelihood estimate exists if the log likelihood function is constant on the boundary of the parameter space and if the negative of the Hessian matrix of the log likelihood, defined as $\boldsymbol{I}_{\mathrm{T}}(\boldsymbol{Y},\boldsymbol{\theta})$, is positive definite at solutions of the likelihood equations.

Theorem 2.1 (Makelainen et al. 1981, Corollary 2.5). *Let Θ be a connected open subset of R^b, $b \geq 1$, and let $\ell(\boldsymbol{\theta})$ be a twice continuously differentiable real-valued function on Θ with $\lim_{\boldsymbol{\theta}\to\partial\Theta} \ell(\boldsymbol{\theta}) = c$, where c is either a real number or $-\infty$. Suppose that $\boldsymbol{I}_{\mathrm{T}}(\boldsymbol{Y},\boldsymbol{\theta}) = -\partial^2\ell(\boldsymbol{\theta})/\partial\boldsymbol{\theta}\,\partial\boldsymbol{\theta}^T$ is positive definite at every point $\boldsymbol{\theta} \in \Theta$ for which $\partial\ell(\boldsymbol{\theta})/\partial\boldsymbol{\theta}^T = \mathbf{0}$. Then $\ell(\boldsymbol{\theta})$ has a unique global maximum and no other critical points. Furthermore, $\ell(\boldsymbol{\theta}) > c$ for every $\boldsymbol{\theta} \in \Theta$.*

The key to this result is the condition that $\boldsymbol{I}_{\mathrm{T}}(\boldsymbol{Y},\boldsymbol{\theta})$ is positive definite at *every* $\boldsymbol{\theta}$ for which $\boldsymbol{S}(\boldsymbol{\theta}) = 0$. This assumption rules out local minima, and multiple local maxima cannot occur without local minima. Applying the result in practice necessitates finding all solutions to $\boldsymbol{S}(\boldsymbol{\theta}) = 0$ and checking that $\boldsymbol{I}_{\mathrm{T}}(\boldsymbol{Y},\boldsymbol{\theta})$ is positive definite at each solution.

Example 2.26 (Uniqueness of binomial MLE). If Y is binomial(n,p), then $\Theta = (0,1)$ and the likelihood $L(p|Y)$ is proportional to $p^Y(1-p)^{n-Y}$ with $\ell(p) =$ constant $+\, Y \log p + (n-Y)\log(1-p)$. If $0 < Y < n$, then $\ell(p) \to -\infty$ as $p \to 0$ or 1 and is thus constant on the boundary. If $Y = 0$ or $Y = n$, then $\ell(p)$ is not constant on the boundary, and Theorem 2.1 does not apply. In that case, the maximum occurs at 0 or 1, respectively, not in our parameter space. Taking derivatives, we have

$$\frac{\partial\ell(p)}{\partial p} = \frac{Y}{p} - \frac{n-Y}{1-p} = 0,$$

which has solution $\widehat{p} = Y/n$ unless $Y = 0$ or $Y = n$. Taking another derivative, we have

$$-\frac{\partial^2\ell(p)}{\partial p^2} = \frac{Y}{p^2} + \frac{n-Y}{(1-p)^2},$$

which is positive for all p and Y, implying that $\ell(p)$ is strictly concave, and a concave function can have only one maxima. Thus, Theorem 2.1 assures us that $\widehat{p} = Y/n$ is the unique maximum likelihood estimator when $0 < Y < n$. ♦

The next theorem strengthens the condition on $\boldsymbol{I}_{\mathrm{T}}(\boldsymbol{Y},\boldsymbol{\theta})$ to be positive definite throughout the parameter space which implies that the function $\ell(\boldsymbol{\theta})$ is strictly concave on Θ. This allows us to remove any conditions about the boundary. But note the requirement that at least one critical point exists.

Theorem 2.2 (Makelainen et al. 1981, Theorem 2.6). *Let $\ell(\boldsymbol{\theta})$ be twice continuously differentiable with $\boldsymbol{\theta}$ varying in a connected open subset $\Theta \subset R^b$. Suppose that* *(i)* *the likelihood equations* $\partial\ell(\boldsymbol{\theta})/\partial\boldsymbol{\theta}^T = \mathbf{0}$ *have at least one solution* $\boldsymbol{\theta} \in \Theta$

and that

(ii) $\boldsymbol{I}_{\mathrm{T}}(\boldsymbol{Y}, \boldsymbol{\theta}) = -\partial^2\ell(\boldsymbol{\theta})/\partial\boldsymbol{\theta}\,\partial\boldsymbol{\theta}^T$ *is positive definite at every point* $\boldsymbol{\theta} \in \Theta$.

Then

(a) $\ell(\boldsymbol{\theta})$ *is a strictly concave function of* $\boldsymbol{\theta}$;
(b) *there is a unique maximum likelihood estimate* $\widehat{\boldsymbol{\theta}}_{\mathrm{MLE}} \in \Theta$;
(c) $\ell(\boldsymbol{\theta})$ *has no other maxima or minima or other stationary points in* Θ.

Recall that a matrix A is positive definite if $\boldsymbol{x}^T A \boldsymbol{x} > 0$ for all $\boldsymbol{x}$. One criterion for positive definiteness of a matrix A is as follows (see Graybill 1988 p. 397). A is positive definite if and only if

$$a_{11} > 0; \quad \begin{vmatrix} a_{11} & a_{12} \\ a_{21} & a_{22} \end{vmatrix} > 0; \quad \begin{vmatrix} a_{11} & a_{12} & a_{13} \\ a_{21} & a_{22} & a_{23} \\ a_{31} & a_{32} & a_{33} \end{vmatrix} > 0; \quad \ldots; \quad |A| > 0.$$

2.7.3 Application of Theorem 2.2 to the Multinomial

Suppose that $(N_1, \ldots, N_k)$ are distributed as multinomial$(n; p_1, p_2, \ldots, p_k)$. The likelihood is

$$L(\boldsymbol{p} \mid N_1, \ldots, N_k) = \frac{n!}{N_1! N_2! \cdots N_k!} p_1^{N_1} p_2^{N_2} \cdots p_k^{N_k},$$

where $\sum_{i=1}^k N_i = n$, $\sum_{i=1}^k p_i = 1$, $0 < p_i < 1$, $i = 1, \ldots, k$. The log likelihood is thus

$$\ell(\boldsymbol{p}) = c_0 + \sum_{i=1}^{k-1} N_i \log p_i + N_k \log\left(1 - \sum_{i=1}^{k-1} p_i\right).$$

Taking derivatives we have

$$\frac{\partial\ell(\boldsymbol{p})}{\partial p_1} = \frac{N_1}{p_1} - \frac{N_k}{p_k} = 0 \quad \Rightarrow \quad \frac{N_1}{p_1} = \frac{N_k}{p_k}$$

$$\vdots \qquad\qquad \vdots \qquad\qquad \vdots$$

$$\frac{\partial\ell(\boldsymbol{p})}{\partial p_{k-1}} = \frac{N_{k-1}}{p_{k-1}} - \frac{N_k}{p_k} = 0 \Rightarrow \frac{N_{k-1}}{p_{k-1}} = \frac{N_k}{p_k}$$

Assume that $0 < N_i < n$, $i = 1, \ldots, k$. Otherwise, $L(\boldsymbol{p})$ is maximized in p_i if $n_i = 0$ by setting $\widehat{p}_i = 0$ which is not in the parameter space. To solve the above system, note that $N_i = p_i(N_k/p_k)$ and summing both sides gives

$$\sum_{i=1}^{k-1} N_i = \left(\sum_{i=1}^{k-1} p_i\right)\left(\frac{N_k}{p_k}\right)$$

or $n - N_k = (1 - p_k)(N_k/p_k)$ or $n = N_k/p_k$, and thus $\widehat{p}_i = N_i/n$, $i = 1, \ldots, k$. Taking another derivative, we find that $\boldsymbol{I}_{\mathrm{T}}(\boldsymbol{Y}, \boldsymbol{p})$ can be written as $\boldsymbol{D} + (N_k/p_k^2)\,\mathbf{1}\mathbf{1}^T$, where D is a diagonal matrix with diagonal elements $d_{ii} = N_i/p_i^2$, $i = 1, \ldots k-1$, and $\mathbf{1}$ is a vector of ones. Thus, $\boldsymbol{I}_{\mathrm{T}}(\boldsymbol{Y}, \boldsymbol{p})$ is positive definite because for any $\boldsymbol{x} \neq \mathbf{0}$

$$\boldsymbol{x}^T \boldsymbol{I}_{\mathrm{T}}(\boldsymbol{Y}, \boldsymbol{p})\boldsymbol{x} = \sum_{i=1}^{k-1} N_i x_i^2/p_i^2 + (N_k/p_k^2)\left(\sum x_i\right)^2 > 0.$$

Then by Theorem 2.2 (p. 91), the maximum likelihood estimators are unique if all the N_i lie between zero and n.

2.7.4 *Uniqueness of the MLE in the Normal location-scale Model*

Suppose that $Y_1, \ldots, Y_n$ are iid $\mathrm{N}(\mu, \sigma^2)$, $-\infty < \mu < \infty$, $0 < \sigma < \infty$, $\theta^T = (\mu, \sigma)$. Then

$$\ell(\mu, \sigma) = c - n\log\sigma - \frac{1}{2\sigma^2}\sum_{i=1}^{n}(Y_i - \mu)^2.$$

$$\frac{\partial \ell(\boldsymbol{\theta})}{\partial \boldsymbol{\theta}^T} = \begin{pmatrix} \dfrac{\partial \ell(\boldsymbol{\theta})}{\partial \mu} \\ \dfrac{\partial \ell(\boldsymbol{\theta})}{\partial \sigma} \end{pmatrix} = \begin{pmatrix} \dfrac{1}{\sigma^2}\displaystyle\sum_{i=1}^{n}(Y_i - \mu) \\ -\dfrac{n}{\sigma} + \dfrac{1}{\sigma^3}\displaystyle\sum_{i=1}^{n}(Y_i - \mu)^2 \end{pmatrix}.$$

Note that the latter partial derivative is with respect to σ.

$$\boldsymbol{I}_{\mathrm{T}} = -\frac{\partial^2 \ell(\boldsymbol{\theta})}{\partial \boldsymbol{\theta}\, \partial \boldsymbol{\theta}^T} = \begin{pmatrix} \dfrac{n}{\sigma^2} & \dfrac{2}{\sigma^3}\displaystyle\sum_{i=1}^{n}(Y_i - \mu) \\ \dfrac{2}{\sigma^3}\displaystyle\sum_{i=1}^{n}(Y_i - \mu) & -\dfrac{n}{\sigma^2} + \dfrac{3}{\sigma^4}\displaystyle\sum_{i=1}^{n}(Y_i - \mu)^2 \end{pmatrix}.$$

The only solution of $\partial\ell(\boldsymbol{\theta})/\partial\boldsymbol{\theta}^T = \mathbf{0}$ is of course $\widehat{\mu} = \overline{Y}$ and $\widehat{\sigma} = s_n$. Moreover,

$$\boldsymbol{I}_{\mathrm{T}}(\overline{Y}, s_n) = \begin{pmatrix} n/s_n^2 & 0 \\ \\ 0 & 2n/s_n^2 \end{pmatrix},$$

which is positive definite since $\boldsymbol{x}^T \boldsymbol{I}_{\mathrm{T}}(\overline{Y}, s_n)\boldsymbol{x} = nx_1^2/s_n^2 + 2nx_2^2/s_n^2 > 0$ for any $\boldsymbol{x}$. Thus, $(\overline{Y}, s_n)$ is a relative maximum of the log likelihood. Is it a global maximum? To use Theorem 2.2 (p. 91), we need to show that $\boldsymbol{I}_{\mathrm{T}}$ is positive definite everywhere in $(-\infty, \infty) \times (0, \infty)$. Unfortunately,

$$|\boldsymbol{I}_{\mathrm{T}}| = \frac{n^2}{\sigma^4}\left[-1 + \frac{3}{\sigma^2}\left\{\frac{1}{n}\sum_{i=1}^{n}(Y_i - \mu)^2\right\} - \frac{4}{\sigma^2}(\overline{Y} - \mu)^2\right]$$

$$= \frac{n^2}{\sigma^6}\left\{3s_n^2 - \sigma^2 - (\overline{Y} - \mu)^2\right\},$$

and this can be negative—take $\mu = \overline{Y}$ and $\sigma^2 = 4s_n^2$. Thus, $\boldsymbol{I}_{\mathrm{T}}$ is not concave over $(-\infty, \infty) \times (0, \infty)$. It is concave in a region around $(\overline{Y}, s_n)$. Since $(\overline{Y}, s_n)$ is the only critical value, to use Theorem 2.1 (p. 91), it would be sufficient to show that $\ell(\mu, \sigma)$ is constant on the boundary of the region $(-\infty, \infty) \times (0, \infty)$. We could argue successfully here, but we take another approach that illustrates advantages of using profile likelihoods.

Theorem 2.3. *Let $f(\mu, \sigma)$ be a function such that $\mu_1(\sigma)$ is a global maximum for each value of σ. If σ_1 is a global maximum of $f^*(\sigma) = f(\mu_1(\sigma), \sigma)$, then $(\mu_1(\sigma_1), \sigma_1)$ is a global maximum of $f(\mu, \sigma)$.*

Proof. $f^*(\sigma_1) = f(\mu_1(\sigma_1), \sigma_1) \geq f(\mu_1(\sigma), \sigma) \geq f(\mu, \sigma)$for all$(\mu, \sigma)$. ∎

For $\mathrm{N}(\mu, \sigma^2)$ data, consider $\ell(\mu, \sigma)$ as a function of only μ. Then

$$\frac{\partial\ell(\mu, \sigma)}{\partial\mu} = \frac{1}{\sigma^2}\sum_{i=1}^{n}(Y_i - \mu) = 0$$

implies that $\mu_1 = \overline{Y}$ is a critical value for each σ. Since

$$-\frac{\partial^2\ell(\mu, \sigma)}{\partial\mu^2} = \frac{n}{\sigma^2} > 0,$$

$f(\mu, \sigma)$ is concave as a function of μ for each σ, and $\mu_1 = \overline{Y}$ is a global maximum of $\ell(\mu, \sigma)$ for each σ. Now we can "concentrate" $\ell(\mu, \sigma)$ to get the profile likelihood

$$\ell_n^*(\sigma) = \ell(\overline{Y}, \sigma) = c - n\log\sigma - \frac{ns_n^2}{2\sigma^2}.$$

Then

$$\frac{\partial \ell_n^*(\sigma)}{\partial \sigma} = -\frac{n}{\sigma} + \frac{ns_n^2}{\sigma^3} = 0$$

implies that $\sigma_1 = s_n$ is a critical value of $\ell_n^*(\sigma)$. Now

$$-\frac{\partial^2 \ell_n^*(\sigma)}{\partial \sigma^2} = -\frac{n}{\sigma^2} + \frac{3ns_n^2}{\sigma^4},$$

and this latter expression is positive at $\sigma_1 = s_n$. Thus, $\sigma_1 = s_n$ is a *local* maximum of $\ell_n^*(\sigma)$. Unfortunately, plugging in $\sigma^2 = 4s_n^2$ in this latter expression shows that $\ell_n^*(\sigma)$ is not concave. However, we can now check the boundary points $\sigma = 0$ and $\sigma = \infty$ and see that

$$\lim_{\sigma \to 0} \ell_n^*(\sigma) = -\infty \qquad \text{and} \qquad \lim_{\sigma \to \infty} \ell_n^*(\sigma) = -\infty.$$

Thus, by Theorem 2.1 (p. 91), $\sigma_1 = s_n$ is a global maximum of ℓ_n^*, and by Theorem 2.3, $(\widehat{\mu}, \widehat{\sigma}) = (\overline{X}, s)$ is a global maximum of $\ell(\mu, \sigma)$.

2.7.5 *Application of Theorems 2.1 and 2.3 to the Exponential Threshold Model*

Suppose that $Y_1, \ldots, Y_n$ are iid from the exponential threshold model,

$$f(y; \mu, \sigma) = \frac{1}{\sigma} \exp\left\{-\frac{(y-\mu)}{\sigma}\right\}, \qquad \mu \le y < \infty, -\infty < \mu < \infty,\ 0 < \sigma < \infty.$$

The likelihood is then

$$L(\mu, \sigma | \boldsymbol{Y}) = \sigma^{-n} \exp\left\{-\sum_{i=1}^{n} \left(\frac{Y_i - \mu}{\sigma}\right)\right\} \prod_{i=1}^{n} I(\mu \le Y_i < \infty).$$

For each value of σ, $L(\mu, \sigma)$ is maximized at $\widehat{\mu} = Y_{(1)}$, the smallest order statistic, and the profile (or concentrated) likelihood is

$$L^*(\sigma) = L(Y_{(1)}, \sigma) = \sigma^{-n} \exp\left\{-\sum_{i=1}^{n} \left(\frac{Y_i - Y_{(1)}}{\sigma}\right)\right\}.$$

Then

$$\log L^*(\sigma) = -n \log \sigma - \frac{1}{\sigma} \sum_{i=1}^{n} \left(Y_i - Y_{(1)}\right),$$

and

$$\frac{\partial}{\partial \sigma} \log L^*(\sigma) = -\frac{n}{\sigma} + \frac{n}{\sigma^2}\left(\overline{Y} - Y_{(1)}\right) = 0,$$

implies that $\widehat{\sigma} = \overline{Y} - Y_{(1)}$ if $\overline{Y} > Y_{(1)}$. Furthermore,

$$-\frac{\partial^2}{\partial \sigma^2} \log L^*(\sigma) = -\frac{n}{\sigma^2} + \frac{2n}{\sigma^3}\left(\overline{Y} - Y_{(1)}\right).$$

Clearly, this second derivative is greater than zero at $\widehat{\sigma}$ if $\overline{Y} > Y_{(1)}$ but not for all values of $\sigma \epsilon (0, \infty)$. Thus, $\widehat{\sigma}$ is a relative maximum of $\log L^*(\sigma)$ but $\log L^*(\sigma)$ is not concave on $(0, \infty)$. Since $\widehat{\sigma}$ is the only critical point, it is a global maximum of $\log L^*(\sigma)$ by Theorem 2.1 (p. 91) because

$$\lim_{\sigma \to 0} \log L^*(\sigma) = \lim_{\sigma \to \infty} \log L^*(\sigma) = -\infty.$$

Finally, Theorem 2.3 (p. 94) yields that $(\widehat{\mu}, \widehat{\sigma}) = (Y_{(1)}, \overline{Y} - Y_{(1)})$ is the unique maximum likelihood estimator of (μ, σ).

2.7.6 *Uniqueness of the MLE for Exponential Families*

The binomial, multinomial, and normal distributions are exponential families. Thus, in these cases the uniqueness of maximum likelihood estimators can be obtained directly from the following corollary to Theorem 2.2 (p. 91).

Theorem 2.4. *Suppose that X is distributed according to a minimal exponential family of form (2.59, p. 98) where $\boldsymbol{g}(\boldsymbol{\theta})$ is one-to-one and twice continuously differentiable in Θ, and Θ is an open subset of R^s. If there is at least one solution to the transformed likelihood equations $E_{\boldsymbol{\theta}}\,\{\boldsymbol{T}(X)\} = \boldsymbol{T}(X)$, then that solution is the unique maximum likelihood estimator.*

Proof. The likelihood equations $\mathrm{E}_{\eta}\,\boldsymbol{T}(X) = \boldsymbol{T}(X)$ in the canonical parameter space $\boldsymbol{\eta} = \boldsymbol{g}(\boldsymbol{\theta})$ have a unique solution $\widehat{\boldsymbol{\eta}}_{\mathrm{MLE}}$ since $I_X(\boldsymbol{\eta})$ is the covariance matrix of $\boldsymbol{T}$ which is positive definite everywhere since $\boldsymbol{T}$ is affinely independent. Then $\widehat{\boldsymbol{\theta}}_{\mathrm{MLE}} = \boldsymbol{g}^{-1}(\widehat{\boldsymbol{\eta}}_{\mathrm{MLE}})$ is also unique. ■

Example 2.27 (Normal location-scale as exponential family). Suppose that $Y_1, \ldots, Y_n$ are iid $\mathrm{N}(\mu, \sigma^2)$ as in Section 2.7.4 (p. 93). From (2.61, p. 100) of Appendix B on exponential families, we see that $\eta_1 = g_1(\boldsymbol{\theta}) = \mu/\sigma^2$ and $\eta_2 = -1/2\sigma^2$, and $T_1 = \sum_{i=1}^n Y_i$ and $T_2 = \sum_{i=1}^n Y_i^2$. The "transformed likelihood equations" are $n\mu = \sum_{i=1}^n Y_i$ and $n(\sigma^2 + \mu^2) = \sum_{i=1}^n Y_i^2$ leading to the usual estimators $\widehat{\mu}_{\mathrm{MLE}} = \overline{Y}$ and $\widehat{\sigma}_{\mathrm{MLE}} = s_n$ (as long as $n \geq 2$ and the Y_i are not all equal). The technical condition of being a "minimal" exponential family is that $T_1 = \sum_{i=1}^n Y_i$ and $T_2 = \sum_{i=1}^n Y_i^2$ do not satisfy a linear constraint like $T_1 + 2T_2 = 4$ and similarly that $g_1(\boldsymbol{\theta})$ and $g_2(\boldsymbol{\theta})$ do not satisfy a linear constraint. Clearly $g(\boldsymbol{\theta})$ is suitably differentiable and one-to-one, and thus we can use Theorem 2.4 to justify the uniqueness of the maximum likelihood estimators. ♦

Finally, we give one last example to show that existence and uniqueness of maximum likelihood estimators is not always to be expected.

Example 2.28 (Mixture of normals). Suppose that $Y_1, \ldots, Y_n$ are iid from the mixture of normals model

$$f(y; \mu, \sigma, p) = p\phi(y - \mu) + (1 - p)\frac{1}{\sigma}\phi\left(\frac{y - \mu}{\sigma}\right),$$

where ϕ is the standard normal density function. The log likelihood with $\mu = Y_1$ is

$$\log\left\{p\phi(0) + (1 - p)\frac{1}{\sigma}\phi(0)\right\} + \sum_{i=2}^{n} \log\left\{p\phi(Y_i - Y_1) + (1 - p)\frac{1}{\sigma}\phi\left(\frac{Y_i - Y_1}{\sigma}\right)\right\}.$$

As $\sigma \to 0$, the first term in this last expression tends to infinity, but each of the individual terms in the second term are bounded. Thus, for some paths to the boundary of the parameter space, the log likelihood tends to infinity, and thus the maximum likelihood estimator does not exist in the strict sense. Moreover, there can be multiple local maxima. Nevertheless, local maxima within the parameter space that satisfy the likelihood equations tend to behave well and are asymptotically normal with the usual asymptotic variance. In Section 2.6.3 (p. 83) we showed how the EM Algorithm can find at least one of these solutions. So, uniqueness of the maximum likelihood estimator is a nice property, but it is not necessary in order to obtain useful parameter estimates. ♦

2.8 Appendix B – Exponential Family Distributions

Fisher (1934b) is often given credit for initiating the study of families of distributions that we now call *exponential families*. In particular he gave the result that the only families of distributions that have a one-dimensional sufficient statistic are one-parameter exponential families. This result was extended to multiparameter families by Darmois (1935), Koopman (1936), and Pitman (1936).

The most thorough treatments of exponential families are found in the books by Barndorff-Nielsen (1978), Johansen (1979), and Brown (1986). Of these, Johansen (1979) is the most accessible for students. Other valuable sources are Lehmann (1983), and Lehmann and Casella (1998), which contain most of the results relevant to statistical inference. The *Encyclopedia of Statistics* entry by Barndorff-Nielsen (1982) is also an accessible exposition of the basic properties.

A family of distributions $\{\mathcal{F}_{\boldsymbol{\theta}}, \boldsymbol{\theta} \in \Omega\}$ is an *exponential family* if densities $f(x;\boldsymbol{\theta})$ exist with respect to a measure ν and have the form

$$f(x;\boldsymbol{\theta}) = h(x)\exp\left\{\sum_{i=1}^{s} g_i(\boldsymbol{\theta})T_i(x) - B(\boldsymbol{\theta})\right\}. \tag{2.59}$$

For those not familiar with measure theory, the only fact you need to know here is that measure theory allows us to write expectations with respect to a density as integrals even when the densities are for discrete random variables. Thus,

$$\begin{aligned} \mathrm{E}\{q(X)\} &= \int q(x)f(x;\boldsymbol{\theta})d\nu(x) \\ &= \begin{cases} \int q(x)f(x;\boldsymbol{\theta})dx, & \text{for continuous random variables;} \\ \sum q(x_i)f(x_i;\boldsymbol{\theta}), & \text{for discrete random variables.} \end{cases} \end{aligned}$$

Both x and $\boldsymbol{\theta}$ can be vectors, but $\boldsymbol{\theta}$ is bold unless it is clearly a scalar, and x is bold only when it helps to emphasize that it is a vector. In general x and $\boldsymbol{\theta}$ take values in subsets of R^k and R^m (often $m = s$), respectively, and ν is Lebesgue measure or a counting measure.

Example 2.29 (Normal distribution with known variance 1**).** The normal(μ, 1) density is

$$\begin{aligned} f(x;\mu) &= \frac{1}{\sqrt{2\pi}}\exp\left\{-\frac{1}{2}(x-\mu)^2\right\} \\ &= \frac{e^{-x^2/2}}{\sqrt{2\pi}}\exp\left\{\mu x - \frac{\mu^2}{2}\right\}, \quad -\infty < x < \infty,\ -\infty < \mu < \infty, \end{aligned}$$

from which we can identify $\theta = \mu$, $\Omega = (-\infty,\infty)$, $g(\mu) = \mu$, $T(x) = x$, $B(\mu) = \mu^2/2$, and $h(x) = e^{-x^2/2}/\sqrt{2\pi}$. Of course ν is Lebesgue measure on $(-\infty,\infty)$. ♦

Example 2.30 (Poisson distribution). The Poisson(λ) density is

$$f(x;\lambda) = \frac{\lambda^x e^{-\lambda}}{x!} = \frac{1}{x!}\exp\{\log(\lambda)x - \lambda\}, \qquad x = 0, 1, \ldots \quad \lambda > 0,$$

from which we see that $\theta = \lambda$, $\Omega = (0,\infty)$, $g(\lambda) = \log(\lambda)$, $T(x) = x$, $B(\lambda) = \lambda$, $h(x) = 1/x!$, and ν may be taken as counting measure on the nonnegative integers. ♦

2.8.1 Canonical Representation

The set of parameter values

$$\Omega_0 = \left\{ \theta : \int h(x) \exp\left(\sum_{i=1}^{s} g_i(\boldsymbol{\theta}) T_i(x) \right) d\nu(x) < \infty \right\}$$

is called the *natural parameter space* of the family (2.59). It is the largest possible parameter space for a specific family type. In a given situation, the actual parameter space Ω may equal Ω_0 or be restricted to a proper subset.

For many purposes it is simpler to reparameterize in terms of $\boldsymbol{\eta} = \boldsymbol{g}(\boldsymbol{\theta})$ so that (2.59) becomes

$$f(x; \boldsymbol{\eta}) = h(x) \exp\left\{ \sum_{i=1}^{s} \eta_i T_i(x) - A(\boldsymbol{\eta}) \right\}. \tag{2.60}$$

This latter expression is called the *canonical representation*, and $\boldsymbol{\eta}$ is the *canonical parameter*. The set

$$\Omega_\eta = \left\{ \boldsymbol{\eta} : \int h(x) \exp\left(\sum_{i=1}^{s} \eta_i T_i(x) \right) d\nu(x) < \infty \right\}$$

is also called the *natural parameter space*.

In the canonical representation, if $\boldsymbol{\eta}$ is in the interior of Ω_η, then the moment generating function of $\boldsymbol{T}$ exists in a neighborhood of the origin and is given by (see, Lehmann and Casella 1998, Theorem 5.10, p. 28)

$$M_T(u) = \mathrm{E} e^{u_1 T_1 + \ldots + u_s T_s} = e^{A(\boldsymbol{\eta}+\boldsymbol{u}) - A(\boldsymbol{\eta})}.$$

The *cumulant generating function* is just

$$K_T(\boldsymbol{u}) = \log\{M_T(\boldsymbol{u})\} = A(\boldsymbol{\eta} + \boldsymbol{u}) - A(\boldsymbol{\eta}).$$

Both functions lead to the following simple expressions for the mean and covariance matrix of $\boldsymbol{T}(X)$:

$$\frac{\partial A(\boldsymbol{\eta})}{\partial \boldsymbol{\eta}^T} = \mathrm{E}\{\boldsymbol{T}(X)\}, \qquad \frac{\partial^2 A(\boldsymbol{\eta})}{\partial \boldsymbol{\eta} \partial \boldsymbol{\eta}^T} = \mathrm{Var}\{\boldsymbol{T}(X)\}.$$

Note also that taking two derivatives of the logarithm of (2.60) leads to $\boldsymbol{I}(\boldsymbol{\eta}) = \mathrm{Var}\{\boldsymbol{T}(X)\}$. The information matrix for representation (2.59) is then $\boldsymbol{I}(\boldsymbol{\theta}) = J^T I(\boldsymbol{g}(\boldsymbol{\theta}))J$, where $J = \partial g_i(\boldsymbol{\theta})/\partial \theta_j$. These expressions are useful if $A(\boldsymbol{\eta})$ is easily manipulated. In Example 2.29 (p. 98) we have $A(\eta) = B(\mu(\eta)) = \eta^2/2$, and $\mathrm{E}(X) = d(\eta^2/2)/d\eta = \eta = \mu$, $\mathrm{Var}(X) = d^2(\eta^2/2)/d\eta^2 = 1$. Similarly, for Example 2.30 (p. 98) we have $A(\eta) = B(\lambda(\eta)) = e^\eta$, and $\mathrm{E}(X) = d(e^\eta)/d\eta = e^\eta = \lambda$, $\mathrm{Var}(X) = d^2(e^\eta)/d\eta^2 = e^\eta = \lambda$. In the next example we see that $A(\boldsymbol{\eta})$ is harder to use than direct calculation.

Example 2.31 (Normal location-scale model). The normal(μ, σ^2) density is

$$
\begin{aligned}
f(x;\boldsymbol{\theta}) &= \frac{1}{\sqrt{2\pi\sigma^2}} \exp\left\{-\frac{1}{2\sigma^2}(x-\mu)^2\right\} \\
&= \frac{1}{\sqrt{2\pi}} \exp\left\{\frac{\mu}{\sigma^2}x - \frac{1}{2\sigma^2}x^2 - \frac{\mu^2}{2\sigma^2} - \log(\sigma)\right\}, \qquad (2.61)
\end{aligned}
$$

for $-\infty < x < \infty$, $-\infty < \mu < \infty$, $0 < \sigma < \infty$. We can identify $\boldsymbol{\theta} = (\mu, \sigma)$, $\Omega_0 = (-\infty, \infty) \times (0, \infty)$, $g_1(\boldsymbol{\theta}) = \mu/\sigma^2$, $g_2(\boldsymbol{\theta}) = -1/2\sigma^2$, $T_1(x) = x$, $T_2(x) = x^2$, $B(\boldsymbol{\theta}) = \mu^2/2\sigma^2 + \log(\sigma)$, and $h(x) = 1/\sqrt{2\pi}$. For computing moments we put $\mu = -\eta_1/2\eta_2$ and $\sigma^2 = -1/2\eta_2$ into $B(\boldsymbol{\theta})$ to get

$$
A(\boldsymbol{\eta}) = \frac{-\eta_1^2}{4\eta_2} - \frac{1}{2}\log(-\eta_2) - \log(\sqrt{2}).
$$

Taking a derivative with respect to η_1, we get

$$
\mathrm{E}\{X\} = \frac{\partial}{\partial \eta_1} A(\boldsymbol{\eta}) = -2\eta_1/(4\eta_2) = \mu.
$$

The second derivative with respect to η_1 is then $= -2/(4\eta_2) = \sigma^2 = \mathrm{Var}\{X\}$. But note that for the standard normal density ϕ, we have $\int_\infty^\infty x\phi(x)dx = -[\phi(\infty) - \phi(-\infty)] = 0$ directly, and $\int_\infty^\infty x^2\phi(x)dx = \int_\infty^\infty \phi(x)dx = 1$ using integration by parts. Thus, $\mathrm{E}\{X\} = \mu$ and $\mathrm{Var}\{X\} = \sigma^2$ follow fairly easily and perhaps more easily than finding $A(\boldsymbol{\eta})$ and taking two derivatives with respect to η_1. ♦

It is not hard to see that the joint distribution of independent and identically distributed (iid) random variables $\boldsymbol{X}^T = (X_1, \ldots, X_n)$ from an exponential family also form an exponential family with the same parameters but with $T_i(x)$ replaced with $T_i^* = \sum_{j=1}^n T_i(x_j)$, $B(\boldsymbol{\theta})$ replaced by $nB(\boldsymbol{\theta})$, and $h(x)$ replaced by $\prod_{j=1}^n h(x_j)$. For example, the joint density of a sample from the normal(μ, σ^2) density is

$$
f(\boldsymbol{x};\boldsymbol{\theta}) = (2\pi)^{-n/2} \exp\left\{\frac{\mu}{\sigma^2}\sum_{j=1}^n x_j - \frac{1}{2\sigma^2}\sum_{j=1}^n x_j^2 - n\frac{\mu^2}{2\sigma^2} - n\log(\sigma)\right\}.
$$

From the factorization theorem we immediately see that $(\sum_{j=1}^n X_j, \sum_{j=1}^n X_j^2)$ is a sufficient statistic and note that it is a one-to-one function of the more familiar sufficient statistic $(\overline{X}, S^2)$. In general

$$(T_1^*, \ldots, T_s^*) = \left(\sum_{j=1}^{n} T_i(X_j), \ldots, \sum_{j=1}^{n} T_s(X_j) \right)$$

is a sufficient statistic based on an iid sample from (2.59, p. 98). In particular problems we may prefer to work with other equivalent sufficient statistics, possibly nonlinear transformed ones such as $(\overline{X}, S^2)$. However, we almost always ought to choose a sufficient statistic from among the minimal sufficient statistics, which leads us to seek a minimal version of (2.59, p. 98).

2.8.2 Minimal Exponential Family

Although the representation (2.59, p. 98) is not unique because of linear transformations, it is possible and usually worthwhile to find a representation based on the smallest s that satisfies (2.59, p. 98). Thus we seek a version of (2.59, p. 98) such that the real-valued components $T_1, \ldots, T_s$ are "affinely independent with respect to ν, " that is,

$$\sum_{i=1}^{s} c_i T_i(x) = c_{s+1} \text{ a.s } \nu \implies c_i = 0, \; i = 1, \ldots s+1. \tag{2.62}$$

In other words, (2.62) says that almost surely ν the T_i do not satisfy a linear constraint. For example, $T_1(x) = x$ and $T_2(x) = 3x$ are not affinely independent (since $3T_1 - T_2 = 0$), but $T_1(x) = x$ and $T_2(x) = 3x^2$ are affinely independent with respect to $\nu =$ Lebesgue measure. Similarly, we want the real-valued components $\eta_1(\theta), \ldots, \eta_s(\theta)$ to be affinely independent, that is, that the η_i do not satisfy a linear constraint. In such a case when both the T_i and $\eta_i(\boldsymbol{\theta})$ are free of linear constraints, then the family is said to be *minimal*. It is not hard to show by a theorem of Lehmann and Scheffé (see Casella and Berger 2002, Theorem 6.2.13, p. 281) that $(T_1, \ldots, T_s)$ is then minimal sufficient.

Example 2.32 (Multinomial distribution). The multinomial$(n; p_1, \ldots, p_k)$ distribution is a naturally occurring family that is not minimal in the usual parameterization. The density is

$$f(x; \boldsymbol{p}) = \frac{n!}{x_1! \ldots x_k!} \prod_{i=1}^{k} p_i^{x_i} = \frac{n!}{x_1! \ldots x_k!} \exp\left\{ \sum_{i=1}^{k} \log(p_i) x_i \right\}, \tag{2.63}$$

where $\sum_{i=1}^{k} p_i = 1$ and $\sum_{i=1}^{k} x_i = n$. This family has the representation (2.59, p. 98) with $g_i = \log(p_i)$ and $T_i = x_i, i = 1, \ldots, k$. Here it is easy to reduce to a minimal representation with $s = k - 1$ by setting $p_k = 1 - \sum_{i=1}^{k-1} p_i$, $x_k = n - \sum_{i=1}^{k-1} x_i$, $\boldsymbol{\theta} = (p_1, \ldots, p_{k-1})$, and $\boldsymbol{g}(\boldsymbol{\theta}) = \{\log(p_1/p_k), \ldots, \log(p_{k-1}/p_k)\}$.

Thus p_k and x_k may still appear in

$$f(\boldsymbol{x};\boldsymbol{\theta}) = \frac{n!}{x_1!\dots x_k!}\exp\left\{\sum_{i=1}^{k-1}\log(p_i/p_k)x_i + n\log(p_k)\right\}$$

so that the expression is not too unwieldy. The natural parameter space here is

$$\Omega_0 = \left\{(p_1,\dots,p_{k-1}) : 0 < p_i,\ \sum_{i=1}^{k-1} p_i < 1\right\}.$$

The canonical representation is

$$f(x;\boldsymbol{\eta}) = \frac{n!}{x_1!\dots x_k!}\exp\left\{\sum_{i=1}^{k-1}\eta_i x_i - n\log\left(1+\sum_{i=1}^{k-1}e^{\eta_i}\right)\right\},$$

with natural parameter space $\Omega_\eta = \{(\eta_1,\dots,\eta_{k-1}) : -\infty < \eta_i < \infty\}$. Taking derivatives of $A(\boldsymbol{\eta}) = n\log\left(1+\sum_{i=1}^{k-1}e^{\eta_i}\right)$ leads quickly to $\mathrm{E}\{X_i\} = np_i$, $\mathrm{Var}\{X_i\} = np_i(1-p_i)$, and $\mathrm{Cov}(X_i, X_j) = -np_ip_j$ for $i \neq j$. ♦

Example 2.33 (Normal with mean = variance). A less natural family is the normal(σ^2,σ^2) density where upon substituting $\mu = \sigma^2$ into (2.61, p. 100) we get

$$\begin{aligned} f(x;\sigma) &= \frac{1}{\sqrt{2\pi}}\exp\left\{\frac{\sigma^2}{\sigma^2}x - \frac{1}{2\sigma^2}x^2 - \frac{\sigma^4}{2\sigma^2} - \log(\sigma)\right\} \\ &= \frac{e^x}{\sqrt{2\pi}}\exp\left\{-\frac{1}{2\sigma^2}x^2 - \frac{\sigma^2}{2} - \log(\sigma)\right\}. \end{aligned}$$

Thus, this family reduces to an $s = 1$ minimal exponential family. ♦

Example 2.34 (Normal with mean = standard deviation). Now consider the normal
(σ,σ^2) density where substituting $\mu = \sigma$ into (2.61, p. 100) yields

$$\begin{aligned} f(x;\sigma) &= \frac{1}{\sqrt{2\pi}}\exp\left\{\frac{\sigma}{\sigma^2}x - \frac{1}{2\sigma^2}x^2 - \frac{\sigma^2}{2\sigma^2} - \log(\sigma)\right\} \\ &= \frac{e^{-1/2}}{\sqrt{2\pi}}\exp\left\{\frac{1}{\sigma}x - \frac{1}{2\sigma^2}x^2 - \log(\sigma)\right\}. \end{aligned}$$

This case is different because it is a minimal $s = 2$ exponential family (x and x^2 are affinely independent and so are $\eta_1 = 1/\sigma$ and $\eta_2 = -1/2\sigma^2$), yet the dimension of $\theta = \sigma$ is 1. This type of exponential family is called a *curved exponential family*.

For an iid sample from the above density, $(\sum_{j=1}^{n} X_j, \sum_{j=1}^{n} X_j^2)$ (or equivalently $(\overline{X}, S^2)$) is still a minimal sufficient statistic, but simple calculations show that it is not complete. ♦

Example 2.35 (Fourth-power curved exponential example). Another example of a curved exponential family is given by Lehmann and Casella (1998, p. 41):

$$f(x;\theta) = Ce^{-(x-\theta)^4} = Ce^{-x^4}e^{4\theta^3 x - 6\theta^2 x^2 + 4\theta x^3 - \theta^4}.$$

This is a minimal $s = 3$ curved exponential family. For an iid sample, $(\sum_{j=1}^{n} X_j, \sum_{j=1}^{n} X_j^2, \sum_{j=1}^{n} X_j^3)$ is minimal sufficient but not complete. ♦

2.8.3 *Completeness*

Recall that a set of statistics $(T_1, \ldots, T_s)$ is *complete* with respect to the family of their induced distributions indexed by $\boldsymbol{\theta}$ if there are no functions $\phi(\boldsymbol{T})$ (other than $\phi = 0$) such that $E_{\boldsymbol{\theta}}\, \phi(\boldsymbol{T}) = 0$ for all $\boldsymbol{\theta} \in \Omega$.

So what restrictions guarantee completeness of the minimal sufficient statistic? Johansen (1979, p. 11), Lehmann and Casella (1998, p. 42), and Brown (1986, p. 43) each give completeness results in terms of the canonical representation (2.60, p. 99). Johansen's sufficient condition is that $\{\mathcal{F}_{\eta}, \boldsymbol{\eta} \in \Omega_{\eta}\}$ be a "regular exponential family," which means that it is a minimal exponential family with the natural parameter space Ω_{η} and that space is an open set in R^s. Lehmann and Casella's Theorem 6.22 seems a little less restrictive because it requires the family to be of *full rank*, which means that the family is minimal and the parameter space contains an s-dimensional rectangle (not concentrated on a subspace of lower dimension than s). For example, the parameter space in Example 2.34 (p. 102) is defined by the curve $\eta_2 = -\eta_1^2/2$ which is of dimension 1. Clearly, this curve does not contain a 2-dimensional rectangle. Lehmann and Casella's result allows the parameter space of $\boldsymbol{\eta}$ to be a subset of the natural parameter space Ω_{η}. We can restate their theorem as follows.

Theorem 2.5 (Lehmann and Casella Theorem 6.22). *The sufficient statistic $(T_1, \ldots, T_s)$ of a canonical exponential family having representation (2.60, p. 99) is complete sufficient provided that the family is minimal and that the parameter space contains an s-dimensional rectangle.*

To translate this result on completeness to a family with the representation (2.59, p. 98), we need to merely verify that the family is minimal and that in terms of the canonical parameters, the parameter space contains an s-dimensional rectangle. Thus, of the above examples, only for random samples drawn from the densities in Examples 2.34 (p. 102) and 2.35 (p. 103) are the T not complete. If the dimension of $\boldsymbol{\theta}$ is less than s, then we can be sure that the parameter space, after conversion to the canonical framework, is concentrated on a subspace of lower dimension than s.

For Example 2.31 (p. 100) with the normal(μ, σ^2) density, Theorem 2.5 says that $(T_1, T_2) = (X, X^2)$ is complete. One might note that $E_{\boldsymbol{\theta}}(T_1^2 - T_2) = 0$ for all $\boldsymbol{\theta}$. This does not contradict completeness because $T_1^2 - T_2 = 0$.

2.8.4 Distributions of the Sufficient Statistics

Several other important facts about exponential families are summarized by the following result from Lehmann (1986, Lemma 8, p. 58).

Theorem 2.6. *Let X be distributed according to an exponential family of form (2.60, p. 99) where $\eta = (\boldsymbol{\zeta}, \boldsymbol{\psi})$*

$$f(x; \boldsymbol{\zeta}, \boldsymbol{\psi}) = h(x) \exp \left\{ \sum_{i=1}^{r} \zeta_i U_i(x) + \sum_{j=1}^{s} \psi_j T_j(x) - A(\boldsymbol{\zeta}, \boldsymbol{\psi}) \right\}. \tag{2.64}$$

Then

(0) the distribution of $(\boldsymbol{U}, \boldsymbol{T})$ is an exponential family
(i) the marginal distribution of $\boldsymbol{T} = (T_1, \ldots, T_s)$ is an exponential family of the form

$$f(\boldsymbol{t}; \boldsymbol{\zeta}, \boldsymbol{\psi}) = q(\boldsymbol{t}) C(\boldsymbol{\zeta}) \exp \left\{ \sum_{j=1}^{s} \psi_j t_j - A(\boldsymbol{\zeta}, \boldsymbol{\psi}) \right\}; \tag{2.65}$$

(ii) the conditional distribution of $\boldsymbol{U} = (U_1, \ldots, U_s)$ given $\boldsymbol{T} = \boldsymbol{t}$ is an exponential family of the form

$$f(\boldsymbol{u} | \boldsymbol{t}; \boldsymbol{\zeta}) = q_{\boldsymbol{t}}(\boldsymbol{u}) \exp \left\{ \sum_{i=1}^{r} \zeta_i u_i - A_{\boldsymbol{t}}(\boldsymbol{\zeta}) \right\}. \tag{2.66}$$

Example 2.36 (Logistic dose-response model). Consider a toxicology study with k groups of animals who are given a drug at dose levels $d_1, \ldots, d_k$, respectively. The animals are monitored for a reaction such as weight loss or death. The result is that X_i of the n_i animals in group i react, $i = 1, \ldots, k$, and we treat the data $X_1, \ldots, X_k$ as a set of independent binomial(n_i, p_i) random variables. The joint density is then

$$\left\{ \prod_{i=1}^{k} \binom{n_i}{x_i} \right\} \exp \left\{ \sum_{i=1}^{k} \log \left(\frac{p_i}{1 - p_i} \right) x_i - \sum_{i=1}^{k} n_i \log(1 - p_i) \right\}.$$

Without any restrictions on the p_i (other than $0 < p_i < 1$), this is an $s = k$ full rank exponential family with $T_i = X_i$ and $\eta_i = \log\{p_i/(1 - p_i)\}$, $i = 1, \ldots, k$.

A common assumption in such studies is that p_i is modeled as $p_i = F(\beta_0 + \beta_1 d_i)$, where F is a distribution function like the standard normal Φ, logistic $F(t) =$

$(1+e^{-t})^{-1}$, or extreme value $F(t) = \exp(-\exp(-t))$. In these models the logistic distribution function has a particularly nice form because $\log\{F(t)/(1-F(t))\} = t$ and thus $\log\{p_i/(1-p_i)\} = \beta_0+\beta_1 d_i$. Plugging into the joint density above, we get

$$\left\{\prod_{i=1}^{k}\binom{n_i}{x_i}\right\}\exp\left\{\beta_0\sum_{i=1}^{k}x_i+\beta_1\sum_{i=1}^{k}d_i x_i-\sum_{i=1}^{k}n_i\log(1-F(\beta_0+\beta_1 d_i))\right\}.$$

Thus, for the logistic model we have an $s = 2$ full rank exponential family with $\boldsymbol{T} = (\sum_{i=1}^{k} X_i, \sum_{i=1}^{k} d_i X_i)$ and $\boldsymbol{\eta} = (\beta_0, \beta_1)$. This model fits nicely into the previous Theorem about the conditional distribution of $\boldsymbol{U}|\boldsymbol{T} = \boldsymbol{t}$, where typically $U = \sum_{i=1}^{k} d_i X_i$ and $T = \sum_{i=1}^{k} X_i$ since β_1 is usually the parameter of interest, and β_0 is a nuisance parameter. Let us spell out the details as taken from Cox and Snell (1989, Ch. 2). The joint distribution of (T, U) is given by

$$P(T=t, U=u) = c(t,u)\exp\left\{\beta_0 t+\beta_1 u-\sum_{i=1}^{k}\log n_i\left(1+e^{\beta_0+\beta_1 d_i}\right)\right\}, \tag{2.67}$$

where $c(t,u)$ is just the constant needed to make (2.67) a discrete density. The conditional density of $U|T = t$ is found in the usual way

$$P(U=u|T=t) = \frac{P(U=u, T=t)}{P(T=t)} = \frac{c(t,u)e^{\beta_1 u}}{\sum_v c(t,v)e^{\beta_1 v}}, \tag{2.68}$$

where cancellations of terms from (2.67) greatly simplify the expression. Even with the simple form in (2.68), calculation of this conditional density usually requires specialized software.

Further results in Lehmann (1986) use this conditional distribution to find UMPU tests concerning β_1. Methods based on the logistic model with conditional likelihoods are often called *conditional logistic regression*.

Looking back at the general form of the likelihood in terms of F, we note that if F is some other distribution function than the logistic, then typically the density does not reduce as it does for the logistic, and we have a minimal exponential family with $s = k$, but $\boldsymbol{\theta} = (\beta_0, \beta_1)$ only has dimension $= 2$. These densities thus form a curved exponential family. Messig and Strawderman (1993) give conditions for $X_1, \ldots, X_k$ to be minimal sufficient. They also assert completeness of the minimal sufficient statistic in certain cases of these curved exponential families (a surprising result). ♦

2.8.5 *Families with Truncation or Threshold Parameters*

In general, the definition (2.59, p. 98) of an exponential family assumes that the support of X does not depend on θ. This seems to eliminate families with truncation or threshold parameters from being able to use exponential family results. Fortunately, a number of families have conditional exponential family representations, as illustrated using the following results from Quesenberry (1975).

Consider a random sample $X_1, \ldots, X_n$ from one of three types of densities defined over an interval (a, b), possibly $(-\infty, \infty)$:

$$f(x; \boldsymbol{\theta}, \mu_1, \mu_2) = c(\mu_1, \mu_2, \boldsymbol{\theta}) d(x, \boldsymbol{\theta}), \qquad a < \mu_1 < x < \mu_2 < b; \tag{2.69}$$

$$f(x; \boldsymbol{\theta}, \mu) = c_1(\mu, \boldsymbol{\theta}) d_1(x, \boldsymbol{\theta}), \qquad a < \mu < x < b; \tag{2.70}$$

$$f(x; \boldsymbol{\theta}, \mu) = c_2(\mu, \boldsymbol{\theta}) d_2(x, \boldsymbol{\theta}), \qquad a < x < \mu < b; \tag{2.71}$$

where $d(x, \boldsymbol{\theta})$, $d_1(x, \boldsymbol{\theta})$, and $d_2(x, \boldsymbol{\theta})$ are positive, continuous, and integrable over the intervals (μ_1, μ_2), (μ, b), and (a, μ), respectively.

Now let us focus on the first density form (2.69). If $X_{(1)}$ is the smallest order statistic and $X_{(n)}$ is the largest, then for $\boldsymbol{\theta}$ fixed, $(X_{(1)}, X_{(n)})$ is minimal sufficient for (μ_1, μ_2). Denote by $Z_1, \ldots, Z_{n-2}$ the other values of the sample that fall between $X_{(1)}$ and $X_{(n)}$. Quesenberry's (1975) Theorem 1 states that conditional on $(X_{(1)}, X_{(n)}) = (x_{(1)}, x_{(n)})$, the random variables $Z_1, \ldots, Z_{n-2}$ are iid and continuous with density function

$$q(z, \boldsymbol{\theta}) = \frac{d(z, \boldsymbol{\theta}) I\left(x_{(1)} < z < x_{(n)}\right)}{\int_{x_{(1)}}^{x_{(n)}} d(z, \boldsymbol{\theta}) dz},$$

where $I(\cdot)$ is the indicator function. Of course a similar result holds for observations from (2.70) and (2.71) that are conditional on being larger than $X_{(1)}$ or smaller than $X_{(n)}$, respectively. Thus if the $d(z, \boldsymbol{\theta})$ in (2.69), (2.70) and (2.71) have an exponential family form, then the conditional densities do also. The simplest example is the exponential

$$f(x; \theta, \mu) = \theta e^{-\theta(x-\mu)} I(\mu < x < \infty),$$

with $\theta > 0$ and $-\infty < \mu < \infty$. Then, conditional on $X_{(1)} = x_{(1)}$, the observations larger than $x_{(1)}$ have density

$$q(z, \theta) = \theta e^{-\theta(z - x_{(1)})} I(x_{(1)} < x < \infty) = e^{-\theta z + \theta x_{(1)} + \log(\theta)} I(x_{(1)} < x < \infty).$$

2.8.6 Exponential Family Glossary

natural parameter space of the family (2.59):

$$\Omega_0 = \{\boldsymbol{\theta} : \int h(x) \exp\left[\sum_{i=1}^{s} \eta_i(\boldsymbol{\theta}) T_i(x)\right] d\nu(x) < \infty\}.$$

canonical representation ($\boldsymbol{\eta}$ is called the *canonical parameter*):

$$f(x; \boldsymbol{\eta}) = h(x) \exp\left[\sum_{i=1}^{s} \eta_i T_i(x) - A(\boldsymbol{\eta})\right].$$

minimal exponential family : $T_1, \ldots, T_s$ are affinely independent with respect to ν, and $g_1(\boldsymbol{\theta}), \ldots, g_s(\boldsymbol{\theta})$ are affinely independent (affinely independent means "do not satisfy a linear constraint").
regular exponential family : minimal exponential family with the canonical representation and Ω_η is an open set in R^s.
full rank exponential family: minimal exponential family and Ω contains an s-dimensional rectangle.
curved exponential family: $\dim(\theta) < s$.

2.9 Problems

2.1. Let $Y_1, \ldots, Y_n$ be iid positive random variables such that $Y^{(\lambda)}$ is assumed to have a normal(μ, σ^2) distribution, where

$$Y^{(\lambda)} = \begin{cases} \dfrac{Y^\lambda - 1}{\lambda} & \text{when } \lambda \neq 0, \\ \log(Y) & \text{when } \lambda = 0. \end{cases}$$

(The normality assumption is actually only possible for $\lambda = 0$, but ignore that detail.) Derive the log likelihood $\ell_n(\mu, \sigma, \lambda \mid \boldsymbol{Y})$ of the observed data $Y_1, \ldots, Y_n$. Note that $y^{(\lambda)}$ is a strictly increasing function of y (the derivative is always positive). It might be easiest to use the "distribution function" method to get the density of Y_i, but feel free to use Jacobians, etc.

2.2. One of the data sets obtained from the 1984 consulting session on max flow of rivers was $n = 35$ yearly maxima from one station displayed in the following R printout.

```
> data.max
 [1]  5550  4380  2370  3220  8050  4560  2100
 [8]  6840  5640  3500  1940  7060  7500  5370
[15] 13100  4920  6500  4790  6050  4560  3210
[22]  6450  5870  2900  5490  3490  9030  3100
[29]  4600  3410  3690  6420 10300  7240  9130
```

a. Find the maximum likelihood estimates for the extreme value location-scale density $f(y; \mu, \sigma) = f_0((y - \mu)/\sigma)/\sigma$, where

$$f_0(t) = e^{-t} e^{-e^{-t}}.$$

b. Draw a QQ plot and the parametrically estimated distribution function overlaid with the empirical distribution function. Here is sample R code.

```
par(mfrow = c(2,1))  # gives two plots per page
qextval<-function(t,mu,sigma){-sigma*log(-log(t))
                              +mu}
pextval<-function(x,mu,sigma){exp(-exp(-(x-mu)/
                              sigma))}
plot(qextval(ppoints(data.max),0,1),sort(data.max))
seq(1900,13200,,100)->x     # a grid of values
pextval(x,muhat,sigmahat)->y # est. cdf for grid
plot(x,y,type="l")   # plots est. ext. value cdf
1:35/35->ht            # heights for empirical cdf
points(sort(data.max),ht) # adds empirical cdf
```

2.3. Recall the ZIP model

$$P(Y=0) = p + (1-p)e^{-\lambda}$$

$$P(Y=y) = (1-p)\frac{\lambda^y e^{-\lambda}}{y!} \quad y = 1, 2, \ldots$$

a. Reparameterize the model by defining

$$\pi \equiv P(Y=0) = p + (1-p)e^{-\lambda}.$$

Solve for p in terms of λ and π, and then substitute so that the density depends only on λ and π.

b. For an iid sample of sample of size n, let n_0 be the number of zeroes in the sample. Assuming that the complete data is available (no grouping), show that the likelihood factors into two pieces and that $\widehat{\pi} = n_0/n$. (This illustrates why we obtained exact fits for the 0 cell in Example 2.1, p. 32.) Also show that the maximum likelihood estimator for λ is the solution to a simple nonlinear equation involving $\overline{Y}_+$ (the average of the nonzero values).

c. Now consider the truncated or conditional sample consisting of the $n-n_0$ nonzero values. Write down the conditional likelihood for these values and obtain the same equation for $\widehat{\lambda}_{\text{MLE}}$ as in a). (First write down the conditional density of Y given $Y > 0$.)

2.4. In sampling land areas for counts of an animal species, we obtain an iid sample of counts $Y_1, \ldots, Y_n$, where each Y_i has a Poisson distribution with parameter λ, $P(Y_1 = y) = e^{-\lambda}\lambda^y/y!$.

a. Derive the maximum likelihood estimator of λ. Call it $\widehat{\lambda}_{\text{MLE},1}$.

b. For simplicity in quadrat sampling, sometimes only the presence or absence of a species is recorded. Let n_0 be the number of Y_i's that are zero. Write down the binomial likelihood based only on n_0. Show that the maximum likelihood estimator of λ based only on n_0 is $\widehat{\lambda}_{\text{MLE},2} = -\log(n_0/n)$.

c. Recall that the asymptotic relative efficiency of $\widehat{\theta}_1$ to $\widehat{\theta}_2$ is the limit of the ratio of the asymptotic variance of $\widehat{\theta}_2$ to the asymptotic variance of $\widehat{\theta}_1$. Use the delta theorem from Ch. 1 to show that the asymptotic relative efficiency of $\widehat{\lambda}_{\text{MLE},1}$ to $\widehat{\lambda}_{\text{MLE},2}$ is $\{\exp(\lambda) - 1\}/\lambda$.

d. The overall goal of the sampling is to estimate the mean number of the species per unit land area. Comment on the use of $\widehat{\lambda}_{\text{MLE},2}$ in place of $\widehat{\lambda}_{\text{MLE},1}$. That is, explain to a researcher for what λ values and under what distributional assumptions is it reasonable?

2.5. Suppose that X and Y are continuous random variables with joint density $f_{X,Y}(x,y)$ and marginal densities $f_X(x)$ and $f_Y(y)$, respectively. Use the $2h$ method to justify the definition of the conditional density of X given Y:

$$f_{X|Y}(x|y) = \frac{f_{X,Y}(x,y)}{f_Y(y)}.$$

2.6 (Continuation). Generalize Problem 2.5 to the case where $\boldsymbol{X}$ is a vector of dimension p and $\boldsymbol{Y}$ is a vector of dimension q. (Hint: The result looks the same but you do not divide by $2h$.)

2.7 (Continuation). Repeat Problem 2.5 for the case that X is continuous but Y is discrete.

2.8. For the rainfall example (2.8, p. 41), with distribution function

$$F_Y(t; p, c, \sigma) = pI(0 \le t) + (1-p)\left\{1 - e^{-(t/\sigma)^c}\right\},$$

find the mean, variance, and median. It helps to know that the rth moment about 0 of a Weibull distribution (the continuous component in the model) is $\Gamma(1 + r/c)\sigma^r$. The mean and variance expressions depend on $\Gamma(\cdot)$. To get the median, just set the distribution function equal to 1/2 and solve for t.

2.9. The sample $Y_1, \ldots, Y_n$ is iid with distribution function

$$F_Y(y; p_0, p_1, \alpha, \beta) = p_0 I(0 \le y) + (1 - p_0 - p_1)F(y; \alpha, \beta) + p_1 I(y \ge 1),$$

where $F(y;\alpha,\beta)$ is the beta distribution. You may recall that the beta density is positive on $0 < y < 1$ so that $F(0;\alpha,\beta) = 0$ and $F(1;\alpha,\beta) = 1$, but otherwise you do not need to use or know its form in the following; just use $F(y;\alpha,\beta)$ or $f(y;\alpha,\beta)$ where needed. The physical situation relates to test scores standardized to lie in [0, 1], but where n_0 of the sample values are exactly 0 (turned in a blank test), n_1 values are 1 (a perfect score), and the rest are in between 0 and 1. Use the $2h$ method to show that the likelihood is

$$p_0^{n_0} p_1^{n_1} (1 - p_0 - p_1)^{n-n_0-n_1} \prod_{0<Y_i<1} f(Y_i; \alpha, \beta).$$

2.10. Suppose that $Y_1, \ldots, Y_n$ are n distinct numbers. Assume that there exists an $h_* > 0$ such that for all $0 < h < h_*$, $\widehat{F}_h(y)$ is a distribution function satisfying

$$\widehat{F}_h(Y_i + h) - \widehat{F}_h(Y_i - h) = \frac{1}{n}, \qquad i = 1, \ldots, n.$$

Prove that $\widehat{F}_h(y) \longrightarrow n^{-1} \sum_{i=1}^{n} I(Y_i \leq y)$ as $h \longrightarrow 0$ at each continuity point of the limit.

2.11. Let $Y_1, \ldots, Y_n$ be an iid sample, each with distribution function F. We make no restrictions on F. Show that the empirical distribution function is the nonparametric maximum likelihood estimator for the case that there are ties in the data. To fix notation, let n_j be the number of sample values at y_j, $j = 1, \ldots, k$, $\sum_{j=1}^{k} n_j = n$. Then start with the approximate likelihood $\prod_{j=1}^{k} [p_{j,h}]^{n_j}$, where $p_{j,h} = F(y_j + h) - F(y_j - h)$, and use an argument similar to that found in Section 2.2.6, p. 45.

2.12. For an iid sample $Y_1, \ldots, Y_n$, Type II censoring occurs when we observe only the smallest r values. For example, in a study of light bulb lifetimes, we might stop the study after the first $r = 10$ bulbs have failed. Assuming a continuous distribution with density $f(y; \boldsymbol{\theta})$, the likelihood is just the joint density of the smallest r order statistics evaluated at those order statistics:

$$L(\boldsymbol{\theta}; Y_{(1)}, \ldots, Y_{(r)}) = \frac{n!}{(n-r)!} \left[\prod_{i=1}^{r} f(Y_{(i)}; \boldsymbol{\theta}) \right] \left[1 - F(Y_{(r)}; \boldsymbol{\theta}) \right]^{n-r}.$$

For this situation, let $f(y; \sigma) = e^{-y/\sigma}/\sigma$ and find the MLE of σ.

2.13. For the random censoring likelihood, p. 49, we expressed $P(Y_i \in (y-h, y+h], \delta_i = 1)$ and $P(Y_i \in (y-h, y+h], \delta_i = 0)$ as double integrals, divided by $2h$ and took limits. Instead of double integrals, try to find the same expressions by conditioning: e.g., start with

$$\begin{aligned} P(Y_i \in (y-h, y+h], \delta_i = 1) &= P(X_i \in (y-h, y+h], X_i \leq R_i) \\ &= P(X_i \leq R_i \mid X_i \in (y-h, y+h]) \\ &\times P(X_i \in (y-h, y+h]). \end{aligned}$$

(Hint: bound $P(X_i \leq R_i \mid X_i \in (y-h, y+h])$ above by $P(y-h \leq R_i)$ and below by $P(y+h \leq R_i)$.)

2.14. Derive the likelihood found in (2.13, p. 51).

2.15. Derive the likelihood for the normal additive errors nonlinear model of Section 2.3.2 (p. 53).

2.16. The standard Box-Cox regression model (Box and Cox 1964) assumes that after transformation of the observed Y_i to $Y_i^{(\lambda)}$ we have the linear model

$$Y_i^{(\lambda)} = \boldsymbol{x}_i^T \boldsymbol{\beta} + e_i, \quad i = 1, \ldots, n,$$

where Y_i is assumed positive and the $\boldsymbol{x}_i$ are known constants, $i = 1, \ldots, n$. In addition assume that $e_1, \ldots, e_n$ are iid normal$(0, \sigma^2)$ errors. Recall that the Box-Cox transformation is defined in Problem 2.1 (p. 107) and is strictly increasing for all λ. Show that the likelihood is

$$L(\beta, \sigma, \lambda \mid \{Y_i, \boldsymbol{x}_i\}_{i=1}^n) = \left(\frac{1}{\sqrt{2\pi}\sigma}\right)^n \exp\left[-\sum_{i=1}^n \frac{\left\{Y_i^{(\lambda)} - \boldsymbol{x}_i^T \boldsymbol{\beta}\right\}^2}{2\sigma^2}\right] \times \prod_{i=1}^n \left| \left.\frac{\partial t^{(\lambda)}}{\partial t}\right|_{t=Y_i} \right|.$$

2.17. Suppose that we have the same situation as the previous problem except that $e_1, \ldots, e_n$ are iid from a density $f_e(z)$. Find the likelihood of the data.

2.18. The "transform-both-sides" (TBS) model (Carroll and Ruppert 1998) assumes that there is a known physical relationship between Y_i and $\boldsymbol{x}_i$, say $E(Y_i \mid x_i) = g(\boldsymbol{x}_i, \boldsymbol{\beta})$ with unknown $\boldsymbol{\beta}$, but the random error structure is unknown. So, to preserve the known physical relationship, it is assumed that after applying the Box-Cox transformation (see Problem 2.1, p. 107) to both sides of the equation we have

$$Y_i^{(\lambda)} = g(\boldsymbol{x}_i, \boldsymbol{\beta})^{(\lambda)} + e_i, \quad i = 1, \ldots, n.$$

We also assume that both Y_i and $g(\boldsymbol{x}_i, \boldsymbol{\beta})$ are positive quantities and that $e_1, \ldots, e_n$ are iid normal$(0, \sigma^2)$ errors. Show that the likelihood is given by

$$L(\beta, \sigma, \lambda \mid \{Y_i, \boldsymbol{x}_i\}_{i=1}^n) = \left(\frac{1}{\sqrt{2\pi}\sigma}\right)^n \exp\left[-\sum_{i=1}^n \frac{\left\{Y_i^{(\lambda)} - g(\boldsymbol{x}_i, \boldsymbol{\beta})^{(\lambda)}\right\}^2}{2\sigma^2}\right] \times \prod_{i=1}^n \left| \left.\frac{\partial t^{(\lambda)}}{\partial t}\right|_{t=Y_i} \right|.$$

2.19. For the Poisson probability mass function $f(y; \lambda)$, put the log density in the generalized linear model form, (2.14, p. 53), identifying $b(\theta)$, etc., and derive the mean and variance of Y, $E(Y) = \mu$, $\text{Var}(Y) = \mu$.

2.20. One version of the negative binomial probability mass function is given by

$$f(y;\mu,k) = \frac{\Gamma(y+k)}{\Gamma(k)\Gamma(y+1)}\left(\frac{k}{\mu+k}\right)^k\left(1-\frac{k}{\mu+k}\right)^y \quad y=0,1\ldots$$

where μ and k are parameters. Assume that k is known and put $f(y;\mu,k)$ in the generalized linear model form, (2.14, p. 53), identifying $b(\theta)$, etc., and derive the mean and variance of Y, $\mathrm{E}(Y)=\mu$, $\mathrm{Var}(Y)=\mu+\mu^2/k$.

2.21. The usual gamma density is given by

$$f(y;\alpha,\beta) = \frac{1}{\Gamma(\alpha)\beta^\alpha}y^{\alpha-1}e^{-y/\beta} \quad 0\le y<\infty, \quad \alpha,\beta>0,$$

and has mean $\alpha\beta$ and variance $\alpha\beta^2$. First reparameterize by letting $\mu=\alpha\beta$ so that the parameter vector is now (μ,α). Now put this gamma family in the form of a generalized linear model, identifying θ, $b(\theta)$, $a_i(\phi)$, and $c(y,\phi)$. Note that α is unknown here and should be related to ϕ. Make sure that $b(\theta)$ is actually a function of θ and take the first derivative to verify that $b(\theta)$ is correct.

2.22. Consider the standard one-way ANOVA situation with Y_{ij} distributed as $\mathrm{N}(\mu_i,\sigma^2)$, $i=1,\ldots,k$, $j=1,\ldots,n_i$, and all the random variables are independent.

a. Form the log likelihood, take derivatives, and show that the MLEs are $\widehat{\mu}_i=\overline{Y}_i, i=1,\ldots,k$, $\widehat{\sigma}^2$ =SSE/N, where SSE=$\sum_{i=1}^k\sum_{j=1}^{n_i}(Y_{ij}-\overline{Y}_i)^2$ and $N=\sum_{i=1}^k n_i$.
b. Now define $V_i^T=(Y_{i1}-\overline{Y}_i,\ldots,Y_{i,n_i-1}-\overline{Y}_i)$. Using standard matrix manipulations with the multivariate normal distribution, the density of V_i is given by

$$(2\pi)^{-(n_i-1)/2}n_i^{1/2}\sigma^{-(n_i-1)}\exp\left(-\frac{1}{2\sigma^2}v_i^T[I_{n_i-1}+J_{n_i-1}]v_i\right),$$

where I_{n_i-1} is the n_i-1 by n_i-1 identity matrix and J_{n_i-1} is an n_i-1 by n_i-1 matrix of 1's. Now form the (marginal) likelihood based on $V_1,\ldots,V_k$ and show that the MLE for σ^2 is now $\widehat{\sigma}^2$ =SSE/$(N-k)$.
c. Finally, let us take a more general approach and assume that $\boldsymbol{Y}$ has an N-dimensional multivariate normal distribution with mean $X\boldsymbol{\beta}$ and covariance matrix $\Sigma=\sigma^2Q(\theta)$, where X is an $N\times p$ full rank matrix of known constants, $\boldsymbol{\beta}$ is a p-vector of regression parameters, and $Q(\boldsymbol{\theta})$ is an $N\times N$ standardized covariance matrix depending on the unknown parameter $\boldsymbol{\theta}$. Typically $\boldsymbol{\theta}$ would consist of variance component and/or spatial correlation parameters. We can concentrate the likelihood by noting that if $Q(\boldsymbol{\theta})$ were known, then the generalized least squares estimator would be $\widehat{\boldsymbol{\beta}}(\boldsymbol{\theta})=(X^TQ(\boldsymbol{\theta})^{-1}X)^{-1}X^TQ(\boldsymbol{\theta})^{-1}\boldsymbol{Y}$. Substituting for $\boldsymbol{\beta}$ yields the profile log likelihood

$$-\frac{N}{2}\log(2\pi)-N\log\sigma-\frac{1}{2}\log|Q(\boldsymbol{\theta})|-\frac{GSSE(\boldsymbol{\theta})}{2\sigma^2},$$

where $GSSE(\boldsymbol{\theta}) = (\boldsymbol{Y} - X\widehat{\boldsymbol{\beta}}(\boldsymbol{\theta}))^T Q(\boldsymbol{\theta})^{-1}(\boldsymbol{Y} - X\widehat{\boldsymbol{\beta}}(\boldsymbol{\theta}))$. To connect with part a), let $Q(\boldsymbol{\theta})$ be the identity matrix (so that $GSSE(\boldsymbol{\theta})$ is just SSE) and find the maximum likelihood estimator of σ^2.

d. Continuing part c), the REML approach is to transform to $\boldsymbol{V} = A^T\boldsymbol{Y}$, where the $N-p$ columns of A are linearly independent and $A^T X = 0$ so that $\boldsymbol{V}$ is $\text{MN}(0, A^T \Sigma A)$. A special choice of A leads to the REML log likelihood

$$-\frac{N-p}{2}\log(2\pi) - (N-p)\log\sigma - \frac{1}{2}\log|X^T X| - \frac{1}{2}\log|X^T Q(\boldsymbol{\theta})^{-1} X|$$
$$-\frac{1}{2}\log|Q(\boldsymbol{\theta})| - \frac{GSSE(\boldsymbol{\theta})}{2\sigma^2}.$$

To connect with part b), let $Q(\boldsymbol{\theta})$ be the identity matrix and find the maximum likelihood estimator of σ^2.

2.23. Suppose that $Y_{i,j}$ are independent for $i = 1, \ldots, n,\ j = 1, \ldots, m$, with $Y_{i,j} \sim \text{N}(\mu_i, \sigma^2)$. This a generalization of Section 2.4.1 (p. 57).

a. Construct the full likelihood for $(\sigma^2, \mu_1, \ldots, \mu_n)$ and obtain the MLEs of $\mu_1, \ldots, \mu_n$ and σ^2. Show that $\widehat{\sigma}^2_{\text{MLE}}$ is not unbiased and not even consistent for σ^2 as $n \to \infty$ with m fixed.

b. Construct a marginal likelihood for σ^2 analogous to that derived in Example 2.4.1 for the case $m = 2$. Find the marginal MLE of σ^2, call it $\widehat{\sigma}^2_{\text{MMLE}}$, and show that it is both unbiased and consistent for σ^2 as $\max(n, m) \to \infty$.

2.24. For the Neyman-Scott solution in Section 2.4.3 (p. 59), use the fact that if $(\boldsymbol{X}_1, \boldsymbol{X}_2)^T$ is distributed as $\text{MN}\{(\boldsymbol{\mu}_1, \boldsymbol{\mu}_2)^T, \boldsymbol{\Sigma}\}$, then $(\boldsymbol{X}_1 \mid \boldsymbol{X}_2)$ is distributed as $\text{MN}\{\boldsymbol{\mu}_1 + \boldsymbol{\Sigma}_{12}\boldsymbol{\Sigma}_{22}^{-1}(\boldsymbol{X}_2 - \boldsymbol{\mu}_2), \boldsymbol{\Sigma}_{11} - \boldsymbol{\Sigma}_{12}\boldsymbol{\Sigma}_{22}^{-1}\boldsymbol{\Sigma}_{21}\}$, to show that $(Y_{i1} \mid Y_{i1} + Y_{i2})$ is distributed as $\text{N}\{(Y_{i1} + Y_{i2})/2, \sigma^2/2\}$.

2.25. If $\boldsymbol{Y}$ is from an exponential family where $(\boldsymbol{W}, \boldsymbol{V})$ are jointly sufficient for $(\boldsymbol{\theta}_1, \boldsymbol{\theta}_2)$, then the conditional density of $\boldsymbol{W}|\boldsymbol{V}$ is free of the nuisance parameter $\boldsymbol{\theta}_2$ and can be used as a conditional likelihood for estimating $\boldsymbol{\theta}_1$. In some cases it may be difficult to find the conditional density. However, from (2.19, p. 57) we have

$$\frac{f_{\boldsymbol{Y}}(\boldsymbol{y}; \boldsymbol{\theta}_1, \boldsymbol{\theta}_2)}{f_{\boldsymbol{V}}(\boldsymbol{v}; \boldsymbol{\theta}_1, \boldsymbol{\theta}_2)} = f_{\boldsymbol{W}|\boldsymbol{V}}(\boldsymbol{w}|\boldsymbol{v}; \boldsymbol{\theta}_1).$$

Thus, if you know the density of $\boldsymbol{Y}$ and of $\boldsymbol{V}$, then you can get a conditional likelihood equation.

a. Now let $Y_1, \ldots, Y_n$ be iid $\text{N}(\mu, \sigma^2)$, $V = \overline{Y}$, and $\boldsymbol{\theta} = (\sigma, \mu)^T$. Form the ratio above and note that it is free of μ. (It helps to remember that $\sum(Y_i - \mu)^2 = \sum(Y_i - \overline{Y})^2 + n(\overline{Y} - \mu)^2$.)

b. Find the conditional maximum likelihood estimator of σ^2.

2.26. Consider the normal theory linear measurement error model

$$Y_i = \alpha + \beta U_i + \sigma_e e_i, \qquad X_i = U_i + \sigma Z_i, \qquad i = 1, \ldots, n,$$

where $e_1, \ldots, e_n, Z_1, \ldots, Z_n$ are iid N(0, 1) random variables, σ^2 is known, and α, β, σ_e and $U_1, \ldots, U_n$ are unknown parameters.

a. Let s_X^2 denote the sample variance of $X_1, \ldots, X_n$. Show that $E(s_X^2) = s_U^2 + \sigma^2$ where s_U^2 is the sample variance of $\{U_i\}_1^n$.

For the remainder of this problem assume that $s_U^2 \to \sigma_U^2$ as $n \to \infty$ and that s_X^2 converges in probability to $\sigma_U^2 + \sigma^2$. (Note that even though $\{U_i\}_1^n$ are parameters it still makes sense to talk about their sample variance, and denoting the limit of s_U^2 as $n \to \infty$ by σ_U^2 is simply a matter of convenience).

b. Show that the estimate of slope from the least squares regression of $\{Y_i\}_1^n$ on $\{X_i\}_1^n$ (call it $\widehat{\beta}_{Y\,|\,X}$) is not consistent for β as $n \to \infty$. This shows that it is not OK to simply ignore the measurement error in the predictor variable.
c. Now construct the full likelihood for $\alpha, \beta, \sigma_\epsilon^2, U_1, \ldots, U_n$ and show that it has no sensible maximum. Do this by showing that the full likelihood diverges to ∞ when $U_i = (Y_i - \alpha)/\beta$ for all i and $\sigma_\epsilon^2 \to 0$. This is another well-known example of the failure of maximum likelihood to produce meaningful estimates.
d. Consider the simple estimator (of β)

$$\widehat{\beta}_{\text{MOM}} = \frac{s_X^2}{s_X^2 - \sigma^2}\widehat{\beta}_{Y\,|\,X},$$

and show that it is a consistent estimator of β. This shows that consistent estimators exist, and thus the problem with maximum likelihood is not intrinsic to the model.
e. Assuming that all other parameters are known, show that $T_i = Y_i\beta/\sigma_\epsilon^2 + X_i/\sigma^2$ is a sufficient statistic for U_i, $i = 1, \ldots, n$.
f. Find the conditional distribution of $Y_i \mid T_i$ and use it to construct a conditional likelihood for α, β and σ_ϵ^2 in a manner similar to that for the logistic regression measurement error model.

2.27. Suppose that $Y_1, \ldots, Y_m$ are independent Bernoulli random variables with $\mathrm{P}(Y_i = 1) = p_i$, where $0 < p_i < 1$ for all $i = 1, \ldots, m$. Let $S = Y_1 + \cdots + Y_m$, $\boldsymbol{Y} = (Y_1, \ldots, Y_m)^T$, $\boldsymbol{y} = (y_1, \ldots, y_m)^T$, $\mathbf{0}_{m\times 1} = (0, \ldots, 0)^T$, $\theta_i = \log(p_i/(1 - p_i))$, and $\boldsymbol{\theta} = (\theta_1, \ldots, \theta_m)^T$. Let $\mathcal{U}_{m,s}$ denote the collection of $m \times 1$ vectors $\boldsymbol{u}$ whose elements are either 1 or 0 and sum to s.

a. Show that the conditional density of $\mathbf{Y}|S$,

$$f_{\mathbf{Y}|S}(\mathbf{y}|s) = \mathrm{P}(Y_1 = y_1, \ldots, Y_m = y_m|S = s),$$

is given by

$$f_{\mathbf{Y}|S}(\mathbf{y}|s) = \frac{\sum \exp(\boldsymbol{\theta}^T \mathbf{u}) I(\mathbf{y} = \mathbf{u})}{\sum \exp(\boldsymbol{\theta}^T \mathbf{u})}, \qquad 0 \le s \le m;$$

where the summations are over all $\mathbf{u} \in \mathcal{U}_{m,s}$ and I is an indicator function.

b. Find the conditional distributions of $Y_i|S = s$, for $i = 1, \ldots, m$, and $s = 0, 1, \ldots, m$.

2.28. Suppose that $Y_{k,i}$ is the ith Bernoulli observation from the kth stratum of a stratified random sample, $i = 1, \ldots, m_k$, $k = 1, \ldots, n$, and that

$$\mathrm{P}(Y_{k,i} = 1|\mathbf{X}_{k,i}) = F(\alpha_k + \boldsymbol{\beta}^T \mathbf{X}_{k,i}),$$

where $\mathbf{X}_{k,i}$ is a vector of explanatory variables associated with $Y_{k,i}$, and $F(\cdot)$ is the logistic distribution function. Consider the case when the number of strata (n) is large and the strata sample sizes (m_k) are small. In this case $\boldsymbol{\beta}$ is the parameter of interest and $\alpha_1, \ldots, \alpha_k$ are nuisance parameters. Explain why the conditional likelihood of the data given the stratum totals, $S_k = Y_{k,1} + \cdots + Y_{k,m_k}$, is free of the nuisance parameters. In particular show that the contribution to the conditional likelihood from the kth stratum is

$$L(\boldsymbol{\beta}; \mathbf{Y}_k|S_k) = \frac{\exp(\boldsymbol{\beta}^T \sum_{i=1}^{m_k} \mathbf{X}_{k,i} Y_{k,i})}{\sum \exp(\boldsymbol{\beta}^T \sum_{i=1}^{m_k} \mathbf{X}_{k,i} u_i)}, \qquad 0 \le S_k \le m_k;$$

where the summations are over all $\mathbf{u} \in \mathcal{U}_{m_k,S_k}$ and u_i denotes the ith element of $\mathbf{u}$. (See Problem 2.27 for the definition of $\mathcal{U}_{m_k,S_k}$.). Explain, in terms of the likelihood, the phrase "there is no information in strata for which $S_k = 0$ or $S_k = m_k$."

2.29. In matched case-control studies the population of subjects is first stratified. Then within each stratum the population is divided between cases (subjects with a disease) and controls (disease-free subjects). A random sample of cases is obtained and a random sample of controls is obtained. This problem addresses only the most common situation in which one case is selected and multiple ($m_k - 1$) controls are selected, so that the total sample size from the kth stratum is m_k.

a. Argue that the sampling described above is probabilistically equivalent to the situation in which: i) a random sample of size m_k is chosen from the kth stratum without regard to cases or controls; ii) the number of cases, S_k, is determined and the sample is accepted if $S_k = 1$, otherwise it is rejected; iii) the process is repeated until a sample is accepted, i.e., $S_k = 1$. The latter description of the sampling makes clear that the distribution of responses, $Y_{k,1}, \ldots, Y_{k,m_k}$ in the observed sample is a conditional distribution given $S_k = 1$.
b. For 1-to-($m_k - 1$) matching it is common to assign the first index ($i = 1$) to the case and the remaining indices to the controls so that $Y_{k,1} = 1$, $Y_{k,2} = \cdots = Y_{k,m_k} = 0$. Show that with this naming convention the contribution to the

conditional likelihood of data from the kth stratum is

$$L(\boldsymbol{\beta}; \mathbf{Y}_k | S_k = 1) = \frac{\exp(\boldsymbol{\beta}^T \mathbf{X}_{k,1})}{\sum_{i=1}^{m_k} \exp(\boldsymbol{\beta}^T \mathbf{X}_{k,i})}.$$

c. Show that for the case of 1-to-1 matching

$$L(\boldsymbol{\beta}; \mathbf{Y}_k | S_k = 1) = F\left(\boldsymbol{\beta}^T (\mathbf{X}_{k,1} - \mathbf{X}_{k,2})\right),$$

where $F(\cdot)$ is the logistic distribution function.

2.30. Instead of the logistic model for the probability of disease assume the more general model,

$$\mathrm{P}(Y_{k,i} = 1 | \mathbf{X}_{k,i}) = \alpha_k \, p(\mathbf{X}_{k,i}, \boldsymbol{\beta}),$$

for a given function $p(\cdot, \cdot)$. Assuming that the probabilities of disease, $\alpha_k \, p(\mathbf{X}_{k,i}, \boldsymbol{\beta})$ are small for all k and i (a reasonable assumption for rare diseases), use the approximation $\log(x/(1-x))/\log(x) \approx 1$ for x near 0, to show that an approximate conditional likelihood for the case of 1-to-$(m_k - 1)$ matching is

$$L(\boldsymbol{\beta}; \mathbf{Y}_k | S_k = 1) = \frac{p(\mathbf{X}_{k,1}, \boldsymbol{\beta})}{\sum_{i=1}^{m_k} p(\mathbf{X}_{k,i}, \boldsymbol{\beta})},$$

where the case has index $i = 1$. Explain the role of the rare-disease assumption in showing that the approximate likelihood does not depend on the nuisance parameters.

2.31. This problem relates to Example 2.12 (p. 63). Figure 2.8 displays plots of the exponential threshold model likelihood in Example 2.12 and the corresponding $2h$-likelihoods for $h = 0.1000, 0.0667, 0.0333, 0^+$. For clarity (and because scaling does not matter), all of the likelihoods are scaled to a common supremum in the plots.

a. The plots suggest that the $2h$-likelihood MLE $\widehat{\mu}_{2h\text{-MLE}}$ converges to $\widehat{\mu}_{\text{MLE}} = Y_{(1)}$ as $h \to 0$. Prove that this is true in general by showing that $\widehat{\mu}_{2h\text{-MLE}} = Y_{(1)} - h$ for h sufficiently small for any random sample from the threshold model in Example 2.12.
b. Explain the solitary point in the fourth panel of Figure 2.8 by calculating $\lim_{h \to 0} L_h(\mu | Y)$ for the threshold model $2h$-likelihood.

2.32. Derive the Fisher information matrix for the Poisson(λ) distribution (actually just an information *number* here).

2.33. Derive the Fisher information matrix for the multinomial(n; $\boldsymbol{p}$) distribution.

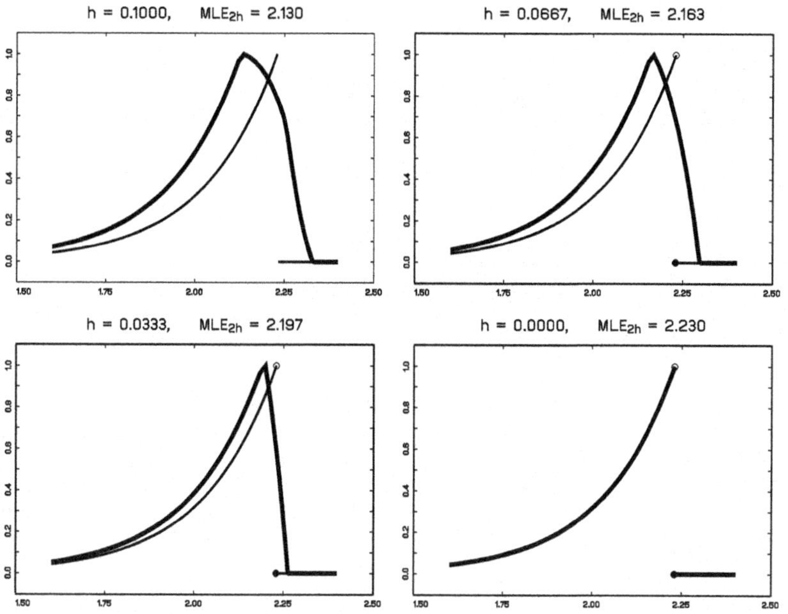

Fig. 2.8 Likelihood (thin line) and $2h$-likelihoods (thick line) for $h = 0.1000, 0.0667, 0.0333, 0^+$ for the sample of size $n = 5$ from Example 2.12. The likelihoods are scaled so that their supremums are all equal

2.34 (Continuation of Problem 2.2, p. 107).

a. Find an estimate of the asymptotic covariance of $(\widehat{\mu}, \widehat{\sigma})$. Here it helps to know that the Fisher information matrix has elements $\boldsymbol{I}(\mu,\sigma)_{11} = 1/\sigma^2$, $\boldsymbol{I}(\mu,\sigma)_{12} = -.423/\sigma^2$, and $\boldsymbol{I}(\mu,\sigma)_{22} = 1.824/\sigma^2$.
b. Estimate the median of the distribution of the largest flow rate in 100 years:

$$\widehat{Q} = \widehat{\sigma}[-\log(-\log(.993))] + \widehat{\mu}.$$

c. Find an estimate of the variance of $\widehat{Q}$.

2.35. Derive the Fisher information matrix $\boldsymbol{I}(\boldsymbol{\theta})$ for the reparameterized ZIP model

$$P(Y = 0) = \pi$$

$$P(Y = y) = \left(\frac{1-\pi}{1-e^{-\lambda}}\right)\frac{\lambda^y e^{-\lambda}}{y!} \quad y = 1, 2, \ldots$$

The (2,2) element of $\boldsymbol{I}(\boldsymbol{\theta})$ is

$$(1-\pi)\left(\frac{1}{\lambda(1-e^{-\lambda})} - \frac{e^{-\lambda}}{(1-e^{-\lambda})^2}\right).$$

2.36. A "multiplicative" form of the ZIP model is

$$P(Y = 0) = \omega e^{-\lambda},$$

$$P(Y = y) = a_{\omega,\lambda} \frac{e^{-\lambda}\lambda^y}{y!} \quad y = 1, 2, \ldots.$$

a. Show that $a_{\omega,\lambda} = \left(1 - \omega e^{-\lambda}\right) / \left(1 - e^{-\lambda}\right)$.
b. Find the element of $\boldsymbol{I}(\boldsymbol{\theta})$ that corresponds to ω.

2.37. Suppose that $Y_1, \ldots, Y_n$ are iid from the one parameter exponential family

$$f(y;\theta) = \exp\{yg(\theta) - b(\theta) + c(y)\}.$$

a. For $g(\theta) = \theta$, find $\overline{I}(Y, \theta)$ (sample version) and explain why it is the same as $I(\theta)$ (Fisher information).
b. Now for general differentiable $g(\theta)$, find $I(\theta)$.

2.38. In the Hardy-Weinberg example, Example 2.2 (p. 33), we have a multinomial problem with $k = 3$ and the probabilities modeled in terms of p_A. Find $I_{\mathrm{T}}(p_A)$ in three ways:

a. Calculate $I_{\mathrm{T}}(p_A)$ directly from the probability mass function.
b. Use the transformation method with $\boldsymbol{I}_{\mathrm{T}}(\boldsymbol{p})$ from the original multinomial parameterization.
c. Find the variance of $\widehat{p}_{A,\mathrm{MLE}} = (2N_{AA} + N_{Aa})/(2n)$ directly from the covariance of the multinomial and take the reciprocal. (Actually, the general method is to take the inverse of the asymptotic variance, but here the variance and asymptotic variance are the same.)

2.39. Recall the Type II censoring in Problem 2.12 where the likelihood is a function of the order statistics. Here it is not as easy to think of the information matrix for a single observation. So for the exponential density $f(y;\sigma) = \sigma^{-1}e^{-y/\sigma}I(y > 0)$, compute the average information by taking the expectation of the negative of the second derivative of the log likelihood function (at the true parameter value) divided by n. It helps to know that the ith order statistic from an exponential has expectation $\sigma \sum_{j=1}^{i}(n - j + 1)^{-1}$.

2.40. a. Ralph decides to start n light bulbs at $t = 0$. At $t = c_1$ hours, he checks and finds r_1 have failed. Then he comes back at $t = c_2$ hours and finds that r_2 more have failed. He ends the experiment with $n - r_1 - r_2$ light bulbs still working. If the lifetime distribution of the light bulbs has density $f(y;\boldsymbol{\theta})$ and distribution function $F(y;\boldsymbol{\theta})$, write down the likelihood function of the observed data (r_1, r_2). (Hint: think "multinomial.")

b. Jordan is a bit more industrious than Ralph. He reruns the experiment and observes failures $Y_{r_1+1}, \ldots, Y_{r_1+r_2}$, and stops the experiment at $t = c_2$ with $n - r_1 - r_2$ light bulbs still working. Write down the likelihood for this case. (Note: he decided to stop at $t = c_2$ before observing the outcomes. So it is a fixed censoring time. The numbers r_1 and r_2 would vary from experiment to experiment.)
c. Sampson is even more industrious. He starts n light bulbs at $t = 0$, observes $Y_1, \ldots, Y_{r_1+r_2}$, and stops the experiment at $t = c_2$ with $n - r_1 - r_2$ light bulbs still working. Write down the likelihood for this case.
d. A section head decides to evaluate the three workers' performances. She has cost figures for light bulbs and hourly rates for the workers' time. But she is not sure how to deal with the results of the experiment. A statistician friend has told her that she can actually make the comparisons before the experiments are even run if she is willing to make some guesses about parameter values. What statistical calculations should she make to complete a cost-efficiency analysis of the three experiments? What guesses would be required? If she waits until after the experiments are completed, what alternative calculations could she make that would be easier and not require guesses?

2.41. Use simulation to verify that the information matrix for Example 2.21 (p. 78) is correct when $\mu = 1$ and $\sigma = 1$. One approach is to generate samples of size $n = 100$ (or larger) from a normal(1,1) distribution and exponentiate to get lognormal data. Then form the log likelihood and use a numerical derivative routine to find the second derivative matrix for each sample. Then average over 1000 replications and compare to the given information matrix.

2.42. Use simulation to verify that the inverse information matrix for Example 2.21 (p. 78) is correct when $\mu = 1$ and $\sigma = 1$. One approach is to generate samples of size $n = 100$ (or larger) from a normal(1,1) distribution and exponentiate to get lognormal data. Then form the log likelihood and find the estimates for (μ, σ^2, λ). Repeat for 1000 replications giving a data matrix of size 1000 by 3 of estimates. Compute the sample covariance matrix of these and compare it to the inverse of the given information matrix.

2.43. For the generalized linear model with link function g, not necessarily the canonical link, write down the likelihood function and show how to obtain the likelihood score equation,

$$S(\boldsymbol{\beta}, \phi) = \sum_{i=1}^{n} \boldsymbol{D}_i \frac{(Y_i - \mu_i)}{\text{Var}(Y_i)} = 0,$$

where $\boldsymbol{D}_i = \boldsymbol{D}_i(\boldsymbol{\beta}) = \partial \mu_i(\boldsymbol{\beta}) / \partial \boldsymbol{\beta}^T$. (In the above expression we have suppressed the dependence of $\boldsymbol{D}_i$ and μ_i on $\boldsymbol{\beta}$.) The key idea used is the chain rule and the fact that the derivative of $\theta_i = b'^{-1}(\mu_i)$ with respect to μ_i is $1/b''(\theta_i)$.

2.44. Continuing the last problem, show that the Fisher information matrix for the $\boldsymbol{\beta}$ part of the generalized linear model is given by

$$\overline{\boldsymbol{I}}(\boldsymbol{\beta}) = \frac{1}{n}\sum_{i=1}^{n} \frac{\boldsymbol{D}_i \boldsymbol{D}_i^T}{\mathrm{Var}(Y_i)}.$$

Here you can use either of two methods: a) take the expectation of the negative of the derivative of $S(\boldsymbol{\beta}, \phi)$, and noting that all the ugly derivatives drop out because $\mathrm{E}(Y_i - \mu_i) = 0$; or b) the individual summed components of $\overline{\boldsymbol{I}}(\boldsymbol{\beta})$ can also be found using the cross-product definition of information in (2.39, p. 67).

2.45. Suppose that X_1 and X_2 are independent and continuous random variables with densities f_1 and f_2, respectively. Z is a Bernoulli(p) random variable and independent of X_1 and X_2. Define Y by $Y = ZX_1 + (1 - Z)X_2$.

a. Use the $2h$ method to show that the joint density of (Y, Z) is given by

$$f_{Y,Z}(y, z) = [pf_1(y)]^z\,[(1 - p)f_2(y)]^{1-z}.$$

b. Use the $2h$ method to show that

$$P(Z = 1|Y = y) = \frac{pf_1(y)}{pf_1(y) + (1 - p)f_2(y)}.$$

2.46. For the mixture of two normals problem that was used to illustrate the EM Algorithm in Section 2.6.3a (p. 84), find the updating formulas for the standard deviations σ_1 and σ_2.

2.47. A mixture of three component densities has the form

$$f(y; \boldsymbol{\theta}, \boldsymbol{p}) = p_1 f_1(y; \boldsymbol{\theta}) + p_2 f_2(y; \boldsymbol{\theta}) + p_3 f_3(y; \boldsymbol{\theta}),$$

where $p_1 + p_2 + p_3 = 1$. We observe an iid sample $Y_1, \ldots, Y_n$ from $f(y; \boldsymbol{\theta}, \boldsymbol{p})$.

a. Show how to define multinomial$(1; p_1, p_2, p_3)$ vectors (Z_{i1}, Z_{i2}, Z_{i3}) to get a representation for the Y_i from $f(y; \boldsymbol{\theta}, \boldsymbol{p})$ based on independent random variables (X_{i1}, X_{i2}, X_{i3}) from the individual components.
b. Give the complete data log likelihood and the function Q to be maximized at the M step.

2.48. Suppose that the data $Y_1, \ldots, Y_n$ are assumed to come from a mixture of two binomial distributions. Thus

$$f(y; p, \theta_1, \theta_2) = p\binom{n}{y}\theta_1^y(1 - \theta_1)^{n-y} + (1 - p)\binom{n}{y}\theta_2^y(1 - \theta_2)^{n-y}.$$

Find $Q(p, \theta_1, \theta_2, p^\nu, \theta_1^\nu, \theta_2^\nu)$ and the updating formulas.

2.49. Recall that the ZIP model is just a mixture of densities $f(y;\lambda,p) = pf_1(y) + (1-p)f_2(y;\lambda)$, where

$$f_1(y) = I(y=0) \qquad\qquad f_2(y;\lambda) = \frac{\lambda^y e^{-\lambda}}{y!}, \quad y = 0,1,2,\ldots$$

Lambert (1992) used it to model product defects as a function of covariates. In the "perfect" state, no defects occur ($P(Y_i = 0) = 0$), whereas in the "imperfect" state, the number of defects Y_i follows a Poisson(λ) distribution. The author used the EM Algorithm as follows (except we won't do the more complicated modeling with covariates.) Let $Z_i = 1$ if the product is in the perfect state and $Z_i = 0$ for the imperfect state. Recall that the contribution to the complete data likelihood for a pair (Y_i, Z_i) is $[pf_1(Y_i)]^{Z_i}\,[(1-p)f_2(Y_i;\lambda)]^{1-Z_i}$ (and here note that the first part reduces to p^{Z_i} because f_1 is a point mass at 0).

a. E step. Write down the complete data log likelihood and find $Q(\lambda, p, \lambda^\nu, p^\nu)$ in terms of $w_i^\nu = \mathrm{E}(Z_i|Y_i, \lambda^\nu, p^\nu)$. (You do not need to give an expression for w_i^ν.)
b. M step. Find expressions for $\lambda^{\nu+1}$ and $p^{\nu+1}$ by maximizing Q from the E step.

2.50. The results of an experiment, $Y_1, \ldots, Y_n$ iid Poisson(λ), were written on slips of paper. Thus, the density of Y_i is

$$f(y;\lambda) = \frac{\lambda^y e^{-\lambda}}{y!} \quad y = 0,1,\ldots$$

Unfortunately, $n-q$ of the slips were lost by accident, completely at random. For notational purposes, suppose that we observe only $Y_1, \ldots, Y_q$.

a. Decide on the vector $\boldsymbol{Z}$ that we would use for the EM Algorithm and write down the complete data log likelihood.
b. Now find the equation giving $\lambda^{\nu+1}$ in terms of λ^ν.
c. Using b., find an explicit representation of the maximum likelihood estimator.

2.51. The lab book containing a student's data had smudges so that for some of the data, the student could not tell whether the data value had a minus sign in front of it. The student's original plan was to assume that the data $Y_1, \ldots, Y_n$ were iid $\mathrm{N}(\mu, \sigma^2)$. Now, she only has q of these Y values; for convenience label them $Y_1, \ldots, Y_q$ although the "smudging" was completely at random. For the smudged values, she only observes $|Y_{q+1}|, \ldots, |Y_n|$. Note that one way to represent Y_i is $[2I(Y_i > 0) - 1]$ times $|Y_i|$ since the sign of Y_i is given by $[2I(Y_i > 0) - 1]$. Also, for the first parts below, assume that you know the function $w_i(\mu,\sigma) = \mathrm{E}\{I(Y_i > 0) \,|\, |Y_i|\} = P(Y_i > 0 \,|\, |Y_i|)$.

a. Give the likelihood function of the observed data $Y_1, \ldots, Y_q, |Y_{q+1}|, \ldots, |Y_n|$.
b. For an EM Algorithm approach, suggest some missing data $\boldsymbol{Z}$, and give the complete data log likelihood function.

c. Now take conditional expectations and find the Q function that should be maximized at the M step.
d. Using the Q function from c), find the update formula for μ.
e. Find $w_i(\mu, \sigma) = P(Y_i > 0 \mid |Y_i|)$.

2.52. For an exponential random variable X with density

$$f(x;\sigma) = \frac{1}{\sigma}e^{-x/\sigma} \qquad x > 0,$$

derive the conditional expectation of X given that $X > r$ for $r > 0$.

2.53. Write down the complete data likelihood for the situation of Example 2.6.3b (p. 87) when we assume that the censoring times $R_1, \ldots, R_n$ are iid with density g and distribution Function G. In this case, we let $\boldsymbol{Z}$ consist of the censored X values plus the unobserved R values, $R_1, \ldots, R_{n_u}$ after relabeling. Now, the complete data are $\{X_i, R_i\}_{i=1}^n$. Explain why the same EM iteration results in this random censoring version of Example 2.6.3b.

2.54. $Y_1, \ldots, Y_n$ are iid continuous random variables with density $f(y;\boldsymbol{\theta})$, but they have been grouped into the k intervals $[a_0, a_1), [a_1, a_2), \ldots, [a_{k-1}, a_k)$, and we only record the number of Y_i in each interval: $N_1, N_2, \ldots, N_k$, respectively.

a. Write down the likelihood for the observed data.
b. For the EM Algorithm, use $Y_1, \ldots, Y_n$ as the complete data and write down the complete data log likelihood.
c. Find the density of Y_1 given $Y_1 \in [a_{i-1}, a_i)$.

2.55. Suppose that $Y_1, \ldots, Y_n$ are iid with density $f(y;\boldsymbol{\theta})$, but we only observe the values of the Y_i's that are greater than some known constant c. Let's say there are n_L values below c (whose actual values we do not know) and label the data above c (for which we know the values) $Y_{n_L+1}, \ldots, Y_n$. For the EM Algorithm, find the function Q to be maximized at the M step. Express any conditional expectations as integrals.

2.56. Suppose that we have a one-way random effects model:

$$Y_{ij} = \mu + \alpha_i + e_{ij}, \quad i = 1, \ldots, k; \; j = 1, \ldots, n_i,$$

where $\alpha_1, \ldots, \alpha_k$ are iid $\mathrm{N}(0, \sigma_\alpha^2)$ and independent of the errors e_{ij} which are all iid $\mathrm{N}(0, \sigma_e^2)$. Set this up as an EM Algorithm, first writing down the complete data log likelihood in terms of the observed Y_{ij} and the unobserved α_i, and then deriving the form of the "Q" function to maximize. Do not evaluate any conditional expectations; just leave them as conditional expectations but be clear about what is being conditioned on, and ignore the M step.

2.57. Consider a situation where the complete data likelihood is from a canonical exponential family of the form

$$L(\theta|\boldsymbol{x}) = h(\boldsymbol{x}) \exp\left\{\sum_{i=1}^{s} \eta_i T_i(\boldsymbol{x}) - A(\boldsymbol{\eta})\right\},$$

where $\boldsymbol{X}^T = (\boldsymbol{Y}^T, \boldsymbol{Z}^T)$. Show that the M step of the EM Algorithm reduces to solving

$$\mathrm{E}_{\eta=\eta^{(k)}}\left[\boldsymbol{T}(\boldsymbol{X})|\boldsymbol{Y}\right] = \mathrm{E}_{\eta} T(\boldsymbol{X}).$$

The following questions are related to Appendix A starting on p. 90.

2.58. If $Y_1, \ldots, Y_n$ are iid from a normal(μ, σ^2) distribution, then the log likelihood is

$$\ell_n(\mu, \sigma) = -\frac{n}{2}\log(2\pi) - n\log\sigma - \frac{1}{2}\sum_{i=1}^{n}\left(\frac{Y_i - \mu}{\sigma}\right)^2.$$

Show that $\ell_n(\mu, \sigma)$ tends to $-\infty$ (and is therefore *constant on the boundary*) as (i) $\sigma \to 0$ and ∞ for μ fixed, (ii) $\mu \to \pm\infty$ for fixed σ, (iii) both parameters are heading towards their extremes.

2.59. Suppose that $Y_1, \ldots, Y_n$ are iid from the truncated (at 0) normal location model. Truncated distributions are different from censored data because for truncated distributions we do not even know about observations below the truncation point (0 in this case). Thus the density is the same as usual but truncated and renormalized to have integral equal to one. In our case the density is $f(y;\mu) = \phi(y-\mu)I(y \geq 0)/(1 - \Phi(-\mu))$, where ϕ and Φ are the density and distribution function of a standard normal random variable, respectively. Show or tell why the maximum likelihood estimator of μ must be unique if it exists. (Hint: exponential family.)

2.60. Suppose that $Y_1, \ldots, Y_n$ are iid from the location-scale family $f(y;\mu,\sigma) = \sigma^{-1} f_0((y-\mu)/\sigma)$. Lehmann (1983, p. 437) says that two different authors have shown in unpublished papers that the MLEs of (μ, σ) are unique if i) $f_0(x)$ is positive and twice differentiable for all $x \in (-\infty, \infty)$ and ii) f_0 is "strongly unimodal," which means that $\log f_0(x)$ is concave on $(-\infty, \infty)$. Let's try to prove this result using Theorem 2.1. First define

$$\psi_0(x) = -\frac{d}{dx}\log f_0(x) = -\frac{f_0'(x)}{f_0(x)}.$$

Condition ii) above is just $\psi_0'(x) > 0$ for all x. Next, note that the MLEs $\widehat{\mu}$ and $\widehat{\sigma}$ must satisfy

$$\sum_{i=1}^{n}\psi_0\left(\frac{Y_i - \widehat{\mu}}{\widehat{\sigma}}\right) = 0 \quad \text{and} \quad \sum_{i=1}^{n}\left[\left(\frac{Y_i - \widehat{\mu}}{\widehat{\sigma}}\right)\psi_0\left(\frac{Y_i - \widehat{\mu}}{\widehat{\sigma}}\right) - 1\right] = 0.$$

Now complete the following steps:

Step 1. Verify that

$$-\frac{\partial^2}{\partial\theta\partial\theta^T}\log f(y;\mu,\sigma) = \frac{1}{\sigma^2}\begin{bmatrix} \psi_0'(x) & \psi_0(x)+x\psi_0'(x) \\ \psi_0(x)+x\psi_0'(x) & 2x\psi_0(x)+x^2\psi_0'(x)-1 \end{bmatrix},$$

where $x = (y-\mu)/\sigma$.

Step 2. Now replace the above matrix by $-(\partial^2/\partial\theta\partial\theta^T)\ell(\widehat{\mu},\widehat{\sigma})$, being sure to take advantage of the fact that certain sums are 0 from the likelihood equations. Verify that this matrix is positive definite. It helps to remember the Cauchy-Schwarz inequality for sums,

$$\left(\sum s_i t_i\right)^2 \le \left(\sum s_i^2\right)\left(\sum t_i^2\right),$$

and this additional hint:

$$\left[\sum \{X_i\psi_0'(X_i)\}\right]^2 = \left[\sum \left\{X_i\sqrt{\psi_0'(X_i)}\sqrt{\psi_0'(X_i)}\right\}\right]^2.$$

Step 3. Now let (μ,σ) go to the boundaries of the parameter space, and show that $l_n(\mu,\sigma)$ always goes to $-\infty$. You may assume that $xf_0(x) \to 0$ as $x \to \pm\infty$ (follows because $\log f_0(x)$ being concave implies that $f_0(x)$ is bounded by $c\exp(-x)$ for some constant c).

2.61. Suppose that $Y_1,\ldots,Y_n$ are iid from the extreme value location-scale family $f(y;\mu,\sigma) = \sigma^{-1}f_0((y-\mu)/\sigma)$, where $f_0(x) = e^{-x}\exp(-e^{-x}), -\infty < x < \infty$. Verify that f_0 satisfies the conditions for the maximum likelihood estimators to be unique (see the previous problem).

Chapter 3
Likelihood-Based Tests and Confidence Regions

There are three asymptotically equivalent testing methods based on the likelihood function: Wald tests, likelihood ratio tests, and score tests also known as *Lagrange multiplier tests* in the econometrics literature. In this chapter we present increasingly more general versions of the three test statistics culminating in the most general forms given in (3.7)–(3.13) starting on p. 136. Under appropriate null hypotheses and regularity conditions, each of these statistics is asymptotically chi-squared distributed. In addition, under "local" alternatives converging to the null hypothesis, they have identical asymptotic noncentral chi-squared distributions. The asymptotic local power results also show that the tests are the best possible asymptotically.

We start with the simplest case of iid data with one unknown real parameter. Then for testing $H_0 : \theta = \theta_0$ versus $H_a : \theta \neq \theta_0$, the three test statistics are:

- Wald —

$$T_{\mathrm{W}} = \frac{(\widehat{\theta}_{\mathrm{MLE}} - \theta_0)^2}{\left\{I_{\mathrm{T}}(\widehat{\theta}_{\mathrm{MLE}})\right\}^{-1}};$$

- Likelihood ratio —

$$T_{\mathrm{LR}} = -2\left\{\ell(\theta_0) - \ell(\widehat{\theta}_{\mathrm{MLE}})\right\},$$

- Score —

$$T_{\mathrm{S}} = \frac{S(\theta_0)^2}{I_{\mathrm{T}}(\theta_0)}.$$

In these definitions, $\ell(\theta)$, $S(\theta)$ and $I_{\mathrm{T}}(\theta)$ are the scalar-parameter versions of the log likelihood and likelihood score functions, and total information matrix (number) appearing in Table 2.1 (p. 69). The asymptotic equivalence of the test statistics results from the close relationships among them. These are easiest to see in the case of a scalar parameter. Relationships among the three test statistics and their relationship to the log likelihood function $\ell(\theta)$ are depicted in Figure 3.1 for the case

D.D. Boos and L.A. Stefanski, *Essential Statistical Inference: Theory and Methods*, Springer Texts in Statistics, DOI 10.1007/978-1-4614-4818-1_3,

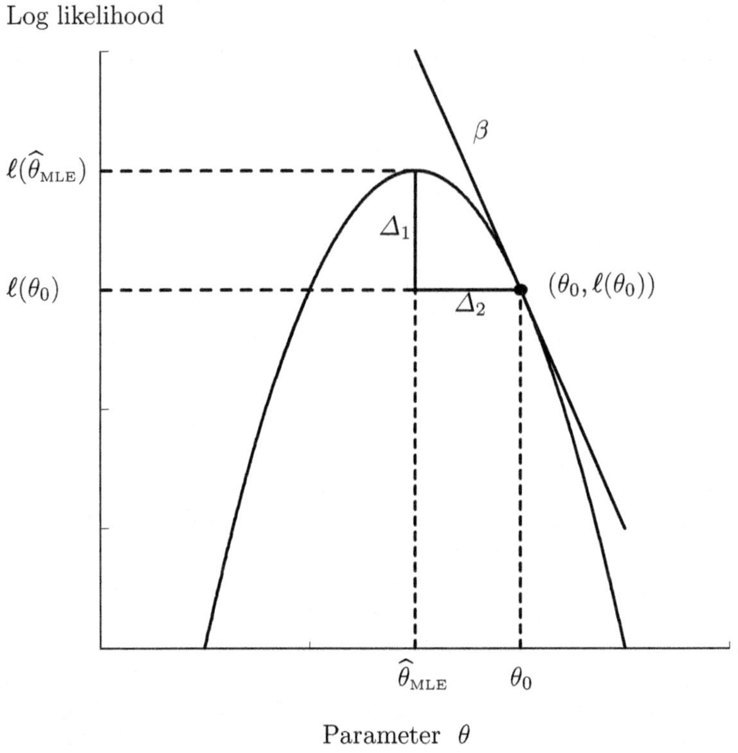

Fig. 3.1 Graphical representation of the relationships between Wald, Score and Likelihood Ratio test statistics. The Likelihood Ratio test statistic is a multiple of the difference, Δ_1; the Wald test statistic is a multiple of the squared difference, Δ_2^2; and the Score test statistic is a multiple of the squared slope, β^2

of a scalar parameter and simple null hypothesis. Visualizing a dynamic graph as θ_0 moves left (right) of its depicted position reveals how all three statistics decrease (increase) in unison in this quadratic likelihood illustration.

The Likelihood Ratio test statistic is a multiple of the difference, Δ_1, between the log likelihood evaluated at $\widehat{\theta}_{\text{MLE}}$ and θ_0. The Wald test statistic is a multiple of the squared difference, Δ_2^2, between $\widehat{\theta}_{\text{MLE}}$ and θ_0. The Score test statistic is a multiple of the squared slope, β^2 $(= S(\theta_0)^2)$, of the line tangent to the log likelihood at θ_0. The multipliers are all different, but are all such that the statistics have equal asymptotic expectations $(= 1)$. Note that the three quantities Δ_1, $|\Delta_2|$ and $|\beta|$ vary together as the location of the point $(\theta_0, \ell(\theta_0))$ on the parabola changes relative to the point $(\widehat{\theta}_{MLE}, \ell(\widehat{\theta}_{MLE}))$.

Wald (1943) is usually credited with the formal derivation of T_{W} in the general multiparameter case and proof of its asymptotic distribution; hence the name

Wald test. Of course, dividing a parameter estimate by a standard error certainly did not begin with Wald (1943). For example, the one-sample t statistic was studied by Gossett ("Student," 1908). The likelihood ratio approach to testing was introduced by Neyman and Pearson (1928, 1933). Wilks (1938) gave the first proof of the asymptotic null distribution. The score statistic and its asymptotic properties were introduced by Rao (1948). A parallel development of the score approach was begun by Aitchison and Silvey (1958) under the name *Lagrange multiplier statistic* still used in the econometrics literature.

Even though these three test statistics are asymptotically equivalent, there are often good reasons to prefer one in certain situations. The Wald statistic is by far the simplest and is available in most standard computing packages. However, it is not invariant to reparameterization, whereas the likelihood ratio and score statistics are. Thus, for example, the Wald test about a scale parameter σ depends on whether the null hypothesis is expressed as $H_0 : \sigma = \sigma_0$ or $H_0 : \sigma^2 = \sigma_0^2$. From a computing standpoint, the likelihood ratio is often more difficult to compute than the Wald or score statistic because it requires $\widehat{\boldsymbol{\theta}}_{\text{MLE}}$ under both the null and alternative hypotheses. In contrast, the Wald statistic only requires $\widehat{\boldsymbol{\theta}}_{\text{MLE}}$ under the alternative hypothesis, and the score statistic only requires $\widehat{\boldsymbol{\theta}}_{\text{MLE}}$ under the null hypothesis. However, for testing a sequence of nested models, likelihood ratio statistics are convenient assuming that all of the log likelihoods for the different models have been computed. For example, consider a parameter $\boldsymbol{\theta}$ with five components and the sequence of possible models and log likelihoods where $\widetilde{\boldsymbol{\theta}}_{i,\text{MLE}}$ denotes the

$$
\begin{array}{ll}
H_1 : \theta_1 = \theta_2 = \theta_3 = \theta_4 = \theta_5 = 0 & \ell(\widetilde{\boldsymbol{\theta}}_{1,\text{MLE}}) \\
H_2 : \theta_2 = \theta_3 = \theta_4 = \theta_5 = 0 & \ell(\widetilde{\boldsymbol{\theta}}_{2,\text{MLE}}) \\
H_3 : \theta_3 = \theta_4 = \theta_5 = 0 & \ell(\widetilde{\boldsymbol{\theta}}_{3,\text{MLE}}) \\
H_4 : \theta_4 = \theta_5 = 0 & \ell(\widetilde{\boldsymbol{\theta}}_{4,\text{MLE}}) \\
H_5 : \theta_5 = 0 & \ell(\widetilde{\boldsymbol{\theta}}_{5,\text{MLE}})
\end{array}
$$

maximum likelihood estimator under the ith hypothesis. There are $\binom{5}{2} = 10$ possible tests, such as H_2 versus $H_4 - H_2$, and all ten Likelihood Ratio test statistics can be obtained simply by subtraction. For example, the statistic for H_2 versus $H_4 - H_2$ is $T_{\text{LR}} = -2\left\{\ell(\widetilde{\boldsymbol{\theta}}_{2,\text{MLE}}) - \ell(\widetilde{\boldsymbol{\theta}}_{4,\text{MLE}})\right\}$. Generally speaking, the Wald statistic with chi-squared critical values does not have as good Type I error probabilities as the likelihood ratio and score statistics in finite samples. From a robustness standpoint, the Wald and score tests are easiest to adjust when the likelihoods are possibly misspecified.

Next we introduce the test statistics in the case of a simple null hypothesis ($\boldsymbol{\theta}$ completely specified) and then proceed to the more interesting composite null hypothesis where $\boldsymbol{\theta}$ is only partially specified under H_0.

3.1 Simple Null Hypotheses

We now consider the simple null hypothesis $H_0 : \boldsymbol{\theta} = \boldsymbol{\theta}_0$ versus $H_a : \boldsymbol{\theta} \neq \boldsymbol{\theta}_0$, where $\boldsymbol{\theta}$ is $b \times 1$ and the data $Y_1, \ldots, Y_n$ are iid from $f(y; \boldsymbol{\theta})$. The three statistics are defined in terms of $\ell(\boldsymbol{\theta})$, $\boldsymbol{S}(\boldsymbol{\theta})$ and $\boldsymbol{I}_{\mathrm{T}}(\boldsymbol{\theta})$ defined in Table 2.1 (p. 69). The asymptotic distribution results for the three statistics stated below hold under suitable regularity conditions. To distinguish it from our notations for information matrices, the identity matrix is denoted by $\mathbb{I}$ in the following.

- The Wald statistic is

$$T_{\mathrm{W}} = (\widehat{\boldsymbol{\theta}}_{\mathrm{MLE}} - \boldsymbol{\theta}_0)^T \left\{\boldsymbol{I}_{\mathrm{T}}(\widehat{\boldsymbol{\theta}}_{\mathrm{MLE}})\right\} (\widehat{\boldsymbol{\theta}}_{\mathrm{MLE}} - \boldsymbol{\theta}_0).$$

  $\widehat{\boldsymbol{\theta}}_{\mathrm{MLE}}$ is $\mathrm{AN}(\boldsymbol{\theta}, \{\boldsymbol{I}_{\mathrm{T}}(\boldsymbol{\theta})\}^{-1})$ or equivalently, $\{\boldsymbol{I}_{\mathrm{T}}(\boldsymbol{\theta})\}^{-1/2}(\widehat{\boldsymbol{\theta}}_{\mathrm{MLE}} - \boldsymbol{\theta}) \xrightarrow{d} N(\boldsymbol{0}, \mathbb{I}_b)$ where $\mathbb{I}_b$ is the b-dimensional identity matrix. It follows that $T_{\mathrm{W}} \xrightarrow{d} \chi^2_b$ as $n \to \infty$ under H_0 provided $\boldsymbol{I}_{\mathrm{T}}(\widehat{\boldsymbol{\theta}}_{\mathrm{MLE}})\{\boldsymbol{I}_{\mathrm{T}}(\boldsymbol{\theta}_0)\}^{-1}$ converges in probability to $\mathbb{I}_b$ as $n \to \infty$ under H_0. In fact, $\boldsymbol{I}_{\mathrm{T}}(\widehat{\boldsymbol{\theta}}_{\mathrm{MLE}})$ can be replaced by $\boldsymbol{I}_{\mathrm{T}}(\boldsymbol{\theta}_0)$ or by the sample information matrix $\boldsymbol{I}_{\mathrm{T}}(\boldsymbol{Y}, \widehat{\boldsymbol{\theta}}_{\mathrm{MLE}})$ in the definition of T_{W} and the asymptotic distribution of the resulting statistic is unchanged. However, if any of these matrices are singular, then some care must be used; see Moore (1977), Andrews (1987), and Hadi and Wells (1990).
- The score test statistic is

$$T_{\mathrm{S}} = \boldsymbol{S}(\boldsymbol{\theta}_0)^T \{\boldsymbol{I}_{\mathrm{T}}(\boldsymbol{\theta}_0)\}^{-1} \boldsymbol{S}(\boldsymbol{\theta}_0). \tag{3.1}$$

  Under H_0, $\boldsymbol{S}(\boldsymbol{\theta}_0)$ has mean $\boldsymbol{0}$, variance $\boldsymbol{I}_{\mathrm{T}}(\boldsymbol{\theta}_0)$, and by The Central Limit Theorem is $\mathrm{AN}(\boldsymbol{0}, \boldsymbol{I}_{\mathrm{T}}(\boldsymbol{\theta}_0))$. Thus $\{\boldsymbol{I}_{\mathrm{T}}(\boldsymbol{\theta}_0)\}^{-1/2}\boldsymbol{S}(\boldsymbol{\theta}_0)$ converges in distribution to $N(\boldsymbol{0}, \mathbb{I}_b)$ under H_0. It follows that $T_{\mathrm{S}} \xrightarrow{d} \chi^2_b$ as $n \to \infty$ under H_0.
- The likelihood ratio statistic is

$$T_{\mathrm{LR}} = -2\log\left\{\frac{\sup_{\boldsymbol{\theta} \in H_0} L(\boldsymbol{\theta} \mid \boldsymbol{Y})}{\sup_{\boldsymbol{\theta} \in \boldsymbol{\Theta}} L(\boldsymbol{\theta} \mid \boldsymbol{Y})}\right\} = -2\left\{\ell(\boldsymbol{\theta}_0) - \ell(\widehat{\boldsymbol{\theta}}_{\mathrm{MLE}})\right\},$$

  where $\boldsymbol{\Theta}$ is the parameter space and sup = *least upper bound* plays the role of "maximum" in cases where the maximum is not attained. The limiting distribution of T_{LR} is not as obvious as it is for T_{W} and T_{S}. However, using a two-term Taylor series expansion, one can show that under H_0, $T_{\mathrm{LR}} = T_{\mathrm{W}} + \delta_n$ where $\delta_n \xrightarrow{p} 0$ as $n \to \infty$. Thus, the convergence of the Wald statistic and Slutsky's Theorem (p. 14) imply that $T_{\mathrm{LR}} \xrightarrow{d} \chi^2_b$ as $n \to \infty$ under H_0.

The two quadratic-form statistics, T_{W} and T_{S}, have an interesting "inverse" relationship to one another in the sense that the components of the defining quadratic forms are inversely related. $\widehat{\boldsymbol{\theta}}_{\mathrm{MLE}}$ on the "wings" of T_{W} is obtained by solving

$\boldsymbol{S}(\boldsymbol{\theta}) = \mathbf{0}$ (an "inverse" operation), whereas $\boldsymbol{S}(\boldsymbol{\theta}_0)$ appears on the wings of T_{S}. Also the product of the middle matrix components, $\boldsymbol{I}_{\mathrm{T}}(\widehat{\boldsymbol{\theta}}_{\mathrm{MLE}})$ and $\{\boldsymbol{I}_{\mathrm{T}}(\boldsymbol{\theta}_0)\}^{-1}$, converge to $\mathbb{I}_b$ and thus are asymptotically matrix inverses of one another.

Example 3.1 (Normal model with known variance). Suppose that $Y_1, \ldots, Y_n$ are iid N$(\mu, 1)$ and $H_0 : \mu = \mu_0$. Then

$$\ell(\mu) = \log L(\mu \mid \boldsymbol{Y}) = -\frac{n}{2}\log(2\pi) - \frac{1}{2}\sum_{i=1}^{n}(Y_i - \mu)^2,$$

$$S(\mu) = \frac{\partial}{\partial \mu}\ell(\mu) = \sum_{i=1}^{n}(Y_i - \mu),$$

and

$$I_{\mathrm{T}}(\boldsymbol{Y}, \mu) = -\frac{\partial}{\partial \mu}S(\mu) = n,$$

so that $\widehat{\mu}_{\mathrm{MLE}} = \overline{Y}$ and $I_{\mathrm{T}}(\mu) = E\{I_{\mathrm{T}}(\boldsymbol{Y}, \mu)\} = n$. It follows that

$$T_{\mathrm{W}} = (\overline{Y} - \mu_0)(n)(\overline{Y} - \mu_0) = n(\overline{Y} - \mu_0)^2,$$

$$T_{\mathrm{S}} = \sum_{i=1}^{n}(Y_i - \mu_0)(n^{-1})\sum_{i=1}^{n}(Y_i - \mu_0) = n(\overline{Y} - \mu_0)^2,$$

and

$$T_{\mathrm{LR}} = -2\left\{-\frac{1}{2}\sum_{i=1}^{n}(Y_i - \mu_0)^2 + \frac{1}{2}\sum_{i=1}^{n}(Y_i - \overline{Y})^2\right\} = n(\overline{Y} - \mu_0)^2,$$

are identical for this model. ◆

Example 3.2 (Bernoulli data). Suppose that $Y_1, \ldots, Y_n$ are iid from a Bernoulli distribution with parameter p and $H_0 : p = p_0$. With $X = \sum_{i=1}^{n} Y_i$,

$$\ell(p) = \log L(p \mid \boldsymbol{Y}) = X \log p + (n - X)\log(1 - p),$$

$$S(p) = \frac{\partial}{\partial p}\ell(p) = \frac{X}{p} - \frac{n - X}{1 - p} = \frac{X - np}{p(1 - p)},$$

and

$$I_{\mathrm{T}}(\boldsymbol{Y}, p) = -\frac{\partial}{\partial p}S(p) = \frac{X}{p^2} + \frac{n - X}{(1 - p)^2}.$$

Thus $\widehat{p}_{\rm MLE} = \widehat{p} = X/n$ and $I_{\rm T}(p) = E\{I_{\rm T}(Y,p)\} = n/p(1-p)$. The test statistics are

$$T_{\rm W} = (\widehat{p} - p_0)\left\{\frac{n}{\widehat{p}(1-\widehat{p})}\right\}(\widehat{p} - p_0) = \frac{n\,(\widehat{p} - p_0)^2}{\widehat{p}(1-\widehat{p})},$$

$$T_{\rm S} = \left\{\frac{X - np_0}{p_0(1-p_0)}\right\}\left\{\frac{n}{p_0(1-p_0)}\right\}^{-1}\left\{\frac{X - np_0}{p_0(1-p_0)}\right\} = \frac{n\,(\widehat{p} - p_0)^2}{p_0(1-p_0)},$$

$$T_{\rm LR} = -2\left[X\log(p_0/\widehat{p}) + (n - X)\log\{(1-p_0)/(1-\widehat{p})\}\right].$$

$T_{\rm LR}$ is not so familiar, but $T_{\rm W}$ and $T_{\rm S}$ are seen in introductory statistics texts and differ only in whether $n\,(\widehat{p} - p_0)^2$ is standardized by $\widehat{p}(1-\widehat{p})$ (Wald) or by $p_0(1-p_0)$ (score). Problem 3.12 relates to the asymptotic equivalence of $T_{\rm W}$, $T_{\rm S}$ and $T_{\rm LR}$ for this example. ♦

3.2 Composite Null Hypotheses

Only rarely are we interested in simple null hypotheses where the entire parameter vector $\boldsymbol{\theta}$ is specified in the null hypothesis. More often interest lies in only certain components of the parameter vector or in constraints on the components. Thus, we now study test statistics appropriate for when the null hypothesis is specified as either $H_0 : \boldsymbol{\theta}_1 = \boldsymbol{\theta}_{10}$ where $\boldsymbol{\theta}^T = (\boldsymbol{\theta}_1^T, \boldsymbol{\theta}_2^T)$, or as $H_0 : h(\boldsymbol{\theta}) = \boldsymbol{0}$, for some function h. For the former we assume that $\boldsymbol{\theta}$ is partitioned as

$$\underset{b\times 1}{\boldsymbol{\theta}} = \begin{pmatrix} \underset{r\times 1}{\boldsymbol{\theta}_1} \\ \\ \underset{(b-r)\times 1}{\boldsymbol{\theta}_2} \end{pmatrix}. \tag{3.2}$$

As in the simple null hypothesis the asymptotic distribution results for the six statistics stated below hold under suitable regularity conditions.

3.2.1 *Wald Statistic – Partitioned* $\boldsymbol{\theta}$

Consider (3.2) and testing $H_0 : \boldsymbol{\theta}_1 = \boldsymbol{\theta}_{10}$ versus $H_a : \boldsymbol{\theta}_1 \neq \boldsymbol{\theta}_{10}$, where no restrictions are made on $\boldsymbol{\theta}_2$. H_0 is a *composite null hypothesis*, and $\boldsymbol{\theta}_2$ is a *nuisance parameter*. For notational simplicity we leave off the subscript MLE and let $\widehat{\boldsymbol{I}}_{\rm T} = \boldsymbol{I}_{\rm T}(\widehat{\boldsymbol{\theta}})$.

Partition the information matrix as

$$\widehat{\boldsymbol{I}}_{\mathrm{T}} = \begin{pmatrix} \widehat{\boldsymbol{I}}_{\mathrm{T},11} & \widehat{\boldsymbol{I}}_{\mathrm{T},12} \\ \widehat{\boldsymbol{I}}_{\mathrm{T},21} & \widehat{\boldsymbol{I}}_{\mathrm{T},22} \end{pmatrix},$$

and denote the upper (1,1) element of its inverse by

$$\widehat{\boldsymbol{I}}_{\mathrm{T}}^{11} = \left(\widehat{\boldsymbol{I}}_{\mathrm{T},11} - \widehat{\boldsymbol{I}}_{\mathrm{T},12}\widehat{\boldsymbol{I}}_{\mathrm{T},22}^{-1}\widehat{\boldsymbol{I}}_{\mathrm{T},21}\right)^{-1}.$$

The Wald test statistic is

$$\begin{aligned} T_{\mathrm{W}} &= (\widehat{\boldsymbol{\theta}}_1 - \boldsymbol{\theta}_{10})^T \left\{\widehat{\boldsymbol{I}}_{\mathrm{T}}^{11}\right\}^{-1} (\widehat{\boldsymbol{\theta}}_1 - \boldsymbol{\theta}_{10}) \\ &= (\widehat{\boldsymbol{\theta}}_1 - \boldsymbol{\theta}_{10})^T \left(\widehat{\boldsymbol{I}}_{\mathrm{T},11} - \widehat{\boldsymbol{I}}_{\mathrm{T},12}\widehat{\boldsymbol{I}}_{\mathrm{T},22}^{-1}\widehat{\boldsymbol{I}}_{\mathrm{T},21}\right)(\widehat{\boldsymbol{\theta}}_1 - \boldsymbol{\theta}_{10}). \end{aligned} \tag{3.3}$$

Similar to the simple null hypothesis case, $T_{\mathrm{W}} \xrightarrow{d} \chi^2_r$ as $n \to \infty$ under H_0, where the degrees of freedom r corresponds to the number of restrictions under H_0.

3.2.2 Score Statistic – Partitioned $\boldsymbol{\theta}$

The score statistic T_{S} for a composite null hypothesis requires the maximum likelihood estimator of $\boldsymbol{\theta}$ under the null hypothesis $H_0 : \boldsymbol{\theta}_1 = \boldsymbol{\theta}_{10}$. We denote this restricted MLE by $\widetilde{\boldsymbol{\theta}} = (\widetilde{\boldsymbol{\theta}}_1, \widetilde{\boldsymbol{\theta}}_2)$, where $\widetilde{\boldsymbol{\theta}}_1 = \boldsymbol{\theta}_{10}$, and $\widetilde{\boldsymbol{\theta}}_2$ maximizes $\ell(\boldsymbol{\theta}_{10}, \boldsymbol{\theta}_2)$ with respect to $\boldsymbol{\theta}_2$. Under the regularity conditions assumed throughout this section, $\widetilde{\boldsymbol{\theta}}$ satisfies $\boldsymbol{S}_2(\widetilde{\boldsymbol{\theta}}) = 0$ where $\boldsymbol{S}_2(\boldsymbol{\theta})$ is from the partitioned score function

$$\boldsymbol{S}(\boldsymbol{\theta}) = \begin{pmatrix} \boldsymbol{S}_1(\boldsymbol{\theta}) \\ \boldsymbol{S}_2(\boldsymbol{\theta}) \end{pmatrix} = \begin{pmatrix} \left\{\dfrac{\partial}{\partial \boldsymbol{\theta}_1}\ell(\boldsymbol{\theta})\right\}^T \\ \left\{\dfrac{\partial}{\partial \boldsymbol{\theta}_2}\ell(\boldsymbol{\theta})\right\}^T \end{pmatrix}.$$

Let $\widetilde{\boldsymbol{I}}_{\mathrm{T}} = \boldsymbol{I}_{\mathrm{T}}(\widetilde{\boldsymbol{\theta}})$. The score test statistic

$$T_{\mathrm{S}} = \boldsymbol{S}(\widetilde{\boldsymbol{\theta}})^T \widetilde{\boldsymbol{I}}_{\mathrm{T}}^{-1} \boldsymbol{S}(\widetilde{\boldsymbol{\theta}}), \tag{3.4}$$

was introduced by Rao (1948) who showed that $T_{\mathrm{S}} \xrightarrow{d} \chi^2_r$ as $n \to \infty$ under H_0. Because $\widetilde{\boldsymbol{I}}_{\mathrm{T}}$ is not an estimator of the asymptotic variance of $\boldsymbol{S}(\widetilde{\boldsymbol{\theta}})$, the asymptotic distribution result is not as obvious as it is for the simple-null score statistic in (3.1). However, because $S_2(\widetilde{\theta}) = 0$, (3.4) is equivalent to

$$T_{\mathrm{S}} = \left(S_1(\widetilde{\theta})^T, \mathbf{0}\right) \begin{pmatrix} \widetilde{I}_{\mathrm{T},11} & \widetilde{I}_{\mathrm{T},12} \\ \widetilde{I}_{\mathrm{T},21} & \widetilde{I}_{\mathrm{T},22} \end{pmatrix}^{-1} \begin{pmatrix} S_1(\widetilde{\theta}) \\ 0 \end{pmatrix}$$

$$= S_1(\widetilde{\theta})^T \left(\widetilde{I}_{\mathrm{T},11} - \widetilde{I}_{\mathrm{T},12}\widetilde{I}_{\mathrm{T},22}^{-1}\widetilde{I}_{\mathrm{T},21}\right)^{-1} S_1(\widetilde{\theta}). \tag{3.5}$$

Comparing (3.5) to (3.3) shows that the composite-null T_{W} and T_{S} share a similar inverse relationship as was pointed out in the simple null hypothesis case (p. 129). Furthermore, it can be shown that $(\widetilde{I}_{\mathrm{T},11} - \widetilde{I}_{\mathrm{T},12}\widetilde{I}_{\mathrm{T},22}^{-1}\widetilde{I}_{\mathrm{T},21})$ is an estimator of the asymptotic variance of $S_1(\widetilde{\theta})$ (see 6.17, p. 288) and thus (3.5) is a natural and intuitive starting point for proving the asymptotic χ^2_r distribution of T_{S}.

3.2.3 Likelihood Ratio Statistic – Partitioned θ

The likelihood ratio statistic for a composite null hypothesis is

$$T_{\mathrm{LR}} = -2\log\left\{\frac{\sup_{\theta \in H_0} L(\theta \mid Y)}{\sup_{\theta \in \Theta} L(\theta \mid Y)}\right\} = -2\left\{\ell(\widetilde{\theta} \mid Y) - \ell(\widehat{\theta} \mid Y)\right\}. \tag{3.6}$$

As in the simple null case, showing that $T_{\mathrm{LR}} \xrightarrow{d} \chi^2_r$ as $n \to \infty$ under H_0 entails using a two-term Taylor series approximation showing that $T_{\mathrm{LR}} = T_{\mathrm{W}} + \delta_n$ where δ_n is asymptotically negligible.

3.2.4 Normal Location-Scale Model

Suppose that $Y_1, \ldots, Y_n$ are iid $\mathrm{N}(\mu, \sigma^2)$. The log likelihood is

$$\ell(\mu, \sigma) = \log L(\mu, \sigma \mid Y) = c - n\log\sigma - \frac{1}{2\sigma^2}\sum_{i=1}^{n}(Y_i - \mu)^2,$$

and the score function components are

$$S_1(\mu, \sigma) = \frac{\partial}{\partial\mu}\ell(\mu, \sigma) = \frac{1}{\sigma^2}\sum_{i=1}^{n}(Y_i - \mu),$$

$$S_2(\mu, \sigma) = \frac{\partial}{\partial\sigma}\ell(\mu, \sigma) = \frac{1}{\sigma^3}\sum_{i=1}^{n}\left\{(Y_i - \mu)^2 - \sigma^2\right\}.$$

Of course, solving $S_j(\mu,\sigma) = 0,\ j = 1,2$ results in $\widehat{\mu} = \overline{Y}$ and $\widehat{\sigma}^2 = s_n^2 = n^{-1}\sum_{i=1}^n (Y_i - \overline{Y})^2$. The information matrix is $\widehat{\boldsymbol{I}}_{\mathrm{T}}(\mu,\sigma) = \mathrm{diag}(n/\sigma^2, 2n/\sigma^2)$.

Consider the composite null hypothesis $H_0 : \mu = \mu_0$ with σ unrestricted. The Wald statistic is

$$T_{\mathrm{W}} = \left(\overline{Y} - \mu_0\right)^T \left(\frac{n}{s_n^2}\right) \left(\overline{Y} - \mu_0\right) = \frac{n\left(\overline{Y} - \mu_0\right)^2}{s_n^2} = \left(\frac{n}{n-1}\right) t^2,$$

where t^2 is the square of the usual Student-t statistic. For T_{S} and T_{LR}, $\widetilde{\mu} = \mu_0$ and $\widetilde{\sigma}$ solves

$$S_2(\widetilde{\mu},\sigma) = \frac{1}{\sigma^3}\sum_{i=1}^n \{(Y_i - \mu_0)^2 - \sigma^2\} = 0,$$

so that

$$\widetilde{\sigma}^2 = \frac{1}{n}\sum_{i=1}^n (Y_i - \mu_0)^2 = s_n^2 + \left(\overline{Y} - \mu_0\right)^2,$$

resulting in

$$\begin{aligned} T_{\mathrm{S}} &= \left\{\frac{1}{\widetilde{\sigma}^2}\sum_{i=1}^n (Y_i - \mu_0)\right\} \left(\frac{\widetilde{\sigma}^2}{n}\right) \left\{\frac{1}{\widetilde{\sigma}^2}\sum_{i=1}^n (Y_i - \mu_0)\right\} \\ &= \frac{n\left(\overline{Y} - \mu_0\right)^2}{\widetilde{\sigma}^2} = \frac{n\left(\overline{Y} - \mu_0\right)^2}{s_n^2 + \left(\overline{Y} - \mu_0\right)^2} = \frac{nT_{\mathrm{W}}}{n + T_{\mathrm{W}}}, \end{aligned}$$

and

$$T_{\mathrm{LR}} = n\log\left\{1 + \frac{\left(\overline{Y} - \mu_0\right)^2}{s_n^2}\right] = n\log\left[1 + \frac{T_{\mathrm{W}}}{n}\right\}.$$

Using the inequalities

$$\frac{x}{1+x} \le \log(1+x) \le x, \qquad x > -1,$$

(which are strict for $x \neq 0$) with $x = T_{\mathrm{W}}/n$, shows that $T_{\mathrm{S}} < T_{\mathrm{LR}} < T_{\mathrm{W}}$. In a more general result for the normal linear model, Berndt and Savin (1977) proved that $T_{\mathrm{S}} \le T_{\mathrm{LR}} \le T_{\mathrm{W}}$ for testing regression parameters. Of course from an exact testing standpoint T_{S}, T_{LR}, T_{W} are all equivalent to using t^2. However, it is instructive to consider the true Type I error probabilities when the asymptotic χ_1^2 critical values are used as this analysis sheds light on the relative performance of T_{W}, T_{S} and T_{LR} when exact distributions are not available.

For T_W

$$P\left\{T_W \geq \chi_1^2(1-\alpha)\right\} = P\left\{\left(\frac{n}{n-1}\right)t^2 \geq \chi_1^2(1-\alpha)\right\}$$
$$= P\left\{F_{1,n-1} \geq \left(\frac{n-1}{n}\right)\chi_1^2(1-\alpha)\right\},$$

where $\chi_1^2(1-\alpha)$ is the upper $1-\alpha$ quantile of the χ_1^2 distribution and $F_{1,n-1}$ denotes an F random variable with 1 and $n-1$ degrees of freedom. For the score statistic

$$P\left\{T_S \geq \chi_1^2(1-\alpha)\right\} = P\left\{F_{1,n-1} \geq \left(\frac{n-1}{n-\chi_1^2(1-\alpha)}\right)\chi_1^2(1-\alpha)\right\},$$

whereas for T_{LR}

$$P\left\{T_{LR} \geq \chi_1^2(1-\alpha)\right\} = P\left\{F_{1,n-1} \geq (n-1)\left[\exp\{\chi_1^2(1-\alpha)/n\} - 1\right]\right\}.$$

The table below displays the exact finite-sample sizes of the T_W, T_S and T_{LR} tests obtained using the formulas above, for $\alpha = .05$ and $n = 10, 20$. The Wald test

Finite-Sample Sizes for Testing μ at Asymptotic Nominal Size = .05

	T_W	T_S	T_{LR}
$n = 10$	.096	.042	.070
$n = 20$	.071	.047	.059

sizes, .096 and .071, illustrate the often liberal tendency of T_W. The score test is slightly conservative having sizes .042 and .047, whereas the likelihood ratio test with sizes .070 and .059. is liberal although not to the extent of the Wald test. For all three tests the improvement in size as n increases from 10 to 20 illustrates the rate of convergence to the asymptotic chi-squared distribution.

If the exact finite-sample first moment of T_W or T_{LR} can be determined, then these statistics can be adjusted to have the same mean as the asymptotic χ_1^2 distribution resulting in what is known as a "Bartlett's correction." The correction generally improves the chi-squared approximation. Applying Bartlett's correction to T_W the Type I error probabilities above improve to .053 and .052, compared to .096 and .071, respectively. Similar improvements are available for T_{LR}. Discussions of Bartlett's corrections are found in Lawley (1956), Barndorff-Nielsen and Cox (1979), and Reid (1988).

3.2.5 Wald, Score, and Likelihood Ratio Tests – $H_0 : h(\theta) = 0$

In many situations, it is not convenient to use the partitioned-vector approach to testing. For example, consider bivariate normal data where μ_1 and μ_2 are the mean parameters, respectively, and $H_0 : \mu_1 = \mu_2$ is of interest. Testing this null hypothesis using the partitioned-vector approach requires reparameterization such that $\theta_1 = \mu_1 - \mu_2$ and $\theta_2 = \mu_2$.

We now consider testing null hypotheses specified as $H_0 : \boldsymbol{h}(\boldsymbol{\theta}) = \boldsymbol{0}$, where $\boldsymbol{h}(\cdot)$ is an $r \times 1$ vector function with $r \times b$ matrix of first partial derivatives $\boldsymbol{H}(\boldsymbol{\theta}) = \partial \boldsymbol{h}(\boldsymbol{\theta})/\partial \boldsymbol{\theta}$. For example, consider testing $\mu_1 = \mu_2$ with bivariate normal data. Set $\boldsymbol{\theta} = (\mu_1, \mu_2, \sigma_1, \sigma_2, \rho)^T$ and $h(\boldsymbol{\theta}) = \mu_1 - \mu_2$. Then $\boldsymbol{H}(\boldsymbol{\theta}) = (1, -1, 0, 0, 0)$. Note that the linear hypotheses $\boldsymbol{K}^T\boldsymbol{\beta} = \boldsymbol{m}$ common in linear models inference is of the form $\boldsymbol{h}(\boldsymbol{\theta}) = \boldsymbol{0}$ with $\boldsymbol{h}(\boldsymbol{\beta}) = \boldsymbol{K}^T\boldsymbol{\beta} - \boldsymbol{m}$. Also the partitioned-vector hypothesis $H_0 : \boldsymbol{\theta}_1 = \boldsymbol{\theta}_{10}$ has the form $\boldsymbol{h}(\boldsymbol{\theta}) = \boldsymbol{0}$ where $\boldsymbol{h}(\boldsymbol{\theta}) = \boldsymbol{\theta}_1 - \boldsymbol{\theta}_{10}$. When testing $H_0 : \boldsymbol{h}(\boldsymbol{\theta}) = \boldsymbol{0}$ we assume that $r \le b$ and that $\boldsymbol{H}(\boldsymbol{\theta})$ is full rank.

- The Wald statistic T_W for $H_0 : \boldsymbol{h}(\boldsymbol{\theta}) = \boldsymbol{0}$ follows from the Delta Theorem (p. 14) applied to $\boldsymbol{h}(\widehat{\boldsymbol{\theta}})$. Since $\widehat{\boldsymbol{\theta}}$ is $\text{AN}(\boldsymbol{\theta}, \boldsymbol{I}_T(\boldsymbol{\theta})^{-1})$, the Delta Theorem implies that $\boldsymbol{h}(\widehat{\boldsymbol{\theta}})$ is $\text{AN}\big[\boldsymbol{h}(\boldsymbol{\theta}), \boldsymbol{H}(\boldsymbol{\theta})\boldsymbol{I}_T(\boldsymbol{\theta})^{-1}\boldsymbol{H}(\boldsymbol{\theta})^T\big]$. Thus
$$T_W = \boldsymbol{h}(\widehat{\boldsymbol{\theta}})^T \left[\boldsymbol{H}(\widehat{\boldsymbol{\theta}})\boldsymbol{I}_T(\widehat{\boldsymbol{\theta}})^{-1}\boldsymbol{H}(\widehat{\boldsymbol{\theta}})^T\right]^{-1} \boldsymbol{h}(\widehat{\boldsymbol{\theta}}).$$
This Wald statistic is intuitive and easily computed. However, it has a double non-invariance: not only does T_W vary with reparameterization, it also varies with the choice of $\boldsymbol{h}$. For example, $H_0 : \mu_1 - \mu_2 = 0$ and $H_0 : \mu_1/\mu_2 - 1 = 0$ result in different Wald statistics (assuming that μ_2 is not zero). The lack of invariance is illustrated in Section 3.2.8 (p. 139).
- The score test statistic given in (3.4, p. 131) is appropriate for testing $H_0 : \boldsymbol{h}(\boldsymbol{\theta}) = \boldsymbol{0}$ provide that $\widetilde{\boldsymbol{\theta}}$ maximizes the likelihood subject to the constraint $\boldsymbol{h}(\boldsymbol{\theta}) = \boldsymbol{0}$. However, an alternative equivalent form is often used. Note that $\widetilde{\boldsymbol{\theta}}$ maximizes the log likelihood subject to the constraint $h(\boldsymbol{\theta}) = \boldsymbol{0}$. Thus $\widetilde{\boldsymbol{\theta}}$ satisfies $S(\widetilde{\boldsymbol{\theta}}) - \boldsymbol{H}(\widetilde{\boldsymbol{\theta}})^T\widetilde{\boldsymbol{\lambda}} = 0$ and $\boldsymbol{h}(\widetilde{\boldsymbol{\theta}}) = \boldsymbol{0}$, where $\widetilde{\boldsymbol{\lambda}}$ is a data-dependent vector of Lagrange multipliers. Substituting $\boldsymbol{H}(\widetilde{\boldsymbol{\theta}})^T\widetilde{\boldsymbol{\lambda}}$ for $S(\widetilde{\boldsymbol{\theta}})$ in (3.4, p. 131), we obtain
$$T_S = \widetilde{\boldsymbol{\lambda}}^T \boldsymbol{H}(\widetilde{\boldsymbol{\theta}})\widetilde{\boldsymbol{I}}_T^{-1}\boldsymbol{H}(\widetilde{\boldsymbol{\theta}})^T\widetilde{\boldsymbol{\lambda}}.$$
This form motivates the name *Lagrange Multiplier Test* used in the econometrics literature.

  To simply illustrate, consider the situation where $\boldsymbol{\theta} = (\theta_1, \theta_2)^T$ and $H_0 : \theta_1 = \theta_2$. Here $h(\theta_1, \theta_2) = \theta_1 - \theta_2$ leads to $\boldsymbol{H}(\theta_1, \theta_2) = (1, -1)$, and the restricted maximum likelihood estimator $\widetilde{\boldsymbol{\theta}}$ is found by solving the augmented set of equations
$$S_1(\boldsymbol{\theta}) - \lambda = 0$$
$$S_2(\boldsymbol{\theta}) + \lambda = 0$$
$$\theta_1 - \theta_2 = 0,$$

where λ is a scalar. Of course, $\widetilde{\boldsymbol{\theta}}$ can also be found by substituting $\theta_1 = \theta_2 = \theta$ into the log likelihood and finding the maximum with respect to θ. The latter method does not use S_1 and S_2 directly, although the resulting test statistics are identical.

- The likelihood ratio test statistic for testing $H_0 : \boldsymbol{h}(\boldsymbol{\theta}) = \mathbf{0}$ is identical to that in (3.6, p. 132), when $\widetilde{\boldsymbol{\theta}}$ maximizes the likelihood subject to the constraint $\boldsymbol{h}(\boldsymbol{\theta}) = \mathbf{0}$., and there are no widely used alternative forms.

3.2.6 Summary of Wald, Score, and LR Test Statistics

We summarize the forms of the three test statistics here. Note that the formulas below are appropriate for general data structure, assuming neither independence nor identical distributions.

- For $H_0 : \boldsymbol{\theta}_1 = \boldsymbol{\theta}_{10}$, the Wald statistic is

$$T_{\mathrm{W}} = (\widehat{\boldsymbol{\theta}}_1 - \boldsymbol{\theta}_{10})^T \left[\left\{\boldsymbol{I}_{\mathrm{T}}(\widehat{\boldsymbol{\theta}})^{-1}\right\}_{11}\right]^{-1} (\widehat{\boldsymbol{\theta}}_1 - \boldsymbol{\theta}_{10}) \tag{3.7}$$

$$= (\widehat{\boldsymbol{\theta}}_1 - \boldsymbol{\theta}_{10})^T \left(\widehat{\boldsymbol{I}}_{\mathrm{T},11} - \widehat{\boldsymbol{I}}_{\mathrm{T},12}\widehat{\boldsymbol{I}}_{\mathrm{T},22}^{-1}\widehat{\boldsymbol{I}}_{\mathrm{T},21}\right) (\widehat{\boldsymbol{\theta}}_1 - \boldsymbol{\theta}_{10}); \tag{3.8}$$

whereas for $H_0 : \boldsymbol{h}(\boldsymbol{\theta}_1) = \mathbf{0}$ with $\boldsymbol{H}(\boldsymbol{\theta}) = \partial \boldsymbol{h}(\boldsymbol{\theta})/\partial\boldsymbol{\theta}$,

$$T_{\mathrm{W}} = \boldsymbol{h}(\widehat{\boldsymbol{\theta}})^T \left\{\boldsymbol{H}(\widehat{\boldsymbol{\theta}})\boldsymbol{I}_{\mathrm{T}}(\widehat{\boldsymbol{\theta}})^{-1}\boldsymbol{H}(\widehat{\boldsymbol{\theta}})^T\right\}^{-1} \boldsymbol{h}(\widehat{\boldsymbol{\theta}}). \tag{3.9}$$

- The score statistic formula,

$$T_{\mathrm{S}} = \boldsymbol{S}(\widetilde{\boldsymbol{\theta}})^T \left\{\boldsymbol{I}_{\mathrm{T}}(\widetilde{\boldsymbol{\theta}})\right\}^{-1} \boldsymbol{S}(\widetilde{\boldsymbol{\theta}}), \tag{3.10}$$

is always correct where $\widetilde{\boldsymbol{\theta}}$ is the MLE under H_0. However, for $H_0 : \boldsymbol{\theta}_1 = \boldsymbol{\theta}_{10}$, T_{S} is equivalent to

$$T_{\mathrm{S}} = \boldsymbol{S}_1(\widetilde{\boldsymbol{\theta}})^T \left(\widetilde{\boldsymbol{I}}_{\mathrm{T},11} - \widetilde{\boldsymbol{I}}_{\mathrm{T},12}\widetilde{\boldsymbol{I}}_{\mathrm{T},22}^{-1}\widetilde{\boldsymbol{I}}_{\mathrm{T},21}\right)^{-1} \boldsymbol{S}_1(\widetilde{\boldsymbol{\theta}}), \tag{3.11}$$

where $\widetilde{\boldsymbol{I}}_{\mathrm{T,ij}}$ denotes the ijth submatrix of $\boldsymbol{I}_{\mathrm{T}}(\widetilde{\boldsymbol{\theta}})$. Also for $H_0 : \boldsymbol{h}(\boldsymbol{\theta}) = \mathbf{0}$, the Lagrange Multiplier form of T_{S} is

$$T_{\mathrm{S}} = \widetilde{\boldsymbol{\lambda}}^T H(\widetilde{\boldsymbol{\theta}}) \left\{\boldsymbol{I}_{\mathrm{T}}(\widetilde{\boldsymbol{\theta}})\right\}^{-1} H(\widetilde{\boldsymbol{\theta}})^T\widetilde{\boldsymbol{\lambda}}, \tag{3.12}$$

which is equivalent to (3.10) since $\boldsymbol{S}(\widetilde{\boldsymbol{\theta}}) = H(\widetilde{\boldsymbol{\theta}})^T\widetilde{\boldsymbol{\lambda}}$.

- Finally, the likelihood ratio test statistic is always given by

$$T_{\mathrm{LR}} = -2\left\{\log L(\widetilde{\boldsymbol{\theta}} \mid \boldsymbol{Y}) - \log L(\widehat{\boldsymbol{\theta}} \mid \boldsymbol{Y})\right\}. \tag{3.13}$$

3.2.7 Score Statistic for Multinomial Data

For a multinomial $(n; p_1, \ldots, p_k)$ vector $(N_1, \ldots, N_k)$, the score statistic T_{S} can always be put in the form of the usual Pearson chi-squared goodness-of-fit statistic, $\sum_{i=1}^{k}(O_i - \widetilde{E}_i)^2/\widetilde{E}_i$. This follows from the fact that the log likelihood is $c - \sum_{i=1}^{k} N_i \log(p_i)$ and the score vector is

$$\boldsymbol{S}(\boldsymbol{p}) = \left(\frac{N_1}{p_1} - \frac{N_k}{p_k}, \ldots, \frac{N_{k-1}}{p_{k-1}} - \frac{N_k}{p_k}\right)^T,$$

where $p_k = 1 - \sum_{i=1}^{k-1} p_i$. Note that we use the full rank parameterization $\boldsymbol{p} = (p_1, \ldots, p_{k-1})^T$ and thus $\boldsymbol{S}$ is a $k-1$ dimensional vector. Use of p_k in place of $1 - \sum_{i=1}^{k-1} p_i$ is for notational purposes only. From Example 2.14 (p. 70), we have the $k-1$ by $k-1$ matrix $\boldsymbol{I}_{\mathrm{T}}(\boldsymbol{p})^{-1} = \{\mathrm{diag}(\boldsymbol{p}) - \boldsymbol{p}\boldsymbol{p}^T\}/n$, and thus

$$\begin{aligned}
T_{\mathrm{S}} &= \boldsymbol{S}(\widetilde{\boldsymbol{p}})^T \{\boldsymbol{I}_{\mathrm{T}}(\widetilde{\boldsymbol{p}})\}^{-1} \boldsymbol{S}(\widetilde{\boldsymbol{p}}) \\
&= \sum_{i=1}^{k}\left(\frac{N_i}{\widetilde{p}_i} - \frac{N_k}{\widetilde{p}_k}\right)^2 \frac{\widetilde{p}_i}{n} - \left\{\sum_{i=1}^{k}\left(\frac{N_i}{\widetilde{p}_i} - \frac{N_k}{\widetilde{p}_k}\right)\frac{\widetilde{p}_i}{n}\right\}^2 \\
&= \sum_{i=1}^{k}\left\{\frac{N_i}{\widetilde{p}_i} - \frac{N_k}{\widetilde{p}_k} - \left(n - \frac{N_k}{\widetilde{p}_k}\right)\right\}^2 \frac{\widetilde{p}_i}{n} \\
&= \sum_{i=1}^{k}\frac{(N_i - n\widetilde{p}_i)^2}{n\widetilde{p}_i}.
\end{aligned} \tag{3.14}$$

In the second equality, the upper summation limit is set to k rather than $k-1$ because the kth summands are identically 0. This allows us to invoke the variance equality $\sum a_i^2 p_i - (\sum a_i p_i)^2 = \sum(a_i - \sum a_j p_j)^2 p_i$ to obtain the third equality.

The same argument applies to sets of independent multinomial vectors. Thus, for all contingency tables the score statistic is the chi-squared goodness-of-fit statistic whenever testing any null hypothesis against the alternative of a full unrestricted model.

Example 3.3 (Testing for Hardy-Weinberg equilibrium). Recall from Example 2.2 (p. 33) that the Hardy-Weinberg model of random mating has $p_1 = p_{AA} = p_A^2$, $p_2 = p_{Aa} = 2p_A(1 - p_A)$, and $p_3 = p_{aa} = (1 - p_A)^2$, where p_A is the

Table 3.1 Data reproduced from Table 2.6 of Agresti (2002)

Victim's Race	Defendant's Race	Proportion Receiving Death Penalty
White	White	53/467 = 0.11
White	Black	11/48 = 0.23
Black	White	0/16 = 0
Black	Black	4/143 = 0.03

Table 3.2 Results of fitting three models to the data in Table 3.1

Model	Log Likelihood	Parameters Fit	df	Deviance	T_{LR}	Pearson $\chi^2 = T_{\text{S}}$	Change in T_{S}
1	−211.99	2	2	5.39		5.81	
2	−209.48	3	1	0.38	5.01	0.20	5.61
3	−209.29	4	0	0	0.38	0	0.20

population frequency of A alleles. For testing the validity of this model based on multinomial $k = 3$ data, (N_{AA}, N_{Aa}, N_{aa}), the restricted estimators are the indicated functions of $\widetilde{p}_A = (2N_{AA} + N_{Aa})/(2n)$, and from (3.14)

$$T_{\text{S}} = \frac{\left(N_{AA} - n\widetilde{p}_A^2\right)^2}{n\widetilde{p}_A} + \frac{\{N_{Aa} - n2\widetilde{p}_A(1-\widetilde{p}_A)\}^2}{n2\widetilde{p}_A(1-\widetilde{p}_A)} + \frac{\left\{N_{aa} - n(1-\widetilde{p}_A)^2\right\}^2}{n(1-\widetilde{p}_A)^2}. \quad \blacklozenge$$

Example 3.4 (Death penalty sentencing and race). Table 2.6 of Agresti (2002) presents results from Radelet and Pierce (1991) on the proportion of defendants given the death penalty, indexed by the race of the victim and race of the defendant. Agresti (2002) presents the data in two 2-by-2 tables, one for white victims and one for black victims, and uses them to illustrate the dangers of Simpson's paradox when pooling two tables. Here we use the data, reproduced in Table 3.1, to illustrate various test statistics. The proportion of defendants who received the death penalty is higher for blacks than for whites, and more so when the victims are white. In the following we give results for three models assuming independent binomial variation in the four rows of Table 3.1. Model 1. The proportions are the same for both defendant races controlling for victim's races (the common odds ratio is 1). Model 2. The common odds ratio for both tables is different from 1. Model 3. The model with no restrictions (also known as the saturated model). These models were fit using logistic regression. We could also fit an equivalent loglinear model based on assuming a multinomial distribution with 8 cells; the only difference from the above results would be that the log likelihoods would differ by a constant, and the number of estimated parameters would be 5, 6, and 7 for Models 1, 2, and 3, respectively.

The deviance is obtained by taking twice the absolute difference between the log likelihood for a given model and the log likelihood for the saturated model, i.e., for the three models in Table 3.2 the deviance in row j is $-2\{\ell(\widehat{\boldsymbol{\theta}}_j) - \ell(\widehat{\boldsymbol{\theta}}_3)\}$. The deviance is a goodness-of-fit statistic similar to the Pearson χ^2.

Comparing the deviance of 5.39 or $T_S = 5.81$ to a χ_2^2 distribution suggests that Model 1 is at best marginally adequate. Because focused alternatives typically lead to more powerful tests than goodness-of-fit tests, we usually are interested in differences of deviances, the column labeled T_{LR} in the results table. For comparing Model 1 to Model 2, $T_{LR} = 5.01 = -2(-211.99 + 209.29)$ with p-value=.025 (based on a χ_1^2 distribution), supports the conclusion that the probability of receiving the death penalty depends on defendant race.

The $T_{LR} = 0.38 = -2(-209.48 + 209.29)$ results for Model 2 versus Model 3 is clearly not significant, suggesting that the odds ratio in the white victim's 2-by-2 table is similar to that in the black victim's table.

The last column displays differences in the Pearson χ^2 values. Such differences of T_S values are not themselves score statistics, but can be used much like the differences of deviances as long as the binomial sample sizes are reasonably large. The general approach justifying use of these differences is explained in Lindsay and Qu (2003) and discussed briefly in Section 8.3.4 (p. 356). The score statistic for comparing Models 1 and 2 is 5.81, quite similar to the 5.01 and 5.61, but it cannot be computed from the Pearson χ^2 statistics (and not to be confused with the 5.81 in the first row above). ♦

3.2.8 Wald Test Lack of Invariance

We now illustrate the lack of invariance of Wald statistics to the choices of parameterization and function $\boldsymbol{h}$ when the null hypothesis is specified as $H_0 : \boldsymbol{h}(\boldsymbol{\theta}) = \boldsymbol{0}$.

In Examples 2.1 (p. 30) and 2.5 (p. 34) we discussed use of the chi-squared goodness-of-fit statistic for revealing deficiencies in the Poisson model for the fetal lamb data. We now use the ZIP model

$$P(Y_1 = y) = \begin{cases} p + (1-p)f(0;\lambda), & y = 0, \\ (1-p)f(y;\lambda), & y = 1, 2, \ldots, \end{cases}$$

for the same purpose by testing $H_0 : p = 0$ (the Poisson model is adequate). First we need the information matrix $\boldsymbol{I}_{\mathrm{T}}(p, \lambda)$ for the ZIP model. Routine (but tedious) calculations yield

$$I_{\mathrm{T},11}(p,\lambda) = n\left\{\frac{1-e^{-\lambda}}{\pi(1-p)}\right\},$$

$$I_{\mathrm{T},12}(p,\lambda) = -n\left\{\frac{e^{-\lambda}}{\pi}\right\},$$

$$I_{\mathrm{T},22}(p,\lambda) = n\left\{\frac{1-p}{\lambda} - \frac{p(1-p)e^{-\lambda}}{\pi}\right\},$$

where $\pi = P(Y_i = 0) = p + (1-p)e^{-\lambda}$. Then, substituting the ZIP model estimates $\widehat{p} = .577$ and $\widehat{\lambda} = .847$ to get $\widehat{\boldsymbol{I}}_{\mathrm{T}} = \boldsymbol{I}_{\mathrm{T}}(\widehat{p}, \widehat{\lambda})$, we obtain

$$T_{\mathrm{W}} = \widehat{p}^2(\widehat{I}_{\mathrm{T},11} - \widehat{I}^2_{\mathrm{T},12}/\widehat{I}_{\mathrm{T},22}) = 71.73, \tag{3.15}$$

to be compared to a χ^2_1 distribution, and thus is highly significant. Recall that in Example (2.5, p. 34) we found $\chi^2_{\mathrm{G}}(\mathrm{GOF}) = 18.4$ (with p-value $= .00010$). One expects that a test of the Poisson assumption directed at p within the ZIP model is more powerful than an omnibus test like the chi-squared goodness-of-fit test, thus the larger value of T_{W} in (3.15).

In the ZIP model, there are certain advantages to changing the parameter from (p, λ) to (π, λ), where $\pi = P(Y_i = 0) = p + (1-p)e^{-\lambda}$. In the (π, λ) parameterization the maximum likelihood estimator of π is simply the sample proportion of Y_i's that are zero, $\widehat{\pi} = n_0/n$ in the notation of Example 2.1 (p. 31). The maximum likelihood estimator for λ is the same as in the (p, λ) parameterization but is now found by solving a simple one-variable equation. Also, the information matrix is diagonal with

$$I_{\mathrm{T},11}(\pi, \lambda) = n\frac{1}{\pi(1-\pi)},$$

$$I_{\mathrm{T},22}(\pi, \lambda) = n(1-\pi)\left\{\frac{1}{\lambda\left(1-e^{-\lambda}\right)} - \frac{e^{-\lambda}}{\left(1-e^{-\lambda}\right)^2}\right\}.$$

Testing $H_0 : p = 0$ is equivalent to testing $H_0 : \pi = e^{-\lambda}$, or $H_0 : h_1(\pi, \lambda) = 0$, with $h_1(\pi, \lambda) = \pi - e^{-\lambda}$. After taking derivatives to get the associated $\boldsymbol{H}$ function, we substitute $\widehat{\pi} = 182/240 = .758$ and $\widehat{\lambda} = .847$ to get $\widehat{\boldsymbol{I}}_{\mathrm{T}} = \boldsymbol{I}_{\mathrm{T}}(\widehat{\pi}, \widehat{\lambda})$ and

$$T_{\mathrm{W}h_1} = \frac{n\left(\widehat{\pi} - e^{-\widehat{\lambda}}\right)^2}{\dfrac{1}{\widehat{I}_{\mathrm{T},11}} + \dfrac{e^{-2\widehat{\lambda}}}{\widehat{I}_{\mathrm{T},22}}} = 21.88, \tag{3.16}$$

to be compared to a χ^2_1 distribution, say $\chi^2_1(.95) = 3.84$. Taking logarithms of both sides of $\pi = e^{-\lambda}$ leads to testing $H_0 : h_2(\pi, \lambda) = 0$ with $h_2(\pi, \lambda) = \log(\pi) + \lambda$ and

$$T_{\mathrm{W}h_2} = \frac{n\left(\log(\widehat{\pi}) + \widehat{\lambda}\right)^2}{\dfrac{1}{\widehat{\pi}^2\widehat{I}_{\mathrm{T},11}} + \dfrac{1}{\widehat{I}_{\mathrm{T},22}}} = 13.44. \tag{3.17}$$

Finally, solving for p in terms of π, we find $p = (\pi - e^{-\lambda})/(1 - e^{-\lambda})$, leading to yet another logically-equivalent hypothesis test $H_0 : h_3(\pi, \lambda) = 0$ with $h_3(\pi, \lambda) = (\pi - e^{-\lambda})/(1 - e^{-\lambda})$ and

$$T_{\mathrm{W}h_3} = \frac{n\left(\widehat{\pi} - e^{-\widehat{\lambda}}\right)^2}{\frac{1}{\widehat{I}_{\mathrm{T},11}} + \frac{e^{-2\widehat{\lambda}}}{\widehat{I}_{\mathrm{T},22}}\left(\frac{1-\widehat{\pi}}{1-e^{-\widehat{\lambda}}}\right)^2} = 71.73. \tag{3.18}$$

First note that the latter Wald statistic is identical to the Wald statistic for the (p, λ) parameterization and $H_0 : p = 0$. This makes sense because $p = h_3(\pi, \lambda) = 0$ in both. Comparing $T_{\mathrm{W}h_1}$ and $T_{\mathrm{W}h_3}$ is instructive because the only difference is the factor $(1 - \widehat{\pi})^2/(1 - e^{-\widehat{\lambda}})^2$ in the denominator of $T_{\mathrm{W}h_3}$. When H_0 is true, this factor should be close to 1; here its value is .18, which greatly reduces the denominator of $T_{\mathrm{W}h_3}$. It is a bit disturbing that these Wald statistics are so different, although they are all much larger than the nominal .05 critical value 3.84 and thus result in identical conclusions at the .05 level of significance. Also it can shown that under the null hypothesis they are likely to be much more similar.

For comparison, we compute T_{S} and T_{LR}. In the original (p, λ) parameterization for the ZIP model,

$$S_1(p, \lambda) = \frac{\partial \log L(p, \lambda \mid \boldsymbol{Y})}{\partial p} = \frac{n_0\left(1 - e^{-\lambda}\right)}{p_0} - \frac{n - n_0}{1 - p},$$

and the maximum likelihood estimators under $H_0 : p = 0$ are $\widetilde{p} = 0$ and $\widetilde{\lambda} = \overline{Y} = .358$ (because under H_0 the data are Poisson). After some algebra we find that

$$T_{\mathrm{S}} = \frac{\left\{S_1(\widetilde{p}, \widetilde{\lambda})\right\}^2}{\left(\widetilde{I}_{\mathrm{T},11} - \widetilde{I}^2_{\mathrm{T},12}/\widetilde{I}_{\mathrm{T},22}\right)} = \frac{n\left(n_0/n - e^{-\widetilde{\lambda}}\right)^2}{e^{-\widetilde{\lambda}}\left(1 - e^{-\widetilde{\lambda}} - \widetilde{\lambda}e^{-\widetilde{\lambda}}\right)} = 23.96.$$

In the (π, λ) parameterization we find

$$S_1(\pi, \lambda) = \frac{n_0 - n\pi}{\pi(1 - \pi)}, \quad S_2(\pi, \lambda) = \pi S_1(\pi, \lambda),$$

and under $H_0 : \pi = e^{-\lambda}$, $\widetilde{\lambda} = \overline{Y} = .358$ and $\widetilde{\pi} = e^{-\widetilde{\lambda}} = .70$. Then,

$$T_{\mathrm{S}} = (\widetilde{S}_1, \widetilde{S}_2)\begin{pmatrix} \widetilde{I}^{-1}_{\mathrm{T},11} & 0 \\ 0 & \widetilde{I}^{-1}_{\mathrm{T},22} \end{pmatrix}\begin{pmatrix} \widetilde{S}_1 \\ \widetilde{S}_2 \end{pmatrix}$$

$$= \frac{\widetilde{S}^2_1}{\widetilde{I}_{\mathrm{T},11}} + \frac{\widetilde{S}^2_2}{\widetilde{I}_{\mathrm{T},22}} = \frac{n(n_0/n - \widetilde{\pi})^2}{\widetilde{\pi}(1 - \widetilde{\pi} - \widetilde{\lambda}\widetilde{\pi})} = 23.96.$$

The two versions of T_{S} are equal because the score test is invariant to reparameterization. Although the $\widetilde{I}_{\mathrm{T,ij}}$ notation is the same in both version of T_{S} these information matrix components do depend on the parameterizations and thus differ between versions.

Finally, we note that the maximized log likelihoods in Example 2.1 (p. 31) are -201.044 and -190.437 for the Poisson models and ZIP models, respectively. Thus

$$T_{\rm LR} = -2\{-201.044 - (-190.437)\} = 21.21,$$

regardless of the parameterization. The score statistic is slightly larger than the likelihood ratio statistic and larger than two of the Wald statistics computed for these data (although in general, the score statistic is often smaller than the other two statistics, as illustrated in Section 3.2.4 (p. 132)).

van den Broek (1995) gives the score test for the Poisson distribution within the ZIP model and mentions that a simulation study by El-Shaarawi (1985) shows that the score test is more powerful than the likelihood ratio test. Ridout et al. (2001) give the score test for testing the appropriateness of a ZIP model within the context of a zero-inflated negative binomial distribution. For the fetal lamb data they obtain $T_{\rm S} = 4.72$ and p-value $= .03$ suggesting that even the ZIP model for a negative binomial distribution is not adequate for these data.

3.2.9 Testing Equality of Binomial Probabilities: Independent Samples

Suppose that Y_j are independent binomial(n_j, p_j), $j = 1, 2$, and the task is to test $H_0 : p_1 = p_2$. With $h(p_1, p_2) = p_1 - p_2$, and $\boldsymbol{I}_{\rm T}(p_1, p_2) = {\rm diag}\,(n_1/\{p_1(1-p_1)\}, n_2/\{p_2(1-p_2)\})$, using (3.9, p. 136) results in

$$T_{\rm W} = \frac{(\widehat{p}_1 - \widehat{p}_2)^2}{\left\{\dfrac{\widehat{p}_1(1-\widehat{p}_1)}{n_1} + \dfrac{\widehat{p}_2(1-\widehat{p}_2)}{n_2}\right\}},$$

where $\widehat{p}_1 = Y_1/n_1$ and $\widehat{p}_2 = Y_2/n_2$. For the score statistic, (3.10) leads to

$$T_{\rm S} = \frac{(\widehat{p}_1 - \widehat{p}_2)^2}{\left\{\dfrac{\widetilde{p}(1-\widetilde{p})}{n_1} + \dfrac{\widetilde{p}(1-\widetilde{p})}{n_2}\right\}},$$

where $\widetilde{p} = (Y_1 + Y_2)/(n_1 + n_2)$ is the pooled estimator of p obtained from the likelihood with $p_1 = p_2 = p$. Note that $T_{\rm S}$ is the same as the chi-squared test statistic for homogeneity or independence in 2-by-2 contingency tables. Finally,

$$T_{\rm LR} = 2\sum_{i=1}^{2}\left\{Y_i \log\left(\frac{\widehat{p}_i}{\widetilde{p}}\right) + (n_i - Y_i)\log\left(\frac{1-\widehat{p}_i}{1-\widetilde{p}}\right)\right\}.$$

For this problem $T_{\rm S}$ is usually the preferred statistic.

3.2.10 Test Statistics for the Behrens-Fisher Problem

Suppose that $Y_1, \ldots, Y_{n_1}$ are iid from a $\mathrm{N}(\mu_1, \sigma_1^2)$ distribution, $X_1, \ldots, X_{n_2}$ are iid from a $\mathrm{N}(\mu_2, \sigma_2^2)$ distribution, and the null hypothesis is $H_0 : \mu_1 = \mu_2$. This is known as the *Behrens-Fisher Problem* and has a long history. Using independence and (2.43, p. 72), $\boldsymbol{I}_{\mathrm{T}}(\boldsymbol{\theta}) = \mathrm{diag}\left(n_1/\sigma_1^2, n_2/\sigma_2^2, 2n_1/\sigma_1^2, 2n_2/\sigma_2^2\right)$, where $\boldsymbol{\theta}^T = (\mu_1, \mu_2, \sigma_1, \sigma_2)$. The Wald statistic is

$$T_{\mathrm{W}} = \frac{\left(\overline{Y} - \overline{X}\right)^2}{\left\{\dfrac{s_{n_1}^2}{n_1} + \dfrac{s_{n_2}^2}{n_2}\right\}},$$

via (3.9, p. 136). Using (3.10, p. 136), the score statistic is

$$T_{\mathrm{S}} = \frac{\left(\overline{Y} - \overline{X}\right)^2}{\left\{\dfrac{\sum (Y_i - \widetilde{\mu})^2}{n_1^2} + \dfrac{\sum (X_i - \widetilde{\mu})^2}{n_2^2}\right\}},$$

where $\widetilde{\mu}$ is a root of a cubic equation that arises from solving simultaneously

$$\sigma_1^2(\mu) = \frac{1}{n_1}\sum (X_i - \mu)^2, \qquad \sigma_2^2(\mu) = \frac{1}{n_2}\sum (Y_i - \mu)^2,$$

and

$$\mu = \left\{\frac{n_1 \overline{X}}{\sigma_1^2(\mu)} + \frac{n_2 \overline{Y}}{\sigma_2^2(\mu)}\right\} \Big/ \left\{\frac{n_1}{\sigma_1^2(\mu)} + \frac{n_2}{\sigma_2^2(\mu)}\right\}.$$

The likelihood ratio statistic is

$$T_{\mathrm{LR}} = n_1 \log\left\{\frac{\sum (Y_i - \widetilde{\mu})^2}{\sum (Y_i - \overline{Y})^2}\right\} + n_2 \log\left\{\frac{\sum (X_i - \widetilde{\mu})^2}{\sum (X_i - \overline{X})^2}\right\}.$$

For this problem the preferred method is a modified version of T_{W} obtained by replacing $s_{n_1}^2$ and $s_{n_2}^2$ by the corresponding degrees-of-freedom-adjusted sample variances. The signed square root of the modified T_{W} is known as *Welch's t*, and is typically compared to a t distribution with an estimated degrees of freedom. See Best and Rayner (1987) for comparison of T_{W}, T_{S}, and T_{LR}.

3.3 Confidence Regions

Suppose that we are interested in a confidence region for the $r \times 1$ subvector $\boldsymbol{\theta}_1$ where $\boldsymbol{\theta}$ has the partition $(\boldsymbol{\theta}_1^T, \boldsymbol{\theta}_2^T)^T$. Confidence procedures are obtained by inverting the various test statistics. That is, we find the values of $\boldsymbol{\theta}_1$, considered as null values, for which the test statistics are less than or equal to the upper $1-\alpha$ of a chi-squared distribution with r degrees of freedom. The confidence region is thus

$$C_{1-\alpha} = \left\{\boldsymbol{\theta}_1 : T(\boldsymbol{\theta}_1) \le \chi_r^2(1-\alpha)\right\}, \tag{3.19}$$

where $\chi_r^2(1-\alpha)$ is the $1-\alpha$ quantile of a χ_r^2 random variable and T denotes any one of T_{W}, T_{S}, or T_{LR}. Writing $T(\boldsymbol{\theta}_1)$ as a function of $\boldsymbol{\theta}_1$ means that the statistic is calculated treating $\boldsymbol{\theta}_1$ as a null value. The proof that this approach is asymptotically valid is very simple. Let $\boldsymbol{\theta}_{10}$ be the true parameter value. Then

$$P(C_{1-\alpha} \ni \boldsymbol{\theta}_{10}) = P\left\{T(\boldsymbol{\theta}_{10}) \le \chi_r^2(1-\alpha)\right\},$$

which converges to $1-\alpha$ as $T(\boldsymbol{\theta}_{10}) \xrightarrow{d} \chi_r^2$. Note that "$C_{1-\alpha} \ni \boldsymbol{\theta}_{10}$" is read as "the (random) set $C_{1-\alpha}$ contains the (nonrandom) point $\boldsymbol{\theta}_{10}$."

3.3.1 *Confidence Interval for a Binomial Probability*

Example 3.2 (p. 129) is easy to use for illustration. In this simple binomial model, there is just one parameter p and thus $b = r = 1$. In the following $z_{1-\alpha/2}$ is the upper $1-\alpha/2$ quantile of the standard normal distribution, and we use the fact that $\chi_1^2(1-\alpha) = z_{1-\alpha/2}^2$.

Solving the inequality

$$\frac{n\left(\widehat{p}-p\right)^2}{\widehat{p}(1-\widehat{p})} \le z_{1-\alpha/2}^2$$

for p leads to

$$\left(\widehat{p} - z_{1-\alpha/2}\sqrt{\frac{\widehat{p}(1-\widehat{p})}{n}},\ \widehat{p} + z_{1-\alpha/2}\sqrt{\frac{\widehat{p}(1-\widehat{p})}{n}}\right),$$

the standard Wald interval taught in introductory classes. Alternatively, solving the inequality

$$\frac{n\left(\widehat{p}-p\right)^2}{p(1-p)} \le z_{1-\alpha/2}^2$$

for p leads to, via the quadratic formula, the score interval with endpoints

$$\left\{\widehat{p}\left(\frac{n}{n+z^2}\right)+\frac{1}{2}\left(\frac{z^2}{n+z^2}\right)\right\}$$
$$\pm \sqrt{\frac{\widehat{p}(1-\widehat{p})}{n}\left\{\frac{n^2z^2}{(n+z^2)^2}\right\}+\frac{1}{4}\left\{\frac{z^4}{(n+z^2)^2}\right\}}, \quad (3.20)$$

where for conciseness $z_{1-\alpha/2}$ is shortened to z. Santner (1998), Agresti and Coull (1998), and Brown et al. (2001), show that the score interval has coverage properties superior to the Wald interval. In fact, the score interval performs well even when $\widehat{p} = 0$ or 1. The interval derived from T_{LR} must be found numerically and has not received much attention in the literature.

3.3.2 Confidence Interval for the Difference of Binomial Probabilities: Independent Samples

Consider the more complicated setup of Section 3.2.9 (p. 142) with two independent binomial random variables and the parameter $\Delta = p_1 - p_2$. The Wald interval is straightforward and familiar because all parameter estimates are computed under the unrestricted model,

$$\widehat{p}_1 - \widehat{p}_2 \pm z_{1-\alpha/2}\sqrt{\frac{\widehat{p}_1(1-\widehat{p}_1)}{n_1}+\frac{\widehat{p}_2(1-\widehat{p}_2)}{n_2}}.$$

The score and likelihood ratio intervals are difficult to compute because they require maximum likelihood estimators for p_1 and p_2 under the restriction $H_0 : p_1 - p_2 = \Delta$ for multiple values of Δ. Denote such estimators by $\widetilde{p}_1(\Delta)$ and $\widetilde{p}_2(\Delta)$. Then the score interval requires finding all values of Δ such that

$$T_{\text{S}}(\Delta) = \frac{\{\widetilde{p}_1(\Delta) - \widetilde{p}_2(\Delta)\}^2}{\left[\dfrac{\widetilde{p}_1(\Delta)\{1-\widetilde{p}_1(\Delta)\}}{n_1}+\dfrac{\widetilde{p}_2(\Delta)\{1-\widetilde{p}_2(\Delta)\}}{n_2}\right]}$$

is less than $\chi^2_1(1-\alpha)$. The likelihood ratio interval is comparably difficult, and hence neither interval is used much in practice.

3.4 Likelihood Testing for Regression Models

3.4.1 Linear Model

For testing linear hypotheses about $\boldsymbol{\beta}$ of the form $H_0 : K^T\boldsymbol{\beta} = \boldsymbol{m}$, the statistics T_{W}, T_{S}, and T_{LR} are all functions of the usual F statistics and thus equivalent if exact critical values are used.

Returning to Example 2.10 (p. 52) dealing with maximum sea levels in Venice, recall that the least squares and extreme value MLE slope estimates are very close (.567 and .563), but the standard errors of the slopes are .177 and .136, respectively. The square root of the Wald test statistic for $H_0 : \beta = 0$ is thus .563/.136=4.14. The square root of the extreme value likelihood ratio statistic is 3.75. Both are highly significant and larger than the corresponding least squares $t = .567/.177 = 3.2$.

3.4.2 Additive Errors Nonlinear Model

Recall the standard nonlinear regression model (2.10, p. 50) where $Y_i = g(\boldsymbol{x}_i, \boldsymbol{\beta}) + e_i$, where g is a known function. For testing $H_0 : \beta_j = 0$, computer packages make Wald tests easy to conduct by printing the standard errors obtained from the estimated information matrix (2.46, p. 74). As in other models, the Wald tests are often liberal, and dependent on the parameterization. Seber and Wild (1989, Ch. 5) give a discussion of the advantages of likelihood ratio and score tests and F approximations to them. The likelihood ratio tests are easily obtained from the output by subtracting sums of squares from full and reduced model fits.

Practitioners often want a statistic comparable to R^2 in linear regression to assess model fit, but most statistical packages do not give an R^2-like statistic for nonlinear least squares. One R^2-like statistic based on the likelihood is $R^2_{\mathrm{LR}} = 1 - \exp(-T_{\mathrm{LR}}/n)$, where T_{LR} tests the hypothesis that all non-intercept slope parameters are zero. For nonlinear least squares this is just 1−SSE/SST, where SST and SSE are the total and error sums of squares. See Magee (1990) for a discussion of alternative definitions of R^2.

3.4.3 Generalized Linear Model

For comparing nested generalized linear models, likelihood ratio tests dominate via their connection to the model *deviance* statistic that is standard output in GLM model fitting software. The deviance of a particular model is just the likelihood ratio statistic for testing that model against the alternative saturated model (essentially one parameter for each data point). Likelihood ratio test statistics for comparing different nested submodels are then readily obtained as the differences of the model deviances because the contribution of the saturated-model log likelihood to the deviance is the same for any two models and thus cancels.

Wald tests are also easy to implement because computer packages usually print standard errors from the estimated information matrix. In the case of logistic regression, Hauck and Donner (1977) show that Wald tests have non-monotone power functions and probably should be avoided. Many commonly used tests turn out to be score tests for logistic models. We describe a few in the following examples.

3.4.3a Logistic Dose-Response Model

Suppose we have data from a dose-response study wherein rats receive different doses of a suspected carcinogen, and the response indicates the occurrence of tumors appear within a two-month followup. Then $\boldsymbol{x}_i^T = (1, d_i)$ for $i = 1, \ldots, n$, where $d_1, \ldots, d_n$ are the dose levels. At the ith dose level suppose that m_i rats are used with Y_i = the number of rats having at least one tumor at the end of two months. Assume that the logit model $\log\{p_i/(1-p_i)\} = \beta_1 + \beta_2 d_i$ holds where p_i is the probability of an individual rat having at least one tumor at the end of two months. The score test for $H_0 : \beta_2 = 0$,

$$T_S = \frac{\left\{\sum_{i=1}^n \left(Y_i - m_i\overline{Y}\right) d_i\right\}^2}{\overline{Y}(1-\overline{Y})\sum_{i=1}^n m_i(d_i - \overline{d})^2},$$

where $\overline{Y} = \sum Y_i / \sum m_i$ and $\overline{d} = \sum m_i d_i / \sum m_i$, is known as the Cochran-Armitage trend test. It turns out that this test statistic is also the score statistic for any model in which $p_i = F(\beta_1 + \beta_2 d_i)$ for some distribution function F, not just the logistic distribution function. (see Tarone and Gart (1980), for details).

3.4.3b Adequacy of Logistic Dose-Response Model

Continuing the previous logistic regression example, suppose that we want to test the adequacy of the logistic model, $H_0 : p_i = \left\{1 + \exp(-\boldsymbol{x}_i^T \boldsymbol{\beta})\right\}^{-1}, i = 1, \ldots, n$ versus the alternative that the p_i are not constrained. The score test turns out to be the usual chi-squared goodness-of-fit test. That is, think of each dose level having two cells consisting of rats with counts Y_i and $m_i - Y_i$. Then:

$$\begin{aligned} T_S &= \sum \frac{(O_i - \widehat{E}_i)^2}{\widehat{E}_i} \\ &= \sum_{i=1}^n \left[\frac{(Y_i - m_i\widetilde{p}_i)^2}{m_i\widetilde{p}_i} + \frac{\{m_i - Y_i - m_i(1-\widetilde{p}_i)\}^2}{m_i(1-\widetilde{p}_i)}\right] \end{aligned}$$

$$= \sum_{i=1}^{n} (Y_i - m_i \widetilde{p}_i)^2 \left\{ \frac{1}{m_i \widetilde{p}_i} + \frac{1}{m_i(1-\widetilde{p}_i)} \right\}$$

$$= \sum_{i=1}^{n} \frac{(Y_i - m_i \widetilde{p}_i)^2}{m_i \widetilde{p}_i (1-\widetilde{p}_i)},$$

where the $\widetilde{p}_i$ are the fitted probabilities from the logistic model; the tilde ˜ is used here because the logistic model is the null hypothesis. T_S should be compared to a chi-squared distribution with $n - 3$ degrees of freedom. This test statistic is also the score test for testing the adequacy of a general logistic model with n distinct $\boldsymbol{x}$ vectors. An alternative statistic used with generalized linear models is the *deviance* for the fitted model as described in earlier in this section. Most software packages print both the score test statistic and the deviance test statistic by default.

3.4.3c Mantel-Haenszel Statistic

Example 3.2.9 (p. 142) gave the score statistic for testing $H_0 : p_1 = p_2$ for two independent binomials. A logistic regression formulation of the same problem is $H_0 : \beta_2 = 0$, where $\text{logit}(p_1) = \log\{p_1/(1-p_1)\} = \beta_1 + \beta_2$ and $\text{logit}(p_2) = \log\{p_2/(1-p_2)\} = \beta_1$ for which β_2 is the log odds ratio. We now consider the generalization to the case of k sets of independent binomials (often arising from k strata or k independent 2-by-2 tables), where it is assumed that each data set has the same log odds ratio β_{k+1}. Thus for the jth data set (or table) we have $\text{logit}\{p_{1+2(j-1)}\} = \beta_j + \beta_{k+1}$ and $\text{logit}\{p_{2+2(j-1)}\} = \beta_j$. A traditional way to display data of this type arising from a clinical trial is, for the jth table,

	Success	Failure	
Treatment	a_j	b_j	n_{1j}
Control	c_j	d_j	n_{2j}
Total	m_{1j}	m_{2j}	t_j

Here $Y_{1+2(j-1)} = a_j$ and $Y_{2+2(j-1)} = c_j$. The score test for testing that the common odds ratio is unity in all tables ($e^{\beta_{k+1}} = 1$) is formulated as $H_0 : \beta_{k+1} = 0$ with test statistic

$$T_S = \frac{\displaystyle\sum_{i=1}^{k} (a_j - m_{1j} n_{1j} / t_j)^2}{\displaystyle\sum_{i=1}^{k} (n_{1j} n_{2j} m_{1j} m_{2j} / t_j^3)}. \tag{3.21}$$

The classical Mantel-Haenszel statistic, say T_{MH}, conditions on the margins of the tables and thus differs slightly from (3.21) by replacing the divisor t_j^3 in the denominator of (3.21) by $t_j^2(t_j - 1)$.

3.5 Model Adequacy

Checking model assumptions is a key component of data analysis. In regression models, most attention is given to plotting residuals versus predicted values and explanatory variables. In all types of parametric models, though, there is concern about whether distributional assumptions are warranted, e.g., whether data are normally distributed, or have an extreme value distribution, or some other distribution.

The chi-squared goodness-of-fit statistic, used to assess whether the Poisson distribution was appropriate in (2.3, p. 31), can be used for discrete-data model checking more generally. Example 3.4.3b also illustrates its use. For continuous data, we might look at histograms or QQ plots. In Figure 2.2 (p. 36) we demonstrated the use of the quantile-quantile (QQ) plot to assess distributional type. A QQ plot is just a plot of the ordered sample values (empirical quantiles) versus the quantiles of the fitted model, and a straight line suggests an adequate distribution choice. For continuous data we can also group the data into cells and use the chi-squared goodness-of-fit statistic as in the following.

Example 3.5 (Grouped likelihood - hurricane data). Recall that in Example 2.6 (p. 36) the QQ plot (Figure 2.2, p. 37) suggested that the gamma model is inappropriate for the hurricane rainfall data. A more objective assessment is obtained by first grouping the data into cells and using the chi-squared statistic along with the grouped maximum likelihood estimators of α and β. Moore (1986) recommends choosing $k \approx 2n^{2/5}$ equiprobable cells for the chi-squared statistic, here resulting in $k = 8$ since $n = 36$. Thus we found the (.125,.25,.375,.5,.625,.75,.875) quantiles (percentiles) of the gamma distribution with the raw maximum likelihood estimators. For example, the .125 quantile of a gamma($\alpha = 2.19, \beta = 3.33$) distribution is 2.39. So the first cell is (0,2.39]. The .25 quantile is 3.67. So the second cell is (2.39,3.67]. Now we form the multinomial likelihood for the counts of a gamma random variable falling in these cells (the probabilities are just the difference of the gamma distribution function at the cell boundaries) and maximize that. The result is $\widehat{\alpha} = 3.37$ and $\widehat{\beta} = 2.03$ which differ from the full maximum likelihood estimates ($\widehat{\alpha} = 2.19, \widehat{\beta} = 3.33$). Perhaps the grouped likelihood is not as sensitive to the two large rainfall values as the likelihood is. Finally, we compute the chi-squared statistic with probabilities given by the estimated gamma distribution: χ^2(GOF)=4.05 with p-value= .54 from a χ^2_5 distribution (df=$8 - 2 - 1 = 5$). Apparently the hurricane rainfall data fit a gamma distribution better than initially thought, at least as indicated by the grouped-data chi-squared statistic. ♦

Although the chi-squared statistic after grouping is simple to understand and use, it is typically not the most powerful method. A class of more powerful statistics is based on a distance between the fitted parametric model and the empirical distribution function $F_n(y) = n^{-1}\sum_{i=1}^{n} I(Y_i \le y)$. For example, weighted Cramer-von-Mises statistics have the form

$$T_{\text{WCVM}} = n \int_{-\infty}^{\infty} \left[F(y;\widehat{\boldsymbol{\theta}}) - F_n(y)\right]^2 w(y,\widehat{\boldsymbol{\theta}}) f(y;\widehat{\boldsymbol{\theta}})\, dy,$$

where $\widehat{\boldsymbol{\theta}}$ is usually $\widehat{\boldsymbol{\theta}}_{\mathrm{MLE}}$, and $w(y, \widehat{\boldsymbol{\theta}})$ is a weight function. A good choice of $w(y, \boldsymbol{\theta})$ is $[F(y; \boldsymbol{\theta})(1 - F(y; \boldsymbol{\theta}))]^{-1}$ resulting in the Anderson-Darling goodness-of-fit statistic. In location-scale families, the null distribution of T_{WCVM} is free of the unknown parameter values, and can thus be tabled. The gamma distribution is not a location-scale family; so the null distribution is harder to calculate. However, the parametric bootstrap described in Ch. 11 (p. 413) can be used to find p-values. For the hurricane data of Example 2.6 (p. 36), the parametric bootstrap p-value for testing adequacy of the gamma distribution is .032. Removing the largest data point 31.00 leads to a p-value of .16. Thus, most of the evidence against the gamma distribution is due to the largest observation being larger than one would expect from a gamma distribution.

In more complicated models, it is not always easy to assess the adequacy of the model. Most data analysts use a variety of techniques, especially residual plots, to check assumptions. After carrying out these checks, we still may be unsure about the model. Thus, Chapter 7 (p. 297), Chapter 8 (p. 339), Chapter 10 (p. 385), Chapter 11 (p. 413), and Chapter 12 (p. 449) describe methods to robustify inference in the face of model inadequacy or to relax assumptions.

Model adequacy is sometimes addressed by considering classes of models differing with respect to flexibility and hence usually complexity as well. Comparisons among model classes are then made with the objective of determining the least complex adequate model. *Model selection* is the name given to formal procedures used to decide on a model. For example, in regression, deciding on what explanatory variables to include in the fitted model is called model selection or variable selection. With multivariate data, deciding on the type of covariance structure is a form of model selection. Historically, the most important methods for model selection are based on the fitted negative log likelihood (times 2) plus a penalty term depending on the number of parameters in the model. For example, Akaike's Information Criterion (AIC, see Akaike, 1973) uses the penalty term $2k$ where k is the number of parameters in the candidate model. The Bayesian Information Criterion (BIC, Schwarz, 1978), uses $k \log(n)$. Surveys of these approaches appear in Hastie et al. (2001) and Rao and Wu (2001).

3.6 Nonstandard Hypothesis Testing Problems

The likelihood-based inference methods of Sections 3.1–3.4 depend on regularity conditions to be discussed in Ch. 4 that guarantee $\widehat{\boldsymbol{\theta}}_{\mathrm{MLE}}$ is asymptotically normal and the associated test statistics have limiting chi-squared distributions. We have already mentioned situations like Example 2.12 (p. 63) where $\widehat{\boldsymbol{\theta}}_{\mathrm{MLE}}$ is not asymptotically normal. Such anomalies can occur when the densities are not suitably smooth or the support of the density depends on the unknown $\boldsymbol{\theta}$. The key condition ensuring regular asymptotic properties is that $\widehat{\boldsymbol{\theta}}_{\mathrm{MLE}}$ is a solution to the likelihood equations asymptotically,

$$P\left(\frac{\partial}{\partial \boldsymbol{\theta}^T} \log L(\widehat{\boldsymbol{\theta}}_{\text{MLE}} \mid \boldsymbol{Y}) = \boldsymbol{0}\right) \longrightarrow 1. \tag{3.22}$$

When (3.22) does not hold, then typically it is no longer true that $\widehat{\boldsymbol{\theta}}_{\text{MLE}} \sim \text{AN}\{\boldsymbol{\theta}, \boldsymbol{I}_{\text{T}}(\boldsymbol{\theta})^{-1}\}$, and in such cases the three likelihood based tests do not always have a limiting χ_r^2 under H_0.

However, test statistics with non-chi-squared limits also arise when the hypotheses are not of the usual form, $\theta = \theta_0$ versus $\theta \neq \theta_0$. In the next two subsections we discuss two interesting situations: extensions of one-sided hypotheses; and cases where the null hypothesis occurs on the boundary of the parameter space. In these nonstandard situations the limiting distribution is not chi-squared but often related to chi-squared distributions.

3.6.1 *One-Sided Hypotheses and Their Extensions*

The hypothesis tests considered so far have been mainly of the form $H_0 : \boldsymbol{\theta}_1 = \boldsymbol{\theta}_{10}$ versus $H_a : \boldsymbol{\theta}_1 \neq \boldsymbol{\theta}_{10}$, often called *two-sided* tests. If θ_1 is a single real parameter, then we usually call $H_0 : \theta_1 \leq \theta_{10}$ versus $H_a : \theta_1 > \theta_{10}$ or $H_0 : \theta_1 \geq \theta_{10}$ versus $H_a : \theta_1 < \theta_{10}$ *one-sided* tests. One-sided tests are common. For example, if Y is binomial(n, p), we might be interested in $H_0 : p \leq 1/2$ versus $H_a : p > 1/2$ and reject H_0 if $Y \geq k$, where $P(Y \geq k \mid p = 1/2) = \alpha$. Or for $Y_1, \ldots, Y_n \sim N(\mu, \sigma^2)$, we might test $H_0 : \mu \geq 2.1$ versus $H_a : \mu < 2.1$ with the rejection region $t = \sqrt{n}(\overline{Y} - 2.1)/s < -t_{\alpha,n-1}$, where $t_{\alpha,n-1}$ is the $1 - \alpha$ quantile from the t distribution with $n - 1$ degrees of freedom. Often in these tests, for simplicity, we write the null hypothesis as $H_0 : \theta = \theta_0$ and let the alternative describe the direction of interesting alternatives. In these cases the two different specifications of H_0 result in equivalent tests because θ_0 is the hardest of the null hypothesis values to differentiate from the possible alternative values.

3.6.1a Isotonic Regression

Consider data consisting of k independent normal samples of size $n_1, \ldots, n_k$, each iid with means $\mu_1, \mu_2, \ldots, \mu_k$, and common variance σ^2, i.e., the usual ANOVA setup. In some cases there is reason to believe a priori that the means are nondecreasing, $\mu_1 \leq \mu_2 \leq \cdots \leq \mu_k$. In this case, the maximum likelihood estimators minimize $\sum_{i=1}^k n_i(\overline{Y}_i - \mu_i)^2$ subject to the constraint $\mu_1 \leq \mu_2 \leq \cdots \leq \mu_k$, a type of order-restricted least squares method known as isotonic regression.

Now consider testing $H_0 : \mu_1 = \mu_2 \cdots = \mu_k$ versus $H_a : \mu_1 \leq \mu_2 \leq \cdots \leq \mu_k$, *with at least one strict inequality*. For this hypothesis test the likelihood ratio test statistic has a non-chi-squared distribution which in the known variance case is called the $\overline{\chi}^2$ distribution, and in the variance unknown case, is called the $\overline{E}^2$ distribution. For example, Barlow et al. (1972, p. 126) prove that

$$P(\overline{\chi}_k^2 \geq C) = \sum_{\ell=1}^{k} P(\ell, k) P(\chi_{\ell-1}^2 \geq C), \tag{3.23}$$

where χ_0^2 denotes a point mass at zero, and $P(\ell, k)$ are the probabilities that the order-restricted estimators assume exactly ℓ distinct values under the null hypothesis. For the case $k = 2$, we have $P(\overline{\chi}_k^2 \geq C) = (1/2)I(C = 0) + (1/2)P(\chi_1^2 \geq C)$ because $\overline{Y}_1 > \overline{Y}_2$ with probability 1/2, in which case the maximum likelihood estimates under the inequality constraint are $\widehat{\mu}_1 = \widehat{\mu}_2 = (n_1\overline{Y}_1 + n_2\overline{Y}_2)/(n_1 + n_2)$. Thus $P(1,2) = P(2,2) = 1/2$, and $\overline{\chi}^2$ has the mixture distribution consisting of a point mass at 0 and a χ_1^2 random variable. The resulting test is equivalent to using a one-sided normal test of the difference of means. For the case of unknown variance, the null distribution, $\overline{E}^2$, is similar to (3.23), with the difference that beta random variables replace the χ_{l-1}^2 random variables in (3.23).

Example 3.6 (Isotonic regression P-values for $k = 3$). Consider one-way ANOVA normally-distributed data with $k = 3$, $n_1 = n_2 = n_3 = 5$, and known variance $\sigma^2 = 14$. In this case, $T_{\text{LR}} = \sum_{i=1}^{3} n_i(\widehat{\mu}_i - \overline{Y})^2/\sigma^2$ for $H_0 : \mu_1 = \mu_2 = \mu_3$ versus $H_a : \mu_1 \leq \mu_2 \leq \mu_3$, where $\widehat{\mu}_i, i = 1, \ldots, 3$ are the maximum likelihood estimators under the H_a constraint, and $\overline{Y}$ is the pooled mean over all observations. In this example we compare the usual ANOVA test of equality of means, the isotonic regression ANOVA p-values, and a simple linear regression p-value. For simplicity, we now restrict attention to data sets for which the sample means are in increasing order so that $\widehat{\mu}_i = \overline{Y}_i, i = 1, \ldots, 3$. For such data sets, the usual ANOVA statistic for H_0 versus the complement of H_0 has exactly the same value as the isotonic-regression likelihood ratio statistic $T_{\text{LR}} = \sum_{i=1}^{3} n_i(\widehat{\mu}_i - \overline{Y})^2/\sigma^2$. However, the former is compared to a χ_2^2 distribution, and the latter to the $\overline{\chi}^2$ distribution in (3.23) with $P(1,3) = 1/3$, $P(2,3) = 1/2$, and $P(3,3) = 1/6$ (Bartholomew 1959). Linear regression is useful for detecting monotone trends whether they are linear or not and so we also consider the simple linear regression test statistic

$$T_{\text{fi}} = \frac{\sqrt{5}\sum_{i=1}^{3}(x_i - \overline{x})\overline{Y}_i}{\sigma\left\{\sum_{i=1}^{3}(x_i - \overline{x})^2\right\}^{1/2}},$$

derived from the working model $\overline{Y}_i \sim N(\gamma + \beta x_i, \sigma^2/5)$ with $x_1 = 1, x_2 = 2, x_3 = 3$ and testing $H_0 : \beta = 0$ versus $H_a : \beta > 0$. The p-values for three different sets of means are displayed in Table 3.3.

These p-values suggest that the isotonic regression likelihood ratio test is more powerful than the usual ANOVA test. The regression slope test is best in these three data sets even though the means are not linear in x_i. Perhaps it is a little unfair to include this slope test because it has power for parameter values not in H_a, for example, when population means increase and then decrease. Nevertheless, the

Table 3.3 P-values for three samples, $n_1 = n_2 = n_3 = 5$, $\sigma^2 = 14$

$\overline{Y}_1$	$\overline{Y}_2$	$\overline{Y}_3$	ANOVA	ISO	Reg
2	6	6	0.149	0.050	0.045
2	6	7	0.082	0.026	0.017
2	6	8	0.036	0.011	0.006

relative ease of using regression is probably why isotonic regression is not used more.

Finally, although the p-values reported in Table 3.3 are for the known-variance case, qualitatively similar results are obtained for the more complicated case of σ unknown. ♦

There is a large literature on order-restricted inference. For testing for ordered alternatives, there has been more emphasis on T_{LR} than on T_{W} and T_{S}. The classic references are Barlow et al. (1972) and Robertson et al. (1988), whereas a more recent account is Silvapulle and Sen (2005).

3.6.2 *Null Hypotheses on the Boundary of the Parameter Space*

When a null hypothesis value, say $\boldsymbol{\theta}_0$ lies on the boundary of the parameter space, then maximum likelihood estimators are often truncated at that boundary because by definition $\widehat{\boldsymbol{\theta}}_{\text{MLE}}$ must lie in the parameter space of $\boldsymbol{\theta}$. Thus $\widehat{\boldsymbol{\theta}}_{\text{MLE}}$ is equal to the boundary value $\boldsymbol{\theta}_0$ with positive probability and correspondingly T_{LR} is zero for those cases. The result is that the limiting distribution of T_{LR} is a mixture of a point mass at zero and a chi-squared distribution. We illustrate first with an artificial example and then consider the one-way random effects model.

3.6.2a Normal Mean with Restricted Parameter Space

Suppose that $Y_1, \ldots, Y_n \sim N(\mu, 1)$. Usually, $\widehat{\mu}_{\text{MLE}} = \overline{Y}$, but suppose that we restrict the parameter space for μ to be $[\mu_0, \infty)$ where μ_0 is some given constant, instead of $(-\infty, \infty)$. Then $\widehat{\mu}_{\text{MLE}} = \overline{Y}$ if $\overline{Y} \geq \mu_0$ and $\widehat{\mu}_{\text{MLE}} = \mu_0$ if $\overline{Y} < \mu_0$. Now suppose that the null hypothesis is $H_0 : \mu = \mu_0$. We first consider the three likelihood-based test statistics, showing that only the score statistic has a limiting χ^2_1 distribution. Then we provide a simple solution to this testing problem.

Under H_0, the Wald statistic is $T_{\text{W}} = n(\widehat{\mu}_{\text{MLE}} - \mu_0)^2$, which is thus $T_{\text{W}} = 0$ if $\widehat{\mu}_{\text{MLE}} = \mu_0$ and $T_{\text{W}} = n(\overline{Y} - \mu_0)^2$ if $\overline{Y} \geq \mu_0$. The score statistic is $T_{\text{S}} = n(\overline{Y} - \mu_0)^2$, and the likelihood ratio statistic is the same as the Wald statistic. Thus, only the score statistic converges to a χ^2_1 distribution under H_0. The Wald and the likelihood ratio statistics converge to a distribution that is an equal mixture of a point mass at 0 and a χ^2_1 distribution, the same distribution as in (3.23) for $k = 2$. In fact the limiting distribution function is similar to the "wind speed" example, given by (2.3, p. 28) with $p = 1/2$ and F_T equal to the χ_1^2 distribution. Note that the limiting distribution is the distribution of the random variable $Z^2 I(Z > 0)$, where Z is a standard normal random variable. This uncommon limiting distribution does not cause problems here because the test is one sided. Thus a suitable level-α test for the Wald and likelihood ratio statistics is to reject H_0 if the test statistic exceeds the $1 - 2\alpha$ quantile of the χ_1^2 distribution. To see that this is correct, note that by independence of Z^2 and $I(Z > 0)$

$$\begin{aligned} P\left\{Z^2 I(Z > 0) > \chi_1^2(2\alpha)\right\} &= P\left\{Z^2 > \chi_1^2(2\alpha), Z > 0\right\} \\ &= P\left\{Z^2 > \chi_1^2(2\alpha)\right\} P(Z > 0) \\ &= (2\alpha)(1/2) = \alpha. \end{aligned}$$

However, there is a problem with the score test. Note that rejecting for large values of T_S is equivalent to conducting a two-sided test of $H_0 : \theta = \theta_0$ using the statistic $\sqrt{n}(\overline{Y} - \mu_0)$ and thus rejects either if $\overline{Y} < \mu_0 - \sqrt{\chi_1^2(\alpha)/n}$ or $\overline{Y} > \mu_0 + \sqrt{\chi_1^2(\alpha)/n}$. However, rejecting for $\overline{Y} < \mu_0 - \sqrt{\chi_1^2(\alpha)/n}$ makes no sense when the parameter space is $[\mu_0, \infty)$. The natural solution is to reject H_0 when $\sqrt{n}(\overline{Y} - \mu_0) > z_\alpha$, where $P(Z > z_\alpha) = \alpha$.

The main point of this artificial example is that the restricted parameter space with μ_0 on the boundary has led to nonstandard limiting distributions for T_W and T_{LR}. A similar problem occurs in real problems as illustrated in the next example.

3.6.2b One-Way Random Effects Model

Following notation in Searle (1971, Ch. 9), consider the one-way balanced random effects model,

$$Y_{ij} = \mu + \alpha_i + e_{ij}, \quad j = 1, \ldots, r, \ i = 1, \ldots, t,$$

where μ is an unknown mean, and the α_i are independent $N(0, \sigma_\alpha^2)$ random variables and independent of the errors e_{ij} that are iid $N(0, \sigma_e^2)$. If $\sigma_\alpha^2 > 0$ and $\sigma_e^2 > 0$, then the maximum likelihood estimators are

$$\widehat{\mu} = \overline{Y}_{..} = \frac{1}{rt} \sum_{i=1}^{t} \sum_{j=1}^{r} Y_{ij},$$

$$\widehat{\sigma}_\alpha^2 = \begin{cases} \dfrac{\text{SSA}}{rt} - \dfrac{\text{MSE}}{r} & \text{when SSA/t} > \text{MSE}, \\ 0 & \text{otherwise,} \end{cases}$$

$$\widehat{\sigma}_e^2 = \begin{cases} \text{MSE} & \text{when SSA/t} > \text{MSE}, \\ \dfrac{\text{SSE} + \text{SSA}}{rt} & \text{otherwise}, \end{cases}$$

where

$$\overline{Y}_{i.} = \frac{1}{r}\sum_{j=1}^{r} Y_{ij}, \quad \text{SSA} = r\sum_{i=1}^{t}\left(\overline{Y}_{i.} - \overline{Y}_{..}\right)^2,$$

$$\text{SSE} = \sum_{i=1}^{t}\sum_{j=1}^{r}\left(Y_{ij} - \overline{Y}_{i.}\right)^2, \quad \text{MSE} = \frac{\text{SSE}}{t(r-1)}.$$

The log likelihood is

$$\ell(\boldsymbol{\theta}) = C - \frac{t(r-1)}{2}\log\sigma_e^2 - \frac{t}{2}\log V$$
$$-\frac{1}{2}\left\{\frac{\text{SSE}}{\sigma_e^2} + \frac{\text{SSA} + rt\left(\overline{Y}_{..} - \mu\right)^2}{V}\right\},$$

where C is a constant and $V = \sigma_e^2 + r\sigma_\alpha^2$. Taking derivatives with respect to $\boldsymbol{\theta}^T = (\sigma_\alpha^2, \sigma_e^2, \mu)$, and noting that SSA and SSE are independent with $\text{E(SSA)} = (t-1)V$ and $\text{E(SSE)} = t(r-1)\sigma_e^2$, we obtain

$$\overline{\boldsymbol{I}}(\boldsymbol{\theta}) = \frac{1}{rt}\begin{pmatrix} \dfrac{r^2t}{2V^2} & \dfrac{rt}{2V^2} & 0 \\ \dfrac{rt}{2V^2} & \left[\dfrac{t(r-1)}{2\sigma_e^4} + \dfrac{t}{2V^2}\right] & 0 \\ 0 & 0 & \dfrac{rt}{V} \end{pmatrix}$$

with inverse

$$\left[\overline{\boldsymbol{I}}(\boldsymbol{\theta})\right]^{-1} = \begin{pmatrix} \dfrac{2}{r}\left[V^2 + \dfrac{\sigma_e^4}{r-1}\right] & \dfrac{-2\sigma_e^4}{r-1} & 0 \\ \dfrac{-2\sigma_e^4}{r-1} & \dfrac{2r\sigma_e^4}{r-1} & 0 \\ 0 & 0 & V \end{pmatrix}.$$

If $\sigma_\alpha^2 > 0$, $\sigma_e^2 > 0$, and $t \to \infty$ while r remains fixed, then $P(\widehat{\sigma}_\alpha^2 = 0) \to 0$ and $P\left\{\widehat{\sigma}_e^2 = (\text{SSE} + \text{SSA})/(rt)\right\} \to 0$ and the maximum likelihood estimators are asymptotically normal with asymptotic covariance $\left[rt\bar{\boldsymbol{I}}(\boldsymbol{\theta})\right]^{-1}$ as dictated by standard asymptotic results. (If in addition $r \to \infty$, then slightly different asymptotic normality results are obtained; see Miller 1977.)

Consider now the hypothesis testing situation $H_0 : \sigma_\alpha^2 = 0$. Under this null hypothesis, $P(\widehat{\sigma}_\alpha^2 = 0) \to 1/2$, somewhat like the artificial example above. The Wald statistic is $T_{\text{W}} = 0$ when $\widehat{\sigma}_\alpha^2 = 0$ (when $\text{SSA}/t \le \text{MSE}$) and

$$T_{\text{W}} = \frac{[\text{SSA}/(rt) - \text{MSE}/r]^2}{\left[rt\bar{\boldsymbol{I}}(\widehat{\boldsymbol{\theta}}_{\text{MLE}})\right]_{11}^{-1}}$$

when $\text{SSA}/t > \text{MSE}$. Thus, in large samples under H_0, T_{W} converges in distribution to $Z^2 I(Z > 0)$ as in the artificial example. We would reject H_0 if the test statistic is greater than the $1 - 2\alpha$ quantile of the χ_1^2 distribution.

Miller (1977, p. 757) gives the likelihood ratio statistic as

$$\begin{aligned} T_{\text{LR}} &= -t\left[(r-1)\log\{t/(t-1)\} + r\log r \right. \\ &\quad \left. + \log F - r\log\{t(r-1)/(t-1) + F\}\right] \end{aligned}$$

when $F > t/(t-1)$ and $T_{\text{LR}} = 0$ otherwise, where F is the usual F statistic $F = \text{MSA}/\text{MSE}$, $\text{MSA} = \text{SSA}/(t-1)$. It can then be shown that as $t \to \infty$, T_{LR} also converges in distribution to $Z^2 I(Z > 0)$.

For computing the score statistic, note that under H_0 the likelihood estimator for μ remains $\overline{Y}_{..}$ and $\widetilde{\sigma}_e^2 = (\text{SSE} + \text{SSA})/(rt)$. Then the score statistic is

$$T_{\text{S}} = \left[\frac{1}{2\widetilde{\sigma}_e^4}\{(r-1)\text{SSA} - \text{SSE}\}\right]^2 \frac{2\widetilde{\sigma}_e^4}{r(r-1)t}. \tag{3.24}$$

Moreover, it can be shown that T_{S} converges under H_0 to a χ_1^2 distribution. Thus, we are in the same situation as the artificial example where the score statistic has the standard limiting distribution, but it does not take into account the one-sided nature of the null hypothesis. Thus we should redefine T_{S} to be zero whenever $\text{SSA}/t \le \text{MSE}$ and use the $1 - 2\alpha$ critical value of the χ_1^2 distribution.

Manipulation of (3.24) shows that T_{S} is a function of the usual F statistic $F = \text{MSA}/\text{MSE}$ and that T_{S} is large if F is large or small (and small means F near 0, which means $\text{SSA}/t \le \text{MSE}$). Thus, an "exact" corrected score procedure is to just carry out the usual F test with the usual F critical values.

3.6.2c General Solution for Parameters on the Boundary

The study of likelihood ratio statistics with parameters on the boundary of the parameter space was begun by Chernoff (1954). Self and Liang (1987) give a fairly complete description of the general problem. They separate the parameter vector into four components:

1. r_1 parameters of interest on the boundary,
2. r_2 parameters of interest not on the boundary,
3. s_1 nuisance parameters on the boundary,
4. s_2 nuisance parameters not on the boundary.

Thus, the b-dimensional vector $\boldsymbol{\theta}$ has sub-dimensions (r_1, r_2, s_1, s_2), where $r_1+r_2+s_1+s_2 = b$. For example, in the one-way mixed model above, $\boldsymbol{\theta}^T = (\sigma_\alpha^2, \sigma_e^2, \mu)$ with dimensions $(r_1 = 1, r_2 = 0, s_1 = 0, s_2 = 2)$. A related example is to test $H_0 : \sigma_\alpha^2 = 0, \mu = 0$ versus $H_a : \sigma_\alpha^2 > 0, \mu \neq 0$ in the one-way mixed model. In this case, μ has joined σ_α^2 to form the parameter vector of interest and the dimensions are $(r_1 = 1, r_2 = 1, s_1 = 0, s_2 = 1)$. The limiting distribution of T_{LR} in this case is $Z_1^2 I(Z_1 > 0) + Z_2^2$, where Z_1 and Z_2 are independent standard normal random variables. Self and Liang (1987) also give a variety of more complicated examples.

Sometimes it is not obvious when we have a boundary problem. Recall our testing the Poisson assumption with a ZIP model

$$P(Y_1 = y) = \begin{cases} p + (1-p)f(0;\lambda), & y = 0; \\ (1-p)f(y;\lambda), & y = 1, 2, \ldots \end{cases}$$

by testing $H_0 : p = 0$. Logically, this is a boundary value (p is a probability) but mathematically it is not. It is possible to let p be less than zero — the only natural restriction on p is that $P(Y_1 = 0) = p + (1-p)e^{-\lambda} \geq 0$. This restriction translates into the parameter space for (p, λ) given by Figure 3.2. Using the full parameter space in Figure 3.2, the maximum likelihood estimators solve the likelihood equations and the test statistics we computed in Section 2.3 are asymptotically χ_1^2 under $H_0 : p = 0$. Philosophically we might prefer a one-sided hypothesis here, $H_0 : p = 0$ versus $H_a : p > 0$, in which case we would need to modify the statistics computed in Section 2.3 to be 0 if the unrestricted maximum likelihood estimator of p is less than 0 and use $1-2\alpha$ quantiles of the χ_1^2 distribution.

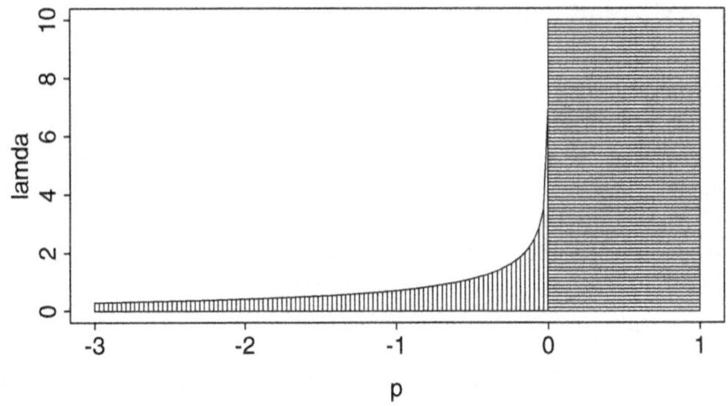

Fig. 3.2 Maximum allowable parameter space for the ZIP density. The space continues to the left and up to ∞ for $p \in [0, 1]$

3.7 Problems

3.1. For a situation where the likelihood is quadratic in a single real parameter θ, i.e.,

$$\log L(\theta \mid \boldsymbol{Y}) = a(\boldsymbol{Y}) + b(\boldsymbol{Y})\theta - c\theta^2,$$

where c is a known constant and $a(\cdot)$ and $b(\cdot)$ are known functions, show that the Wald, score, and likelihood ratio statistics are all the same for $H_0 : \theta = \theta_0$.

3.2. Suppose that we have an iid sample $Y_1, \ldots, Y_n$ from $f(y; \mu, \sigma) = \sigma^{-1} f_0((y - \mu)/\sigma)$, where f_0 is the Laplace density $f_0(y) = (1/2)\exp(-|y|)$. The maximum likelihood estimators are the sample median $\widehat{\mu}$ and the mean deviation from the median $\widehat{\sigma} = n^{-1} \sum |Y_i - \widehat{\mu}|$. Even though the Laplace density doesn't quite have the smoothness conditions that we usually require, most results work as long as you use $d|y|/dy = \text{sgn}(y)$ when $y \neq 0$ and $= 0$ otherwise, and compute the Fisher information matrix from the product of first partial derivatives rather than try to get the matrix of second partial derivatives. (To save you from computations, the result turns out to be $I_{\mu,\sigma} = \sigma^{-2}\text{diag}[1, 1]$. Also $\text{E}|Y_1 - \mu| = \sigma$ and $\text{Var}(Y_1) = 2\sigma^2$.) Find the Wald, score, and likelihood ratio statistics T_{W}, T_{S}, and T_{LR}, for $H_0 : \mu = \mu_0$ versus $H_a : \mu \neq \mu_0$.

3.3. Verify the value 71.73 (3.15, p. 140) for the Wald statistic in the (p, λ) parameterization.

3.4. Verify the values 21.88, 13.44, 71.73 in equations (3.16)–(3.18) starting on p. 140, in Section 3.2.8 for the (π, λ) parameterization.

3.5. For the two independent binomials problem, Section 3.2.9 (p. 142), derive the three statistics T_{W}, T_{S}, and T_{LR} for testing $H_0 : p_1 = p_2$.

3.6. Assume that Y_1 and Y_2 are independent with respective geometric probability mass functions,

$$f(y; p_i) = p_i(1-p_i)^{y-1} \quad y = 0, 1, \ldots \quad 0 \le p_i \le 1, \; i = 1, 2.$$

Recall that the mean and variance of a geometric random variable with parameter p are $1/p$ and $(1-p)/p^2$, respectively. For $H_0 : p_1 = p_2$ versus $H_a : p_1 \neq p_2$ show that the score statistic is

$$T_S = \frac{\widetilde{p}^2}{2(1-\widetilde{p})}(Y_1 - Y_2)^2,$$

where $\widetilde{p} = 2/(Y_1 + Y_2)$.

3.7. For the setting of the previous problem (Y_1 and Y_2 are independent geometric random variables), show that for $h_1(p_1, p_2) = p_1 - p_2$, the Wald statistic is

$$T_{W,1} = (Y_1 - Y_2)^2 \Big/ \left(\frac{1-\widehat{p}_1}{\widehat{p}_2^2} + \frac{1-\widehat{p}_2}{\widehat{p}_1^2} \right),$$

where $\widehat{p}_i = 1/Y_i, i = 1, 2$. Next, show that for $h_2(p_1, p_2) = 1/p_1 - 1/p_2$, the Wald statistic is

$$T_{W,2} = (Y_1 - Y_2)^2 \Big/ \left(\frac{1-\widehat{p}_1}{\widehat{p}_1^2} + \frac{1-\widehat{p}_2}{\widehat{p}_2^2} \right).$$

Which version seems more appropriate?

3.8. Suppose that $Y_1, \ldots, Y_{n_1}$ are iid from a $N(\mu_1, \sigma^2)$ distribution, $X_1, \ldots, X_{n_2}$ are iid from a $N(\mu_2, \sigma^2)$ distribution, the samples are independent of each other, and we desire to test $H_0 : \mu_1 = \mu_2$.

a. Derive the Wald and score tests and express them as a function of the square of the usual two-sample pooled t. Also, show that the likelihood ratio statistic is $T_{LR} = (n_1 + n_2)\log\{1 + t^2/(n_1 + n_2 - 2)\}$.
b. Let $n_1 = n_2 = 5$. These tests reject H_0 at approximate level .05 if they are larger than 3.84. Find exact expressions for $P(T_W \ge 3.84)$, $P(T_S \ge 3.84)$, and $P(T_{LR} \ge 3.84)$ using the fact that t^2 has an $F(1, n_1 + n_2 - 2)$ under H_0.

3.9. Suppose that $Y_1, \ldots, Y_n$ are independently distributed as Poisson random variables with means $\lambda_1, \ldots, \lambda_n$, respectively. Thus,

$$P(Y_i = y) = f(y; \lambda_i) = \frac{\lambda_i^y e^{-\lambda_i}}{y!}, \quad y = 0, 1, 2, \ldots.$$

Show that the score statistic for testing $H_0 : \lambda_1 = \lambda_2 = \cdots = \lambda_n = \lambda$ (i.e., that the Y_i's all have the same distribution) is given by

$$T_S = \sum_{i=1}^{n} \frac{\left(Y_i - \overline{Y}\right)^2}{\overline{Y}}.$$

3.10. Derive expression (3.20, p. 145) for the confidence interval for p.

3.11. In the case that we observe all successes in n Bernoulli(p) trials, an exact $1-\alpha$ lower bound for p is $\alpha^{1/n}$, obtained by solving $p^n(1-p)^0 = \alpha$ (see Example 9.2.5 on p. 425–427 of Casella and Berger 2002). Use the left-hand part of the interval (3.20, p. 145) with $z = z_{1-\alpha}$ to get an approximate lower bound in this case. Compare the two lower bounds in terms of coverage for $\alpha = .05$ and $n = 10$, $n = 20$, and $n = 100$.

3.12. Show that the T_W, T_S and T_{LR} statistics defined in Example 3.2 (p. 129) are asymptotically equivalent under H_0 by showing their differences converge to 0 in probability.

3.13. Consider a dose-response situation similar to Section 3.4.3a (p. 147) with k dose levels. At the ith dose d_i we observe $(Y_{ij}, n_{ij}, j = 1, \ldots, m_i)$, where the n_{ij} are fixed constants. A common assumption is that the Y_{ij} are all independent and distributed binomial$(n_{ij}, p_i(\boldsymbol{\beta}) = F(\boldsymbol{x}_i^T \boldsymbol{\beta}))$, where F is a distribution function like the standard normal or logistic distribution function and typically $\boldsymbol{x}_i^T = (1, d_i)$ or $\boldsymbol{x}_i^T = (1, d_i, d_i^2)$. Then the log of the likelihood is

$$\log L(\boldsymbol{\beta}) = \text{constant} + \sum_{i=1}^{k} \sum_{j=1}^{m_i} \left[Y_{ij} \log\{p_i(\boldsymbol{\beta})\} + (n_{ij} - Y_{ij}) \log\{1 - p_i(\boldsymbol{\beta})\} \right].$$

a. Find the score function $S(\boldsymbol{\beta})$.
b. Find the total sample information $\boldsymbol{I}_T(\boldsymbol{Y}, \boldsymbol{\beta})$ and its expected value $\boldsymbol{I}_T(\boldsymbol{\beta})$.
c. Show that for the logistic distribution function, $F(x) = (1 + \exp(-x))^{-1}$,

$$\boldsymbol{I}_T(\boldsymbol{Y}, \boldsymbol{\beta}) = \boldsymbol{I}_T(\boldsymbol{\beta}) = \sum_{i=1}^{k} \sum_{j=1}^{m_i} n_{ij}\, p_i(\boldsymbol{\beta}) \{1 - p_i(\boldsymbol{\beta})\} \boldsymbol{x}_i \boldsymbol{x}_i^T.$$

3.14. Derive the Cochran-Armitage trend test in Section 3.4.3a (p. 147) as a score test.

3.15. Derive the goodness-of-fit score statistic in Section 3.4.3b (p. 147) assuming that the Y_i are independent binomial(m_i, p_i) random variables, and the parameter vector is $\boldsymbol{\theta} = (p_1, \ldots, p_n)^T$. The only place the logistic parameterization enters is that under H_0, $\widetilde{p}_i = F(\boldsymbol{x}_i^T \widetilde{\boldsymbol{\beta}})$, where F is the logistic distribution function and $\widetilde{\boldsymbol{\beta}}$ is just the logistic model maximum likelihood estimator. Thus, the form of F never enters the derivation since we just substitute $\widetilde{p}_i$ wherever necessary in the score test definition.

3.16. Derive the score statistic T_S in (3.21, p. 148)

3.17. For the lung cancer data of Glasser (1965), Example 2.11 (p. 56), fit lognormal and Weibull models using the SAS program `lung.sas.txt` on the course website http://www4.stat.ncsu.edu/~boos/Essential.Statistical.Inference

a. Get the p-values for Wald and likelihood ratio tests of age, and then after deleting age, get the p-values for performance status (ps).
b. For one of the models with only ps in the model, invert the negative of the Hessian (SAS calls -Hessian the Hessian), and check that the reported standard errors actually come from it.
c. One might like to choose between the normal and Weibull models. Can you think of a way to do this?

3.18. Consider having two independent iid samples, the first with a normal(μ_1,1) distribution and sample size n_1, the second with a normal(μ_2,1) distribution and sample size n_2. For $H_0 : \mu_1 = \mu_2$ versus $H_a : \mu_1 < \mu_2$, find T_{LR} and the testing procedure at $\alpha = .05$. Note that under H_a : the maximum likelihood estimators are the usual ones if $\overline{Y}_1 \leq \overline{Y}_2$, but $\widehat{\mu}_1 = \widehat{\mu}_2 = (n_1\overline{Y}_1 + n_2\overline{Y}_2)/(n_1 + n_2)$ if $\overline{Y}_1 > \overline{Y}_2$. Also, note that $P(2, 2) = 1/2$ in (3.23, p. 152) for this case since the probability is 1/2 that the restricted estimators are the usual sample means with $l = 2$ distinct values.

3.19. Find the .90, .95, and .99 quantiles of the distribution of $Z^2 I(Z > 0)$, where Z is a standard normal random variable.

3.20. Consider testing $H_0 : \mu_1 = \mu_2 = \mu_3 = \mu_4$ versus $H_a : \mu_1 \leq \mu_2 \leq \mu_3 \leq \mu_4$ in the one-way ANOVA context with known variance. Suppose that the observed sample means are in increasing order so that the usual ANOVA T_{LR} statistic and the isotonic regression T_{LR} are exactly the same and equal to 7.5. Find the p-values for both procedures using the χ^2_3 distribution for the usual ANOVA test and (3.23, p. 152) for the isotonic regression test assuming equal sample sizes. In this case, the necessary constants for (3.23, p. 152) are $P(1, 4) = 1/4$, $P(2, 4) = 11/24$, $P(3, 4) = 6/24$, and $P(4, 4) = 1/24$, from Bartholomew (1959).

3.21. Verify the accuracy of the statement near the end of Section 3.6.2b that starts on p. 154, "Manipulation of (3.24) shows that T_S is a function of the usual F statistic $F = \mathrm{MSA}/\mathrm{MSE}$ and that T_S is large if F is large or small (and small means F near 0, which means $\mathrm{SSA}/t \leq \mathrm{MSE}$). Thus, an 'exact' corrected score procedure is to just carry out the usual F test with the usual F critical values."

Chapter 4
Bayesian Inference

4.1 Introduction

The majority of this book is concerned with frequentist inference about an unknown parameter $\boldsymbol{\theta}$ arising in a statistical model, typically a parametric model where data $\boldsymbol{Y}$ are assumed to have density $f(\boldsymbol{y};\boldsymbol{\theta})$. The classical tools are estimation $\boldsymbol{\theta}$ by maximum likelihood, hypothesis testing via T_{W}, T_{S}, or T_{LR} tests, and confidence interval estimation obtained by inverting test statistics. Throughout, the parameter $\boldsymbol{\theta}$ is viewed as a fixed constant, and only $\boldsymbol{Y}$ is random.

The Bayesian approach is a bit different, starting with the assumption that $\boldsymbol{\theta}$ is random with population or prior density $\pi(\boldsymbol{\theta})$. Then Nature chooses a particular value $\boldsymbol{\theta}$ from $\pi(\boldsymbol{\theta})$ and generates data $\boldsymbol{Y}$ from $f(\boldsymbol{y};\boldsymbol{\theta})$, but written as $f(\boldsymbol{y} \mid \boldsymbol{\theta})$ to emphasize that it is a density of $\boldsymbol{Y}$ conditional on the value of $\boldsymbol{\theta}$. At this point, the goals of frequentist and Bayesian inference are the same: there is one value of $\boldsymbol{\theta}$ that generated the data, and our goal is to learn about that value based on $\boldsymbol{Y}$. However, the Bayesian approach is different because it can take advantage of the original random nature of $\boldsymbol{\theta}$ and use the joint distribution of $\boldsymbol{\theta}$ and $\boldsymbol{Y}$ to get the conditional density of $\boldsymbol{\theta}$ given $\boldsymbol{Y}$,

$$\pi(\boldsymbol{\theta} \mid \boldsymbol{Y} = \boldsymbol{y}) = \frac{f(\boldsymbol{y} \mid \boldsymbol{\theta})\pi(\boldsymbol{\theta})}{\displaystyle\int f(\boldsymbol{y} \mid \boldsymbol{\theta})\pi(\boldsymbol{\theta})d\boldsymbol{\theta}}. \tag{4.1}$$

This conditional density is called the *posterior density* of $\boldsymbol{\theta}$ and is the basis for making inferences about $\boldsymbol{\theta}$. For example, the standard point estimator of the value of $\boldsymbol{\theta}$ that generated $\boldsymbol{Y}$ is the mean of the posterior, $\int \boldsymbol{\theta}\pi(\boldsymbol{\theta} \mid \boldsymbol{Y})d\boldsymbol{\theta}$. Analogues of $(1-\alpha)$ confidence regions, called credible regions, are regions of the parameter space that have posterior probability $1-\alpha$.

The density in the denominator of (4.1), $m(\boldsymbol{y}) = \int f(\boldsymbol{y} \mid \boldsymbol{\theta})\pi(\boldsymbol{\theta})d\boldsymbol{\theta}$, is the *marginal density* of $\boldsymbol{Y}$. Before data has been collected, $m(\boldsymbol{y})$ can, in principle, be used to predict a future value of $\boldsymbol{Y}$. When used in this fashion, $m(\boldsymbol{y})$ is usually called

D.D. Boos and L.A. Stefanski, *Essential Statistical Inference: Theory and Methods*, Springer Texts in Statistics, DOI 10.1007/978-1-4614-4818-1_4,

the *prior predictive density*. Note that frequentist inference has no parallel method for predicting before observing data except perhaps to guess a value of $\boldsymbol{\theta}$ and use $f(\boldsymbol{y} \mid \boldsymbol{\theta})$.

A more important forecast density is the marginal density of a future value of $\boldsymbol{Y}$, say $\boldsymbol{Y}_{new}$,

$$m(\boldsymbol{y}_{new} \mid \boldsymbol{Y}) = \int f(\boldsymbol{y}_{new} \mid \boldsymbol{\theta}, \boldsymbol{Y})\pi(\boldsymbol{\theta} \mid \boldsymbol{Y})d\boldsymbol{\theta}.$$

We call $m(\boldsymbol{y}_{new} \mid \boldsymbol{Y})$ the *posterior predictive density*. If $\boldsymbol{Y}_{new}$ is independent of $\boldsymbol{Y}$, then $f(\boldsymbol{y}_{new} \mid \boldsymbol{\theta}, \boldsymbol{Y})$ reduces to $f(\boldsymbol{y}_{new} \mid \boldsymbol{\theta})$ and $m(\boldsymbol{y}_{new} \mid \boldsymbol{Y})$ is defined in exactly the same fashion as $m(\boldsymbol{y})$ except that the integration over $\boldsymbol{\theta}$ is with respect to the posterior instead of the prior. For a specific predicted value, one might use the mean of $m(\boldsymbol{y}_{new} \mid \boldsymbol{Y})$, but $m(\boldsymbol{y}_{new} \mid \boldsymbol{Y})$ allows us to assess the uncertainty about the prediction, perhaps giving a prediction interval.

Here we briefly discuss notation specific for this chapter. Typically, we prefer to use capital letters for random quantities and lowercase letters for arguments of functions, like X and $g(x)$. However, we break our convention here because θ and Θ are hard to distinguish, and standard Bayesian notation is to use lowercase θ everywhere and let the context dictate the usage. Within functions, θ refers to arguments, and $f(y \mid \theta)$ means $f(y \mid \Theta = \theta)$. When referring to densities about θ like priors and posteriors, we use π and let the arguments show whether it is a prior or posterior.

The new quantity introduced in Bayesian analysis is the prior $\pi(\boldsymbol{\theta})$. Where does it come from? For the *subjective* Bayesian, $\pi(\boldsymbol{\theta})$ reflects personal uncertainty about $\boldsymbol{\theta}$ before the data are collected. For other Bayesians, $\pi(\boldsymbol{\theta})$ incorporates all previous information known about $\boldsymbol{\theta}$, perhaps from historical records, and Bayesian analysis is used to combine previous information with current data. (Frequentists can also include historical information for making inferences but not in such a straightforward way.) For others, $\pi(\boldsymbol{\theta})$ may be a convenient technical density that allows them to use the Bayesian machinery easily (conjugate priors) or in an objective fashion often resulting in inference close to frequentist inference (vague or non-informative priors). In any case, $\pi(\boldsymbol{\theta})$ is a crucial quantity in Bayesian inference, and often a stumbling block for non-Bayesians. We will not spend time reviewing debates about $\pi(\boldsymbol{\theta})$, but merely assume that it has been given.

We start with a few simple examples.

Example 4.1 (Binomial(n; p)). For Y that is binomial(n; p),

$$f(y \mid p) = \binom{n}{y} p^y (1-p)^{n-y} \quad y = 0, 1, \ldots n.$$

The usual prior is a beta(α, β) density

$$\pi(p) = \frac{\Gamma(\alpha+\beta)}{\Gamma(\alpha)\Gamma(\beta)} p^{\alpha-1}(1-p)^{\beta-1},$$

where $\Gamma()$ is the Gamma function. Multiplying the two densities results in the joint density

$$\binom{n}{y} p^y (1-p)^{n-y} \frac{\Gamma(\alpha+\beta)}{\Gamma(\alpha)\Gamma(b)} p^{\alpha-1}(1-p)^{\beta-1}$$
$$= c(\alpha,\beta,y)p^{y+\alpha-1}(1-p)^{n-y+\beta-1}.$$

Formally the posterior is obtained by dividing the joint density by the marginal density of Y

$$m(y) = \int_0^1 c(\alpha,\beta,y)p^{y+\alpha-1}(1-p)^{n-y+\beta-1}dp$$
$$= \frac{B(y+\alpha, n-y+\beta)\binom{n}{y}}{B(\alpha,\beta)}, \tag{4.2}$$

where $B(\alpha,\beta) = \Gamma(\alpha)\Gamma(\beta)/\Gamma(\alpha+\beta)$ is the beta function. This beta-binomial density arises in other contexts where one averages over the p of binomial densities.

In this case, we really do not need $m(y)$ because we can recognize from the kernel $p^{y+\alpha-1}(1-p)^{n-y+\beta-1}$ of the joint density that the posterior must be a beta($y+\alpha, n-y+\beta$) density. Whenever the posterior is from the same family as the prior, we say the prior is a *conjugate* prior. Historically, conjugate priors were very important because they simplified finding posteriors. With the advent of modern computing and Markov chain Monte Carlo methods, conjugate priors are not as essential as in the past, but they still make inference very straightforward.

Because the mean of a beta(α,β) density is $\alpha/(\alpha+\beta)$, the mean of the posterior is

$$E(p \mid Y) = \frac{Y+\alpha}{Y+\alpha+n-Y+\beta} = \frac{Y+\alpha}{\alpha+\beta+n}$$
$$= \left(\frac{\alpha+\beta}{\alpha+\beta+n}\right)\left(\frac{\alpha}{\alpha+\beta}\right) + \left(\frac{n}{\alpha+\beta+n}\right)\left(\frac{Y}{n}\right). \tag{4.3}$$

Here we have used upper case Y to emphasize the random nature of the estimator. We see that the posterior mean is a weighted average of the prior mean and the unbiased estimator Y/n. Bayesian point estimators are usually of this general form, shrinking maximum likelihood estimators toward prior values.

For a specific numerical example with (4.3), let $n = 10$ and suppose $Y = 2$ is observed. Then a uniform(0,1) prior ($\alpha = \beta = 1$) gives posterior mean .25 compared to the unbiased estimator .20. Jeffreys's noninformative prior, $\alpha = \beta = 1/2$ from (4.12, p. 175), gives 5/22=.23, and a more informative prior, $\alpha = \beta = 5$ gives equal weighting and posterior mean (1/2)(1/2)+(1/2)(.20)=.35. All these Bayes estimators are drawn toward the prior mean of 1/2 (when $\alpha = \beta$), but the latter was the only one exerting much influence away from the unbiased estimator .20.

Lehmann and Casella (1998, p. 311–312) show that $\alpha = \beta = \sqrt{n}/2$ is a special choice for which the Bayes estimator has lower mean squared error than Y/n for small n over most of the parameter space (0,1). This illustrates the general tendency that Bayes estimators compare favorable in terms of mean squared error to frequentist unbiased estimators. Of course, the choice $\alpha = \beta = 0$ leads to Y/n even though the prior for these values is not a true density because $p^{-1}(1-p)^{-1}$ is not integrable (hence the term *improper prior*). The beta$(Y, n-Y)$ posterior is proper, however, as long as $1 \le Y \le n-1$.

One arena where Bayesian inference is very appealing is when trying to make personal decisions about health care procedures (surgeries, correct drug usage, etc.). Here is one example that is quite real for a lot of folks with heart problems.

Should Stent Patients Stop Taking Blood Thinners Before Surgery?

When a heart artery gets clogged up, the standard procedure is to open up the artery with angioplasty using an inflatable balloon at the end of a catheter that is usually inserted in the leg and guided up to the clogged artery. An important advance in the use of angioplasty is to leave a bare metal stent (thin hollow mesh) in the artery to keep the artery from clogging back up. These stents are effective but still have a rate of clogging up (called restenosis) of 30% in the first year. The next generation of stents are called drug-eluting because they are inserted with a drug coating that keeps the restenosis rate under 10%, a major improvement. However, it has been discovered that these drug-eluting stents have an increased risk for a blood clot to form on their surface if blood thinners, notably clopidogrel (brand name Plavix) plus aspirin, have been stopped. In one study, it was found that a certain percentage of people with drug-eluting stents who stopped clopidogrel at the end of 6 months experienced some kind of event linked to a blood clot in the following 18 months. We shall assume for illustration that the percentage is 5%, but it is likely smaller. So, it is now being recommended that people with drug-eluting stents remain on blood thinners like clopidogrel for much longer than the initial 6 month period after receiving their stents, perhaps indefinitely.

The dilemma for these patients with drug-eluting stents is what to do when they need surgery or even something as simple as a colonoscopy where it is recommended that they stop all blood thinners before the operation. The reason for stopping the blood thinners is to reduce the risk of internal bleeding during and after the operation or procedure. However, stopping the blood thinners yields the increased risk of blood clots on the stent and a possible heart attack.

We first think about a person with a drug-eluting stent who is considering an operation for which he needs to stop taking blood thinners. What is his risk of a blood clot if he stops taking blood thinners? Ignoring the difference between an 18 month window (from the study) and a 7 day window (for his operation), he desires to use the 5% result to discern a relevant prior. The 5% study result can be viewed as arising from a sample of people having the marginal beta-binomial distribution (4.1, p. 164). Each of the persons in the study can be viewed as obtaining a p from the

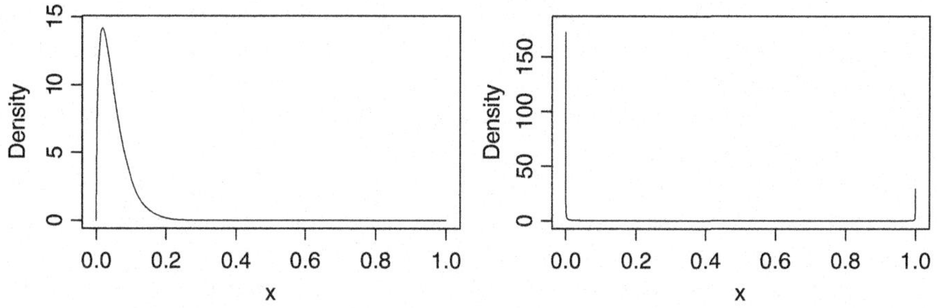

Fig. 4.1 Possible priors for the stent example. Left panel: Beta(1.5,28.5) density. Right panel: Beta(.01,.19) density

population distribution of p's, then experiencing a Bernoulli event during the next 18 months with probability p. Presumably, different people have different risks of having a blood clot once they stop blood thinners leading to the distribution of p's. The mean and variance of the beta-binomial in (4.1, p. 164) with $n = 1$ are $\alpha/(\alpha + \beta)$ and $\alpha\beta/(\alpha + \beta)^2$, respectively. Thus, it makes sense to set $.05 = \alpha/(\alpha + \beta)$ or $\beta = 19\alpha$, but it is not clear how to set their actual values because $(\alpha, \beta)^T$ is not estimable from a sample of beta-binomial random variables, each with $n = 1$. Fortunately, this is not required here. Using (4.1, p. 164) with $Y = 1$ and $n = 1$, the prior predicted probability that a person will have a blood clot is

$$P(Y = 1) = \frac{B(1 + \alpha, \beta)}{B(\alpha, \beta)} = \frac{\alpha}{\alpha + \beta} = .05$$

regardless of the actual values of α and β. Here we have used the fact that $\Gamma(\alpha + 1) = \alpha\Gamma(\alpha)$. Thus, the patient should evaluate his risk at 1 in 20 if he stops taking blood thinners.

We actually know one person who stopped the blood thinners for a colonoscopy and had a blood clot during the procedure. He was restented and his dilemma now is the same: should he stop blood thinners in the future for an important surgery? His situation is a bit different, though, because he actually has data, $Y_1 = 1$ for $n = 1$, leading to his personal posterior beta$(1 + \alpha, \beta)$ and forecast density

$$P(Y_2 = 1 \mid Y_1 = 1) = \frac{B(2 + \alpha, \beta)}{B(1 + \alpha, \beta)} = \frac{\alpha + 1}{\alpha + \beta + 1}.$$

Substituting $\beta = 19\alpha$, this probability is $(\alpha + 1)/(20\alpha + 1)$. Figure 4.1 shows two possibilities, $\alpha = 1.5$ on the left and $\alpha = .01$ on the right. With $\alpha = 1.5$, the predicted probability of an event is only .08, barely up from .05 because the prior is unimodal above .05. However, with $\alpha = .01$ the prior is bimodal putting .95 of the mass near 0 and .05 near 1. Then, his risk is $(.01+1)/(20(.01)+1) = .84$, and with $\alpha = .001$ it is .98. These bimodal priors reflect a belief that there are essentially two types of individuals, those who do not need blood thinners beyond 6 months and those that do. And for those that do need blood thinners, the risk is almost 1 if they stop taking blood thinners.

Thus, we have a situation that a person without any data has a 1 in 20 risk, but a person who has already experienced a blood clot, the risk could be anywhere between .05 and 1. In other words, without further information about the shape of the population of p's, the second person cannot really evaluate his risk. ♦

Example 4.2 (Normal(θ, σ_Y^2)). Suppose that $Y_1, \ldots, Y_n$ are iid $N(\theta, \sigma_Y^2)$, where σ_Y^2 is known, and the prior density for θ is $N(\mu_0, \sigma_0^2)$, where μ_0 and σ_0^2 are known. Straightforward calculations (Problem 4.1) reveal that the posterior of θ given $\boldsymbol{Y}$ (or equivalently, given the sufficient statistic $\overline{Y}$) is normal with mean

$$\frac{\tau_0\mu_0 + \tau_n\overline{Y}}{\tau_0 + \tau_n} \tag{4.4}$$

and variance $(\tau_0+\tau_n)^{-1}$, where $\tau_0 = 1/\sigma_0^2$ is the prior *precision* of θ and $\tau_n = n/\sigma_Y^2$ is the *precision* of $\overline{Y}$. Thus, the posterior mean is again a weighted average of the prior mean and the unbiased estimator.

A $1-\alpha$ credible interval or region is a set of θ values that has posterior probability $1-\alpha$, much like a confidence interval or region. For the above normal posterior, the 95% credible interval for θ is

$$\left(\frac{\tau_0\mu + \tau_n\overline{Y}}{\tau_0 + \tau_n} - \frac{1.96}{(\tau_0 + \tau_n)^{1/2}}, \frac{\tau_0\mu + \tau_n\overline{Y}}{\tau_0 + \tau_n} + \frac{1.96}{(\tau_0 + \tau_n)^{1/2}}\right). \tag{4.5}$$

One advantage of these intervals is that people can say naturally "I believe with probability .95 that θ lies in the interval." Frequentists, on the other hand, need to explain that 95% confidence intervals are intervals such that in repeated sampling, 95% of them on average contain the true θ.

The scope of application for this simple normal example is quite large if one is willing to ignore the difference between estimating variability and assuming it known. For example, we know maximum likelihood estimators are approximately normal with asymptotic variances given by the elements of the inverse of the total information matrix. Thus, suppose that $\boldsymbol{\theta}$ is $b \times 1$ but $\widehat{\theta}_1$ is a single parameter of interest with variance estimated by $\widehat{\sigma}_1^2 = \left\{\boldsymbol{I}_{\mathrm{T}}(\widehat{\boldsymbol{\theta}})^{-1}\right\}_{11}$. Then, using a $N(\mu y_{01}, \sigma_{01}^2)$ prior for θ_1, the posterior is approximately normal with mean

$$\frac{\mu_{01}/\sigma_{01}^2 + \overline{Y}/\widehat{\sigma}_1^2}{1/\sigma_{01}^2 + 1/\widehat{\sigma}_1^2}$$

and variance $\left(1/\sigma_{01}^2 + 1/\widehat{\sigma}_1^2\right)^{-1}$. ♦

A nice feature of independent data is that one can sequentially update the prior for each additional datum, or all at once. For example, suppose $Y_1, \ldots, Y_n$ are independent with respective densities $f_i(y_i \mid \boldsymbol{\theta}), i = 1, \ldots, n$. Then, the posterior is

$$\pi(\boldsymbol{\theta} \mid \boldsymbol{Y}) = \frac{\pi(\boldsymbol{\theta}) \prod_{i=1}^{n} f_i(Y_i \mid \boldsymbol{\theta})}{\int \pi(\boldsymbol{\theta}) \prod_{i=1}^{n} f_i(Y_i \mid \boldsymbol{\theta})\, d\boldsymbol{\theta}}$$

$$= \frac{\pi(\boldsymbol{\theta} \mid Y_1) \prod_{i=2}^{n} f_i(Y_i \mid \boldsymbol{\theta})}{\int \pi(\boldsymbol{\theta} \mid Y_1) \prod_{i=2}^{n} f_i(Y_i \mid \boldsymbol{\theta})\, d\boldsymbol{\theta}}$$

$$= \frac{\pi(\boldsymbol{\theta} \mid Y_1, \ldots Y_k) \prod_{i=k+1}^{n} f_i(Y_i \mid \boldsymbol{\theta})}{\int \pi(\boldsymbol{\theta} \mid Y_1, \ldots, Y_k) \prod_{i=k+1}^{n} f_i(Y_i \mid \boldsymbol{\theta})\, d\boldsymbol{\theta}}. \tag{4.6}$$

Here, $\pi(\boldsymbol{\theta} \mid Y_1)$ is the posterior from using the prior and Y_1. It then is used as the prior for the remaining data. In the simple normal example above, the intermediate posterior after observing the first k Y's has mean

$$\frac{\tau_0 \mu_0 + k\overline{Y}_k/\sigma_0^2}{\tau_0 + k/\sigma_0^2},$$

where $\overline{Y}_k = k^{-1} \sum_{i=1}^{k} Y_i$.

Sufficient statistics often make calculations easier. Because of the factorization theorem, if a sufficient statistic for $\boldsymbol{\theta}$ exists, then the posterior depends on the data only through the sufficient statistic. This is clear in the simple normal example, where the posterior depends only on $\overline{Y}$. In addition, one can replace the likelihood in Bayesian analysis by the likelihood based only on the sufficient statistic. For the simple normal example, we could use the $\mathrm{N}(\theta, \sigma_0^2/n)$ density of $\overline{Y}$ in place of the full likelihood.

There are large philosophical differences between frequentist and Bayesian approaches. The frequentist uses the sampling distribution of statistics to define estimators, confidence regions, and hypothesis tests. The Bayesian relies on the posterior density for inference. However, the details of the difference in approaches often reduces to how nuisance parameters are handled in each approach.

Suppose that $\boldsymbol{\theta}$ is partitioned into $(\boldsymbol{\theta}_1^T, \boldsymbol{\theta}_2^T)^T$, where $\boldsymbol{\theta}_1$ is the parameter of interest and $\boldsymbol{\theta}_2$ is often called a nuisance parameter. The Bayesian approach is to integrate $\boldsymbol{\theta}_2$ out of the posterior density, yielding the marginal posterior density for $\boldsymbol{\theta}_1$. The frequentist uses a variety of techniques to deal with $\boldsymbol{\theta}_2$: (i) use a conditional or marginal likelihood that does depend on $\boldsymbol{\theta}_2$; (ii) plug in the maximum likelihood estimator for $\boldsymbol{\theta}_2$ and try to handle the resulting sampling distributions; (iii) use pivotal quantities like the t statistic; or (iv) take the maximum value of p-values over the nuisance parameter space. The frequentist approaches often seem ad hoc compared to the Bayesian approach. However, assigning meaningful priors to the nuisance parameters is often much harder than for the parameters of interest. Integration to get the marginal posterior might be daunting, but computing strategies are available. Thus, there are tradeoffs in both Bayesian and frequentist approaches, and neither is uniformly easier to use than the other. In this chapter we try to give an overview of the main ideas of Bayesian inference, and include comparisons to frequentist procedures when appropriate.

4.2 Bayes Estimators

Although $\boldsymbol{\theta}$ is random in Bayesian analysis, once Nature has chosen a particular value of $\boldsymbol{\theta}$ and data $\boldsymbol{Y}$ is generated, then $\boldsymbol{\theta}$ is fixed, and it makes sense to define point estimators of $\boldsymbol{\theta}$. Bayes estimators are straightforward minimizers of Bayes risk, which we now define after first defining risk.

A loss function $L(\boldsymbol{\theta}, \boldsymbol{\delta}(\boldsymbol{Y}))$ is a nonnegative function of the true parameter value $\boldsymbol{\theta}$ and an estimator $\boldsymbol{\delta}(\boldsymbol{Y})$. The most common loss function is squared error, $L(\boldsymbol{\theta}, \boldsymbol{\delta}(\boldsymbol{Y})) = \sum\{\theta_i - \delta_i(\boldsymbol{Y})\}^2$. The risk is the expectation of $L(\boldsymbol{\theta}, \boldsymbol{\delta}(\boldsymbol{Y}))$ with respect to the conditional distribution of $\boldsymbol{Y}$ given $\boldsymbol{\theta}$,

$$R(\boldsymbol{\theta}, \boldsymbol{\delta}) = \int L(\boldsymbol{\theta}, \boldsymbol{\delta}(\boldsymbol{y})) f(\boldsymbol{y} \mid \boldsymbol{\theta})\, d\boldsymbol{y}.$$

At this point there are several standard frequentist approaches. The first approach is to restrict the class of estimators, say to unbiased estimators, and then to minimize $R(\boldsymbol{\theta}, \boldsymbol{\delta})$ over the restricted class. In the case of unbiased estimators and squared error loss, the resulting optimal estimator, if it exists, is called the *minimum variance unbiased estimator* (MVUE). The second general frequentist approach is to maximize the risk $R(\boldsymbol{\theta}, \boldsymbol{\delta})$ over $\boldsymbol{\theta}$ and then seek an estimator that minimizes the maximum risk. Estimators $\boldsymbol{\delta}_{mm}$ that satisfy

$$\inf_{\boldsymbol{\delta}} \sup_{\boldsymbol{\theta}} R(\boldsymbol{\theta}, \boldsymbol{\delta}) = \sup_{\boldsymbol{\theta}} R(\boldsymbol{\theta}, \boldsymbol{\delta}_{mm}),$$

are called *minimax* estimators.

Instead of maximizing the risk, the Bayes approach is to average the risk over $\boldsymbol{\theta}$ leading to the Bayes risk

$$R_{\text{Bayes}}(\pi, \boldsymbol{\delta}) = \int R(\boldsymbol{\theta}, \boldsymbol{\delta}) \pi(\boldsymbol{\theta})\, d\boldsymbol{\theta}. \tag{4.7}$$

An estimator δ_{Bayes} that minimizes $R_{\text{Bayes}}(\pi, \delta)$ is called a Bayes estimator. For squared error loss, the Bayes estimator is the posterior mean $\int \theta\, \pi(\theta \mid y)\, d\theta$. For absolute error loss in the case of a scalar parameter, $L(\theta, \delta(y)) = |\theta - \delta(y)|$, the Bayes estimator is a median of the posterior $\pi(\theta \mid y)$ and is not necessarily unique.

Some authors (e.g., Robert, 2001, p. 62–63) call (4.7) the *integrated risk* and its value at the minimizer δ_{min}, $R_{\text{Bayes}}(\pi, \delta_{min})$, the Bayes risk. If this latter quantity is finite, then the Bayes estimator also minimizes at each y the *posterior expected loss*

$$\int L(\theta, \delta(y))\pi(\theta \mid y)\, d\theta.$$

Bayes estimators with proper priors are generally not unbiased in the frequentist sense. The weighted means of Examples 4.1 (p. 164) and 4.2 (p. 168) are clearly biased. However, they typically have good risk behavior in the frequentist sense. Interestingly, a key technique for finding minimax estimators starts with a Bayes estimator (see Lehmann and Casella, 1998, Ch. 5). Moreover, an *admissible estimator* (an estimator not uniformly larger in risk compared to any other estimator) must be a Bayes estimator or the limit of Bayes estimators. Thus, Bayes estimators are not only good in terms of Bayes risk but are often of interest to frequentists willing to sacrifice unbiasedness.

4.3 Credible Intervals

A $1 - \alpha$ credible interval or region is just a region of θ values with posterior probability $1 - \alpha$. When the region volume is minimized, the region is called a highest posterior density (HPD) region. For example, the normal interval (4.5, p. 168) is a 95% HPD interval due to the symmetry and unimodality of the normal distribution. Because Monte Carlo methods (Section 4.8, p. 193) are often used to estimate posteriors, simple equal tail area intervals are the most commonly used type of credible intervals. That is, given a set of N replications from the end of a Markov chain Monte Carlo run, one just orders the values, and the empirical .025 and .975 quantiles are the endpoints of the 95% credible interval.

Often credible intervals are called Bayesian confidence intervals. And in some cases they correspond exactly to frequentist confidence intervals such as the usual t interval for the case of a single normal mean with unknown variance and improper priors (see the end of Section 4.6.1, p. 176). In general, however, credible intervals do not have the frequentist property of including the true parameter with probability $1-\alpha$ when repeated sampling with the same θ value. Rather, for given data $Y = y$, they have probability $1-\alpha$ of containing the true parameter value. As an illustration, Agresti and Min (2005) investigated the frequentist coverage of Bayesian credible intervals for $p_1 - p_2$, p_1/p_2, and odds ratios for two independent binomials. They found that the frequentist coverage of the Bayesian intervals was not as good as the best of the frequentist intervals.

Recall that even though θ is random, when Nature chooses a particular θ_0 and generates data $\boldsymbol{Y}$ from $f(y \mid \theta_0)$, for that data $\boldsymbol{Y}$, θ is fixed at θ_0 and at least Nature knows whether θ_0 lies in the interval. In fact, on the computer we can play the role of Nature and check the validity of Bayesian specifications by generating $\theta_k, k = 1, \ldots, K$, from a continuous prior $\pi(\theta)$ and for each θ_k, generate data $\boldsymbol{Y}_k$, form the posterior and credible interval and see if the proportion of intervals that contain the associated θ_i is close to $1 - \alpha$. A related approach is to compute

$$H_k = \int_{-\infty}^{\theta_k} \pi(t \mid \boldsymbol{Y}_k)dt, \quad k = 1, \ldots, K.$$

By the probability integral transformation, H_k should be a uniform random variable, and $H_1, \ldots, H_K$ can be checked for uniformity using a statistic like the Anderson-Darling goodness-of-fit statistic. Lack of uniformity suggests the presence of mistakes in the mathematics or the computations. Examples may be found in Monahan and Boos (1992). Cook et al. (2006) give a similar proposal for testing validity of Bayesian posteriors and computations.

4.4 Conjugate Priors

When the data $\boldsymbol{Y}$ has density $f(\boldsymbol{y} \mid \boldsymbol{\theta})$ and the prior and posterior are from the same family of densities, we say that the prior is *conjugate*. More specifically, if the conjugate prior has density $\pi(\boldsymbol{\theta} \mid \boldsymbol{\gamma}_{\text{prior}})$, where $\boldsymbol{\gamma}_{\text{prior}}$ is a vector of fixed hyperparameters, then the posterior density has the form $\pi(\boldsymbol{\theta} \mid \boldsymbol{\gamma}_{\text{post}})$, where the updated hyperparameter $\boldsymbol{\gamma}_{\text{post}}$ is a known function of $\boldsymbol{\gamma}_{\text{prior}}$ and the data $\boldsymbol{Y}$.

In Examples 4.1 (p. 164) and 4.2 (p. 168) we have already seen conjugate priors. The beta(α, β) prior for binomial(n, p) data Y leads to the beta$(Y + \alpha, n - Y + \beta)$ posterior. The $\text{N}(\mu_0, \sigma_0^2)$ prior for a sample of $\text{N}(\theta, \sigma_Y^2)$ data leads to a $\text{N}\{\tau_0\mu_0 + \tau_n\overline{Y})/(\tau_0 + \tau_n), 1/(\tau_0 + \tau_n)\}$ posterior, where $\tau_0 = 1/\sigma_0^2$ and $\tau_n = n/\sigma_Y^2$. Other examples illustrated in problems are: a beta(α, β) prior is conjugate for the negative binomial family; a gamma prior is conjugate for Poisson data and also for gamma data; a Pareto prior is conjugate for uniform$(0,\theta)$ data.

The examples mentioned above are for one real θ. An interesting vector case is the multinomial$(n; p_1, \ldots, p_k)$. Here the conjugate prior is a Dirichlet$(\alpha_1, \ldots, \alpha_k)$ distribution with density

$$\pi(p_1, \ldots, p_k \mid \alpha_1, \ldots, \alpha_k) = \frac{1}{B(\boldsymbol{\alpha})} \prod_{i=1}^{k} p_i^{\alpha_i - 1}, \tag{4.8}$$

where $B(\boldsymbol{\alpha}) = \prod_{i=1}^{k} \Gamma(\alpha_i)/\Gamma(\sum_{i=1}^{k} \alpha_i)$. Recall that the likelihood for multinomial data $(N_1, \ldots, N_k)$ is proportional to $\prod_{i=1}^{k} p_i^{N_i}$. Thus the posterior must be proportional to

$$\prod_{i=1}^{k} p_i^{N_i} \prod_{i=1}^{k} p_i^{\alpha_i - 1} = \prod_{i=1}^{k} p_i^{\alpha_i + N_i - 1},$$

which we recognize as Dirichlet$(\alpha_1 + N_1, \ldots, \alpha_k + N_k)$.

DeGroot (1970, Sect. 9.3) shows that any data density that has a sufficient statistic of fixed dimension for all $n \geq 1$ has a conjugate prior. The most useful densities of this type belong to the exponential family. For simplicity we use the natural parameter version given in (2.60, p. 99)

$$f(\boldsymbol{y};\boldsymbol{\eta}) = h(\boldsymbol{y}) \exp\left\{\sum_{i=1}^{s} \eta_i T_i(\boldsymbol{y}) - A(\boldsymbol{\eta})\right\}. \tag{4.9}$$

The conjugate prior has density

$$\pi(\boldsymbol{\eta} \mid \boldsymbol{\gamma}, \lambda) = K(\boldsymbol{\gamma}, \lambda) \exp\left\{\sum_{i=1}^{s} \gamma_i \eta_i - \lambda A(\boldsymbol{\eta})\right\}. \tag{4.10}$$

The posterior is $\pi(\boldsymbol{\eta} \mid \boldsymbol{\gamma} + \boldsymbol{T}(\boldsymbol{Y}), \lambda + 1)$ as can be seen by multiplying likelihood and prior

$$\begin{aligned} h(\boldsymbol{Y}) \exp&\left\{\sum_{i=1}^{s} \eta_i T_i(\boldsymbol{Y}) - A(\boldsymbol{\eta})\right\} K(\boldsymbol{\gamma}, \lambda) \exp\left\{\sum_{i=1}^{s} \gamma_i \eta_i - \lambda A(\boldsymbol{\eta})\right\} \\ \propto \quad \exp&\left\{\sum_{i=1}^{s} \eta_i (T_i(\boldsymbol{Y}) + \gamma_i) - (\lambda + 1) A(\boldsymbol{\eta})\right\}. \end{aligned}$$

This natural exponential family form is often not the easiest (or most "natural") form to work with. The binomial density $\binom{n}{y} p^y (1-p)^{n-y}$ is an exponential family density with $h(y) = \binom{n}{y}$, $T(y) = y$, $\eta = \log\{p/(1-p)\}$, and $A(\eta) = n \log(1+e^\eta)$. From (4.10) the conjugate prior is

$$\pi(\eta \mid \gamma, \lambda) \propto \exp\{\gamma\eta - \lambda n \log(1 + e^\eta)\} = \frac{e^{\gamma\eta}}{(1 + e^\eta)^{\lambda n}}.$$

Using the change-of-variables formula to transform back to the density of $p = e^\eta(1 + e^\eta)^{-1}$, we find $\pi(p \mid \gamma, \lambda)$ is beta($\alpha = \gamma + 2, \beta = \lambda n + 2 - \gamma$). The update from prior $\pi(\eta \mid \gamma, \lambda)$ to posterior $\pi(\eta \mid \gamma + Y, \lambda + 1)$ is parallel to the usual update of beta(α, β) to beta($\alpha + Y, n - Y + \beta$) of Example 4.1 (p. 164), but the latter is much simpler to use.

As mentioned in the chapter Introduction use of conjugate priors are not as important as in the past, but they are straightforward. Moreover, one might argue philosophically that updating the prior with data should not change the form of the density but just change the hyperparameters. More examples of conjugate priors for normal data appear in Section 4.6 (p. 176).

4.5 Noninformative Priors

The search for truly noninformative priors has been long and extensive, but it does not seem possible to have a general rule for specifying ignorance about $\boldsymbol{\theta}$ in the prior. Kass and Wasserman (1996) give an extensive discussion, and here we give briefly a few ideas from that paper, but most of the ideas originate with Jeffreys (1961).

Consider first the case of a discrete finite set for θ, say θ can only take the values $1, 2, 3$. Then it seems quite plausible that a non-informative prior give equal weight to each value, $\pi(\theta) = 1/3$ for $\theta = 1, 2, 3$. One criticism is for the more general case of events A, B, C where $C = C_1 \cup C_2$. Then, we might specify $P(A) = P(B) = P(C) = 1/3$ or $P(A) = P(B) = P(C_1) = P(C_2) = 1/4$. So there is not a clear definition of how to proceed in such cases.

For the case of a location parameter μ taking values on the whole real line $(-\infty, \infty)$, the improper prior $\pi(\mu) = 1$ giving equal weight to all values can be justified on a variety of grounds. An improper prior does not have a finite integral. However, there seems to be no philosophical problem with improper priors as long as they lead to proper posteriors. From Example 4.2 (p. 168), the posterior using $\pi(\mu) = 1$ (equivalent to taking $\mu_0 = 0$ and $\sigma_0 = \infty$) is normal with mean $\overline{Y}$ and variance σ_Y^2/n.

For the case of a scale parameter σ taking values in $(0, \infty)$, Jeffreys suggests $\pi(\sigma) = 1/\sigma$. His invariance argument is that any power transformation of σ, say $\gamma = \sigma^a$, has via a change-of-variables, the improper density

$$\pi(\gamma) = \frac{1}{\gamma^{1/a}} \left| \frac{\gamma^{1/a-1}}{a} \right| = \frac{1}{a\gamma}, \quad 0 < \gamma < \infty,$$

which is similar in form to $1/\sigma$.

Combining these last two improper priors, a location-scale family with $(\mu, \sigma) \in (-\infty, \infty) \times (0, \infty)$, suggests using the improper prior V

$$\pi(\mu, \sigma) \propto \frac{1}{\sigma}. \tag{4.11}$$

Now, moving to the case of general parameters on continuous parameter spaces, let us first consider the binomial(n, p) case. It seems quite natural to think first of $\pi(p) = 1$, the uniform prior on $(0, 1)$. However, transformations of the parameter no longer have the uniform property. For example, the odds parameter $\theta = p/(1-p)$ has density

$$\pi(\theta) = 1 \left| \frac{d(\theta/(\theta+1))}{d\theta} \right| = \frac{1}{(1+\theta)^2}, \quad 0 < \theta < \infty.$$

For the one-parameter case, Jeffreys (1961) proposed the general noninformative prior for θ as

$$\pi(\theta) \propto I(\theta)^{1/2}, \tag{4.12}$$

where $I(\theta)$ is the usual information number. Recall from (2.50, p. 79) that the transformed parameter $\gamma = g(\theta)$ has information

$$I(\gamma) = \frac{1}{\{g'(\theta)\}^2} I(\theta) = \frac{1}{\{g'(g^{-1}(\gamma))\}^2} I(g^{-1}(\gamma)). \tag{4.13}$$

Now notice that using the change-of-variable formula for densities with $\gamma = g(\theta)$ applied to (4.12) yields

$$\pi(\gamma) \propto \frac{1}{|g'(g^{-1}(\gamma))|} I(g^{-1}(\gamma))^{1/2} = I(\gamma)^{1/2},$$

the same as the square root of (4.13). Thus, (4.12) leads to the same form in terms of information regardless of parameterization.

In the binomial case, $I(p) = n/\{p(1-p)\}$, and thus Jeffreys's prior is proportional to $\{p(1-p)\}^{-1/2}$, a beta$(1/2, 1/2)$ density. The odds parameter $\gamma = p/(1-p)$ has Jeffreys's prior

$$\pi(\theta) \propto \frac{1}{\gamma^{1/2}(1+\gamma)}, \quad 0 < \theta < \infty. \tag{4.14}$$

Further transforming to the log odds, $\eta = \log\{p/(1-p)\}$, gives

$$\pi(\eta) \propto \frac{e^{\eta/2}}{1+e^{\eta}}, \quad -\infty < \eta < \infty. \tag{4.15}$$

Moving to the multiparameter case, Jeffreys gives

$$\pi(\boldsymbol{\theta}) \propto |\boldsymbol{I}(\boldsymbol{\theta})|^{1/2}, \tag{4.16}$$

where $|\boldsymbol{I}(\boldsymbol{\theta})|$ is the determinant of $\boldsymbol{I}(\boldsymbol{\theta})$. As an example, consider the multinomial $(n; p_1, \ldots, p_k)$ distribution where $\sum_{i=1}^k p_i = 1$. Recall from Chapter 2 Example 2.13 (p. 70) that $\boldsymbol{I}_{\mathrm{T}}(\boldsymbol{p}) = n\{\mathrm{diag}(1/p_1, \ldots, 1/p_{k-1}) + \boldsymbol{1}\boldsymbol{1}^T/p_k\}$ leading to

$$|\boldsymbol{I}_{\mathrm{T}}(\boldsymbol{p})| = \frac{n}{p_1 p_2 \cdots p_k},$$

and Jeffreys's prior

$$\pi(\boldsymbol{p}) \propto \frac{1}{\sqrt{p_1 p_2 \cdots p_k}},$$

a Dirichlet$(1/2, 1/2, \ldots, 1/2)$ distribution. Recall that the Dirichlet distribution is the conjugate prior for the multinomial distribution.

For the $N(\mu, \sigma^2)$ model with $\theta = (\mu, \sigma)$, the information matrix is Diag $(\sigma^{-2}, 2\sigma^{-2})$ and Jeffreys's prior would be $\pi(\mu, \sigma) \propto \sigma^{-2}$ in contrast to σ^{-1} of (4.11, p. 174). Thus, Jeffreys modified his original proposal (4.16) in the presence of location parameters, say $\mu_1, \ldots, \mu_k$, to

$$\pi(\mu_1, \ldots, \mu_k, \boldsymbol{\theta}) \propto |\boldsymbol{I}(\boldsymbol{\theta})|^{1/2}, \tag{4.17}$$

where $\boldsymbol{I}(\boldsymbol{\theta})$ is the information matrix with $\mu_1, \ldots, \mu_k$ held fixed (treated as known). With this modification, Jeffreys's prior for the $N(\mu, \sigma^2)$ case is σ^{-1}. Moreover, the extension to the normal linear model would yield the same prior.

Many noninformative priors are improper. Often, they result in proper and useful posteriors. But in multiparameter problems, it is not as easy to assess the effect of using improper priors. As an alternative, very diffuse proper priors are often used in practice. But such priors, as they approximate improper priors in a limiting sense, sometimes share problems similar to those of improper priors. Kass and Wasserman (1996, p. 1358) conclude after reviewing a number of examples: "The message from this and similar examples is that improper priors must be used with care when the dimension of the parameter space is large."

There are many other approaches for obtaining noninformative priors as surveyed in Kass and Wasserman (1996). However, they conclude that Jeffreys's priors, (4.11, p. 174), (4.12, p. 175), (4.16, p. 175), and (4.17, p. 176) are the "default among the defaults" for noninformative priors.

4.6 Normal Data Examples

4.6.1 One Normal Sample with Unknown Mean and Variance

In Example 4.2 (p. 168) we made the unrealistic assumption that the variance of the normal data is known. Here we assume that $Y_1, \ldots, Y_n$ are iid $N(\mu, \sigma^2)$, where both μ and $\tau = 1/\sigma^2$ are unknown, $\boldsymbol{\theta} = (\mu, \tau)^T$. Use of the precision τ in place of σ^2 makes the calculations easier.

The prior is usually given in two steps: $\pi(\mu \mid \tau)$ is $N(\mu_0, (\tau n_0)^{-1})$ and $\pi(\tau)$ is gamma$(\alpha_0, 1/\beta_0)$ with density $\pi(\tau) = \Gamma(\alpha_0)^{-1}\beta_0^{\alpha_0}\tau^{\alpha_0-1}e^{-\tau\beta_0}$. The use of $1/\beta_0$ in the gamma prior instead of β_0 also makes the computations easier. The full prior is then $\pi(\mu, \tau) = \pi(\mu \mid \tau)\pi(\tau)$. Because the gamma prior mean is α_0/β_0, we are roughly thinking that the prior variance is like $(\beta_0/\alpha_0)n_0^{-1}$. Another way to assess the strength of the prior is to note that the marginal prior distribution of μ is t with center μ_0, $scale^2 = (\beta_0/\alpha_0)/n_0$, and degrees of freedom $2\alpha_0$. For further calibration, recall that a t distribution with k degrees of freedom and $scale = \sigma$ has variance $\sigma^2 k/(k-2)$ provided $k > 2$.

The likelihood is

$$(2\pi\sigma^2)^{-n/2}\exp\left[-\frac{1}{2\sigma^2}\left\{\sum(Y_i-\overline{Y})^2+n(\overline{Y}-\mu)^2\right\}\right].$$

Multiplying the likelihood and the prior, squaring and regrouping (Problem 4.10, page 202, asks you to fill in the steps), the joint distribution and posterior have the form

$$\pi(\mu,\tau\mid \boldsymbol{Y})=c_1\tau^{1/2}\exp\left\{-\frac{\tau n'}{2}(\mu-\mu')^2\right\}\tau^{\alpha'-1}e^{-\tau\beta'}, \tag{4.18}$$

where $\mu'=(n_0\mu_0+n\overline{Y})/(n_0+n)$, $n'=n_0+n$, $\alpha'=n/2+\alpha_0$, and

$$\beta'=\beta_0+\frac{\sum(Y_i-\overline{Y})^2}{2}+\frac{1}{2}\left(\frac{n_0n}{n_0+n}\right)(\mu_0-\overline{Y})^2.$$

We recognize that (4.18) is the product of the $\mathrm{N}(\mu',(\tau n')^{-1})$ density and the $\mathrm{Gamma}(\alpha',1/\beta')$ density, and thus the prior is conjugate for the normal family. For inference on μ, we integrate with respect to τ leading to the marginal t posterior for μ,

$$\pi(\mu\mid \boldsymbol{Y})=c(df)scale^{-1}\left\{1+\frac{(\mu-\mu')^2}{(scale)^2df}\right\}^{-\left[\frac{df+1}{2}\right]}, \tag{4.19}$$

where $df=2\alpha'=n+2\alpha_0$, $scale^2=(\beta'/\alpha')/(n_0+n)$, and $c(df)$ is the constant for a t density. The posterior mean is $E(\mu\mid \boldsymbol{Y})=\mu'$ and the posterior variance is $\mathrm{Var}(\mu\mid \boldsymbol{Y})=scale^2\{df/(df-2)\}$ as long as $df>2$.

For forecasting or predicting a new observation Y_{n+1}, the *posterior predictive density* is

$$m(y_{n+1}\mid Y_1,\ldots,Y_n)=\int f(y_{n+1}\mid\theta)\pi(\theta\mid Y_1,\ldots,Y_n)d\theta, \tag{4.20}$$

the marginal density of Y_{n+1} based on the posterior of θ given the data. Starting with the full posterior, multiplying by the density of Y_{n+1}, and integrating with respect to μ and then τ gives that $m(y_{n+1}\mid Y_1,\ldots,Y_n)$ is a t density with mean μ', degrees of freedom $2\alpha'$, and $scale^2=(\beta'/\alpha')(1+1/n')$.

If $\alpha_0\to-1/2$, $\beta_0\to0$, and $n_0\to0$ such that $(n_0/\beta_0)^{1/2}\to1$, then the limit of $\pi(\mu,\tau)$ is proportional to $1/\tau$, an improper prior on $(-\infty,\infty)\times(0,\infty)$. Plugging into the marginal posterior for μ, we find it has a t distribution with mean $\overline{Y}$, $df=n-1$, $\alpha'=(n-1)/2$, $\beta'=\sum(Y_i-\overline{Y})^2/2$, and $scale^2=\sum(Y_i-\overline{Y})^2/\{n(n-1)\}=s^2_{n-1}/n$. Thus, the Bayes estimator for this improper prior is $\overline{Y}$, and the credible interval is identical to the usual t interval of classical frequentist inference.

Summarizing, the marginal posterior for μ using the informative prior has a t distribution with mean $\mu' = (n_0\mu_0 + n\overline{Y})/(n_0 + n)$ and

$$scale^2 = \frac{2\beta_0 + \sum(Y_i - \overline{Y})^2 + \left(\frac{n_0 n}{n_0 + n}\right)(\mu_0 - \overline{Y})^2}{(2\alpha_0 + n)(n_0 + n)}. \tag{4.21}$$

The marginal prior for μ has a t distribution with mean μ_0 and

$$scale^2 = \frac{\beta_0}{\alpha_0 n_0}. \tag{4.22}$$

The marginal posterior for μ using the non-informative improper prior has a t distribution with mean $\overline{Y}$ and

$$scale^2 = \frac{s_{n-1}^2}{n}. \tag{4.23}$$

These latter two $scale^2$ values, (4.22) and (4.23), assess the weight of the prior and data, respectively. Then, (4.21) relates to the combination of those two in the posterior. Interestingly, the posterior $scale$ does not have to be smaller than the $scale$ of either the prior or the data because the term $(\mu_0 - \overline{Y})^2$ can be made arbitrarily large by choosing μ_0 sufficiently far from $\overline{Y}$. This counter-intuitive result contrasts with the scale-known case of Example 4.2 (p. 168) where the posterior variance of μ, $(\tau_0 + \tau_n)^{-1}$, is always smaller than τ_0^{-1} or τ_n^{-1}.

4.6.2 Two Normal Samples

Here we have $X_i, \ldots, X_m$ iid $N(\mu_1, \sigma^2)$ and $Y_1, \ldots, Y_n$ iid $N(\mu_2, \sigma^2)$, where $\boldsymbol{\theta} = (\mu_1, \mu_2, \tau = 1/\sigma^2)^T$ and $\Delta = \mu_1 - \mu_2$ is the parameter of interest. We assume the usual Normal-gamma priors, $\pi(\mu_1 \mid \tau)$ is $N(\mu_{10}, (\tau m_0)^{-1})$, $\pi(\mu_2 \mid \tau)$ is $N(\mu_{20}, (\tau n_0)^{-1})$, and $\pi(\tau)$ is gamma$(\alpha_0, 1/\beta_0)$.

Following the procedure as in the previous section, the joint posterior of $\boldsymbol{\theta}$ given $(\boldsymbol{X}, \boldsymbol{Y})$ is the product of the $N(\mu_1', (\tau m')^{-1})$, $N(\mu_2', (\tau n')^{-1})$, and Gamma$(\alpha', 1/\beta')$ densities, where $\mu_1' = (m_0\mu_{10} + m\overline{X})/(m_0 + m)$, $m' = m_0 + m$, $\mu_2' = (n_0\mu_{20} + m\overline{Y})/(n_0 + n)$, $n' = n_0 + n$, $\alpha' = \alpha_0 + m/2 + n/2$, and V

$$\beta' = \beta_0 + \frac{\sum(X_i - \overline{X})^2}{2} + \frac{1}{2}\left(\frac{m_0 m}{m_0 + m}\right)(\mu_{10} - \overline{X})^2$$
$$+ \frac{\sum(Y_i - \overline{Y})^2}{2} + \frac{1}{2}\left(\frac{n_0 n}{n_0 + n}\right)(\mu_{20} - \overline{Y})^2.$$

Because we really want the posterior of $\Delta = \mu_1 - \mu_2$, we need to make a change of variables, say from (μ_1, μ_2) to (Δ, μ_2) and integrate out μ_2 and then τ. The first step is easy—because μ_1 and μ_2 have conditionally independent normal distributions, their difference is normal with mean $\mu_1' - \mu_2'$ and variance $1/(\tau m') + 1/(\tau n')$. Thus, the posterior of (Δ, τ) given $(\boldsymbol{X}, \boldsymbol{Y})$ is the product of $\mathrm{N}\{\Delta', 1/(\tau m') + 1/(\tau n')\}$ and the Gamma$(\alpha', 1/\beta')$ density. Finally, integrating out τ leads to a marginal t posterior for Δ with center $\Delta' = \mu_1' - \mu_2'$, degrees of freedom $2\alpha'$, and $scale^2 = (\beta'/\alpha')(m' + n')/(m'n')$.

The improper prior $\pi(\boldsymbol{\theta}) = 1/\tau$ obtained by letting $\alpha_0 \to -1, \beta_0 \to 0, m_0 \to 0$, and $n_0 \to 0$ such that $(m_0 n_0)^{1/2}/\beta_0 \to 1$, leads to a marginal t posterior for Δ with mean $\overline{X} - \overline{Y}$, degrees of freedom $m + n - 2$, and $scale^2 = s_p^2(1/m + 1/n)$, where s_p^2 is the usual pooled estimate of variance.

4.6.3 Normal Linear Model

Suppose that $\boldsymbol{Y}$ follows the usual normal linear model with mean $\boldsymbol{X\Delta}$ and covariance matrix $\sigma^2 \boldsymbol{I}_n$, that is,

$$\boldsymbol{Y} = \boldsymbol{X\Delta} + \boldsymbol{e},$$

and $e_1, \ldots, e_n$ are iid $\mathrm{N}(0, \sigma^2)$. The dimension of $\boldsymbol{\Delta}$ is $p \times 1$, and $\boldsymbol{X}$ is an $n \times p$ matrix of constants. We assume again a Normal-gamma prior, $\boldsymbol{\Delta}$ given $\tau = 1/\sigma^2$ is $\mathrm{N}(\Delta_0, \boldsymbol{\Sigma}_0^{-1}/\tau)$, where $\boldsymbol{\Sigma}_0$ is a positive definite matrix, and τ is gamma$(\alpha_0, 1/\beta_0)$. (The use of $\boldsymbol{\Delta}$ as the coefficient parameter is so that we can keep using β_0 as a gamma hyperparameter.)

Following the style of the previous sections, we find the posterior also has the Normal-gamma form, where $\boldsymbol{\Delta}$ has posterior mean

$$\boldsymbol{\Delta}' = (\boldsymbol{X}^T\boldsymbol{X} + \boldsymbol{\Sigma})^{-1}(\boldsymbol{X}^T\boldsymbol{Y} + \boldsymbol{\Sigma}_0\boldsymbol{\Delta}_0)$$

and covariance matrix $(\boldsymbol{X}^T\boldsymbol{X} + \boldsymbol{\Sigma}_0)^{-1}/\tau, \alpha' = \alpha_0 + n/2$, and

$$\beta' = \beta_0 + \frac{1}{2}\left\{\left(\boldsymbol{Y} - \boldsymbol{X\Delta}'\right)^T \boldsymbol{Y} + \left(\boldsymbol{\Delta}_0 - \boldsymbol{\Delta}'\right)^T \boldsymbol{\Sigma}_0\boldsymbol{\Delta}_0\right\}. \tag{4.24}$$

We can also reexpress the posterior mean $\boldsymbol{\Delta}'$ as a weighted average of the least squares estimator $\widehat{\boldsymbol{\Delta}} = (\boldsymbol{X}^T\boldsymbol{X})^{-1}\boldsymbol{X}^T\boldsymbol{Y}$ and the prior mean $\boldsymbol{\Delta}_0$,

$$\boldsymbol{\Delta}' = (\boldsymbol{X}^T\boldsymbol{X} + \boldsymbol{\Sigma}_0)^{-1}(\boldsymbol{X}^T\boldsymbol{X}\widehat{\boldsymbol{\Delta}} + \boldsymbol{\Sigma}_0\boldsymbol{\Delta}_0).$$

Note that if $\boldsymbol{\Delta}_0 = 0$ and $\boldsymbol{\Sigma}_0 = d\boldsymbol{I}_p$, where d is a constant and $\boldsymbol{I}_p$ is the p-dimensional identity matrix, then the posterior mean $\boldsymbol{\Delta}' = (\boldsymbol{X}^T\boldsymbol{X} + d\boldsymbol{I}_p)^{-1}\boldsymbol{X}^T\boldsymbol{Y}$

Table 4.1 Weight Gain of Cotton Rats by Litter Size (from Randolph et al. 1977)

Litter Size (X)	2	3	4	4	4	4	5	5
Weight Gain (Y)	31.1	36.9	41.6	46.1	48.4	48.4	30.1	44.4
Litter Size (X)	5	5	6	6	6	6	6	7
Weight Gain (Y)	46.8	54.0	48.9	50.1	51.2	56.5	68.4	77.1

is the *ridge regression* estimator of Hoerl and Kennard (1970), originally proposed to deal with multicollinearity and near singularity of $\boldsymbol{X}^T\boldsymbol{X}$.

Integrating out τ, we find the marginal posterior of $\boldsymbol{\Delta}$ is multivariate t with center $\boldsymbol{\Delta}'$, $2\alpha' = 2\alpha_0 + n$ degrees of freedom, and scale matrix $(\beta'/\alpha')(\boldsymbol{X}^T\boldsymbol{X} + \boldsymbol{\Sigma}_0)^{-1}$. The p-dimensional multivariate t with k degrees of freedom, center μ and scale matrix $\boldsymbol{S}$, has density

$$c\left\{1 + \frac{1}{k}(\boldsymbol{x} - \mu)^T \boldsymbol{S}^{-1}(\boldsymbol{x} - \mu)\right\}^{-(k+p)/2},$$

where $c = \Gamma(k/2 + p/2)|\boldsymbol{\Sigma}|^{-1}/\left\{\Gamma(k/2)(k\pi)^{p/2}\right\}$.

Suppose that we are interested in only a subset of the $\boldsymbol{\Delta}$ vector, say $\boldsymbol{\Delta}_1$ of dimension p_1, where $\boldsymbol{\Delta} = (\boldsymbol{\Delta}_1^T, \boldsymbol{\Delta}_2^T)^T$. Then the marginal posterior distribution of $\boldsymbol{\Delta}_1$ is the p_1 dimensional multivariate t with $2\alpha_0 + n$ degrees of freedom, center $\boldsymbol{\Delta}_1'$ and scale matrix $(\beta'/\alpha')\left\{(\boldsymbol{X}^T\boldsymbol{X} + \boldsymbol{\Sigma}_0)^{-1}\right\}_{11}$.

If $\boldsymbol{X}^T\boldsymbol{X}$ is nonsingular and we let $\boldsymbol{S} \to \boldsymbol{0}$, $\alpha_0 \to -p/2$, and $\beta_0 \to 0$ such that $\pi(\boldsymbol{\Delta}, \tau) \to 1/\tau$, then $\boldsymbol{\Delta}'$ converges to the least squares estimator $\boldsymbol{X}^T\boldsymbol{X}^{-1}\boldsymbol{X}^T\boldsymbol{Y}$, and the marginal posterior of $\boldsymbol{\Delta}$ is a p-dimensional t with degrees of freedom $n - p$ and scale matrix $s^2\boldsymbol{X}^T\boldsymbol{X}^{-1}$, where $s^2 = (n-p)^{-1}\sum(Y_i - \boldsymbol{x}_i^T\boldsymbol{\Delta}')^2$.

Example 4.3 (Simple linear regression). Randolph et al. (1977) studied energy use in cotton rats and presented a scatter plot of weight gain in the 12 days after birth. Boos and Monahan (1986) extracted the data in Table 4.1 from the scatter plot and used it to illustrate Bayesian analysis using robust regression. Here we use the standard conjugate prior Bayesian analysis for normal data presented above.

Figure 4.2 (p. 181) gives the scatter plot along with the least squares lines, $E(Y \mid X) = 15.6 + 6.8X$ for all 16 data points (solid line), and $E(Y \mid X) = 25.6 + 4.2X$ for the first 14 data points (dashed line). Randolph et al. (1977) reported both lines because they were worried that the last two points were unusual.

Boos and Monahan (1986) used a previous study of cotton rats, Kilgore (1970), to build a normal-gamma informative prior: given τ, $\boldsymbol{\Delta}$ is normal with mean $\boldsymbol{\Delta}_0 = (20.4, 8.01)^T$ and covariance

$$\frac{\boldsymbol{\Sigma}_0^{-1}}{\tau} = \frac{1}{\tau}\begin{pmatrix} 3.39 & -0.56 \\ -0.56 & 0.10 \end{pmatrix},$$

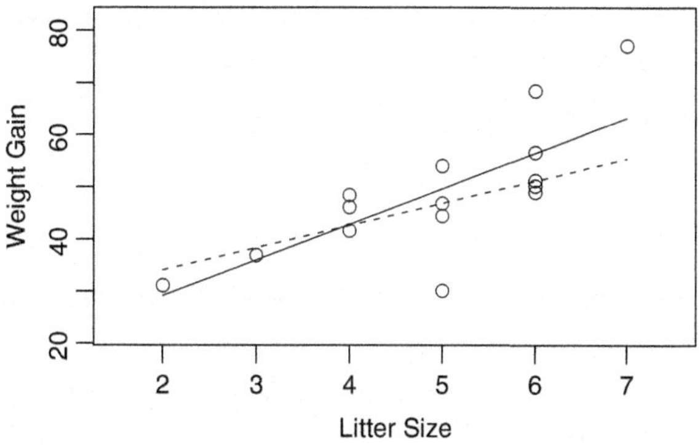

Fig. 4.2 Scatter plot of Table 4.1 (p. 180) data with least squares lines (full data: solid, without last two points: dashed)

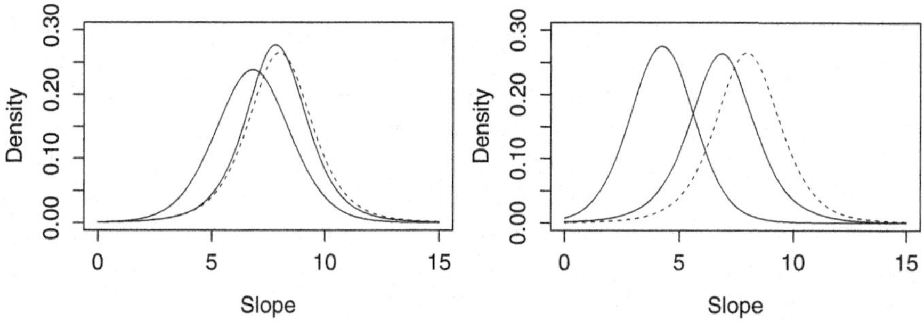

Fig. 4.3 Marginal prior and posterior densities for the slope Δ_2 for the cotton rat data. Left panel is for the full data, and right panel is for data without last two points. Prior density is dashed line. Solid lines are posteriors for noninformative prior (on left of each panel) and for informative prior (center densities)

and τ is gamma$(\alpha_0, 1/\beta_0)$ with $\alpha_0 = 2.5$ and $\beta_0 = 51.3$. Figure 4.3 (p. 181) gives the marginal prior (dashed lines) and two posteriors (solid lines): from the noninformative prior on the left of each graph and from the informative prior in the middle of each graph. The left panel is for the full data, and the right panel is for the data reduced by deleting the last two data points.

The posteriors for the noninformative priors can be viewed as representing the data; thus the right panel has a taller density on the left because deleting the last two points gives a slope estimate with smaller standard error. The posterior for the informative prior combines information from the data and from the prior. Our intuition is that this latter posterior should be narrower and taller than either of the other two densities. This intuition holds in the left panel although the posterior $scale$ is barely smaller than the prior $scale$. In the right panel, the posterior $scale$ is smaller than the noninformative prior posterior but larger than the prior $scale$. This non-intuitive result is due to the fact that in the right panel, the prior mean of 8.01 is far away from the least squares estimate 4.2. It is not as clear from the expression (4.24, p. 179) that the $scale$ might be inflated by a large difference between prior mean and the data, but basically it is a generalization of (4.21, p. 178) for the one sample case. There is an additional factor, the correlation between the slope and the intercept in the prior and also in the data. Although not presented here, we found that changing the correlation in the prior to be smaller also reduces the size of the scale of the posterior for Δ_2. ♦

4.7 Hierarchical Bayes and Empirical Bayes

The hierarchical Bayes approach is to specify the prior distribution $\pi(\boldsymbol{\theta})$ in two stages. The usual prior has parameters $\boldsymbol{\alpha}$ that need to be specified exactly, say $\pi(\boldsymbol{\theta}) = \pi(\boldsymbol{\theta} \mid \boldsymbol{\alpha}_0)$. For example, in the normal problem with known variance (Example 4.2, p. 168), the normal prior for μ has a mean and variance that need to be given. In the unknown normal case (Section 4.6.1, p. 176) there are four parameters $(\mu_0, n_0, \alpha_0, \beta_0)$ to be set, although the normal-gamma prior has a hierarchical flavor because the normal prior for μ is conditional on τ.

In the hierarchical approach, the hyperparameter $\boldsymbol{\alpha}$ is not specified but rather a prior density $h(\boldsymbol{\alpha})$ is given for $\boldsymbol{\alpha}$. Typically, $h(\boldsymbol{\alpha}) = h(\boldsymbol{\alpha} \mid \boldsymbol{\gamma}_0)$ has a hyperparameter $\boldsymbol{\gamma}$ that needs to be given. However, specifying $\boldsymbol{\gamma}_0$ is felt to be more robust and less subjective than specifying $\boldsymbol{\alpha}_0$. In a sense, pushing the choice of specific parameters higher in the hierarchy makes the resulting prior

$$\pi(\boldsymbol{\theta}) = \pi(\boldsymbol{\theta} \mid \boldsymbol{\gamma}_0) = \int \pi(\boldsymbol{\theta} \mid \boldsymbol{\alpha}) h(\boldsymbol{\alpha} \mid \boldsymbol{\gamma}_0)\, d\boldsymbol{\alpha} \tag{4.25}$$

less sensitive to specification. That is, $\pi(\boldsymbol{\theta} \mid \boldsymbol{\gamma}_0)$ is less sensitive to assigning a specific value to $\boldsymbol{\gamma}_0$ via $h(\boldsymbol{\alpha} \mid \boldsymbol{\gamma}_0)$ than $\pi(\boldsymbol{\theta} \mid \boldsymbol{\alpha})$ is to assigning a specific value $\boldsymbol{\alpha}_0$. Note that hierarchical Bayes can be viewed as single prior Bayes with prior $\pi(\boldsymbol{\theta})$ in (4.25). In fact, in simple problems like Example 4.2 (p. 168) with just one parameter of interest, there seems to be little interest in using a hierarchy because finite mixtures of priors can approximate (4.25) easily. The most important use of hierarchical modeling is in more complicated situations like one-way random effects ANOVA and meta-analysis discussed below where the Bayesian approach leads to natural pooling and shrinkage so that estimators of individual effects can borrow strength from the whole data set. The Bayesian approach also leads to parametric empirical Bayes methods where specification of the hyperprior $h(\boldsymbol{\alpha})$ can be avoided.

We can write the posterior in the hierarchical case using (4.25) (but suppressing dependence on $\boldsymbol{\gamma}_0$) as

$$\pi(\boldsymbol{\theta} \mid \boldsymbol{Y}) = \frac{f(\boldsymbol{Y} \mid \boldsymbol{\theta}) \int \pi(\boldsymbol{\theta} \mid \boldsymbol{\alpha}) h(\boldsymbol{\alpha})\, d\boldsymbol{\alpha}}{\int\int f(\boldsymbol{Y} \mid \boldsymbol{\theta}) \pi(\boldsymbol{\theta} \mid \boldsymbol{\alpha}) h(\boldsymbol{\alpha})\, d\boldsymbol{\alpha}\, d\boldsymbol{\theta}}.$$

Within the numerator integral, if we multiply and divide by the marginal $m(y \mid \boldsymbol{\alpha}) = \int f(y \mid \boldsymbol{\theta})\pi(\boldsymbol{\theta} \mid \boldsymbol{\alpha})\, d\boldsymbol{\theta}$, then we are led to another representation of the posterior,

$$\pi(\boldsymbol{\theta} \mid \boldsymbol{Y}) = \int \pi(\boldsymbol{\theta} \mid \boldsymbol{\alpha}, \boldsymbol{Y})\pi(\boldsymbol{\alpha} \mid \boldsymbol{Y})\, d\boldsymbol{\alpha},$$

where

$$\pi(\boldsymbol{\theta} \mid \boldsymbol{\alpha}, \boldsymbol{Y}) = \frac{f(\boldsymbol{Y} \mid \boldsymbol{\theta})\pi(\boldsymbol{\theta} \mid \boldsymbol{\alpha})}{m(\boldsymbol{Y} \mid \boldsymbol{\alpha})} \tag{4.26}$$

and

$$\pi(\boldsymbol{\alpha} \mid \boldsymbol{Y}) = \frac{m(\boldsymbol{Y} \mid \boldsymbol{\alpha})h(\boldsymbol{\alpha})}{\int\int f(\boldsymbol{Y} \mid \boldsymbol{\theta})\pi(\boldsymbol{\theta} \mid \boldsymbol{\alpha})h(\boldsymbol{\alpha})\, d\boldsymbol{\alpha}\, d\boldsymbol{\theta}}.$$

In certain situations, we can estimate $\boldsymbol{\alpha}$ by treating $m(\boldsymbol{Y} \mid \boldsymbol{\alpha})$ as a marginal likelihood. Then, supposing further that $\pi(\boldsymbol{\alpha} \mid \boldsymbol{Y})$ is highly peaked around the maximum likelihood estimator $\widehat{\boldsymbol{\alpha}}$, we have

$$\pi(\boldsymbol{\theta} \mid \boldsymbol{Y}) \approx \pi(\boldsymbol{\theta} \mid \widehat{\boldsymbol{\alpha}}, \boldsymbol{Y}), \tag{4.27}$$

and the mean of this approximate posterior is called an *empirical Bayes estimator*. Usually, the mean of (4.27) is close to the mean of a full posterior with $h(\boldsymbol{\alpha})$ specified, but the variability from (4.27) does not fully incorporate the uncertainty about $\boldsymbol{\alpha}$ as does a full Bayesian approach.

4.7.1 One-Way Normal Random Effects Model

We consider data in a one-way design, $Y_{ij}, i = 1, \ldots, k; j = 1, \ldots, n_i$. The standard frequentist ANOVA models are either fixed effects or random effects. The Bayesian analogue of the fixed effects model is given by the normal linear model (Section 4.6.3, p. 179) with $\boldsymbol{X}^T\boldsymbol{X} = \text{diag}(n_1, \ldots, n_k)$. In that model, the prior covariance $\boldsymbol{\Sigma}_0^{-1}/\tau$ can be used to relate the means to one another, but it does not have the flavor of a random effects model.

A simple Bayesian random effects ANOVA model is to assume that Y_{ij} given θ_i and σ_e^2 has a normal density with mean θ_i and variance σ_e^2. Given $\theta_1, \ldots, \theta_k$ and σ_e^2, the Y_{ij} are all independent. Note the abuse of notation here because $\boldsymbol{\theta}$ of the previous section here includes both $\theta_1, \ldots, \theta_k$ and σ_e^2. Next, we let $\theta_1, \ldots, \theta_k$ given $\boldsymbol{\alpha} = (\mu, \sigma_a)$ be iid from a $N(\mu, \sigma_a^2)$ density, and the hyperparameters have density $h(\mu, \sigma_a)$. The parameter σ_e^2 might also be drawn from a distribution with parameters added to $\boldsymbol{\alpha}$. However, a typical specification is as follows: σ_e^2 has a Jeffreys's noninformative prior $\propto 1/\sigma_e^2$ on $(0, \infty)$, σ_a has a flat noninformative prior $\propto 1$ on $(0, \infty)$, and μ has a flat noninformative prior on $(-\infty, \infty)$. These choices are nontrivial because other seemingly natural noninformative priors can cause the posterior to be improper (see, for example, Hobert and Casella, 1996).

Table 4.2 One-Way Random Effects Data

	Group 1	Group 2	Group 3	Group 4	Group 5	Group 6
	4.9	8.2	8.1	7.4	22.1	5.7
	4.1	4.8	3.0	13.5	15.7	8.1
	9.1	9.0	13.9	13.4	19.3	12.8
			12.6	6.8	19.3	7.6
					11.4	13.9
Means	6.0	7.3	9.4	10.3	17.6	9.6
SD's	2.7	2.2	4.9	3.7	4.1	3.5

Table 4.3 Posterior and Frequentist Estimates for the Table 4.2 Data

	Posterior Summaries					EBLUP & REML		
	mean	sd	2.5%	50%	97.5%	EST	SE	SE2
θ_1	7.0	2.2	2.8	7.1	11.1	7.1	1.9	2.1
θ_2	8.1	2.1	3.8	8.1	12.2	8.1	1.9	2.1
θ_3	9.6	1.8	6.0	9.6	13.2	9.6	1.7	1.9
θ_4	10.2	1.8	6.8	10.2	13.7	10.3	1.7	1.9
θ_5	16.2	2.0	11.7	16.4	19.9	16.2	1.6	1.7
θ_6	9.8	1.6	6.4	9.8	12.9	9.7	1.6	1.7
μ	10.1	2.6	5.0	10.2	14.9	10.2	1.7	1.7
σ_a^2	36.2	103.4	2.3	18.8	152.3	13.1		
σ_e^2	16.2	6.1	8.0	15.0	31.3	14.2		

The posterior results are from `WinBUGS` via `R2WinBUGS`. Monte Carlo standard errors of the entries are in the second decimal place except for σ_a^2, where for example, it is 3.2 for the mean. The SE are estimates of (4.30), and SE2 are bias-corrected versions.

A key focus here is the random effects population $N(\mu, \sigma_a^2)$, which is the prior $\pi(\theta_i \mid \mu, \sigma_a)$, and therefore we focus on the posterior densities of μ and σ_a^2. We also may be interested in the posterior densities of the individual θ_i and forecasts of new individuals from these populations.

To illustrate, Table 4.2 gives data generated from the above model with $\mu = 10$, $\sigma_e = 5$, and $\sigma_a = 4$. We analyzed these data with Gibbs sampling using the `R` package `R2WinBUGS`; computational details are given in Example 4.6 (p. 197). Table 4.3 gives summary statistics for the posterior distributions of all the parameters. Note that the table entries are MCMC estimates (Section 4.8.2, p. 195) and generally have Monte Carlo standard errors in the second decimal place. The posterior mean and median are essentially the same for μ and $\boldsymbol{\theta}$ but differ somewhat for the variance parameters because of their skewed distributions. The posterior means for the individual θ_i are essentially weighted means of the individual sample means given in Table 4.2, and the posterior mean of $\mu = 10.1$.

The frequentist formulation is to let $Y_{ij} = \mu + a_i + e_{ij}$, where the e_{ij} are iid $\mathrm{N}(0, \sigma_e^2)$ and independent of the a_i that are iid $\mathrm{N}(0, \sigma_a^2)$. If these variances were known, then the best linear unbiased estimator (BLUE) of μ would be $\widehat{\mu} = \sum_{i=1}^k w_i \overline{Y}_{i.} / \sum_{i=1}^k w_i$, where $w_i^{-1} = \mathrm{Var}(\overline{Y}_{i.}) = \sigma_a^2 + \sigma_e^2/n_i$. Estimating the variances, say by REML, leads to estimated weights $\widehat{w}_i$ and the empirical best linear unbiased estimator (EBLUE) $\widehat{\mu} = \sum_{i=1}^k \widehat{w}_i \overline{Y}_{i.} / \sum_{i=1}^k \widehat{w}_i$ with estimated variance $1/\sum_{i=1}^k \widehat{w}_i$. In Table 4.3 we see that $\widehat{\mu} = 10.2$ is close to the Bayes posterior mean 10.1, but with smaller standard error (1.7 versus 2.6).

For estimating the random effects, a_i or $\mu + a_i$, best linear unbiased prediction (BLUP) proceeds by noting

$$E(\mu + a_i \mid \boldsymbol{Y}) = \mu + E(a_i \mid \overline{Y}_{i.}) = \mu + \frac{\sigma_a^2}{\sigma_a^2 + \sigma_e^2/n_i}(\overline{Y}_{i.} - \mu).$$

Here we have used joint normality of $(a_i, \overline{Y}_{i.})$ and the standard result that if $(\boldsymbol{X}_1, \boldsymbol{X}_2)^T$ is distributed as $\mathrm{MN}\{(\boldsymbol{\mu}_1, \boldsymbol{\mu}_2)^T, \boldsymbol{\Sigma}\}$, then $(\boldsymbol{X}_1 \mid \boldsymbol{X}_2)$ is distributed as $\mathrm{MN}\{\boldsymbol{\mu}_1 + \boldsymbol{\Sigma}_{12}\boldsymbol{\Sigma}_{22}^{-1}(\boldsymbol{X}_2 - \boldsymbol{\mu}_2), \boldsymbol{\Sigma}_{11} - \boldsymbol{\Sigma}_{12}\boldsymbol{\Sigma}_{22}^{-1}\boldsymbol{\Sigma}_{21}\}$. The conditional expectation above suggests inserting $\widehat{\mu}$ for μ yielding

$$\widehat{\theta}_i^* = \widehat{\mu} + \frac{\sigma_a^2}{\sigma_a^2 + \sigma_e^2/n_i}(\overline{Y}_{i.} - \widehat{\mu}). \tag{4.28}$$

Harville (1976, Theorem 1) verifies that (4.28) is indeed the BLUP for $\theta_i = \mu + a_i$, that is, it minimizes the mean squared error between θ_i and any unbiased estimator that is linear in the Y_{ij}. Finally, inserting estimators for σ_a^2 and σ_e^2, we find the empirical best linear unbiased predictor (EBLUP) of $\theta_i = \mu + a_i$ is

$$\widehat{\theta}_i = (1 - \widehat{B}_i)\overline{Y}_{i.} + \widehat{B}_i\widehat{\mu}, \quad \widehat{B}_i = \frac{\widehat{\sigma}_e^2/n_i}{\widehat{\sigma}_a^2 + \widehat{\sigma}_e^2/n_i}. \tag{4.29}$$

The mean squared error of prediction for the BLUP version in (4.28) is

$$\begin{aligned} E(\widehat{\theta}_i^* - \theta_i)^2 &= (1 - B_i)^2 \frac{\sigma_e^2}{n_i} + B_i^2 \left(\frac{1}{\sum w_i} + \sigma_a^2 - \frac{2 w_i \sigma_a^2}{\sum w_i} \right) \\ &\quad + 2(1 - B_i) B_i \left(\frac{w_i}{\sum w_i} \right) \frac{\sigma_e^2}{n_i} \\ &= \left(\frac{1}{\sigma_a^2} + \frac{n_i}{\sigma_e^2} \right)^{-1} \left\{ 1 + \left(\frac{w_i}{\sum w_i} \right) \frac{(\sigma_e^2/n_i)}{\sigma_a^2} \right\}. \end{aligned} \tag{4.30}$$

Kackar and Harville (1984) and Kenward and Roger (1997) discuss how to adjust this calculation to take into account the estimation of σ_a^2 and σ_e^2; basically they estimate (4.30) plus a bias term from Taylor series expansions.

The last three columns of Table 4.3 contain estimates from (4.29), standard errors from (4.30), $\widehat{\mu}$, and REML estimates of σ_a^2 and σ_e^2 for the data in Table 4.2. These frequentist estimates are close to the Bayesian counterparts except, perhaps, for σ_a^2. The next-to-last column of Table 4.3 contains estimated standard errors based on estimating the mean squared error of prediction in (4.30). These standard errors are all less than the Bayesian posterior standard deviations, reflecting the fact that estimation of σ_a^2 and σ_e^2 is ignored in the mean squared error of prediction calculation. The last column shows the adjusted standard errors of Kackar and Harville (1984) and Kenward and Roger (1997). They are closer to the Bayesian standard deviations.

For the empirical Bayes approach, we go back to the Bayesian formulation but simplify the problem by assuming that σ_e^2 is known. Then $\pi(\boldsymbol{\theta} \mid \boldsymbol{\alpha}, \boldsymbol{Y})$ of (4.26, p. 183) is direct from (4.4, p. 168) of Example 4.2 (p. 168) just a product of k independent normal densities, where the ith density has mean

$$\frac{n_i \overline{Y}_{i.}/\sigma_e^2 + \mu/\sigma_a^2}{n_i/\sigma_e^2 + 1/\sigma_a^2} \tag{4.31}$$

and variance $(n_i/\sigma_e^2 + 1/\sigma_a^2)^{-1}$. The empirical Bayes approach is to approximate the posterior by $\pi(\boldsymbol{\theta} \mid \widehat{\boldsymbol{\alpha}}, \boldsymbol{Y})$ where $\widehat{\boldsymbol{\alpha}}$ are obtained from the marginal likelihood $m(\boldsymbol{y} \mid \boldsymbol{\alpha}) = \int f(\boldsymbol{y} \mid \boldsymbol{\theta})\pi(\boldsymbol{\theta} \mid \boldsymbol{\alpha})\, d\boldsymbol{\theta}$, which is the same likelihood as in the frequentist analysis for estimating $(\mu, \sigma_a^2, \sigma_e^2)$. If we use the same estimation method and insert those estimates in (4.31), then the empirical Bayes estimators of θ_i are exactly the EBLUPS of (4.29). Also, the posterior variances $(n_i/\sigma_e^2 + 1/\sigma_a^2)^{-1}$ are exactly the first term of (4.30) and typically only slightly less than (4.30) because the second term of (4.30) is small unless σ_e^2 is much larger than σ_a^2. For example, compared to the next-to-last column of Table 4.3 (p. 184), the estimated empirical Bayes standard errors are only .03 to .05 less than the EBLUP standard errors.

4.7.2 *James-Stein Estimation*

In a famous paper, James and Stein (1961) proved that the estimator (4.32) of a multivariate normal mean vector dominates the maximum likelihood estimator in terms of mean squared error as long as the number of components is three or greater. Here we briefly describe this result and its empirical Bayes interpretation.

Suppose that $Y_1, \ldots, Y_b$ are independent with Y_i distributed as $N(\theta_i, \sigma_0^2)$, where σ_0^2 is known. In other words, $\boldsymbol{Y}$ is $\text{MN}(\boldsymbol{\theta}, \sigma_0^2 \boldsymbol{I}_b)$, where $\boldsymbol{I}_b$ is the b-dimension identity matrix. The James-Stein estimator is

$$\widehat{\boldsymbol{\theta}}_{\text{JS}} = \left\{1 - (b-2)\sigma_0^2 \left(\sum_{i=1}^{b} Y_i^2\right)^{-1}\right\} \boldsymbol{Y}. \tag{4.32}$$

Note that $\widehat{\boldsymbol{\theta}}_{\mathrm{JS}}$ shrinks $\boldsymbol{Y}$ towards 0. Lehmann and Casella (1998, p. 276) give the risk or expected squared error loss of $\widehat{\boldsymbol{\theta}}_{\mathrm{JS}}$ as

$$\sum_{i=1}^{b} E(\widehat{\theta}_{\mathrm{JS,i}} - \theta_i)^2 = b\sigma_0^2 - (b-2)^2\sigma_0^4\, E\left(\sum_{i=1}^{b} Y_i^2\right)^{-1}, \tag{4.33}$$

which is clearly less than the risk $b\sigma_0^2$ for $\widehat{\boldsymbol{\theta}}_{\mathrm{MLE}} = \boldsymbol{Y}$.

For the empirical Bayes interpretation, suppose that the prior distribution for $\boldsymbol{\theta}$ is $\mathrm{MN}(\mathbf{0}, \sigma_a^2 \boldsymbol{I}_b)$. If σ_a^2 is known, then the Bayes posterior for $\boldsymbol{\theta}|\boldsymbol{Y}$ is normal with mean

$$\left(1 - \frac{\sigma_0^2}{\sigma_0^2 + \sigma_a^2}\right)\boldsymbol{Y}. \tag{4.34}$$

The marginal distribution of $\boldsymbol{Y}$ is $\mathrm{MN}\{\mathbf{0}, (\sigma_a^2 + \sigma_0^2)\boldsymbol{I}_b\}$, and under this marginal distribution, $(b-2)\sigma_0^2/\sum_{i=1}^{b} Y_i^2$ is an unbiased estimator of $\sigma_0^2/(\sigma_0^2 + \sigma_a^2)$. Substituting this estimator in (4.34) leads to (4.32), and thus $\widehat{\boldsymbol{\theta}}_{\mathrm{JS}}$ may be viewed as an empirical Bayes estimator. This empirical Bayes characterization of the James-Stein estimator was originally due to Efron and Morris (1972).

Efron and Morris (1973, p. 126) noted that the prior distribution $\mathrm{MN}(\mathbf{0}, \sigma_a^2 \boldsymbol{I}_b)$ for $\boldsymbol{\theta}$ could be replaced by $\mathrm{MN}(\mu, \sigma_a^2 \boldsymbol{I}_b)$ resulting in the Bayes posterior mean

$$\frac{\sigma_0^2}{\sigma_0^2 + \sigma_a^2}\mu + \left(1 - \frac{\sigma_0^2}{\sigma_0^2 + \sigma_a^2}\right)\boldsymbol{Y}. \tag{4.35}$$

The marginal distribution of $\boldsymbol{Y}$ is now $\mathrm{MN}\{\mu, (\sigma_a^2 + \sigma_0^2)\boldsymbol{I}_b\}$, leading to unbiased estimators $\widehat{\mu} = \overline{Y}$ and $\widehat{B} = (b-3)\sigma_0^2/\sum_{i=1}^{b}(Y_i - \overline{Y})^2$ for $B = \sigma_0^2/(\sigma_0^2 + \sigma_a^2)$. Substituting these estimators into (4.35) yields

$$\widehat{B}\overline{Y} + (1 - \widehat{B})\boldsymbol{Y}, \tag{4.36}$$

very similar to (4.29, p. 185), each component of $\boldsymbol{Y}$ is shrunk towards the sample mean $\overline{Y}$. The estimator (4.36) is also called a James-Stein estimator, and it can be shown to have expected loss similar to (4.33) that is less than $\boldsymbol{Y}$ as long as $b \geq 4$ (Lehmann and Casella, 1998, p. 367). A detailed account of James-Stein estimation and related minimax results may be found in Chapters 4 and 5 of Lehmann and Casella (1998).

4.7.3 Meta-Analysis Applications of Hierarchical and Empirical Bayes

Meta-analysis refer to statistical analysis of a group of studies related to the same basic question of interest. For example, one might be interested in the effect of zinc lozenges on preventing colds. There might be k studies that provide estimates of the treatment effect, for example, the difference between success rates or odds ratio of people taking zinc lozenges versus those taking a placebo. Each study would also provide a standard error for the estimate.

The hierarchical Bayes framework for use in meta-analysis is often very similar to the random effects model of the last section. We have the results of k studies, each one providing information on effect parameters θ_i for the ith study, $i = 1, \ldots, k$. The data from the ith study is Y_i, and $Y_i \mid \theta_i$ has density $f_i(y_i \mid \theta_i)$, and all the Y_i's are independent given $\theta_1, \ldots, \theta_k$. Compared to the previous section, here Y_i plays the role of $\overline{Y}_{i.}$.

We assume that the θ_i are drawn from the density $\pi(\theta \mid \boldsymbol{\alpha})$, where $\boldsymbol{\alpha}$ is unknown and comes from a prior density $h(\boldsymbol{\alpha})$. Note that in this formulation we do not have any nuisance parameters to deal with in the distribution of $Y_i \mid \theta_i$; for example, unknown variances in normal problems. Typically, one just assumes that these variance parameters are estimated well-enough to be considered known. In the binomial models discussed below, however, there is a nuisance control group probability that is given a prior distribution. Ignoring any first stage nuisance parameters, the three levels of the model are

$$f(\boldsymbol{y} \mid \boldsymbol{\theta}) = f(y_1, \ldots, y_k \mid \theta_1, \ldots, \theta_k) = \prod_{i=1}^{k} f(y_i \mid \theta_i),$$

$$\pi(\boldsymbol{\theta} \mid \alpha) = \pi(\theta_1, \ \ldots, \theta_k \mid \boldsymbol{\alpha}) = \prod_{i=1}^{k} \pi(\theta_i \mid \boldsymbol{\alpha}),$$

$$h(\boldsymbol{\alpha}).$$

The second and third stages of the hierarchy thus provide a prior for $\boldsymbol{\theta}$ of the form (4.25, p. 182)

$$\pi(\boldsymbol{\theta}) = \pi(\theta_1, \ldots, \theta_k) = \int \prod_{i=1}^{k} \pi(\theta_i | \boldsymbol{\alpha}) h(\boldsymbol{\alpha}) d\boldsymbol{\alpha}.$$

In the empirical Bayes framework where $\boldsymbol{\alpha}$ is estimated by $\hat{\boldsymbol{\alpha}}$, this results in a simple estimated prior for θ_i of the form $\pi(\theta_i | \hat{\boldsymbol{\alpha}})$ and an estimated posterior of the form $\pi(\theta_i | \hat{\boldsymbol{\alpha}}, y_i)$. The mean of this estimated posterior (the empirical Bayes *estimate*) is usually close to the posterior mean of the full Bayesian hierarchical model. However, the variance of the estimated posterior tends to be too small when compared to the variance of the full Bayesian posterior (because the variability in α is ignored). Kass and Steffey (1989) provides corrections for this underestimation.

4.7.3a Meta-Analysis using Normal Models with Known Variance

At minimum, most study summaries include an effect estimate and standard error. The effect estimate could be the difference of sample means or a coefficient in a linear or logistic regression. Thus, the data for a meta-analysis of k studies often consists of $(Y_1, V_1), \ldots, (Y_k, V_k)$, where the Y_i are at least approximately normally distributed with variance estimated by V_i but treated as a known variance in the analysis.

A fixed effect approach would assume that all the Y_i are approximately unbiased estimators of the true effect μ. An approximately optimal estimator is then $\overline{Y} = \sum_{i=1}^{k} V_i^{-1} Y_i / \sum_{i=1}^{k} V_i^{-1}$. Because of study heterogeneity caused by differences in study protocols, investigators, and subject populations, it is unrealistic to think that each Y_i is actually estimating the same μ. Thus, it makes sense to assume that the ith study is estimating a θ_i that is drawn from a population with center μ, typically a normal population. This random effects model is popular in both frequentist and Bayesian approaches.

The Bayesian approach here is very similar to the random effects approach in Section 4.7.1 (p. 183). The only difference here is that Y_i given θ_i is $N(\theta_i, V_i)$ with V_i assumed known. A standard specification follows Section 4.7.1 (p. 183): θ_i given (μ, σ_a) is $N(\mu, \sigma_a^2)$, σ_a has a flat noninformative prior $\propto 1$ on $(0, \infty)$, and μ has a flat noninformative prior on $(-\infty, \infty)$. Although V_i is treated as known, the posterior densities are nontrivial and we resort to Markov chain Monte Carlo to estimate them (see Section 4.8.2, p. 195).

For frequentist and empirical Bayes approaches, the likelihood of $\boldsymbol{Y}$ used for estimating $\boldsymbol{\alpha} = (\mu, \sigma_a)$ is the product of independent $N(\mu, \sigma_a^2 + V_i)$ densities, $i = 1, \ldots, k$. Given an estimator $\widehat{\sigma}_a^2$, an approximately optimal estimator of μ is given by $\widehat{\mu} = \sum_{i=1}^{k} \widehat{w}_i Y_i / \sum_{i=1}^{k} \widehat{w}_i$, where $\widehat{w}_i = (\widehat{\sigma}_a^2 + V_i)^{-1}$. Typically $\sum_{i=1}^{k} \widehat{w}_i$ is used to estimate $\text{Var}(\widehat{\mu})$ and 95% confidence limits are $\widehat{\mu} \pm 1.96(\sum_{i=1}^{k} \widehat{w}_i)^{1/2}$.

A variety of estimators for σ_a^2 are possible. The most popular is a moment estimator due to DerSimonian and Laird (1986),

$$\widehat{\sigma}_{a,DL}^2 = \max\left\{0, \frac{Q_w - (k-1)}{\sum_{i=1}^{k} V_i^{-1} - \sum_{i=1}^{k} V_i^{-2} / \sum_{i=1}^{k} V_i^{-1}}\right\}, \qquad (4.37)$$

where $Q_w = \sum_{i=1}^{k} V_i^{-1}(Y_i - \overline{Y})^2$ and $\overline{Y} = \sum_{i=1}^{n} V_i^{-1} Y_i / \sum_{i=1}^{k} V_i^{-1}$. The second argument of the $\max(\cdot, \cdot)$ function is the method-of-moments estimator of σ_a^2 obtained by equating the Q_w to its expectation and solving for σ_a^2.

In a simulation study of meta-analyses for log odds ratios, Sidik and Jonkman (2007) found that $\widehat{\sigma}_{a,DL}^2$ and the maximum likelihood estimator and the restricted maximum likelihood estimator (REML) are all biased too small. They proposed a less biased moment estimator

$$\widehat{\sigma}^2_{a,SJ} = \frac{1}{k-1}\sum_{i=1}^{k}\widehat{v}_i^{-1}(Y_i - \overline{Y}_{\widehat{v}})^2, \tag{4.38}$$

where $\overline{Y}_{\widehat{v}} = \sum_{i=1}^{k}\widehat{v}_i^{-1}Y_i/\sum_{i=1}^{k}\widehat{v}_i^{-1}$, $\widehat{v}_i = \widehat{r}_i + 1$, $\widehat{r}_i = V_i/s_k^2$, and s_k^2 is the sample variance of the Y_i with divisor k. The estimator is motivated by rewriting $\text{Var}(Y_i) = \sigma_a^2(1 + r_i)$, $r_i = V_i/\sigma_a^2$, and has the added advantage that it is automatically nonnegative.

Although estimation of μ and σ_a^2 is the main focus of meta-analyses, empirical Bayes offers automatic improved estimators of the θ_i by borrowing strength from all the data. Because $\theta_i|\boldsymbol{Y}, \mu, \sigma_a^2$ is normal with mean $B_i\mu + (1 - B_i)Y_i$ and variance $(V_i^{-1} + \sigma_a^{-2})^{-1}$ where $B_i = V_i/(\sigma_a^2 + V_i)$, the empirical Bayes estimator of θ_i is

$$\widehat{\theta}_{i,EB} = \widehat{B}_i\,\widehat{\mu} + (1 - \widehat{B}_i)Y_i, \tag{4.39}$$

where $\widehat{B}_i = V_i/(\widehat{\sigma}_a^2 + V_i)$. Similar to the one-way random effects case, the estimated posterior variance $(V_i^{-1} + \widehat{\sigma}_a^{-2})^{-1}$ is too small because it does not take into account the estimation of μ and σ_a^2, and a variety of authors have suggested improvements (e.g., Kass and Steffey, 1989).

Example 4.4 (Meta-Analysis of hernia repair). Sidik and Jonkman (2007) reanalyzed results from 29 studies analyzed in Memon et al. (2003) that compared two types of surgical techniques for hernia repair. The endpoint postoperative complication is binary, and here we use the 29 estimates of log odds ratios along with the usual asymptotic standard errors for V_i, $(n_{11}^{-1} + n_{12}^{-1} + n_{21}^{-1} + n_{22}^{-1})^{1/2}$ in 2×2 table notation.

Here we focus on the results given in the top half of Table 4.4. The estimates of μ are similar for all four methods. The fixed effect standard error for $\widehat{\mu}$ is much smaller than for the three random effect approaches as one might expect. The DerSimonian and Laird estimator $\widehat{\sigma}^2_{a,DL} = 0.43$ appears too small when compared to the Bayes estimator and to the Sidik and Jonkman estimator. The Bayes estimate 0.65 reported in Table 4.4 is the posterior median. In addition, the Bayes posterior standard deviation for σ_a^2 is 0.32, suggesting that even with 29 studies, there is quite a bit of uncertainty in the random effects population variability. ♦

4.7.3b Meta-Analysis Using the Binomial Model

Here we want to consider meta-analysis of studies with two groups to compare based on a binary endpoint. The data consists of $(Y_{1c}, n_{1c}, Y_{1t}, n_{1t})$, ..., $(Y_{kc}, n_{kc}, Y_{kt}, n_{kt})$, where the control group Y_{ic} is binomial(n_{ic}, p_{ic}) and independent of the treatment group Y_{it} that is binomial(n_{it}, p_{it}), $i = 1, \ldots, k$. Example 4.4 (p. 190) is this type study, but there we used the approximate normality of the log odds ratio to

Table 4.4 Meta-Analysis of 29 Studies on Hernia Repair from Memon et al. (2003)

	Approximate Normal Results Based on Log Odds Ratio							
	Log Odds Ratio Scale					Odds Ratio Scale		
	$\widehat{\sigma}_a^2$	$\widehat{\mu}$	SE($\widehat{\mu}$)	L95%	U95%	exp($\widehat{\mu}$)	L95%	U95%
Fixed		−0.49	0.07	−0.63	−0.35	0.61	0.53	0.70
DL	0.43	−0.48	0.16	−0.78	−0.17	0.62	0.46	0.84
SJ	0.82	−0.47	0.20	−0.85	−0.08	0.63	0.43	0.92
Bayes	0.65	−0.47	0.18	−0.83	−0.10	0.63	0.43	0.90
	Binomial Assumption Results							
MH-Fixed		−0.53	0.07	−0.67	−0.40	0.59	0.51	0.67
GLMM	0.64	−0.47		−0.84	−0.11	0.62	0.43	0.89
Bayes	0.77	−0.48	0.20	−0.87	−0.09	0.63	0.42	0.91

L95% and U95% refer to 95% confidence or credibility intervals. For Bayes estimates, the Monte Carlo standard errors are in the range 0.001−0.007. DL uses (4.37, p. 189). SJ uses (4.38, p. 190). MH = Mantel-Haenszel. GLMM = binomial mixed model. Bayes estimates for σ_a^2 are posterior medians.

carry out the analysis. Here we work directly with the binomial outcomes but still focus on the log odds ratio.

Bayes random effects approaches are given in Smith et al. (1995) and Warn et al. (2002). We prefer the model on p. 2687 of Smith et al. (1995). As in the previous section, we assume that the logs odds ratios θ_i are drawn from a random effects population (prior) that is $N(\mu, \sigma_a^2)$. The θ_i are related to the binomial parameters by $\text{logit}(p_{ic}) = \gamma_i - \theta_i/2$ and $\text{logit}(p_{it}) = \gamma_i + \theta_i/2$, where $\text{logit}(p) = \log(p/(1-p))$, and $\gamma_i = \{\text{logit}(p_{ic}) + \text{logit}(p_{it})\}/2$ is an average logit for the ith study. Then, Y_{ic} given p_{ic} is binomial(n_{ic}, p_{ic}), and Y_{it} given p_{it} is binomial(n_{it}, p_{it}). As in previous specifications, σ_a has a flat noninformative prior $\propto 1$ on $(0, \infty)$, and μ has a flat noninformative prior on $(-\infty, \infty)$. Also, the prior for γ_i is N(0, 10).

The last row of Table 4.4 (p. 191) give results for Example 4.4 (p. 190). The Bayes posterior median for σ_a^2 is a little higher than the corresponding median for the approximate normal approach, but otherwise the two Bayes models yield similar results.

For a frequentist comparison in Table 4.4 (p. 191), we present the standard fixed effects Mantel-Haenszel approach and a generalized linear mixed model (GLMM) that is a logistic regression with random intercept and random slope coefficient for the treatment indicator variable. This random slope coefficient is specified to be $N(\mu, \sigma_a^2)$ and thus corresponds directly to the other random effects analyses in Table 4.4 (p. 191). The GLMM estimate of σ_a^2 is 0.64, somewhat smaller than the Bayes estimate. In general, all the random effect approaches in Table 4.4 (p. 191) are fairly close and distinct from the fixed effect approaches.

Table 4.5 Affect of Magnesium on Short Term Mortality of Heart Patients, from Teo et al. (1991)

	Mag. Group		Control Group			
Study	Died	Alive	Died	Alive	Log Odds Ratio	SE
1	1	39	2	34	−0.83	1.25
2	9	126	23	112	−1.06	0.41
3	2	198	7	193	−1.28	0.81
4	1	47	1	45	−0.04	1.43
5	10	140	8	140	0.22	0.49
6	1	58	9	47	−2.41	1.07
7	1	24	3	20	−1.28	1.19

Table 4.6 Meta-Analysis of Table 4.5 Data

	Approximate Normal Results Based on Log Odds Ratio							
	Log Odds Ratio Scale					Odds Ratio Scale		
	$\widehat{\sigma}_a^2$	$\widehat{\mu}$	SE($\widehat{\mu}$)	L95%	U95%	exp($\widehat{\mu}$)	L95%	U95%
Fixed		−0.75	0.27	−1.27	−0.23	0.47	0.28	0.79
DL	0.17	−0.80	0.33	−1.46	−0.15	0.45	0.23	0.86
SJ	0.34	−0.84	0.38	−1.59	−0.09	0.43	0.20	0.91
Bayes	0.43	−0.85	0.48	−1.84	0.08	0.48	0.16	1.08
	Binomial Assumption Results							
MH-Fixed		−0.83		−1.32	−0.34	0.44	0.27	0.71
GLMM	0.78	−0.74	0.11	−1.01	−0.47	0.47	0.36	0.62
Bayes	0.50	−0.91	0.48	−1.93	0.06	0.45	0.14	1.06

L95% and U95% refer to 95% confidence or credibility intervals. For Bayes estimates, the Monte Carlo standard errors are in the range 0.001–0.03. DL uses (4.37, p. 189). SJ uses (4.38, p. 190). MH = Mantel-Haenszel. GLMM = binomial mixed model. Bayes estimates for σ_a^2 are posterior medians.

Example 4.5 (Meta-Analysis on the use of magnesium in heart attack patients). Brockwell and Gordon (2007) report on a meta-analysis by Teo et al. (1991) of seven small clinical trials that studied the effect of injecting heart attack patients with magnesium. The endpoint was death while in the hospital (5 studies) or within one month after treatment (one study). The data are in Table 4.5.

Table 4.6 gives the results of the meta-analyses in the same format as Table 4.4. The DerSimonian and Laird estimator of σ_a^2 is much smaller than any of the others, and recall that the Bayes estimators of σ_a^2 are posterior medians. The GLMM standard error for $\widehat{\mu}$ appears far too small and results in a short and likely overoptimistic confidence interval. The Bayes results are again quite close to one another.

Brockwell and Gordon (2007) note that a later large clinical trial resulted in an odds ratio 95% interval of (1.00, 1.12), creating a controversy because of the discrepancy with the earlier studies. From that perspective, the Bayes intervals (0.16,1.08) and (0.14,1.06) may be preferred because they are the only ones casting doubt on the effectiveness of using magnesium with these patients. ♦

4.8 Monte Carlo Estimation of a Posterior

The main technical problem in Bayesian analysis is obtaining the posterior density for the parameters of interest and associated summary quantities like the posterior mean, standard deviation, and quantiles. Recall the simple form for a posterior

$$\pi(\theta \mid Y = y) = \frac{f(y \mid \theta)\pi(\theta)}{\int f(y \mid \theta)\pi(\theta)d\theta}.$$

and the form for hierarchical models

$$\pi(\theta \mid Y = y) = \frac{f(y \mid \theta) \int \pi(\theta \mid \alpha)h(\alpha)\,d\alpha}{\int\int f(y \mid \theta)\pi(\theta \mid \alpha)h(\alpha)\,d\alpha\,d\theta}.$$

In either case, we see the need to calculate integrals. Moreover, we usually need to integrate out the nuisance parameters and calculate the posterior mean leading to more integration. In the case of conjugate priors, all these calculations can be done analytically. However, as soon as we move away from conjugate priors and/or to hierarchical models, computing difficult integrals is routinely needed.

Historically, there have been a number of important techniques to do numerical integration such as Simpson's rule, the mid-point rule, and Gaussian quadrature. Moreover, modern techniques like Laplace approximation and Monte Carlo integration are available and still very important. Monahan (2001, Ch. 12) gives a good introduction to these approaches.

The main focus of modern Bayesian computing, however, has been on Markov chain Monte Carlo (MCMC) methods whereby a dependent sequence of random variables are obtained with the property that in the limit these random variables have the posterior distribution. The end of these sequences can be used as a sample to approximate the posterior density via a histogram or kernel density estimator. Moreover, the posterior mean and variance can be estimated by the sample mean and variance, and posterior quantiles are estimated by sample quantiles. In fact, in many ways it is simpler to have a sample from the posterior than it is to have a complex density function.

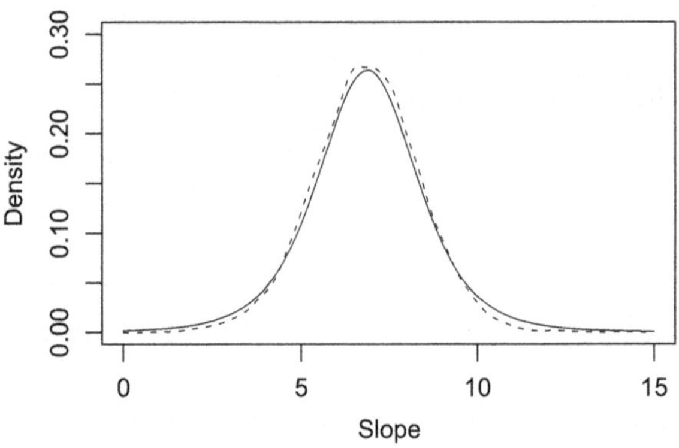

Fig. 4.4 Posterior (t with mean 6.88 and scale 1.44) for the slope of the cotton rat regression without last two points (solid) and kernel estimate from $N = 10,000$ Monte Carlo variates from the posterior (dashed)

We first illustrate Monte Carlo sampling in a case where the random variables are independent and then move to the more general case of Gibbs sampling resulting in dependent sequences.

4.8.1 Direct Monte Carlo Sampling from a Posterior

Recall from Section 4.6 (p. 176) that a standard approach for normal data leads to the Normal-gamma form for the posterior. Here it is very easy to sample from the posterior because the posterior distribution of $\sigma^2 = 1/\tau$ does not depend on the mean parameters, and the mean parameters given σ^2 have a normal distribution (multivariate if $\boldsymbol{\mu}$ is a vector). So the plan is to simply generate τ_1 from the posterior, then generate a $\boldsymbol{\mu}_1$ from the appropriate normal distribution with $\tau = \tau_1$. Then repeat this a total of N times leading to $(\boldsymbol{\mu}_1, \tau_1), \ldots, (\boldsymbol{\mu}_N, \tau_N)$ where the pairs are independent because the τ_i are generated independently (but $\boldsymbol{\mu}_i$ and τ_i are dependent).

To illustrate with a real data example, we use the simple linear regression Example 4.3 (p. 180) and focus on the posterior in the middle of the right panel of Figure 4.3 (p. 181) for the case where two data points were dropped and the informative prior was used. In this case the posterior mean of the intercept and slope is $\boldsymbol{\Delta}' = (16.30, 6.88)^T$, $\alpha' = 9.5$, $\beta' = 630.0$, and the posterior covariance matrix of $\boldsymbol{\Delta}$ given τ is

$$\tau^{-1}\begin{pmatrix} 0.79 & -0.15 \\ -0.56 & 0.03 \end{pmatrix}.$$

The marginal posterior density of the slope is a t density with mean 6.88, $scale$ $(\beta'(0.03)/\alpha')^{1/2} = 1.44$ and standard deviation $scale\{2\alpha'/(2\alpha' - 2)\}^{1/2} = 1.52$. Figure 4.4 (p. 194) plots this density.

The dashed line in Figure 4.4 (p. 194) is for a kernel density estimator based on data generated from the posterior and plotted using the following code

```
N<-10000
set.seed(373)
tau<-rgamma(N,shape=alpha1,rate=beta1)
Deltas<-rnorm(N,mean=Delta1[2],sd=sqrt(COV[2,2]/tau))
lines(density(Deltas,from=0,to=15),lty=2)
```

In this code, we have generated only the second component of the mean parameter, that is, the slope. For the purposes of looking at the marginal posterior density of the slope, it does not matter whether we generated from the marginal normal via

```
Deltas<-rnorm(N,mean=Delta1[2],sd=sqrt(COV[2,2]/tau))
```

or from the bivariate normal with `mean=Delta1` and `cov=COV/tau` and take the second component. The beauty of the Monte Carlo approach is that once we generate a sample of vectors from the joint posterior, we can analyze each component separately for marginal densities or analyze groups of parameters together for joint densities. That may not seem important until one tries analytically to find a marginal density from a complicated joint density.

It is clear in Figure 4.4 (p. 194) that the Monte Carlo approach is closely reproducing the density. Estimated summary descriptive statistics are also close to the true values:

```
> round(mean(Deltas),2)
[1] 6.85
> round(sd(Deltas),2)
[1] 1.52
> round(quantile(Deltas,c(.025,.5,.975)),2)
 2.5%   50% 97.5%
 3.79  6.87  9.78
```

Because these samples are independent, the Monte Carlo standard error of the mean estimate is $1.518/(10000)^{1/2} = .015$. Thus the difference between the true mean 6.88 and the estimate is a little less than two Monte Carlo standard errors. The estimated standard deviation is very close to the true standard deviation 1.52. The true .025 and .975 quantiles are (3.87,9.89), reasonably close to the estimated values 3.79 and 9.78, respectively. Of course, if better accuracy is desired, then N can be made larger.

4.8.2 Markov chain Monte Carlo Sampling from a Posterior

In the previous section, Monte Carlo sampling from the Normal-gamma posterior was simple because the posterior of σ^2 did not depend on other parameters, and we could generate a σ^2 and then a $\boldsymbol{\Delta}$ given that σ^2. Repeating N times leads to an iid sample of vectors from the posterior.

In general, though, the posterior does not factor into such a nice form. Suppose that $\boldsymbol{\theta}$ is b-dimensional and we would like to estimate a complicated posterior

$\pi(\boldsymbol{\theta} \mid \boldsymbol{Y})$. Suppose further that we are able to generate random variables from each of the *full* univariate conditional posterior densities

$$\pi(\theta_1 \mid \theta_2, \theta_3, \ldots, \theta_b, \boldsymbol{Y})$$
$$\pi(\theta_2 \mid \theta_1, \theta_3, \ldots, \theta_b, \boldsymbol{Y})$$
$$\ldots$$
$$\pi(\theta_b \mid \theta_1, \theta_2, \ldots, \theta_{b-1}, \boldsymbol{Y}).$$

Then given starting values $\theta_1^{(0)}, \ldots, \theta_b^{(0)}$:

1. Generate $\theta_1^{(i)}$ from $\pi(\theta_1 \mid \theta_2^{(i-1)}, \theta_3^{(i-1)}, \ldots, \theta_b^{(i-1)}, \boldsymbol{Y})$;
2. Generate $\theta_2^{(i)}$ from $\pi(\theta_2 \mid \theta_1^{(i)}, \theta_3^{(i-1)}, \ldots, \theta_b^{(i-1)}, \boldsymbol{Y})$;

   ...

b. Generate $\theta_b^{(i)}$ from $\pi(\theta_b \mid \theta_1^{(i)}, \theta_2^{(i)}, \ldots, \theta_{b-1}^{(i)}, \boldsymbol{Y})$;

repeating for $i = 1, \ldots, N$.

In contrast to the previous section, these N vectors are not iid. However, under suitable regularity conditions, the limiting joint distribution of $\boldsymbol{\theta}^{(N)} = (\theta_1^{(N)}, \theta_2^{(N)}, \ldots, \theta_b^{(N)})$ as $N \to \infty$ is exactly the joint posterior distribution of $\boldsymbol{\theta}$ given $\boldsymbol{Y}$. Moreover, for any integrable function $g()$,

$$\frac{1}{N}\sum_{i=1}^{N} g\left(\theta_1^{(i)}, \theta_2^{(i)}, \ldots, \theta_b^{(i)}\right) \xrightarrow{wp1} E\{g(\theta_1, \ldots, \theta_b)\} \quad \text{as } N \to \infty.$$

The above sampling scheme is called *Gibbs sampling* and was pioneered by Geman and Geman (1984), Tanner and Wong (1987), and Gelfand and Smith (1990).

The sequences generated are Markov chains, and in the limit ($N \to \infty$), any collection of k generated vectors $(\boldsymbol{\theta}^{(N+1)}, \boldsymbol{\theta}^{(N+2)}, \ldots, \boldsymbol{\theta}_{(N+k)})$ has a stationary distribution. Also each $\boldsymbol{\theta}^{(N+j)}$ is distributed approximately according to the marginal posterior distribution of $(\boldsymbol{\theta} \mid \boldsymbol{Y})$, but $\boldsymbol{\theta}^{(N+j_1)}$ and $\boldsymbol{\theta}^{(N+j_2)}$ are not independent. Thus some care must be taken when using Monte Carlo means to approximate expectations.

In practice, one typically generates a sequence of length N but discards the first N_0 *burn-in* elements as a means of assuring convergence to the stationary distribution of the Markov chain. Then, the remaining $N - N_0$ elements of the sequence are used to estimate functionals of the posterior distribution such as the mean, standard deviation, quantiles, etc. However the lack of independence among the $\boldsymbol{\theta}^{(N+j)}$ means that assessing the Monte Carlo variability in these Monte Carlo estimates must account for the dependence among the random variables. For example, time series models and methods are sometimes used. Alternatively, one can generate k independent sequences, estimate the posterior mean for each, then use the sample standard deviation of these posterior means divided by $\sqrt{k}$ as a standard error for the mean of the means. We illustrate with the data from Table 4.2 (p. 184).

Example 4.6 (Gibbs sampling in the normal random effects model). Recall the one-way random effect model with data $Y_{ij}, i = 1, \ldots, k; j = 1, \ldots, n_i$. We assume that all random quantities on the same level are conditionally independent with distributions

$$Y_{ij} \mid \theta_i, \sigma_e^2 \sim N(\theta_i, \sigma_e^2)$$
$$\theta_i \mid \mu, \sigma_a^2 \sim N(\mu, \sigma_a^2) \quad \text{and} \quad \tau_e = 1/\sigma_e^2 \sim \text{gamma}(a_2, 1/b_2)$$
$$\mu \sim N(\mu_0, 1/\tau_0) \quad \text{and} \quad \tau_a = 1/\sigma_a^2 \sim \text{gamma}(a_1, 1/b_1).$$

In Section 4.7.1 (p. 183), we gave noninformative priors on τ_e and τ_a, but here we want to use gamma priors so that we can write out all the full conditionals. Following Jones and Hobert (2004), the posterior conditional distributions are

$$\mu \mid \boldsymbol{\theta}, \tau_a, \tau_e, \boldsymbol{Y} \sim N\left(\frac{\tau_0\mu_0 + k\tau_a\overline{\theta}}{\tau_0 + k\tau_a}, \frac{1}{\tau_0 + k\tau_a}\right)$$

$$\theta_i \mid \boldsymbol{\theta}_{[i]}, \mu, \tau_a, \tau_e, \boldsymbol{Y} \sim N\left(\frac{\tau_a\mu + n_i\tau_e\overline{Y}_{i.}}{\tau_a + n_i\tau_e}, \frac{1}{\tau_a + n_i\tau_e}\right)$$

where $\overline{\theta} = k^{-1}\sum_{i=1}^k \theta_i$, and $\boldsymbol{\theta}_{[i]} = (\theta_1, \ldots, \theta_{i-1}, \theta_{i+1}, \ldots, \theta_k)$, and

$$\tau_a \mid \boldsymbol{\theta}, \mu, \tau_e, \boldsymbol{Y} \sim \text{gamma}\left(\frac{k}{2} + a_1, \frac{\sum(\theta_i - \mu)^2}{2} + b_1\right)$$

$$\tau_e \mid \boldsymbol{\theta}, \mu, \tau_a, \boldsymbol{Y} \sim \text{gamma}\left(\frac{\sum n_i}{2} + a_2, \frac{\sum n_i(\theta_i - \overline{Y}_{i.})^2 + \text{SSE}}{2} + b_2\right),$$

where SSE$= \sum_{i=1}^k \sum_{j=1}^{n_i}(Y_{ij} - \overline{Y}_{i.})^2$.

We choose $\mu_0 = 0$, $\tau_0 = .000001$ to approximate the usual noninformative prior on $(-\infty, \infty)$ for μ. For τ_e we choose $(a_2 = b_2 = .001)$, which is a standard approximation to Jeffreys's noninformative prior $\propto 1/\sigma_e^2$ on $(0, \infty)$. For τ_a, we choose $(a_1 = b_1 = .01)$ (very mildly informative) because the prior $\propto 1/\sigma_a^2$ on $(0, \infty)$ yields an improper posterior. Here is part of the R code to generate N random vectors

```
# starting values
mu[1]<-rnorm(1,mean=0,sd=1/sqrt(mu0))
theta[,1]<-rnorm(k,mean=mean(ybar),sd=2*sd(ybar))
tau.e[1]<-rgamma(1,shape=a2,rate=b2)
tau.a[1]<-rgamma(1,shape=a1,rate=b1)
# start Gibbs loop
for(i in 2:N){
mu[i]<-rnorm(1,mean=(tau.0*mu0+k*tau.a[i-1]
  *mean(theta[,i-1]))/(tau.0+k*tau.a[i-1]),
                sd=1/sqrt(tau.0+k*tau.a[i-1]))
```

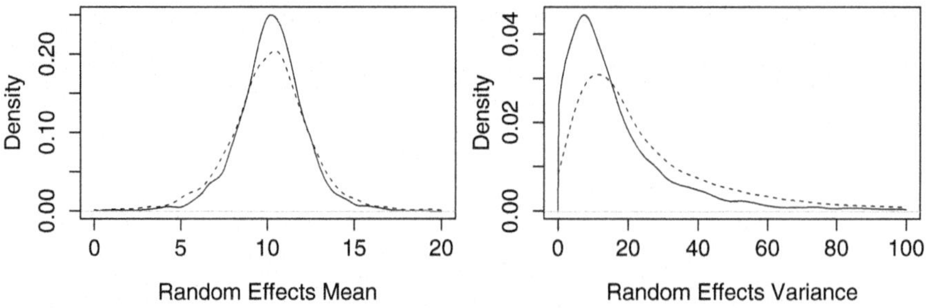

Fig. 4.5 Posterior densities for μ and σ^2 of the random effects $N(\mu, \sigma^2)$ distribution for the Table 4.2 (p. 184) data. Solid lines are for the R code implementation, and dashed lines are for the WinBUGS implementation with flat prior on σ_a

```
theta[,i]<-rnorm(k,mean=(tau.a[i-1]*mu[i]+ni
  *tau.e[i-1]*y)/(tau.a[i-1]+ni*tau.e[i-1]),
             sd=1/sqrt(tau.a[i-1]+ni*tau.e[i-1]))
tau.a[i]<-rgamma(1,shape=k/2+a1,
          rate=(sum((theta[,i]-mu[i])^2))/2+b1)
tau.e[i]<-rgamma(1,shape=ntot/2+a2,
       rate=(sum(ni*(theta[,i]-ybar)^2)+SSE)/2+b2)
} # ends update loop
```

We then ran this code 5 separate times with $N = 1500$ each time, retaining the last 1000 from each run, resulting in 5000 total vectors of $(\boldsymbol{\theta}, \mu, \sigma_e^2, \sigma_e^2)$ from the posterior. The solid lines in Figure 4.5 (p. 198) are kernel density estimates of the posteriors for μ and σ^2.

A simpler way to do the Gibbs sampling for this problem is to use the BUGS software; BUGS stands for **B**ayesian inference **U**sing **G**ibbs Sampling. Actually we used the R package `R2WinBUGS` to call WinBUGS. The BUGS model statement used to duplicate the above R code is given by

```
model{
  for(i in 1:N) {y[i]~dnorm(theta[group[i]],tau.e)}
      tau.e~dgamma(0.001,0.001)
      sig2.e<-1/tau.e
  for(j in 1:k){theta[j]~dnorm(mu,tau.a)}
      mu~dnorm(0.0,1.0E-6)
      tau.a~dgamma(0.01,0.01)
      sig2.a<-1/tau.a
}
```

The code is very similar to R, but note that the normal densities are specified with precisions, that is, `dnorm(mu,tau.a)` is equivalent to the R code `rnorm(1,mean=mu,sd=1/sqrt(tau.a))`, and `dgamma(a1,b1)` is equivalent to the R code `rgamma(1,shape=a1,rate=b1)` or `rgamma(1,shape=a1,scale=1/a1)`.

In Section 4.7.1 (p. 183) we used a flat prior on σ_a, which seems to be standard now that it is well known that Jeffreys's prior $\sigma_a^2 \propto 1/\sigma_a^2$ on $(0, \infty)$ yields an improper posterior. The BUGS model code for this specification is the same as above except that

```
sig.a~dunif(0,100)
sig2.a<-pow(sig.a,2)
tau.a<-1/sig2.a
```

replaces the `tau.a dgamma(0.01,0.01)`, where `pow` is the power function. In Figure 4.5 (p. 198) the dashed lines are for this specification. Basically, the posterior densities are a little more spread out, but the quantiles (see Table 4.3, p. 184) are not much different. ♦

There are many practical details in using Gibbs sampling and checking correctness of the output. Typically, one looks at plots over time of the generated values and at a variety of diagnostics to make sure that convergence has been reached. It can be difficult to detect an improper posterior, and there is a lot of active research on implementation of Gibbs sampling. There is an older sampling method, the *Metropolis* algorithm, which is more general than Gibbs sampling and also results in MCMC dependent samples. It is often used when Gibbs sampling in not available or easy to carry out, but we do not discuss it here.

4.8.3 *Why Does Gibbs Sampling Work?*

At first blush, Gibbs sampling is both mysterious and amazing. It seems magical that one can generate sequences from conditional distributions and end up with random variables having the desired posterior joint distribution. A rigorous explanation of how Gibbs sampling works requires Markov chain theory, which is beyond the level of this book. However, the following example sheds light on the process in a special case.

Example 4.7 (Gibbs sampling in the bivariate normal model). Suppose that the data consists of one pair (X, Y) from the bivariate normal with mean $\boldsymbol{\theta} = (\theta_X, \theta_Y)^T$, and covariance matrix

$$\begin{pmatrix} 1 & \rho \\ \rho & 1 \end{pmatrix},$$

where $-1 < \rho < 1$ is known. If the (improper) prior on $\boldsymbol{\theta}$ consists of independent uniform densities on $(-\infty, \infty)$, then the posterior for $\boldsymbol{\theta}$ is bivariate normal with mean (X, Y) and the same covariance matrix. This is easy to see after writing down the bivariate normal density. Of course, because we know the joint posterior, there is no reason to use Gibbs, but it is easy to follow the convergence. The conditional densities required for Gibbs sampling are

$$\theta_X \mid \theta_Y, X, Y \sim N(X + \rho(\theta_Y - Y), 1 - \rho^2)$$

$$\theta_Y \mid \theta_X, X, Y \sim N(Y + \rho(\theta_X - X), 1 - \rho^2).$$

Starting with a nonrandom $\theta_Y^{(0)}$, the first step is to calculate

$$\begin{aligned}
\theta_X^{(1)} &= Z_{X1}\sqrt{1-\rho^2} + X + \rho(\theta_Y^{(0)} - Y) \\
\theta_Y^{(1)} &= Z_{Y1}\sqrt{1-\rho^2} + Y + \rho(\theta_X^{(1)} - X) \\
&= Z_{Y1}\sqrt{1-\rho^2} + Y + \rho Z_{X1}\sqrt{1-\rho^2} + \rho^2(\theta_Y^{(0)} - Y),
\end{aligned}$$

where the random variables $Z_{X1}, Z_{X2}, \dots$ and $Z_{Y1}, Z_{Y2}, \dots$ are mutually independent, standard normal random variables. Continuing the iteration, results in

$$\begin{aligned}
\theta_X^{(k)} &= X + \sqrt{1-\rho^2}\left\{Z_{Xk} + \rho^{2k-1}(\theta_Y^{(0)} - Y)\right\} \\
&\quad + \sqrt{1-\rho^2}\left\{\sum_{i=1}^{k-1}\rho^{2i} Z_{X(k-i)} + \sum_{i=1}^{k-1}\rho^{2i-1} Z_{Y(k-i)}\right\}.
\end{aligned}$$

$$\begin{aligned}
\theta_Y^{(k)} &= Y + \sqrt{1-\rho^2}\left\{Z_{Xk} + \rho^{2k}(\theta_Y^{(0)} - Y)\right\} \\
&\quad + \sqrt{1-\rho^2}\left\{\sum_{i=1}^{k}\rho^{2i-1} Z_{X(k+1-i)} + \sum_{i=1}^{k-1}\rho^{2i} Z_{Y(k-i)}\right\}.
\end{aligned}$$

Note that $(\theta_X^{(k)}, \theta_Y^{(k)})$ are linear functions of iid(μ, σ^2) normal random variables and thus have a bivariate normal distribution. Taking expectations shows that

$$E\{\theta_X^{(k)} \mid X, Y\} = X + \sqrt{1-\rho^2}\rho^{2k-1}(\theta_Y^{(0)} - Y) \to X \quad \text{as} \quad k \to \infty.$$

Similarly, $E\{\theta_Y^{(k)} \mid X, Y\}$ converges to Y. The variances and covariances are

$$\operatorname{Var}\{\theta_X^{(k)} \mid X, Y\} = (1-\rho^2)\left\{1 + \sum_{i=1}^{k-1}(\rho^{4i} + \rho^{4i-2})\right\}$$

$$\operatorname{Var}\{\theta_Y^{(k)} \mid X, Y\} = (1-\rho^2)\left\{1 + \rho^{4k-2} + \sum_{i=1}^{k-1}(\rho^{4i} + \rho^{4i-2})\right\}$$

$$\operatorname{Cov}\{\theta_X^{(k)}, \theta_Y^{(k)} \mid X, Y\} = (1-\rho^2)\left\{\sum_{i=1}^{k}(\rho^{4i-3} + \rho^{4i-5}) - \rho^{-1}\right\}.$$

Problem 4.19 (p. 203) is to verify that the above variances converge to 1, and that the covariance converges to ρ as $k \to \infty$. Thus, because the bivariate normal is characterized by these moments, $(\theta_X^{(k)}, \theta_Y^{(k)})$ converges in distribution to the bivariate normal posterior of $\boldsymbol{\theta}$. ◆

4.9 Problems

4.1. Fill in the details to get the posterior mean (4.4, p. 168).

4.2. Give details to show that (4.6, p. 169) is correct.

4.3. For the credible interval in (4.5, p. 168), show that in repeated sampling from a normal distribution with mean μ' and variance σ_Y^2, the coverage probability is

$$\Phi\left(1.96\left(\frac{\tau_0+\tau_n}{\tau_n}\right)^{1/2}-\frac{\tau_0(\mu_0-\mu')}{\tau_n^{1/2}}\right)-\Phi\left(-1.96\left(\frac{\tau_0+\tau_n}{\tau_n}\right)^{1/2}-\frac{\tau_0(\mu_0-\mu')}{\tau_n^{1/2}}\right).$$

Set $n = 10$ and $\sigma_Y^2 = 1$ and compute this probability for several combinations of the other parameters.

4.4. The negative binomial density is

$$f(y \mid r, p) = \binom{r+y-1}{y} p^r (1-p)^y \quad y = 0, 1, \ldots;\ 0 \le p \le 1.$$

Show that beta (α, β) is a conjugate prior for p with beta $(\alpha + r, \beta + Y)$ posterior.

4.5. The Poisson density is

$$f(y \mid \lambda) = \frac{e^{-\lambda}\lambda^y}{y!} \quad y = 0, 1, \ldots;\ 0 \le p \le 1.$$

Show that gamma (α, β) is a conjugate prior for λ with gamma $(\alpha + Y, \beta/(\beta+1))$ posterior.

4.6. Suppose that Y has a gamma$(\alpha_0, 1/\theta)$ density,

$$f(y \mid \theta) = \frac{1}{\Gamma(\alpha_0)}\theta^{\alpha_0} y^{\alpha_0-1} e^{-y\theta} \quad 0 < y < \infty;\ 0 < \theta < \infty.$$

Show that gamma$(\alpha, 1/\beta)$ is a conjugate prior for θ with gamma$(\alpha+\alpha_0, 1/(\beta+Y))$ posterior. What is the posterior if instead of a single Y, we have a sample $Y_1, \ldots, Y_n$ from the gamma$(\alpha_0, 1/\theta)$?

4.7. Suppose that data $\boldsymbol{Y}$ with likelihood $f(\boldsymbol{Y} \mid \boldsymbol{\theta})$ has conjugate prior $\pi(\boldsymbol{\theta} \mid \boldsymbol{\gamma})$ and posterior $\pi(\boldsymbol{\theta} \mid q(\boldsymbol{Y}, \boldsymbol{\gamma}))$ for known function q. Show that the mixture prior,

$$\pi(\boldsymbol{\theta} \mid \boldsymbol{\gamma}_1, \ldots, \boldsymbol{\gamma}_k) = \sum_{i=1}^{k} w_i \pi(\boldsymbol{\theta} \mid \boldsymbol{\gamma}_i),$$

with know weights $w_1, \ldots, w_k$, is also conjugate, and give the form of the posterior.

4.8. Show that Jeffreys's prior for the odds parameter $\gamma = p/(1-p)$ is given by (4.14, p. 175).

4.9. Show that Jeffreys's prior for the log odds parameter $\gamma = \log\{p/(1-p)\}$ is given by (4.15, p. 175).

4.10. Fill in the details to get the posterior (4.18, p. 177) .

4.11. Show the step that leads from (4.18, p. 177) to (4.19, p. 177).

4.12. (Use results from Section 4.6.1, p. 176) Joe and Mary are in their 40's and have four boys (now in their 20's) whose heights are 68 inches, 72 inches, 74 inches, and 69 inches, respectively. Recently, Joe and Mary had a fifth son. They would like to get a posterior density for the adult height of their fifth boy. They decide to assume that the four observations they have on their first four sons have come from a normal distribution with mean μ and variance $1/\tau$. Note also that Joe is himself 68 inches tall and Mary is 62 inches tall.

a. First use the improper prior $\pi(\mu, \tau) = 1/\tau$ and find the marginal posterior of μ given the four data points. Also find the mean and standard deviation of this posterior.
b. Now elicit from yourselves a "normal-gamma" conjugate prior based on any thoughts you might have from the above description. Try to put yourself in Joe and Mary's shoes. This really ought to be done without knowledge of the heights of the children since those values are the data.
c. Find the marginal posterior of μ and its mean and standard deviation based on b. and the sample.
d. The parents actually want the Bayesian forecast (prediction) density of the adult height of their fifth son. Thus they could assume that Y_5 is the adult height of their fifth son and that $Y_5 \mid \mu, \tau$ is normal$(\mu, 1/\tau)$. Recall from Section 4.6.1 (p. 176) that the prediction density is a t density with center=μ', $scale^2 = (\beta'/\alpha')(1 + (n_0+n)^{-1})$, and degrees of freedom=$2\alpha'$.

   i. Find the mean and standard deviation of this t distribution using the actual values from a) and c) in order to compare the effect of uninformative and informative priors.
   ii. They would like their fifth son to be a basketball player. Using the predictive density, find the probability that their fifth son will be taller than 78 inches using both the a) and c) values.

4.13. Verify that the expression (4.30, p. 185) follows from the expression above it.

4.14. For the random effects example (Section 4.7.1, p. 183), verify that the marginal density of the data $(Y_{ij}, i = 1, \ldots, k; j = 1, \ldots, n_i)$ in the Bayesian specification, formally given by $m(\boldsymbol{y} \mid \boldsymbol{\alpha}) = \int f(\boldsymbol{y} \mid \boldsymbol{\theta})\pi(\boldsymbol{\theta} \mid \boldsymbol{\alpha})\, d\boldsymbol{\theta}$, is the same as the joint density for the frequentist specification (but with $\theta_i = \mu + a_i$) by noting that in the Bayes specification, each Y_{ij} given (θ_i, σ_e^2) is $N(\theta_i, \sigma_e^2)$ and all independent and each θ_i given (μ, σ_a^2) is $N(\mu, \sigma_a^2)$ and all independent, implies $Y_{ij} = \theta_i + e_{ij}$, where the e_{ij} are iid $N(0, \sigma_e^2)$ and independent of $\theta_1, \ldots, \theta_k$.

4.15. From Section 4.7.2 (p. 186), if $\boldsymbol{Y}$ is MN$\{\boldsymbol{0}, (\sigma_a^2 + \sigma_0^2)\boldsymbol{I}_b\}$, show that $(b - 2)\sigma_0^2 / \sum_{i=1}^{b} Y_i^2$ is an unbiased estimator of $\sigma_0^2/(\sigma_0^2 + \sigma_a^2)$.

4.16. Derive $\widehat{\sigma}^2_{a,DL}$ of (4.37, p. 189).

4.17. Derive $\widehat{\sigma}^2_{a,SJ}$ of (4.38, p. 190) by assuming that the variance ratios $r_i = V_i/\sigma_a^2$ are known.

4.18. Higgins et al. (2009) extracted effect sizes and associated standard errors from Roberts et al. (2007) meta analysis of 14 studies on set shifting ability in people with eating disorders:

```
size = 0.38  0.07  0.52  0.85  0.45  0.01 -0.58
       0.44  0.46  0.93  0.28  0.20  0.46  0.59
se   = 0.40  0.21  0.29  0.25  0.29  0.35  0.36
       0.25  0.22  0.47  0.24  0.28  0.23  0.36
```

a. Use the DerSimonian and Laird approach to estimate μ and σ_a^2 of the random effects distribution, and give a 95% confidence interval for μ.
b. Run a full Bayesian analysis using MCMC and report the (2.5%,50%,97.5%) percentiles of the posterior distribution of μ and σ_a^2.

4.19. First verify that $\sum_{i=1}^{k-1} \rho^{4i} = (\rho^4 - \rho^{4k})/(1 - \rho^4)$. Then give exact expressions for the variances and covariances of $(\theta_X^{(k)}, \theta_Y^{(k)})$ in Example 4.7 (p. 199) and show that they converge to the appropriate quantities.

4.20. Generate $N = 200$ iterations of the Gibbs sequence for Example 4.7 (p. 199) using $(X = 1, Y = 2, \rho = .9, \theta_Y^{(0)} = 2$. Make separate scatter plots of the first 100 pairs and the second 100 pairs.

Part III
Large Sample Approximations in Statistics

Chapter 5
Large Sample Theory: The Basics

5.1 Overview

A fundamental problem in inferential statistics is to determine, either exactly or approximately, the distribution of a statistic calculated from a probability sample of data. The statistic is usually a parameter estimate, in which case the distribution characterizes the sampling variability of the estimate, or a test statistic, in which case the distribution provides the critical values of the test and also is useful for power calculations. Because the number and type of inference problems admitting statistics for which exact distributions can be determined is limited, approximating the distribution of a statistic is usually necessary. The well-known results that $\overline{Y}$ and $\sqrt{n}(\overline{Y} - \mu)/s$ have $N(\mu, \sigma^2/n)$ and t_{n-1} distributions, respectively, when $Y_1, \ldots, Y_n$ are independently and identically distributed $N(\mu, \sigma^2)$ and similar results for normal linear models, are exceptions rather than the norm. If the population distribution is not normal, then neither distribution is exact in general. Yet the distributions of $\overline{Y}$ and $\sqrt{n}(\overline{Y} - \mu)/s$ are approximately $N(\mu, \sigma^2/n)$ and t_{n-1} respectively, provided $Y_1, \ldots, Y_n$ are independently and identically distributed from a distribution with mean μ and variance $\sigma^2 < \infty$. The larger n is, the better the approximation.

Approximate inference, i.e., inference based on an approximation to the distribution of a statistic, is essential for most modern statistical methods. Even though it is usually the case that the exact distribution of a statistic cannot be obtained, it is generally possible to obtain an approximation to the distribution of sufficient accuracy to be useful in applications. Most often, although not exclusively, approximations are derived via determination of the asymptotic distribution (as sample size n increases to ∞) of the statistic of interest. The asymptotic distribution forms the basis for determining a useful approximate distribution of a statistic calculated from a sample of size n.

The modern statistical researcher should know how to determine when a large sample approximation to the distribution of a statistic is appropriate, how to derive the approximation, and how to use it for inference in applications. The fact that

D.D. Boos and L.A. Stefanski, *Essential Statistical Inference: Theory and Methods*, Springer Texts in Statistics, DOI 10.1007/978-1-4614-4818-1_5,

only a few types of asymptotic distributions arise in the vast majority of statistical applications greatly simplifies the task of developing a working knowledge of large sample theory. Statistics that are averages, or are asymptotically equivalent to averages, generally have normal asymptotic distributions. An important corollary is that quadratic forms of such statistics have asymptotic chi-squared distributions. Statistics that are extreme values, e.g., maximum or minimum order statistics, have one of three extremal-type asymptotic distributions. Most statistics arising in practice are either average-like or extreme value-like, and the statistician who is familiar with the asymptotic theory of averages and extreme values can handle most statistical asymptotic problems. Of the two types of statistics, those that are average-like are far more common in applications and are studied first and in greater detail throughout this chapter. We cover extremal-type asymptotic distributions as a special case of convergence in distribution in Section 5.2.3 (p. 220).

5.1.1 Statistics Approximated by Averages

Large-sample theory for statistics that are approximated by averages consists largely of applications of the *Laws of Large Numbers* and of the *Central Limit Theorem*. If $\overline{X}$ is the sample mean of n independent and identically distributed (iid) random variables $X_1, \ldots, X_n$, with mean $\mu = E(X_1)$, then the Laws of Large Numbers guarantee that $\overline{X}$ is close to μ in a probabilistic sense when n is large. Specifically, the Strong Law of Large Numbers states that

$$P\left(\lim_{n\to\infty} |\overline{X} - \mu| < \epsilon, \text{for every } \epsilon > 0\right) = 1,$$

and the Weak Law of Large Numbers states that

$$\lim_{n\to\infty} P(|\overline{X} - \mu| < \epsilon) = 1, \text{for every } \epsilon > 0.$$

The Central Limit Theorem guarantees that the distribution of $\overline{X}$ is approximately the same as a normal distribution with mean μ and variance σ^2/n, as $n \to \infty$, provided the population variance $\sigma^2 = \text{Var}(X_1)$ is finite. Specifically, the Central Limit Theorem for iid summands states that

$$\lim_{n\to\infty} P\left(\frac{\overline{X} - \mu}{\sigma/\sqrt{n}} \le t\right) = \Phi(t),$$

where Φ is the standard normal cumulative distribution function.

The Laws of Large Numbers and the Central Limit Theorem describe the asymptotic behavior of arithmetic averages. Most statistics are not arithmetic averages, but are asymptotically equivalent to arithmetic averages. The key to deriving the asymptotic distribution of a statistic is identifying the appropriate approximating average, which is not always immediately apparent. We now describe some classes of statistics that are approximately averages.

5.1.1a Averages

First, of course, is the class of statistics that are exactly equal to arithmetic averages. It appears that we are stating the obvious here, but we do so to emphasize the fact that the class of averages includes averages of functions of the sample values. Thus statistics like $n^{-1}\sum X_i^2, n^{-1}\sum \log(X_i), n^{-1}\sum I(X_i \le a)$ and $n^{-1}\sum \exp(tX_i)$ are averages to which the Laws of Large Numbers and Central Limit Theorem apply provided the required means and variances exist. Note also that random quantities like $n^{-1}\sum(X_i - \mu)^2$ are also in this class even though they depend on possibly unknown parameters like μ. The Laws of Large Numbers and the Central Limit Theorem apply to such quantities in exactly the same way as *statistics*, which by definition are not a function of unknown parameters.

5.1.1b Functions of Averages

Functions of averages are approximately averages (or quadratic functions of averages). Suppose that $\overline{\boldsymbol{Y}}_{m\times 1}$ is a multivariate sample mean and f is a smooth scalar valued function. By the multivariate Laws of Large Numbers we know that $\overline{\boldsymbol{Y}}$ is close to the population mean $\boldsymbol{\mu}$ and a Taylor Series approximation is justified. Letting $f'(\boldsymbol{y})$ denote the row vector $\partial f(\boldsymbol{y})/\partial \boldsymbol{y}$,

$$\begin{aligned} f(\overline{\boldsymbol{Y}}) &\approx f(\boldsymbol{\mu}) + f'(\boldsymbol{\mu})(\overline{\boldsymbol{Y}} - \boldsymbol{\mu}) \\ &= n^{-1}\sum \{f(\boldsymbol{\mu}) + f'(\boldsymbol{\mu})(\boldsymbol{Y}_i - \boldsymbol{\mu})\}. \end{aligned} \tag{5.1}$$

Thus $f(\overline{\boldsymbol{Y}}) \approx \overline{W} = n^{-1}\sum W_i$ where $W_i = f(\boldsymbol{\mu}) + f'(\boldsymbol{\mu})(\boldsymbol{Y}_i - \boldsymbol{\mu})$.

For example, consider an iid sample of pairs $(Y_{1i}, Y_{2i}), \ldots, (Y_{1n}, Y_{2n})$ and the ratio estimator $f(\overline{\boldsymbol{Y}}) = \overline{Y}_1/\overline{Y}_2$, where $\mu_2 \neq 0$. In this case $f'(\boldsymbol{\mu}) = (1/\mu_2, -\mu_1/\mu_2^2)$ and thus $\overline{Y}_1/\overline{Y}_2 \approx n^{-1}\sum W_i$ where

$$W_i = \frac{\mu_1}{\mu_2} + \frac{1}{\mu_2}(Y_{1i} - \mu_1) - \frac{\mu_1}{\mu_2^2}(Y_{2i} - \mu_2). \tag{5.2}$$

Some statistics are functions of averages even though they are not commonly expressed as such. For example, consider $X_1, \ldots, X_n$ iid(μ, σ^2) and the variance estimator $s_n^2 = n^{-1}\sum_{i=1}^n (X_i - \overline{X})^2$. Define $\overline{Y}_1 = \overline{X}$, $\overline{Y}_2 = n^{-1}\sum X_i^2$ and

$f(\boldsymbol{y}) = y_2 - y_1^2$ for which $f'(\boldsymbol{y}) = (-2y_1, 1)$. Then $s_n^2 = f(\overline{\boldsymbol{Y}})$ and a Taylor Series approximation yields $s_n^2 \approx n^{-1} \sum W_i$ where $W_i = (X_i - \mu)^2$.

For applying the Laws of Large Numbers to $f(\overline{\boldsymbol{Y}})$, we can just use continuity of f (see Theorem 5.9, p. 227) rather than (5.1). For applying the Central Limit Theorem to $f(\overline{\boldsymbol{Y}})$, the approximation of $f(\overline{\boldsymbol{Y}})$ by an arithmetic average is the key step in the application of the Delta Theorem (p. 238). In some cases $f'(\boldsymbol{\mu}) = 0$, in which case we need to replace (5.1) by a second order Taylor expansion and the approximation is no longer an average but rather a quadratic form, $f(\overline{\boldsymbol{Y}}) \approx f(\boldsymbol{\mu}) + (\overline{\boldsymbol{Y}} - \boldsymbol{\mu})^T \boldsymbol{B}(\boldsymbol{\mu})(\overline{\boldsymbol{Y}} - \boldsymbol{\mu})$, where $2\boldsymbol{B}(\boldsymbol{\mu})$ is the matrix of second partial derivatives of f. Thus, there are functions of averages that are not approximately averages but rather approximately quadratic functions of averages.

5.1.1c Statistics Implicitly Defined by Averages

A wide class of statistics that are approximated by averages are defined by systems of equations that are themselves averages. In the one-dimensional case, the statistic $\widehat{\theta}$ is defined by solving $G_n(\theta) = 0$, i.e., $G_n(\widehat{\theta}) = 0$, where $G_n(\theta) = n^{-1} \sum g(Y_i, \theta)$ for some function g specific to the estimator. Supposing that $\widehat{\theta}$ is consistent for θ justifies the following Taylor Series expansion when n is large,

$$\begin{aligned} 0 = G_n(\widehat{\theta}) &\approx G_n(\theta) + G_n'(\theta)(\widehat{\theta} - \theta) \\ &= n^{-1} \sum g(Y_i, \theta) + \left\{ n^{-1} \sum g'(Y_i, \theta) \right\} (\widehat{\theta} - \theta), \end{aligned}$$

where $g'(y, \theta) = \partial g(y, \theta)/\partial \theta$. Solving for $\widehat{\theta} - \theta$ yields

$$\widehat{\theta} - \theta \approx -\frac{n^{-1} \sum g(Y_i, \theta)}{n^{-1} \sum g'(Y_i, \theta)}. \tag{5.3}$$

The right-hand side of (5.3) is a ratio estimator which, as shown in (5.2, p. 209), is approximated by an average and thus so too is $\widehat{\theta}$. For an example, suppose that $X_1, \ldots, X_n$ are iid with density $f(x; \theta) = \theta \exp(-\theta x) I(x \geq 0)$, with $\theta > 0$. Then the maximum likelihood estimator of θ is found by solving $n^{-1} \sum g(X_i, \theta) = 0$, where $g(x, \theta) = \theta^{-1} - x$.

Estimators implicitly defined by averages are also known as M-estimators. M-estimators are an important and ubiquitous class of statistics and are studied in greater detail in Chapter 5.

One purpose for describing the three classes of statistics above is to emphasize the simple yet important fact that

Most common statistics are approximately averages (or quadratic functions of averages).

Of course, not all statistics can be approximated by averages. As alluded to previously, extreme order statistics such as the minimum $X_{(1)}$ and the maximum $X_{(n)}$ can not be approximated by averages. Asymptotic theory for extreme value-like statistics differs from that of average-like statistics.

When a statistic can be approximated by an average, the approximating average is usually an average of some function of the sample values. The general form of the approximation for a statistic $T = T(X_1, \ldots, X_n)$ can be written in the form

$$T - T_\infty = \frac{1}{n}\sum_{i=1}^{n} h(X_i) + R_n, \tag{5.4}$$

where T_∞ is the large sample stochastic limit of T, h is a function (often called the *influence function* or *influence curve*), and the remainder R_n is negligibly small as n increases. This expansion with remainder term can be easily derived for the sample variance $s_n^2 = n^{-1}\sum_{i=1}^n (X_i - \overline{X})^2$, where substituting $\sum_{i=1}^n (X_i - \overline{X})^2 = \sum_{i=1}^n (X_i - \mu)^2 - n(\overline{X} - \mu)^2$ leads to

$$s_n^2 - \sigma^2 = \frac{1}{n}\sum_{i=1}^{n}[(X_i - \mu)^2 - \sigma^2] - (\overline{X} - \mu)^2. \tag{5.5}$$

The remainder term $R_n = -(\overline{X} - \mu)^2$ can be shown to be suitably small by using the Markov inequality given in Section 5.5.1 (p. 226). Thus we can see that $s_n^2 - \sigma^2$ is approximated by an average of the transformed random variables $h(X_i) = [(X_i - \mu)^2 - \sigma^2]$. Note that this is the same approximation obtained via the function-of-averages approach.

The strategy of much of large sample theory is then to verify the approximation of $T - T_\infty$ by an average and to use the Laws of Large Numbers and the Central Limit Theorem to deduce the asymptotic properties of the approximating average $n^{-1}\sum_{i=1}^n h(X_i)$. In this chapter we present the tools (theorems and techniques) used to approximate statistics by averages and to derive their asymptotic distribution properties.

We start with the basic notions of convergence of random quantities and then move to the approximations and tools. The knowledgeable reader will notice a strong similarity between the notation and language in these notes and of that found in the classic reference by Serfling (1980, *Approximation Theorems of Mathematical Statistics*). That book is still a very useful reference and we encourage all students to buy it and study it. This chapter is more concise and focuses mainly on the above approximation (5.4, p. 211) as the guiding principle for understanding and proving useful large sample results.

5.2 Types of Stochastic Convergence

Recall that a nonrandom sequence $y_1, y_2, \ldots$ converges to a limiting value c if every neighborhood of c contains all but a finite number of the full sequence, i.e., for each $\epsilon > 0$ there exists an n_ϵ such that for all $n \geq n_\epsilon$, $|y_n - c| < \epsilon$. Common notation for indicating convergence are

$$\lim_{n\to\infty} y_n = c, \qquad \text{and} \qquad y_n \to c \text{ as } n \to \infty.$$

An example of a convergent sequence is $y_n = 1/2^n$ where $c = 0$. An example of a nonconvergent sequence is $y_n = 1/2^n$ for odd values of n, and $y_n = 10$ for even values. In this latter sequence, an infinite number of the y_n remain bounded away from 0.

Convergence of sequences of random variables is complicated by the fact that there is no single unique value to assign to a random variable Y_n, i.e., there is a distribution of possible values. Thus any useful definition of convergence must involve probabilistic concepts. There are three useful modes of convergence: convergence *with probability 1*, *in probability*, and *in distribution*. The sequence Y_n *converges with probability 1* to Y if

$$P\left(\lim_{n\to\infty} Y_n = Y\right) = 1;$$

Y_n *converges in probability* to Y provided that for each $\epsilon > 0$,

$$\lim_{n\to\infty} P(|Y_n - Y| < \epsilon) = 1;$$

and Y_n *converges in distribution* to Y (with distribution function F) provided

$$\lim_{n\to\infty} P(Y_n \leq y) = P(Y \leq y) = F(y),$$

for all y such that $F(y) - F(y^-) = 0$ (points y where F is continuous). Convergence with probability 1 is the strongest form of convergence followed by convergence in probability. Convergence in distribution is the weakest form of convergence but also the most important in large sample inference. It provides the justification for basing standard errors and confidence intervals on the limiting distribution (usually normal) of a parameter estimate and for using critical values from the limiting distribution (usually normal or chi-squared) of a test statistic. However, in deriving the asymptotic distribution of a statistic, it is often essential to use convergence in probability and convenient to use convergence almost surely in intermediate steps of the analysis. Thus it is important to have a good working knowledge of all three modes of convergence.

In the sections that follow we study the three modes of convergence. Throughout these sections we assume a sequence of random variables $Y_1, Y_2, \ldots$ and a limiting random variable Y (a constant c in many applications) all defined on some probability space $(\Omega, \mathcal{A}, P)$. Here Ω is a set of points, $\mathcal{A}$ is a σ-field of sets in Ω closed under complements and countable unions, and P is a probability measure (distribution). Random variables Y are functions on Ω into the real line such that $Y^{-1}(B) \in \mathcal{A}$ for Borel sets B. Important examples are often based on having a sample of independent and identically distributed (iid) random variables. To denote iid random variables with mean μ and finite variance $\sigma^2 > 0$, we use the notation iid(μ, σ^2).

5.2.1 *Convergence with Probability 1 (Almost Sure Convergence)*

The natural extension of convergence of nonrandom sequences to convergence of sequences of random variables gives rise to what is called convergence *with probability 1* (wp1), also known as convergence *almost surely* (a.s.). The key idea is the same as in convergence of nonrandom sequences — that only a finite number of random variables in the sequence can be bounded away from their limit — except that we require the probability of this event to be one. That is, Y_n converges to Y wp1 provided

$$P(\text{for each } \epsilon > 0, \text{ there exists } n_\epsilon \text{ such that for all } n > n_\epsilon, \ |Y_n - Y| < \epsilon) = 1.$$

Example 5.1 (Uniform random variables scaled by n). For a simple example of an almost sure convergent sequence, suppose that $Y_n = U_n/n$ where $U_1, U_2, \ldots$ are iid Uniform(0, 1) random variables. Now consider that for any $\epsilon > 0$, if $n_\epsilon = [1/\epsilon] + 1$ where $[\cdot]$ is the greatest integer function, then for any $n > n_\epsilon$, $|u/n| < 1/n < 1/n_\epsilon < \epsilon$, for any number u, $0 \le u \le 1$. Because $P(\text{all } |U_n| \le 1) = 1$, it follows that for the sequence $Y_1, Y_2, \ldots$,

$$P(\text{for each } \epsilon > 0, \ |Y_n| < \epsilon \text{ for all } n > [1/\epsilon] + 1) = 1,$$

that is, Y_n converges to 0 wp1. Because it uses a construction based on uniform random variables, this example is almost trivial. Nevertheless it illustrates the concept of almost sure convergence. ♦

Convergence wp1 is conceptually similar to convergence of nonrandom sequences because it is based on viewing each realization of an infinite sequence of random variables as a sequence of real numbers. That is, suppose that the random variables $Y_1, Y_2, \ldots$ and Y are all defined on the same probability space $(\Omega, \mathcal{A}, P)$. Then for each $\omega \in \Omega$, $Y_1(\omega), Y_2(\omega), \ldots$ is a sequence of real numbers and we can use the definition of convergence of a sequence of real numbers to describe convergence of that sequence to $Y(\omega)$. If the set of ω's where this convergence takes place has probability 1 with respect to the underlying probability space $(\Omega, \mathcal{A}, P)$, then we say that Y_n converges with probability 1 to Y. Formally we state Y_n *converges with probability 1* to Y if

$$P\left(\lim_{n\to\infty} Y_n = Y\right) = 1.$$

By this we mean

$$P\left(\omega \in \Omega : \lim_{n\to\infty} Y_n(\omega) = Y(\omega)\right) = 1.$$

That is, there exists a subset Ω_0 of Ω with $P(\Omega_0) = 1$, such that for each $\omega \in \Omega_0$, the real sequence $Y_1(\omega), Y_2(\omega), \ldots$ converges to the real number $Y(\omega)$, i.e.,

$$\lim_{n\to\infty} Y_n(\omega) = Y(\omega).$$

Our notation for this convergence is

$$Y_n \xrightarrow{wp1} Y \qquad \text{as } n \to \infty.$$

The common alternative expression *almost sure convergence* uses the notation $Y_n \xrightarrow{a.s.} Y$.

5.2.1a Strong Law of Large Numbers

The Strong Law of Large Numbers (SLLN) for iid random variables states that if $X_1, \ldots, X_n$ are iid with finite mean $\mathrm{E}(X_1) = \mu$, then the sample mean $\overline{X}$ converges with probability 1 to μ. For reference we state

Theorem 5.1 (Strong Law of Large Numbers). *If $X_1, \ldots, X_n$ are iid with finite mean $E(X_1) = \mu$, then $\overline{X} \xrightarrow{wp1} \mu$ as $n \to \infty$.*

In some ways *with probability 1* convergence is easier to use than *in probability* convergence, discussed in the next section, because the connection to convergence of nonrandom sequences is closer. That is, if you know how to prove a result for nonrandom sequences, then it is often clear how to proceed for proving with probability 1 convergence for random sequences as long as you pay some attention to the sets of probability 1 on which the convergence take place.

5.2.2 Convergence in Probability

We now study convergence in probability in greater detail. The sequence Y_n *converges in probability* to Y if for each $\epsilon > 0$,

$$\lim_{n\to\infty} P(|Y_n - Y| < \epsilon) = 1;$$

or equivalently

$$\lim_{n\to\infty} P(|Y_n - Y| > \epsilon) = 0.$$

For proofs, it is sometimes useful to note an alternate more technical definition: for each $\epsilon > 0$, there exists n_ϵ such that for all $n \geq n_\epsilon$,

$$P(|Y_n - Y| < \epsilon) > 1 - \epsilon.$$

Shorthand notation is

$$Y_n \xrightarrow{p} Y \qquad \text{as } n \to \infty.$$

Example 5.2 (Binary sequence). Let us start with a simple example:

$$\begin{aligned} Y_n &= \frac{1}{2^n} && \text{with probability } p_n; \\ &= 10 && \text{with probability } 1 - p_n. \end{aligned}$$

Then for $n > -\log \epsilon / \log 2$,

$$P(|Y_n - 0| > \epsilon) = 1 - p_n,$$

and thus $Y_n \xrightarrow{p} 0$ if $p_n \to 1$. An advantage of convergence in probability is that a small and diminishing amount of "bad behavior" (here $Y_n = 10$) is allowed for a random sequence, and convergence can still follow. ♦

The sequence in the next example converges in probability but does not converge wp1.

Example 5.3 (Bernoulli random variables). Suppose that $Y_1, Y_2, \ldots$ are independent Bernoulli trials with $P(Y_n = 1) = E(Y_n) = 1/n$. The mean and variance of this sequence converge to 0. Thus if the sequence converges in any sense, the only sensible limit is 0. We now show that this sequence does not converge to 0 wp1. Because the only possible values for Y_n are 0 and 1, $|Y_n - 0| < \epsilon$ if and only if $Y_n = 0$ for $\epsilon < 1$. Thus the condition for convergence to 0 wp1 is that $P(\text{there exists } m, \text{ such that all } Y_n = 0 \text{ for } n > m) = 1$. However,

$$\begin{aligned} P(\text{all } Y_n = 0 \text{ for } n > m) &< P(Y_{m+1} = 0, Y_{m+2} = 0, \ldots, Y_{m+B} = 0) \\ &= \left(\frac{m+1-1}{m+1}\right)\left(\frac{m+2-1}{m+2}\right)\cdots\left(\frac{m+B-1}{m+B}\right) \\ &= \frac{m}{m+B} \end{aligned}$$

for every positive integer B. Letting $B \to \infty$ shows that $P(\text{all } Y_n = 0 \text{ for } n > m) = 0$, and thus the sequence does not converge to 0 wp1. Nevertheless, if one had to predict the value of Y_n for large n, the prediction $\widehat{Y}_n = 0$ is correct with probability $P(Y_n = \widehat{Y}_n) = (n-1)/n \to 1$, showing that $\widehat{Y}_n \xrightarrow{p} 0$. Even though the sequence does not converge wp1, the asymptotic behavior of the sequence ($\widehat{Y}_n \xrightarrow{p} 0$) is regular enough to be useful for statistical inference. ◆

5.2.2a Weak Law of Large Numbers

Many important applications of convergence in probability are connected to the Weak Law of Large Numbers (WLLN). For $X_1, \ldots, X_n$ that are iid with finite mean $E(X_1) = \mu$, the WLLN states that $\overline{X}$ converges in probability to μ as $n \to \infty$. For reference, we state

Theorem 5.2 (Weak Law of Large Numbers). *If $X_1, \ldots, X_n$ are iid with finite mean $\mu = E(X_1)$, then*

$$\overline{X} \xrightarrow{p} \mu \qquad \text{as } n \to \infty.$$

If we assume that X_1 also has a finite variance σ^2, then the WLLN follows directly from Chebychev's inequality:

$$P(|\overline{X} - \mu| > \epsilon) \leq \frac{E(\overline{X} - \mu)^2}{\epsilon^2} = \frac{\text{Var}(\overline{X})}{n\epsilon^2} = \frac{\sigma^2}{n\epsilon^2}.$$

Other versions of the weak law of large numbers are available that allow weakening of the iid assumptions in Theorem 5.2 but add stronger moment assumptions. The following theorem is an example, directly requiring $\text{Var}(\overline{X}) \to 0$.

Theorem 5.3. *If $X_1, \ldots, X_n$ are random variables with finite means $\mu_i = E(X_i)$, variances $E(X_i - \mu_i)^2 = \sigma_i^2$, and covariances $E(X_i - \mu_i)(X_j - \mu_j) = \sigma_{ij}$ such that*

$$Var(\overline{X}) = \frac{1}{n^2}\left[\sum_{i=1}^{n} \sigma_i^2 + 2\sum_{i=1}^{n-1}\sum_{j=i+1}^{n} \sigma_{ij}\right] \to 0 \qquad \text{as } n \to \infty,$$

then

$$\overline{X} - \overline{\mu} \xrightarrow{p} 0 \qquad \text{as } n \to \infty,$$

where $\overline{\mu} = n^{-1}\sum_{i=1}^{n} \mu_i$.

Example 5.4 (Autoregressive Time Series). Consider the first-order autoregressive time series

$$X_i - \mu = \rho(X_{i-1} - \mu) + e_i, \qquad i = 1, 2, \ldots,$$

where the e_i are iid $\mathrm{N}(0, \sigma_e^2)$, X_0 is distributed as $\mathrm{N}\left\{\mu, \sigma_e^2/(1-\rho^2)\right\}$ and independent of the e_i, and $0 < |\rho| < 1$. Then

$$\mathrm{Cov}(X_i, X_{i+h}) = \sigma_{i,i+h} = \frac{\rho^h}{1-\rho^2}\sigma_e^2, \qquad h = 0, 1, \ldots$$

and $\sigma_i^2 = \sigma_{ii} = [\sigma_e^2/(1-\rho^2)]$. Thus,

$$\begin{aligned}\mathrm{Var}(\overline{X}) &= \frac{1}{n^2}\frac{\sigma_e^2}{1-\rho^2}\left[n + 2\sum_{i=1}^{n-1}\sum_{j=i+1}^{n}\rho^{j-i}\right] \\ &= \frac{1}{n^2}\frac{\sigma_e^2}{1-\rho^2}\left[n + 2(n-1)\left(\frac{\rho}{1-\rho}\right) - 2\frac{(\rho^2-\rho^{n+1})}{(1-\rho)^2}\right] \to 0\end{aligned}$$

as $n \to \infty$ and $\overline{X} \xrightarrow{p} \mu$ by Theorem 5.3. ♦

Example 5.5 (Equicorrelation). In the previous example it is crucial for the correlations between members of the sequence to die out as the distance between their indices grows. A simple illustration of what happens otherwise is as follows. Suppose that $Z_0, Z_1, \ldots$ is a sequence of independent standard normal random variables. Define

$$X_i = Z_0 + Z_i, \qquad i = 1, 2, \ldots$$

Then $\mathrm{Cov}(X_i, X_j) = 1$ for $i \neq j$ and $\mathrm{Var}(\overline{X}) = n^{-2}[2n + n(n-1)] = 1 + n^{-1}$, and clearly the conditions of Theorem 5.3 do not hold. It is easy to see that the conditions of Theorem 5.2 or Theorem 5.3 hold for $\overline{X} - Z_0$ and thus $\overline{X} \xrightarrow{p} Z_0$ as $n \to \infty$. This example also illustrates a case where the stochastic limit is a random variable rather than a constant. ♦

5.2.3 *Convergence in Distribution*

Suppose that the random variables $Y, Y_1, Y_2, \ldots$ have distribution functions $F(y), F_1(y), F_2(y), \ldots$ respectively. Let C_F be the subset of $(-\infty, \infty)$ where F is continuous. Then Y_n *converges in distribution* to Y if F_n converges to F pointwise on C_F, i.e.,

$$\lim_{n\to\infty} F_n(y) = F(y) \quad \text{for each} \quad y \in C_F.$$

For this convergence our shorthand notation is

$$Y_n \xrightarrow{d} Y \qquad \text{as } n \to \infty.$$

An alternative expression is *convergence in law* with the related notation

$$Y_n \xrightarrow{L} Y \qquad \text{as } n \to \infty.$$

The shorthand notations are useful and concise, but it is important to remember that the convergence is for the distribution functions and not for the random variables themselves. For example, if Z is a standard normal random variable and $Y_n \xrightarrow{d} Z$, then it is also true that $Y_n \xrightarrow{d} -Z$ even though it seems contradictory.

The sequence in the following example converges in distribution but not wp1 or in probability.

Example 5.6 (Convergence in distribution but not in probability). Suppose that $Y_1, Y_2, \ldots$ are independent random variables with $Y_n \sim \mathrm{N}(\mu, \sigma_n^2)$, where the sequence $\sigma_n^2 \to \sigma^2 > 0$ as $n \to \infty$. This sequence does not converge in probability (see Problem 5.2, p. 263). However,

$$\begin{aligned} P(a < Y_n \leq b) &= \Phi\{(b-\mu)/\sigma_n\} - \Phi\{(a-\mu)/\sigma_n\} \\ &\to \Phi\{(b-\mu)/\sigma\} - \Phi\{(a-\mu)/\sigma\} = P(a < Y \leq b), \end{aligned}$$

for all $a \leq b$, where $Y \sim \mathrm{N}(\mu, \sigma^2)$ and Φ is the distribution function of a standard normal random variable.

Thus $Y_n \xrightarrow{d} Y$ and the $\mathrm{N}(\mu, \sigma^2)$ distribution provides a good approximation to the distribution of Y_n for large n even though Y_n does not converge in probability or wp1. ♦

Because convergence in distribution is about convergence of distribution functions, there is no need for the random variables to be defined on the same sample space. However, in many applications the random variables in the sequence are defined on the same probability space, but the limiting random variable need not be. In fact, we do not even need a limiting random variable, but merely a limiting distribution function. Nevertheless, it is convenient to use the language of limiting random variables as in the Central Limit Theorem discussed next.

5.2.3a Central Limit Theorem

Consider the standardized sample mean from a sample of iid(μ, σ^2) random variables, $Y_n = n^{1/2}(\overline{X} - \mu)/\sigma$. The Central Limit Theorem (CLT) states that $P(Y_n \leq y) \to \Phi(y)$ for each $y \in (-\infty, \infty)$, where Φ is the distribution function of a standard normal random variable. We now state the Lindeberg-Levy version of the CLT.

Theorem 5.4 (Central Limit Theorem). *If $X_1, \ldots, X_n$ are iid with mean $E(X_1) = \mu$ and variance $Var(X_1) = \sigma^2 < \infty$, then*

$$\frac{\overline{X} - \mu}{\sigma/\sqrt{n}} \xrightarrow{d} Z \qquad as \ n \to \infty,$$

where Z is a standard normal random variable.

Example 5.7 (Normal approximation to the binomial). If Y is a binomial(n, p) random variable, then Y has the same distribution as $\sum_{i=1}^{n} X_i$, where $X_1, \ldots, X_n$ are iid Bernoulli(p) random variables, $\mathrm{E}(X_i) = p$, $\mathrm{Var}(X_i) = p(1-p)$. The normal approximation for calculations like $P(Y \geq k)$ comes from the CLT,

$$\begin{aligned} P(Y \geq k) = P(\overline{X} \geq k/n) &= P\left\{\frac{\sqrt{n}(\overline{X} - p)}{\sqrt{p(1-p)}} \geq \frac{\sqrt{n}(k/n - p)}{\sqrt{p(1-p)}}\right\} \\ &\approx 1 - \Phi\left\{\frac{(k - np)}{\sqrt{np(1-p)}}\right\}, \end{aligned}$$

where Φ is the standard normal distribution function. Because of the discreteness of the binomial distribution, continuity corrections such as subtracting 1/2 from k in the above formula, can improve this approximation. ♦

5.2.3b Sample Size and the Central Limit Theorem

Students at all levels often ask, "How large does n have to be for the normal approximation to be useful?" The answer to that question is not quite as simple as the $n \geq 30$ often given in introductory courses. The CLT says that $P(Z_n \leq t) \approx \Phi(t)$, where $Z_n = \sqrt{n}(\overline{X} - \mu)/\sigma$. An improved approximation, called an *Edgeworth Expansion*, states that if X_1 has a finite third moment with skewness coefficient $\mathrm{Skew}(X_1) = E(X_1 - \mu)^3/\sigma^3$, then

$$P(Z_n \leq t) \approx \Phi(t) - \frac{\mathrm{Skew}(X_1)}{\sqrt{n}} \frac{(t^2 - 1)}{6} \phi(t), \tag{5.6}$$

where ϕ is the standard normal density (see, e.g., Feller 1966, p. 539). The key feature of this improved approximation is that $\mathrm{Skew}(X_1)/\sqrt{n}$ is the main quantity regulating how fast the distribution of $\overline{X}$ (or equivalently of Z_n) approaches a normal distribution. For a symmetric distribution $\mathrm{Skew}(X_1) = 0$, and the CLT approximation is very fast.

Example 5.8 (Convergence to normality for the binomial). As an illustration, consider the previous binomial example where simple calculations show that the skewness coefficient of a Bernoulli random variable is $\text{Skew}(X_1) = (1 - 2p)/\sqrt{p(1-p)}$. Thus for $p = 1/2$ the skewness is zero suggesting fast convergence and a good approximation for small n, whereas at $p = .10$, $\text{Skew}(X_1) = 2.67$ and a much larger n is required. In this case, formal rules such as $\min\{np, n(1-p)\} \geq 10$ are often found in introductory texts. Here, for $p = .10$ that rule leads to $n = 100$ with the corresponding $\text{Skew}(X_1)/\sqrt{100} = .267$, whereas at $p = 1/2$, it leads to $n = 20$. ♦

In general, one should pay attention to $\text{Skew}(X_1)/\sqrt{n}$ instead of simple rules such as $n \geq 30$. To explore this further, note that the skewness coefficient of $\overline{X}$ is given by

$$\text{Skew}(\overline{X}) = \frac{\text{Skew}(X_1)}{\sqrt{n}}.$$

This expression follows from several simple facts: $\text{Skew}(a + bX_1) = \text{Skew}(X_1)$ and $\text{E}(\overline{X} - \mu)^3 = n^{-3}\sum_{i=1}^{n} \text{E}(X_i - \mu)^3$. Now suppose that we have a sample $X_1, \ldots, X_{30}$ from some distribution with skewness coefficient $\text{Skew}(X_1)$. Let us define new random variables $Y_1 = (X_1 + X_2)/2, Y_2 = (X_3 + X_4)/2, \ldots, Y_{15} = (X_{29} + X_{30})/2$. Then $\overline{Y} = \overline{X}$ and

$$\text{Skew}(\overline{Y}) = \frac{\text{Skew}(Y_1)}{\sqrt{15}} = \frac{\text{Skew}(X_1)/\sqrt{2}}{\sqrt{15}} = \frac{\text{Skew}(X_1)}{\sqrt{30}} = \text{Skew}(\overline{X}).$$

Clearly, the normal approximation for $\overline{Y}$ must be the same as for $\overline{X}$ but the sample sizes are $n^* = 15$ for the Y sample and $n = 30$ for the original sample. Thus, for thinking about the quality of the normal approximation in the CLT, we must look at $\text{Skew}(X_1)/\sqrt{n}$ and not simply at n or $\text{Skew}(X_1)$ by themselves.

5.2.3c Convergence in Distribution for the Sample Extremes

We will see that most convergence in distribution results follow from some version of the CLT coupled with an approximation-by-averages representation. As mentioned in the chapter introduction, though, sample extremes have a different limiting behavior from the CLT. In this section we describe some limit results for the sample extremes. When working with sample extremes, it is often best to directly use the definition of convergence of distribution functions. We illustrate the approach in a simple example and then give the main classical theorem about the limiting distribution of the sample extremes.

Example 5.9 (Maximum of an exponential sample). Suppose that $X_1, \ldots, X_n$ are a random sample from the standard exponential distribution with distribution function $F(x) = 1 - \exp(-x)$, and consider the largest sample value $X_{(n)}$. We take advantage of the fact that if the largest value is less than x, then so are each of the sample values and vice versa. Thus for each $x \in (-\infty, \infty)$, as $n \to \infty$, we have

$$
\begin{aligned}
F_n(x) = P\left(X_{(n)} - \log n \le x\right) &= P\left(X_1 - \log n \le x, \ldots, X_n - \log n \le x\right) \\
&= \{1 - \exp(-x - \log n)\}^n \\
&= \left\{1 - \frac{\exp(-x)}{n}\right\}^n \\
&\to \exp\{-\exp(-x)\} = G(x).
\end{aligned}
$$

♦

As you might recall from previous chapters, $G(x)$ is called the extreme value or Gumbel distribution function. A famous result due to Gnedenko (1943) says that if $(X_{(n)} - a_n)/b_n$ converges in distribution for some real sequences a_n and $b_n > 0$, then the limiting distribution must be one of only three types. The distribution function $G(x) = \exp(-e^{-x})$ in the previous example is the best-known of the three types. Here is a statement of Gnedenko's Theorem (see, for example, Leadbetter et al. 1983, p. 10–12).

Theorem 5.5 (Extremal Limiting distributions). *Let $X_1, \ldots, X_n$ be iid with maximum $X_{(n)}$. If for some constants a_n, $b_n > 0$,*

$$
P\left\{\frac{X_{(n)} - a_n}{b_n} \le x\right\} \to G(x),
$$

for some nondegenerate distribution function $G(x)$, then $G(x)$ must be of the form $F_0\{(x - \mu)/\sigma\}$, $-\infty < \mu < \infty$, $\sigma > 0$, where F_0 is one of the following three types:

$$
\textit{Type I:} \quad F_0(y) = \exp\left(-e^{-y}\right) \qquad -\infty < y < \infty,
$$

$$
\textit{Type II:} \quad F_0(y) = \begin{cases} 0 & y \le 0, \\ \exp\left(-y^{-\alpha}\right) & y > 0, \textit{ for some } \alpha > 0; \end{cases}
$$

$$
\textit{Type III:} \quad F_0(y) = \begin{cases} \exp\{-(-y)^{\alpha}\} & y \le 0, \textit{ for some } \alpha > 0, \\ 1 & y > 0. \end{cases}
$$

As mentioned above, Type I is the most common limiting distribution for $X_{(n)}$ and occurs when X_i has normal distributions, gamma distributions, and many more. If X_i has a Pareto distribution, then Type II is the correct limiting form. If X_i has a uniform distribution on any interval, then Type III is the correct form. Figure 5.1 shows the densities of these distributions; for Types II and III the displayed densities use $\alpha = 1$. Note that if Y has a Type II density, then $1/Y$ has a Weibull density with shape parameter α. If Y has a Type III density, then $-Y$ has a Weibull density with shape parameter α.

For statistical inference where the data are sample extremes, it is common to put all three Types together with scale and location parameters, and estimate the three parameters by maximum likelihood. For example, Coles and Dixon (1999) give the

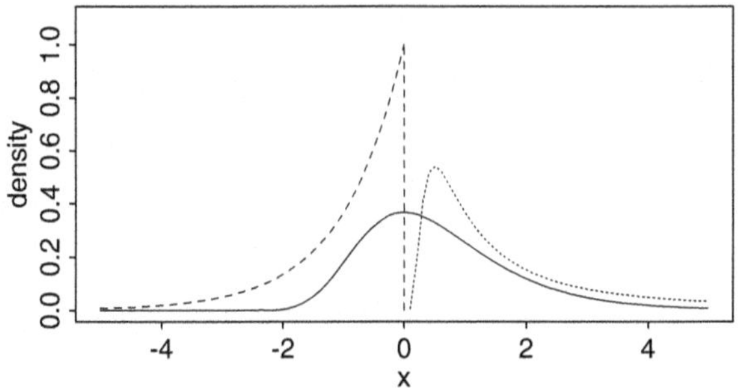

Fig. 5.1 Densities of Extremal Types I (solid), II with $\alpha = 1$ (dotted), and III with $\alpha = 1$ (dashed).

following parameterization of the resulting *Generalized Extreme Value* distribution function:

$$F(x; \mu, \sigma, \eta) = \exp\left[-\{1 + \eta(x - \mu)/\sigma\}^{-1/\eta}\right], \quad \eta \neq 0$$
$$= \exp[-\exp\{-(x - \mu)/\sigma\}] \qquad \eta = 0.$$

defined on $\{x : 1 + \eta(x - \mu)/\sigma > 0\}$ for $\mu \in (-\infty, \infty)$, $\sigma > 0$, and $\eta \in (-\infty, \infty)$, where $\eta = 0$ corresponds to Type I, $\eta > 0$ corresponds to Type II, and $\eta < 0$ corresponds to Type III.

Theorem 5.5 only states what the limiting distribution is when $X_{(n)}$ has a limiting distribution; it may not have a limiting distribution. For example, if the X_i have a Poisson distribution, then there are no constants $a_n, b_n > 0$ such that $(X_{(n)} - a_n)/b_n$ has a nondegenerate limiting distribution. Theorem 5.5 also covers the smallest value, $X_{(1)}$, by noting that $X_{(1)} = -\max\{-X_1, \ldots, -X_n\}$. See Leadbetter et al. (1983, Ch. 1) for more details and examples.

5.2.3d Uniform Convergence in Distribution

The most common limit distributions encountered in statistical inference are normal, chi-squared , or extreme value and therefore have continuous distribution functions. Thus the definition of convergence in distribution usually entails pointwise convergence of distribution functions at each $y \in (-\infty, \infty)$ or $y \in (0, \infty)$. In this case we automatically have *uniform convergence* of the distribution functions as provided by the following classical theorem attributed to Pólya:

Theorem 5.6 (Pólya). *If the sequence of distribution functions F_n converges pointwise to F, where F is continuous on $(-\infty, \infty)$, then*

$$\lim_{n\to\infty} \sup_{y\in(-\infty,\infty)} |F_n(y) - F(y)| = 0.$$

5.3 Relationships between Types of Convergence

The most important relationships between the modes of convergence are that with probability 1 convergence is the strongest and implies convergence in probability which in turn implies convergence in distribution. In symbols we have

$$Y_n \xrightarrow{wp1} Y \quad \Longrightarrow \quad Y_n \xrightarrow{p} Y \quad \Longrightarrow \quad Y_n \xrightarrow{d} Y.$$

Recall that Example 5.3 (p. 215) and Example 5.6 (p. 218) showed that the arrows cannot be reversed without some additional conditions. In the following we give just a few more details about the relationships.

1. $Y_n \xrightarrow{wp1} Y$ implies $Y_n \xrightarrow{p} Y$. This follows directly from the following equivalent condition for $\xrightarrow{wp1}$:

$$Y_n \xrightarrow{wp1} Y \Longleftrightarrow \lim_{n\to\infty} P(|Y_m - Y| < \epsilon, \text{ all } m \geq n) = 1 \quad \text{for each } \epsilon > 0. \tag{5.7}$$

   See Serfling (1980, p. 7) for a proof. This latter condition also gives a way to obtain $\xrightarrow{wp1}$ from probability calculations used for convergence in probability:

$$\sum_{n=1}^{\infty} P(|Y_n - Y| > \epsilon) < \infty \quad \text{for each } \epsilon > 0 \text{ implies } \; Y_n \xrightarrow{wp1} Y \quad \text{as } n \to \infty.$$

   Also because of the Markov inequality $P(|Y_n - Y| > \epsilon) \leq E|Y_n - Y|^r/\epsilon^r$, we can replace $P(|Y_n - Y| > \epsilon)$ by $E|Y_n - Y|^r$ in the above condition and obtain $\xrightarrow{wp1}$.
2. $Y_n \xrightarrow{p} Y$ implies $Y_n \xrightarrow{d} Y$. This is not obvious, but one method of proof follows from Slutsky's Theorem (p. 241) after noting that $Y_n = (Y_n - Y) + Y$.
3. For a constant c, $Y_n \xrightarrow{p} c$ is equivalent to $Y_n \xrightarrow{d} c$.

Serfling (1980, p. 6–11) gives more relationships among the types of convergence, but the ones above are the most important.

5.4 Extension of Convergence Definitions to Vectors

Now we extend the definitions of the three types of convergence to random vectors.

5.4.1 Convergence wp1 and in Probability for Random Vectors

From real analysis we have that a k-vector sequence $\boldsymbol{y}_n$ converges to $\boldsymbol{c}$ if the Euclidean distance between $\boldsymbol{y}_n$ and $\boldsymbol{c}$ converges to zero, that is,

$$\| \boldsymbol{y}_n - \boldsymbol{c} \| \to 0 \qquad \text{as } n \to \infty,$$

where $\| x \| = \left(\sum_{i=1}^k x_i^2\right)^{1/2}$. This norm convergence is equivalent to the convergence of the coordinates y_{in} to c_i, respectively, $i = 1, \ldots, k$. The generalization of convergence with probability 1 and in probability for random vectors $\boldsymbol{Y}_n$ (and matrices) to $\boldsymbol{Y}$ is similar. We say that $\boldsymbol{Y}_n$ converges wp1 to $\boldsymbol{Y}$ if

$$\|\boldsymbol{Y}_n - \boldsymbol{Y} \| \xrightarrow{wp1} 0 \qquad \text{as } n \to \infty,$$

and $\boldsymbol{Y}_n$ converges in probability to $\boldsymbol{Y}$ if

$$\| \boldsymbol{Y}_n - \boldsymbol{Y} \| \xrightarrow{p} 0 \qquad \text{as } n \to \infty.$$

Similar to the case of nonrandom sequences, the stochastic convergence of this Euclidean distance $\|\boldsymbol{Y}_n - \boldsymbol{Y} \|$ is equivalent to the componentwise convergence. That is, a sequence of vectors (or matrices) $\boldsymbol{Y}_n$ converges with probability 1 to $\boldsymbol{Y}$ if and only if each component of $\boldsymbol{Y}_n$ converges with probability 1 to the corresponding component of $\boldsymbol{Y}$. The same statement holds if we replace "with probability 1" by "in probability."

Thus, for example, the Strong Law of Large Numbers (Theorem 5.1, p. 214) and Weak Law of Large Numbers (Theorem 5.2, p. 216) also hold immediately for multivariate sample means $\overline{\boldsymbol{X}}$ because the sample mean of $\boldsymbol{X}_1, \ldots, \boldsymbol{X}_n$

$$\overline{\boldsymbol{X}} = \frac{1}{n}\sum_{i=1}^{n} \boldsymbol{X}_i = \left(\frac{1}{n}\sum_{i=1}^{n} X_{1i}, \ldots, \frac{1}{n}\sum_{i=1}^{n} X_{ki}\right)$$

is equal to a vector of sample means.

5.4.2 Convergence in Distribution for Vectors

Because convergence in distribution refers to pointwise convergence of distribution functions, it is quite natural for convergence in distribution of random vectors to mean convergence of the associated distribution functions. Thus, we say $\boldsymbol{Y}_n \xrightarrow{d} \boldsymbol{Y}$ as $n \to \infty$ if the distribution function of $\boldsymbol{Y}_n$ converges to the distribution function of $\boldsymbol{Y}$ at all points $\boldsymbol{y} \in R^k$ where the distribution function of the limiting random vector

Y, $F(y) = P(Y \le y)$, is continuous. Note that for k-vectors x and y, $x \le y$ means $x_1 \le y_1, \ldots, x_k \le y_k$.

This natural definition of convergence in distribution to random k-vectors is quite different from the extensions of the last section for wp1 and in probability convergence. Moreover, because the marginal distributions of a random vector are not sufficient to determine its joint distribution, we shall see that convergence in distribution of a vector is not equivalent to convergence of its components (an exception is when the components are independent).

But first we now state the simplest vector version of the CLT; the proof is left for Problem 5.48 (p. 272).

Theorem 5.7 (Multivariate CLT). *If $\boldsymbol{X}_1, \ldots, \boldsymbol{X}_n$ are iid random k-vectors with finite mean $E(\boldsymbol{X}_1) = \boldsymbol{\mu}$ and covariance matrix $\boldsymbol{\Sigma}$, then*

$$\sqrt{n}(\overline{\boldsymbol{X}} - \boldsymbol{\mu}) \xrightarrow{d} \boldsymbol{Y} \qquad \text{as } n \to \infty,$$

where $\boldsymbol{Y}$ is a $N_k(\boldsymbol{0}, \boldsymbol{\Sigma})$ random vector.

Example 5.10 (Central Limit Theorem for multinomial vector). Suppose that $\boldsymbol{X}_1, \ldots, \boldsymbol{X}_n$ are iid multinomial$(1; p_1, \ldots, p_k)$ random vectors. Then the sum of these vectors is multinomial$(n; p_1, \ldots, p_k)$ and we get directly from Theorem 5.7 that

$$\sqrt{n}\,(\widehat{p}_1 - p_1, \ldots, \widehat{p}_k - p_k)^T \xrightarrow{d} \boldsymbol{Y} \qquad \text{as } n \to \infty,$$

where $\widehat{\boldsymbol{p}} = n^{-1}\sum_{i=1}^n \boldsymbol{X}_i$ and $\boldsymbol{Y}$ is a normal random vector with mean $\boldsymbol{0}$ and $\text{Cov}(Y_i, Y_j) = -p_i p_j$ for $i \ne j$ and $\text{Cov}(Y_i, Y_j) = p_i(1 - p_i)$ for $i = j$. ♦

As mentioned earlier, vector convergence in distribution does not follow from convergence of the individual components. Here is a simple example to illustrate.

Example 5.11 (Marginal convergence but not joint convergence). Suppose that Z_1, Z_2, Z_3 are independent standard normal random variables, and let $\boldsymbol{X}_n = (X_{1n}, X_{2n})^T$ be $(Z_1 + Z_3, Z_2 + Z_3)^T$ for odd values of n and $(Z_1 + Z_3, Z_2 - Z_3)^T$ for even values of n. Then, clearly the individual components X_{1n} and X_{2n} converge in distribution to N$(0, 2)$ random variables because they have that distribution for each n. But X_n has a bivariate normal distribution with mean $\boldsymbol{0}$, variances = 2, and correlation = 1/2 for odd n, and correlation = $-1/2$ for even n. Thus the distribution of $\boldsymbol{X}_n$ does not converge to any distribution. ♦

Note that this counterexample depends on the components being related. However, if the components are **independent** and converge in distribution individually, then the vector converges in distribution. We state this as

Theorem 5.8 (Joint convergence under independence). *Suppose that the components of $\boldsymbol{X}_n$ are independent and converge in distribution to the components of $\boldsymbol{X}$ that are also independent. Then $\boldsymbol{X}_n \xrightarrow{d} \boldsymbol{X}$ as $n \to \infty$.*

Proof. Assume that $P(X_i \le x)$ is continuous at $x = x_i, i = 1, \ldots, k$.

$$P(\boldsymbol{X}_n \le \boldsymbol{x}) = P(X_{1n} \le x_1, \ldots, X_{kn} \le x_k)$$
$$= \prod_{i=1}^{k} P(X_{in} \le x_i) \longrightarrow \prod_{i=1}^{k} P(X_i \le x_i) = P(\boldsymbol{X} \le \boldsymbol{x}).$$

5.5 Tools for Proving Large Sample Results

We have stated the most important definitions and relationships between modes of convergence for random quantities. Now we are ready to discuss the tools that make asymptotic statistical analysis fairly routine.

5.5.1 *Markov Inequality*

An important tool used for proving convergence in probability is the basic Markov inequality: for $r > 0$

$$P(|Y_n - Y| > \epsilon) \le \frac{E|Y_n - Y|^r}{\epsilon^r}.$$

For $r = 2$ and $Y = c$, this becomes Chebychev's inequality (already mentioned in connection with the WLLN, p. 216)

$$P(|Y_n - c| > \epsilon) \le \frac{E(Y_n - c)^2}{\epsilon^2} = \frac{\text{Var}(Y_n) + \{E(Y_n) - c\}^2}{\epsilon^2}.$$

Use of Chebychev's inequality often requires stronger moment conditions than are necessary. This was seen in the WLLN where a proof using Chebychev's inequality requires a finite second moment, but the WLLN only requires a finite first moment (but a harder proof that we have not given). A second illustration of this fact is the sample variance as follows.

Example 5.12 (Convergence in probability of the sample variance). For an iid sample $X_1, \ldots, X_n$ with finite fourth moment, we have

$$P(|\,s_n^2 - \sigma^2\,| > \epsilon) \le \frac{E(s_n^2 - \sigma^2)^2}{\epsilon^2} = \frac{\sigma^4}{\epsilon^2}\left(\frac{n}{n-1}\right)^2\left(\frac{2}{n-1} + \frac{\text{Kurt}(X_1) - 3}{n}\right)$$
$$+ \frac{\sigma^4}{\epsilon^2}\left(\frac{n}{n-1} - 1\right)^2,$$

where $\text{Kurt}(X_1) = \mu_4/\sigma^4$. Thus $s_n^2 \xrightarrow{p} \sigma^2$ as $n \to \infty$. We see in the next section that only a finite second moment is required for this convergence. ♦

5.5.2 Continuous Functions of Convergent Sequences

This subsection gives powerful results for proving convergence of functions $\boldsymbol{g}(\boldsymbol{Y}_n)$ of random sequences $\boldsymbol{Y}_n$. This approach usually gives the best results for proving wp1 and convergence in probability. Before stating the main theorem, we first review briefly the definition of a continuous function. Here we state these definitions for real-valued functions of a real sequence, but they are also true for vector-valued functions of vector sequences.

A real function f is continuous at a point c if

$$\lim_{y \to c} f(y) = f(c).$$

A lesser known equivalent definition is: if

$$\lim_{y_n \to c} f(y_n) = f(c)$$

for all sequences y_n that converge to c, then f is continuous at c. The following theorem is analogous to this latter definition but allows some "bad behavior" on sets in the sample space of the limiting random variable having probability 0. For convergence with probability 1 or in probability, the limiting random variable is often a constant c. Then the assumption of the following theorem is merely that g is continuous at c. Note also that the theorem is stated for vector functions $\boldsymbol{g}$ defined on vectors $\boldsymbol{Y}_n$.

Theorem 5.9 (Continuity). *Let $\boldsymbol{g}$ be a vector-valued function defined on R^k that is continuous except for a set A_g where $P(Y \in A_g) = 0$. Then as $n \to \infty$*

(i)

$$\boldsymbol{Y}_n \xrightarrow{wp1} \boldsymbol{Y} \quad \Longrightarrow \quad g(\boldsymbol{Y}_n) \xrightarrow{wp1} g(\boldsymbol{Y}).$$

(ii)

$$\boldsymbol{Y}_n \xrightarrow{p} \boldsymbol{Y} \quad \Longrightarrow \quad g(\boldsymbol{Y}_n) \xrightarrow{p} g(\boldsymbol{Y}).$$

(iii)

$$\boldsymbol{Y}_n \xrightarrow{d} \boldsymbol{Y} \quad \Longrightarrow \quad g(\boldsymbol{Y}_n) \xrightarrow{d} g(\boldsymbol{Y}).$$

Proof of (i). Let Ω_0 be the set with $P(\Omega_0) = 1$ where $\boldsymbol{Y}_n$ converges. Then for $\omega \in \Omega_0 \cap Y^{-1}(A_g^c)$, $\boldsymbol{Y}_n(\omega) \to \boldsymbol{Y}(\omega)$ and $g(\boldsymbol{Y}_n(\omega)) \to g(\boldsymbol{Y}(\omega))$, and $P\left\{\Omega_0 \cap Y^{-1}(A_g^c)\right\} = 1$. ■

Proof of (ii). For simplicity we deal only with real Y_n and $Y = c$. By the definition of continuity, for given $\epsilon > 0$, there exists δ_ϵ such that $|y_n - c| < \delta_\epsilon$ implies that

$|g(y_n) - g(c)| < \epsilon$. Let $n_{\epsilon,\delta}$ be such that for $n \geq n_{\epsilon,\delta}$, $P(|Y_n - c| < \delta_\epsilon) > 1 - \epsilon$. Then, for $n \geq n_{\epsilon,\delta}$,

$$P(|g(Y_n) - g(c)| < \epsilon) \geq P(|Y_n - c| < \delta_\epsilon) > 1 - \epsilon.$$

For the proof of part (iii), see Serfling (1980, p. 25).

Two useful corollaries, combined in the following theorem, are obtained by taking g to be the sum and product of the k components of $\boldsymbol{y}$, $g(\boldsymbol{y}) = \sum_{j=1}^k y_j$ and $g(\boldsymbol{y}) = \prod_{j=1}^k y_j$.

Theorem 5.10 (Continuity: products and sums). *If* $\boldsymbol{Y}_n \xrightarrow{wp1} \boldsymbol{Y}$, *then* $\sum_{j=1}^k Y_{jn} \xrightarrow{wp1} \sum_{j=1}^k Y_j$ *and* $\prod_{j=1}^k Y_{jn} \xrightarrow{wp1} \prod_{j=1}^k Y_j$. *The result also holds for* $\xrightarrow{p}$ *and* $\xrightarrow{d}$.

Now we give a variety of examples of the use of Theorems 5.9 and 5.10.

5.5.2a Examples for wp1 and in Probability Convergence

Example 5.13 (Simple continuous functions). Let $g(x) = x^2$ or $g(x) = \exp(x)$. Then for any c such that $Y_n \xrightarrow{p} c$ we have $g(Y_n) \xrightarrow{p} g(c)$ since these g's are continuous at all points. ♦

Example 5.14 (Convergence of the sample variance). The variance s_n^2 of an iid(μ, σ^2) sample converges almost surely to σ^2. Consider a proof using the approximation-by-averages representation

$$s_n^2 - \sigma^2 = \frac{1}{n}\sum_{i=1}^n \{(X_i - \mu)^2 - \sigma^2\} - (\overline{X} - \mu)^2.$$

By the SLLN, $\overline{X} - \mu$ converges to 0 wp1. The function $g(y) = y^2$ is continuous at $y = 0$ and thus by Theorem 5.9, $(\overline{X} - \mu)^2$ converges to 0 wp1. A second appeal to the SLLN, assuming a finite second moment, shows that

$$\frac{1}{n}\sum_{i=1}^n \{(X_i - \mu)^2 - \sigma^2\} \xrightarrow{wp1} 0.$$

Finally an appeal to Theorem 5.10 (p. 228) shows that $s_n^2 - \sigma^2 \xrightarrow{wp1} 0$, or equivalently that $s_n^2 \xrightarrow{wp1} \sigma^2$. Because $n/(n-1) \xrightarrow{wp1} 1$, another appeal to Theorem 5.10 shows that the unbiased sample variance $\{n/(n-1)\}s_n^2$ also converges to σ^2 wp1. Finally, we note the advantage of the above approach because it requires only a second finite moment as compared to using Chebychev's inequality (Example 5.12, p. 226) that requires a finite fourth moment. ♦

Example 5.15 (Poisson distribution function). Suppose that $X_1, \ldots, X_n$ are iid Poisson (μ) with μ unknown and we want to estimate the probability that a "new" Poisson (μ) random variable X_{n+1} is less than its mean, i.e.,

$$\theta = P(X_{n+1} \le \mu) = \sum_{i=1}^{[\mu]} \frac{e^{-\mu}\mu^i}{i!} = g(\mu),$$

where $[\mu]$ is the greatest integer part of μ. The natural estimator of θ is

$$\widehat{\theta} = \sum_{i=1}^{[\overline{X}]} \frac{e^{-\overline{X}}\overline{X}^i}{i!} = g(\overline{X}).$$

Theorem 5.9 (p. 227) tells us that $g(\overline{X}) \xrightarrow{p} g(\mu)$ as long as μ is not one of the points of discontinuity $0, 1, 2, \ldots$ that make up the set A_g. What happens when μ is a positive integer, say $\mu = 2$? Then $\overline{X}$ is close to 2 as n gets large but randomly varies on either side of 2 with probability $\approx 1/2$. Thus $g(\overline{X})$ randomly jumps between $g(2)$ and $g(2^-) = g(1)$ with probability approaching $1/2$, and never converges.

For a slightly different application of Theorem 5.9 (p. 227) but with the same g, look back to Example 5.5 (p. 217) where $\overline{X} \xrightarrow{p} Z_0$ and Z_0 is a standard normal random variable. Theorem 5.9 would apply since $P(Z_0 \in A_g) = P(Z_0 = 0 \text{ or } 1 \text{ or } 2 \text{ or } \ldots) = 0$, and thus $g(\overline{X}) \xrightarrow{p} g(Z_0)$. ♦

5.5.2b Examples for Convergence in Distribution

Example 5.16 (Function with discontinuity at zero, continuous Y). Suppose that $Y_n \xrightarrow{d} Y$ and $P(Y = 0) = 0$. Then

$$\frac{1}{Y_n} \xrightarrow{d} \frac{1}{Y}.$$ ♦

Example 5.17 (Function with discontinuity at zero, discrete Y). Suppose that $Y_n \xrightarrow{d} Y$, where Y has a Poisson(λ) distribution. Then $Y_n^{1/2} \xrightarrow{d} Y^{1/2}$, but a similar convergence does not hold for $1/Y_n$. ♦

Example 5.18 (Quadratic forms). Suppose that $\boldsymbol{Y}_n \xrightarrow{d} \boldsymbol{Y}$, where $\boldsymbol{Y}_n$ and $\boldsymbol{Y}$ are k-vectors and $\boldsymbol{C}$ is a k by k matrix. Then $\boldsymbol{Y}_n^T \boldsymbol{C} \boldsymbol{Y}_n \xrightarrow{d} \boldsymbol{Y}^T \boldsymbol{C} \boldsymbol{Y}$. ♦

Example 5.19 (Chi-squared goodness-of-fit statistic). An important case of the previous example is to the chi-squared goodness-of-fit statistic

$$\sum_{i=1}^{k} \frac{(O_i - E_i)^2}{E_i} = \sum_{i=1}^{k} \frac{(n\widehat{p}_i - np_i)^2}{np_i} = \sum_{i=1}^{k} \frac{1}{p_i} n(\widehat{p}_i - p_i)^2 = \boldsymbol{Y}_n^T \boldsymbol{C} \boldsymbol{Y}_n,$$

where from Example 5.10 (p. 225), $\widehat{\boldsymbol{p}} = n^{-1}\sum_{i=1}^{n} \boldsymbol{X}_i$, $\boldsymbol{C} = \text{Diag}[1/p_1, \ldots, 1/p_k]$, and $\boldsymbol{Y}_n = \sqrt{n}(\widehat{\boldsymbol{p}}_n - \boldsymbol{p})$. ♦

Example 5.20 (One-way ANOVA F). Another important case of Example 5.18 is to the one-way ANOVA F statistic based on k independent samples, each composed of n_i iid(μ, σ^2) random variables under the null hypothesis of equal means. If for simplicity we let $n_i = n$, then

$$F = \frac{1}{k-1}\sum_{i=1}^{k} n(\overline{X}_i - \overline{\overline{X}})^2/s_p^2 = \frac{\boldsymbol{Y}_n^T \boldsymbol{C} \boldsymbol{Y}_n}{(k-1)s_p^2},$$

where the $\overline{X}_i$ are the individual sample means, $\overline{\overline{X}}$ is the overall mean, s_p^2 is the pooled variance estimate, $\boldsymbol{Y}_n^T = \sqrt{n}(\overline{X}_1 - \mu, \ldots, \overline{X}_k - \mu)$, $\boldsymbol{C} = (\boldsymbol{I}_k - k^{-1}\boldsymbol{1}_k\boldsymbol{1}_k^T)$, $\boldsymbol{I}_k$ is the k-dimensional identity matrix, and $\boldsymbol{1}_k$ is a vector of ones. If $n \to \infty$ and k remains fixed, then the numerator of the F statistic converges to a quadratic form $\boldsymbol{Y}^T\boldsymbol{C}\boldsymbol{Y}$ whose distribution can be shown to be the same as $\sigma^2\chi^2_{k-1}$ using the following results. ♦

To aid in using these results, we now state several standard results on quadratic forms in normal random variables. The first result, from Graybill (1976, p. 135-136), is for nonsingular covariance matrices. Although the middle matrix $\boldsymbol{C}$ is assumed symmetric, the results apply to nonsymmetric $\boldsymbol{C}$ because $\boldsymbol{Y}^T\boldsymbol{C}\boldsymbol{Y}$ is a scalar and thus $\boldsymbol{Y}^T\boldsymbol{C}\boldsymbol{Y} = \boldsymbol{Y}^T\boldsymbol{C}^T\boldsymbol{Y} = \boldsymbol{Y}^T(\boldsymbol{C} + \boldsymbol{C}^T)\boldsymbol{Y}/2$. Note that we use the version of the noncentral chi-squared distribution with mean equal to the degrees of freedom plus the noncentrality parameter (as found in the computing packages R and SAS).

Theorem 5.11. *Suppose that the $k \times 1$ vector $\boldsymbol{Y}$ is distributed as $MN_k(\boldsymbol{\mu}, \boldsymbol{\Sigma})$ where $\boldsymbol{\Sigma}$ has rank k. Then for symmetric $\boldsymbol{C}$ the quadratic form $\boldsymbol{Y}^T\boldsymbol{C}\boldsymbol{Y}$ has a noncentral chi-squared distribution with noncentrality parameter $\lambda = \boldsymbol{\mu}^T\boldsymbol{C}\boldsymbol{\mu}$ and r degrees freedom if and only if one of the following three conditions are satisfied:*

1. *$\boldsymbol{C}\boldsymbol{\Sigma}$ is an idempotent matrix of rank tr$(\boldsymbol{C}\boldsymbol{\Sigma}) = r$;*
2. *$\boldsymbol{\Sigma}\boldsymbol{C}$ is an idempotent matrix of rank tr$(\boldsymbol{\Sigma}\boldsymbol{C}) = r$;*
3. *$\boldsymbol{\Sigma}$ is a generalized inverse of $\boldsymbol{C}$ and $\boldsymbol{C}$ has rank r.*

This next theorem, from Styan (1970, Theorem 4) who gives credit to Khatri (1963), allows $\boldsymbol{\Sigma}$ to be singular.

Theorem 5.12. *Suppose that the $k \times 1$ vector $\boldsymbol{Y}$ is distributed as $MN_k(\boldsymbol{\mu}, \boldsymbol{\Sigma})$ where $\boldsymbol{\Sigma}$ has rank $r \le k$. Then for symmetric $\boldsymbol{C}$ the quadratic form $\boldsymbol{Y}^T\boldsymbol{C}\boldsymbol{Y}$ has a noncentral chi-squared distribution with noncentrality parameter $\lambda = \boldsymbol{\mu}^T\boldsymbol{C}\boldsymbol{\mu}$ and r degrees freedom if and only if all of the following four conditions are satisfied:*

1. *$\boldsymbol{\Sigma}\boldsymbol{C}\boldsymbol{\Sigma}\boldsymbol{C}\boldsymbol{\Sigma} = \boldsymbol{\Sigma}\boldsymbol{C}\boldsymbol{\Sigma}$;*
2. *rank$(\boldsymbol{\Sigma}\boldsymbol{C}\boldsymbol{\Sigma}) = tr(\boldsymbol{C}\boldsymbol{\Sigma}) = r$;*
3. *$\boldsymbol{\mu}^T\boldsymbol{C}\boldsymbol{\Sigma}\boldsymbol{C}\boldsymbol{\Sigma} = \boldsymbol{\mu}^T\boldsymbol{C}\boldsymbol{\Sigma}$;*
4. *$\boldsymbol{\mu}^T\boldsymbol{C}\boldsymbol{\Sigma}\boldsymbol{C}\boldsymbol{\mu} = \boldsymbol{\mu}^T\boldsymbol{C}\boldsymbol{\mu}$.*

When $\mu = 0$, considerable simplification occurs, and in that case Shanbhag (1968) shows that 1. and 2. of Theorem 5.12 are equivalent to $(\boldsymbol{C}\boldsymbol{\Sigma})^2 = (\boldsymbol{C}\boldsymbol{\Sigma})^3$ and $\text{tr}(\boldsymbol{C}\boldsymbol{\Sigma}) = r$.

Continuation of Example 5.20 (p. 230). Theorem 5.11 applies directly to Example 5.20 (p. 230) because by the multivariate Central Limit Theorem $\boldsymbol{Y}_n{}^T = \sqrt{n}(\overline{X}_1 - \mu, \ldots, \overline{X}_k - \mu) \xrightarrow{d} \boldsymbol{Y}$, where $\boldsymbol{Y}$ is $MN_k(\boldsymbol{0}, \sigma^2\boldsymbol{I}_k)$. Moreover, $\boldsymbol{C} = (\boldsymbol{I}_k - k^{-1}\boldsymbol{1}_k\boldsymbol{1}_k^T)$ is idempotent and $\text{tr}(\boldsymbol{C}\boldsymbol{\Sigma}) = \text{tr}(\boldsymbol{C}) = k-1$. Therefore $\boldsymbol{Y}^T\boldsymbol{C}\boldsymbol{Y}/\sigma^2$ is distributed as χ^2_{k-1}. Asymptotic power approximations can be made by letting the ith mean in Example 5.20 (p. 230) have the form $\mu + d_i/\sqrt{n}$. A strengthened version of the multivariate Central Limit Theorem yields in this case that the limiting random variable $\boldsymbol{Y}$ has a $MN_k(\boldsymbol{d}, \sigma^2\boldsymbol{I}_k)$ distribution. Then Theorem 5.11 yields that $\boldsymbol{Y}^T\boldsymbol{C}\boldsymbol{Y}/\sigma^2$ is distributed as noncentral chi-squared with $r = k-1$ degrees of freedom and noncentrality parameter $\lambda = \boldsymbol{d}^T(\boldsymbol{I}_k - k^{-1}\boldsymbol{1}_k\boldsymbol{1}_k^T)\boldsymbol{d}/\sigma^2 = \sum_{i=1}^k (d_i - \overline{d})^2/\sigma^2$. Therefore, the approximate power for the one-way ANOVA F is the probability that this latter noncentral chi-squared random variable is larger than the $1-\alpha$ quantile of a χ^2_{k-1} distribution. In Problem 5.11 (p. 265) we see that the approximation is fairly crude for small n since it is comparable to using the standard normal to approximate a t distribution.

Continuation of Example 5.19 (p. 229). Theorem 5.12 is required for this application because $\boldsymbol{Y}$ from the multinomial distribution has a singular normal distribution. It is not hard to verify that $\boldsymbol{C}\boldsymbol{\Sigma} = \boldsymbol{I} - \boldsymbol{p}\boldsymbol{1}^T$ is idempotent and therefore $(\boldsymbol{C}\boldsymbol{\Sigma})^2 = (\boldsymbol{C}\boldsymbol{\Sigma})^3$ and $\text{tr}(\boldsymbol{C}\boldsymbol{\Sigma}) = k-1$.

5.5.3 *Order Notation*

In working with both nonrandom sequences and random sequences, it helps to be able to have simple rules and notation to bound sums and products of sequences with possibly different rates of convergence. We first give the deterministic versions of "order" notation, and then discuss analogues for random sequences.

5.5.3a Nonrandom Order Relations

For real sequences $\{u_n\}$ and $\{v_n\}$ "$u_n = O(v_n)$ as $n \to \infty$" means that there exist a positive constant M and an integer n_0 such that

$$\left|\frac{u_n}{v_n}\right| \le M \quad \text{for all } n \ge n_0.$$

The phrase "$u_n = o(v_n)$ as $n \to \infty$" means that

$$\lim_{n\to\infty} \left| \frac{u_n}{v_n} \right| = 0,$$

and "$u_n \sim v_n$ as $n \to \infty$" means that

$$\lim_{n\to\infty} \left| \frac{u_n}{v_n} \right| = 1.$$

Example 5.21 (Polynomial in n). Let $u_n = 2n^3 + 6n^2 + 1$. Then $u_n = O(n^3)$, where $(M = 4, n_0 = 6)$ is one pair which meets the bounding requirements. Of course $u_n = O(4n^3)$ and $u_n = O(n^4)$, etc., are true as well. Also $u_n = o(n^4)$, $u_n = o(n^3 log(n))$, and $u_n \sim 2n^3$. ♦

Rules and Relationships of Deterministic Order Relations

1. If $u_n = o(v_n)$, then $u_n = O(v_n)$.
2. If $u_{1n} = O(v_{1n})$ and $u_{2n} = O(v_{2n})$, then $u_{1n}u_{2n} = O(v_{1n}v_{2n})$.
3. If $u_{1n} = o(v_{1n})$ and $u_{2n} = O(v_{2n})$, then $u_{1n}u_{2n} = o(v_{1n}v_{2n})$.
4. If $u_{1n} = o(v_{1n})$ and $u_{2n} = o(v_{2n})$, then $u_{1n}u_{2n} = o(v_{1n}v_{2n})$.
5. If $u_{n_j} = o(v_n)$ for $j = 1, \ldots, m$, then $u_{n_1} + \cdots + u_{n_m} = o(v_n)$ for any finite m.
6. If $u_{n_j} = O(v_n)$ for $j = 1, \ldots, m$, then $u_{n_1} + \cdots + u_{n_m} = O(v_n)$ for any finite m.
7. If $u_{n_1} = o(v_n)$ and $u_{n_2} = O(v_n)$, then $u_{n_1} + u_{n_2} = O(v_n)$.

These rules are simple known facts about sequences. But to illustrate, consider the following proof of 3. By assumption

$$\frac{u_{1n}}{v_{1n}} \to 0 \qquad \text{as } n \to \infty,$$

and there exist $M > 0$ and n_0 such that

$$\left| \frac{u_{2n}}{v_{2n}} \right| \le M \quad \text{for all } n \ge n_0.$$

Then

$$\left| \frac{u_{1n}u_{2n}}{v_{1n}v_{2n}} \right| \le \left| \frac{u_{1n}}{v_{1n}} \right| M \quad \text{for all } n \ge n_0$$
$$\to 0 \qquad \text{as } n \to \infty.$$

■

5.5.3b "With Probability 1" Order Relations

If the sequences $\{U_n\}$ and $\{V_n\}$ are random variables, then we can add "wp1" to each of the above order relations, and we get meaningful statements. For example, "$U_n = O(V_n)$ $wp1$" means that there exists $\Omega_0 \subset \Omega$ such that $P(\Omega_0) = 1$ and for each $\omega \in \Omega_0$, $U_n(\omega) = O(V_n(\omega))$. Note that the constants do not need to be uniform in ω. For example, a pair $(M(\omega), n_0(\omega))$ can depend on ω. For completeness, we restate the rules and relationships.

Rules and Relationships of wp1 Order Relations

1. If $U_n = o(V_n)$ wp1, then $U_n = O(V_n)$ wp1.
2. If $U_{1n} = O(V_{1n})$ wp1 and $U_{2n} = O(V_{2n})$ wp1, then $U_{1n}U_{2n} = O(V_{1n}V_{2n})$ wp1.
3. If $U_{1n} = o(V_{1n})$ wp1 and $U_{2n} = O(V_{2n})$ wp1, then $U_{1n}U_{2n} = o(V_{1n}V_{2n})$ wp1.
4. If $U_{1n} = o(V_{1n})$ wp1 and $U_{2n} = o(V_{2n})$ wp1, then $U_{1n}U_{2n} = o(V_{1n}V_{2n})$ wp1.
5. If $U_{n_j} = o(V_n)$ wp1 for $j = 1, \ldots, m$, then $U_{n_1} + \cdots + U_{n_m} = o(V_n)$ wp1 for any finite m.
6. If $U_{n_j} = O(V_n)$ wp1 for $j = 1, \ldots, m$, then $U_{n_1} + \cdots + U_{n_m} = O(V_n)$ wp1 for any finite m.
7. If $U_{n_1} = o(V_n)$ wp1 and $U_{N_2} = O(V_n)$ wp1, then $U_{n_1} + U_{n_2} = O(V_n)$ wp1.

To further illustrate, let us prove 3. above. By assumption there exists Ω_{10} with $P(\Omega_{10}) = 1$ such that for $\omega \in \Omega_{10}$

$$\frac{U_{1n}(\omega)}{V_{1n}(\omega)} \to 0 \qquad \text{as } n \to \infty,$$

and there exists Ω_{20} with $P(\Omega_{20}) = 1$ such that for $\omega \in \Omega_{20}$ there exist $M(\omega) > 0$ and $n_0(\omega)$ such that

$$\left| \frac{U_{2n}(\omega)}{V_{2n}(\omega)} \right| \le M(\omega) \quad \text{for all } n \ge n_0(\omega).$$

Then for $\omega \in \Omega_{10} \cap \Omega_{20}$

$$\left| \frac{U_{1n}(\omega)U_{2n}(\omega)}{V_{1n}(\omega)V_{2n}(\omega)} \right| \le \left| \frac{U_{1n}(\omega)}{V_{1n}(\omega)} \right| M(\omega) \quad \text{for all } n \ge n_0(\omega)$$
$$\to 0 \qquad \text{as } n \to \infty.$$

Moreover, $P(\Omega_{10} \cap \Omega_{20}) = 1 - P(\Omega_{10}^C \cup \Omega_{20}^C) \ge 1 - \{P(\Omega_{10}^C) + P(\Omega_{20}^C)\} = 1.$ ■

5.5.3c "In Probability" Order Relations

For a sequence of random variables $\{Y_n\}$ we say that the sequence is *bounded in probability* if for each $\epsilon > 0$ there exist M_ϵ and n_ϵ such that

$$P(|Y_n| > M_\epsilon) < \epsilon \quad \text{for all } n \ge n_\epsilon,$$

or equivalently

$$P(|Y_n| \le M_\epsilon) > 1 - \epsilon \text{ for all } n \ge n_\epsilon.$$

If $F_n(y) = P(Y_n \le y)$, then an equivalent condition is

$$F_n(M_\epsilon) - F_n(-M_\epsilon) > 1 - \epsilon \text{ for all } n \ge n_\epsilon.$$

Shorthand notation is

$$Y_n = O_p(1) \text{ as } n \to \infty.$$

When Y_n is bounded in probability, then the sequence of probability measures associated with Y_n is said to be *tight* (c.f., Billingsley, 1999, Ch. 1). Whatever the language used, the importance of $Y_n = O_p(1)$ is that probability mass is not allowed to escape to $\pm\infty$. Thus it should make intuitive sense that

$$Y_n \xrightarrow{d} Y \quad \text{implies} \quad Y_n = O_p(1),$$

which can be easily verified.

Now given the definition of $O_p(1)$ we can define for random sequences $\{U_n\}$ and $\{V_n\}$ that "$U_n = O_p(V_n)$" means that

$$\left|\frac{U_n}{V_n}\right| = O_p(1) \quad \text{as } n \to \infty.$$

Similarly, "$U_n = o_p(V_n)$" means that

$$\left|\frac{U_n}{V_n}\right| \xrightarrow{p} 0 \quad \text{as } n \to \infty.$$

Rules and Relationships of In Probability Order Relations

1. If $U_n = o_p(V_n)$, then $U_n = O_p(V_n)$.
2. If $U_{1n} = O_p(V_{1n})$ and $U_{2n} = O_p(V_{2n})$, then $U_{1n}U_{2n} = O_p(V_{1n}V_{2n})$.
3. If $U_{1n} = o_p(V_{1n})$ and $U_{2n} = O_p(V_{2n})$, then $U_{1n}U_{2n} = o_p(V_{1n}V_{2n})$.
4. If $U_{1n} = o_p(V_{1n})$ and $U_{2n} = o_p(V_{2n})$, then $U_{1n}U_{2n} = o_p(V_{1n}V_{2n})$.
5. If $U_{n_j} = o_p(V_n)$ for $j = 1, \ldots, m$, then $U_{n_1} + \cdots + U_{n_m} = o_p(V_n)$ for any finite m.
6. If $U_{n_j} = O_p(V_n)$ for $j = 1, \ldots, m$, then $U_{n_1} + \cdots + U_{n_m} = O_p(V_n)$ for any finite m.
7. If $U_{n_1} = o_p(V_n)$ and $U_{N_2} = O_p(V_n)$, then $U_{n_1} + U_{n_2} = O_p(V_n)$.

Notice that 4. is direct from Theorem 5.10 (p. 228). The other extensions are straightforward but take a little more work to prove than the previous versions. For example, let us prove 2.

By assumption there exists $M_{1\epsilon} > 0$, $M_{2\epsilon} > 0$, $n_{1\epsilon}$, and $n_{2\epsilon}$ such that

$$P\left(\left|\frac{U_{in}}{V_{in}}\right| \leq M_{i\epsilon}\right) > 1 - \epsilon/2 \quad \text{for all } n \geq n_{i\epsilon},\ i = 1, 2.$$

Then for all $n \geq \max(n_{1\epsilon}, n_{2\epsilon})$,

$$\begin{aligned} P\left(\left|\frac{U_{1n}U_{2n}}{V_{1n}V_{2n}}\right| \leq M_{1\epsilon}M_{2\epsilon}\right) &\geq P\left(\left|\frac{U_{1n}}{V_{1n}}\right| \leq M_{1\epsilon}, \left|\frac{U_{2n}}{V_{2n}}\right| \leq M_{2\epsilon}\right) \\ &\geq 1 - (\epsilon/2 + \epsilon/2) = 1 - \epsilon. \end{aligned}$$

We have used the fact that for $a > 0, b > 0$, if $|A| \leq a$ and $|B| \leq b$, then $|AB| \leq ab$. Thus, the event $\{|A| \leq a, |B| \leq b\}$ is a subset of $\{|AB| \leq ab\}$, and $P(|A| \leq a, |B| \leq b) \leq P(|AB| \leq ab)$. ■

Note that $S_n = O(1)$ implies $S_n = O_p(1)$ and $S_n = o(1)$ implies $S_n = o_p(1)$.

Example 5.22 (Adding and subtracting). Often we have a quantity of interest for which we want to prove convergence in probability, say $\widehat{\theta}$, and a related quantity, say $\widehat{\theta}^*$, where we know that $\widehat{\theta}^* \xrightarrow{p} \theta$ and $\widehat{\theta} - \widehat{\theta}^* \xrightarrow{p} 0$ as $n \to \infty$. Then, the method of proof is simply to subtract and add,

$$\begin{aligned} \widehat{\theta} - \theta &= \widehat{\theta} - \widehat{\theta}^* + \widehat{\theta}^* - \theta \\ &= o_p(1) + o_p(1) = o_p(1). \end{aligned}$$

♦

Example 5.23 (Convergence of sample variance). We close this subsection on order relations with one of our main examples, the sample variance, written as

$$s_n^2 - \sigma^2 = \frac{1}{n}\sum_{i=1}^{n}[(X_i - \mu)^2 - \sigma^2] + R_n,$$

where $R_n = -(\overline{X} - \mu)^2$. To prove $s_n^2 \xrightarrow{p} \sigma^2$, we needed to show that $R_n \xrightarrow{p} 0$, or $R_n = o_p(1)$. But this follows from the WLLN and 4. above with $U_{1n} = U_{2n} = \overline{X} - \mu$ and $V_{1n} = V_{2n} = 1$. For asymptotic normality of s_n^2, we need $\sqrt{n}R_n \xrightarrow{p} 0$ or $R_n = o_p(n^{-1/2})$. This follows from 3. above with the same U_{in} and with $U_{1n} = o_p(1)$ and $U_{2n} = O_p(n^{-1/2})$, which follows from the Central Limit Theorem. ♦

5.5.4 Asymptotic Normality and Related Results

A useful language convention (actually introduced in Chapter 1) is to say that Y_n is *asymptotically normal* with "mean" μ_n and "variance" σ_n^2 if

$$\frac{Y_n - \mu_n}{\sigma_n} \xrightarrow{d} N(0, 1) \quad \text{as} \quad n \to \infty.$$

Our shorthand notation is Y_n is AN(μ_n, σ_n^2). Note that μ_n and σ_n^2 need not be the mean and variance of Y_n, and Y_n itself may not converge in distribution. Still we may approximate Y_n by a N(μ_n, σ_n^2) random variable as seen by using Theorem 5.6 (p. 222):

$$\begin{aligned}
&\sup_{y\in(-\infty,\infty)} |P(Y_n \le y) - P(\sigma_n Z + \mu_n \le y)| \\
&\quad = \sup_{y\in(-\infty,\infty)} \left| P\left(\frac{Y_n - \mu_n}{\sigma_n} \le \frac{y-\mu_n}{\sigma_n}\right) - P\left(Z \le \frac{y-\mu_n}{\sigma_n}\right)\right| \\
&\quad = \sup_{t\in(-\infty,\infty)} \left| P\left(\frac{Y_n - \mu_n}{\sigma_n} \le t\right) - P\left(Z \le t\right)\right| \to 0,
\end{aligned}$$

where Z is a standard normal random variable.

Example 5.24 (Normal approximation to a chi-squared random variable). If Y_n is a χ_n^2 random variable, then by the Central Limit Theorem Y_n is AN$(n, 2n)$ (since Y_n is equal in distribution to a sum of n independent χ_1^2 random variables). Certainly Y_n does not converge in distribution; in fact $Y_n \xrightarrow{d} \infty$. However, $\sqrt{2n}(1.645) + n$ is a reasonable approximation to the 90th percentile of Y_n as can be checked for various values of n. ♦

Here are three useful theorems taken from Serfling (1980, p. 20) associated with asymptotic normality.

Theorem 5.13. *If Y_n is AN(μ, σ_n^2), then $Y_n \xrightarrow{p} \mu$ if and only if $\sigma_n \to 0$ as $n \to \infty$.*

Theorem 5.14. *If Y_n is AN(μ_n, σ_n^2), then Y_n is AN(μ_n^*, σ_n^{*2}) if and only if*

$$\lim_{n\to\infty} \frac{\sigma_n^*}{\sigma_n} = 1 \quad \text{and} \quad \lim_{n\to\infty} \frac{\mu_n^* - \mu_n}{\sigma_n} = 0.$$

Theorem 5.15. *If Y_n is AN(μ_n, σ_n^2), then $a_n Y_n + b_n$ is AN(μ_n, σ_n^2) if and only if*

$$\lim_{n\to\infty} a_n = 1 \quad \text{and} \quad \lim_{n\to\infty} \frac{\mu_n(a_n - 1) + b_n}{\sigma_n} = 0.$$

Example 5.25 (Versions of the sample variance). Let s_n^2 be the "$1/n$" version of the sample variance from an iid(μ, σ^2) sample with finite fourth central moment $\mu_4 = E(X_1 - \mu)^4$. We show in the next section that s_n^2 is AN$(\sigma^2, (\mu_4 - \sigma^4)/n)$. The last theorem above tells us that the same asymptotic normality applies to the "$1/(n-1)$" version of the sample variance using $a_n = n/(n-1)$. ♦

Generalization to random vectors: $\{\boldsymbol{Y}_n\}$ is *asymptotically multivariate normal* with "mean" $\boldsymbol{\mu}_n$ and "covariance matrix" $\boldsymbol{\Sigma}_n$ if for every $\boldsymbol{c} \in R^k$ such that

$c^T \Sigma_n c > 0$ for all n sufficiently large, $c^T Y_n$ is $\text{AN}(c^T \mu_n, c^T \Sigma_n c)$. Shorthand notation is Y_n is $\text{AMN}(\mu_n, \Sigma_n)$ or $\text{AN}_k(\mu_n, \Sigma_n)$. Many cases of interest are covered by the following result for a simpler form for Σ_n.

Theorem 5.16. *Y_n is $AN_k(\mu_n, b_n^2 \Sigma)$ if and only if*

$$\frac{Y_n - \mu_n}{b_n} \xrightarrow{d} MN(\mathbf{0}, \Sigma).$$

Analogous to part of Theorem 5.13 (p. 236), we also have

Theorem 5.17. *If Y_n is $AN_k(\mu, \Sigma_n)$ with $\Sigma_n \longrightarrow 0$, then $Y_n \xrightarrow{p} \mu$ as $n \to \infty$.*

Recall that $\Sigma_n \longrightarrow 0$ means pointwise convergence of the elements of Σ_n.

5.5.5 The Delta Theorem

We introduced the Delta Theorem in Chapter 1 (p. 14), but for completeness we repeat it here in three versions; the last version allows vector functions $g(\theta)^T = (g_1(\theta), \ldots, g_s(\theta))$ of a k-dimensional estimator $\widehat{\theta}$. The reason for the different versions is that in simple cases, the vector versions seem like overkill and obscure the basic Taylor expansion $g(\widehat{\theta}) \approx g(\theta) + g'(\theta)(\widehat{\theta} - \theta)$ that lies behind the method. The proof of these theorems may be found in Serfling (1980, Ch. 3).

Theorem 5.18 (Delta Method - Scalar Case). *Suppose that $\widehat{\theta}$ is $AN(\theta, \sigma_n^2)$ with $\sigma_n \to 0$ and g is a real-valued function differentiable at θ with $g'(\theta) \neq 0$. Then as $n \to \infty$*

$$g(\widehat{\theta}) \text{ is } AN\left[g(\theta), \{g'(\theta)\}^2 \sigma_n^2\right].$$

Example 5.26 (Variance stabilizing transformation for the sample variance). "Variance stabilizing" transformations are usually applied to sample statistics whose asymptotic variance depends on their asymptotic mean. The purpose is to convert them into new statistics for which the asymptotic variance does not depend on the asymptotic mean. This can be useful for constructing confidence intervals or using the statistics in an analysis of variance type setting. One of our key examples has been the sample variance, s_n^2, based on an iid sample $X_1, \ldots, X_n$, which is AN $\left[\sigma^2, \sigma^4 \{\text{Kurt}(X_1) - 1\} / n\right]$, where $\text{Kurt}(X_1) = \mu_4/\sigma^4$ is the kurtosis. Note that the asymptotic variance depends directly on σ^2, the asymptotic mean. By Theorem 5.18, however, since the derivative of $\log(x)$ is $1/x$ and

$$(1/\sigma^2)^2 \left[\sigma^4 \{\text{Kurt}(X_1) - 1\} / n\right] = \{\text{Kurt}(X_1) - 1\} / n,$$

then $\log(s_n^2)$ is $\text{AN}[\log(\sigma^2), \{\text{Kurt}(X_1) - 1\} / n\}$. Thus, $\log(y)$ is a variance stabilizing transformation for the sample variance because $\{\text{Kurt}(X_1) - 1\} / n$ does not depend on $\log(\sigma^2)$. ♦

Example 5.27 (Square of the sample mean). Suppose that $X_1, \ldots, X_n$ are iid(μ, σ^2) and we are interested in the asymptotic distribution of $\overline{X}^2$. If $\mu \neq 0$, then $g'(x) = 2x$ and $\overline{X}^2$ is AN$(\mu^2, 4\mu^2\sigma^2/n)$. However, if $\mu = 0$, then Theorem 5.18 cannot be used because the derivative of x^2 at $\mu = 0$ is $2(0) = 0$. In this case $n\overline{X}^2 \xrightarrow{d} (\sigma Z)^2$, where Z is a standard normal random variable, using Theorem 5.9 (p. 227). Special theorems to handle cases where derivatives are zero are given on p. 119 and 124 of Serfling (1980). ♦

Example 5.28 (Reciprocal of the sample mean). Suppose that $X_1, \ldots, X_n$ are iid(μ, σ^2) and we are interested in the asymptotic distribution of $1/\overline{X}$. The derivative of g is $g'(x) = -1/x^2$ and if $\mu \neq 0$, then $1/\overline{X}$ is AN$\{1/\mu, (\sigma^2/\mu^4)/n\}$. Theorem 5.18, however, cannot be used for $1/\overline{X}$ when $\mu = 0$ because then the derivative does not exist at $\mu = 0$. In this case, Theorem 5.9 (p. 227) gives $1/(\sqrt{n}\overline{X}) \xrightarrow{d} 1/N(0, \sigma^2)$. ♦

The next version of the Delta Method is for real-valued functions g defined on a vector $\boldsymbol{\theta}$. Here we allow the asymptotic variance to be of the form $b_n^2 \boldsymbol{\Sigma}$, where often $b_n^2 = 1/n$.

Theorem 5.19 (Delta Method - Real-Valued Function of a Vector). *Suppose that $\widehat{\boldsymbol{\theta}}$ is $AN_k(\boldsymbol{\theta}, b_n^2 \boldsymbol{\Sigma})$ with $b_n \to 0$ and that g is a real-valued function with partial derivatives existing in a neighborhood of $\boldsymbol{\theta}$ and continuous at $\boldsymbol{\theta}$ with $g'(\boldsymbol{\theta}) = \partial g(\boldsymbol{\theta})/\partial \boldsymbol{\theta}$ not identically zero. Then as $n \to \infty$*

$$g(\widehat{\boldsymbol{\theta}}) \text{ is } AN\left[g(\boldsymbol{\theta}), b_n^2 g'(\boldsymbol{\theta}) \boldsymbol{\Sigma} g'(\boldsymbol{\theta})^T\right].$$

Example 5.29 (Risk difference for multinomial 2×2 table). Consider a 2×2 table of counts in a medical setting given by

	Cured	Not Cured
Group 1	N_{11}	N_{12}
Group 2	N_{21}	N_{22}

where $(N_{11}, N_{12}, N_{21}, N_{22})^T$ has a multinomial$(n; p_{11}, p_{12}, p_{21}, p_{22})$ distribution, $n = N_{11} + N_{12} + N_{21} + N_{22}$. The *risk difference* of the proportion cured for the two groups is

$$g(\boldsymbol{p}) = g(p_{11}, p_{12}, p_{21}, p_{22}) = \frac{p_{11}}{p_{11} + p_{12}} - \frac{p_{21}}{p_{21} + p_{22}},$$

and $\widehat{\theta}$ is obtained by substituting $\widehat{p}_{ij} = N_{ij}/n$ for p_{ij} in this latter expression. For use with Theorem 5.19 (p. 238), we have

$$g'(\boldsymbol{p}) = \left[\frac{p_{12}}{(p_{11} + p_{12})^2}, \frac{-p_{11}}{(p_{11} + p_{12})^2}, \frac{-p_{22}}{(p_{21} + p_{22})^2}, \frac{p_{21}}{(p_{21} + p_{22})^2}\right].$$

From Example 5.10 (p. 225) the asymptotic covariance times n of $\widehat{\boldsymbol{p}} = (\widehat{p}_{11}, \widehat{p}_{12}, \widehat{p}_{21}, \widehat{p}_{22})$ is given by

$$\boldsymbol{\Sigma} = \begin{pmatrix} p_{11}(1-p_{11}) & -p_{11}p_{12} & -p_{11}p_{21} & -p_{11}p_{22} \\ -p_{12}p_{11} & p_{12}(1-p_{12}) & -p_{12}p_{21} & -p_{12}p_{22} \\ -p_{21}p_{11} & -p_{21}p_{12} & p_{21}(1-p_{21}) & -p_{21}p_{22} \\ -p_{22}p_{11} & -p_{22}p_{12} & -p_{22}p_{21} & p_{22}(1-p_{22}) \end{pmatrix}$$

Multiplying $g'(\boldsymbol{p})^T \boldsymbol{\Sigma} g'(\boldsymbol{p})$ (in Maple, for example) yields

$$\frac{p_{11}p_{12}}{(p_{11}+p_{12})^3} + \frac{p_{21}p_{22}}{(p_{21}+p_{22})^3} \tag{5.8}$$

for n times the asymptotic variance of $g(\widehat{\boldsymbol{p}})$. If one makes the substitutions $p_1 = p_{11}/(p_{11}+p_{12})$, $p_2 = p_{21}/(p_{21}+p_{22})$, $n_1 = n(p_{11}+p_{12})$, and $n_2 = n(p_{21}+p_{22})$ that would be appropriate for the independent binomials version of the 2×2 table (see Example 5.31, p. 240), then this asymptotic variance has the usual form $p_1(1-p_1)/n_1 + p_2(1-p_2)/n_2$. ♦

Example 5.30 (General Multinomial). Now consider a real-valued function $g(\widehat{\boldsymbol{p}})$ of $\widehat{\boldsymbol{p}} = (N_1/n, \ldots, N_k/n)^T$ in the general multinomial case and using the standard notation that $(N_1, \ldots, N_k)$ is multinomial$(n; p_1, \ldots, p_k)$. Certainly this covers the previous example, but the general case is easier to handle because the covariance matrix (also the asymptotic covariance matrix) of $\widehat{\boldsymbol{p}}$ may be written as $\boldsymbol{\Sigma}/n = \left[\text{Diag}(\boldsymbol{p}) - \boldsymbol{p}\boldsymbol{p}^T\right]/n$. Then

$$\begin{aligned} g'(\boldsymbol{p})\boldsymbol{\Sigma} g'(\boldsymbol{p})^T &= g'(\boldsymbol{p})\left[\text{Diag}(\boldsymbol{p}) - \boldsymbol{p}\boldsymbol{p}^T\right]g'(\boldsymbol{p})^T \\ &= \sum_{i=1}^k p_i \left\{\frac{\partial g(\boldsymbol{p})}{\partial p_i}\right\}^2 - \left\{\sum_{i=1}^k p_i \frac{\partial g(\boldsymbol{p})}{\partial p_i}\right\}^2. \end{aligned} \tag{5.9}$$

Looking at the previous example, and making the notation conversion, the second term of (5.9) is zero and the first term is easily found to yield (5.8).

Our last version of the Delta Method allows vector-valued functions g of dimension s.

Theorem 5.20 (Delta Method - Vector-Valued Function of a Vector). *Suppose that $\widehat{\boldsymbol{\theta}}$ is $AN_k(\boldsymbol{\theta}, b_n^2\boldsymbol{\Sigma})$ with $b_n \to 0$ and that $\boldsymbol{g}$ is an s-valued function on $\mathbf{R}^k$ possessing an $s \times k$ matrix of partial derivatives*

$$\boldsymbol{g}'(\boldsymbol{x}) = \frac{\partial \boldsymbol{g}(\boldsymbol{x})}{\partial \boldsymbol{x}} = \begin{pmatrix} \frac{\partial g_1(\boldsymbol{x})}{\partial x_1} & \frac{\partial g_1(\boldsymbol{x})}{\partial x_2} & \cdots & \frac{\partial g_1(\boldsymbol{x})}{\partial x_k} \\ \frac{\partial g_2(\boldsymbol{x})}{\partial x_1} & \frac{\partial g_2(\boldsymbol{x})}{\partial x_2} & \cdots & \frac{\partial g_2(\boldsymbol{x})}{\partial x_k} \\ \cdot & \cdot & \cdots & \cdot \\ \cdot & \cdot & \cdots & \cdot \\ \frac{\partial g_s(\boldsymbol{x})}{\partial x_1} & \frac{\partial g_s(\boldsymbol{x})}{\partial x_2} & \cdots & \frac{\partial g_s(\boldsymbol{x})}{\partial x_k} \end{pmatrix}, \qquad (5.10)$$

in a neighborhood of $\boldsymbol{\theta}$, and each partial derivative is continuous at $\boldsymbol{\theta}$ with at least one component nonzero at $\boldsymbol{\theta}$ in every row.

Then as $n \to \infty$

$$\boldsymbol{g}(\widehat{\boldsymbol{\theta}}) \text{ is } AN\left[\boldsymbol{g}(\boldsymbol{\theta}), b_n^2 \boldsymbol{g}'(\boldsymbol{\theta})\boldsymbol{\Sigma}\boldsymbol{g}'(\boldsymbol{\theta})^T\right].$$

Example 5.31 (Product binomial 2×2 table). Let X_1 be binomial(n_1, p_1) and X_2 be binomial(n_2, p_2) with associated estimators $\widehat{p}_1 = X_1/n_1$ and $\widehat{p}_2 = X_2/n_2$. The setup is similar to Example 5.29 (p. 238) but the distributions and notation are different:

	Cured	Not Cured	Sample Size
Group 1	X_1	$n_1 - X_1$	n_1
Group 2	X_2	$n_2 - X_2$	n_2

Here we want to consider the *risk difference* $p_1 - p_2$ along with the *risk ratio* p_1/p_2. To obtain the joint asymptotic distribution of $(\widehat{p}_1 - \widehat{p}_2, \widehat{p}_1/\widehat{p}_2)^T$ via Theorem 5.20, we run into a problem because we actually have two sample sizes n_1 and n_2 and no "n." The standard approach in this type of situation is to define $n = n_1 + n_2$ and $\lambda_n = n_1/n$, and assume that $\lambda_n \to \lambda$, $\lambda \in (0, 1)$, as $\min(n_1, n_2) \to \infty$. Then we can replace n_1 by $n\lambda_n$ and n_2 by $n(1 - \lambda_n)$ and say that $\widehat{p}_1$ is AN$\{p_1, p_1(1 - p_1)/\lambda_n n\}$ and by Theorem 5.14 (p. 236) that $\widehat{p}_1$ is AN$(p_1, p_1(1 - p_1)/\lambda n)$. Similarly, $\widehat{p}_2$ is AN$\{p_2, p_2(1 - p_2)/(1 - \lambda)n\}$. Because of independence, we can assert that $(\widehat{p}_1, \widehat{p}_2)^T$ are jointly asymptotically normal and use Theorem 5.20 to get that $(\widehat{p}_1 - \widehat{p}_2, \widehat{p}_1/\widehat{p}_2)^T$ is jointly asymptotically normal with asymptotic mean $(p_1 - p_2, p_1/p_2)^T$ and asymptotic variance times n given by

$$\begin{pmatrix} 1 & -1 \\ \frac{1}{p_2} & \frac{-p_1}{p_2^2} \end{pmatrix} \begin{pmatrix} \frac{p_1(1-p_1)}{\lambda} & 0 \\ 0 & \frac{p_2(1-p_2)}{1-\lambda} \end{pmatrix} \begin{pmatrix} 1 & \frac{1}{p_2} \\ -1 & \frac{-p_1}{p_2^2} \end{pmatrix}.$$

Problem 5.26 (p. 267) shows that the asymptotic correlation is one when $p_1 = p_2$.

5.5.6 Slutsky's Theorem

The next theorem (actually first given in Chapter 1) lets us use results proved for one sequence to be applied to other closely related sequences. It is one of the most important tools of asymptotic analysis.

Theorem 5.21 (Slutsky). *If $X_n \xrightarrow{d} X$, $Y_n \xrightarrow{p} a$, and $Z_n \xrightarrow{p} b$ as $n \to \infty$, then*

$$Y_n X_n + Z_n \xrightarrow{d} aX + b \qquad as\ n \to \infty.$$

Proof. First note that $Y_n X_n + Z_n = (Y_n - a)X_n + aX_n + (Z_n - b) + b$, so that $Y_n X_n + Z_n = W_n + R_n$, where $R_n = (Y_n - a)X_n + (Z_n - b) = o_p(1)$ and $W_n = aX_n + b \xrightarrow{d} aX + b$ as $n \to \infty$. Then, for arbitrary t and ϵ chosen so that $t + \epsilon$ and $t - \epsilon$ are continuity points of the distribution function of $aX + b$,

$$\begin{aligned} P(W_n + R_n \le t) &= P(W_n + R_n \le t, |\, R_n \,| > \epsilon) \\ &\quad + P(W_n + R_n \le t, |\, R_n \,| \le \epsilon) \\ &\le P(|\, R_n \,| > \epsilon) + P(W_n \le t + \epsilon), \end{aligned}$$

and similarly

$$P(W_n + R_n \le t) \ge P(W_n \le t - \epsilon).$$

This leads to

$$\lim_{n\to\infty} |\, F_n(t) - F(t) \,| \le \max\{F(t) - F(t - \epsilon), F(t + \epsilon) - F(t)\},$$

where $F_n(t) = P(Y_n X_n + Z_n \le t)$ and $F(t) = P(aX + b \le t)$. Since $\epsilon > 0$ is arbitrary, the result follows. Note that we have used the fact that distribution functions have only a countable number of jump discontinuities. ■

Example 5.32 (One sample t statistic). The standard first example for Slutsky's Theorem is the one sample t statistic $t = \sqrt{n}(\overline{X} - \mu)/s_{n-1}$ where the numerator converges in distribution by the CLT and the denominator converges in probability by Example 5.14 (p. 228) and Theorem 5.9 (p. 227). ♦

Slutsky's Theorem along with the CLT provide asymptotic normality for many estimators and statistics via the approximation by averages Theorem 5.23 of the next subsection. We can also give a multivariate version of Slutsky's Theorem as follows.

Theorem 5.22 (Multivariate Slutsky). *Let $\boldsymbol{X}_n$ and $\boldsymbol{X}$ be random k-vectors such $\boldsymbol{X}_n \xrightarrow{d} \boldsymbol{X}$. Let Y_n be an $m \times k$ random matrix such that $\boldsymbol{Y}_n \xrightarrow{p} \boldsymbol{A}$, where $\boldsymbol{A}$ is an $m \times k$ constant matrix. Let $\boldsymbol{Z}_n$ be a random m-vector such that $\boldsymbol{Z}_n \xrightarrow{p} \boldsymbol{B}$, where $\boldsymbol{B}$ is a constant m-vector. Then*

$$\boldsymbol{Y}_n \boldsymbol{X}_n + \boldsymbol{Z}_n \xrightarrow{d} \boldsymbol{A}\boldsymbol{X} + \boldsymbol{B} \qquad as\ n \to \infty.$$

5.5.7 *Approximation by Averages*

Theorem 5.23 (Approximation by Averages). *Suppose that $X_1, \ldots, X_n$ are iid and the statistic $T = T(X_1, \ldots, X_n)$ has the following approximation,*

$$T - T_\infty = \frac{1}{n}\sum_{i=1}^{n} h(X_i) + R_n,$$

where $E\{h(X_1)\} = 0$, $\text{Var}\{h(X_1)\} = \sigma_h^2$, and $\sqrt{n}R_n \xrightarrow{p} 0$ as $n \to \infty$. Then

$$\sqrt{n}(T - T_\infty) \xrightarrow{d} N(0, \sigma_h^2) \qquad \text{as } n \to \infty.$$

Proof. First apply the CLT to $n^{-1/2}\sum_{i=1}^{n} h(X_i)$ which is playing the role of X_n in Slutsky's Theorem. Then apply Slutsky's Theorem with $\sqrt{n}\,R_n$ playing the role of Z_n.

Example 5.33 (Sample variance). Let $X_1, \ldots, X_n$ be iid(μ, σ^2) with finite fourth central moment $\mu_4 = E(X_1 - \mu)^4$. Consider the "$1/n$" version of the sample variance $s_n^2 = n^{-1}\sum_{i=1}^{n}(X_i - \overline{X})^2$. Some algebra leads to

$$s_n^2 - \sigma^2 = \frac{1}{n}\sum_{i=1}^{n}\{(X_i - \mu)^2 - \sigma^2\} - (\overline{X} - \mu)^2.$$

The remainder term $R_n = -(\overline{X} - \mu)^2$ is $o_p(n^{-1/2})$ because $\sqrt{n}R_n = -\sqrt{n}(\overline{X} - \mu)(\overline{X} - \mu)$, and $-\sqrt{n}(\overline{X} - \mu)$ is $O_p(1)$ by the CLT, and $\overline{X} - \mu$ is $o_p(1)$ by the WLLN. Now apply Slutsky's Theorem with $h(X_i) = [(X_i - \mu)^2 - \sigma^2]$ to get $\sqrt{n}(s^2 - \sigma^2) \xrightarrow{d} N(0, \mu_4 - \sigma^4)$ as $n \to \infty$. ♦

5.5.7a Sample Central Moments

So far we have covered the sample mean and the sample second central moment s_n^2. Now we give results for all the sample central moments. Let $X_1, \ldots, X_n$ be iid(μ, σ^2) with finite $2k$th moment μ_{2k}, where $\mu_k = E(X_1 - \mu)^k$. Note that $\mu = E(X_1)$, $\mu_1 = E(X_1 - \mu) = 0$, and $\mu_2 = \sigma^2$. Let the sample kth central moment be given by $m_k = n^{-1}\sum_{i=1}^{n}(X_i - \overline{X})^k$. Then we have the following approximation theorem which verifies the conditions of Theorem 5.23 for the sample kth central moments. A direct proof is given in Serfling (1980, p. 72), but it also follows from Theorem 5.28 (p. 249).

Theorem 5.24 (Lemma 2.2.3, p. 72, of Serfling, 1980). *If $X_1, \ldots, X_n$ are iid with mean μ and $\mu_{2k} < \infty$, then*

$$m_k - \mu_k = \frac{1}{n}\sum_{i=1}^{n}\left[(X_i-\mu)^k - \mu_k - k\mu_{k-1}(X_i-\mu)\right] + R_n,$$

where $\sqrt{n}R_n \xrightarrow{p} 0$ as $n \to \infty$.

Noting that for $h_k(X_i) = \{(X_i-\mu)^k - \mu_k - k\mu_{k-1}(X_i-\mu)\}$, $Eh_k(X_i) = 0$ and

$$\sigma^2_{h,k} = \operatorname{Var} h_k(X_i) = \mu_{2k} - \mu_k^2 - 2k\mu_{k-1}\mu_{k+1} + k^2\mu_2\mu_{k-1}^2, \tag{5.11}$$

then Theorem 5.23 tells us that m_k is $\mathrm{AN}(\mu_k, \sigma^2_{h,k}/n)$. For example, $k = 3$ corresponds to the central third moment m_3, and $h_3(X_i) = (X_i-\mu)^3 - \mu_3 - 3\mu_2(X_i-\mu)$ and $\sigma^2_{h,3} = \mu_6 - \mu_3^2 - 6\mu_2\mu_4 + 9\mu_2^3$.

5.5.7b Sample Percentiles/Quantiles

Let $X_1, \ldots, X_n$ be iid with distribution function F and inverse distribution function (or quantile function)

$$\eta_p = F^{-1}(p) = \inf\{x : F(x) \geq p\}, \quad 0 < p < 1\,. \tag{5.12}$$

The simplest definition of a sample pth quantile (or 100pth percentile) is $\widehat{\eta}_p = F_n^{-1}(p)$, where F_n is the empirical distribution function with jumps of $1/n$ at the X_i's. The sample pth quantile $\widehat{\eta}_p = F_n^{-1}(p)$ is then the npth order statistic if np is an integer and the $([np]+1)$th order statistic otherwise, where $[\cdot]$ is the greatest integer function. Other common definitions of sample pth quantiles are weighted averages of the two order statistics. Virtually all definitions have the same asymptotic normal limiting distribution whose proof was available as early as the 1940's and likely earlier. Modern proofs of the asymptotic normality of $\widehat{\eta}_p = F_n^{-1}(p)$ depend on Theorem 5.23 (p. 242) via a proof of the basic approximation which originated with Bahadur (1966). In fact any approximation of a statistic by an average plus a negligible remainder is often called a "Bahadur representation." The best version of that approximation for use with Theorem 5.23 (p. 242) is due to Ghosh (1971):

Theorem 5.25. *Suppose that $X_1, \ldots, X_n$ are iid with distribution function F and $F'(\eta_p)$ exists and is positive. Then*

$$\widehat{\eta}_p - \eta_p = \frac{1}{n}\sum_{i=1}^{n}\left[\frac{p - I(X_i \leq \eta_p)}{F'(\eta_p)}\right] + R_n,$$

where $\sqrt{n}R_n \xrightarrow{p} 0$ as $n \to \infty$.

Because

$$h_p(X_i) = \left[\frac{p - I(X_i \le \eta_p)}{F'(\eta_p)}\right]$$

has mean 0 and variance $\sigma^2_{h,p} = p(1-p)/[F'(\eta_p)]^2$, Theorem 5.23 and Theorem 5.25 give that $\widehat{\eta}_p$ is $\mathrm{AN}(\eta_p, \sigma^2_{h,p}/n)$ as $n \to \infty$. For example, when $p = 0.5$, we have the sample median, and $h(X_i) = \{1/2 - I(X_i \le \eta_{0.5})\}/F'(\eta_{0.5})$ with asymptotic variance $1/\left[4nF'(\eta_{0.5})^2\right]$.

5.5.8 Finding h in the Approximation by Averages

The approximation-by-averages representation

$$T - T_\infty = \frac{1}{n}\sum_{i=1}^{n} h_{\mathrm{T}}(X_i) + R_n,$$

of Theorem 5.23 (p. 242) is a key tool for finding the asymptotic distribution of random quantities. In this section we begin adding a subscript T to h in order to distinguish it from various other h functions that might be involved in its definition. So far we have illustrated this representation for sample central moments (Theorem 5.24, p. 243) including the sample variance (Example 5.33, p. 242) and sample percentiles (Theorem 5.25, p. 243). In each of these examples, we merely presented the appropriate h_{T}. But where does h_{T} come from? How can we find h_{T} in a new situation?

A general answer can be given for functional statistics. A functional statistic is one that can be represented in terms of the empirical distribution function $F_n(t) = n^{-1}\sum_{i=1}^{n} I(X_i \le t)$. For example, the sample mean $\overline{X} = \int y\, dF_n(y)$ is a functional statistic. In this case we write $\overline{X} = T(F_n)$, where the functional $T(\cdot)$ is defined as $T(F) = \int y\, dF(y)$. For functional statistics,

$$h_{\mathrm{T}}(x) = \frac{\partial}{\partial \epsilon} T(F + \epsilon(\delta_x - F))\bigg|_{\epsilon=0+} = \lim_{\epsilon \to 0+} \frac{T(F + \epsilon(\delta_x - F)) - T(F)}{\epsilon} \tag{5.13}$$

is called the *influence curve* as introduced by Hampel (1974). Note that $\delta_x(t) = I(x \le t)$, viewed as a function of t, is the distribution function of a constant random variable with value x.

However, there are many situations where we can motivate and give h_{T} without resorting to this functional derivative approach. Thus, in this subsection, we list a number of general forms for random quantities and their associated h_{T} functions. At the end of the subsection, we return to the functional definition of the influence curve given in (5.13).

For all of the following situations, we are thinking of a sample $X_1, \ldots, X_n$ of iid random variables.

5.5.8a Averages of Functions

The asymptotic properties of $T = n^{-1}\sum_{i=1}^{n} q(X_i)$ are easily handled directly by the laws of large numbers and the Central Limit Theorem, but it is still worth pointing out the approximation-by-averages representation

$$\frac{1}{n}\sum_{i=1}^{n} q(X_i) - E\{q(X_1)\} = \frac{1}{n}\sum_{i=1}^{n} [q(X_i) - E\{q(X_1)\}] + 0. \tag{5.14}$$

Here the definition $h_{\mathrm{T}}(x) = q(x) - E\{q(X_1)\}$ is obvious, but it helps reinforce that h_{T} is always standardized to have mean 0. The proof that $\sqrt{n}R_n \xrightarrow{d} 0$ is automatic since $R_n = 0$. The asymptotic variance is obtained from $\sigma^2_{h_{\mathrm{T}}} = \mathrm{Var}\{q(X_1)\}$. Examples are $q(x) = \log(x)$, $q(x) = x^2$, $q(x) = I(x \le a)$, etc.

5.5.8b Functions of Averages

This class refers to Delta Theorem examples where the quantity of interest, $T = g(\overline{q}_1, \ldots, \overline{q}_k)$, is a real-valued function of averages of the type in (5.14),

$$\overline{q}_j = \frac{1}{n}\sum_{i=1}^{n} q_j(X_i).$$

The representation then follows from a first order Taylor expansion

$$\begin{aligned} g(\overline{q}_1, \ldots, \overline{q}_k) - g(\boldsymbol{\mu}) &= \sum_{j=1}^{k} g'_j(\boldsymbol{\mu})\left[\overline{q}_j - \mu_j\right] + R_n \\ &= \frac{1}{n}\sum_{i=1}^{n}\left[\sum_{j=1}^{k} g'_j(\boldsymbol{\mu})\left\{q_j(X_i) - \mu_j\right\}\right] + R_n, \end{aligned} \tag{5.15}$$

where $\overline{q}_j = n^{-1}\sum_{i=1}^{n} q_j(X_i)$, $\boldsymbol{\mu} = [E\{q_1(X_1)\}, \ldots, E\{q_k(X_1)\}]^T$, and

$$g'_j(\boldsymbol{\mu}) = \left.\frac{\partial}{\partial x_j} g(\boldsymbol{X})\right|_{\boldsymbol{X}=\boldsymbol{\mu}}.$$

Thus $h_{\mathrm{T}}(x) = \sum_{j=1}^{k} g'_j(\boldsymbol{\mu})\left\{q_j(x) - \mu_j\right\}$.

Example 5.34 (Sample central third moment). The sample central third moment may be expressed as

$$T = m_3 = n^{-1}\sum_{i=1}^{n}(X_i - \overline{X})^3 = n^{-1}\sum_{i=1}^{n}\left(X_i^3 - 3X_i^2\overline{X} + 3X_i\overline{X}^2 - \overline{X}^3\right)$$

$$= m_3' - 3m_2'\overline{X} + 2\overline{X}^3,$$

which is a function of the raw sample moments $\overline{q}_1 = \overline{X}$, $\overline{q}_2 = m_2'$, and $\overline{q}_3 = m_3'$. Thus $T = g(\overline{X}, m_2', m_3')$, where

$$g(y_1, y_2, y_3) = y_3 - 3y_2y_1 + 2y_1^3.$$

Taking partial derivatives and substituting $\boldsymbol{\mu} = (E\{X_1\}, E\{X_1^2\}, E\{X_1^3\})^T = (\mu, \mu_2', \mu_3')^T$, we have

$$h_{\mathrm{T}}(X_i) = (6\mu^2 - 3\mu_2')(X_i - \mu) - 3\mu(X_i^2 - \mu_2') + (X_i^3 - \mu_3').$$

This can be compared to the representation of $T = m_3$ in terms of centered quantities as given in Theorem 5.24 (p. 243), $h_{\mathrm{T}}(X_i) = (X_i - \mu)^3 - \mu_3 - 3\mu_2(X_i - \mu)$. This latter h_{T} from Theorem 5.24 is preferable because then the asymptotic variance is given in terms of central moments rather than raw moments. ♦

Note that this class of statistics, $T = g(\overline{q}_1, \ldots, \overline{q}_k)$, is a subset of the class handled by the Delta Theorem (Theorem 5.19, p. 238). Moreover, the main reason to get an approximation-by-averages representation is to obtain asymptotic normality that Theorem 5.19 (p. 238) already gives in more generality. Nevertheless, (5.15, p. 245) reveals how the form of h_{T} for this case comes from Taylor expansion. As in Example 5.34 above, $(\overline{q}_1, \ldots, \overline{q}_k)$ are typically subsets of the raw sample moments $\overline{X}, m_2', m_3', \ldots$, and functions of them of interest include s_n^2, m_3, estimates of Skew(X_1) and Kurt(X_1) given by $m_3/m_2^{3/2}$ and m_4/m_2^2, respectively, and the sample correlation coefficient. The fact that $\sqrt{n}R_n \xrightarrow{p} 0$ as $n \to \infty$ is given in the following theorem which has nearly the same conditions as the Delta Theorem (Theorem 5.19, p. 238).

We state this theorem in a slightly more general way in terms of an iid sample of vectors $\boldsymbol{Y}_1, \ldots, \boldsymbol{Y}_n$. The most important applications are of the type mentioned above, where $\boldsymbol{Y}_i = \{q_1(X_i), \ldots, q_k(X_i)\}^T$, and $X_1, \ldots, X_n$ is an iid sample. In this case (5.16) below reduces to (5.15, p. 245).

Theorem 5.26 (Approximation-by-Averages Representation for Functions of Means). *Suppose that $\boldsymbol{Y}_1, \ldots, \boldsymbol{Y}_n$ are iid vectors with finite mean $E(\boldsymbol{Y}_1) = \boldsymbol{\mu}$ and covariance $\boldsymbol{\Sigma}$. If g is a real-valued function such that g' exists in a neighborhood of $\boldsymbol{\mu}$ and is continuous at $\boldsymbol{\mu}$, and $g'(\boldsymbol{\mu}) \neq \boldsymbol{0}$, then*

$$g(\overline{\boldsymbol{Y}}) - g(\boldsymbol{\mu}) = \frac{1}{n}\sum_{i=1}^{n} g'(\boldsymbol{\mu})(\boldsymbol{Y}_i - \boldsymbol{\mu}) + R_n, \tag{5.16}$$

where $\sqrt{n}R_n \xrightarrow{p} 0$ as $n \to \infty$.

Proof. By Young's form of Taylor's expansion (see, for example, p. 45 of Serfling, 1980), (5.16) holds where

$$\frac{\sqrt{n}R_n}{\|\sqrt{n}(\overline{\boldsymbol{Y}} - \boldsymbol{\mu})\|} \xrightarrow{p} 0,$$

and $\|\cdot\|$ is the usual Euclidean norm. But by the Central Limit Theorem, the denominator is converging to a random variable and is thus $O_p(1)$. Multiplying $o_p(1)$ and $O_p(1)$ results then yields an $o_p(1)$ result for $\sqrt{n}R_n$. ■

Note that the asymptotic variance of $g(\overline{\boldsymbol{Y}})$ that follows from Theorem 5.26 is similar to the one given by the Delta Theorem (Theorem 5.19, p. 238), $g'(\boldsymbol{\mu})\boldsymbol{\Sigma}g'(\boldsymbol{\mu})^T/n$, although with slightly different notation, $\boldsymbol{\theta} = \boldsymbol{\mu}$ and $b_n^2 = n$.

5.5.8c Functions of Statistics with Approximation-by-Averages Representation

We hope to convince the reader that approximation by averages are ubiquitous. Moreover, they build on one another. The next result says that a smooth function of statistics that have an approximation-by-averages representation also has an approximation-by-averages representation. This is a generalization of (5.15, p. 245) and essentially gives approximation-by-averages representation for statistics that are covered by the Delta Theorems. We let $\boldsymbol{\theta}$ have dimension k in accordance with Theorem 5.26 (p. 246) although $\boldsymbol{\theta}$ usually has dimension b.

Theorem 5.27. *For an iid sample $X_1, \ldots, X_n$, suppose that each component of $\widehat{\boldsymbol{\theta}}$ has the approximation-by-averages representation given by*

$$\widehat{\theta}_j - \theta_j = \frac{1}{n}\sum_{i=1}^{n} h_j(X_i) + R_{nj}, \quad j = 1, \ldots, k, \tag{5.17}$$

where $\sqrt{n}R_{nj} \xrightarrow{p} 0$ as $n \to \infty$, $E\{h_j(X_1)\} = 0$, and $\text{Var}\{h_j(X_1)\}$ is finite, $j = 1, \ldots, k$. Also assume that g is a real-valued function with partial derivatives existing in a neighborhood of the true value $\boldsymbol{\theta}$ and continuous at $\boldsymbol{\theta}$. Then

$$g(\widehat{\boldsymbol{\theta}}) - g(\boldsymbol{\theta}) = \frac{1}{n}\sum_{i=1}^{n}\left[\sum_{j=1}^{k} g_j'(\boldsymbol{\theta})h_j(X_i)\right] + R_n, \tag{5.18}$$

where $\sqrt{n}R_n \xrightarrow{p} 0$ as $n \to \infty$.

Proof. The proof is nearly identical to the proof of Theorem 5.26 (p. 246) except that $\widehat{\theta}_j - \theta_j$ plays the role of $\overline{q}_j - \mu_j$ in (5.15, p. 245) or $\overline{Y}_j - \mu_j$ in (5.16) and is in turn replaced by $\sum_{i=1}^{n} h_j(X_i) + R_{nj}$. ■

5.5.8d Averages of Functions with Estimated Parameters

Often we need asymptotic normality of a sum with estimated parameters in the summands, $T = n^{-1}\sum_{i=1}^{n} q(X_i, \widehat{\boldsymbol{\theta}})$. A simple example would be the sample central moments, $m_k = n^{-1}\sum_{i=1}^{n}(X_i - \overline{X})^k$, for which an approximation by averages result has already been given (Theorem 5.24, p. 243). A more difficult statistic to handle arises from replacing $\overline{X}$ in m_k by the sample median $\widehat{\eta}_{1/2}$, resulting in $n^{-1}\sum_{i=1}^{n}(X_i - \widehat{\eta}_{1/2})^k$.

In the general case considered here, we need each component of the $b \times 1$ estimator $\widehat{\boldsymbol{\theta}}$ to have an approximation-by-averages representation

$$\widehat{\theta}_j - \theta_j = \frac{1}{n}\sum_{i=1}^{n} h_j(X_i) + R_{nj}, \quad j = 1, \ldots, b, \tag{5.19}$$

where $\sqrt{n}R_{nj} \xrightarrow{p} 0$ as $n \to \infty$, $E\{h_j(X_1)\} = 0$, and $\mathrm{Var}\{h_j(X_1)\}$ is finite, $j = 1, \ldots, b$. Note that (5.19) is the same (5.17, p. 247) except b replaces k as the dimension. In the following we put the above h_j functions into a $b\times 1$ vector $\boldsymbol{h}$. Note also that the vector of partial derivatives of q with respect to $\boldsymbol{\theta}$, q', is also a $b \times 1$ vector. Then, Taylor expansion along with some adding and subtracting leads to

$$\frac{1}{n}\sum_{i=1}^{n} q(X_i, \widehat{\boldsymbol{\theta}}) - E\{q(X_1, \boldsymbol{\theta})\} = \frac{1}{n}\sum_{i=1}^{n} h_{\mathrm{T}}(X_i) + R_n,$$

where

$$h_{\mathrm{T}}(X_i) = q(X_i, \boldsymbol{\theta}) - E\{q(X_1, \boldsymbol{\theta})\} + \left[E\{q'(X_1, \boldsymbol{\theta})\}\right]\boldsymbol{h}(X_i), \tag{5.20}$$

and for some $\widehat{\boldsymbol{\theta}}^*$ lying between $\widehat{\boldsymbol{\theta}}$ and $\boldsymbol{\theta}$

$$\begin{aligned} R_n = &\left[\frac{1}{n}\sum_{i=1}^{n} q'(X_i, \boldsymbol{\theta}) - E\{q'(X_1, \boldsymbol{\theta})\}\right](\widehat{\boldsymbol{\theta}} - \boldsymbol{\theta}) \\ &+ E\{q'(X_1, \boldsymbol{\theta})\}\left\{\widehat{\boldsymbol{\theta}} - \boldsymbol{\theta} - \frac{1}{n}\sum_{i=1}^{n} \boldsymbol{h}(X_i)\right\} \\ &+ \frac{1}{2}(\widehat{\boldsymbol{\theta}} - \boldsymbol{\theta})^T\left\{\frac{1}{n}\sum_{i=1}^{n} q''(X_i, \widehat{\boldsymbol{\theta}}^*)\right\}(\widehat{\boldsymbol{\theta}} - \boldsymbol{\theta}). \end{aligned} \tag{5.21}$$

Before showing that $\sqrt{n}R_n \xrightarrow{p} 0$ as $n \to \infty$ in Theorem 5.28 below, let us look at $h_{\mathrm{T}}(X_i)$ from (5.20) for a few examples:

1. The kth sample central moment has $q(X_i, \mu) = (X_i - \mu)^k$ and $\widehat{\mu} = \overline{X}$. Thus $q'(X_i, \mu) = -k(X_i - \mu)^{k-1}$, $h_1(X_i) = X_i - \mu$, and

$$h_{\mathrm{T}}(X_i) = (X_i - \mu)^k - \mu_k - k\mu_{k-1}(X_i - \mu),$$

which is the same as given in Theorem 5.24 (p. 243).

2. The second central moment about the median has $q(X_i, \eta_{0.5}) = (X_i - \eta_{0.5})^2$ and for the sample median $h_1(X_i) = \{1/2 - I(X_i \le \eta_{0.5})\}/F'(\eta_{0.5})$. Thus

$$h_{\mathrm{T}}(X_i) = (X_i - \eta_{0.5})^2 - E(X_1 - \eta_{0.5})^2 - 2(\mu - \eta_{0.5}) \left[\frac{1/2 - I(X_i \le \eta_{0.5})}{F'(\eta_{0.5})} \right].$$

   If the population mean and medians are the same, then this second central moment about the median has the same asymptotic behavior as the sample variance $m_2 = s_n^2$.
3. A more standard dispersion estimator used with the sample median is the mean absolute deviation from the sample median,

$$T = \frac{1}{n} \sum_{i=1}^{n} |X_i - \widehat{\eta}_{0.5}|.$$

   Unfortunately, the results in this subsection do not apply to this statistic because the absolute value function is not differentiable at 0. Nevertheless, T has an approximation-by-averages representation, with

$$h_{\mathrm{T}}(x) = |x - \eta_{0.5}| - E|X_1 - \eta_{0.5}|, \tag{5.22}$$

   as can be verified by using (5.13, p. 244).

The following theorem formally states the approximation-by-averages representation for $T = n^{-1} \sum_{i=1}^{n} q(X_i, \widehat{\boldsymbol{\theta}})$. A simple corollary is that T is AN$(E\{q(X_1, \boldsymbol{\theta})\}$, $\mathrm{Var}\{h_{\mathrm{T}}(X_1)\}/n)$ where h_{T} is given in (5.20, p. 248).

Theorem 5.28. *For an iid sample $X_1, \ldots, X_n$, suppose that each component of $\widehat{\boldsymbol{\theta}}$ has the approximation-by-averages representation given by*

$$\widehat{\theta}_j - \theta_j = \frac{1}{n} \sum_{i=1}^{n} h_j(X_i) + R_{nj}, \quad j = 1, \ldots, b, \tag{5.23}$$

where $\sqrt{n} R_{nj} \xrightarrow{p} 0$ as $n \to \infty$, $E\{h_j(X_1)\} = 0$, and $\mathrm{Var}\{h_j(X_1)\}$ is finite, $j = 1, \ldots, b$. Also assume that the real-valued function $q(X_i, \boldsymbol{\theta})$ has two partial derivatives with respect to $\boldsymbol{\theta}$, and

1. *$\mathrm{Var}\{q(X_1, \boldsymbol{\theta})\}$ and $E\{q'(X_1, \boldsymbol{\theta})\}$ are finite;*
2. *there exists a function $M(x)$ such that for all $\boldsymbol{\theta}^*$ in a neighborhood of the true value $\boldsymbol{\theta}$ and all $j, k \in \{1, \ldots, b\}$, $|q''(x, \boldsymbol{\theta}^*)_{jk}| \le M(x)$, where $E\{M(X_1)\} < \infty$.*

Then,

$$\frac{1}{n} \sum_{i=1}^{n} q(X_i, \widehat{\boldsymbol{\theta}}) - E\{q(X_1, \boldsymbol{\theta})\} = \frac{1}{n} \sum_{i=1}^{n} h_{\mathrm{T}}(X_i) + R_n,$$

where

$$h_{\mathrm{T}}(X_i) = q(X_i, \boldsymbol{\theta}) - E\{q(X_1, \boldsymbol{\theta})\} + \left[E\{q'(X_1, \boldsymbol{\theta})\}\right] \boldsymbol{h}(X_i),$$

and $\sqrt{n}R_n \xrightarrow{p} 0$ *as* $n \to \infty$.

Proof. The first two terms of (5.21) are clearly $o_p(n^{-1/2})$ due to the approximation-by-averages representation for $\widehat{\boldsymbol{\theta}}$ and the WLLN applied to $n^{-1}\sum_{i=1}^n q'(X_i, \boldsymbol{\theta})$. For the last term of (5.21) we use the fact that $n^{-1}\sum_{i=1}^n q''(X_i, \widehat{\boldsymbol{\theta}}^*)$ is ultimately bounded by $n^{-1}\sum_{i=1}^n M(X_i)$ because $\widehat{\boldsymbol{\theta}} \xrightarrow{p} \boldsymbol{\theta}$ and $n^{-1}\sum_{i=1}^n M(X_i) = O_p(1)$ by the WLLN. ■

A version of Theorem 5.28 is given in Presnell and Boos (2004). Randles (1982) presents more general results on statistics with estimated parameters. An important application of Theorem 5.28 is found in the proof of Theorem 6.9 (p. 288).

5.5.8e Maximum Likelihood Estimators

Consider $X_1, \ldots, X_n$ iid from a parametric model density $f(x; \boldsymbol{\theta})$. Under suitable regularity conditions to be given in Chapter 6,

$$\widehat{\boldsymbol{\theta}}_{\mathrm{MLE}} - \boldsymbol{\theta} = \frac{1}{n}\sum_{i=1}^{n} \boldsymbol{I}(\boldsymbol{\theta})^{-1}\left[\frac{\partial}{\partial \boldsymbol{\theta}} \log f(X_i; \boldsymbol{\theta})\right] + R_n,$$

where $\sqrt{n}R_n \xrightarrow{p} 0$ as $n \to \infty$.

5.5.8f M-Estimators

In Chapter 7 we generalize maximum likelihood estimators to $\widehat{\boldsymbol{\theta}}$ that satisfy $\sum_{i=1}^n \boldsymbol{\psi}(X_i, \widehat{\boldsymbol{\theta}}) = \boldsymbol{0}$, where $\boldsymbol{\psi}$ is a known $(b \times 1)$-function that does not depend on i or n. In the Appendix of Chapter 7, we show under suitable regularity conditions that

$$\widehat{\boldsymbol{\theta}} - \boldsymbol{\theta} = \frac{1}{n}\sum_{i=1}^{n} \boldsymbol{A}(\boldsymbol{\theta})^{-1}\boldsymbol{\psi}(X_i, \boldsymbol{\theta}) + R_n,$$

where $\sqrt{n}R_n \xrightarrow{p} 0$ as $n \to \infty$ and $\boldsymbol{A}(\boldsymbol{\theta}) = E\left\{-\boldsymbol{\psi}'(X_1, \boldsymbol{\theta})\right\}$.

5.5.8g U-Statistics

U-statistics were introduced by Hoeffding (1948) and are a large class of statistics that generalize the sample mean $\overline{X}$. They are statistics that can be represented as arithmetic averages of functions of subsets of the original sample. Specifically they are statistics that can be written as

$$T = \frac{1}{\binom{n}{m}} \sum_{c} v(X_{i_1}, X_{i_2}, \ldots, X_{i_m}), \tag{5.24}$$

where v is a function called a *kernel* that is symmetric in its arguments, and $\sum_c$ denotes the m-fold summation over all $\binom{n}{m}$ distinct subsets of $\{i_1, \ldots, i_m\}$ from $\{1, \ldots, n\}$. The U in U-statistics is for "unbiased," as the kernel v is such that $E(T) = E\{v(X_1, \ldots, X_m)\} = \theta$, the parameter that T is estimating. The form of T in (5.24) looks foreboding, but usually $m = 2$ or $m = 3$, and the resulting expression is relatively simple. For example, the unbiased version of the sample variance is seen to be a U-statistic upon algebraic manipulation,

$$s_{n-1}^2 = \frac{1}{n-1} \sum_{i=1}^{n} (X_i - \overline{X})^2 = \frac{2}{n(n-1)} \sum_{i<j} \frac{(X_i - X_j)^2}{2} = T, \tag{5.25}$$

where $m = 2$, $v = (x_1, x_2) = (x_1 - x_2)^2/2$, and c is the set of all $\binom{n}{2}$ sets of observations $\{X_i, X_j\}$.

Although U-statistics have identically distributed summands, many of the summands are correlated due to overlapping X_i values; for example, $v(X_1, X_2)$ and $v(X_1, X_3)$ are correlated due to X_1 appearing in both. Thus, an approximation-by-averages representation is important because the iid Central Limit Theorem does not apply to sums of dependent summands. Many results about U-statistics may be found in Serfling (1980, Ch. 5), but here we want to merely mention the approximation by averages result that follows from Theorem 5.3.2, p. 188, of that text:

Theorem 5.29. *For an iid sample $X_1, \ldots, X_n$, if $E\{v(X_1, \ldots, X_m)\}^2 < \infty$, then*

$$T - \theta = \frac{1}{n} \sum_{i=1}^{n} h(X_i) + R_n,$$

where

$$h(x) = m\left[E\left\{v(X_1, X_2, \ldots, X_m) \mid X_1 = x\right\} - \theta\right],$$

and $\sqrt{n} R_n \xrightarrow{p} 0$ as $n \to \infty$.

In the U-statistic context, the average $n^{-1} \sum_{i=1}^{n} h(X_i)$ above is often called the *projection* of $T - \theta$ onto the space of averages of mean zero random variables because of its conditional expectation origin.

Example 5.35 (Sample variance). To find the h for the sample variance given in (5.24), we note that $v(x_1, x_2) = (x_1 - x_2)^2/2$ and compute

$$E\left\{(X_1 - X_2)^2/2 \mid X_1 = x\right\} = E(x - X_2)^2/2$$
$$= E\left\{(x - \mu) - (X_2 - \mu)\right\}^2/2$$
$$= \frac{(x-\mu)^2 + \sigma^2}{2}.$$

Finally, we subtract $\theta = E(X_1 - X_2)^2/2 = \sigma^2$ from $\left\{(x-\mu)^2 + \sigma^2\right\}/2$ and multiply by $m = 2$ to obtain $h(x) = (x-\mu)^2 - \sigma^2$, a form we have derived before in other ways.

5.5.8h The Influence Curve

The approximation-by-averages representation of most statistics in use can be found from one of the above general classes of random quantities. The maximum likelihood and M-estimator classes are extremely relevant and by themselves cover most situations of interest, at least in combination with one of the general results of Theorems 5.26 (p. 246) through 5.27 (p. 247). Several notable exceptions are the class of linear combinations of order statistics, *L-statistics*, and the class of estimators obtained by inverting rank tests, *R-statistics*, discussed in Chs. 8 and 9 of Serfling, 1980.

If we can obtain h of the approximation by averages essentially by Taylor expansion in all of the above classes, why do we need the complicated functional derivative

$$h(x) = \frac{\partial}{\partial \epsilon} T(F + \epsilon(\delta_x - F))\bigg|_{\epsilon=0+} \tag{5.26}$$

of (5.13, p. 244)? The reason is that this h (typically called the *Influence Curve*) unifies all of the approximation by averages results. That is, all of the above approximation-by-averages representations can be derived via (5.26). Our goal here is to merely show why this is the case. One small caveat: equation (5.26) only shows how to obtain h, not how to prove $\sqrt{n}R_n \xrightarrow{p} 0$ as $n \to \infty$.

For a functional $T(\cdot)$ defined on distribution functions $F(t)$, the Gateaux derivative in the "direction" $\Delta(t)$ is given by

$$T(F; \Delta) = \frac{\partial}{\partial \epsilon} T(F + \epsilon\Delta)\bigg|_{\epsilon=0+} \tag{5.27}$$

The Gateaux derivative has a long history in variational calculus and in the definition of derivatives for functionals. von Mises (1947) introduced the Gateaux derivative into the statistics literature and used a Taylor expansion in ϵ of $T(F + \epsilon\Delta)$ with $\Delta = F_n - F$ to get asymptotic results for functional statistics $T(F_n)$, where F_n is the empirical distribution function $F_n(t) = n^{-1}\sum_{i=1}^{n} I(X_i \le t)$. Specifically von Mises (1947) used Taylor expansion about $\epsilon = 0$ to get

$$T(F + \epsilon\Delta) = T(F) + T(F;\Delta)\epsilon + R(F,\Delta).$$

Now letting $\Delta = F_n - F$ and $\epsilon = 1$ in the previous expression, we get

$$T(F_n) = T(F) + T(F; F_n - F) + R_n, \tag{5.28}$$

where we have changed $R(F,\Delta)$ to R_n to reflect dependence on n. Note that

$$\begin{aligned} F_n(t) - F(t) &= \frac{1}{n}\sum_{i=1}^{n}\{I(X_i \le t) - F(t)\} \\ &= \frac{1}{n}\sum_{i=1}^{n}\{\delta_{X_i}(t) - F(t)\}. \end{aligned} \tag{5.29}$$

The key to understanding why the Influence Curve of (5.26) provides the h of the approximation by averages is the following. Starting with von Mises' first order Taylor expansion (5.28) and substituting (5.29), we have

$$T(F_n) - T(F) = T(F; F_n - F) + R_n \tag{5.30}$$

$$= T\left\{F; \frac{1}{n}\sum(\delta_{X_i} - F)\right\} + R_n \tag{5.31}$$

$$= \frac{1}{n}\sum T(F; \delta_{X_i} - F) + R_n, \tag{5.32}$$

where in this last step we have used linearity of the Gateaux derivative, $T(F; \sum a_i F_i) = \sum a_i T(F; F_i)$. Linearity is not guaranteed by definition (5.27), but it usually holds. Thus, simple Taylor expansion of $T(F + \epsilon(F_n - F))$ and linearity leads to the approximation-by-averages representation with h given by (5.13, p. 244) or (5.26, p. 252). Now we give several examples to illustrate.

5.5.8i Influence Curve Derivation for Sample Mean and Variance

The mean functional is $T_1(F) = \int t\,dF(t)$ and the variance functional is $T_2(F) = \int (t-\mu)^2 dF(t)$. To find the influence curve of the mean functional,

$$T_1(F + \epsilon\Delta) = \int t\,d[F(t) + \epsilon\Delta(t)] = \int t\,dF(t) + \epsilon\int t\,d\Delta(t).$$

Then the Gateaux derivative is

$$T_1(F;\Delta) = \frac{\partial}{\partial\epsilon}T_1(F + \epsilon\Delta)\bigg|_{\epsilon=0} = \int t\,d\Delta(t),$$

and setting $\Delta(t) = \delta_x(t) - F(t)$, we have

$$h(x) = \int t\, d\,[\delta_x(t) - F(t)] = x - \int t\, dF(t) = x - \mu,$$

where we have put in μ for the mean functional in the last step, and used the fact that $\int t\, d\delta_x(t) = x$.

For the variance functional, let us use the alternate expression $\mathrm{Var}(X) = E(X^2) - \mu^2$ and substitute $T_1(F) = \mu$ to get

$$T_2(F + \epsilon\Delta) = \int t^2\, dF(t) + \epsilon \int t^2\, d\Delta(t) - \{T_1(F + \epsilon\Delta)\}^2.$$

Then the Gateaux derivative is

$$\begin{aligned} T_2(F;\Delta) &= \frac{\partial}{\partial\epsilon} T_2(F + \epsilon\Delta)\bigg|_{\epsilon=0} \\ &= \int t^2\, d\Delta(t) - 2T_1(F)T_1(F;\Delta). \end{aligned}$$

Setting $\Delta(t) = \delta_x(t) - F(t)$ and using the results for the mean, we have

$$\begin{aligned} h(x) &= \int t^2\, d\,[\delta_x(t) - F(t)] - 2\mu(x - u) \\ &= x^2 - (\sigma^2 + \mu^2) - 2\mu x + 2\mu^2 = (x - \mu)^2 - \sigma^2, \end{aligned}$$

where we have used the fact that $\int t^2\, dF(t) = \sigma^2 + \mu^2$.

5.5.8j Influence Curve Derivation for the Median

The sample median provides a good example of using the functional approach to obtain the appropriate h function without getting bogged down by mathematical rigor. Recall that the definition of the pth quantile (5.12, p. 243) was already in a functional form, $\eta_p = F^{-1}(p)$. We use the fact that when F is continuous at η_p, $F(F^{-1}(p)) = p$. Then, for $p = 0.5$, we have for Δ continuous,

$$F\left\{(F + \epsilon\Delta)^{-1}(0.5)\right\} + \epsilon\Delta\left\{(F + \epsilon\Delta)^{-1}(0.5)\right\} = 0.5.$$

Using implicit differentiation, we take the partial derivative with respect to ϵ,

$$\begin{aligned} &F'\left\{(F + \epsilon\Delta)^{-1}(0.5)\right\} \frac{\partial}{\partial\epsilon}(F + \epsilon\Delta)^{-1}(0.5) \\ &\quad + \Delta\left\{(F + \epsilon\Delta)^{-1}(0.5)\right\} + \epsilon\frac{\partial}{\partial\epsilon}\Delta\left\{(F + \epsilon\Delta)^{-1}(0.5)\right\} = 0. \quad (5.33) \end{aligned}$$

Note that it is not clear that we can differentiate Δ, but we ignore that detail because setting $\epsilon = 0$ gets rid of the last term. Thus, setting $\epsilon = 0$ and substituting $\eta_{0.5} = F^{-1}(0.5)$, we have

$$F'(\eta_{0.5})T(F;\Delta) + \Delta(\eta_{0.5}) = 0,$$

and solving for $T(F;\Delta)$, the Gateaux derivative is

$$T(F;\Delta) = -\frac{\Delta(\eta_{0.5})}{F'(\eta_{0.5})}.$$

Although we derived $T(F;\Delta)$ under the assumption that Δ was continuous, we now set $\Delta(t) = \delta_x(t) - F(t) = I(x \le t) - F(t)$ in the formula above, and using the fact that $F(\eta_{0.5}) = 0.5$, obtain

$$h(x) = \frac{0.5 - I(x \le 0.5)}{F'(\eta_{0.5})}.$$

5.5.9 *Proving Convergence in Distribution of Random Vectors*

Previously we noted that convergence in distribution of vectors is harder to prove than the vector versions of wp1 convergence and convergence in probability. The key reason is that convergence in distribution of the components of a vector, i.e., convergence of the marginal distribution functions, does not entail joint convergence of the vector except in the special case that the components are independent (Theorem 5.8, p. 225).

The main tool for proving convergence in distribution of vectors is the Cramer-Wold device given in Theorem 5.31 (p. 256) below. In order to prove Theorem 5.31, we need the following classical result giving the equivalence between convergence in distribution and pointwise convergence of characteristic functions. The characteristic function $\phi(\boldsymbol{t})$ of a random vector $\boldsymbol{Y}$ is defined to be the complex-valued function

$$\phi(\boldsymbol{t}) = \mathrm{E}\left\{e^{i\boldsymbol{t}^T\boldsymbol{Y}}\right\} = \mathrm{E}\left\{\cos(\boldsymbol{t}^T\boldsymbol{Y})\right\} + i\mathrm{E}\left\{\sin(\boldsymbol{t}^T\boldsymbol{Y})\right\},$$

where $\boldsymbol{t} \in R^k$ and $i = \sqrt{-1}$. The characteristic function uniquely determines the distribution function and vice versa. When the moment generating function $m(\boldsymbol{t}) = E\left(e^{\boldsymbol{t}^T\boldsymbol{Y}}\right)$ exists, the characteristic function and moment generating function are related via $m(\boldsymbol{t}) = \phi(-i\boldsymbol{t})$. The characteristic function always exists and thus is more useful than the moment generating function for proving convergence results.

Theorem 5.30. *Let $\boldsymbol{Y}_n$ be a sequence of random k-vectors with associated characteristic functions ϕ_n. Then*

$$\boldsymbol{Y}_n \xrightarrow{d} \boldsymbol{Y} \iff \lim_{n\to\infty} \phi_n(\boldsymbol{t}) = \phi(\boldsymbol{t}) \quad \textit{for all } \boldsymbol{t} \in R^k.$$

Although Theorem 5.30 is used to prove Theorem 5.31, its most important application is to prove the Central Limit Theorem. Now we are ready to state the following general result concerning the equivalence of convergence in distribution of random vectors and the convergence in distribution of all linear combinations of the vector components. Its proof is straightforward using characteristic functions (see Serfling, 1980, p. 18).

Theorem 5.31 (Cramér-Wold).

$$\boldsymbol{Y}_n \xrightarrow{d} \boldsymbol{Y} \iff \boldsymbol{c}^T \boldsymbol{Y}_n \xrightarrow{d} \boldsymbol{c}^T \boldsymbol{Y} \quad \textit{for all } \boldsymbol{c} \in R^k.$$

Recall Example 5.11 (p. 225) where convergence in distribution of the components did not entail joint convergence in distribution. The Cramér-Wold condition picks this up: $(c_1, c_2)(X_{1n}, X_{2n})^T = c_1 Z_1 + c_2 Z_2 + (c_1 + c_2) Z_3$ for n odd is distributed as $\mathrm{N}(0, c_1^2 + c_2^2 + (c_1 + c_2)^2)$, whereas $(c_1, c_2)(X_{1n}, X_{2n})^T = c_1 Z_1 + c_2 Z_2 + (c_1 - c_2) Z_3$ for n even is distributed as $\mathrm{N}(0, c_1^2 + c_2^2 + (c_1 - c_2)^2)$ and thus $c^T \boldsymbol{X}_n$ does not converge in distribution. Thus by Theorem 5.31, $\boldsymbol{X}_n$ does not converge in distribution.

In many applications, we verify the Cramér-Wold condition by approximating all the components by averages. The next example illustrates the type of problem that can be handled by such an approximation. Then the next subsection generalizes the result into a theorem.

Example 5.36 (Joint convergence in distribution: sample mean and variance). Suppose that $X_1, \ldots, X_n$ are iid(μ, σ^2) and μ_4 exists, where $\mu_k = E(X_1 - \mu)^k$. Because of the separate approximation by averages results for both $\overline{X}$ and s_n^2, we might guess that

$$\sqrt{n}(\overline{X} - \mu, s_n^2 - \sigma^2)^T \xrightarrow{d} BN(\boldsymbol{0}, \boldsymbol{\Sigma}) \text{ as } n \to \infty, \tag{5.34}$$

where $\boldsymbol{\Sigma}_{11} = \sigma^2$, $\boldsymbol{\Sigma}_{12} = \mu_3$, and $\boldsymbol{\Sigma}_{22} = \mu_4 - \sigma^4$. Perhaps the off-diagonal element μ_3 is not totally obvious, but it makes sense to use the covariance of the two h functions from their approximation-by-averages representations, and that turns out to be μ_3.

To use Theorem 5.31 in showing this joint convergence, we need to show that $c_1[\sqrt{n}(\overline{X} - \mu)] + c_2[\sqrt{n}(s^2 - \sigma^2)] \xrightarrow{d} N\left\{0, (c_1, c_2)\boldsymbol{\Sigma}(c_1, c_2)^T\right\}$. Note that using the representation in Example 5.33 (p. 242) for $s_n^2 - \sigma^2$, we have

$$c_1(\overline{X} - \mu) + c_2(s^2 - \sigma^2) = \frac{1}{n}\sum_{i=1}^{n}[c_1(X_i - \mu) + c_2\{(X_i - \mu)^2 - \sigma^2\}] + R_n,$$

where $R_n = -c_2(\overline{X} - \mu)^2$, which is an approximation-by-averages representation. Then by Theorem 5.23 (p. 242) we have the appropriate convergence to $\mathrm{N}\{0, (c_1, c_2)\boldsymbol{\Sigma}(c_1, c_2)^T\}$ since $\mathrm{Var}[c_1(X_i - \mu) + c_2\{(X_i - \mu)^2 - \sigma^2\}] = c_1^2\sigma^2 + c_2^2(\mu_4 - \sigma^4) + 2c_1c_2\mu_3 = (c_1, c_2)\boldsymbol{\Sigma}(c_1, c_2)^T$. ♦

5.5.10 Multivariate Approximation by Averages

Theorem 5.32 (Multivariate Approximation by Averages). *For an iid sample $X_1, \ldots, X_n$, suppose that each component of $\boldsymbol{T}$ has the approximation-by-averages representation given by*

$$T_j - T_{j\infty} = \frac{1}{n}\sum_{i=1}^{n} h_j(X_i) + R_{nj}, \quad j = 1, \ldots, k, \tag{5.35}$$

where $\sqrt{n}R_{nj} \xrightarrow{p} 0$ as $n \to \infty$, $E\{h_j(X_1)\} = 0$, and the covariance of $(h_1(X_1), \ldots, h_k(X_1))^T$ is $\boldsymbol{\Sigma}$. Then

$$\sqrt{n}(T_1 - T_{1\infty}, \ldots, T_k - T_{k\infty})^T \xrightarrow{d} N(0, \boldsymbol{\Sigma}).$$

Theorem 5.32 follows easily from Theorem 5.22 (p. 241) and the multivariate Central Limit Theorem (Theorem 5.7, p. 225). Theorem 5.32 tells us that any statistics having component-by-component approximation by averages can be put together to get joint asymptotic normality. For example, any group of sample central moments and any group of sample percentiles have joint asymptotic normal distributions. These convergence are illustrated below. Moreover, one can combine any set of sample moments with any set of sample percentiles to get joint asymptotic normality of the whole set. This can be useful if one is interested in the asymptotic correlations between these statistics or if one has a statistic which is a function of both sample moments and sample percentiles.

Recall that joint convergence in probability and with probability 1 for random k-vectors is equivalent to convergence of the individual components. This equivalence is not true for convergence in distribution (except for the case where all the components are independent). Theorem 5.32 is an equivalence result in that statistics that are component-by-component approximated by averages and thus are univariately asymptotically normal are also automatically jointly asymptotically normal.

Example 5.37 (Sample moments). Consider the vector composed of the sample mean $\overline{X}$ and the second through the kth sample central moments $(\overline{X}, m_2, \ldots, m_k)^T$. Theorem 5.24 (p. 243) and Theorem 5.32 (p. 257) yield that this vector is asymptotically normal with mean $(\mu, \mu_2, \ldots, \mu_k)^T$ and covariance matrix $\boldsymbol{\Sigma}/n$, where the ijth element of $\boldsymbol{\Sigma}$ for $i \geq 2$ and $j \geq 2$ is given by

$$\begin{aligned} &\mathrm{E}\left[(X_1-\mu)^i - \mu_i - i\mu_{i-1}(X_1-\mu)\right]\left[(X_1-\mu)^j - \mu_j - j\mu_{j-1}(X_1-\mu)\right] \\ &= \mu_{i+j} - \mu_i\mu_j - i\mu_{i-1}\mu_{j+1} - j\mu_{i+1}\mu_{j-1} + ij\mu_{i-1}\mu_{j-1}\sigma^2. \end{aligned}$$

♦

Example 5.38 (Sample quantiles). Suppose that $F'(\eta_{p_i}) > 0, i = 1, \ldots, k$, where $0 < p_1 < \cdots < p_k < 1$. Then by Theorem 5.24 (p. 243) and Theorem 5.32 (p. 257), the vector $(\widehat{\eta}_{p_1}, \ldots, \widehat{\eta}_{p_k})^T$ is asymptotically normal with mean $(\eta_{p_1}, \ldots, \eta_{p_k})^T$ and covariance matrix $\boldsymbol{\Sigma}/n$, where the ijth element of $\boldsymbol{\Sigma}$ for $i \le j$ is given by

$$E\left[\left\{\frac{p_i - I(X_1 \le \eta_{p_i})}{F'(\eta_{p_i})}\right\}\left\{\frac{p_j - I(X_1 \le \eta_{p_j})}{F'(\eta_{p_j})}\right\}\right]$$
$$= \frac{p_i(1-p_j)}{F'(\eta_{p_i})F'(\eta_{p_j})}. \qquad \blacklozenge$$

5.6 Summary of Methods for Proving Convergence in Distribution

We have covered a lot of ground in this chapter. Here we remind the reader of the most important approaches for proving convergence in distribution.

1. **Averages:**

$$\frac{1}{n}\sum_{i=1}^{n} X_i \text{ is AN}\{\text{E}(X_1), \text{Var}(X_1)/n\} \qquad \text{as } n \to \infty$$

directly from the CLT (p. 219).

2. **Averages of Functions:**

$$\frac{1}{n}\sum_{i=1}^{n} h(X_i) \text{ is AN}\{\text{E}\, h(X_1), \text{Var}\, h(X_1)/n\} \qquad \text{as } n \to \infty$$

directly from the CLT (p. 219).

3. **Functions of Averages** (or of Asymptotically Normal Y_n)

   a) Continuous Functions of Standardized Y_n:

$$Y_n \text{ is AN}(\mu_n, \sigma_n^2) \implies g\left(\frac{Y_n - \mu_n}{\sigma_n}\right) \xrightarrow{d} g(Z) \qquad \text{as } n \to \infty,$$

   where Z is a standard normal random variable. Of course, Theorem 5.9 (p. 227) is a bit more general since it applies to any Y_n converging in distribution.

   b) Differentiable Functions of Asymptotically Normal Y_n:

$$Y_n \text{ is AN}(\theta, \sigma^2/n) \implies g(Y_n) \text{ is AN}\left[g(\theta), \{g'(\theta)\}^2\sigma^2/n\right]$$

   as $n \to \infty$. The third Delta Theorem, Theorem 5.20 (p. 239), is more general since it applies to vector-valued functions of a vector $\boldsymbol{Y}_n$.

4. **Approximation by Averages**:

$$T - T_\infty = \frac{1}{n}\sum_{i=1}^{n} h(X_i) + R_n,$$

where $\mathrm{E}\, h(X_1) = 0$, $\mathrm{Var}\, h(X_1) = \sigma_h^2$, and $R_n = o_p\left(n^{-1/2}\right)$ implies

$$T \text{ is } \mathrm{AN}(T_\infty, \sigma_h^2/n) \qquad \text{as } n \to \infty.$$

This is Theorem 5.23 (p. 242) with the more general version found in Theorem 5.32 (p. 257).

5.7 Appendix – Central Limit Theorem for Independent Non-Identically Distributed Summands

A Central Limit Theorem gives conditions under which an appropriately standardized (or normalized) sum of random variables converges in distribution to a standard normal distribution. The most well known Central Limit Theorem is the one for independent and identically distributed (iid) summands, Theorem 5.4 (p. 219). However, there are Central Limit Theorems for summands that are neither independent nor identically distributed, and certain of these are useful for large sample inference, especially for regression settings.

The general nature of the conditions under which a standardized sum converges to a standard normal are that: (a) the summands are not too dependent; and (b) all summands contribute more or less equally to the sum in the sense that no single summand contributes significantly more variability to the sum than any other summand. Independent summands satisfy (a) in the strongest possibly sense, and identically distributed summands satisfy (b) in the strongest possible sense. Hence the well-known version of the Central Limit Theorem for iid random variables follows.

However, not all sums are asymptotically normal. Counter examples are useful for illustrating the necessity of certain conditions in Central Limit Theorems. We start with two such examples. The first shows that some conditions on the dependence structure are necessary; the second illustrates the necessity for conditions ensuring that no random variable contributes significantly to the variance of the sum.

Example 5.39 (Persistent correlations). Suppose that T has a non-normal density f_T, with $E(T) = 0$ and $\mathrm{Var}(T) = \sigma_T^2 > 0$, and let $Z_1, Z_2, \ldots,$ be iid standard normal random variables that are independent of T. Define $X_i = T + Z_i$ for $i = 1, 2, \ldots$. Note that the X_i are identically distributed but not independent. The correlation between X_i and X_j for $i \neq j$ is $\sigma_T^2/(\sigma_T^2 + 1) > 0$. Note that the pairwise correlation can be made arbitrarily small by taking σ_T^2 sufficiently small. Now set $S_n = X_1 + X_2 + \cdots + X_n = nT + Z_1 + \cdots + Z_n$. The sum S_n has mean 0 and variance $v_n^2 = n^2\sigma_T^2 + n$, so that the standardized sum is $S_n/v_n = Y_{1,n} + Y_{2,n}$ where $Y_{1,n} = nT/v_n$ and $Y_{2,n} = (Z_1 + \cdots + Z_n)/v_n$. It is easily verified that $Y_{1,n}$ converges in distribution to T/σ_T and that $Y_{2,n}$ converges in probability to 0. Thus the standardized sum converges in distribution to T/σ_T which is non-normal. This example shows that a standardized sum of dependent summands need not converge to a standard normal distribution, even when the components are identically distributed and the pairwise correlations are small. The key feature of this example is that the correlation between X_i and X_j does not decay to zero as $|i - j| \to \infty$ for all i and j, and thus T dominates the sum. ♦

Next we present an example of summands that are mutually independent, but whose distributions differ sufficiently that their standardized sum does not converge to a standard normal.

Example 5.40 (Components contribute unequally). Suppose that $B, B_1, B_2, \ldots$ are iid random variables with $P(B = -1) = P(B = 1) = 1/2$. Note that $E(B) = 0$ and $\text{Var}(B) = 1$. Define $X_i = 2^{i/2}B_i$. The X_i are mutually independent and have common mean 0, but different variances, $\text{Var}(X_i) = 2^i$. The sum $S_n = X_1 + \cdots + X_n$ has mean 0 and variance $\text{Var}(S_n) = v_n^2 = \sum_{i=1}^n 2^i = 2^{n+1} - 2$ so the standardized sum is $Y_n = S_n/v_n$. We show that the standardized sum does not converge in distribution to a standard normal by using a proof by contradiction. Assume that Y_n does converge in distribution to a standard normal. It follows that Y_{n-1} must also converge in distribution to N(0, 1). Write $Y_n = S_{n-1}/v_n + X_n/v_n = (v_{n-1}/v_n)Y_{n-1} + (2^{n/2}/v_n)B_n$. Note that v_{n-1}/v_n and $(2^{n/2}/v_n)$ both converge to $1/\sqrt{2}$. So using the facts that $Y_{n-1} \xrightarrow{d} \text{N}(0, 1)$ and that B_n and B have the same distribution, it follows that $Y_n \xrightarrow{d} Z/\sqrt{2} + B/\sqrt{2}$ where Z and B are independent and Z is standard normal. However, the sum of a normal random variable $(Z/\sqrt{2})$ and a discrete random variable $(B/\sqrt{2})$ does not have a normal distribution, contradicting the assumption that $Y_n \xrightarrow{d} \text{N}(0, 1)$. Note that in this example $\text{Var}(X_n)/\text{Var}(S_n) \to 1/2$, so that one summand (X_n) contributes one-half of the total variability to sum (S_n) asymptotically. The fact that $\text{Var}(X_n)$ is increasing with n is not the key feature of this example. The same phenomenon holds if the variance decreases with n as illustrated in Problem 5.51 (p. 272). ♦

Central Limit Theorems for dependent summands play an important role in the analysis of dependent data, e.g., time series and spatial statistics. Our concern in this text is primarily with independent, but not necessarily identically distributed, such as arises in regression modeling and in data from designed experiments. Accordingly we limit further discussion to Central Limit Theorems for independent but non-identically distributed random variables.

5.7.1 Double Arrays

In statistical applications we are usually interested in standardized sums of the form $Y_n = (S_n - a_n)/b_n$ where $S_n = X_1 + \cdots + X_n$. We assume that the X_i are mutually independent. Reexpression results in

$$Y_n = \sum_{i=1}^{n}\left(\frac{X_i}{b_n} - \frac{a_n}{nb_n}\right) = \sum_{i=1}^{n} X_{n,i}$$

where $X_{n,i} = X_i/b_n - a_n/nb_n$. The $X_{n,i}$ can be arranged in a *triangular array*

$$\begin{array}{llll} X_{1,1}, & & & \\ X_{2,1} & X_{2,2} & & \\ X_{3,1} & X_{3,2} & X_{3,3} & \\ \vdots & \vdots & \vdots & \\ X_{n,1} & X_{n,2} & \ldots & X_{n,n} \end{array}$$

Note that the random variables in each row are independent, and that the standardized sum Y_n is just the nth row sum. The triangular array above is a special case of a *double array* in which the number of entries in each row is arbitrary,

$$\begin{array}{llll} X_{1,1}, & X_{1,2} & \ldots & X_{1,k_1} \\ X_{2,1}, & X_{2,2} & \ldots & X_{2,k_2} \\ \vdots & \vdots & \vdots & \vdots \\ X_{n,1} & X_{n,2} & \ldots & X_{n,k_n}. \end{array} \tag{5.36}$$

Central Limit Theorems are usually stated in terms of the row sums of a double array. This is convenient for statistical applications as it sometimes is the case that the summands in a statistic depend on sample size. For example, in some regression problems with experimental data, the response might be measured at equally-spaced time points $t_{n,i} = i/(n+1)$, $i = 1, 2, \ldots, n$, giving rise to observed response $X_{n,i}$. Even in standard linear regression problems with data $\{X_i, Y_i\}$, the least squares estimator of slope is

$$\widehat{\beta} = \frac{\sum_{i=1}^{n}(X_i - \overline{X})Y_i}{\sum_{i=1}^{n}(X_i - \overline{X})^2} = \sum_{i=1}^{n} Y_{n,i}$$

where $Y_{n,i} = Y_i(X_i - \overline{X})/\sum_{i=1}^{n}(X_i - \overline{X})^2$. Thus we see that often the data themselves, or statistics calculated from data, are row sums of double arrays. So it is natural to formulate and state Central Limit Theorems in terms of such arrays.

5.7.2 *Lindeberg-Feller Central Limit Theorem*

Consider a double array as in (5.36, p. 261), and set $S_n = X_{n,1} + \cdots + X_{n,k_n}$. Define $\mu_{n,i} = E(X_{n,i})$, $\sigma^2_{n,i} = \text{Var}(X_{n,i})$, $\mu_n = \sum_{i=1}^{k_n} \mu_{n,i}$, $\sigma^2_n = \sum_{i=1}^{k_n} \sigma^2_{n,i}$. The standardized sum is $(S_n - \mu_n)/\sigma_n$. Note that $(S_n - \mu_n)/\sigma_n = X'_{n,1} + \cdots + X'_{n,k_n}$ where $X'_{n,i} = (X_{n,i} - \mu_{n,i})/\sigma_n$, where $E(X'_{n,i}) = 0$ and $\sum_{i=1}^{k_n} \text{Var}(X'_{n,i}) = 1$. Thus the standardized row sums from a double array $\{X_{n,i}\}$ are equal to the row sums of double array $\{X'_{n,i}\}$ for which $\mu'_{n,i} = E(X'_{n,i}) = 0$, and $\sigma'^2_{n,i} = \text{Var}(X'_{n,i})$ are such that $\sigma'^2_n = \sum_{i=1}^{k_n} \sigma'^2_{n,i} = 1$. It follows that there is no loss in generality in assuming that the double array is such that the entries have mean zero and the row sum variances are equal to 1. Thus we drop the $'$ notation and simply assume that the double array under study is such that $\mu_{n,i} = 0$ for all i and n, and that $\sigma^2_n = 1$ for all n.

As alluded to in the opening remarks and examples, a Central Limit Theorem requires that the summands all contribute only a small amount asymptotically. The appropriate condition is known as *uniform asymptotic negligibility*. The double array (5.36) is said to be uniformly asymptotically negligible if and only if

$$\lim_{n\to\infty} \max_{1\le i\le k_n} P(|X_{n,i}| > \epsilon) = 0. \tag{5.37}$$

Note that (5.37) implies that $\lim_{n\to\infty} P(|X_{n,i}| > \epsilon) = 0$ for each i, and thus each component $X_{n,i}$ must converge in probability to 0.

Theorem 5.33 (Lindeberg-Feller). *Consider the double array in (5.36, p. 261) with independence in each row, $\mu_{n,i} = 0$ and $\sigma^2_{n,i} < \infty$ for all n and i, and $\sigma^2_n = 1$. Then in order that*

(i) $S_n \xrightarrow{d} N(0,1)$

(ii) *the double array is uniformly asymptotical negligible*

both hold as $n \to \infty$, it is necessary and sufficient that for each $\delta > 0$

$$\lim_{n\to\infty} \sum_{i=1}^{k_n} E\left\{X^2_{n,i} I(|\, X_{n,i} \,| > \delta)\right\} = 0. \tag{5.38}$$

Equation (5.38) is called the "Lindeberg Condition." Theorem 5.33 says that the desired asymptotic normality of S_n and the uniform asymptotically negligible condition (5.37) are together equivalent to the Lindeberg Condition. Thus using Theorem 5.33 requires that we know how to verify (5.38). Here we illustrate the basic approach by proving the iid Central Limit Theorem, Theorem 5.4 (p. 219).

The assumptions of Theorem 5.4 are that $X_1, \ldots, X_n$ are iid with mean $E(X_1) = \mu$ and variance $\mathrm{Var}(X_1) = \sigma^2 < \infty$. For Theorem 5.33, $k_n = n$ and define

$$X_{n,i} = \frac{X_i - \mu}{\sigma\sqrt{n}}$$

so that $E(X_{n,i}) = 0$, $\mathrm{Var}(X_{n,i}) = 1/n$, and $\sum_{i=1}^{n} \mathrm{Var}(X_{n,i}) = 1$. Then

$$\begin{aligned}\sum_{i=1}^{n} E\left\{X_{n,i}^2\, I(X_{n,i} > \delta)\right\} &= \frac{1}{\sigma^2} E\left\{(X_1 - \mu)^2 I(|\,X_1 - \mu\,| > \delta\sqrt{n})\right\} \\ &= \frac{1}{\sigma^2}\int (x-\mu)^2 I(|\,x-\mu\,| > \delta\sqrt{n})\, dF(x),\end{aligned}$$

where F is the distribution function of X_1. Now notice that

$$\int (x-\mu)^2 I(|\,x-\mu\,| > \delta\sqrt{n})\, dF(x) = \sigma^2 - \int_{[\mu-\delta\sqrt{n},\mu+\delta\sqrt{n}]} (x-\mu)^2\, dF(x).$$

The integral on the right-hand side above increases monotonically to σ^2 as $n \to \infty$ and consequently

$$\lim_{n\to\infty} \int (x-\mu)^2 I(|\,x-\mu\,| > \delta\sqrt{n})\, dF(x) = 0,$$

thus verifying (5.38, p. 262).

So $S_n = \sum_{i=1}^{n} X_{n,i} = \sqrt{n}(\overline{X} - \mu)/\sigma$ is AN(0,1) as stated by Theorem 5.4 (p. 219). A variety of problems given at the end of the chapter require verification of the Lindeberg Condition, and this basic type of proof is usually adequate.

5.8 Problems

5.1. Suppose that $Y_1, \ldots, Y_n$ are identically distributed with mean $\mathrm{E}(Y_1) = \mu$, $\mathrm{var}(Y_1) = \sigma^2$, and covariances given by

$$\begin{aligned}\mathrm{Cov}(Y_i, Y_{i+j}) &= \rho\sigma^2, \qquad |j| \le 2 \\ &= 0 \qquad\qquad |j| > 2.\end{aligned}$$

Prove that $\overline{Y} \xrightarrow{p} \mu$ as $n \to \infty$.

5.2. Suppose that $Y_1, Y_2, \ldots$ are independent random variables with $Y_n \sim \mathrm{N}(\mu, \sigma_n^2)$, where the sequence $\sigma_n^2 \to \sigma^2 > 0$ as $n \to \infty$. Prove that there is no random variable Y such that $Y_n \xrightarrow{p} Y$. (Hint: Assume there is such a Y and obtain a contradiction from $|Y_n - Y_{n+1}| \le |Y_n - Y| + |Y_{n+1} - Y|$.)

5.3. Show that $Y_n \xrightarrow{d} c$ for some constant c implies $Y_n \xrightarrow{p} c$ by directly using the definitions of convergence in probability and in distribution. Start with $P(|Y_n - c| > \epsilon)$.

5.4. In Section 5.2.3b (p. 219) the sample size discussion refers to convergence in the Central Limit Theorem, but in many practical situations one may be more interested in convergence of the one sample t statistic, $t = \sqrt{n}(\overline{X} - \mu)/s_{n-1}$. An Edgeworth expansion (Hall, 1987) for the t statistic similar to (5.6, p. 219) shows that $\sqrt{\beta_1}(X_1)/\sqrt{n}$ is still the important quantity but that t is skewed in the direction **opposite** to $Z_n = \sqrt{n}(\overline{X} - \mu)/\sigma$. Thus Boos and Hughes-Oliver (2000) developed the following equation

$$\alpha + [.19 + .026\log(\alpha)]\frac{\sqrt{\beta_1}(X_1)}{\sqrt{n}}$$

for the probability that the $1-\alpha$ one-sided t bound $(-\infty, \overline{X} + t_{\alpha,n-1}s_{n-1}/\sqrt{n})$ fails to include the mean μ.

a. Find how large n needs to be for a nominal 95% upper bound for μ to have a "miss rate" of at most .06 if $\sqrt{\beta_1}(X_1) = .80$. (In other words, to be sure the coverage is at least .94.)
b. Working backwards in the above equation, find an adjusted α so that the miss rate would be .05.

5.5. Consider the simple linear regression setting,

$$Y_i = \alpha + \beta x_i + e_i, \quad i = 1, \ldots, n,$$

where the x_i are known constants, and $e_1, \ldots, e_n$ are iid with mean 0 and finite variance σ^2. After a little algebra, the least squares estimator has the following representation,

$$\widehat{\beta} - \beta = \frac{\sum_{i=1}^{n}(x_i - \overline{x})e_i}{\sum_{i=1}^{n}(x_i - \overline{x})^2}.$$

Using that representation, prove that $\widehat{\beta} \xrightarrow{p} \beta$ as $n \to \infty$ if $\sum_{i=1}^{n}(x_i - \overline{x})^2 \to \infty$.

5.6. Let $X_1, \ldots, X_n$ be iid with finite fourth moment. Show that the sample kurtosis $b_2 = m_4/(m_2)^2$ is weakly consistent.

5.7. Let $(X_1, Y_1), \ldots, (X_n, Y_n)$ be iid bivariate random vectors with finite (s,t)th moment, $\mathrm{E}|X_1|^s|Y_1|^t < \infty$, where s and t are positive integers. Consider the sample covariance

$$m_{11} = \frac{1}{n}\sum_{i=1}^{n}(X_i - \overline{X})(Y_i - \overline{Y}).$$

Prove that $m_{11} \xrightarrow{p} \mu_{11}$, where μ_{11} is the population covariance. Find the smallest values of (s, t) that you can.

5.8. Suppose that $Y_1, \ldots, Y_n$ are iid random variables with $P(Y_1 > 0) = 1$. Define the arithmetic ($\widehat{\theta}_A$), geometric ($\widehat{\theta}_G$), and harmonic ($\widehat{\theta}_H$) *sample* means as follows:

$$\widehat{\theta}_A = \frac{1}{n}\sum_{i=1}^{n} Y_i$$

$$\widehat{\theta}_G = \left(\prod_{i=1}^{n} Y_i\right)^{1/n}$$

$$\widehat{\theta}_H = \left(\frac{1}{n}\sum_{i=1}^{n}\frac{1}{Y_i}\right)^{-1}.$$

Give some simple conditions under which each of these estimators converges in probability, and describe the limit for each in terms of expectations of functions of Y_1.

5.9. Let $X_1, \ldots, X_n$ be iid from a distribution with mean μ, variance σ^2, and finite central moments μ_3 and μ_4. Consider $\widehat{\theta} = \overline{X}/s_n$, a measure of "effect size" used in meta-analysis. Prove that $\widehat{\theta} \xrightarrow{p} \theta = \mu/\sigma$ as $n \to \infty$.

5.10. Prove the following extension of Example 5.18 (p. 229): Suppose that $\boldsymbol{Y}_n \xrightarrow{d} \boldsymbol{Y}$, where $\boldsymbol{Y}_n$ and $\boldsymbol{Y}$ are k-vectors and $\boldsymbol{C}_n$ is a k by k matrix converging in probability to the k by k matrix $\boldsymbol{C}$. Then $\boldsymbol{Y}_n^T\boldsymbol{C}_n\boldsymbol{Y}_n \xrightarrow{d} \boldsymbol{Y}^T\boldsymbol{C}\boldsymbol{Y}$.

5.11. In the continuation of Example 5.20 (p. 230) on 231, it follows that under the local alternative $\mu_i = \mu + d_i/\sqrt{n}$, $(k-1)F = \sum_{i=1}^{k} n(\overline{X}_i - \overline{\overline{X}})^2/s_p^2$ converges in distribution to a noncentral chi-squared distribution with $k-1$ degrees of freedom and noncentrality parameter $\lambda = \boldsymbol{d}^T(\boldsymbol{I}_k - k^{-1}\boldsymbol{1}_k\boldsymbol{1}_k^T)\boldsymbol{d}/\sigma^2 = \sum_{i=1}^{k}(d_i - \overline{d})^2/\sigma^2$. Now, if the data in a one-way ANOVA setup such as Example 5.20 (p. 230) are normally distributed with means $\mu_1, \ldots, \mu_k$ and common variance σ^2, then the exact power of the ANOVA F statistic is the probability that a noncentral F distribution with $k-1$ and $k(n-1)$ degrees of freedom and noncentrality parameter $\lambda = \sum_{i=1}^{k} n(\mu_i - \overline{\mu})^2/\sigma^2$ is larger than a percentile of the central version of that F distribution. In R, that would be `1-pf(qf(.95,k-1,k*(n-1)),k-1,k*(n-1),nc)` , where nc=λ and the level is .05. Show that if $\mu_i = \overline{\mu} + d_i/\sqrt{n}$, then $\sum_{i=1}^{k}(d_i - \overline{d})^2/\sigma^2 = \sum_{i=1}^{k} n(\mu_i - \overline{\mu})^2/\sigma^2$, that is, the asymptotic chi-squared power approximation uses the same noncentrality parameter. In R that would be `1-pchisq(qchisq(.95,k-1),2,nc)`. Compare the approximation for $k = 3$ and nc=10 to the true power for $n = 5$, 10, and 20.

5.12. Prove that $Y_n \xrightarrow{d} Y$ implies that $Y_n = O_p(1)$.

5.13. Suppose that $\widehat{\theta}_1$ is $\mathrm{AN}(\theta_0, A_1/n)$ as $n \to \infty$.

a. What can we say about consistency of $\widehat{\theta}_1$? Why?
b. If $\widehat{\theta}_1 - \widehat{\theta}_2 \xrightarrow{p} 0$, what can we say about consistency of $\widehat{\theta}_2$? Why?
c. If $\sqrt{n}(\widehat{\theta}_1 - \widehat{\theta}_2) \xrightarrow{p} b$, what can we say about asymptotic normality of $\widehat{\theta}_2$? Why?

5.14. Prove: if $U_{1n} = O(V_{1n})$ wp1 and $U_{2n} = O(V_{2n})$ wp1, then $U_{1n}U_{2n} = O(V_{1n}V_{2n})$ wp1.

5.15. Prove: if $U_{1n} = o_p(V_{1n})$ and $U_{2n} = O_p(V_{2n})$, then $U_{1n}U_{2n} = o_p(V_{1n}V_{2n})$.

5.16. Show that $S_n = O(1)$ implies $S_n = O_p(1)$ and $S_n = o(1)$ implies $S_n = o_p(1)$.

5.17. Consider the convergence concepts

(i) $O(1)$
(ii) $o(1)$
(iii) $O_p(1)$
(iv) $o_p(1)$
(v) $\xrightarrow{d}$
(vi) $\xrightarrow{p}$

For each of the following quantities, list which of (i)-(vi) apply directly without any standardization. For (v) and/or (vi), consider only cases where the limit is a random variable or a finite constant (not $\pm\infty$). In those cases, give the limit.

a. $X_n = n + 5$.
b. $Y_n = -1/n$ for n odd and $Y_n = 1/n$ for n even.
c. $Z_n = n^{-1}\sum_{i=1}^{n} U_i$, where $U_1, \ldots, U_n$ are iid, $E(U_1) = 0$, $\mathrm{Var}(U_1) = 1$.
d. X_nY_n, where X_n and Y_n are defined in a. and b.
e. $\sqrt{X_n}Z_n$, where X_n and Y_n are defined in a. and b.

5.18. For proving convergence wp1, it is sometimes useful to use the following simple result: If events $A_1, \ldots, A_k$ jointly imply the event B and $P(A_1) = P(A_2) = \cdots = P(A_k) = 1$, then $P(B) = 1$. Prove this result.

5.19. For proving convergence in probability, it is sometimes useful to use the following simple result: If events $A_{1\epsilon}, \ldots, A_{k\epsilon}$ jointly imply the event B_ϵ and $P(A_{i\epsilon}) \geq 1 - \epsilon$, $i = 1, \ldots, k$, then $P(B_\epsilon) \geq 1 - k\epsilon$. Prove this result.

5.20. Show that $Y_n \xrightarrow{p} Y$ implies $Y_n \xrightarrow{d} Y$. Hint: use Slutsky's Theorem and the representation $Y_n = (Y_n - Y) + Y$.

5.21. From Theorem 5.13 (p. 236) we obtain the useful result: If Y_n is $\mathrm{AN}(\mu, \sigma_n^2)$ and $\sigma_n \to 0$, then $Y_n \xrightarrow{p} \mu$. Prove this using Slutsky's Theorem.

5.22. For the statistic $\widehat{\theta} = \overline{X}/s_n$ in Problem 5.9 (p. 265), show that the asymptotic variance times n is given by

$$1 - \frac{\mu\mu_3}{\sigma^4} + \frac{\mu^2\mu_4}{4\sigma^6} - \frac{\mu^2}{4\sigma^2}$$

using Theorem 5.19 (p. 238) and (5.34, p. 256).

5.23. In Problem 5.8 (p. 265), you were asked to work on the consistency of $\widehat{\theta}_A$, $\widehat{\theta}_G$, and $\widehat{\theta}_H$. Here we focus on asymptotic normality. For these estimators, state why each is asymptotically normal, any conditions needed for asymptotic normality, and the explicit form of the asymptotic mean and variance.

5.24. When two independent binomials, X_1 is binomial (n_1, p_1) and X_2 is binomial (n_2, p_2), are put in the form of a 2×2 table (see Example 5.31, p. 240), then one often estimates the odds ratio

$$\theta = \frac{\dfrac{p_1}{1-p_1}}{\dfrac{p_2}{1-p_2}} = \frac{p_1(1-p_2)}{p_2(1-p_1)}.$$

The estimate $\widehat{\theta}$ is obtained by inserting $\widehat{p}_1 = X_1/n_1$ and $\widehat{p}_2 = X_2/n_2$ in the above expression. Show that $\log(\widehat{\theta})$ has asymptotic variance

$$\frac{1}{n_1 p_1(1-p_1)} + \frac{1}{n_2 p_2(1-p_2)}.$$

5.25. Using the multinomial form of the 2×2 table found in Example 5.29 (p. 238), show that for the odds ratio defined in Problem 5.24, $\log(\widehat{\theta})$ has asymptotic variance

$$\frac{1}{np_{11}} + \frac{1}{np_{12}} + \frac{1}{np_{21}} + \frac{1}{np_{22}}.$$

5.26. Multiply the matrices in Example 5.31 (p. 240) to obtain the asymptotic covariance matrix. Then, find an expression for the asymptotic correlation between the risk difference and risk ratio, and show that this correlation is one when $p_1 = p_2$.

5.27. For an iid sample $Y_1, \ldots, Y_n$, consider finding the asymptotic joint distribution of $(\overline{Y}, s_n, s_n/\overline{Y})$ using Theorem 5.20 (p. 239) and (5.34, p. 256).

a. Find the matrices $\boldsymbol{g}'(\boldsymbol{\theta})$ and $\boldsymbol{\Sigma}$ used to compute the asymptotic covariance $\boldsymbol{g}'(\boldsymbol{\theta})\boldsymbol{\Sigma}\boldsymbol{g}'(\boldsymbol{\theta})^T$.
b. Multiply the matrices in a. to get the asymptotic covariance. (It might help to use Maple or Mathematica.)

5.28. Suppose that $(\widehat{\theta}_1, \widehat{\theta}_2)^T$ are jointly asymptotically normal $((\theta_1, \theta_2)^T, \Sigma/n)$ as $n \to \infty$. Let $g_1(x_1)$ and $g_2(x_2)$ be increasing functions with non-zero derivatives at θ_1 and θ_2, respectively. Show that the asymptotic correlation between $g_1(\widehat{\theta}_1)$ and $g_2(\widehat{\theta}_2)$ is the same as the asymptotic correlation between $\widehat{\theta}_1$ and $\widehat{\theta}_2$.

5.29. In most of Chapter 5 we have dealt with iid samples of size n of either univariate or multivariate random variables. Another situation of interest is when we have a number of different independent samples of different sizes. For simplicity, consider the case of two iid samples, $X_1, \ldots, X_m$ and $Y_1, \ldots, Y_n$, with common variance σ^2 and under a null hypothesis they have a common mean, say μ. Then the two-sample pooled t statistic is

$$t_p = \frac{\overline{X} - \overline{Y}}{\sqrt{s_p^2\left(\frac{1}{m} + \frac{1}{n}\right)}},$$

where

$$s_p^2 = \frac{(m-1)s_{\rm X}^2 + (n-1)s_{\rm Y}^2}{m+n-2},$$

and

$$s_{\rm X}^2 = \frac{1}{m-1}\sum_{i=1}^{m}(X_i - \overline{X})^2, \quad s_{\rm Y}^2 = \frac{1}{n-1}\sum_{i=1}^{n}(Y_i - \overline{Y})^2.$$

It can be shown that $t_p \xrightarrow{d} N(0,1)$ as $\min(m,n) \to \infty$. However, the proof is fairly tricky. Instead it is common to assume that both sample sizes go to ∞ at a similar rate, i.e., $\lambda_{m,n} = m/(m+n) \to \lambda > 0$ as $\min(m,n) \to \infty$. Under this assumption prove that $t_p \xrightarrow{d} N(0,1)$. Hint: show that $t_p = \left\{\sqrt{1-\lambda_{m,n}}\sqrt{m}\left(\overline{X} - \mu\right) - \sqrt{\lambda_{m,n}}\sqrt{n}\left(\overline{Y} - \mu\right)\right\}/s_p$.

5.30. For the setting of Problem 5.7 (p. 264) prove that m_{11} is $AN(\mu_{11}, \sigma_h^2)$ and find σ_h^2 by showing that the conditions of (an extended version for vectors of) Theorem 5.23 (p. 242) holds with

$$h(X_i, Y_i) = (X_i - \mu_X)(Y_i - \mu_Y) - \mu_{11}.$$

Give the best moment conditions that you can for the asymptotic normality result.

5.31. For the sample third moment m_3, show directly that the result of Theorem 5.24 (p. 243) holds. That is, add and subtract μ inside the definition of m_3, and expand in terms of $(X_i - \mu)$ and $(\overline{X} - \mu)$. Then subtract μ_3 and the appropriate $(1/n)\sum h(X_i)$ for m_3. This defines R_n, and you are to show that $\sqrt{n}R_n \xrightarrow{p} 0$.

5.32. Suppose that $\widehat{\theta}_1, \ldots, \widehat{\theta}_k$ each satisfy the assumptions of Theorem 5.23 (p. 242):

$$\widehat{\theta}_i - \theta_i = \frac{1}{n}\sum_{j=1}^{n} h_i(X_j) + R_{in},$$

where $\sqrt{n}R_{in} \xrightarrow{p} 0$ as $n \to \infty$ and $\mathrm{E}\, h_i(X_1) = 0$ and $\mathrm{var}\, h_i(X_1) = \sigma^2_{hi} < \infty$. Let $T = \sum_{i=1}^k c_i \widehat{\theta}_i$ for any set of constants $c_1, \ldots, c_k$. Find the correct approximating function h_T for T, show that Theorem 5.23 may be used (verify directly without using later theorems), and find the limiting distribution of T.

5.33. Let $(X_1, Y_1), \ldots, (X_n, Y_n)$ be iid pairs with $\mathrm{E}\, X_1 = \mu_1, \mathrm{E}\, Y_1 = \mu_2$, $\mathrm{Var}\, X_1 = \sigma_1^2$, $\mathrm{Var}\, Y_1 = \sigma_2^2$, and $\mathrm{Cov}\,(X_1, Y_1) = \sigma_{12}$.

a. What can we say about the asymptotic distribution of $(\overline{X}, \overline{Y})^T$?
b. Suppose that $\mu_1 = \mu_2 = 0$ and let $T = (\overline{X})(\overline{Y})$. Show that

$$nT \xrightarrow{d} Q \qquad \text{as } n \to \infty,$$

and describe the random variable Q.
c. Suppose that $\mu_1 = 0, \mu_2 \neq 0$ and let $T = (\overline{X})(\overline{Y})$. Show that

$$\sqrt{n}T \xrightarrow{d} R \qquad \text{as } n \to \infty,$$

and describe the random variable R.

5.34. Prove the result of Example 5.20 (p. 230) for the case of unequal sample sizes and $\lambda_{iN} = n_i/N \to \lambda_i > 0, i = 1, \ldots, k, n_1 + \cdots + n_k = N$. That is, show that $(k-1)s_p^2 F = \sum_{i=1}^k n_i(\overline{X}_i - \overline{\overline{X}})^2$ can be written as a quadratic form and converges in distribution to $\sigma^2\chi^2_{k-1}$ under the null hypothesis of equal means and common variance σ^2.

5.35. One of the uses of asymptotic normality is to compare asymptotic variances of estimators. In Ch. 1 we defined the asymptotic relative efficiency (ARE) of two estimators of the same quantity to be the ratio of their asymptotic variances:

$$\mathrm{ARE}(\widehat{\theta}_1, \widehat{\theta}_2) = \frac{\mathrm{Avar}(\widehat{\theta}_2)}{\mathrm{Avar}(\widehat{\theta}_1)}.$$

a. For an iid sample $X_1, \ldots, X_n$ from a distribution with finite second moment and positive density f at the population median $\eta_{1/2} = F^{-1}(1/2)$, $f(\eta_{1/2}) > 0$, find an expression for the asymptotic relative efficiency of the sample mean to the sample median, $\mathrm{ARE}(\overline{X}, \widehat{\eta}_{1/2})$.
b. Now we want to consider location scale models

$$f(x; \mu, \sigma) = \frac{1}{\sigma} f_0\left(\frac{x-\mu}{\sigma}\right).$$

Show that the expression derived in a. does not depend on μ or σ.

c. Compute $\text{ARE}(\overline{X}, \widehat{\eta}_{1/2})$ for the following families:

(i) uniform(0,1): $f_0(x) = I(0 < x < 1)$;
(ii) normal(0,1): $f_0(x) = (2\pi)^{-1/2} \exp(-x^2/2)$;
(iii) logistic: $f_0(x) = e^{-x}/[1 + e^{-x}]^2$;
(iv) Laplace: $f_0(x) = (1/2)e^{-|x|}$.

The variances for these four f_0 are 1/12, 1, $\pi^2/3$, and 2, respectively. You might note that it makes sense to compute ARE's here because both estimators are estimating the same parameter, the center of symmetry μ, which is both the population mean and the population median.

5.36. Find the approximating h_T function for the sample coefficient of variation $T = s_n/\overline{X}$.

5.37. Find the approximating h_T function for the sample skewness $T = m_3/(m_2)^{3/2}$.

5.38. Find the approximating h_T function for the sample kurtosis $T = m_4/(m_2)^2$.

5.39. Find the approximating h_T function for the sample interquartile range $T = \widehat{\eta}_{3/4} - \widehat{\eta}_{1/4}$.

5.40. Formulate an extension of Theorem 5.27 (p. 247) for the situation of two independent samples $X_1, \ldots, X_m$ and $Y_1, \ldots, Y_n$, where $\widehat{\theta}_1$ is a function $X_1, \ldots, X_m$, and $\widehat{\theta}_2$ is a function $Y_1, \ldots, Y_n$. The statistic of interest is $T = g(\widehat{\theta}_1, \widehat{\theta}_2)$, and the conclusion is

$$g(\widehat{\theta}_1, \widehat{\theta}_2) - g(\theta_1, \theta_2) = \frac{1}{m}\sum_{i=1}^{m} g_1'(\boldsymbol{\theta})h_1(X_i) + \frac{1}{n}\sum_{i=1}^{n} g_2'(\boldsymbol{\theta})h_2(Y_i) + R_{mn},$$

where $\sqrt{\max(m,n)}R_{mn} \xrightarrow{p} 0$ as $n \to \infty$ and g_1' and g_2' are the first partial derivatives.

5.41. For the situation of the previous problem give the approximating sum of averages for $T = s_m^2/s_n^2$, the ratio of the sample variances.

5.42. Thinking of the kth central moment as a functional,

$$T_k(F) = \int (t - \mu)^k \, dF(t),$$

show that the Gateaux derivative is given by

$$T_k(F; \Delta) = \int \{t - T_1(F)\}^k \, d\Delta(t) - T_1(F; \Delta) \int k \, \{t - T_1(F)\}^{k-1} \, dF(t),$$

where $T_1(F;\Delta) = \int t\, d\Delta(t)$ is the Gateaux derivative for the mean functional given in Example 5.5.8i (p. 253). Then, substitute $\Delta(t) = \delta_x(t) - F(t)$ and obtain h_k given in Theorem 5.24 (p. 243).

5.43. Show that (5.22, p. 249) may be obtained by using (5.13, p. 244).

5.44. A location M-estimator (to be studied formally in Chapter 7, p. 297) may be represented as $T(F_n)$, where $T(\cdot)$ satisfies

$$\int \psi(t - T(F))\, dF(t) = 0,$$

and ψ is a known differentiable function. Using implicit differentiation, show that the Gateaux derivative is

$$T(F;\Delta) = \frac{\int \psi(t - T(F))\, d\Delta(t)}{\int \psi'(t - T(F))\, d\Delta(t)}.$$

Then substitute $\Delta(t) = \delta_x(t) - F(t)$ and obtain influence function $h(x)$.

5.45. One representation of a "smooth" linear combination of order statistics is $T(F_n)$, where $T(F) = \int_0^1 J(p)F^{-1}(p)dp$, and J is a weighting function. Using the results in Example 5.5.8j (p. 254), find the influence function $h(x)$.

5.46. Suppose that $Y_1, \ldots Y_n$ are iid from the Weibull distribution with distribution function $F(t) = 1 - \exp\{-(t/\sigma)^c\}$, $t \geq 0$. Dubey (1967, *Technometrics*) proposed the following "method of percentile" estimator of c based on two sample quantiles, $\widehat{\eta}_{p_1}$ and $\widehat{\eta}_{p_2}$,

$$\widehat{c} = \frac{\log\{-\log(1-p_1)\} - \log\{-\log(1-p_2)\}}{\log(\widehat{\eta}_{p_1}) - \log(\widehat{\eta}_{p_2})}, \qquad 0 < p_1 < p_2 < 1.$$

He minimized the asymptotic variance of these estimators and found that $p_1 = .17$ and $p_2 = .97$ give the most efficient estimator which is 66% efficient relative to the maximum likelihood estimator. (The point is that it is very simple to compute such an estimator.) Justify that $\widehat{c}$ is asymptotically normal and show what calculations would be necessary to get the asymptotic variance (but do not do the calculations, just set things up.)

5.47. One feature of the boxplot that stands out is the distance from the top of the box to the bottom of the box, the so-called interquartile range $\widehat{\eta}_{3/4} - \widehat{\eta}_{1/4}$. The asymptotic variance of the interquartile range is

$$\frac{1}{16n}\left[\frac{3}{f^2(\eta_{1/4})} - \frac{2}{f(\eta_{1/4})f(\eta_{3/4})} + \frac{3}{f^2(\eta_{3/4})}\right].$$

Here we are using the density f in place of F', and of course we are assuming that the density is positive at the first and third quartiles. We would like to compare the efficiency of the interquartile range to that of other scale estimators, but the problem is that they aren't estimating the same quantities. To take that into account we define the standardized asymptotic relative efficiency (SARE) of two scale estimators, $\widehat{\sigma}_1 \xrightarrow{p} \sigma_1$ and $\widehat{\sigma}_2 \xrightarrow{p} \sigma_2$, by

$$\text{SARE}(\widehat{\sigma}_1, \widehat{\sigma}_2) = \frac{\text{Avar}(\widehat{\sigma}_2)/\sigma_2^2}{\text{Avar}(\widehat{\sigma}_1)/\sigma_1^2}.$$

Now compute the SARE of the sample standard deviation s_n to the interquartile range for the four f_0 given in c) of Problem 5.35. You may use the fact that the denominator of the SARE here is $(\text{Kurt} - 1)/(4n)$ (you could verify that using the delta theorem), and Kurt for the four f_0 are 1.8, 3.0, 4.2, 6.0, respectively. (You may be surprised at how inefficient the interquartile range is.)

5.48. Use Theorem 5.4 (p. 219), the univariate CLT and Theorem 5.31 (p. 256) to prove Theorem 5.7 (p. 225, the multivariate CLT). Perhaps it is easier to use an alternate statement of the conclusion of the univariate CLT than given in Theorem 5.4: $\sqrt{n}(\overline{X} - \mu) \xrightarrow{d} Y$, where Y is a normal$(0, \sigma^2)$ random variable.

5.49. Use Theorem 5.30 (p. 255) to prove the following result. If $\boldsymbol{X}_n \xrightarrow{d} \boldsymbol{X}$, $\boldsymbol{Y}_n \xrightarrow{d} \boldsymbol{Y}$, $\boldsymbol{X}_n$ and $\boldsymbol{Y}_n$ are random vectors defined on the same probability space and are independent, and $\boldsymbol{X}$ and $\boldsymbol{Y}$ are defined on the same probability space and are independent, then

$$\boldsymbol{X}_n + \boldsymbol{Y}_n \xrightarrow{d} \boldsymbol{X} + \boldsymbol{Y} \quad \text{as } n \to \infty.$$

5.50. For an iid sample $X_1, \ldots, X_n$ with finite fourth moments, show that the asymptotic correlation between the sample mean $\overline{X}$ and the sample standard deviation s_n is $\text{Skew}(X_1)/\sqrt{\text{Kurt}(X_1) - 1}$.

5.51. Suppose that $B, B_1, B_2, \ldots$ are iid random variables with $P(B = -1) = P(B = 1) = 1/2$. Define $X_i = 2^{-i/2} B_i$ and $S_n = X_1 + \cdots + X_n$. Prove that the standardized sum $S_n/\sqrt{\text{Var}(S_n)}$ does not converge in distribution to a standard normal. In doing so show that $\text{Var}(X_1)$ contributes to one-half of the variability of S_n asymptotically. Hint: Show that S_{n-1} and $\sqrt{2}(X_2 + \cdots + X_n)$ are identically distributed.

5.52. Consider a Gauss-Markov linear model

$$\boldsymbol{Y} = \boldsymbol{X}\boldsymbol{\beta} + \boldsymbol{\epsilon},$$

where $\boldsymbol{Y}$ is $n\times 1$, the components of $\boldsymbol{e} = (e_1, \ldots, e_n)^T$ are iid$(0, \sigma^2)$, and $\boldsymbol{X}$ is $n\times p_n$. Note that the number of predictors p_n depends on n. Let $\boldsymbol{H} = \boldsymbol{X}\left(\boldsymbol{X}^T\boldsymbol{X}\right)^{-1}\boldsymbol{X}^T$ denote the projection (or "hat") matrix with entries $h_{i,j}$. Note that the $h_{i,j}$ also depend on n. We are interested in the asymptotic properties ($n \to \infty$) of the ith residual, $Y_i - \widehat{Y}_i$, from this regression model, for a fixed i. Prove that if n and $p_n \to \infty$ such that

$$h_{i,i} \to c_i \quad \text{for some} \quad 0 \le c_i < 1, \qquad \text{and} \qquad \max_{\substack{1\le j\le n \\ j\neq i}} |h_{i,j}| \to 0,$$

then

$$Y_i - \widehat{Y}_i \xrightarrow{d} (1 - c_i)e_i + \{(1 - c_i)c_i\}^{1/2}\sigma Z,$$

where Z is a standard normal random variable independent of e_i. Hint: First prove that $\text{Var}\left(\sum_{j=1,j\neq i}^{n} h_{i,j}e_j\right) = (h_{i,i} - h_{i,i}^2)\sigma^2$. Then formulate and verify the appropriate Lindeberg condition. (The result illustrates that least squares residuals tend to have a distribution closer to a normal distribution than that of the original error distribution when that distribution is not normal. Thus, tests for nonnormality may have little power when many predictors are used.)

5.53. Consider the simple linear regression model with design points depending on sample size,

$$Y_{i,n} = \beta_1 + \beta_x X_{i,n} + e_i, \quad i = 1, \ldots, n \ge 2,$$

where $e_1, \ldots, e_n$ are iid$(0, \sigma^2)$.

a. Prove that if

$$\lim_{n\to\infty} \frac{\max_{1\le i\le n}(X_{i,n} - \overline{X})^2}{\sum_1^n (X_{i,n} - \overline{X})^2} = 0,$$

then the least squares slope estimator is asymptotically normal.

b. Determine whether the above condition for asymptotic normality holds for the following cases:

   i. $X_{i,n} = i$;
   ii. $X_{i,n} = \sqrt{i}$;
   iii. $X_{i,n} = 1/\sqrt{i}$;
   iv. $X_{i,n} = 1/i$;
   v. $X_{i,n}$ are iid $N(0, \sigma^2)$;
   vi. $X_{i,n}$ are iid $N(0, n\sigma^2)$.

5.54. Consider the simple linear regression model with design points depending on sample size,

$$Y_{i,n} = \beta_1 + \beta_x X_{i,n} + e_i, \quad i = 1, \ldots, n \ge 2,$$

where $e_1, \ldots, e_n$ are iid$(0, \sigma^2)$, and

$$X_{i,n} = \begin{cases} i/n, & i = 1, \ldots, n-1, \\ n, & i = n. \end{cases}$$

Determine the asymptotic properties (consistency and asymptotic distribution) of the least squares, estimators $\widehat{\beta}_1$ and $\widehat{\beta}_x$, and of the ith residual $Y_i - \widehat{Y}_i$ for the two cases: a) e_1 is normally distributed; b) e_1 is not normally distributed. Relate the findings to the applied regression practice of examining high-leverage points.

Chapter 6
Large Sample Results for Likelihood-Based Methods

6.1 Introduction

Most large sample results for likelihood-based methods are related to asymptotic normality of the maximum likelihood estimator $\widehat{\theta}_{\text{MLE}}$ under standard regularity conditions. In this chapter we discuss these results. If consistency of $\widehat{\theta}_{\text{MLE}}$ is assumed, then the proof of asymptotic normality of $\widehat{\theta}_{\text{MLE}}$ is straightforward. Thus we start with consistency and then give theorems for asymptotic normality of $\widehat{\theta}_{\text{MLE}}$ and for the asymptotic chi-squared convergence of the likelihood-based tests T_{W}, T_{S}, and T_{LR}. Recall that *Strong consistency* of $\widehat{\theta}_{\text{MLE}}$ means $\widehat{\theta}_{\text{MLE}}$ converges with probability one to the true value, and *weak consistency* of $\widehat{\theta}_{\text{MLE}}$ refers to $\widehat{\theta}_{\text{MLE}}$ converging in probability to the true value.

6.2 Approaches to Proving Consistency of $\widehat{\theta}_{\text{MLE}}$

The maximum likelihood estimator is defined to be the parameter value that maximizes the likelihood (2.24, p. 62), but in regular cases $\widehat{\theta}_{\text{MLE}}$ also solves the likelihood equations (2.25, p. 62). These two descriptions have led to two different methods for proving consistency of $\widehat{\theta}_{\text{MLE}}$. Wald (1949) is generally credited with starting the method associated with the maximizing definition. The conditions are fairly strong and restrictive but do not involve second or higher derivatives of the likelihood function. Huber (1967, Theorem 1) generalized these results to misspecified models and to the general estimating equations situation. Haberman (1989, Theorem 5.1) gives further results for the special case when the log likelihood is concave as a function of the parameter.

The second general approach to consistency is related to showing the existence of a consistent sequence of estimators defined by the likelihood equations. This classical approach is characterized by Cramér (1946, p. 500–503) and Serfling (1980, Ch. 4) and involves second and possibly higher derivatives of the log

D.D. Boos and L.A. Stefanski, *Essential Statistical Inference: Theory and Methods*, Springer Texts in Statistics, DOI 10.1007/978-1-4614-4818-1_6,

likelihood function. If additional conditions that imply uniqueness of the estimator are added, then true consistency of the estimator, not just the existence of a consistent solution, follows immediately. Huber (1967, Theorem 2) gives conditions for the latter in a more general setting.

Most rigorous proofs of consistency involve showing that the estimator is ultimately trapped in a compact neighborhood of the true value. Many authors elect to avoid this problem by merely assuming that the parameter space is itself compact (e.g., White, 1981, Gallant, 1987). Assuming a compact parameter space approach avoids almost all problems but is not as theoretically appealing. We give this approach in Section 6.4 (p. 282).

After introducing some notation, we give some "existence of a consistent root" arguments in the next section. Here it helps to separate the case of a real parameter θ from the more difficult vector $\boldsymbol{\theta}$ case.

As usual let $Y_1, \ldots, Y_n$ be iid $f(y; \boldsymbol{\theta})$ where $\boldsymbol{\theta} = (\theta_1, \ldots, \theta_b)^T$ belongs to the parameter space Θ (a subset of R^b), $f(y; \boldsymbol{\theta})$ is a density or probability mass function, and $F(y; \boldsymbol{\theta})$ is the associated distribution function. The random variables Y_i may be real or vector-valued. Often we denote the "true" value of $\boldsymbol{\theta}$ by $\boldsymbol{\theta}_0$. The log likelihood is $l_n(\boldsymbol{\theta}) = \sum_{i=1}^n \log f(Y_i; \boldsymbol{\theta})$, and the average log likelihood is

$$\bar{l}_n(\boldsymbol{\theta}) = n^{-1} \sum_{i=1}^{n} \log f(Y_i; \boldsymbol{\theta}).$$

The expected average log likelihood under the "true" $\boldsymbol{\theta}_0$ is

$$\bar{l}(\boldsymbol{\theta}, \boldsymbol{\theta}_0) = E_{\boldsymbol{\theta}_0} \log f(Y_1; \boldsymbol{\theta}) = \int \log f(y; \boldsymbol{\theta})\, dF(y; \boldsymbol{\theta}_0),$$

where recall that the notation $\int q(y) dF(y)$ means either an integral $\int q(y) f(y)\, dy$ for the continuous case or $\sum_{j=1}^{\infty} q(y_j)[F(y_j) - F(y_j^-)]$ for the discrete case.

Example 6.1 (Normal scale log likelihood). If $Y_1, \ldots, Y_n$ are iid $\mathrm{N}(0, \sigma^2)$, then the average log likelihood is

$$\bar{l}_n(\sigma) = -\frac{1}{2} \log(2\pi) - \log \sigma - \frac{1}{n} \sum_{i=1}^{n} \frac{Y_i^2}{2\sigma^2}.$$

Taking the expected value under σ_0, we have

$$\bar{l}(\sigma, \sigma_0) = -\frac{1}{2} \log(2\pi) - \log \sigma - \frac{\sigma_0^2}{2\sigma^2}. \tag{6.1}$$

In Figure 6.1, the top curve is $\bar{l}(\sigma, \sigma_0)$ for the case $\sigma_0 = 3$. The three vertical lines show values that are $\pm .01$ from the curve $\bar{l}(\sigma, \sigma_0)$ at the points $(\sigma_0 - \delta, \sigma_0, \sigma_0 + \delta) = (2.5, 3.0, 2.5)$ for $\delta = .5$. The second (lower) curve is $\bar{l}_n(\sigma)$ for $n = 10{,}000$. The figure illustrates the generally true behavior of log likelihoods, namely that $\overline{l}_n(\theta)$ converges to $\overline{l}(\theta, \theta_0)$ uniformly in a neighborhood of the true parameter. The figure also suggests and illustrates a strategy for proving the existence of a consistent root of the likelihood equations in the case of a one-dimensional parameter. The argument presented in the next section uses the fact that any smooth curve intersecting the three vertical lines must have a local maximum (flat spot) in the interval $(\sigma_0 - \delta, \sigma_0 + \delta)$. Moreover, because $\overline{l}_n(\sigma)$ is guaranteed to intersect the vertical lines with probability one for sufficiently large n (by the SLLN), there must exist a solution to the likelihood equation, $\partial \overline{l}_n(\sigma)/\partial\sigma = 0$, in $(\sigma_0 - \delta, \sigma_0 + \delta)$ with probability one for sufficiently large n. ♦

6.3 Existence of a Consistent Root of the Likelihood Equations

The inequality in the next theorem shows that any $\overline{l}(\boldsymbol{\theta}, \boldsymbol{\theta}_0)$ has a local maximum near $\boldsymbol{\theta}_0$. Figure 6.1 is an example, but in other situations $\overline{l}(\boldsymbol{\theta}, \boldsymbol{\theta}_0)$ could be much less smooth.

The conditions of the theorem and more generally, conditions for consistency and asymptotic normality of maximum likelihood estimates, involve the *support* of the distributions involved. A point y is said to be in the support of a distribution F if and only if every open neighborhood of y has strictly positive probability assigned to it by F. The set of all such points is called the support of F. The support of F is the smallest closed set of y whose complement has probability 0 assigned to it by F. If F is a distribution on the real line with density $f(y) = dF(y)/dy$, then the

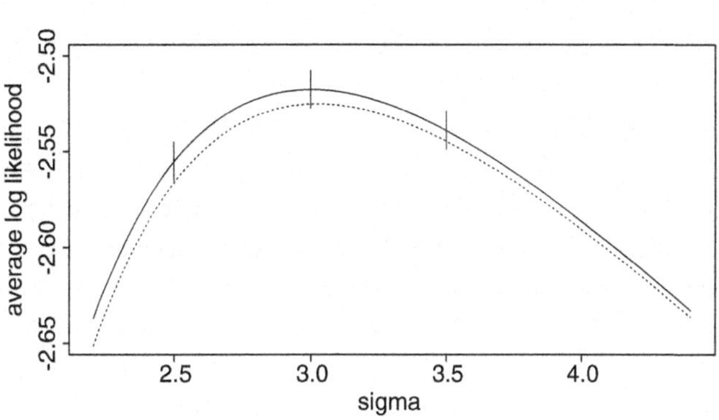

Fig. 6.1 Average log likelihood ($n = 10,000$) and expected average log likelihood for a $N(0, \sigma_0^2)$ sample, $\sigma_0 = 3$.

support of F is the closure of the set $\{y : f(y) > 0\}$; if F is a discrete distribution with probability mass function $f(y)$, then the support of F is the set of all points y such that $f(y) > 0$.

Theorem 6.1 (following Rao, 1973, p. 59). *Let $f(y)$ and $g(y)$ be densities or probability mass functions and let S be the support of f. Then*

$$\int_S \log f(y)\, dF(y) \geq \int_S \log g(y)\, dF(y),$$

with equality only when $F(y) = G(y)$ for all points in S.

Proof. Note that g can be zero on a subset of S. In such cases we interpret the right hand integral as $-\infty$. Now apply Jensen's inequality in the form: if $H_1(y)$ is strictly convex, then $H_1\left[\mathrm{E}\{H_2(Y)\}\right] < \mathrm{E}\left[H_1\{H_2(Y)\}\right]$ unless $H_2(y)$ is constant almost everywhere. Let Y have density $f(y)$, $H_1(y) = -\log y$ which is strictly convex, and $H_2(y) = g(y)/f(y)$. The theorem follows. ■

6.3.1 Real-Valued $\boldsymbol{\theta}$

We now apply Theorem 6.1 and the Strong Law of Large Numbers (SLLN) to show existence of a consistent solution of the likelihood equations $\partial l_n(\boldsymbol{\theta})/\partial\boldsymbol{\theta} = 0$. The case when $\boldsymbol{\theta}$ is real-valued is simpler than the general b-dimensional case and is considered first. Define conditions (**A**), (**B**), and (**C**) by

(**A**) Identifiability: $\boldsymbol{\theta}_1 \neq \boldsymbol{\theta}_2$ implies that $F(y;\boldsymbol{\theta}_1) \neq F(y;\boldsymbol{\theta}_2)$ for at least one y.
(**B**) $|\bar{l}(\boldsymbol{\theta},\boldsymbol{\theta}_0)| < \infty$ for $\boldsymbol{\theta}$ in a neighborhood of $\boldsymbol{\theta}_0$.
(**C**) $\log f(y;\boldsymbol{\theta})$ has a continuous derivative with respect to $\boldsymbol{\theta}$ in a neighborhood of $\boldsymbol{\theta}_0$ for each y in the support of $F(y;\boldsymbol{\theta}_0)$.

Note that these conditions are stated for vector $\boldsymbol{\theta}$ so that we can use them later as well. Basically, (**A**) says that different values of $\boldsymbol{\theta}$ uniquely define different distributions in the parametric family. (**B**) is a very minimal assumption required for the Strong Law of Large Numbers, and (**C**) guarantees that $\widehat{\boldsymbol{\theta}}_{\text{MLE}}$ is a root of the likelihood equations. Note that we do not explicitly rule out densities whose support depends on $\boldsymbol{\theta}$, but (**B**) often will. For example, the gamma threshold model of Problem 6.5 (p. 293) is

$$f(y;\theta) = \frac{1}{\Gamma(\alpha)}(y-\theta)^{\alpha-1}e^{-(y-\theta)}, \quad \theta < y < \infty, \tag{6.2}$$

which includes the exponential threshold model when $\alpha = 1$. This family does not satisfy (**B**) because $\bar{l}(\boldsymbol{\theta},\boldsymbol{\theta}_0) = -\infty$ for values of θ to the right of θ_0.

Here we give a few densities to illustrate condition (**C**).

1. The "Huber" density (Huber, 1981, p. 86), for $k > 0$,

$$f(y;\theta) = \begin{cases} \dfrac{1-\epsilon}{\sqrt{2\pi}} e^{-(y-\theta)^2/2} & \text{for } |y-\theta| \le k \\ \dfrac{1-\epsilon}{\sqrt{2\pi}} e^{k^2/2-k|y-\theta|} & \text{for } |y-\theta| > k, \end{cases} \tag{6.3}$$

satisfies (**C**) where the continuous derivative of $\log f(y;\theta)$ with respect to θ is $\max[-k, \min(k, y-\theta)]$, that is,

$$\frac{\partial}{\partial\theta} \log f(y;\theta) = \begin{cases} k & \text{for } \theta < y-k \\ y-\theta & \text{for } y-k \le \theta \le y+k \\ -k & \text{for } \theta > y+k. \end{cases} \tag{6.4}$$

2. The Laplace location density $f(y;\theta) = (1/2)\exp(-|y-\theta|)$ is continuous at $\theta = y$ but is not differentiable there and therefore does not satisfy (**C**).

Theorem 6.2. *If $Y_1, \ldots, Y_n$ are iid from a density with a real parameter θ and (**A**), (**B**), and (**C**) hold, then there exists a strongly consistent solution $\widehat{\theta}_{\text{MLE}}$ of the likelihood equation $\partial l_n(\theta)/\partial\theta = 0$.*

The following proof uses the SLLN to show that for any $\delta > 0$, $\overline{l}_n(\theta)$ is arbitrarily close to $\overline{l}(\theta, \theta_0)$ at $\theta_0 - \delta$, θ_0, and $\theta_0 + \delta$ for n sufficiently large; for example, within the vertical bars of Figure 6.1. This implies that for n sufficiently large, $\overline{l}_n(\theta)$ increases somewhere in the interval $(\theta_0 - \delta, \theta_0)$ and decreases somewhere in the interval $(\theta_0, \theta_0 + \delta)$, and hence must have a critical point in $(\theta_0 - \delta, \theta_0 + \delta)$.

Proof. If we set $f(y) = f(y;\theta_0)$ and $g(y) = f(y;\theta)$ in Theorem 6.1 above, then under (**A**) and (**B**) we have $\overline{l}(\theta_0, \theta_0) > \overline{l}(\theta, \theta_0)$ if $\theta \ne \theta_0$. Let $\delta > 0$ be given. By Theorem 6.1 there exists $\epsilon > 0$ such that

$$\overline{l}(\theta_0, \theta_0) - \overline{l}(\theta_0 - \delta, \theta_0) > \epsilon \tag{6.5}$$

and

$$\overline{l}(\theta_0, \theta_0) - \overline{l}(\theta_0 + \delta, \theta_0) > \epsilon. \tag{6.6}$$

Now by the SLLN using (**B**), we have

$$\overline{l}_n(\theta) \xrightarrow{wp1} \overline{l}(\theta, \theta_0) \quad \text{as } n \to \infty,$$

for each θ in the neighborhood where (**B**) holds. Let Ω_1 be the subset of the underlying probability space with $P(\Omega_1) = 1$ and such that this latter convergence takes place for both $\theta = \theta_0 - \delta$ and $\theta = \theta_0 + \delta$. For each $\omega \in \Omega_1$ we can choose $n(\epsilon, \omega)$ such that for all $n > n(\epsilon, \omega)$,

$$-\frac{\epsilon}{2} < \overline{l}_n(\theta_0 - \delta) - \overline{l}(\theta_0 - \delta, \theta_0) < \frac{\epsilon}{2}, \tag{6.7}$$

$$-\frac{\epsilon}{2} < \overline{l}_n(\theta_0 + \delta) - \overline{l}(\theta_0 + \delta, \theta_0) < \frac{\epsilon}{2}, \tag{6.8}$$

and

$$-\frac{\epsilon}{2} < \overline{l}_n(\theta_0) - \overline{l}(\theta_0, \theta_0) < \frac{\epsilon}{2}. \tag{6.9}$$

(Note that we have not put ω in the above expressions, but $\{\overline{l}_n\}_{i=1}^{\infty}$ is actually a different sequence for each ω of the underlying probability space. In a proof of this type it is common to ignore all such references to ω or Ω_1 and just say "with probability one.") Putting (6.5) with (6.7) and (6.9) we have for all $n > n(\epsilon, \omega)$,

$$\begin{aligned}\overline{l}_n(\theta_0) - \overline{l}_n(\theta_0 - \delta) &= \overline{l}_n(\theta_0) - \overline{l}(\theta_0, \theta_0) + \overline{l}(\theta_0, \theta_0) - \overline{l}(\theta_0 - \delta, \theta_0) \\ &\quad + \overline{l}(\theta_0 - \delta, \theta_0) - \overline{l}_n(\theta_0 - \delta) \\ &> -\frac{\epsilon}{2} + \epsilon - \frac{\epsilon}{2} = 0.\end{aligned}$$

Thus, $\overline{l}_n(\theta_0) > \overline{l}_n(\theta_0 - \delta)$. Similarly, we can show that $\overline{l}_n(\theta_0) > \overline{l}_n(\theta_0 + \delta)$. Because $\overline{l}_n(\theta)$ is continuously differentiable in θ by (**C**), a solution of $\partial \overline{l}_n(\theta)/\partial\theta = 0$ exists in $(\theta_0 - \delta, \theta_0 + \delta)$ for each $\omega \in \Omega_1$ and for all $n > n(\epsilon, \omega)$. Because $\delta > 0$ is arbitrary, Theorem 6.2 follows. ■

6.3.2 *Vector θ*

For the general case where $\boldsymbol{\theta}$ is b-dimensional, the previous method breaks down because $\overline{l}_n(\boldsymbol{\theta}_0) > \overline{l}_n(\boldsymbol{\theta}_\delta)$ for any finite number of points $\boldsymbol{\theta}_\delta$ a distance δ from $\boldsymbol{\theta}_0$ (i.e., $||\boldsymbol{\theta}_\delta - \boldsymbol{\theta}_0|| = \delta$) does not guarantee existence of a solution to $\partial l_n(\boldsymbol{\theta})/\partial \boldsymbol{\theta}^T = \mathbf{0}$ in $\{\boldsymbol{\theta} : |\boldsymbol{\theta}_0 - \boldsymbol{\theta}| < \delta\}$. Figure 6.2 shows a $\delta = .5$ neighborhood of a two-dimensional parameter $\boldsymbol{\theta} = (\mu, \sigma)$ with $\sigma_0 = 3$ and $\mu_0 = 2$ to illustrate this point. It is possible for $\overline{l}_n(\boldsymbol{\theta})$ to have values of $\overline{l}_n(\boldsymbol{\theta})$ on the four points of the circle that are smaller than $\overline{l}_n(\boldsymbol{\theta}_0)$, yet there need not be a critical point inside the circle. This type of situation is unusual, but to guarantee that a solution exists inside the circle, we need $\overline{l}_n(\boldsymbol{\theta}_0) > \overline{l}_n(\boldsymbol{\theta})$ for all $\boldsymbol{\theta}$ on the boundary of a δ-neighborhood of $\boldsymbol{\theta}_0$. However, the usual SLLN only allows us to conclude that $\overline{l}_n(\boldsymbol{\theta}_0) > \overline{l}_n(\boldsymbol{\theta})$ with probability one, all $n > n(\epsilon)$, for at most a finite number of $\boldsymbol{\theta}$ simultaneously. (The reason is that we can use the SLLN for each of a finite number of $\boldsymbol{\theta}$ values and intersect the associated underlying sets Ω_1, Ω_2, etc., and still have a set with probability one. But we cannot intersect an infinite number and be sure the resulting intersection has probability one.) Since we need the $\overline{l}_n(\boldsymbol{\theta}_0) > \overline{l}_n(\boldsymbol{\theta})$ to hold for all points such that $|\boldsymbol{\theta}_0 - \boldsymbol{\theta}| = \delta$, an uncountable infinity of $\boldsymbol{\theta}$ values if $b \geq 2$, we appeal to the following uniform SLLN.

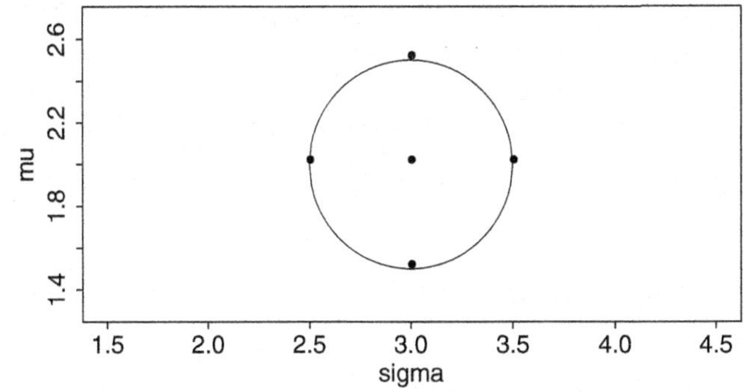

Fig. 6.2 $\delta = .5$ neighborhood of $(\sigma_0 = 3, \mu_0 = 2)$.

Theorem 6.3 (Jennrich, 1969). *Let g be a function on $\mathcal{Y} \times \Theta^*$ where $\mathcal{Y}$ is a Euclidean space and Θ^* is a compact subset of a Euclidean space. Let $g(y, \boldsymbol{\theta})$ be a continuous function of $\boldsymbol{\theta}$ for each y and a measurable function of y for each $\boldsymbol{\theta}$. Assume also that $|g(y, \boldsymbol{\theta})| \le h(y)$ for all y and $\boldsymbol{\theta}$, where h is integrable with respect to a distribution function F on $\mathcal{Y}$. If $Y_1, \ldots, Y_n$ is a random sample from F, then as $n \to \infty$*

$$\frac{1}{n}\sum_{i=1}^{n} g(Y_i, \boldsymbol{\theta}) \xrightarrow{wp1} \int g(y, \boldsymbol{\theta})dF(y) \qquad \textit{uniformly for all } \boldsymbol{\theta} \in \Theta^*.$$

We can now state a result similar to the $b = 1$ case using (**A**), (**B**), (**C**), and

(**D**) $g(y, \boldsymbol{\theta}) = \log f(y; \boldsymbol{\theta})$ satisfies the assumptions of Theorem 6.3 with $\Theta^* = \{\boldsymbol{\theta} : \|\boldsymbol{\theta} - \boldsymbol{\theta}_0\| \le \delta\}$ for some $\delta > 0$, and $F(x) = F(x; \boldsymbol{\theta}_0)$.

Theorem 6.4. *If $Y_1, \ldots, Y_n$ are iid from a density $f(y; \boldsymbol{\theta})$ that satisfies (**A**), (**B**), (**C**), and (**D**), then there exists a strongly consistent solution $\widehat{\boldsymbol{\theta}}_{\text{MLE}}$ of the likelihood equations $\partial l_n(\boldsymbol{\theta})/\partial \boldsymbol{\theta}^T = \mathbf{0}$.*

We now discuss conditions (**A**) and (**D**) for several multiparameter families of densities.

1. Consider the mixture density $f(y; \mu_1, \mu_2, p) = p\phi(y-\mu_1)+(1-p)\phi(y-\mu_2)$, where ϕ is the standard normal density. Condition (**A**) (p. 278) is violated unless some restrictions are placed on the parameters. For example, $(\mu_1 = 1, \mu_2 = 2, p = .4)$ yields the same density as $(\mu_1 = 2, \mu_2 = 1, p = .6)$, and $p = 0$ or $p = 1$ or $\mu_1 = \mu_2$ each violate Condition (**A**) because they render one of the remaining parameters nonidentifiable. Thus, if $p = 0$, then $f(y; \mu_1 = 3, \mu_2 = 5, p = 0)$ is identical to $f(y; \mu_1 = 4, \mu_2 = 5, p = 0)$ or any other $f(y; \mu_1, \mu_2 = 5, p = 0)$. One solution is to require $-\infty < \mu_1 < \mu_2 < \infty$ and $0 < p < 1$. We might also suggest looking back to Example 2.28 (p. 97), where for a similar mixture density, it was noted that the maximum likelihood

estimator does not exist in the strict sense since the likelihood tends to infinity along certain paths in the parameter space. Thus, Theorem 6.4 above is a good result for mixture densities.

2. Condition (**D**) is a technical condition that we generally expect to be satisfied when Condition (**B**) is satisfied. To illustrate how to verify condition (**D**), consider the normal (μ, σ^2) density, where

$$\log f(y; \mu, \sigma) = -\log(\sqrt{2\pi}) - \log\sigma - \frac{(y-\mu)^2}{2\sigma^2}.$$

For any Euclidean neighborhood $\{\boldsymbol{\theta} : \|\boldsymbol{\theta} - \boldsymbol{\theta}_0\| \le \delta\}$, we can find δ_1 and δ_2 such that the neighborhood is contained by the box $\{(\mu, \sigma) : |\mu - \mu_0| \le \delta_1, |\sigma - \sigma_0| \le \delta_2|\}$. For any (μ, σ) in this latter set,

$$|\log f(y; \mu, \sigma)| \le |\log(\sqrt{2\pi})| + |\log(\sigma_0 + \delta_2)| \\ + \frac{1}{2(\sigma_0 - \delta_2)^2}\left[y^2 + 2(|\mu_0| + \delta_1)|y| + (|\mu_0| + \delta_1)^2\right].$$

Since the right-hand side of the above inequality does not depend on (μ, σ) and is integrable with respect to any normal distribution function, we may take the right-hand side as the bounding function $h(y)$ in Theorem 6.3, and thus (**D**) is satisfied.

As mentioned in the Introduction to this chapter, the results of Theorems 6.2 (p. 279) and 6.4 (p. 281) are not fully satisfying because they only ensure that at least one of the roots of the likelihood equation is strongly consistent. Of course, if $\widehat{\boldsymbol{\theta}}_{\text{MLE}}$ is unique using theorems from the appendix to Chapter 2, then consistency follows naturally. For compact parameter spaces we now show consistency for the $\widehat{\boldsymbol{\theta}}_{\text{MLE}}$ defined as the maximizer of the likelihood.

6.4 Compact Parameter Spaces

As mentioned in the Introduction, rigorous consistency proofs typically consist of showing that the estimator is ultimately trapped in a compact set and then using a form of Theorem 6.3 (p. 281) to get the result. We now give a consistency result for compact parameter spaces that often can be combined with other methods of proof to get more general results for noncompact parameter spaces. A similar theorem appears in White (1981, JASA, p. 420, Theorem 2.1).

Theorem 6.5. *Let $Y_1, \ldots, Y_n$ be iid with density $f(y; \boldsymbol{\theta}_0)$, where $\boldsymbol{\theta}_0 \in \Theta$, where Θ is a compact subset of R^b. Suppose that assumptions* (**A**) *and* (**B**) *on page 278 are* *satisfied and that* $g(y, \boldsymbol{\theta}) = \log f(y; \boldsymbol{\theta})$ *satisfies the assumptions of Theorem 6.3 (p. 281) with* $F(y) = F(y; \boldsymbol{\theta}_0)$. *Then for* $\widehat{\boldsymbol{\theta}}_n$ *= the maximizer of* $l_n(\boldsymbol{\theta})$ *on* Θ, *we have*

$$\widehat{\boldsymbol{\theta}}_n \xrightarrow{wp1} \boldsymbol{\theta}_0 \qquad as\ n \to \infty.$$

Proof. Since $l_n(\boldsymbol{\theta})$ is continuous on Θ, it has at least one maximizing value. The following argument may be applied to each $\widehat{\boldsymbol{\theta}}_n$ maximizing $l_n(\boldsymbol{\theta})$ if there are more than one. Let Ω_1 be the subset of the underlying probability space on which the convergence in Theorem 6.3 (p. 281) occurs. For $\omega \in \Omega_1$, suppose that $\widehat{\boldsymbol{\theta}}_n(\omega) \not\to \boldsymbol{\theta}_0$. Because of compactness there exists a subsequence n_k and a value $\boldsymbol{\theta}_1 \neq \boldsymbol{\theta}_0$ such that $\widehat{\boldsymbol{\theta}}_{n_k}(\omega) \to \boldsymbol{\theta}_1$. Now

$$|\bar{l}_{n_k}(\widehat{\boldsymbol{\theta}}_{n_k}(\omega)) - \bar{l}(\boldsymbol{\theta}_1, \boldsymbol{\theta}_0)| \le |\bar{l}_{n_k}(\widehat{\boldsymbol{\theta}}_{n_k}(\omega)) - \bar{l}(\widehat{\boldsymbol{\theta}}_{n_k}(\omega), \boldsymbol{\theta}_0)| \\ + |\bar{l}(\widehat{\boldsymbol{\theta}}_{n_k}(\omega), \boldsymbol{\theta}_0) - \bar{l}(\boldsymbol{\theta}_1, \boldsymbol{\theta}_0)|.$$

By the uniform convergence of Theorem 6.3 (p. 281), the first term on the right-hand side of (6.10) converges to zero. The second term is bounded by

$$\int f(y; \boldsymbol{\theta}_0) \,|\log f(y; \widehat{\boldsymbol{\theta}}_{n_k}(\omega)) - \log f(y; \boldsymbol{\theta}_1)| dy.$$

The latter term converges to zero by the dominated convergence theorem because $\log f(y; \boldsymbol{\theta})$ is continuous and

$$|\log f(x; \widehat{\boldsymbol{\theta}}_{n_k}(\omega)) - \log f(x; \boldsymbol{\theta}_1)| \le 2h(x)$$

and $h(x)$ is integrable with respect to $f(y; \boldsymbol{\theta}_0)$. Thus $\bar{l}_{n_k}(\widehat{\boldsymbol{\theta}}_{n_k}(\omega)) \longrightarrow \bar{l}(\boldsymbol{\theta}_1, \boldsymbol{\theta}_0) < \bar{l}(\boldsymbol{\theta}_0, \boldsymbol{\theta}_0)$. But this latter convergence and inequality (from Theorem 6.1, p. 278) contradict

$$\bar{l}_{n_k}(\widehat{\boldsymbol{\theta}}_{n_k}(\omega)) \ge \bar{l}_{n_k}(\boldsymbol{\theta}_0) \longrightarrow \bar{l}(\boldsymbol{\theta}_0, \boldsymbol{\theta}_0),$$

where the inequality is from the definition of $\widehat{\boldsymbol{\theta}}_n$ as the maximizer of $\bar{l}_n(\boldsymbol{\theta})$. Thus $\widehat{\boldsymbol{\theta}}_n(\omega) \to \boldsymbol{\theta}_0$ for each $\omega \in \Omega_1$, i.e., $\widehat{\boldsymbol{\theta}}_n \xrightarrow{wp1} \boldsymbol{\theta}_0$. ■

6.5 Asymptotic Normality of Maximum Likelihood Estimators

This account of conditions for the asymptotic normality of maximum likelihood estimators is similar to accounts in Ch. 4 of Serfling (1980) and Ch. 6 of Lehmann and Casella (1998). For simplicity we first give a theorem and proof for a single real parameter θ; the multiparameter case follows. The parameter space is denoted by Θ. Under a similar set of conditions to those below, Bahadur (1964) has given the result that if t_n is any estimator such that $n^{1/2}(t_n - \theta) \xrightarrow{d} N(0, v(\theta))$ for each $\theta \in \Theta$, then $v(\theta) \geq 1/I(\theta)$ except perhaps on a set of θ of Lebesgue measure zero. Thus, the maximum likelihood estimator with asymptotic variance $1/I(\theta)$ achieves a type of asymptotic Cramer-Rao lower. Estimators such as $\widehat{\boldsymbol{\theta}}_{\text{MLE}}$ achieving this bound asymptotically are often called *Best Asymptotically Normal* (BAN). The multiparameter extension of this result is also discussed in Bahadur (1964), where $\boldsymbol{v}(\boldsymbol{\theta})$ and $\boldsymbol{I}(\boldsymbol{\theta})$ are now matrices: $\boldsymbol{v}(\boldsymbol{\theta}) - \boldsymbol{I}(\boldsymbol{\theta})^{-1}$ is positive semi-definite and the asymptotic variance of any scalar $\boldsymbol{c}^T\boldsymbol{t}_n$ satisfies $\boldsymbol{c}^T\boldsymbol{v}(\boldsymbol{\theta})\boldsymbol{c} \geq \boldsymbol{c}^T\boldsymbol{I}(\boldsymbol{\theta})^{-1}\boldsymbol{c}$.

6.5.1 Real-Valued θ

Theorem 6.6. *Let $Y_1, \ldots, Y_n$ be iid with density $f(y;\theta)$, where θ is an interior point of Θ, and $f(y;\theta)$ satisfies the following conditions.*

1. *Identifiability: $\theta_1 \neq \theta_2$ implies that $F(y;\theta_1) \neq F(y;\theta_2)$ for at least one y.*
2. *For each $\theta \in \Theta$, $F(y;\theta)$ has the same support not depending on θ.*
3. *For each $\theta \in \Theta$, the first three partial derivatives of $\log f(y;\theta)$ with respect to θ exist for y in the support of $F(y;\theta)$.*
4. *For each $\theta_0 \in \Theta$, there exists a function $g(y)$ (possibly depending on θ_0), such that for all θ in a neighborhood of the given θ_0, $|\partial^3 \log f(y;\theta)/\partial\theta^3| \leq g(y)$ for all y and where $\int g(y)\, dF(y;\theta_0) < \infty$.*
5. *For each $\theta \in \Theta$, $E\{\partial \log f(Y_1;\theta)/\partial\theta\} = 0$, $I(\theta) = E\{\partial \log f(Y_1;\theta)/\partial\theta\}^2 = E\{-\partial^2 \log f(Y_1;\theta)/\partial\theta^2\}$, $0 < I(\theta) < \infty$.*

If $\widehat{\theta}$ satisfies $S(\widehat{\theta}) = \sum_{i=1}^n \partial \log f(Y_i;\widehat{\theta})/\partial\theta = 0$ and $\widehat{\theta} \xrightarrow{p} \theta$ as $n \to \infty$, then

$$\sqrt{n}(\widehat{\theta} - \theta) \xrightarrow{d} N(0, I(\theta)^{-1}) \qquad \text{as } n \to \infty.$$

Before giving the proof, we comment briefly on the above Conditions 1–5. In the proof we also demonstrate where the conditions are needed. Condition 1 is the usual identifiability condition (**A**), p. 278. Condition 2 is a classical condition used to eliminate nonstandard densities like the exponential threshold model. Under Conditions 3-5 and assuming $\widehat{\theta} \xrightarrow{p} \theta$, we do not actually need Conditions 1 and 2, but we include them because Condition 1 is necessary for consistency and Condition 2 indicates when to expect problems with Conditions 3 and 5. Smith (1985) gives a variety of different results for the maximum likelihood estimators in cases that violate Condition 2 (see also Problem 6.5, p. 293). Condition 3 requires appropriate smoothness of the log density to allow Taylor expansion in the proof of Theorem 6.6. Condition 3 is not necessary for asymptotic normality, however. This is illustrated by the Laplace location density $f(y;\theta) = (1/2)\exp(-|y-\theta|)$, where we know that $\widehat{\theta}_{\text{MLE}}$ =sample median is asymptotically normal as proved in Ch. 3. Condition 4 allows us to bound the third derivative of $l_n(\theta)$ that resulted from the Taylor expansion. Condition 5 assumes that the information $I(\theta)$ exists in both forms. Other authors give bounds on first and second derivatives that imply parts of Condition 5 rather than assume it (see Problems 6.3 and 6.4, p. 293).

The proof below proceeds by Taylor expansion of the likelihood equation $S(\widehat{\theta}) = \partial l_n(\widehat{\theta})/\partial\theta = 0$, resulting in $\widehat{\theta} - \theta_0 \approx I(\theta_0)^{-1}S(\theta_0)$. Then the CLT gives the result.

Proof. Let θ_0 be the true parameter value. Let $S'(\theta)$ and $S''(\theta)$ denote the first two derivatives of $S(\theta)$. Taylor expansion of $S(\widehat{\theta})$ around θ_0 yields

$$\begin{aligned} 0 = S(\widehat{\theta}) &= S(\theta_0) + S'(\theta_0)(\widehat{\theta} - \theta_0) + \frac{1}{2}S''(\widehat{\theta}^*)(\widehat{\theta} - \theta_0)^2 \\ &= S(\theta_0) + (\widehat{\theta} - \theta_0)\left\{S'(\theta_0) + \frac{1}{2}S''(\widehat{\theta}^*)(\widehat{\theta} - \theta_0)\right\} \end{aligned}$$

where $\widehat{\theta}^*$ is between θ_0 and $\widehat{\theta}$. Rearranging the last equation yields

$$\sqrt{n}(\widehat{\theta} - \theta_0) = \frac{-S(\theta_0)/\sqrt{n}}{\left\{\frac{1}{n}S'(\theta_0) + \frac{1}{2n}S''(\widehat{\theta}^*)(\widehat{\theta} - \theta_0)\right\}}. \tag{6.10}$$

The numerator of (6.10) converges in distribution to $N(0, I(\theta_0))$ by the Central Limit Theorem and the fact that $\partial \log f(Y_i;\theta)/\partial\theta|_{\theta=\theta_0}$ has mean zero and variance $I(\theta_0)$.

The first term in the denominator of (6.10) converges in probability to $-I(\theta_0)$ by the Weak Law of Large Numbers and Condition 5. Let R_n denote the second term in the denominator of (6.10), i.e.,

$$\begin{aligned} R_n &= \frac{1}{2n}S''(\widehat{\theta}^*)(\widehat{\theta} - \theta_0) \\ &= \frac{1}{2n}\sum_{i=1}^{n}\left\{\frac{\partial^3}{\partial\theta^3}\log f(Y_i;\theta)\bigg|_{\theta=\widehat{\theta}^*}\right\}(\widehat{\theta} - \theta_0). \end{aligned}$$

The proof is completed by showing that $R_n = o_p(1)$ and appealing to Slutsky's Theorem. To this end define

$$R_n^* = \frac{1}{2n}\sum_{i=1}^{n} g(Y_i)(\widehat{\theta} - \theta_0),$$

and note that there exists a $\delta > 0$ such that $|R_n| \le |R_n^*|$ when $|\widehat{\theta} - \theta_0| < \delta$ by Assumption 4. Also by Assumption 4 and the Weak Law of Large Numbers,

$n^{-1}\sum_{i=1}^{n} g(Y_i)$ converges in probability to $Eg(Y_1) < \infty$; and thus $R_n^* = o_p(1)$ because $\widehat{\theta} \xrightarrow{p} \theta_0$. Finally, note that for any $0 < \epsilon < \delta$,

$$\begin{aligned} P(|R_n| > \epsilon) &= P(|R_n| > \epsilon, |\widehat{\theta} - \theta_0| > \epsilon) + P(|R_n| > \epsilon, |\widehat{\theta} - \theta_0| \le \epsilon) \\ &\le P(|\widehat{\theta} - \theta_0| > \epsilon) + P(|R_n^*| > \epsilon, |\widehat{\theta} - \theta_0| \le \epsilon) \\ &\le P(|\widehat{\theta} - \theta_0| > \epsilon) + P(|R_n^*| > \epsilon). \end{aligned}$$

Because both $|\widehat{\theta} - \theta_0|$ and $|R_n^*|$ are $o_p(1)$, it follows that $P(|R_n| > \epsilon) \to 0$ as $n \to \infty$ and hence $R_n = o_p(1)$, thus concluding the proof. ∎

6.5.2 Vector θ

Now we give a theorem on asymptotic normality of $\widehat{\boldsymbol{\theta}}_{\text{MLE}}$ in the multiparameter case where $\boldsymbol{\theta}$ is b dimensional. The proof is similar to the proof for Theorem 6.6; details may be found in Lehmann and Casella (1998, Ch. 6).

Theorem 6.7. *Let $Y_1, \ldots, Y_n$ be iid with density $f(y;\boldsymbol{\theta})$, where $\boldsymbol{\theta}$ is an interior point of Θ, and $f(y;\theta)$ satisfies the following conditions.*

1. *Identifiability: $\boldsymbol{\theta}_1 \ne \boldsymbol{\theta}_2$ implies that $F(y;\boldsymbol{\theta}_1) \ne F(y;\boldsymbol{\theta}_2)$ for at least one y.*
2. *For each $\boldsymbol{\theta} \in \Theta$, $F(y;\boldsymbol{\theta})$ has the same support not depending on $\boldsymbol{\theta}$.*
3. *For each $\boldsymbol{\theta} \in \Theta$, the first three partial derivatives of* $\log f(y;\boldsymbol{\theta})$ *with respect to $\boldsymbol{\theta}$ exist for y in the support of $F(y;\boldsymbol{\theta})$.*
4. *For each $\boldsymbol{\theta}_0 \in \Theta$, there exists a function $g(y)$ (possibly depending on $\boldsymbol{\theta}_0$), such that in a neighborhood of the given $\boldsymbol{\theta}_0$ and for all $j,k,l \in \{1,\ldots,b\}$,*

$$\left|\frac{\partial^3}{\partial\theta_j\,\partial\theta_k\,\partial\theta_l}\log f(y;\boldsymbol{\theta})\right| \le g(y)$$

   for all y and where $\int g(y)\,dF(y;\boldsymbol{\theta}_0) < \infty$.
5. *For each $\boldsymbol{\theta} \in \Theta$, $E[\partial \log f(Y_1;\boldsymbol{\theta})/\partial\boldsymbol{\theta}] = \mathbf{0}$,*

$$\boldsymbol{I}(\boldsymbol{\theta}) = E\left\{\frac{\partial}{\partial\boldsymbol{\theta}}\log f(Y_1;\boldsymbol{\theta})\frac{\partial}{\partial\boldsymbol{\theta}^T}\log f(Y_1;\boldsymbol{\theta})\right\} = E\left\{-\frac{\partial^2}{\partial\boldsymbol{\theta}\,\partial\boldsymbol{\theta}^T}\log f(Y_1;\boldsymbol{\theta})\right\},$$

   and $\boldsymbol{I}(\boldsymbol{\theta})$ is nonsingular.

If $\widehat{\boldsymbol{\theta}}$ satisfies $\boldsymbol{S}(\widehat{\boldsymbol{\theta}}) = \sum_{i=1}^{n} \partial \log f(Y_i;\widehat{\boldsymbol{\theta}})/\partial\boldsymbol{\theta}^T = \mathbf{0}$ and $\widehat{\boldsymbol{\theta}} \xrightarrow{p} \boldsymbol{\theta}$ as $n \to \infty$, then

$$\sqrt{n}(\widehat{\boldsymbol{\theta}} - \boldsymbol{\theta}) \xrightarrow{d} N(\mathbf{0}, \boldsymbol{I}(\boldsymbol{\theta})^{-1}) \qquad \text{as } n \to \infty.$$

6.6 Asymptotic Null Distribution of Likelihood-Based Tests

6.6.1 Wald Tests

Recall from Chapter 3 that composite null hypotheses are typically specified by either of two forms. The first is the partitioned-vector form,

$$H_0 : \boldsymbol{\theta}_1 = \boldsymbol{\theta}_{10}, \tag{6.11}$$

where $\boldsymbol{\theta} = (\boldsymbol{\theta}_1^T, \boldsymbol{\theta}_2^T)^T$ is $b \times 1$ and $\boldsymbol{\theta}_1$ is $r \times 1$, $r \leq b$. The second is the constraint form,

$$H_0 : \boldsymbol{h}(\boldsymbol{\theta}) = \boldsymbol{0}, \tag{6.12}$$

where $\boldsymbol{h}$ is $r \times 1$ and differentiable with $\boldsymbol{H}(\boldsymbol{\theta}) = \partial \boldsymbol{h}(\boldsymbol{\theta})/\partial \boldsymbol{\theta}^T$ having dimension $r \times b$. The two forms of the Wald statistic are then

$$T_{\mathrm{W}} = n(\widehat{\boldsymbol{\theta}}_1 - \boldsymbol{\theta}_{10})^T \left[\left\{ \boldsymbol{I}(\widehat{\boldsymbol{\theta}})^{-1} \right\}_{11} \right]^{-1} (\widehat{\boldsymbol{\theta}}_1 - \boldsymbol{\theta}_{10}) \tag{6.13}$$

and

$$T_{\mathrm{W}} = n\boldsymbol{h}(\widehat{\boldsymbol{\theta}})^T \left\{ \boldsymbol{H}(\widehat{\boldsymbol{\theta}}) \boldsymbol{I}(\widehat{\boldsymbol{\theta}})^{-1} \boldsymbol{H}(\widehat{\boldsymbol{\theta}})^T \right\}^{-1} \boldsymbol{h}(\widehat{\boldsymbol{\theta}}). \tag{6.14}$$

Since (6.12) is more general, including the first form by using $\boldsymbol{h}(\boldsymbol{\theta}) = \boldsymbol{\theta}_1 - \boldsymbol{\theta}_{10}$, the following theorem is stated for the second form, but only for the iid case.

Theorem 6.8. *Suppose that $Y_1, \ldots, Y_n$ are iid with density $f(y; \boldsymbol{\theta})$, $\boldsymbol{\theta} \in \Theta$. Assume that all the conditions of Theorem 6.7 (p. 286) hold and that $\boldsymbol{H}(\boldsymbol{\theta})$ and $\boldsymbol{I}(\boldsymbol{\theta})$ are continuous and $\boldsymbol{H}(\boldsymbol{\theta})\boldsymbol{I}(\boldsymbol{\theta})^{-1}\boldsymbol{H}(\boldsymbol{\theta})^T$ is nonsingular. Then under (6.12), $T_{\mathrm{W}} \xrightarrow{d} \chi^2_r$ as $n \to \infty$.*

Proof. Letting the solution of (6.12) be denoted $\boldsymbol{\theta}_0$, by the Delta Theorem and the asymptotic normality of the maximum likelihood estimator, $\boldsymbol{h}(\widehat{\boldsymbol{\theta}})$ is $AN(\boldsymbol{0}, \boldsymbol{H}(\boldsymbol{\theta}_0) \boldsymbol{I}(\boldsymbol{\theta}_0)^{-1}\boldsymbol{H}(\boldsymbol{\theta}_0)^T/n)$. Then

$$T_{\mathrm{W}} = \sqrt{n}\boldsymbol{h}(\widehat{\boldsymbol{\theta}})^T \left\{ \boldsymbol{H}(\widehat{\boldsymbol{\theta}}) \boldsymbol{I}(\widehat{\boldsymbol{\theta}})^{-1} \boldsymbol{H}(\widehat{\boldsymbol{\theta}})^T \right\}^{-1} \sqrt{n}\boldsymbol{h}(\widehat{\boldsymbol{\theta}})$$

converges in distribution to a χ^2_r distribution by Problem 5.10, p. 265. ■

Any consistent estimator of $\boldsymbol{I}(\boldsymbol{\theta}_0)$ could be used in defining T_{W}, for example, $\overline{\boldsymbol{I}}(\boldsymbol{Y}, \widehat{\boldsymbol{\theta}})$, and Theorem 6.8 still follows. More general versions of Theorem 6.8 can be given for regression situations where some assumption about the convergence of $\boldsymbol{I}(\widehat{\boldsymbol{\theta}})$ is needed. Moore (1977), Hadi and Wells (1990), and Andrews (1987) give generalizations that allow generalized inverses of singular matrices in the middle of Wald statistics.

6.6.2 Score Tests

The score test for either (6.11) or (6.12) has the general form $T_S = \boldsymbol{S}(\widetilde{\boldsymbol{\theta}})^T \boldsymbol{I}_n(\widetilde{\boldsymbol{\theta}})^{-1}\boldsymbol{S}(\widetilde{\boldsymbol{\theta}})/n$, where $\widetilde{\boldsymbol{\theta}}$ is the maximum likelihood estimator under the null hypothesis. Here we give a theorem that covers the iid case for (6.11). In this case, T_S has the form

$$\begin{aligned} T_S &= \boldsymbol{S}_1(\widetilde{\boldsymbol{\theta}})^T \frac{1}{n}\left[I(\widetilde{\boldsymbol{\theta}})^{-1}\right]_{11} \boldsymbol{S}_1(\widetilde{\boldsymbol{\theta}}) \\ &= \boldsymbol{S}_1(\widetilde{\boldsymbol{\theta}})^T \frac{1}{n}\left(\widetilde{\boldsymbol{I}}_{11} - \widetilde{\boldsymbol{I}}_{12}\widetilde{\boldsymbol{I}}_{22}^{-1}\widetilde{\boldsymbol{I}}_{21}\right)^{-1} \boldsymbol{S}_1(\widetilde{\boldsymbol{\theta}}), \end{aligned} \tag{6.15}$$

where $\widetilde{\boldsymbol{I}}_{ij}$ is the i, j submatrix of $\boldsymbol{I}(\widetilde{\boldsymbol{\theta}})$ corresponding to the partition of $\boldsymbol{\theta} = (\boldsymbol{\theta}_1^T, \boldsymbol{\theta}_2^T)^T$.

Theorem 6.9. *Suppose that $Y_1, \ldots, Y_n$ are iid with density $f(y; \boldsymbol{\theta})$, $\boldsymbol{\theta} \in \Theta$. Assume that all the conditions of Theorem 6.7 (p. 286) hold and that $\boldsymbol{I}(\boldsymbol{\theta})$ is continuous. Then under (6.11), $T_S \xrightarrow{d} \chi_r^2$ as $n \to \infty$.*

Proof. From the proof of Theorem 6.7 (p. 286), $\widetilde{\boldsymbol{\theta}}_2$ has the approximation-by-averages representation

$$\widetilde{\boldsymbol{\theta}}_2 - \boldsymbol{\theta}_{20} = \frac{1}{n}\sum_{i=1}^{n} \{I(\boldsymbol{\theta}_0)_{22}\}^{-1} \frac{\partial}{\partial \boldsymbol{\theta}_2} \log f(Y_i; \boldsymbol{\theta}_0) + R_n, \tag{6.16}$$

where $\sqrt{n} R_n \xrightarrow{p} \boldsymbol{0}$ as $n \to \infty$. Using (6.16) and Condition 4 of Theorem 6.7 (p. 286) with Theorem 5.28 (p. 249), we have that

$$\frac{1}{\sqrt{n}} \boldsymbol{S}_1(\widetilde{\boldsymbol{\theta}}) \text{ is AN}\left(\boldsymbol{0}, \left[\{\boldsymbol{I}(\boldsymbol{\theta}_0)^{-1}\}_{11}\right]^{-1}\right). \tag{6.17}$$

Then T_S in (6.15) converges to a χ_r^2 random variable by Problem 5.10 (p. 265). ■

6.6.3 Likelihood Ratio Tests

The first proof of convergence in distribution of T_{LR} to a chi-squared random variable under the composite null hypothesis (6.11) is due to Wilks (1938). Our proof is similar to many others in the literature and follows naturally from Taylor expansion and Theorem 6.7.

Theorem 6.10. *Suppose that $Y_1, \ldots, Y_n$ are iid with density $f(y;\boldsymbol{\theta})$, $\boldsymbol{\theta} \in \Theta$. Assume that all the conditions of Theorem 6.7 (p. 286) hold. Then under (6.11), $T_{\rm LR} \xrightarrow{d} \chi^2_r$ as $n \to \infty$.*

Proof. First expand $l_n(\widetilde{\boldsymbol{\theta}})$ about $\widehat{\boldsymbol{\theta}}$ to obtain

$$l_n(\widetilde{\boldsymbol{\theta}}) = l_n(\widehat{\boldsymbol{\theta}}) + \boldsymbol{S}(\widehat{\boldsymbol{\theta}})^T(\widetilde{\boldsymbol{\theta}} - \widehat{\boldsymbol{\theta}}) - \sqrt{n}(\widetilde{\boldsymbol{\theta}} - \widehat{\boldsymbol{\theta}})^T \frac{1}{2} \boldsymbol{I}_n(\boldsymbol{Y}, \widehat{\boldsymbol{\theta}}^*)\sqrt{n}(\widetilde{\boldsymbol{\theta}} - \widehat{\boldsymbol{\theta}}), \quad (6.18)$$

where $\widehat{\boldsymbol{\theta}}^*$ lies between $\widetilde{\boldsymbol{\theta}}$ and $\widehat{\boldsymbol{\theta}}$. Note that $\boldsymbol{S}(\widehat{\boldsymbol{\theta}}) = 0$ by definition of $\widehat{\boldsymbol{\theta}}$. Then rearranging (6.18), we have

$$T_{\rm LR} = -2[l_n(\widetilde{\boldsymbol{\theta}}) - l_n(\widehat{\boldsymbol{\theta}})] = \sqrt{n}(\widetilde{\boldsymbol{\theta}} - \widehat{\boldsymbol{\theta}})^T \boldsymbol{I}_n(\boldsymbol{Y}, \widehat{\boldsymbol{\theta}}^*)\sqrt{n}(\widetilde{\boldsymbol{\theta}} - \widehat{\boldsymbol{\theta}}). \quad (6.19)$$

Using the convergence in probability of $\widehat{\boldsymbol{\theta}}$ and Condition 4 of Theorem 6.7 (p. 286), $\boldsymbol{I}_n(\boldsymbol{Y}, \widehat{\boldsymbol{\theta}}^*) \xrightarrow{p} \boldsymbol{I}(\boldsymbol{\theta}_0)$ as $n \to \infty$. Thus to get the convergence in distribution of $T_{\rm LR}$, we only need to get the asymptotic distribution of $\sqrt{n}(\widetilde{\boldsymbol{\theta}} - \widehat{\boldsymbol{\theta}})$. To that end, we seek an expansion of $\boldsymbol{S}(\widetilde{\boldsymbol{\theta}})$ about $\widehat{\boldsymbol{\theta}}$ of the form

$$\boldsymbol{S}(\widetilde{\boldsymbol{\theta}}) = \boldsymbol{S}(\widehat{\boldsymbol{\theta}}) - n\widetilde{\boldsymbol{I}}^*_n(\widetilde{\boldsymbol{\theta}} - \widehat{\boldsymbol{\theta}}), \quad (6.20)$$

where $\widetilde{\boldsymbol{I}}^*_n \xrightarrow{p} \boldsymbol{I}(\boldsymbol{\theta}_0)$ as $n \to \infty$. For now assume that such an expansion exists. Then using the results that $\boldsymbol{S}(\widehat{\boldsymbol{\theta}}) = 0$, $S_2(\widetilde{\boldsymbol{\theta}}) = 0$, and that $S_1(\widetilde{\boldsymbol{\theta}})/n$ is $\text{AN}(\boldsymbol{0}, \left[\{\boldsymbol{I}(\boldsymbol{\theta}_0)^{-1}\}_{11}\right]^{-1}/n)$ (see 6.17, p. 288), we have that

$$\sqrt{n}(\widetilde{\boldsymbol{\theta}} - \widehat{\boldsymbol{\theta}}) \xrightarrow{d} \boldsymbol{I}(\boldsymbol{\theta}_0)^{-1}\begin{pmatrix} \boldsymbol{Z} \\ \boldsymbol{0} \end{pmatrix} \qquad \text{as } n \to \infty,$$

where $\boldsymbol{Z}$ is $\text{MN}_r(\boldsymbol{0}, \left[\{\boldsymbol{I}(\boldsymbol{\theta}_0)^{-1}\}_{11}\right]^{-1})$. Finally, putting this last result along with (6.19) and Slutsky's theorem yields that

$$T_{\rm LR} \xrightarrow{d} \boldsymbol{Z}^T \left\{\boldsymbol{I}(\boldsymbol{\theta}_0)^{-1}\right\}_{11} \boldsymbol{Z} \qquad \text{as } n \to \infty,$$

which is clearly distributed as χ^2_r because the covariance matrix of $\boldsymbol{Z}$ is the inverse of the middle matrix of the quadratic form.

The proof is completed by showing that an expansion of the form (6.20) exists. This step is complicated by the fact that there is no Mean Value Theorem for multidimensional-valued functions. The key is to consider the components of $\boldsymbol{S}(\boldsymbol{\theta})$ separately. Suppose that $S_j(\boldsymbol{\theta})$ is the jth component of $\boldsymbol{S}(\boldsymbol{\theta})$. The Mean Value Theorem applied to $S_j(\boldsymbol{\theta})$ results in

$$S_j(\widetilde{\boldsymbol{\theta}}) = S_j(\widehat{\boldsymbol{\theta}}) + S'_j(\widehat{\boldsymbol{\theta}}^{**}_j)(\widetilde{\boldsymbol{\theta}} - \widehat{\boldsymbol{\theta}}), \quad (6.21)$$

where $\widehat{\boldsymbol{\theta}}_j^{**}$ is on the line segment joining $\widetilde{\boldsymbol{\theta}}$ and $\widehat{\boldsymbol{\theta}}$. The system of equations (6.21) for $j = 1, \ldots, b$ is equivalent to the equation in (6.20) with $\widetilde{\boldsymbol{I}}_n^*$ defined as the matrix whose jth row is equal to $-S_j'(\widehat{\boldsymbol{\theta}}_j^{**})$. Because $\widehat{\boldsymbol{\theta}}_j^{**} \xrightarrow{p} \boldsymbol{\theta}_0$ as $n \to \infty$ for each j, it follows that $S_j'(\widehat{\boldsymbol{\theta}}_j^{**}) \xrightarrow{p} E\{S_j'(\boldsymbol{\theta}_0)\}$ and hence that $\widetilde{\boldsymbol{I}}_n^* \xrightarrow{p} \boldsymbol{I}(\boldsymbol{\theta}_0)$ as $n \to \infty$. ■

6.6.4 Local Asymptotic Power

For a fixed $\boldsymbol{\theta}$ in the alternative hypothesis, most reasonable tests have the property that the probability of rejection converges to 1 as n increases. Thus, the standard approach in the testing situation $H_0 : \boldsymbol{\theta}_1 = \boldsymbol{\theta}_{10}$ versus $H_a : \boldsymbol{\theta}_1 \neq \boldsymbol{\theta}_{10}$ (and $\boldsymbol{\theta}_2$ unrestricted with true value $\boldsymbol{\theta}_{20}$) is to consider local alternatives $\boldsymbol{\theta}_n$ where the first r elements have the form

$$\boldsymbol{\theta}_{10} + \frac{\boldsymbol{d}}{\sqrt{n}},$$

where $\boldsymbol{d}$ is an arbitrary $r \times 1$ vector. These are called *local* alternatives because they are close to the null value for n large. They are also called Pitman alternatives, named after Pitman who popularized a measure of asymptotic relative efficiency based on these alternatives (to be discussed more fully in Chapter 12 in the context of rank tests). Assume that we can strengthen the conditions of Theorem 6.7 (p. 286) so that for $\boldsymbol{\theta} = \boldsymbol{\theta}_n$

$$\sqrt{n}(\widehat{\boldsymbol{\theta}}_{\text{MLE}} - \boldsymbol{\theta}_0) \xrightarrow{d} N\left[\begin{pmatrix} \boldsymbol{d} \\ \boldsymbol{0} \end{pmatrix}, \boldsymbol{I}(\boldsymbol{\theta}_0)^{-1}\right] \qquad \text{as } n \to \infty. \tag{6.22}$$

Then, one can show that T_{W}, T_{S}, and T_{LR} each converge in distribution to a noncentral $\chi_r^2(\lambda)$ with noncentrality parameter

$$\lambda = \boldsymbol{d}^T \left[\{\boldsymbol{I}(\boldsymbol{\theta}_0)^{-1}\}_{11}\right]^{-1} \boldsymbol{d}. \tag{6.23}$$

Thus, in terms of these local alternatives, T_{W}, T_{S}, and T_{LR} are asymptotically equivalent. If $r = 1$, then these tests are the best asymptotically possible, but in the general $r > 1$ case, no such assertion is possible (see, e.g., van der Vaart 1998, Chs. 15 and 16).

A second use of these local alternatives is to justify the approximate power calculation

$$P\left\{\chi_r^2(\lambda) > \chi_r^2(1-\alpha)\right\}, \tag{6.24}$$

where $\boldsymbol{d} = n^{1/2}(\boldsymbol{\theta}_{1a} - \boldsymbol{\theta}_{10})$ and $\boldsymbol{\theta}_{1a}$ is the alternative of interest, and λ is given by (6.23). Such power approximations can be used to plan studies, find sample sizes necessary for a given power and alternative $\boldsymbol{\theta}_{1a}$, etc. Since simulation methods are now so easy to use (see Chapter 9), these power approximations are less important in practice.

6.6.5 Nonstandard Situations

We have previously mentioned a number of nonstandard situations where the limiting distributions of one or more of our statistics is not chi-squared with r degrees of freedom. Here we try to summarize some of these cases, but we do not claim to be exhaustive in our coverage.

1. **The alternative hypothesis is order-restricted such as $H_a : \mu_1 \leq \mu_2 \leq \mu_3$.** As mentioned in Section 3.6 (p. 150), these *isotonic regression* cases lead to T_{LR} having an asymptotic distribution such as given by (3.23). The underlying densities here are regular, but the constrained alternative causes the nonstandard asymptotics. These cases are not specifically eliminated by the conditions of Chapter 6 theorems, but the estimators and test statistics in this chapter were not constructed for these alternatives. Key references here are Barlow et al. (1972) and Robertson et al. (1988).
2. **The null hypothesis is on the boundary of the parameter space.** This case was also discussed in Section 3.6 (p. 150). The key example there was the one-way random effects model where the test of $H_0 : \sigma_\alpha^2 = 0$ versus $H_0 : \sigma_\alpha^2 > 0$ leads to the nonstandard limiting distribution $Z^2 I(Z > 0)$ for T_{LR}. In Theorems 6.6 (p. 284) and 6.7 (p. 286), these cases are eliminated by requiring the true value to be an interior point of the parameter space Θ. The original reference for these types of problems is Chernoff (1954). and Self and Liang (1987) describe the general form of the limiting distribution. Verbeke and Molenberghs (2003) discuss the use of the score statistic T_{S} in variance component testing problems.
3. **Densities with threshold parameters like the exponential of Example 2.12 (p. 63).** The maximum likelihood estimator in the exponential case, the sample minimum, has an extreme value limit distribution and is not a solution of the likelihood equations. Thus, we have ruled out these densities in most of this chapter. However, Smith (1985) gives situations where the usual limiting distributions hold. Problem 6.5 (p. 293) is about one of his examples. A more recent account with different types of problems is found in Dubinin and Vardeman (2003).
4. **Regression models with linear or nonlinear models pieced together at a number of join-points or change-points.** For example, the mean model might be $\beta_1 + \beta_2 x$ on $x \leq x_0$ and $\beta_3 + \beta_4 x$ on $x > x_0$. The model may or may not be constrained to be continuous at the change-point x_0, and typically x_0 would be unknown. A variety of nonstandard asymptotic distributions arise in these problems. See Chapter 9 of Seber and Wild (1989) for an overview.
5. **Densities with loss of identifiability under the null hypothesis.** These problems are also described as situations where the nuisance parameters are only present under the alternative hypothesis. An important class of these problems is testing for the number of components in a mixture distribution. For example, consider

$$f(y; \Delta, p) = p\phi(y) + (1 - p)\phi(y - \Delta), \tag{6.25}$$

where ϕ is the standard normal density function. Under $H_0 : \Delta = 0$, the parameter p does not appear in the density, and under $H_0 : p = 1$, the density does not depend on Δ. Either of these H_0 cause nonidentifiability of the remaining parameter, and the Fisher information matrix is singular. Thus, conditions 1. and 5. of Theorem 6.7 (p. 286) are not satisfied. Recent papers on the asymptotic distribution of T_{LR} in such cases include Dacunha-Castelle and Gassiat (1999) and Liu and Shao (2003). For example, the latter paper describes the limit distribution as the supremum of $[\max(W_{\mathrm{S}}, 0)]^2$, where W_{S} is a special mean zero Gaussian stochastic process. Note that this limit distribution is reminiscent of the $Z^2 I(Z > 0)$ limit distribution found in other nonstandard problems. In some problems found in genetics, the parameter p may be known, nevertheless, the limiting distribution is still nonstandard. For example, in the model (6.25) above, with p fixed and known but not $0, 1$, or $1/2$, Goffinet et al. (1992) give $Z^2 I(Z > 0)/2$ as the limiting distribution of T_{LR} under $H_0 : \Delta = 0$. Other problems in this general class include testing for the order of an ARMA(p, q) time series and some of the problems with change-points mentioned above. Davies (1977, 1987) proposed a solution to these problems based on score statistics.

6. **Dimension of $\boldsymbol{\theta}$ grows with sample size.** Throughout this chapter we have implicitly assumed that b is fixed. However, Example 2.4.1 (p. 57) showed that consistency of the $\widehat{\boldsymbol{\theta}}_{\mathrm{MLE}}$ can be defeated by letting b grow with n. In fact, much of the motivation for marginal and conditional likelihoods arises from these type examples. Huber (1973, 1981) shows in the linear model that $c^T \widehat{\boldsymbol{\beta}}$ can still be asymptotically normal if b grows slowly enough. Portnoy (1988) gives a general treatment for regular exponential families including conditions for T_{LR} to be asymptotically normal. In the one-way ANOVA situation Brownie and Boos (1994) show that for the usual F statistic for comparing means, $k^{1/2}(F - 1) \xrightarrow{d} N(0, 2n/(n-1))$ as the number of groups k goes to ∞ while the sample size in each group n remains fixed (similar to the Neyman-Scott situation). Then, T_{W}, T_{S}, and T_{LR} are also asymptotically normal since they are functions of F.

6.7 Problems

6.1. Verify that the "Huber" density in (6.3, p. 279), satisfies condition (**B**), p. 278.

6.2. Consider the density

$$f(y; \sigma) = \frac{2y}{\sigma^2} \exp\left(-\frac{y^2}{\sigma^2}\right) \qquad y > 0,\ \sigma > 0.$$

Verify that the regularity conditions 1. to 5. of Theorem 6.6 (p. 284) hold for the asymptotic normality of the maximum likelihood estimator of σ.

6.3. In Condition 5 of Theorem 6.6 (p. 284) we have the assumption that $\mathrm{E}[\partial \log f(Y_1;\theta)/\partial\theta] = 0$. For continuous distributions this mean zero assumption follows if

$$\int \left[\frac{\partial}{\partial\theta} f(y;\theta)\right] dy = \frac{\partial}{\partial\theta}\left[\int f(y;\theta)\, dy\right]$$

because this latter integral is one by the definition of a density function. The typical proof that this interchange of differentiability and integration is allowed assumes that for each $\theta_0 \in \Theta$, there is a bounding function $g_1(y)$ (possibly depending on θ_0) and a neighborhood of θ_0 such that for all y and for all θ in the neighborhood $|\partial f(y;\theta)/\partial\theta| \le g_1(y)$ and $\int g_1(y)\, dy < \infty$. Use the dominated convergence theorem to show that this condition allows the above interchange.

6.4. Similar to the last problem, show that

$$\mathrm{E}\,[\partial \log f(Y_1;\theta)/\partial\theta]^2 = \mathrm{E}\left[-\partial^2 \log f(Y_1;\theta)/\partial\theta^2\right]$$

follows if in addition to the condition in the last problem there is a similar bounding function $g_2(y)$ for the second derivative of $f(y;\theta)$.

6.5. Smith (1985) considers densities of the form $f(y;\theta) = f_0(y-\theta)$ for $\theta < y < \infty$, where $f_0(y) \sim \alpha c y^{\alpha-1}$ as $y \downarrow 0$. For simplicity, we assume one particular type of such densities, the gamma threshold model,

$$f(y;\theta) = \frac{1}{\Gamma(\alpha)}(y-\theta)^{\alpha-1}e^{-(y-\theta)}, \quad \theta < y < \infty,$$

where α is assumed known. Show that both forms of the information $I(\theta)$ exist and are equal for $\alpha > 2$ and neither exist for $\alpha \le 2$.

6.6. The proof of the asymptotic normality of the maximum likelihood estimator does not use an approximation by averages. Show, however, that one can extend the proof to obtain an approximation by averages result for the maximum likelihood estimator. Hint: add and subtract the numerator of (6.10, p. 285) divided by the probability limit of the denominator of (6.10, p. 285).

6.7. Using (6.16, p. 288) and Condition 4 of Theorem 6.7 (p. 286) with Theorem 5.28 (p. 249), show that (6.17, p. 288) holds.

6.8. Assuming (6.17, p. 288) and (6.20, p. 289) hold, give the details that finish the proof of Theorem 6.10 (p. 288).

6.9. Using (6.22, p. 290) and assuming continuity of $\boldsymbol{I}(\boldsymbol{\theta})$, show that under the Pitman alternatives $\boldsymbol{\theta}_{10} + \boldsymbol{d}/\sqrt{n}$, T_{W} converges in distribution to a noncentral $\chi^2_r(\lambda)$ with noncentrality parameter $\lambda = \boldsymbol{d}^T\left\{\left[\boldsymbol{I}(\boldsymbol{\theta}_0)^{-1}\right]_{11}\right\}^{-1}\boldsymbol{d}$.

Part IV
Methods for Misspecified Likelihoods and Partially Specified Models

Chapter 7
M-Estimation (Estimating Equations)

7.1 Introduction

In Chapter 1 we made the distinction between the parts of a fully specified statistical model. The primary part is the part that is most important for answering the underlying scientific questions. The secondary part consists of all the remaining details of the model. Usually the primary part is the mean or systematic part of the model, and the secondary part is mainly concerned with the distributional assumptions about the random part of the model. The full specification of the model is important for constructing the likelihood and for using the associated classical methods of inference as spelled out in Chapters 2 and 3 and supported by the asymptotic results of Chapter 6.

However, we are now ready to consider robustifying the inference so that misspecification of some secondary assumptions does not invalidate the resulting inferential methods. Basically this robustified inference relies on replacing the information matrix inverse $I(\boldsymbol{\theta})^{-1}$ in the asymptotic normality result for $\widehat{\boldsymbol{\theta}}_{\text{MLE}}$ by a generalization $I(\boldsymbol{\theta})^{-1}B(\boldsymbol{\theta})I(\boldsymbol{\theta})^{-1}$ called the sandwich matrix. In correctly specified models, $I(\boldsymbol{\theta}) = B(\boldsymbol{\theta})$, and the sandwich matrix just reduces to the usual $I(\boldsymbol{\theta})^{-1}$. When the model is not correctly specified, $I(\boldsymbol{\theta}) \neq B(\boldsymbol{\theta})$, and the sandwich matrix is important for obtaining approximately valid inference. Thus, use of this more general result accommodates misspecification but is still appropriate in correctly specified models although its use there in small samples can entail some loss of efficiency relative to standard likelihood inference.

Development of this robustified inference for likelihood-based models leads to a more general context. As discussed in Chapter 6, the asymptotic normal properties of $\widehat{\boldsymbol{\theta}}_{\text{MLE}}$ follow from Taylor expansion of the likelihood equation $\boldsymbol{S}(\boldsymbol{\theta}) = \sum_{i=1}^{n} \partial \log f(Y_i;\boldsymbol{\theta})/\partial \boldsymbol{\theta}^T = \mathbf{0}$. The more general approach is then to define an estimator of interest as the solution of an estimating equation but without the equation necessarily coming from the derivative of a log likelihood. For historical reasons and for motivation from maximum likelihood, this more general approach

D.D. Boos and L.A. Stefanski, *Essential Statistical Inference: Theory and Methods*, Springer Texts in Statistics, DOI 10.1007/978-1-4614-4818-1_7,

is called *M-estimation*. In recent years the approach is often referred to loosely as *estimating equations*. This chapter borrows heavily from the systematic description of M-estimation in Stefanski and Boos (2002).

M-estimators are solutions of the vector equation $\sum_{i=1}^{n} \boldsymbol{\psi}(\boldsymbol{Y}_i, \boldsymbol{\theta}) = \mathbf{0}$. That is, the M-estimator $\widehat{\boldsymbol{\theta}}$ satisfies

$$\sum_{i=1}^{n} \boldsymbol{\psi}(\boldsymbol{Y}_i, \widehat{\boldsymbol{\theta}}) = \mathbf{0}. \tag{7.1}$$

Here we are assuming that $\boldsymbol{Y}_1, \ldots, \boldsymbol{Y}_n$ are independent but not necessarily identically distributed, $\boldsymbol{\theta}$ is a b-dimensional parameter, and $\boldsymbol{\psi}$ is a known $(b \times 1)$-function that does not depend on i or n. In this description $\boldsymbol{Y}_i$ represents the ith datum. In some applications it is advantageous to emphasize the dependence of $\boldsymbol{\psi}$ on particular components of $\boldsymbol{Y}_i$. For example, in a regression problem $\boldsymbol{Y}_i = (\boldsymbol{x}_i, Y_i)$ and (7.1) would typically be written

$$\sum_{i=1}^{n} \boldsymbol{\psi}(Y_i, \boldsymbol{x}_i, \widehat{\boldsymbol{\theta}}) = \mathbf{0}. \tag{7.2}$$

where $\boldsymbol{x}_i$ is the ith regressor.

Huber (1964,1967) introduced M-estimators and their asymptotic properties, and they were an important part of the development of modern robust statistics. Liang and Zeger (1986) helped popularize M-estimators in the biostatistics literature under the name *generalized estimating equations* (GEE). Obviously, many others have made important contributions. For example, Godambe (1960) introduced the concept of an *optimum estimating function* in an M-estimator context, and that paper could be called a forerunner of the M-estimator approach.

There is a large literature on M-estimation and estimating equations. We will not attempt to survey this literature or document its development. Rather we want to show that the M-estimator approach is simple, powerful, and widely applicable. We especially want students to feel comfortable finding and using the asymptotic approximations that flow from the method.

One key advantage of the approach is that a very large class of asymptotically normal statistics including delta method transformations can be put in the general M-estimator framework. This unifies large sample approximation methods, simplifies analysis, and makes computations routine although sometimes tedious. Fortunately, the tedious derivative and matrix calculations often can be performed symbolically with programs such as Maple and Mathematica.

Many estimators not typically thought of as M-estimators can be written in the form of M-estimators. Consider as a simple example the mean deviation from the sample mean

$$\widehat{\theta}_1 = \frac{1}{n} \sum_{i=1}^{n} | Y_i - \overline{Y} |.$$

Is this an M-estimator? There is certainly no single equation of the form

$$\sum_{i=1}^{n} \psi(Y_i, \theta) = 0$$

that yields $\widehat{\theta}_1$. Moreover, there is no family of densities $f(y; \boldsymbol{\theta})$ such that $\widehat{\theta}_1$ is a component of the maximum likelihood estimator of $\boldsymbol{\theta}$. But if we let $\psi_1(y, \theta_1, \theta_2) = |\, y - \theta_2 \,| - \theta_1$ and $\psi_2(y, \theta_1, \theta_2) = y - \theta_2$, then

$$\sum_{i=1}^{n} \boldsymbol{\psi}(Y_i, \widehat{\theta}_1, \widehat{\theta}_2) = \begin{pmatrix} \sum_{i=1}^{n} \left(|\, Y_i - \widehat{\theta}_2 \,| - \widehat{\theta}_1 \right) \\ \sum_{i=1}^{n} \left(Y_i - \widehat{\theta}_2 \right) \end{pmatrix} = \begin{pmatrix} 0 \\ 0 \end{pmatrix}$$

yields $\widehat{\theta}_2 = \overline{Y}$ and $\widehat{\theta}_1 = (1/n) \sum_{i=1}^{n} |Y_i - \overline{Y}|$. We like to use the term "partial M-estimator" for an estimator that is not naturally an M-estimator until additional ψ functions are added. The key idea is simple: any estimator that would be an M-estimator if certain parameters were known, is a partial M-estimator because we can "stack" ψ functions for each of the unknown parameters. This aspect of M-estimators is related to the general approach of Randles (1982) for replacing unknown parameters by estimators.

From the above example it should be obvious that we can replace $\widehat{\theta}_2 = \overline{Y}$ by any other estimator defined by an estimating equation; for example, the sample median. Moreover, we can also add ψ functions to give delta method asymptotic results for transformations of parameters, for example, $\widehat{\theta}_3 = \log(\widehat{\theta}_1)$; see Examples 7.2.3 (p. 304) and 7.2.4 (p. 305) and also Benichou and Gail (1989).

The combination of "approximation by averages" and "delta theorem" methodology from Chapter 5 can handle a larger class of problems than the enhanced M-estimation approach described in this chapter. However, enhanced M-estimator methods, implemented with the aid of symbolic mathematics software (for deriving analytic expressions) and standard numerical routines for derivatives and matrix algebra (for obtaining numerical estimates) provide a unified approach that is simple in implementation, easily taught, and applicable to a broad class of complex problems.

A description of the basic approach is given in Section 7.2 along with a few examples. Connections to the influence curve are given in Section 7.3 and then extensions for nonsmooth ψ functions are given in Section 7.4. Extensions for regression are given in Section 7.5. A discussion of a testing problem is given in Section 7.6, and Section 7.7 summarizes the key features of the M-estimator method. The Appendix gives theorems for the consistency and asymptotic normality of $\widehat{\boldsymbol{\theta}}$ as well as Weak Laws of Large Numbers for averages of summands with estimated parameters.

7.2 The Basic Approach

M-estimators solve (7.1, p. 298), where the vector function $\boldsymbol{\psi}$ must be a known function that does not depend on i or n. For regression situations, the argument of $\boldsymbol{\psi}$ is expanded to depend on regressors $\boldsymbol{x}_i$, but the basic $\boldsymbol{\psi}$ still does not depend on i. For the moment we confine ourselves to the iid case where $Y_1, \ldots, Y_n$ are iid (possibly vector-valued) with distribution function F. The true parameter value $\boldsymbol{\theta}_0$ is defined by

$$\mathrm{E}_F \boldsymbol{\psi}(Y_1, \boldsymbol{\theta}_0) = \int \boldsymbol{\psi}(y, \boldsymbol{\theta}_0)\, dF(y) = \mathbf{0}. \tag{7.3}$$

For example, if $\psi(Y_i, \theta) = Y_i - \theta$, then clearly the population mean $\theta_0 = \int y\, dF(y)$ is the unique solution of $\int (y - \theta)\, dF(y) = 0$.

If there is one unique $\boldsymbol{\theta}_0$ satisfying (7.3), then in general there exists a sequence of M-estimators $\widehat{\boldsymbol{\theta}}$ such that the Weak Law of Large Numbers leads to $\widehat{\boldsymbol{\theta}} \xrightarrow{p} \boldsymbol{\theta}_0$ as $n \to \infty$. These type results are similar to the consistency results discussed in Chapter 6. Theorem 7.1 (p. 327) in this chapter gives one such result for compact parameter spaces. Furthermore, if $\boldsymbol{\psi}$ is suitably smooth, then Taylor expansion of $\boldsymbol{G}_n(\boldsymbol{\theta}) = n^{-1} \sum_{i=1}^n \boldsymbol{\psi}(Y_i, \boldsymbol{\theta})$ gives

$$\mathbf{0} = \boldsymbol{G}_n(\widehat{\boldsymbol{\theta}}) = \boldsymbol{G}_n(\boldsymbol{\theta}_0) + \boldsymbol{G}_n'(\boldsymbol{\theta}_0)(\widehat{\boldsymbol{\theta}} - \boldsymbol{\theta}_0) + \boldsymbol{R}_n,$$

where $\boldsymbol{G}_n'(\boldsymbol{\theta}) = \partial \boldsymbol{G}_n(\boldsymbol{\theta})/\partial \boldsymbol{\theta}$. For n sufficiently large, we expect $\boldsymbol{G}_n'(\boldsymbol{\theta}_0)$ to be nonsingular so that upon rearrangement

$$\sqrt{n}(\widehat{\boldsymbol{\theta}} - \boldsymbol{\theta}_0) = \left\{-\boldsymbol{G}_n'(\boldsymbol{\theta}_0)\right\}^{-1} \sqrt{n}\boldsymbol{G}_n(\boldsymbol{\theta}_0) + \sqrt{n}\boldsymbol{R}_n^*. \tag{7.4}$$

Define $\boldsymbol{\psi}'(y, \boldsymbol{\theta}) = \partial \boldsymbol{\psi}(y, \boldsymbol{\theta})/\partial \boldsymbol{\theta}$ and

$$\boldsymbol{A}(\boldsymbol{\theta}_0) = \mathrm{E}_F \left\{-\boldsymbol{\psi}'(Y_1, \boldsymbol{\theta}_0)\right\}, \tag{7.5}$$

$$\boldsymbol{B}(\boldsymbol{\theta}_0) = \mathrm{E}_F \left\{\boldsymbol{\psi}(Y_1, \boldsymbol{\theta}_0)\boldsymbol{\psi}(Y_1, \boldsymbol{\theta}_0)^T\right\}. \tag{7.6}$$

Under suitable regularity conditions as $n \to \infty$,

$$-\boldsymbol{G}_n'(\boldsymbol{\theta}_0) = \frac{1}{n}\sum_{i=1}^n \left\{-\boldsymbol{\psi}'(Y_i, \boldsymbol{\theta}_0)\right\} \xrightarrow{p} \boldsymbol{A}(\boldsymbol{\theta}_0), \tag{7.7}$$

$$\sqrt{n}\boldsymbol{G}_n(\boldsymbol{\theta}_0) \xrightarrow{d} N\left\{\mathbf{0}, \boldsymbol{B}(\boldsymbol{\theta}_0)\right\}, \tag{7.8}$$

and

$$\sqrt{n}\boldsymbol{R}_n^* \xrightarrow{p} \mathbf{0}. \tag{7.9}$$

Putting (7.1) and (7.4)–(7.9) together with Slutsky's Theorem, we have that

$$\widehat{\boldsymbol{\theta}} \text{ is AN}\left(\boldsymbol{\theta}_0, \frac{\boldsymbol{V}(\boldsymbol{\theta}_0)}{n}\right) \quad \text{as } n \to \infty, \tag{7.10}$$

where $\boldsymbol{V}(\boldsymbol{\theta}_0) = \boldsymbol{A}(\boldsymbol{\theta}_0)^{-1}\boldsymbol{B}(\boldsymbol{\theta}_0)\{\boldsymbol{A}(\boldsymbol{\theta}_0)^{-1}\}^T$. The limiting covariance $V(\boldsymbol{\theta}_0)$ is called the sandwich matrix because the "meat" $\boldsymbol{B}(\boldsymbol{\theta}_0)$ is placed between the "bread" $\boldsymbol{A}(\boldsymbol{\theta}_0)^{-1}$ and $\{\boldsymbol{A}(\boldsymbol{\theta}_0)^{-1}\}^T$.

If $\boldsymbol{A}(\boldsymbol{\theta}_0)$ exists, the Weak Law of Large Numbers gives (7.7). If $\boldsymbol{B}(\boldsymbol{\theta}_0)$ exists, then (7.8) follows from the Central Limit Theorem. The hard part to prove is (7.9). Huber (1967) was the first to give general results for (7.9), but there have been many others since then (see e.g., Serfling 1980 Ch.7). Theorem 7.2 (p. 328) in the Appendix to this chapter gives conditions for (7.10), and a by-product of its proof is verification of (7.9).

Extension. Suppose that instead of (7.1, p. 298), $\widehat{\boldsymbol{\theta}}$ satisfies

$$\sum_{i=1}^{n} \boldsymbol{\psi}(Y_i, \widehat{\boldsymbol{\theta}}) = \boldsymbol{c}_n, \tag{7.11}$$

where $\boldsymbol{c}_n/\sqrt{n} \xrightarrow{p} \boldsymbol{0}$ as $n \to \infty$. Following the above arguments and noting that $\boldsymbol{c}_n/\sqrt{n}$ is absorbed in $\sqrt{n}\boldsymbol{R}_n^*$ of (7.4), we can see that as long as (7.11), (7.4), and (7.7)–(7.9) hold, then (7.10) also holds. This extension allows us to cover a much wider class of statistics including empirical quantiles, estimators whose $\boldsymbol{\psi}$ function depends on n, and Bayesian estimators.

7.2.1 *Estimators for A, B, and V*

For maximum likelihood estimation, $\boldsymbol{\psi}(y, \boldsymbol{\theta}) = \partial \log f(y; \boldsymbol{\theta})/\partial \boldsymbol{\theta}^T$ is often called the score function. If the data truly come from the assumed parametric family $f(y; \boldsymbol{\theta})$, then $\boldsymbol{A}(\boldsymbol{\theta}_0) = \boldsymbol{B}(\boldsymbol{\theta}_0) = I(\boldsymbol{\theta}_0)$, the information matrix. Note that $\boldsymbol{A}(\boldsymbol{\theta}_0)$ is Definition 2 of $I(\boldsymbol{\theta}_0)$ in (2.33, p. 66), and $\boldsymbol{B}(\boldsymbol{\theta}_0)$ is Definition 1 of $I(\boldsymbol{\theta}_0)$ in (2.29, p. 64). In this case the sandwich matrix $V(\boldsymbol{\theta}_0)$ reduces to the usual $I(\boldsymbol{\theta}_0)^{-1}$. One of the key contributions of M-estimation theory has been to point out what happens when the assumed parametric family is not correct. In such cases there is often a well-defined $\boldsymbol{\theta}_0$ satisfying (7.3, p. 300) and $\widehat{\boldsymbol{\theta}}$ satisfying (7.1, p. 298) but $\boldsymbol{A}(\boldsymbol{\theta}_0) \neq \boldsymbol{B}(\boldsymbol{\theta}_0)$, and valid inference should be carried out using the correct limiting covariance matrix $V(\boldsymbol{\theta}_0) = \boldsymbol{A}(\boldsymbol{\theta}_0)^{-1}\boldsymbol{B}(\boldsymbol{\theta}_0)\{\boldsymbol{A}(\boldsymbol{\theta}_0)^{-1}\}^T$, not $I(\boldsymbol{\theta}_0)^{-1}$.

Using the left-hand side of (7.7, p. 300), we define the empirical estimator of $\boldsymbol{A}(\boldsymbol{\theta}_0)$ by

$$\boldsymbol{A}_n(\boldsymbol{Y}, \widehat{\boldsymbol{\theta}}) = -\boldsymbol{G}_n'(\widehat{\boldsymbol{\theta}}) = \frac{1}{n}\sum_{i=1}^{n}\left\{-\boldsymbol{\psi}'(Y_i, \widehat{\boldsymbol{\theta}})\right\}.$$

Note that for maximum likelihood estimation, $A_n(Y, \widehat{\theta})$ is the average observed information matrix $\overline{I}(Y, \widehat{\theta})$ (see 2.34, p. 66). Similarly, the empirical estimator of $B(\theta_0)$ is

$$B_n(Y, \widehat{\theta}) = \frac{1}{n} \sum_{i=1}^{n} \psi(Y_i, \widehat{\theta}) \psi(Y_i, \widehat{\theta})^T.$$

The sandwich matrix of these matrix estimators yields the empirical sandwich variance estimator

$$V_n(Y, \widehat{\theta}) = A_n(Y, \widehat{\theta})^{-1} B_n(Y, \widehat{\theta}) \{A_n(Y, \widehat{\theta})^{-1}\}^T. \tag{7.12}$$

$V_n(Y, \widehat{\theta})$ is generally consistent for $V(\theta_0)$ under mild regularity conditions (see Theorem 7.3, p. 329, and Theorem 7.4, p. 330, in the Appendix to this chapter).

Calculation of $V_n(Y, \widehat{\theta})$ requires no analytic work beyond specifying ψ. In some problems, it is simpler to work directly with the limiting form $V(\theta_0) = A(\theta_0)^{-1} B(\theta_0) \{A(\theta_0)^{-1}\}^T$, plugging in estimators for θ_0 and any other unknown quantities in $V(\theta_0)$. The notation $V(\theta_0)$ suggests that θ_0 is the only unknown quantity in $V(\theta_0)$, but in reality $V(\theta_0)$ often involves higher moments or other characteristics of the distribution function F of Y_i. In fact there is a range of possibilities for estimating $V(\theta_0)$ depending on what model assumptions are used. For simplicity, we use the notation $V_n(Y, \widehat{\theta})$ for the purely empirical estimator and $V(\widehat{\theta})$ for any of the versions based on expected value plus model assumptions.

For maximum likelihood estimation with a correctly specified family, the three competing estimators for $I(\theta)^{-1}$ are $V_n(Y, \widehat{\theta})$, $\overline{I}(Y, \widehat{\theta})^{-1} = A_n(Y, \widehat{\theta})^{-1}$, and $I(\widehat{\theta})^{-1} = V(\widehat{\theta})$. In this case the standard estimators $\overline{I}(Y, \widehat{\theta})^{-1}$ and $I(\widehat{\theta})^{-1}$ are generally more efficient than $V_n(Y, \widehat{\theta})$ for estimating $I(\theta)^{-1}$. Clearly, for maximum likelihood estimation with a correctly specified family, no estimator can have smaller asymptotic variance for estimating $I(\theta)^{-1}$ than $I(\widehat{\theta}_{\text{MLE}})^{-1}$.

Now we illustrate these ideas with examples.

7.2.2 *Sample Mean and Variance*

Let $\widehat{\theta} = (\overline{Y}, s_n^2)^T$, the sample mean and variance. Here

$$\psi(Y_i, \theta) = \begin{pmatrix} Y_i - \theta_1 \\ \\ (Y_i - \theta_1)^2 - \theta_2 \end{pmatrix}.$$

The first component, $\widehat{\theta}_1 = \overline{Y}$, satisfies $\sum(Y_i - \widehat{\theta}_1) = 0$, and is by itself an M-estimator. The second component $\widehat{\theta}_2 = s_n^2 = n^{-1} \sum(Y_i - \overline{Y})^2$, when considered by itself, is not an M-estimator. However, when combined with $\widehat{\theta}_1$, the pair $(\widehat{\theta}_1, \widehat{\theta}_2)^T$ is a 2×1 M-estimator so that $\widehat{\theta}_2$ satisfies our definition of a partial M-estimator.

Now let us calculate $\boldsymbol{A}(\boldsymbol{\theta}_0)$ and $\boldsymbol{B}(\boldsymbol{\theta}_0)$ where $\boldsymbol{\theta}_0^T = (\theta_{10}, \theta_{20})$:

$$\boldsymbol{A}(\boldsymbol{\theta}_0) = \mathrm{E}\left\{-\boldsymbol{\psi}'(Y_1, \boldsymbol{\theta}_0)\right\} = \mathrm{E}\begin{pmatrix} 1 & 0 \\ 2(Y_1 - \theta_{10}) & 1 \end{pmatrix} = \begin{pmatrix} 1 & 0 \\ 0 & 1 \end{pmatrix} \tag{7.13}$$

$$\boldsymbol{B}(\boldsymbol{\theta}_0) = \mathrm{E}\left\{\boldsymbol{\psi}(Y_1, \boldsymbol{\theta}_0)\boldsymbol{\psi}(Y_1, \boldsymbol{\theta}_0)^T\right\}$$

$$= \mathrm{E}\begin{pmatrix} (Y_1 - \theta_{10})^2 & (Y_1 - \theta_{10})\left[(Y_1 - \theta_{10})^2 - \theta_{20}\right] \\ (Y_1 - \theta_{10})\left[(Y_1 - \theta_{10})^2 - \theta_{20}\right] & \left[(Y_1 - \theta_{10})^2 - \theta_{20}\right]^2 \end{pmatrix}$$

$$= \begin{pmatrix} \theta_{20} & \mu_3 \\ \mu_3 & \mu_4 - \theta_{20}^2 \end{pmatrix} = \begin{pmatrix} \sigma^2 & \mu_3 \\ \mu_3 & \mu_4 - \sigma^4 \end{pmatrix}, \tag{7.14}$$

where μ_k is our notation for the kth central moment of Y_1 and the more familiar notation $\sigma^2 = \theta_{20}$ has been substituted at the end. In this case, since $\boldsymbol{A}(\boldsymbol{\theta}_0)$ is the identity matrix, $V(\boldsymbol{\theta}_0) = \boldsymbol{B}(\boldsymbol{\theta}_0)$. To estimate $\boldsymbol{B}(\boldsymbol{\theta}_0)$, we may use

$$\boldsymbol{B}_n(\boldsymbol{Y}, \widehat{\boldsymbol{\theta}}) = \frac{1}{n}\sum_{i=1}^{n}\begin{pmatrix} (Y_i - \overline{Y})^2 & (Y_i - \overline{Y})\left[(Y_i - \overline{Y})^2 - s_n^2\right] \\ (Y_i - \overline{Y})\left[(Y_i - \overline{Y})^2 - s_n^2\right] & \left[(Y_i - \overline{Y})^2 - s_n^2\right]^2 \end{pmatrix}$$

$$= \begin{pmatrix} s_n^2 & m_3 \\ m_3 & m_4 - s_n^4 \end{pmatrix},$$

where the m_k are sample kth moments. Looking back at the form for $\boldsymbol{V}(\boldsymbol{\theta}_0)$ and plugging in empirical moment estimators leads to equality of the empirical estimator and the expected value estimator: $\boldsymbol{V}(\widehat{\boldsymbol{\theta}}) = \boldsymbol{V}_n(\boldsymbol{Y}, \widehat{\boldsymbol{\theta}})$ in this case.

Note that $\widehat{\boldsymbol{\theta}}$ is a maximum likelihood estimator for the normal model density $f(y; \boldsymbol{\theta}) = (2\pi\theta_2)^{-1/2}\exp\left\{-(y - \theta_1)^2/2\theta_2\right\}$, but $\psi_1 = Y_i - \theta_1$ and $\psi_2 = (Y_i - \theta_1)^2 - \theta_2$ are not the score functions that come from this normal density. The partial derivative of this normal log density yields $\psi_1 = (Y_i - \theta_1)/\theta_2$ and $\psi_2 = (Y_i - \theta_1)^2/(2\theta_2^2) - 1/(2\theta_2)$. Thus $\boldsymbol{\psi}$ functions are not unique—many different ones can lead to the same estimator. However, different $\boldsymbol{\psi}$ functions associated with the same estimator yield different $\boldsymbol{A}$ and $\boldsymbol{B}$ but the same $\boldsymbol{V}$. For example, using these latter two ψ functions, the resulting $\boldsymbol{A}$ and $\boldsymbol{B}$ matrices are

$$\boldsymbol{A}(\boldsymbol{\theta}_0) = \begin{pmatrix} \dfrac{1}{\sigma^2} & 0 \\ 0 & \dfrac{1}{2\sigma^4} \end{pmatrix} \qquad \boldsymbol{B}(\boldsymbol{\theta}_0) = \begin{pmatrix} \dfrac{1}{\sigma^2} & \dfrac{\mu_3}{2\sigma^6} \\ \dfrac{\mu_3}{2\sigma^6} & \dfrac{\mu_4 - \sigma^4}{4\sigma^8} \end{pmatrix}.$$

Note that the sandwich matrix of these matrices is the same as the sandwich matrix of the matrices in (7.13) and (7.14). If we further assume that the data truly are normally distributed, then $\mu_3 = 0$ and $\mu_4 = 3\sigma^4$ resulting in $\boldsymbol{A}(\boldsymbol{\theta}_0) = \boldsymbol{B}(\boldsymbol{\theta}_0) = \boldsymbol{I}(\boldsymbol{\theta}_0) = \text{Diag}(1/\sigma^2, 1/2\sigma^4)$. Here the expected value model-based covariance estimator would be $\boldsymbol{V}(\widehat{\boldsymbol{\theta}}) = \text{Diag}(s_n^2, 2s_n^4)$.

Note that the likelihood score $\boldsymbol{\psi}$ functions, $\widehat{\boldsymbol{\psi}}_{\text{MLE}}$, are related to the original $\boldsymbol{\psi}$ functions by $\boldsymbol{\psi}_{\text{MLE}} = \boldsymbol{c}\boldsymbol{\psi}$, where $\boldsymbol{c} = \text{diag}(1/\theta_{20}, 1/2\theta_{20}^2)$. A little algebra shows that all $\boldsymbol{\psi}$ of the form $\boldsymbol{c}\boldsymbol{\psi}$, where $\boldsymbol{c}$ is nonsingular (but possibly depending on $\boldsymbol{\theta}_0$), lead to an equivalence class having the same estimator and asymptotic covariance matrix $\boldsymbol{V}(\boldsymbol{\theta}_0)$.

7.2.3 *Ratio Estimator*

Let $\widehat{\theta} = \overline{Y}/\overline{X}$, where $(Y_1, X_1), \ldots, (Y_n, X_n)$ is an iid sample of pairs with means $\text{E}Y_1 = \mu_Y$ and $\text{E}X_1 = \mu_X \neq 0$, variances $\text{Var}(Y_1) = \sigma_Y^2$ and $\text{Var}(X_1) = \sigma_X^2$, and covariance $\text{cov}(Y_1, X_1) = \sigma_{YX}$. A ψ function for $\widehat{\theta} = \overline{Y}/\overline{X}$ is $\psi(Y_i, X_i, \theta) = Y_i - \theta X_i$ leading to $\boldsymbol{A}(\boldsymbol{\theta}_0) = \mu_X$, $\boldsymbol{B}(\boldsymbol{\theta}_0) = \text{E}(Y_1 - \theta_0 X_1)^2$, $\boldsymbol{V}(\boldsymbol{\theta}_0) = \text{E}(Y_1 - \theta_0 X_1)^2/\mu_X^2$, $\boldsymbol{A}_n(\boldsymbol{Y}, \widehat{\theta}) = \overline{X}$, and

$$\boldsymbol{B}_n(\boldsymbol{Y}, \widehat{\theta}) = \frac{1}{n}\sum_{i=1}^{n}\left(Y_i - \frac{\overline{Y}}{\overline{X}}X_i\right)^2,$$

and

$$\boldsymbol{V}_n(\boldsymbol{Y}, \widehat{\theta}) = \frac{1}{\overline{X}^2}\frac{1}{n}\sum_{i=1}^{n}\left(Y_i - \frac{\overline{Y}}{\overline{X}}X_i\right)^2.$$

This variance estimator is often encountered in finite population sampling contexts.

Now consider the following $\boldsymbol{\psi}$ of dimension 3 that yields $\widehat{\theta}_3 = \overline{Y}/\overline{X}$ as the third component of $\widehat{\boldsymbol{\theta}}$:

$$\boldsymbol{\psi}(Y_i, X_i, \boldsymbol{\theta}) = \begin{pmatrix} Y_i - \theta_1 \\ X_i - \theta_2 \\ \theta_1 - \theta_3\theta_2 \end{pmatrix}.$$

This $\boldsymbol{\psi}$ function is interesting because the third component does not depend on the data. Nevertheless, this $\boldsymbol{\psi}$ satisfies all the requirements and illustrates how to implement the delta method via M-estimation. The A and B matrices are

$$\boldsymbol{A}(\boldsymbol{\theta}_0) = \begin{pmatrix} 1 & 0 & 0 \\ 0 & 1 & 0 \\ -1 & \theta_{30} & \theta_{20} \end{pmatrix} \quad \boldsymbol{B}(\boldsymbol{\theta}_0) = \begin{pmatrix} \sigma_Y^2 & \sigma_{YX} & 0 \\ \sigma_{YX} & \sigma_X^2 & 0 \\ 0 & 0 & 0 \end{pmatrix}.$$

This example illustrates the fact that $\boldsymbol{B}(\boldsymbol{\theta}_0)$ can be singular (although $\boldsymbol{A}(\boldsymbol{\theta}_0)$ generally cannot). In fact whenever a $\boldsymbol{\psi}$ function has components that involve no data, then the resulting B matrix is singular. Using Maple we computed $V(\boldsymbol{\theta}_0) = \boldsymbol{A}(\boldsymbol{\theta}_0)^{-1}\boldsymbol{B}(\boldsymbol{\theta}_0)\{\boldsymbol{A}(\boldsymbol{\theta}_0)^{-1}\}^T$, and obtained for the (3,3) element

$$v_{33} = \frac{1}{\theta_{20}^2}\left[\sigma_Y^2 - 2\theta_{30}\sigma_{YX} + \theta_{30}^2\sigma_X^2\right].$$

This latter expression for the asymptotic variance of $\sqrt{n}\widehat{\theta}_3$ can be shown to be the same as $\mathrm{E}(Y_1 - \theta_{30}X_1)^2/\mu_X^2$ obtained earlier upon noting that $\theta_{20} = \mu_X$.

Sample Maple Program

```
with(linalg):                                Brings in the
                                               linear
                                             algebra package
vA:=[1,0,0,0,1,0,-1,theta[3],theta[2]]; Makes a vector of
                                             the entries of A
A:=matrix(3,3,vA);                           Creates A from vA
Ainv:=inverse(A);
vB:=[sigma[y]^2,sigma[xy],0,sigma[xy],sigma[x]^2,0,0,0,0];
B:=matrix(3,3,vB);
V:=multiply(Ainv,B,transpose(Ainv));
simplify(V[3,3]);
```

$$\frac{{\sigma_y}^2 - 2\,\theta_3\,\sigma_{xy} + {\theta_3}^2\,{\sigma_x}^2}{{\theta_2}^2}$$

The above display is what appears on the Maple window for the last command.

7.2.4 Delta Method Via M-Estimation

In the context of Section 7.2.2 (p. 302), suppose we are interested in $s_n = \sqrt{s_n^2}$ and $\log(s_n^2)$. We could of course just redefine θ_2 in Example 7.2.2 to be θ_2^2 and $\exp(\theta_2)$, respectively. Instead, we prefer to add $\psi_3(Y_i, \boldsymbol{\theta}) = \sqrt{\theta_2} - \theta_3$ and $\psi_4(Y_i, \boldsymbol{\theta}) = \log(\theta_2) - \theta_4$ because it seems conceptually simpler and it also gives the joint asymptotic distribution of all quantities. Now we have

$$\boldsymbol{A}(\boldsymbol{\theta}_0) = \begin{pmatrix} 1 & 0 & 0 & 0 \\ 0 & 1 & 0 & 0 \\ 0 & -\dfrac{1}{2\sqrt{\theta_{20}}} & 1 & 0 \\ 0 & -\dfrac{1}{\theta_{20}} & 0 & 1 \end{pmatrix} \qquad \boldsymbol{B}(\boldsymbol{\theta}_0) = \begin{pmatrix} \theta_{20} & \mu_3 & 0 & 0 \\ \mu_3 & \mu_4 - \theta_{20}^2 & 0 & 0 \\ 0 & 0 & 0 & 0 \\ 0 & 0 & 0 & 0 \end{pmatrix}$$

and $\boldsymbol{V}(\boldsymbol{\theta}_0) = \boldsymbol{A}(\boldsymbol{\theta}_0)^{-1}\boldsymbol{B}(\boldsymbol{\theta}_0)\{\boldsymbol{A}(\boldsymbol{\theta}_0)^{-1}\}^T$ is

$$\boldsymbol{V}(\boldsymbol{\theta}_0) = \begin{pmatrix} \theta_{20} & \mu_3 & \dfrac{\mu_3}{2\sqrt{\theta_{20}}} & \dfrac{\mu_3}{\theta_{20}} \\ \mu_3 & \mu_4 - \theta_{20}^2 & \dfrac{\mu_4 - \theta_{20}^2}{2\sqrt{\theta_{20}}} & \dfrac{\mu_4 - \theta_{20}^2}{\theta_{20}} \\ \dfrac{\mu_3}{2\sqrt{\theta_{20}}} & \dfrac{\mu_4 - \theta_{20}^2}{2\sqrt{\theta_{20}}} & \dfrac{\mu_4 - \theta_{20}^2}{4\theta_{20}} & \dfrac{\mu_4 - \theta_{20}^2}{2\theta_{20}^{3/2}} \\ \dfrac{\mu_3}{\theta_{20}} & \dfrac{\mu_4 - \theta_{20}^2}{\theta_{20}} & \dfrac{\mu_4 - \theta_{20}^2}{2\theta_{20}^{3/2}} & \dfrac{\mu_4 - \theta_{20}^2}{\theta_{20}^2} \end{pmatrix}.$$

Thus the asymptotic variance of s_n is $(\mu_4 - \theta_{20}^2)/(4\theta_{20}) = (\mu_4 - \sigma^4)/4\sigma^2$, and the asymptotic variance of $\log(s_n^2)$ is $(\mu_4 - \theta_{20}^2)/\theta_{20}^2 = \mu_4/\sigma^4 - 1$. We might point out that the rank of a product of matrices is less than or equal to the minimum rank of the matrices in the product. For this example, the rank of the above $\boldsymbol{V}(\boldsymbol{\theta}_0)$ matrix is 2, the same as the rank of $\boldsymbol{B}$. In fact the correlations in the bottom 3×3 portion of $\boldsymbol{V}(\theta_0)$ are all 1, which makes intuitive sense because $\widehat{\theta}_2$, $\widehat{\theta}_3$, and $\widehat{\theta}_4$ are all monotonic functions of one another.

The results above can be formalized into an M-estimation version of the delta method, actually Theorem 5.20 (p. 239) with $\boldsymbol{g}$ vector-valued. In problem 7.6 (p. 332), we give the M-estimation version of Theorem 5.19 (p. 238), where g is real-valued.

7.2.5 *Posterior Mode*

Consider the standard Bayesian model in an iid framework where the posterior density is proportional to

$$\pi(\boldsymbol{\theta}) \prod_{i=1}^{n} f(Y_i; \boldsymbol{\theta}),$$

and π is the prior density. Posterior *mode* estimators satisfy (7.11, p. 301) with $\boldsymbol{\psi}(y, \boldsymbol{\theta}) = \partial \log f(y; \boldsymbol{\theta})/\partial \boldsymbol{\theta}^T$ the same as for maximum likelihood and $\boldsymbol{C}_n = -\pi'(\widehat{\boldsymbol{\theta}})/\pi(\widehat{\boldsymbol{\theta}})$. Thus, as long as $\boldsymbol{c}_n/\sqrt{n} \xrightarrow{p} \boldsymbol{0}$, the Bayesian mode estimator has the same asymptotic covariance matrix as maximum likelihood estimators.

7.2.6 *Instrumental Variable Estimation*

Instrumental variable estimation is a method for estimating regression parameters when predictor variables are measured with error (Fuller 1987; Carroll et al. 2006).

We use a simple instrumental variable model to illustrate some features of the M-estimation approach. Suppose that triples (Y_i, W_i, T_i) are observed such that

$$\begin{aligned} Y_i &= \alpha + \beta X_i + \sigma_e e_{1,i} \\ W_i &= X_i + \sigma_U e_{2,i} \\ T_i &= \gamma + \delta X_i + \sigma_\tau e_{3,i} \end{aligned}$$

where $e_{j,i}$ are mutually independent random errors with common mean 0 and variance 1. For simplicity also assume that $X_1, \ldots, X_n$ are iid, independent of the $\{e_{j,i}\}$ and have finite variance. In the language of measurement error models, W_i is a measurement of X_i, and T_i is an instrumental variable for X_i (for estimating β), provided that $\delta \neq 0$ which we now assume. Note that $X_1, \ldots, X_n$ are latent variables and not observed. Let σ_S^2 and $\sigma_{S,T}$ denote variances and covariances of any random variables S and T.

The least squares estimator of slope obtained by regressing Y on W, $\widehat{\beta}_{Y|W}$, converges in probability to $\{\sigma_X^2/(\sigma_X^2 + \sigma_U^2)\}\,\beta$, and thus is not consistent for β when the measurement error variance $\sigma_U^2 > 0$. However, the instrumental variable estimator,

$$\widehat{\beta}_{IV} = \frac{\widehat{\beta}_{Y|T}}{\widehat{\beta}_{W|T}},$$

where $\widehat{\beta}_{Y|T}$ and $\widehat{\beta}_{W|T}$ are the slopes from the least squares regressions of Y on T and W on T, respectively, is a consistent estimator of β under the stated assumptions regardless of σ_U^2.

The instrumental variable estimator, $\widehat{\beta}_{\text{IV}}$ is a partial M-estimator as defined in the Introduction, and there are a number of ways to complete the ψ function in this case. Provided interest lies only in estimation of the β, a simple choice is

$$\psi(Y, W, T, \theta) = \begin{pmatrix} \theta_1 - T \\ (Y - \theta_2 W)(\theta_1 - T) \end{pmatrix},$$

with associated M-estimator,

$$\widehat{\theta}_1 = \overline{T}, \qquad \widehat{\theta}_2 = \widehat{\beta}_{\text{IV}}.$$

The $\boldsymbol{A}$ and $\boldsymbol{B}$ matrices calculated at the true parameters assuming the instrumental variable model are

$$\boldsymbol{A} = \begin{pmatrix} 1 & 0 \\ \alpha & \sigma_{X,T} \end{pmatrix} \quad \text{and} \quad \boldsymbol{B} = \begin{pmatrix} \sigma_T^2 & \alpha\sigma_T^2 \\ \alpha\sigma_T^2 & \sigma_T^2(\alpha^2 + \sigma_e^2 + \beta^2\sigma_U^2) \end{pmatrix},$$

which yield the asymptotic covariance matrix

$$\boldsymbol{A}^{-1}\boldsymbol{B}\left(\boldsymbol{A}^{-1}\right)^{T} = \begin{pmatrix} \sigma_T^2 & 0 \\ 0 & \sigma_T^2(\sigma_e^2 + \beta^2\sigma_U^2)/\sigma_{X,T}^2 \end{pmatrix}.$$

Under the stated assumptions the instrumental variable estimator and the naive estimator are both consistent for β when $\sigma_U^2 = 0$, yet have different asymptotic means when $\sigma_U^2 > 0$. Thus for certain applications their joint asymptotic distribution is of interest, e.g., for inference about the difference $\widehat{\beta}_{\mathrm{IV}} - \widehat{\beta}_{Y|W}$. The M-estimator approach easily accommodates such calculations. For this task consider the $\boldsymbol{\psi}$ function

$$\boldsymbol{\psi}(Y, W, T, \theta) = \begin{pmatrix} \theta_1 - T \\ \theta_2 - W \\ (Y - \theta_3 W)(\theta_2 - W) \\ (Y - \theta_4 W)(\theta_1 - T) \end{pmatrix}.$$

Note the change in the definitions of θ_2 and the ordering of the components of this $\boldsymbol{\psi}$ function. The configuration is primarily for convenience as it leads to a triangular $\boldsymbol{A}$ matrix. In general when the kth component of $\boldsymbol{\psi}$ depends only on $\theta_1, \ldots, \theta_k$, $k = 1, 2, \ldots$, the partial derivative matrix $\partial\boldsymbol{\psi}/\partial\boldsymbol{\theta}^T$ is lower triangular and so too is the $\boldsymbol{A}$ matrix.

The M-estimator associated with this $\boldsymbol{\psi}$ function is

$$\widehat{\theta}_1 = \overline{T}, \quad \widehat{\theta}_2 = \overline{W}, \quad \widehat{\theta}_3 = \widehat{\beta}_{Y|W}, \quad \widehat{\theta}_4 = \widehat{\beta}_{\mathrm{IV}}.$$

The $\boldsymbol{A}$ matrix calculated at the true parameters assuming the instrumental variable model is

$$\begin{pmatrix} 1 & 0 & 0 & 0 \\ 0 & 1 & 0 & 0 \\ 0 & \alpha + \beta\mu_X\sigma_U^2/\sigma_W^2 & \sigma_W^2 & 0 \\ \alpha & 0 & 0 & \sigma_{X,T} \end{pmatrix}.$$

The expression for the $\boldsymbol{B}$ matrix is unwieldy. However, primary interest lies in the lower 2×2 submatrix of the asymptotic variance matrix $\boldsymbol{A}^{-1}\boldsymbol{B}\left(\boldsymbol{A}^{-1}\right)^{T}$. We used Maple to calculate this submatrix and to substitute expressions for the various mixed moments of (Y, W, T) under the assumption of joint normality, resulting in the asymptotic covariance matrix for $(\widehat{\theta}_3, \widehat{\theta}_4)$,

$$\begin{pmatrix} (\sigma_e^2\sigma_W^2 + \beta^2\sigma_U^2\sigma_X^2)/\sigma_W^4 & \left\{\sigma_e^2\sigma_W^2 + \beta^2(\sigma_U^2\sigma_X^2 - \sigma_U^4)\right\}/\sigma_W^4 \\ \left\{\sigma_e^2\sigma_W^2 + \beta^2(\sigma_U^2\sigma_X^2 - \sigma_U^4)\right\}/\sigma_W^4 & \sigma_T^2(\sigma_e^2 + \beta^2\sigma_U^2)/\sigma_{X,T}^2 \end{pmatrix}.$$

The covariance formula given above assumes normality of the errors $\varepsilon_{j,i}$ and the X_i in the model. Instrumental variable estimation works more generally, and in the absence of distributional assumptions (beyond those of lack of correlation), estimated variances can be obtained using the sandwich formula. We illustrate the calculations with data from the Framingham Heart Study. For this illustration Y and W are systolic blood pressure and serum cholesterol respectively measured at the third exam, and T is serum cholesterol respectively measured at the second exam. The data include measurements on $n = 1615$ males aged 45 to 65.

The 4×1 $\boldsymbol{\psi}$ function was used to determine the estimates (standard errors in parentheses)

$$\widehat{\theta}_1 = \overline{T} = 227.2(1.1), \quad \widehat{\theta}_2 = \overline{W} = 228.4(1.0),$$
$$\widehat{\theta}_3 = \widehat{\beta}_{Y|W} = 0.042(0.011), \quad \widehat{\theta}_4 = \widehat{\beta}_{\text{IV}} = 0.065(0.015).$$

The empirical sandwich variance estimate (direct computer output) is

1785.8453	1291.8722	-1.3658812	-3.8619519
1291.8722	1718.3129	-1.1578449	-2.6815324
-1.3658812	-1.1578449	0.20737770	0.19878711
-3.8619519	-2.6815324	0.19878711	0.35584612

The estimated contrast $\widehat{\beta}_{\text{IV}} - \widehat{\beta}_{Y|W} = 0.023$ has standard error 0.010, resulting in the test statistic $t = 2.29$. The test statistic is consistent with the hypothesis that serum cholesterol is measured with non-negligible error.

7.3 Connections to the Influence Curve (Approximation by Averages)

The Influence Curve (Hampel, 1974) of an estimator $\widehat{\boldsymbol{\theta}}$ based on an iid sample may be defined as the "$\boldsymbol{h}$" function from Theorem 5.23 (p. 242) that satisfies

$$\widehat{\boldsymbol{\theta}} - \boldsymbol{\theta}_0 = \frac{1}{n}\sum_{i=1}^{n} \boldsymbol{h}(Y_i, \boldsymbol{\theta}_0) + \boldsymbol{R}_n,$$

where $\sqrt{n}\boldsymbol{R}_n \xrightarrow{p} 0$ as $n \to \infty$, or by the technical definition in (5.13, p. 244) for functional statistics. Note that here we are allowing $\boldsymbol{h}$ to be a vector and have added the dependence on $\boldsymbol{\theta}_0$. If $\mathrm{E}[\boldsymbol{h}(Y_1, \boldsymbol{\theta}_0)] = 0$ and $\mathrm{E}[\boldsymbol{h}(Y_1, \boldsymbol{\theta}_0)\boldsymbol{h}(Y_1, \boldsymbol{\theta}_0)^T] = \boldsymbol{\Sigma}$ exists, then by Slutsky's Theorem and the CLT, $\widehat{\boldsymbol{\theta}}$ is $\mathrm{AN}(\boldsymbol{\theta}, \boldsymbol{\Sigma}/n)$. It is easy to verify that $\boldsymbol{h}(y; \boldsymbol{\theta}_0) = \boldsymbol{A}(\boldsymbol{\theta}_0)^{-1}\boldsymbol{\psi}(y, \boldsymbol{\theta}_0)$ for M-estimators and thus $\boldsymbol{\Sigma}$ is

$$\begin{aligned}\mathrm{E}\left[\boldsymbol{h}(Y_1, \boldsymbol{\theta}_0)\boldsymbol{h}(Y_1, \boldsymbol{\theta}_0)^T\right] &= \mathrm{E}\left[\boldsymbol{A}(\boldsymbol{\theta}_0)^{-1}\boldsymbol{\psi}(Y_1, \boldsymbol{\theta}_0)\{\boldsymbol{\psi}(Y_1, \boldsymbol{\theta}_0)\}^T\{\boldsymbol{A}(\boldsymbol{\theta}_0)^{-1}\}^T\right] \\ &= \boldsymbol{A}(\boldsymbol{\theta}_0)^{-1}\boldsymbol{B}(\boldsymbol{\theta}_0)\{\boldsymbol{A}(\boldsymbol{\theta}_0)^{-1}\}^T = V(\boldsymbol{\theta}_0).\end{aligned}$$

The Influence Curve or "approximation by averages" approach is more general than the M-estimator approach; however, for many problems they are equivalent. Our experience teaching both methods indicates that students more readily learn the M-estimator approach and are therefore more likely to use it (and use it correctly) in their research and work. Especially in messy problems with a large number of parameters, it appears easier to stack $\boldsymbol{\psi}$ functions and compute $\boldsymbol{A}$ and $\boldsymbol{B}$ matrices than it is to compute and stack influence curves and then compute $\boldsymbol{\Sigma}$.

If the influence curve is known, then defining $\boldsymbol{\psi}(Y_i, \boldsymbol{\theta}) = \boldsymbol{h}(Y_i, \boldsymbol{\theta}_0) - (\boldsymbol{\theta} - \boldsymbol{\theta}_0)$ allows one to use the M-estimator approach even when $\widehat{\boldsymbol{\theta}}$ is not an M-estimator. In this case $\boldsymbol{A}(\boldsymbol{\theta}_0)$ is the identity matrix and $\boldsymbol{B}(\boldsymbol{\theta}_0) = \boldsymbol{\Sigma}$. (A minor modification is that for the empirical variance estimators we need to define $\boldsymbol{\psi}(Y_i, \widehat{\boldsymbol{\theta}}) = \boldsymbol{h}(Y_i, \widehat{\boldsymbol{\theta}})$; that is, plugging in $\widehat{\boldsymbol{\theta}}$ for both $\boldsymbol{\theta}$ and $\boldsymbol{\theta}_0$.) More importantly, this fact allows one to combine M-estimators with estimators that may not be M-estimators but for which we have already computed influence curves. The next example illustrates this.

Example 7.1 (Hodges-Lehmann location estimator). Hodges and Lehmann (1963)suggested that estimators could be obtained by inverting rank tests, and the class of such estimators is called R-estimators. One of the most interesting R-estimators is called the Hodges-Lehmann location estimator

$$\widehat{\theta}_{HL} = \text{median}\left\{\frac{X_i + X_j}{2}, 1 \le i \le j \le n\right\}.$$

It is not clear how to put this estimator directly in the M-estimator framework, but for distributions symmetric around θ_0, that is having $F(y) = F_0(y - \theta_0)$ for a distribution F_0 symmetric about 0, Huber (1981, p. 64) gives

$$h(y;\theta_0) = \frac{F_0(y-\theta_0) - \frac{1}{2}}{\int f_0^2(y)dy},$$

where $f_0(y)$ is the density function of $F_0(y)$. The variance of this influence curve is

$$\frac{1}{12\left[\int f_0^2(y)dy\right]^2},$$

which is easily obtained after noting that $F_0(Y_1 - \theta_0)$ has a uniform distribution.

Now for obtaining the asymptotic joint distribution of $\widehat{\theta}_{HL}$ and any set of M-estimators, we can stack $\psi(Y_i, \theta) = h(y;\theta_0) - (\theta - \theta_0)$ with the $\boldsymbol{\psi}$ functions of the M-estimators. The part of the $\boldsymbol{A}$ matrix associated with $\widehat{\theta}_{HL}$ is all zeroes except for the diagonal element that is a one. The diagonal element of the $\boldsymbol{B}$ matrix is the asymptotic variance given above, but one still needs to compute correlations of $h(Y_1, \theta_0)$ with the other ψ functions to get the off-diagonal elements of the $\boldsymbol{B}$ matrix involving $\widehat{\theta}_{HL}$. ♦

7.4 Nonsmooth $\boldsymbol{\psi}$ Functions

In some situations the $\boldsymbol{\psi}$ function may not be differentiable everywhere, thus invalidating the definition of the $\boldsymbol{A}$ matrix as the expected value of a derivative. The appropriately modified definition of $\boldsymbol{A}$ interchanges the order of differentiation and expectation,

$$A(\boldsymbol{\theta}_0) \equiv -\frac{\partial}{\partial \boldsymbol{\theta}} \{\mathrm{E}_F \boldsymbol{\psi}(Y_1, \boldsymbol{\theta})\}\bigg|_{\boldsymbol{\theta}=\boldsymbol{\theta}_0}. \tag{7.15}$$

The expectation is with respect to the true distribution of the data (denoted E_F), but $\boldsymbol{\theta}$ within $\boldsymbol{\psi}$ varies freely with respect to differentiation, after which the true parameter value $\boldsymbol{\theta}_0$ replaces $\boldsymbol{\theta}$.

7.4.1 *Robust Location Estimation*

Huber (1964) proposed estimating the center of symmetry of symmetric distributions using $\widehat{\theta}$ that satisfies $\sum \psi_k(Y_i - \widehat{\theta}) = 0$, where

$$\psi_k(x) = \begin{cases} -k & \text{when } x < -k, \\ x & \text{when } |x| \le k, \\ k & \text{when } x > k. \end{cases}$$

This ψ function is continuous everywhere but not differentiable at $\pm k$. By definition (7.15, p. 311), and assuming that F has density f,

$$\begin{aligned} A(\theta_0) = -\frac{\partial}{\partial \theta} \{\mathrm{E}_F \psi_k(Y_1 - \theta)\}\bigg|_{\theta=\theta_0} &= -\frac{\partial}{\partial \theta} \left\{ \int \psi_k(y - \theta) dF(y) \right\} \bigg|_{\theta=\theta_0} \\ &= \int \left\{ -\frac{\partial}{\partial \theta} \psi_k(y - \theta) \right\} \bigg|_{\theta=\theta_0} dF(y) \\ &= \int \psi_k'(y - \theta_0) dF(y). \end{aligned}$$

The notation ψ_k' inside the integral stands for the derivative of ψ_k where it exists. Verifying the second equality above is an instructive calculus exercise.

For $B(\theta_0)$ we have $B(\theta_0) = \mathrm{E}\psi_k^2(Y_1 - \theta_0) = \int \psi_k^2(y - \theta_0) dF(y)$, and thus

$$V(\theta_0) = \frac{\int \psi_k^2(y - \theta_0) dF(y)}{\left[\int \psi_k'(y - \theta_0) dF(y)\right]^2}.$$

For estimating $A(\theta_0)$ and $B(\theta_0)$, our usual estimators are

$$A_n(\boldsymbol{Y},\widehat{\theta}) = n^{-1}\sum_{i=1}^{n}\left[-\psi_k'(Y_i-\widehat{\theta})\right]$$

and

$$B_n(\boldsymbol{Y},\widehat{\theta}) = n^{-1}\sum_{i=1}^{n}\psi_k^2(Y_i-\widehat{\theta}).$$

Sometimes n is replaced by $n-1$ in this last expression. Here we can use the notation $\psi_k'(Y_i-\widehat{\theta})$ because we expect to have data at $Y_i-\widehat{\theta}=\pm k$ with probability 0.

7.4.2 *Quantile Estimation*

The sample pth quantile $\widehat{\theta}=\widehat{\eta}_p=F_n^{-1}(p)$ satisfies

$$\sum\left[p-I(Y_i\le\widehat{\theta})\right]=c_n,$$

where $|\,c_n\,| = n|\,F_n(\widehat{\theta})-p\,|\le 1$, and F_n is the empirical distribution function. Thus using the extended definition (7.11, p. 301), the ψ function is $\psi(Y_i,\theta)=p-I(Y_i\le\theta)$. This ψ function is discontinuous at $\theta=Y_i$, and its derivative with respect to θ vanishes almost everywhere. However, definition (7.15, p. 311) of $A(\theta_0)$ continues to give us the correct asymptotic results:

$$A(\theta_0) = -\frac{\partial}{\partial\theta}\left[\mathrm{E}_F\{p-I(Y_1\le\theta)\}\right]\bigg|_{\theta=\theta_0} = -\frac{\partial}{\partial\theta}\{p-F(\theta)\}\bigg|_{\theta=\theta_0} = f(\theta_0).$$

$$B(\theta_0)=\mathrm{E}\{p-I(Y_1\le\theta_0)\}^2=p(1-p).$$

$$V(\theta_0)=\frac{p(1-p)}{f^2(\theta_0)}.$$

Also, we could easily stack any finite number of quantile ψ functions together to get the joint asymptotic distribution of $(\widehat{\eta}_{p_1},\ldots,\widehat{\eta}_{p_k})$. There is a cost, however, for the jump discontinuities in these ψ functions: we no longer can use $A_n(\boldsymbol{Y},\widehat{\theta})$ to estimate $\boldsymbol{A}(\theta_0)$. In fact, the derivative of the pth quantile ψ function is zero everywhere except at the location of the jump discontinuity. There are several options for estimating $\boldsymbol{A}(\theta_0)$. One is to use a smoothing technique to estimate f (kernel density estimators, for example). Another is to approximate ψ by a smooth ψ function and use the $\boldsymbol{A}_n(\boldsymbol{Y},\widehat{\theta})$ from this smooth approximation.

7.4.3 Positive Mean Deviation

The positive mean deviation from the median is defined to be

$$\widehat{\theta}_1 = \frac{2}{n}\sum_{i=1}^{n}(Y_i - \widehat{\theta}_2)I(Y_i > \widehat{\theta}_2),$$

where $\widehat{\theta}_2$ is the sample median. Coupled with a similar definition of a negative mean deviation using values less than the median, these "signed distances" from the median form the basis for an alternative to the usual box plot (see Wilson 2002). The two-dimensional $\boldsymbol{\psi}$ to handle $\widehat{\theta}_1$ is

$$\boldsymbol{\psi}(Y_i, \boldsymbol{\theta}) = \begin{pmatrix} 2(Y_i - \theta_2)I(Y_i > \theta_2) - \theta_1 \\ \\ \dfrac{1}{2} - I(Y_i \le \theta_2) \end{pmatrix}.$$

Notice that the first component of $\boldsymbol{\psi}$ is continuous everywhere but not differentiable at $\theta_2 = Y_i$. The second component has a jump discontinuity at $\theta_2 = Y_i$. To get $\boldsymbol{A}(\boldsymbol{\theta}_0)$, we first calculate the expected value of $\boldsymbol{\psi}(Y_1, \boldsymbol{\theta})$ (note that $\boldsymbol{\theta}$ is not $\boldsymbol{\theta}_0$):

$$\mathrm{E}_F\boldsymbol{\psi}(Y_1, \boldsymbol{\theta}) = \begin{pmatrix} 2\int_{\theta_2}^{\infty}(y - \theta_2)dF(y) - \theta_1 \\ \\ \dfrac{1}{2} - F(\theta_2) \end{pmatrix}.$$

To take derivatives of the first component, let us write $dF(y)$ as $f(y)dy$ and expand it out to get

$$2\int_{\theta_2}^{\infty} yf(y)\,dy - 2\theta_2\int_{\theta_2}^{\infty} f(y)\,dy - \theta_1 = 2\int_{\theta_2}^{\infty} yf(y)\,dy - 2\theta_2\,[1 - F(\theta_2)] - \theta_1.$$

The derivative of this latter expression with respect to θ_1 is of course -1. The derivative with respect to θ_2 is $-2\theta_2 f(\theta_2) - 2\,[1 - F(\theta_2)] + 2\theta_2 f(\theta_2) = -2\,[1 - F(\theta_2)]$ (using the Fundamental Theorem of Calculus to get the first term). Setting $\boldsymbol{\theta} = \boldsymbol{\theta}_0$ means that $F(\theta_{20}) = 1/2$ because θ_{20} is the population median. Thus the derivative of the first component with respect to θ_2 and evaluated at $\boldsymbol{\theta} = \boldsymbol{\theta}_0$ is just -1. The partial derivatives of the second component evaluated at $\boldsymbol{\theta} = \boldsymbol{\theta}_0$ are 0 and $-f(\theta_{20})$, respectively. Thus

$$\boldsymbol{A}(\boldsymbol{\theta}_0) = \begin{pmatrix} 1 & 1 \\ 0 & f(\theta_{20}) \end{pmatrix}.$$

Straightforward calculations for $\boldsymbol{B}(\boldsymbol{\theta}_0)$ yield

$$\boldsymbol{B}(\boldsymbol{\theta}_0) = \begin{pmatrix} b_{11} & \dfrac{\theta_{10}}{2} \\ \dfrac{\theta_{10}}{2} & \dfrac{1}{4} \end{pmatrix},$$

where $b_{11} = 4\int_{\theta_{20}}^{\infty}(y - \theta_{20})^2 f(y)\,dy - \theta_{10}^2$. Finally, $\boldsymbol{V}(\boldsymbol{\theta}_0)$ is given by

$$\boldsymbol{V}(\boldsymbol{\theta}_0) = \begin{pmatrix} b_{11} - \dfrac{\theta_{10}}{f(\theta_{20})} + \dfrac{1}{4f^2(\theta_{20})} & \dfrac{\theta_{10}}{2f(\theta_{20})} - \dfrac{1}{4f^2(\theta_{20})} \\ \dfrac{\theta_{10}}{2f(\theta_{20})} - \dfrac{1}{4f^2(\theta_{20})} & \dfrac{1}{4f^2(\theta_{20})} \end{pmatrix}.$$

7.5 Regression M-Estimators

Regression M-estimators are a natural extension of location M-estimators. Although a number of different regression M-estimators have been proposed and studied, the fundamental ideas were established by Huber (1973, 1981, Ch. 7).

There are two situations of interest for M-estimator analysis of regression estimators. The first is where the independent variables are random variables and we can think in terms of iid $(Y, \boldsymbol{X})$ pairs. This situation fits into our basic theory developed in Section 5.2 for iid sampling. The second situation is where the independent variables are fixed constants. This covers standard regression models as well as multi-sample problems like the one-way analysis of variance setup. For this second regression situation we need to introduce new notation to handle the non-iid character of the problem.

7.5.1 Linear Model with Random X

We start with the linear model and least squares estimates for the case of random pairs $(Y_1, \boldsymbol{X}_1), \ldots, (Y_n, \boldsymbol{X}_n)$. The model for the conditional mean is

$$\mathrm{E}(Y_i \mid \boldsymbol{X}_i = \boldsymbol{x}_i) = \boldsymbol{x}_i^T \boldsymbol{\beta}_0 \qquad i = 1, \ldots, n. \tag{7.16}$$

Here $\boldsymbol{\beta} = (\beta_1, \ldots, \beta_p)^T$ and $\boldsymbol{\beta}_0$ is the true value of $\boldsymbol{\beta}$ (not the intercept), and note that we have switched from b to p for the parameter dimension. The $\boldsymbol{X}_i$ are $p \times 1$ random vectors and $\boldsymbol{X}$ is the usual $n \times p$ matrix constructed from these vectors and $\boldsymbol{X}^T\boldsymbol{X} = \sum_{i=1}^n \boldsymbol{X}_i \boldsymbol{X}_i^T$. The ordinary least squares estimator $\widehat{\boldsymbol{\beta}} = (\boldsymbol{X}^T\boldsymbol{X})^{-1}\boldsymbol{X}^T\boldsymbol{Y}$ minimizes $\sum_{i=1}^n (Y_i - \boldsymbol{X}_i^T\boldsymbol{\beta})^2$ and satisfies

$$\sum_{i=1}^n (Y_i - \boldsymbol{X}_i^T\widehat{\boldsymbol{\beta}})\boldsymbol{X}_i = 0.$$

Thus $\widehat{\boldsymbol{\beta}}$ is an M-estimator with $\boldsymbol{\psi}(Y_i, \boldsymbol{X}_i, \boldsymbol{\beta}) = (Y_i - \boldsymbol{X}_i^T\boldsymbol{\beta})\boldsymbol{X}_i$. The conditional expectation (7.16) also shows that $\boldsymbol{\beta}_0$ satisfies the M-estimator defining equation $\mathrm{E}\boldsymbol{\psi}(Y_1, \boldsymbol{X}_1, \boldsymbol{\beta}_0) = 0$:

$$\begin{aligned}\mathrm{E}(Y_1 - \boldsymbol{X}_1^T\boldsymbol{\beta}_0)\boldsymbol{X}_1 &= \mathrm{E}\left\{\mathrm{E}(Y_1 - \boldsymbol{X}_1^T\boldsymbol{\beta}_0)\boldsymbol{X}_1 \mid \boldsymbol{X}_1\right\} \\ &= \mathrm{E}\left\{(\boldsymbol{X}_1^T\boldsymbol{\beta}_0 - \boldsymbol{X}_1^T\boldsymbol{\beta}_0)\boldsymbol{X}_1\right\} = 0.\end{aligned}$$

Now moving to the asymptotic distribution, we have

$$\boldsymbol{A}(\boldsymbol{\beta}_0) = \mathrm{E}\left\{-\frac{\partial}{\partial\boldsymbol{\beta}}(Y_1 - \boldsymbol{X}_1^T\boldsymbol{\beta})\boldsymbol{X}_1\Big|_{\boldsymbol{\beta}=\boldsymbol{\beta}_0}\right\} = \mathrm{E}\left(\boldsymbol{X}_1\boldsymbol{X}_1^T\right) \equiv \boldsymbol{\mu}_{\boldsymbol{X}\boldsymbol{X}^T},$$

and

$$\begin{aligned}\boldsymbol{B}(\boldsymbol{\beta}_0) &= \mathrm{E}\left(Y_1 - \boldsymbol{X}_1^T\boldsymbol{\beta}_0\right)\boldsymbol{X}_1\left\{(Y_1 - \boldsymbol{X}_1^T\boldsymbol{\beta}_0)\boldsymbol{X}_1\right\}^T \\ &= \mathrm{E}\left\{(Y_1 - \boldsymbol{X}_1^T\boldsymbol{\beta}_0)^2\boldsymbol{X}_1\boldsymbol{X}_1^T\right\}.\end{aligned}$$

If we make the additional homoscedastic assumption that $\mathrm{E}\left[(Y_i - \boldsymbol{X}_i^T\boldsymbol{\beta}_0)^2 \mid \boldsymbol{X}_i\right] = \sigma^2$, then $\boldsymbol{B}(\boldsymbol{\beta}_0)$ is just $\boldsymbol{\mu}_{\boldsymbol{X}\boldsymbol{X}^T}\sigma^2$. In this case, the asymptotic variance of $\widehat{\boldsymbol{\beta}}$ is $\boldsymbol{\mu}_{\boldsymbol{X}\boldsymbol{X}^T}^{-1}\sigma^2/n$. The natural estimates of $\boldsymbol{A}(\boldsymbol{\beta}_0)$ and $\boldsymbol{B}(\boldsymbol{\beta}_0)$ without assuming homogeneity of variances are

$$\boldsymbol{A}_n(\boldsymbol{X}, \widehat{\boldsymbol{\beta}}) = \frac{1}{n}\sum_{i=1}^{n}\boldsymbol{X}_i\boldsymbol{X}_i^T = \frac{\boldsymbol{X}^T\boldsymbol{X}}{n},$$

and

$$\boldsymbol{B}_n(\boldsymbol{X}, \boldsymbol{Y}, \widehat{\boldsymbol{\beta}}) = \frac{1}{n-p}\sum_{i=1}^{n}(Y_i - \boldsymbol{X}_i^T\widehat{\boldsymbol{\beta}})^2\boldsymbol{X}_i\boldsymbol{X}_i^T,$$

where we have used the denominator $n - p$ in analogy with the usual estimate of variance.

7.5.2 Linear Model for Nonrandom X

The above estimates fit into our standard iid theory developed in Section 7.2. Now let us move on to allow nonrandom $\boldsymbol{x}_i$. For this we specify the model as

$$Y_i = \boldsymbol{x}_i^T\boldsymbol{\beta} + e_i \qquad i = 1, \ldots, n, \tag{7.17}$$

where $e_1, \ldots, e_n$ are independent each with mean 0 but variance possibly changing, $\mathrm{Var}(e_i) = \sigma_i^2$, $i = 1, \ldots, n$, and the $\boldsymbol{x}_1, \ldots, \boldsymbol{x}_n$ are known constant vectors.

The least squares estimator and associated ψ function are the same as in the previous subsection, but now the defining concept for $\boldsymbol{\beta}_0$ is

$$\mathrm{E}\boldsymbol{\psi}(Y_i, \boldsymbol{x}_i, \boldsymbol{\beta}_0) = \mathrm{E}(Y_i - \boldsymbol{x}_i^T \boldsymbol{\beta}_0)\boldsymbol{x}_i = 0 \qquad i = 1, \ldots, n. \tag{7.18}$$

This of course follows from (7.17), but notice that we require this expectation to be zero for $i = 1, \ldots, n$. In the iid case, it was sufficient to just consider $i = 1$. For the asymptotic properties of $\widehat{\boldsymbol{\beta}} = (\boldsymbol{X}^T\boldsymbol{X})^{-1}\boldsymbol{X}^T\boldsymbol{Y}$, note that from (7.17), we have

$$\sqrt{n}(\widehat{\boldsymbol{\beta}} - \boldsymbol{\beta}_0) = \sqrt{n}(\boldsymbol{X}^T\boldsymbol{X})^{-1}\boldsymbol{X}^T\boldsymbol{e} = \left(\frac{\boldsymbol{X}^T\boldsymbol{X}}{n}\right)^{-1} \frac{1}{\sqrt{n}} \sum_{i=1}^{n} e_i \boldsymbol{x}_i.$$

Thus, we assume the existence of

$$\boldsymbol{A}(\boldsymbol{\beta}_0) = \lim_{n\to\infty} \frac{\boldsymbol{X}^T\boldsymbol{X}}{n} \qquad \text{and} \qquad \boldsymbol{B}(\boldsymbol{\beta}_0) = \lim_{n\to\infty} \frac{1}{n} \sum_{i=1}^{n} \sigma_i^2 \boldsymbol{x}_i \boldsymbol{x}_i^T. \tag{7.19}$$

Then the Lindeberg-Feller version of the Central Limit Theorem (Theorem 5.33, p. 262) leads to

$$\widehat{\boldsymbol{\beta}} \ \text{ is } \ \mathrm{AMN}(\boldsymbol{\beta}_0, \boldsymbol{A}(\boldsymbol{\beta}_0)^{-1}\boldsymbol{B}(\boldsymbol{\beta}_0)\boldsymbol{A}(\boldsymbol{\beta}_0)^{-1}/n) \quad \text{as} \quad n \to \infty.$$

We could also write this in terms of the exact variance,

$$\widehat{\boldsymbol{\beta}} \ \text{ is } \ \mathrm{AMN}\left(\boldsymbol{\beta}_0, (\boldsymbol{X}^T\boldsymbol{X})^{-1}\left(\sum_{i=1}^{n} \sigma_i^2 \boldsymbol{x}_i \boldsymbol{x}_i^T\right)(\boldsymbol{X}^T\boldsymbol{X})^{-1}\right).$$

The estimators for $\boldsymbol{A}(\boldsymbol{\beta}_0)$ and $\boldsymbol{B}(\boldsymbol{\beta}_0)$ are the same as in the random $\boldsymbol{X}_i$ case. If we assume that the variances of the e_i are homogeneous with common value σ^2, then we have the standard result that $\widehat{\boldsymbol{\beta}}$ is $\mathrm{AMN}(\boldsymbol{\beta}_0, (\boldsymbol{X}^T\boldsymbol{X})^{-1}\sigma^2)$. Recall that in this situation, an unbiased estimator of $\mathrm{Var}(\widehat{\boldsymbol{\beta}})$ is $(\boldsymbol{X}^T\boldsymbol{X})^{-1}\widehat{\sigma}^2$, where

$$\widehat{\sigma}^2 = \frac{1}{n-p} \sum_{i=1}^{n} (y_i - \boldsymbol{x}_i^T \widehat{\boldsymbol{\beta}})^2 = \frac{1}{n-p} \sum_{i=1}^{n} \widehat{e}_i^2. \tag{7.20}$$

So, for the fixed-constants version of regression we have expanded the defining equation to (7.18), added assumptions about the constants $\boldsymbol{x}_i$ in (7.19), and used a more general version of the Central Limit Theorem. For more complicated regression models, we need additional assumptions and notation. But first let us consider an example with variance heterogeneity.

Example 7.2 (Linear regression with heteroscedastic errors). For the linear model (7.17, p. 315) with heteroscedastic errors, we consider the covariance estimator

$$A_n(X,\widehat{\beta})^{-1}B_n(X,Y,\widehat{\beta})A_n(X,\widehat{\beta})^{-1}/n$$

$$= (X^T X)^{-1}\left(\frac{n}{n-p}\sum_{i=1}^{n}(Y_i - x_i^T\widehat{\beta})^2 x_i x_i^T\right)(X^T X)^{-1} \qquad (7.21)$$

in place of $(X^T X)^{-1}\widehat{\sigma}^2$, where $\widehat{\sigma}^2$ is given by (7.20, p. 316). How important is it to use this more complicated form?

Consider the simple linear model

$$Y_i = \beta_1 + \beta_2 i + i e_i, \qquad i = 1,\ldots,n_0, \qquad (7.22)$$

where the $e_1,\ldots,e_{n_0}$ are iid with mean zero and variance σ^2. For general n and asymptotic purposes, consider full replicates of (7.22); that is, we repeat (7.22) k times so that $n = kn_0$. Here, for the first n_0, $x_i = (1,i)^T$ and

$$X^T X = \begin{pmatrix} kn_0 & k\sum_{i=1}^{n_0} i \\ k\sum_{i=1}^{n_0} i & k\sum_{i=1}^{n_0} i^2 \end{pmatrix}$$

with

$$\frac{X^T X}{n} = \begin{pmatrix} 1 & n_0^{-1}\sum_{i=1}^{n_0} i \\ n_0^{-1}\sum_{i=1}^{n_0} i & n_0^{-1}\sum_{i=1}^{n_0} i^2 \end{pmatrix}.$$

The variance of the least squares estimator is

$$(X^T X)^{-1}\left(k\sum_{i=1}^{n_0} i^2\sigma^2 x_i x_i^T\right)(X^T X)^{-1}, \qquad (7.23)$$

which is consistently estimated by (7.21). On the other hand, the usual variance estimator $(X^T X)^{-1}\widehat{\sigma}^2$ is estimating $(X^T X)^{-1}\sigma^2 n_0^{-1}\sum_{i=1}^{n_0} i^2$. For $k = 1$ and $n_0 = 10$, this last quantity is

$$\sigma^2\begin{pmatrix} 17.97 & -2.57 \\ -2.57 & .47 \end{pmatrix}$$

whereas the true asymptotic variance (7.23) is

$$\sigma^2\begin{pmatrix} 8.21 & -1.89 \\ -1.89 & .54 \end{pmatrix}.$$

Thus, in this case, the usual variance estimator is overestimating the variance of the intercept estimator and underestimating the variance of the slope estimator. ♦

7.5.3 Nonlinear Model for Nonrandom X—Extended Definitions of A and B

In a number of ways the least squares estimate for linear models is too simple to illustrate the full range of possibilities for M-estimation analysis. A moderately more complicated situation is the additive error nonlinear model

$$Y_i = g(\boldsymbol{x}_i, \boldsymbol{\beta}) + e_i \qquad i = 1, \ldots, n, \tag{7.24}$$

where g is a known differentiable function and $e_1, \ldots, e_n$ are independent with mean 0 and possibly unequal variances $\text{Var}(e_i) = \sigma_i^2$, $i = 1, \ldots, n$, and the $\boldsymbol{x}_1, \ldots, \boldsymbol{x}_n$ are known constant vectors. As usual we put the vectors together and define $\boldsymbol{X} = (\boldsymbol{x}_1, \ldots, \boldsymbol{x}_n)^T$. The least squares estimator satisfies

$$\sum_{i=1}^{n} \left\{ Y_i - g(\boldsymbol{x}_i, \widehat{\boldsymbol{\beta}}) \right\} g'(\boldsymbol{x}_i, \widehat{\boldsymbol{\beta}}) = 0,$$

where $g'(\boldsymbol{x}_i, \widehat{\boldsymbol{\beta}})$ means the partial derivative with respect to $\boldsymbol{\beta}$ and evaluated at $\widehat{\boldsymbol{\beta}}$. Expanding this equation about the true value and rearranging, we get

$$\sqrt{n}(\widehat{\boldsymbol{\beta}} - \boldsymbol{\beta}_0) = \left\{ \frac{1}{n} \sum_{i=1}^{n} -\boldsymbol{\psi}'(Y_i, \boldsymbol{x}_i, \boldsymbol{\beta}_0) \right\}^{-1} \frac{1}{\sqrt{n}} \sum_{i=1}^{n} \boldsymbol{\psi}(Y_i, \boldsymbol{x}_i, \boldsymbol{\beta}_0) + \sqrt{n} R_n^*, \tag{7.25}$$

where of course $\boldsymbol{\psi}(Y_i, \boldsymbol{x}_i, \boldsymbol{\beta}_0) = \{Y_i - g(\boldsymbol{x}_i, \boldsymbol{\beta}_0)\} g'(\boldsymbol{x}_i, \boldsymbol{\beta}_0)$. We now give general definitions for a number of quantities followed by the result for the least squares estimator.

$$\begin{aligned} \boldsymbol{A}_n(\boldsymbol{X}, \boldsymbol{Y}, \boldsymbol{\beta}_0) &= \frac{1}{n} \sum_{i=1}^{n} \left[-\boldsymbol{\psi}'(Y_i, \boldsymbol{x}_i, \boldsymbol{\beta}_0) \right] \\ &= \frac{1}{n} \sum_{i=1}^{n} \left[g'(\boldsymbol{x}_i, \boldsymbol{\beta}_0) g'(\boldsymbol{x}_i, \boldsymbol{\beta}_0)^T - \{Y_i - g(\boldsymbol{x}_i, \boldsymbol{\beta}_0)\} g''(\boldsymbol{x}_i, \boldsymbol{\beta}_0) \right]. \end{aligned} \tag{7.26}$$

The notation principle is the same as before: all arguments of a quantity are included in its name if those quantities are required for calculation. Now taking expectations with respect to the true model, define

$$\begin{aligned} \boldsymbol{A}_n(\boldsymbol{X}, \boldsymbol{\beta}_0) &= \frac{1}{n} \sum_{i=1}^{n} \text{E} \left\{ -\boldsymbol{\psi}'(Y_i, \boldsymbol{x}_i, \boldsymbol{\beta}_0) \right\} \\ &= \frac{1}{n} \sum_{i=1}^{n} g'(\boldsymbol{x}_i, \boldsymbol{\beta}_0) g'(\boldsymbol{x}_i, \boldsymbol{\beta}_0)^T. \end{aligned} \tag{7.27}$$

Notice that we have dropped out the $\boldsymbol{Y}$ from this quantity's name because the expectation eliminates dependence on the Y_i. Also note that the second term for the least squares estimator drops out because of the modeling assumption (7.24). Finally, assuming that the limit exists, we define

$$\begin{aligned} \boldsymbol{A}(\boldsymbol{\beta}_0) &= \lim_{n\to\infty} \frac{1}{n} \sum_{i=1}^{n} \mathrm{E}\left\{-\boldsymbol{\psi}'(Y_i, \boldsymbol{x}_i, \boldsymbol{\beta}_0)\right\} \\ &= \lim_{n\to\infty} \frac{1}{n} \sum_{i=1}^{n} g'(\boldsymbol{x}_i, \boldsymbol{\beta}_0) g'(\boldsymbol{x}_i, \boldsymbol{\beta}_0)^T. \end{aligned} \tag{7.28}$$

In the linear regression case, note that $\boldsymbol{A}(\boldsymbol{\beta}_0) = \lim_{n\to\infty} \boldsymbol{X}^T\boldsymbol{X}/n$. This limit need not exist for the least squares estimator to be consistent and asymptotically normal, but its existence is a typical assumption leading to those desired results. Definition (7.26) leads to the purely empirical estimator of $\boldsymbol{A}(\boldsymbol{\beta}_0)$:

$$\begin{aligned} \boldsymbol{A}_n(\boldsymbol{X}, \boldsymbol{Y}, \widehat{\boldsymbol{\beta}}) &= \frac{1}{n} \sum_{i=1}^{n} \left[-\boldsymbol{\psi}'(Y_i, \boldsymbol{x}_i, \widehat{\boldsymbol{\beta}})\right] \\ &= \frac{1}{n} \sum_{i=1}^{n} \left[g'(\boldsymbol{x}_i, \widehat{\boldsymbol{\beta}}) g'(\boldsymbol{x}_i, \widehat{\boldsymbol{\beta}})^T - (Y_i - g(\boldsymbol{x}_i, \widehat{\boldsymbol{\beta}})) g''(\boldsymbol{x}_i, \widehat{\boldsymbol{\beta}})\right]. \end{aligned} \tag{7.29}$$

Since this is the negative of the Hessian in a final Newton iteration, this is sometimes preferred on computational grounds. But the estimated expected value estimator based on (7.27) is typically simpler:

$$\begin{aligned} \boldsymbol{A}_n(\boldsymbol{X}, \widehat{\boldsymbol{\beta}}) &= \frac{1}{n} \sum_{i=1}^{n} \left[\mathrm{E}\left\{-\boldsymbol{\psi}'(Y_i, \boldsymbol{x}_i, \boldsymbol{\beta}_0)\right\}\right]\Big|_{\boldsymbol{\beta}_0 = \widehat{\boldsymbol{\beta}}} \\ &= \frac{1}{n} \sum_{i=1}^{n} g'(\boldsymbol{x}_i, \widehat{\boldsymbol{\beta}}) g'(\boldsymbol{x}_i, \widehat{\boldsymbol{\beta}})^T. \end{aligned} \tag{7.30}$$

For the "$\boldsymbol{B}$" matrices, we have in this expanded notation

$$\begin{aligned} \boldsymbol{B}_n(\boldsymbol{X}, \boldsymbol{\beta}_0) &= \frac{1}{n} \sum_{i=1}^{n} \mathrm{E}\boldsymbol{\psi}(Y_i, \boldsymbol{x}_i, \boldsymbol{\beta}_0) \boldsymbol{\psi}(Y_i, \boldsymbol{x}_i, \boldsymbol{\beta}_0)^T \\ &= \frac{1}{n} \sum_{i=1}^{n} \sigma_i^2 g'(\boldsymbol{x}_i, \boldsymbol{\beta}_0) g'(\boldsymbol{x}_i, \boldsymbol{\beta}_0)^T. \end{aligned} \tag{7.31}$$

and $\boldsymbol{B}(\boldsymbol{\beta}_0)$ is just the limit of $\boldsymbol{B}_n(\boldsymbol{X}, \boldsymbol{\beta}_0)$ as $n \to \infty$. A natural estimator of $\boldsymbol{B}(\boldsymbol{\beta}_0)$ is

$$\begin{aligned}\boldsymbol{B}_n(\boldsymbol{X}, \boldsymbol{Y}, \widehat{\boldsymbol{\beta}}) &= \frac{1}{n-p}\sum_{i=1}^{n} \boldsymbol{\psi}(Y_i, \boldsymbol{x}_i, \widehat{\boldsymbol{\beta}})\boldsymbol{\psi}(Y_i, \boldsymbol{x}_i, \widehat{\boldsymbol{\beta}})^T \\ &= \frac{1}{n-p}\sum_{i=1}^{n}(Y_i - g(\boldsymbol{x}_i, \widehat{\boldsymbol{\beta}}))^2 g'(\boldsymbol{x}_i, \widehat{\boldsymbol{\beta}}) g'(\boldsymbol{x}_i, \widehat{\boldsymbol{\beta}})^T. \qquad (7.32)\end{aligned}$$

7.5.4 Robust regression

Huber (1973) discussed robust regression alternatives to least squares in the linear regression context. As a specific example, consider the model (7.24, p. 318) with $g(\boldsymbol{x}_i, \boldsymbol{\beta}) = \boldsymbol{x}^T\boldsymbol{\beta}$ and estimator of $\boldsymbol{\beta}$ satisfying

$$\sum_{i=1}^{n} \psi_k(Y_i - \boldsymbol{x}_i^T\widehat{\boldsymbol{\beta}})\boldsymbol{x}_i = 0, \qquad (7.33)$$

where ψ_k is the "Huber" ψ function defined in Example 7.4.1 (p. 311). This is a slight abuse of notation since the official $\boldsymbol{\psi}(Y_i, \boldsymbol{x}_i, \boldsymbol{\beta}) = \psi_k(Y_i - \boldsymbol{x}_i^T\boldsymbol{\beta})\boldsymbol{x}_i$; i.e., ψ is being used as both the original Huber function ψ_k and also as the generic estimating equation function. Since ψ_k is an odd function about zero, the defining equations $\mathrm{E}\psi_k(Y_i - \boldsymbol{x}_i^T\boldsymbol{\beta}_0)\boldsymbol{x}_i = 0$ are satisfied if the e_i have a symmetric distribution about zero. If the e_i are not symmetrically distributed and the X matrix contains a column of ones, then the intercept estimated by $\widehat{\boldsymbol{\beta}}$ is different from that estimated by least squares, but this is the only component of $\boldsymbol{\beta}_0$ affected by asymmetry.

Differentiating results in

$$\boldsymbol{A}_n(\boldsymbol{X}, \boldsymbol{Y}, \boldsymbol{\beta}_0) = \frac{1}{n}\sum_{i=1}^{n}[-\boldsymbol{\psi}'(Y_i, \boldsymbol{x}_i, \boldsymbol{\beta}_0)] = \frac{1}{n}\sum_{i=1}^{n}\psi_k'(e_i)\boldsymbol{x}_i\boldsymbol{x}_i^T$$

and

$$\boldsymbol{A}_n(\boldsymbol{X}, \boldsymbol{\beta}_0) = n^{-1}\sum_{i=1}^{n}\mathrm{E}\psi_k'(e_i)\boldsymbol{x}_i\boldsymbol{x}_i^T.$$

Also,

$$\boldsymbol{B}_n(\boldsymbol{X}, \boldsymbol{\beta}_0) = n^{-1}\sum_{i=1}^{n}\mathrm{E}\psi_k(e_i)^2\boldsymbol{x}_i\boldsymbol{x}_i^T.$$

If we make the homogeneity assumption that the errors $e_1, \ldots, e_n$ all have the same distribution, then $\boldsymbol{A}_n(\boldsymbol{X}, \boldsymbol{\beta}_0) = \mathrm{E}\psi_k'(e_1)\boldsymbol{X}^T\boldsymbol{X}/n$, $\boldsymbol{B}_n(\boldsymbol{X}, \boldsymbol{\beta}_0) = \mathrm{E}\psi_k(e_1)^2\boldsymbol{X}^T\boldsymbol{X}/n$, and

$$V(\boldsymbol{X}, \boldsymbol{\beta}_0) = \left(\frac{\boldsymbol{X}^T\boldsymbol{X}}{n}\right)^{-1} \frac{\mathrm{E}\{\psi_k(e_1)\}^2}{\{\mathrm{E}\psi_k'(e_1)\}^2},$$

a well-known form in robust regression.

7.5.5 Generalized Linear Models

Generalized linear models have score equations

$$\sum_{i=1}^{n} \boldsymbol{D}_i(\boldsymbol{\beta}) \frac{(Y_i - \mu_i(\boldsymbol{\beta}))}{V_i(\boldsymbol{\beta})a_i(\phi)} = 0, \tag{7.34}$$

where $\mu_i(\boldsymbol{\beta}_0) = \mathrm{E}(Y_i) = g^{-1}(\boldsymbol{x}_i^T\boldsymbol{\beta}_0)$, $\boldsymbol{D}_i(\boldsymbol{\beta}) = \partial\mu_i(\boldsymbol{\beta})/\partial\boldsymbol{\beta}^T$, $V_i(\boldsymbol{\beta}_0)a_i(\phi_0) = \mathrm{Var}(Y_i)$, g is the link function, and ϕ is an additional variance parameter. Taking expectations of the negative of the derivative with respect to $\boldsymbol{\beta}$ of the above sum evaluated at $\boldsymbol{\beta}_0$ leads to the average Fisher information matrix

$$\bar{\boldsymbol{I}}(\boldsymbol{\beta}_0) = \frac{1}{n}\sum_{i=1}^{n} \frac{\boldsymbol{D}_i(\boldsymbol{\beta}_0)\boldsymbol{D}_i(\boldsymbol{\beta}_0)^T}{V_i(\boldsymbol{\beta}_0)a_i(\phi_0)}.$$

Note that the second term involving derivatives of $\boldsymbol{D}_i/V_i$ drops out due to the assumption that $\mu_i(\boldsymbol{\beta}_0) = \mathrm{E}(Y_i)$. Now for certain misspecification of densities, the generalized linear model framework allows for estimation of ϕ and approximately correct inference as long as the mean is modeled correctly and the mean-variance relationship is specified correctly. Details of this robustified inference may be found in McCullagh (1983) under the name "quasi-likelihood." Note, though, that only one extra parameter ϕ is used to make up for possible misspecification. Instead, Liang and Zeger (1986) noticed that the M-estimator approach could be used here without ϕ and with only the mean correctly specified (and with V_i absorbing any weight factor from a_i):

$$\boldsymbol{A}_n(\boldsymbol{X}, \widehat{\boldsymbol{\beta}}) = \frac{1}{n}\sum_{i=1}^{n} \frac{\boldsymbol{D}_i(\widehat{\boldsymbol{\beta}})\boldsymbol{D}_i(\widehat{\boldsymbol{\beta}})^T}{V_i(\widehat{\boldsymbol{\beta}})}.$$

$$\boldsymbol{B}_n(\boldsymbol{X}, \boldsymbol{Y}, \widehat{\boldsymbol{\beta}}) = \frac{1}{n-p}\sum_{i=1}^{n} \frac{\left\{Y_i - \mu_i(\widehat{\boldsymbol{\beta}})\right\}^2 \boldsymbol{D}_i(\widehat{\boldsymbol{\beta}})\boldsymbol{D}_i(\widehat{\boldsymbol{\beta}})^T}{V_i^2(\widehat{\boldsymbol{\beta}})}.$$

7.5.6 Generalized Estimating Equations (GEE)

The landmark paper of Liang and Zeger (1986) actually introduced much more than the use of the sandwich matrix to the generalized linear model framework. Liang and Zeger were concerned about allowing dependence in nonnormal response variables Y_i in order to handle longitudinal and repeated measures situations and clustering. For example, suppose that data were available on families where Y_{ij} is the absence or presence of respiratory disease (binary) or the severity of respiratory disease (1-5, ordered categorical) for the jth family member of the ith family. Examples of possible covariates x are nutritional status, age, sex, and family income. Note that age and sex change with j but nutritional status and family income would be the same for all members of the family. The key point, though, is that families may often be taken as independent but members within a family would have correlated respiratory health. Thus, the response data from k families could be put into vectors $\boldsymbol{Y}_1, \ldots, \boldsymbol{Y}_k$ of varying dimensions $n_1, \ldots, n_k$. The covariance matrix of all the small vectors strung together would be block diagonal with covariance matrices $\boldsymbol{V}_1, \ldots, \boldsymbol{V}_k$.

For normally distributed Y_{ij}, standard models and methods existed for this situation (e.g., Laird and Ware 1982). However, for binary and categorical Y_{ij}, very few methods were available because there are not convenient joint likelihoods for such data. Thus, Liang and Zeger (1986) proposed extending (7.34, p. 321) to

$$\sum_{i=1}^{k} \boldsymbol{D}_i(\boldsymbol{\beta})^T V_i(\boldsymbol{\alpha}, \boldsymbol{\beta})^{-1} \{\boldsymbol{Y}_i - \boldsymbol{\mu}_i(\boldsymbol{\beta})\} = 0, \tag{7.35}$$

where $\boldsymbol{\mu}_i(\boldsymbol{\beta})$ is the $n_i \times 1$ modeled mean of $\boldsymbol{Y}_i$, $\boldsymbol{D}_i(\boldsymbol{\beta}) = \partial \boldsymbol{\mu}_i(\boldsymbol{\beta})/\partial \boldsymbol{\beta}^T$ is $n_i \times p$, and $n_i \times n_i$ $V_i(\boldsymbol{\alpha}, \boldsymbol{\beta})$ is an assumed model for the covariance matrix of $\boldsymbol{Y}_i$. Thus, the estimators from (7.35) are M-estimators, and the associated matrices are

$$\boldsymbol{A}_n(\boldsymbol{X}, \boldsymbol{\beta}_0) = \frac{1}{k} \sum_{i=1}^{k} \boldsymbol{D}_i(\boldsymbol{\beta}_0)^T V_i(\boldsymbol{\alpha}, \boldsymbol{\beta})^{-1} \boldsymbol{D}_i(\boldsymbol{\beta}_0),$$

and

$$\boldsymbol{B}_n(\boldsymbol{X}, \boldsymbol{\beta}_0) = \frac{1}{k} \sum_{i=1}^{k} \boldsymbol{D}_i(\boldsymbol{\beta}_0)^T V_i(\boldsymbol{\alpha}, \boldsymbol{\beta})^{-1} \mathrm{Cov}(\boldsymbol{Y}_i) V_i(\boldsymbol{\alpha}, \boldsymbol{\beta})^{-1} \boldsymbol{D}_i(\boldsymbol{\beta}_0).$$

Typically $V_i(\boldsymbol{\alpha}, \boldsymbol{\beta})$is called the "working" covariance matrix to denote that it is just a guess at the true covariance. Simple models for $V_i(\boldsymbol{\alpha}, \boldsymbol{\beta})$may be written in the form $V_i(\boldsymbol{\alpha}, \boldsymbol{\beta}) = \boldsymbol{C}(\boldsymbol{\beta})^{1/2} \boldsymbol{R}(\boldsymbol{\alpha}) \boldsymbol{C}(\boldsymbol{\beta})^{1/2}$, where $\boldsymbol{C}(\boldsymbol{\beta})^{1/2}$ is a diagonal matrix with the modeled standard deviations of Y_{ij} on the diagonal and $\boldsymbol{R}(\boldsymbol{\alpha})$ is a correlation matrix. Popular forms for $\boldsymbol{R}(\boldsymbol{\alpha})$ are independence ($\boldsymbol{R}(\boldsymbol{\alpha})$=identity matrix), equicorrelated (all off-diagonal elements are the same), and autoregressive (correlation of the form $\rho^{|h|}$ between Y_{ij_1} and Y_{ij_2} for $h = j_1 - j_2$).

An attractive feature of the GEE method is that although one guesses the form of $\text{Cov}(\boldsymbol{Y}_i)$, usually we estimate $\boldsymbol{B}_n(\boldsymbol{X}, \boldsymbol{\beta}_0)$ with the empirical form

$$\boldsymbol{B}_n(\boldsymbol{X}, \boldsymbol{Y}, \widehat{\boldsymbol{\beta}}) = \frac{1}{k-p} \sum_{i=1}^{k} \boldsymbol{D}_i(\widehat{\boldsymbol{\beta}})^T V_i(\widehat{\boldsymbol{\alpha}}, \widehat{\boldsymbol{\beta}})^{-1} \widehat{\boldsymbol{e}}_i \widehat{\boldsymbol{e}}_i^T V_i(\widehat{\boldsymbol{\alpha}}, \widehat{\boldsymbol{\beta}})^{-1} \boldsymbol{D}_i(\widehat{\boldsymbol{\beta}}),$$

where $\widehat{\boldsymbol{e}}_i = \boldsymbol{Y}_i - \boldsymbol{\mu}_i(\widehat{\boldsymbol{\beta}})$. Putting $\boldsymbol{B}_n(\boldsymbol{X}, \boldsymbol{Y}, \widehat{\boldsymbol{\beta}})$ together with $\boldsymbol{A}_n(\boldsymbol{X}, \widehat{\boldsymbol{\beta}})$ gives the sandwich estimator $V_n(\boldsymbol{X}, \widehat{\boldsymbol{\beta}})$ that is robust to misspecification of the covariance structure. Equation (7.35) requires an estimate of $\boldsymbol{\alpha}$, which may be found by simple direct use of $\widehat{\boldsymbol{e}}_i$ or by a second set of estimating equations (Prentice, 1988, Zhao and Prentice, 1990). The efficiency of estimating $\boldsymbol{\beta}$ is not affected much by the working covariance model or by the method of estimating $\boldsymbol{\alpha}$.

Dunlop (1994) gives a simple introduction to these models and methods, and Wu et al. (2001) make comparisons with normal theory methods when the covariance structure is of particular interest. Diggle et al. (2002) is an up-to-date book on GEE and related topics. In time series and spatial analyses, there is often correlation among all the Y_i with no independent replication. In such cases the $\boldsymbol{A}$ matrix estimates from the independent case are still consistent, but more complicated methods must be used in estimating the $\boldsymbol{B}$ matrix; see Heagerty and Lumley (2000) and Kim and Boos (2004).

7.6 Application to a Testing Problem

Although we discuss test statistics based on M-estimators in the next chapter, here we present an application of M-estimation to a score statistic in a standard likelihood situation. The main contribution of M-estimation is to analyze the effect of a parameter estimate on the score statistic.

Example 7.3 (Shaq's free throws). In the 2000 National Basketball Association (NBA) playoffs, Los Angeles Lakers star player Shaquille O'Neal played in 23 games. Table 7.1 gives his game-by-game free throw outcomes and Figure 7.1 displays the results.

It is often conjectured that players have streaks where they shoot better or worse. One way to think about that is to assume that the number of free throws made in the ith game, Y_i, is binomial (n_i, p_i) conditional on n_i, the number of free throws attempted in the ith game, and p_i, the probability of making a free throw in the ith game. Having streaks might correspond to some games having high or low p_i values. Thus, a statistical formulation of the problem might be

Table 7.1 Shaquille O'Neal Free Throws in 2000 NBA Playoffs

Game Number	1	2	3	4	5	6	7	8	9	10	11	
FTs Made	4	5	5	5	2	7	6	9	4	1	13	
FTs Attempted	5	11	14	12	7	10	14	15	12	4	27	
Prop. Made	.80	.45	.36	.42	.29	.70	.43	.60	.33	.25	.48	
Game Number	12	13	14	15	16	17	18	19	20	21	22	23
FTs Made	5	6	9	7	3	8	1	18	3	10	1	3
FTs Attempted	17	12	9	12	10	12	6	39	13	17	6	12
Prop. Made	.29	.50	1.0	.58	.30	.67	.17	.46	.23	.59	.17	.25

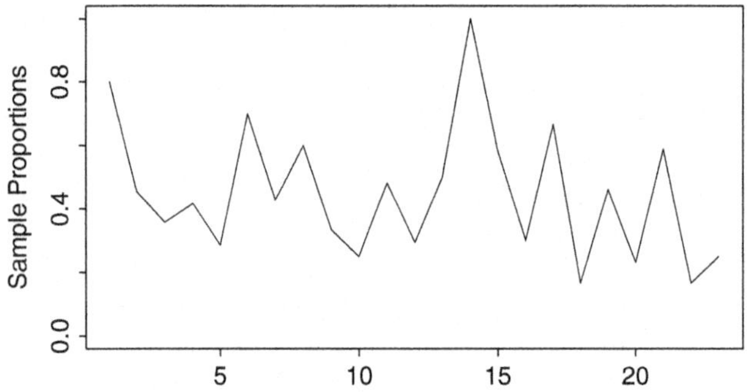

Fig. 7.1 Shaq's Free Throw Percentages in the 2000 NBA Playoffs

"Can the above observed game-to-game variation in sample proportions be explained by binomial variability with a common p?"

Note that the apparent downward trend in sample proportions is not significant; the simple linear regression p-value = .24.

For generality let k be the number of games. The score statistic (see Example 3.4.3b, p. 147) for testing a common binomial proportion versus some differences

$$H_0 : p_1 = p_2 = \cdots = p_k = p \quad \text{vs.} \quad H_1 : p_i \neq p_j \quad \text{for at least one pair } i \neq j$$

is given by

$$T_S = \sum_{i=1}^{k} (Y_i - n_i \widetilde{p})^2 / n_i \widetilde{p}(1 - \widetilde{p}),$$

where $\widetilde{p} = \sum Y_i / \sum n_i$ is the estimate of the common value of p under the null hypothesis. The sample sizes $n_1, \ldots, n_k$ were assumed fixed for this derivation (they aren't really; so this is a conditional approach). T_S is also the simple chi-squared goodness-of-fit statistic with the $2k$ cell expected values $n_1\widetilde{p}, n_1(1 - \widetilde{p}), \ldots, n_k\widetilde{p}, n_k(1 - \widetilde{p})$.

Using the above data, we find $\widetilde{p} = .456$, $T_S = 35.51$ and the p-value is .034 based on a chi-squared distribution with $k - 1 = 22$ degrees of freedom. But the chi-squared approximation is based on each n_i going to infinity, and most of the n_i in our data set are quite small. Since T_S is just a sum $\sum_{i=1}^{k} \widetilde{Q}_i$, $\widetilde{Q}_i = (Y_i - n_i\widetilde{p})^2/n_i\widetilde{p}(1-\widetilde{p})$, we might consider using a normal approximation based on $k \to \infty$ and the Central Limit Theorem. If we assume that p is known, then using moments of the binomial distribution (assuming the n_i are fixed constants), we have $\mathrm{E}(Q_i) = 1$ and $\mathrm{Var}(Q_i) = 2 + (1 - 6p + 6p^2) / \{n_i p(1-p)\}$. Averaging $\mathrm{Var}(Q_i)$, we get a standard error for T_S/k of $1.8035/\sqrt{23}$ using $k^{-1}\sum_{i=1}^{k}(1/n_i) = 0.0998$. The approximate normal test statistic is then $Z = 1.94$ leading to a p-value of .026.

We have ignored estimation of p in the above normal approximation. Does that matter? To find the asymptotic variance of T_S using the M-estimator approach, we need to treat the expected value of T_S/k as a parameter θ_1, and p as θ_2, and form two ψ functions:

$$\psi_1(Y_i, n_i, \theta_1, p) = \frac{(Y_i - n_i p)^2}{n_i p(1-p)} - \theta_1 \qquad \psi_2(Y_i, n_i, \theta_1, p) = Y_i - n_i p.$$

For calculating the $\boldsymbol{A}$ and $\boldsymbol{B}$ matrices we can treat the n_i like fixed constants in regression or as random variables with some distribution. Taking the latter approach and noting that $\theta_1 = 1$ under H_0, we get $\boldsymbol{A}_{11} = 1, \boldsymbol{A}_{12} = (1-2p)/[p(1-p)], \boldsymbol{A}_{21} = 0, \boldsymbol{A}_{22} = \mathrm{E}(n_i) = \mu_n$,

$$\boldsymbol{B}_{11} = 2 + \frac{(1 - 6p + 6p^2)}{p(1-p)} \mathrm{E}\left(\frac{1}{n_i}\right),$$

$\boldsymbol{B}_{12} = (1-2p), \boldsymbol{B}_{22} = \mu_n p(1-p)$. We have used the assumption that conditionally under H_0, $Y_i | n_i$ is binomial(n_i, p). Note that $\boldsymbol{B}_{11}$ corresponds to the average of the $\mathrm{Var}(Q_i)$ in the earlier derivation. The asymptotic variance of interest is then

$$\begin{aligned} \boldsymbol{V}_{11} = \left[\boldsymbol{A}^{-1}\boldsymbol{B}\{\boldsymbol{A}^{-1}\}^T\right]_{11} &= \boldsymbol{B}_{11} - \frac{2\boldsymbol{A}_{12}\boldsymbol{B}_{12}}{\boldsymbol{A}_{22}} + \frac{\boldsymbol{A}_{12}^2\boldsymbol{B}_{22}}{\boldsymbol{A}_{22}^2} \\ &= 2 + \frac{(1 - 6p + 6p^2)}{p(1-p)} \mathrm{E}\left(\frac{1}{n_i}\right) - \frac{(1-2p)^2}{\mu_n p(1-p)}. \end{aligned}$$

Plugging in $k^{-1}\sum(1/n_i)$ for $\mathrm{E}(1/n_i)$ as above and $k^{-1}\sum n_i = 12.87$ for μ_n gives a standard error of $1.8011/\sqrt{23}$ and a p-value of .026 just like the above case where we assumed p is known. Thus, in this case, because $\widetilde{p} = .456$ makes the correction term $(1 - 2\widetilde{p})^2/\{\mu_n\widetilde{p}(1-\widetilde{p})\} = .0024$ negligible, the estimation of p has little effect on the standard error. The effect would be noticeable if $p = .8$, for example. We also ran two parametric bootstraps with 100,000 resamples: conditional on $(n_1, \ldots, n_{23})$ yielding p-value=.028 and also with the n_i drawn with replacement from $(n_1, \ldots, n_{23})$ yielding p-value=.028. So the normal M-estimation approximation .026 to the p-value seems better here than the chi-squared approximation .034. We might add that the results are very sensitive to game 14 where Shaq made 9 free throws out of 9. Also, the related score statistic derived by Tarone (1979) from the beta-binomial model is weighted differently and results in a p-value of .25. ♦

7.7 Summary

M-estimators represent a very large class of statistics, including for example, maximum likelihood estimators and basic sample statistics like sample moments and sample quantiles as well as complex functions of these. The approach we have summarized makes standard error estimation and asymptotic analysis routine regardless of the complexity or dimension of the problem. In summary we would like to bring together the key features of M-estimators:

1. An M-estimator $\widehat{\boldsymbol{\theta}}$ satisfies (7.1, p. 298): $\sum_{i=1}^{n} \boldsymbol{\psi}(\boldsymbol{Y}_i, \widehat{\boldsymbol{\theta}}) = \mathbf{0}$, where $\boldsymbol{\psi}$ is a known function not depending on i or n. See also the extensions (7.2, p. 298) and (7.11, p. 301).
2. Many estimators that do not satisfy (7.1, p. 298) or the extensions (7.2, p. 298) and (7.11, p. 301) are components of higher-dimensional M-estimators and thus are amenable to M-estimator techniques using the method of stacking. Such estimators are called *partial M-estimators.*
3. $\boldsymbol{A}(\boldsymbol{\theta}_0) = \mathrm{E}\{-\partial\boldsymbol{\psi}(Y_1, \boldsymbol{\theta}_0)/\partial\boldsymbol{\theta}\}$ is the Fisher information matrix in regular iid parametric models when $\boldsymbol{\psi}$ is the log likelihood score function. More generally $\boldsymbol{A}(\boldsymbol{\theta}_0)$ must have an inverse but need not be symmetric. See also the extension (7.15, p. 311) for non-differentiable $\boldsymbol{\psi}$.
4. $\boldsymbol{B}(\boldsymbol{\theta}_0) = \mathrm{E}\left\{\boldsymbol{\psi}(Y_1, \boldsymbol{\theta}_0)\boldsymbol{\psi}(Y_1, \boldsymbol{\theta}_0)^T\right\}$ is also the Fisher information matrix in regular iid parametric models when $\boldsymbol{\psi}$ is the log likelihood score function. $\boldsymbol{B}(\boldsymbol{\theta}_0)$ always has the properties of a covariance matrix but is singular when at least one component of $\widehat{\boldsymbol{\theta}}$ is a non-random function of the other components of $\widehat{\boldsymbol{\theta}}$. In general, $\boldsymbol{A}(\boldsymbol{\theta}_0) \neq \boldsymbol{B}(\boldsymbol{\theta}_0)$; the noteworthy exception arises when a parametric model is assumed, and the $\boldsymbol{\psi}$ function results in the maximum likelihood estimates.
5. Under suitable regularity conditions, $\widehat{\boldsymbol{\theta}}$ is AMN $(\boldsymbol{\theta}_0, V(\boldsymbol{\theta}_0)/n)$ as $n \to \infty$, where $V(\boldsymbol{\theta}_0) = \boldsymbol{A}(\boldsymbol{\theta}_0)^{-1}\boldsymbol{B}(\boldsymbol{\theta}_0)\{\boldsymbol{A}(\boldsymbol{\theta}_0)^{-1}\}^T$ is the sandwich matrix.
6. One generally applicable estimator of $V(\boldsymbol{\theta}_0)$ for differentiable $\boldsymbol{\psi}$ functions is the empirical sandwich estimator

$$\boldsymbol{V}_n(\boldsymbol{Y}, \widehat{\boldsymbol{\theta}}) = \boldsymbol{A}_n(\boldsymbol{Y}, \widehat{\boldsymbol{\theta}})^{-1}\boldsymbol{B}_n(\boldsymbol{Y}, \widehat{\boldsymbol{\theta}})\{\boldsymbol{A}_n(\boldsymbol{Y}, \widehat{\boldsymbol{\theta}})^{-1}\}^T.$$

7.8 Appendix – Some Theorems for M-Estimation

In this appendix we give some theorems for consistency and asymptotic normality of M-estimators. In addition we give a WLLN for averages with estimated parameters in order to justify consistency of the sandwich matrix estimators. Standard references for these types of results include Huber (1964, 1967, 1981), Serfling (1980, Ch. 7), Randles (1982), and van der Vaart (1998, Ch. 5).

7.8.1 *Consistency of M-Estimators*

There are a variety of methods for proving consistency of M-estimators. We follow Theorem 6.5 (p. 282) which assumes a compact parameter space Θ and uses Theorem 6.3 (p. 281) to get uniform strong consistency of $\boldsymbol{G}_n(\boldsymbol{\theta})$,

$$\boldsymbol{G}_n(\boldsymbol{\theta}) = \frac{1}{n}\sum_{i=1}^{n} \boldsymbol{\psi}(Y_i, \boldsymbol{\theta}) \xrightarrow{wp1} \boldsymbol{G}_{\mathrm{F}}(\boldsymbol{\theta}) = \int \boldsymbol{\psi}(y, \boldsymbol{\theta})\, dF(y)$$

uniformly on Θ.

Theorem 7.1. *Suppose that $Y_1, \ldots, Y_n$ are iid with distribution function $F(y)$, and $\boldsymbol{\theta}$ is a $b \times 1$ parameter lying in Θ, a compact subset of R^b. Assume that*

1. *(Uniqueness) $\boldsymbol{G}_F(\boldsymbol{\theta})$ exists for all $\boldsymbol{\theta} \in \Theta$, $\boldsymbol{G}_F(\boldsymbol{\theta}_0) = 0$, and $\boldsymbol{G}_F(\boldsymbol{\theta}) \neq 0$ for all other values of $\boldsymbol{\theta}$ in Θ.*
2. *Each component of $\boldsymbol{\psi}(y, \boldsymbol{\theta})$ satisfies the assumptions of Theorem 6.3 (p. 281): $\psi_j(y, \boldsymbol{\theta})$ is continuous in $\boldsymbol{\theta}$ and bounded by an integrable function of y that does not depend on $\boldsymbol{\theta}$, $j = 1, \ldots, b$.*

If $\boldsymbol{G}_n(\widehat{\boldsymbol{\theta}}_n) \xrightarrow{wp1} 0$, then $\widehat{\boldsymbol{\theta}}_n \xrightarrow{wp1} \boldsymbol{\theta}_0$ as $n \to \infty$.

Proof. The proof is almost exactly the same as the proof of Theorem 6.5 (p. 282) with $\boldsymbol{G}_n$ in the place $\bar{l}_n$ and $\boldsymbol{G}_{\mathrm{F}}$ in place of $\bar{l}$. Similarly we assume, for contradiction purposes, existence of a subsequence $\{\widehat{\boldsymbol{\theta}}_{n_k}\}$ such that $\widehat{\boldsymbol{\theta}}_{n_k} \xrightarrow{wp1} \boldsymbol{\theta}_1 \neq \boldsymbol{\theta}_0$. Then

$$|\, \boldsymbol{G}_n(\widehat{\boldsymbol{\theta}}_{n_k}) - \boldsymbol{G}_{\mathrm{F}}(\boldsymbol{\theta}_1)\,| \leq |\, \boldsymbol{G}_n(\widehat{\boldsymbol{\theta}}_{n_k}) - \boldsymbol{G}_{\mathrm{F}}(\widehat{\boldsymbol{\theta}}_{n_k})\,| + |\, \boldsymbol{G}_{\mathrm{F}}(\widehat{\boldsymbol{\theta}}_{n_k}) - \boldsymbol{G}_{\mathrm{F}}(\boldsymbol{\theta}_1)\,|.$$

The first term on the right-hand side converges to 0 by the uniform strong consistency of $\boldsymbol{G}_n(\boldsymbol{\theta})$. The second term converges to 0 by continuity of $\boldsymbol{G}_{\mathrm{F}}$, which easily follows from Condition 2. Thus $\boldsymbol{G}_n(\widehat{\boldsymbol{\theta}}_{n_k}) \xrightarrow{wp1} \boldsymbol{G}_{\mathrm{F}}(\boldsymbol{\theta}_1) \neq 0$, and this contradicts $\boldsymbol{G}_n(\widehat{\boldsymbol{\theta}}_n) \xrightarrow{wp1} 0$. Note that Theorem 7.1 also holds if we change the "wp1" statements to "in probability." ∎

7.8.2 Asymptotic Normality of M-Estimators

Next we give an asymptotic normality theorem that is the direct generalization of Theorem 6.7 (p. 286).

Theorem 7.2. *Let $Y_1, \ldots, Y_n$ be iid with distribution function $F(y)$. Assume that*

1. *$\psi(y, \theta)$ and its first two partial derivatives with respect to θ exist for all y in the support of F and for all θ in a neighborhood of θ_0, where $G_F(\theta_0) = 0$.*
2. *For each θ in a neighborhood of θ_0, there exists a function $g(y)$ (possibly depending on θ_0) such that for all j, k and $l \in \{1, \ldots, b\}$,*

$$\left| \frac{\partial^2}{\partial \theta_j \partial \theta_k} \psi_l(y, \theta) \right| \leq g(y)$$

for all y and where $\int g(y)\, dF(y) < \infty$.

3. *$A(\theta_0) = E\left\{-\psi'(Y_1, \theta_0)\right\}$ exists and is nonsingular.*
4. *$B(\theta_0) = E\left\{\psi(Y_1, \theta_0)\psi(Y_1, \theta_0)^T\right\}$ exists and is finite.*

If $G_n(\widehat{\theta}) = o_p(n^{-1/2})$ and $\widehat{\theta} \xrightarrow{p} \theta_0$, then

$$\sqrt{n}(\widehat{\theta} - \theta_0) \xrightarrow{d} N\left[0, A(\theta_0)^{-1} B(\theta_0) \left\{A(\theta_0)^{-1}\right\}^T\right] \qquad as\ n \to \infty.$$

Proof. The proof uses a component-wise expansion of $G_n(\widehat{\theta})$ similar to that in (6.21, p. 289) used in the proof of Theorem 6.10 (p. 288). By assumption $G_n(\widehat{\theta}) = o_p(n^{-1/2})$ and thus a Taylor series expansion of the jth component of $G_n(\widehat{\theta})$ results in

$$\begin{aligned} o_p(n^{-1/2}) &= G_{n,j}(\widehat{\theta}) \\ &= G_{n,j}(\theta_0) + G'_{n,j}(\theta_0)(\widehat{\theta} - \theta_0) + \frac{1}{2}(\widehat{\theta} - \theta_0)^T G''_{n,j}(\widetilde{\theta}_j^*)(\widehat{\theta} - \theta_0) \\ &= G_{n,j}(\theta_0) + \left\{ G'_{n,j}(\theta_0) + \frac{1}{2}(\widehat{\theta} - \theta_0)^T G''_{n,j}(\widetilde{\theta}_j^*) \right\} (\widehat{\theta} - \theta_0), \end{aligned}$$

where $\widetilde{\theta}_j^*$ is on the line segment joining $\widehat{\theta}$ and θ_0, $j = 1, \ldots, b$. Writing these b equations in matrix notation we have

$$o_p(n^{-1/2}) = G_n(\theta_0) + \left\{ G'_n(\theta_0) + \frac{1}{2}\widetilde{Q}^* \right\} (\widehat{\theta} - \theta_0),$$

where $\widetilde{Q}^*$ is the $b \times b$ matrix with jth row given by $(\widehat{\theta} - \theta_0)^T G''_{n,j}(\widetilde{\theta}_j^*)$. Note that under Condition 2, each entry in $\widetilde{Q}^*$ is bounded by $||\widehat{\theta} - \theta_0|| n^{-1} \sum g(Y_i) = o_p(1)$, and thus $\widetilde{Q}^* = \boldsymbol{o}_p(1)$. By the WLLN $\boldsymbol{G}'_n(\boldsymbol{\theta}_0) \xrightarrow{p} -\boldsymbol{A}(\boldsymbol{\theta}_0)$ which is nonsingular under Condition 3. Thus for n sufficiently large, the matrix in brackets above is nonsingular with probability approaching 1. On the set where the matrix in brackets is nonsingular (call that set S_N) we have

$$\sqrt{n}(\widehat{\boldsymbol{\theta}} - \boldsymbol{\theta}_0) = -\left\{\boldsymbol{G}'_n(\boldsymbol{\theta}_0) + \frac{1}{2}\widetilde{Q}^*\right\}^{-1} \{\sqrt{n}\boldsymbol{G}_n(\boldsymbol{\theta}_0) + \boldsymbol{o}_p(1)\}.$$

Slutsky's Theorem and the CLT then give the result when we note that $P(S_N) \rightarrow 1$. As in Problem 6.6 (p. 293), we could also add and subtract terms to give an approximation-by-averages representation, where $h_F(Y_i, \boldsymbol{\theta}_0) = \boldsymbol{A}(\boldsymbol{\theta}_0)^{-1} \boldsymbol{\psi}(Y_i, \boldsymbol{\theta}_0)$. ■

7.8.3 *Weak Law of Large Numbers for Averages with Estimated Parameters*

One of the most useful aspects of the M-estimator approach is the availability of the empirical sandwich estimator (7.12, p. 302). Thus, it is important that the pieces of this estimator, $\boldsymbol{A}_n(\boldsymbol{Y}, \widehat{\boldsymbol{\theta}})$ and $\boldsymbol{B}_n(\boldsymbol{Y}, \widehat{\boldsymbol{\theta}})$, converge in probability to $\boldsymbol{A}(\boldsymbol{\theta}_0)$ and $\boldsymbol{B}(\boldsymbol{\theta}_0)$, respectively. But note that this convergence would follow immediately from the WLLN except for the presence of $\widehat{\boldsymbol{\theta}}$. Thus, the next two theorems give conditions for the WLLN to hold for averages whose summands are a function of $\widehat{\boldsymbol{\theta}}$ (and thus dependent). The first theorem assumes differentiability and a bounding function similar to Theorem 5.28 (p. 249). The second uses monotonicity.

Theorem 7.3. *Suppose that $Y_1, \ldots, Y_n$ are iid with distribution function F and assume that the real-valued function $q(Y_i, \boldsymbol{\theta})$ is differentiable with respect to $\boldsymbol{\theta}$, $E_F|\, q'(Y_1, \boldsymbol{\theta}_0)\,| < \infty$, and there exists a function $M(y)$ such that for all $\boldsymbol{\theta}$ in a neighborhood of $\boldsymbol{\theta}_0$ and all $j \in \{1, \ldots, b\}$,*

$$\left| \frac{\partial}{\partial \theta_j} q(y, \boldsymbol{\theta}) \right| \leq M(y),$$

where $E_F\{M(Y_1)\} < \infty$. If $\widehat{\boldsymbol{\theta}} \xrightarrow{p} \boldsymbol{\theta}_0$, then $n^{-1}\sum_{i=1}^n q(Y_i, \widehat{\boldsymbol{\theta}}) \xrightarrow{p} E_F q(Y_1, \boldsymbol{\theta}_0)$ as $n \rightarrow \infty$.

Proof.

$$\left| \frac{1}{n}\sum_{i=1}^n q(Y_i, \widehat{\boldsymbol{\theta}}) - E_F q(Y_1, \boldsymbol{\theta}_0) \right| \leq \left| \frac{1}{n}\sum_{i=1}^n q(Y_i, \widehat{\boldsymbol{\theta}}) - \frac{1}{n}\sum_{i=1}^n q(Y_i, \boldsymbol{\theta}_0) \right| + \left| \frac{1}{n}\sum_{i=1}^n q(Y_i, \boldsymbol{\theta}_0) - E_F q(Y_1, \boldsymbol{\theta}_0) \right|$$

The second term on the right-hand side converges to 0 in probability by the WLLN. For the first term, we have for some $\widehat{\boldsymbol{\theta}}^*$ between $\widehat{\boldsymbol{\theta}}$ and $\boldsymbol{\theta}_0$

$$\left|\frac{1}{n}\sum_{i=1}^{n} q(Y_i,\widehat{\boldsymbol{\theta}}) - \frac{1}{n}\sum_{i=1}^{n} q(Y_i,\boldsymbol{\theta}_0)\right| \leq \left|\frac{1}{n}\sum_{i=1}^{n} q'(Y_i,\widehat{\boldsymbol{\theta}}^*)\right| \left|\widehat{\boldsymbol{\theta}} - \boldsymbol{\theta}_0\right|$$

$$\leq \frac{1}{n}\sum_{i=1}^{n} M(Y_i)\left|\widehat{\boldsymbol{\theta}} - \boldsymbol{\theta}_0\right|.$$

This latter bound is $o_p(1)$ by the WLLN and the convergence of $\widehat{\boldsymbol{\theta}}$. ∎

This next theorem is only for a real parameter θ and assumes monotonicity of q with respect to θ. A similar, but more general theorem is given by Theorem 2.9 of Iverson and Randles (1989).

Theorem 7.4. *Let $Y_1,\ldots,Y_n$ be iid with distribution function F. Assume that*

1. *There exists a neighborhood, $N(\theta_0)$, of θ_0 such that the real-valued function $q(\boldsymbol{y},\theta)$ is a monotone function of θ for $\theta \in N(\theta_0)$ and for all $\boldsymbol{y}$ in the support of F.*
2. *$H(\theta) = E\{q(\boldsymbol{Y}_1,\theta)\}$ exists in a neighborhood of θ_0 and is continuous at θ_0.*

If $\widehat{\theta} \xrightarrow{p} \theta_0$ as $n \to \infty$, then

$$\frac{1}{n}\sum_{i=1}^{n} q(\boldsymbol{Y}_i,\widehat{\theta}) \xrightarrow{p} H(\theta_0) \qquad \text{as } n \to \infty.$$

Proof. Assume without loss of generality that q is nondecreasing.

Let $\widehat{T}_n = n^{-1}\sum_{i=1}^{n} q(Y_i,\widehat{\theta})$ and $\widetilde{T}_n = n^{-1}\sum_{i=1}^{n} q(Y_i,\theta_0)$. Then for $\delta > 0$

$$|\widehat{T}_n - \widetilde{T}_n| = R_{n,1} + R_{n,2},$$

where $R_{n,1} = |\widehat{T}_n - \widetilde{T}_n| I(|\widehat{\theta} - \theta_0| > \delta)$ and $R_{n,2} = |\widehat{T}_n - \widetilde{T}_n| I(|\widehat{\theta} - \theta_0| \leq \delta)$. For any $\epsilon > 0$,

$$P(|R_{n,1}| > \epsilon) \leq P(|\widehat{\theta} - \theta_0| > \delta) \to 0 \qquad \text{as } n \to \infty$$

since $\widehat{\theta} \xrightarrow{p} \theta_0$. Now for $R_{n,2}$ choose $\delta = \delta_\epsilon > 0$ such that $|H(\theta_0 + \delta_\epsilon) - H(\theta_0 - \delta_\epsilon)| < \epsilon^2$. Then

$$R_{n,2} = \left|\frac{1}{n}\sum_{i=1}^{n} q(Y_i,\widehat{\theta}) - \frac{1}{n}\sum_{i=1}^{n} q(Y_i,\theta_0)\right| I(|\widehat{\theta} - \theta_0| \leq \delta_\epsilon)$$

$$\leq \frac{1}{n}\sum_{i=1}^{n} \left|q(Y_i,\widehat{\theta}) - q(Y_i,\theta_0)\right| I(|\widehat{\theta} - \theta_0| \leq \delta_\epsilon)$$

$$\leq \frac{1}{n}\sum_{i=1}^{n}\left[q(Y_i,\theta_0+\delta_\epsilon)-q(Y_i,\theta_0-\delta_\epsilon)\right] I(|\widehat{\theta}-\theta_0|\leq\delta_\epsilon)$$

$$\leq \frac{1}{n}\sum_{i=1}^{n}\left[q(Y_i,\theta_0+\delta_\epsilon)-q(Y_i,\theta_0-\delta_\epsilon)\right].$$

Monotonicity of q was used to get the next to the last inequality. Call the last expression above W_n and note that it is nonnegative. Then by the above bound and the Markov inequality, we have

$$P(R_{n,2}>\epsilon)\leq P(W_n>\epsilon)\leq\frac{\mathrm{E}W_n}{\epsilon}=\frac{H(\theta_0+\delta_\epsilon)-H(\theta_0-\delta_\epsilon)}{\epsilon}<\frac{\epsilon^2}{\epsilon}=\epsilon.$$

Thus we have shown that $R_{n,1}$ and $R_{n,2}$ each converge in probability to zero. ∎

7.9 Problems

7.1. Suppose that the $b\times b$ matrix $\boldsymbol{C}(\boldsymbol{\theta})$ is nonsingular at $\boldsymbol{\theta}=\boldsymbol{\theta}_0$. Show that the sandwich matrix $\boldsymbol{V}(\boldsymbol{\theta}_0)$ is the same for the $b\times 1$ $\boldsymbol{\psi}(Y_i,\boldsymbol{\theta})$ as it is for $\boldsymbol{\psi}^*(Y_i,\boldsymbol{\theta})=\boldsymbol{C}(\boldsymbol{\theta})\boldsymbol{\psi}(Y_i,\boldsymbol{\theta})$.

7.2. Let $Y_1,\ldots,Y_n$ be iid from some distribution with finite fourth moment. The coefficient of variation is $\widehat{\theta}_3=s_n/\overline{Y}$.

a. Define a three dimensional ψ so that $\widehat{\theta}_3$ is defined by summing the third component. Find A, B, and V, where

$$V_{33}=\frac{\sigma^4}{\mu^4}-\frac{\mu_3}{\mu^3}+\frac{\mu_4-\sigma^4}{4\mu^2\sigma^2}.$$

b. Now recompute the asymptotic distribution of $\widehat{\theta}_3$ using the delta method applied to $(\overline{Y},s_n^2)$. (I used Maple to do this, but do it by hand if you like.)

7.3. Suppose that $Y_1,\ldots,Y_n$ are iid from a gamma(α,β) distribution.

a. One version of the method of moments is to set $\overline{Y}$ equal to $\mathrm{E}(Y_1)=\alpha\beta$ and $n^{-1}\sum_{i=1}^{n}Y_i^2$ equal to $\mathrm{E}(Y_1^2)=\alpha\beta^2+(\alpha\beta)^2$ and solve for the estimators. Use Maple (at least it's much easier if you do) to find $V=A^{-1}B\{A^{-1}\}^T$. Here it helps to know that $\mathrm{E}(Y_1^3)=[\alpha(1+\alpha)(2+\alpha)]\beta^3$ and $\mathrm{E}(Y_1^4)=[\alpha(1+\alpha)(2+\alpha)(3+\alpha)\beta^4]$. Show your derivation of A and B and attach Maple output.
b. The second version of the method of moments (and perhaps the easier method) is to set $\overline{Y}$ equal to $\mathrm{E}(Y_1)=\alpha\beta$ and s^2 equal to $\mathrm{var}(Y_1)=\alpha\beta^2$ and solve for the estimators. You could use either the "$n-1$" or "n" version of s^2, but here we want to use the "n" version in order to fit into the M-estimator theory. Compute V as in a) except that the second component of the ψ function is different from a) (but V should be same). Here it helps to know that $\mu_3=2\alpha\beta^3$ and $\mu_4=3[\alpha^2+2\alpha]\beta^4$.

c. The asymptotic variance of the MLEs for α and β are $\text{Avar}(\widehat{\alpha}_{\text{MLE}}) = 1.55/n$ for $\alpha = 1.0$ and $\text{Avar}(\widehat{\alpha}_{\text{MLE}}) = 6.90/n$ for $\alpha = 2.0$. Similarly, $\text{Avar}(\widehat{\beta}_{\text{MLE}}) = 2.55\beta^2/n$ for $\alpha = 1.0$ and $\text{Avar}(\widehat{\beta}_{\text{MLE}}) = 3.45\beta^2/n$ for $\alpha = 2.0$. Now calculate the asymptotic relative efficiencies of the MLEs to the method of moment estimators for $\alpha = 1.0$ and $\alpha = 2.0$ using results from a.

7.4. Suppose that $Y_1, \ldots, Y_n$ are iid and $\widehat{\boldsymbol{\theta}}$ satisfies

$$\sum_{i=1}^{n} \boldsymbol{\psi}(Y_i, \widehat{\boldsymbol{\theta}}) = \boldsymbol{c}_n,$$

where we assume

i) $\widehat{\boldsymbol{\theta}} \xrightarrow{p} \boldsymbol{\theta}_0$
ii) $\boldsymbol{c}_n/\sqrt{n} \xrightarrow{p} 0$
iii) The remainder term $\boldsymbol{R}_n$ from the expansion

$$G_n(\widehat{\boldsymbol{\theta}}) = n^{-1} \sum_{i=1}^{n} \boldsymbol{\psi}(Y_i, \widehat{\boldsymbol{\theta}}) = G_n(\boldsymbol{\theta}_0) + G_n'(\boldsymbol{\theta}_0)(\widehat{\boldsymbol{\theta}} - \boldsymbol{\theta}_0) + \boldsymbol{R}_n,$$

satisfies $\sqrt{n}\boldsymbol{R}_n \xrightarrow{p} 0$.

Show that $\widehat{\boldsymbol{\theta}}$ is $\text{AN}(\boldsymbol{\theta}_0, V(\boldsymbol{\theta}_0)/n)$, i.e., the same result as for the usual case when $\boldsymbol{c}_n = 0$.

7.5. Repeat the calculations in Section 7.2.4 (p. 305) with $\widehat{\theta}_3 = \Phi\left\{(a - \overline{Y})/s_n\right\}$ replacing s_n and $\widehat{\theta}_4 = \Phi^{-1}(p)s_n + \overline{Y}$ replacing $\log(s_n^2)$. Note that $\widehat{\theta}_3$ and $\widehat{\theta}_4$ are the maximum likelihood estimators of $P(Y_1 \le a)$ and the pth quantile of Y_1, respectively, under a normal distribution assumption.

7.6 (Delta Theorem via M-estimation). Suppose that $\widehat{\boldsymbol{\theta}}$ is a b-dimensional M-estimator with defining function $\boldsymbol{\psi}(y, \boldsymbol{\theta})$ and such that the usual quantities $\boldsymbol{A}$ and $\boldsymbol{B}$ exist. Here we want to essentially reproduce Theorem 5.19 (p. 238) for $g(\widehat{\boldsymbol{\theta}})$, where g satisfies the assumptions of Theorem 5.19 and $b_n^2\boldsymbol{\Sigma} = n^{-1}\boldsymbol{V}(\boldsymbol{\theta})$, where $\boldsymbol{V}(\boldsymbol{\theta}) = \boldsymbol{A}(\boldsymbol{\theta})^{-1}\boldsymbol{B}(\boldsymbol{\theta})\{\boldsymbol{A}(\boldsymbol{\theta})^{-1}\}^T$. So add the ψ function $g(\boldsymbol{\theta}) - \theta_{b+1}$ to $\boldsymbol{\psi}(y, \boldsymbol{\theta})$, compute the relevant matrices, say $\boldsymbol{A}^*$, $\boldsymbol{B}^*$, and $\boldsymbol{V}^*$, and show that the last diagonal element of $\boldsymbol{V}^*$ is $g'(\boldsymbol{\theta})\boldsymbol{V}(\boldsymbol{\theta})g'(\boldsymbol{\theta})^T$.

7.7. The generalized method of moments (GMM) is an important estimation method found mainly in the econometrics literature and closely related to M-estimation. Suppose that we have iid random variables $Y_1, \ldots, Y_n$ and a p dimensional unknown parameter $\boldsymbol{\theta}$. The key idea is that there are a set of $q \ge p$ possible estimating equations

$$\frac{1}{n} \sum_{i=1}^{n} \psi_j(Y_i; \boldsymbol{\theta}) = 0, \quad j = 1, \ldots, q$$

motivated by the fact that $\mathrm{E}\psi_j(Y_1; \boldsymbol{\theta}_0) = 0$, $j = 1, \ldots, q$, where $\boldsymbol{\theta}_0$ is the true value. These motivating zero expectations come from the theory in the subject area being studied. But notice that if $q > p$, then we have too many equations. The GMM approach is to minimize the objective function

$$T = \left[\frac{1}{n}\sum_{i=1}^{n} \boldsymbol{\psi}(Y_i; \boldsymbol{\theta})\right]^T \boldsymbol{W} \left[\frac{1}{n}\sum_{i=1}^{n} \boldsymbol{\psi}(Y_i; \boldsymbol{\theta})\right],$$

where $\boldsymbol{\psi} = (\psi_1, \ldots, \psi_q)^T$ and $\boldsymbol{W}$ is a matrix of weights. Now let's simplify the problem by letting $q = 2$, $p = 1$ so that θ is real-valued, and $\boldsymbol{W} = \mathrm{diag}(w_1, w_2)$. Then T reduces to

$$T = w_1 \left[\frac{1}{n}\sum_{i=1}^{n} \psi_1(Y_i; \theta)\right]^2 + w_2 \left[\frac{1}{n}\sum_{i=1}^{n} \psi_2(Y_i; \theta)\right]^2.$$

To find $\widehat{\theta}$ we just take the partial derivative of T with respect to θ and set it equal to 0:

$$S(\boldsymbol{Y}; \theta) = 2w_1 \left[\frac{1}{n}\sum_{i=1}^{n} \psi_1(Y_i; \theta)\right]\left[\frac{1}{n}\sum_{i=1}^{n} \psi_1'(Y_i; \theta)\right]$$
$$+2w_2 \left[\frac{1}{n}\sum_{i=1}^{n} \psi_2(Y_i; \theta)\right]\left[\frac{1}{n}\sum_{i=1}^{n} \psi_2'(Y_i; \theta)\right] = 0.$$

a. Prove that $S(\boldsymbol{Y}; \theta_0)$ converges to 0 in probability as $n \to \infty$ making any moment assumptions that you need. (This should suggest to you that the solution of the equation $S(\boldsymbol{Y}; \theta) = 0$ above is consistent.)
b. To get asymptotic normality for $\widehat{\theta}$, a direct approach is to expand $S(\boldsymbol{Y}; \widehat{\theta})$ around θ_0 and solve for $\widehat{\theta} - \theta_0$.

$$\widehat{\theta} - \theta_0 = \left[-\frac{\partial S(\boldsymbol{Y}; \theta_0)}{\partial \theta^T}\right]^{-1} S(\boldsymbol{Y}; \theta_0) + \left[-\frac{\partial S(\boldsymbol{Y}; \theta_0)}{\partial \theta^T}\right]^{-1} R_n.$$

Then one ignores the remainder term and uses Slutsky's Theorem along with asymptotic normality of $S(\boldsymbol{Y}; \theta_0)$. But how to get the asymptotic normality of $S(\boldsymbol{Y}; \theta_0)$? Find $h(Y_i; \theta_0)$ such that

$$S(\boldsymbol{Y}; \theta_0) = \frac{1}{n}\sum_{i=1}^{n} h(Y_i; \theta_0) + R_n^*.$$

No proofs are required.

c. The equation $S(\boldsymbol{Y};\theta) = 0$ is not in the form for using M-estimation results (because the product of sums is not a simple sum). Show how to get it in M-estimation form by adding two new parameters, θ_2 and θ_3, and two new equations so that the result is a system of three equations with three ψ functions; call them ψ_1^*, ψ_2^*, and ψ_3^* because ψ_1^* is actually a function of the original ψ_1 and ψ_2.

7.8. In Example 7.4.1 (p. 311) we discussed the well-known Huber location estimator defined by $\sum_{i=1}^n \psi_k(Y_i - \widehat{\theta}) = 0$. For real data situations, the scaling is not known, and this simple equation is replaced by two equations,

$$\sum_{i=1}^{n} \psi_k\left(\frac{Y_i - \widehat{\theta}}{\widehat{\sigma}}\right) = 0,$$

$$\sum_{i=1}^{n} \chi\left(\frac{Y_i - \widehat{\theta}}{\widehat{\sigma}}\right) = 0,$$

where $\chi(x)$ is an even function about zero. A typical choice of χ is $\chi(x) = \psi_k^2 - c$, where $c = E\psi_k^2(Z)$ and Z is a standard normal random variable. This constant $E\psi_k^2(Z)$ forces σ_0 to be the standard deviation when the data come from a normal distribution. But use the general $\chi(x)$ for the following parts.

a. Find $\boldsymbol{A}$ and $\boldsymbol{B}$ for $(\widehat{\mu}, \widehat{\sigma})$.
b. Simplify the expressions in a. when $Y_1, \ldots, Y_n$ are iid from a location-scale density $\sigma^{-1} f_0((y - \mu)/\sigma)$ and $f_0(x)$ is symmetric about zero. In addition, compute $\boldsymbol{V}$ because the off-diagonal elements of $\boldsymbol{V}$ should be zero.

7.9. Let $Y_1, \ldots, Y_n$ be iid from some continuous distribution with nonzero density at the population mean, $f(\mu_0) > 0$. For defining a "positive" standard deviation similar to the positive mean deviation from the median, one needs to estimate the proportion of the population that lies above the mean, $1 - p_0 = 1 - F(\mu_0)$, using $1 - F_n(\overline{Y})$, where F_n is the empirical distribution function. Letting $\boldsymbol{\psi}(Y_i; \mu, p)^T = (Y_i - \mu, I(Y_i \le \mu) - p)$, find $A(\boldsymbol{\theta}_0)$, $B(\boldsymbol{\theta}_0)$, and $V(\boldsymbol{\theta}_0)$. Note that we have to use the nonsmooth definition of $A(\boldsymbol{\theta}_0)$:

$$A(\boldsymbol{\theta}_0) = -\frac{\partial}{\partial \theta}\{\mathrm{E}_F \boldsymbol{\psi}(Y_1; \mu, p)\}\bigg|_{\mu=\mu_0, p=p_0}.$$

The off-diagonal element of B turns out to be $b_{12} = \int_{-\infty}^{\mu_0} y f(y) dy - p_0\mu_0$. In your calculations, just let this be called b_{12}.

7.10. Let $Y_1, \ldots, Y_n$ be iid from some continuous distribution with nonzero density at the population mean, $f(\mu_0) > 0$. Add $\psi_3(Y_i, \boldsymbol{\theta}) = 2(\theta_2 - Y_i)I(Y_i < \theta_2) - \theta_3$ to the $\boldsymbol{\psi}$ function of Section 7.4.3 (p. 313) and find $\boldsymbol{A}$, $\boldsymbol{B}$, and $\boldsymbol{V}$. You may leave some elements of $\boldsymbol{B}$ in the form b_{ij} but give their integral expression as was done for b_{11}. (This ψ_3 is for the "negative" mean deviation from the median. If the underlying density is symmetric about the true median, then the asymptotic correlation between $\widehat{\theta}_1$ and $\widehat{\theta}_3$ should be 0.)

7.11. Consider the simple no intercept model

$$Y_i = \beta x_i + e_i, \quad i = 1, \ldots, n,$$

where the $x_1, \ldots, x_n$ are fixed constants and $e_1, \ldots, e_n$ are independent with $\mathrm{E}\psi^*(e_i) = 0$ and $\mathrm{E}\psi^*(e_i)^2 < \infty$ with values possibly depending on i, and ψ^* is a real-valued function (with superscript "*" to differentiate from our usual $\psi(Y_i, \boldsymbol{x}_i, \boldsymbol{\beta})$).

a. A classical robust M-estimator $\widehat{\beta}$ satisfies

$$\sum_{i=1}^{n} \psi^*(Y_i - \widehat{\beta} x_i) x_i = 0.$$

Write down the asymptotic variance of $\widehat{\beta}$ and an estimator of this asymptotic variance. You may assume that the derivative $\psi^{*\prime}$ of ψ^* exists everywhere and that $\mathrm{E}|\psi^{*\prime}(e_i)| < \infty$ for all i.

b. Let's simplify and assume that $e_1, \ldots, e_n$ are iid. How does the asymptotic variance expression from a) change? Find an improved asymptotic variance estimator in this situation.

7.12. Suppose that we think our true model is

$$Y_i = \boldsymbol{X}_i^T \boldsymbol{\beta}_0 + g(\boldsymbol{X}_i) e_i,$$

where g is a known function and $e_1, \ldots, e_n$ are iid with mean 0 and variance σ^2.

a. Using this model, write down the weighted least squares estimator of $\boldsymbol{\beta}$.
b. Even though we conjecture that we have the right variance model, and are willing to bet that the weighted least squares estimator has lower variance than the unweighted least squares estimator, we might still want our variance estimate to be robust to misspecification of the variance. Thus, find the sandwich variance estimator for the estimator in a.

(Both a. and b. are trivial if you take the transformation approach to weighted least squares.)

7.13. Consider the nonlinear least squares model

$$Y_i = \exp(\beta X_i) + e_i, \qquad i = 1, \ldots, n,$$

where β is a scalar parameter, $X_1, \ldots, X_n$ are iid random predictor variables, and $e_1, \ldots, e_n$ are iid random errors with mean 0 and variance σ^2 and are independent of $X_1, \ldots, X_n$. The parameter β is estimated by the nonlinear least squares estimator, $\widehat{\beta}_{\mathrm{LS}}$, found by minimizing

$$\sum_{i=1}^{n} \{Y_i - \exp(\beta X_i)\}^2 .$$

The variance parameter, σ^2, is estimated by

$$\widehat{\sigma}^2 = \frac{1}{n}\sum_{i=1}^{n}\left\{Y_i - \exp(\widehat{\beta}_{\mathrm{LS}}X_i)\right\}^2.$$

a. Show that $\widehat{\sigma}^2$ is a partial M-estimator by finding an appropriate ψ function (necessarily of dimension >1) defining $\widehat{\sigma}^2$.
b. Without making any distributional assumptions (you may assume various expectations exist as needed) determine the asymptotic distribution of $\widehat{\sigma}^2$ and derive an estimator of its asymptotic variance.

7.14. Suppose that rat mothers are fed a possible carcinogen after impregnation, and we let Y_i = the number of baby rats born with malformations. For deriving estimators, assume that conditional on $n_1, \ldots, n_k$, $Y_1, \ldots, Y_k$ are independent with Y_i having a binomial (n_i, p) distribution.

a. Write down the likelihood and show that the maximum likelihood estimator is $\widehat{p} = \sum Y_i / \sum n_i$. Also find $\overline{\boldsymbol{I}}(p)$ and the usual estimator of $\mathrm{Var}(\widehat{p})$.
b. Now suppose that the Y_i are no longer binomial but still independent and $\mathrm{E}(Y_i|n_i) = n_i p$. Find $A_k(\widehat{p})$, $B_k(\widehat{p})$, and the sandwich estimator of $\mathrm{Var}(\widehat{p})$. (One simple model for this is the beta-binomial: first you select a rat mother from a population of mothers where the ith mother has p_i drawn from a beta distribution with mean p. Thus $Y_i|n_i, p_i$ is assumed to be binomial (n_i, p_i), and $\mathrm{E}(Y_i|n_i)=\mathrm{E}\mathrm{E}(Y_i|n_i, p_i)=\mathrm{E}(n_i p_i|n_i) = n_i p$. In this case the variance of Y_i is larger than the binomial variance $n_i p(1-p)$.)

7.15. Consider a dose-response situation with k dose levels $d_1, \ldots, d_k$, where at the ith dose we observe $\left\{Y_{ij}, n_{ij}, j = 1, \ldots, m_i\right\}$, and we assume that

$$\mathrm{E}\left(Y_{ij}/n_{ij}|n_{ij}\right) = p_i(\boldsymbol{\beta}) = [1 + \exp(-\beta_1 - \beta_2 d_i)]^{-1}.$$

That is the primary part of the model. To derive an estimator, we make the assumption that the $Y_{ij}|n_{ij}$ have independent binomial$(p_i(\boldsymbol{\beta}), n_{ij})$ distributions. (We may not really believe that assumption.) The log likelihood is thus

$$l_n(\beta) = c + \sum_{i=1}^{k}\sum_{j=1}^{m_i}\left[Y_{ij}\log p_i(\boldsymbol{\beta})(n_{ij} - Y_{ij})\log(1 - p_i(\boldsymbol{\beta}))\right].$$

a. Take derivatives with respect to $\boldsymbol{\beta}$ to get the likelihood equations and the two-dimensional ψ function. Verify that each summand of the likelihood equation has mean 0 using the primary model assumption.
b. Derive the sandwich estimator of $\mathrm{Var}(\widehat{\boldsymbol{\beta}})$. You need to treat the double sum as a single sum and treat the n_{ij} as fixed constants.

7.16. The data `tcdd.dat.txt` on the course website

http://www4.stat.ncsu.edu/~boos/Essential.Statistical.Inference

are from a study on the teratogenic effects of 2,3,7,8-tetrachlorodibenzo-*p*-dioxin (TCDD) in mice (Birmbaum et al. 1991). The third column in the data set is the number of baby mice having cleft palates out of a litter size given by the fourth column. The second column is the dose (the units of measurement are not important). Each row of the data set relates to a different pregnant mice mother given TCDD on gestation day 10. The purpose of this problem is to get familiar with how to use GEE in SAS and to get sandwich variance estimates. Save this data file.

Now get into SAS and run the program `tcdd.sas1.txt` (Note that you have to modify location of the data set in the `infile` statement.) It gives a standard logistic regression of the sample proportions on the logarithm of dose. Next, take the "*" off the repeated statement. This gives a GEE analysis and standard errors from the sandwich matrix. But it does not actually display the sandwich matrix nor does it allow you to try some other "working" correlation structures. (The reason is that genmod actually needs the 0-1 binary Y's instead of the proportions.) The "working" correlation structure is just a guess at the correlation structure within a cluster. So now get the file `tcdd.sas2.txt` from the website. In the data step of this program the individual binary responses for each litter are constructed. You might run a `PROC PRINT` to see what the created data actually looks like. Submit this program and you get the same answers as in the previous program but now you get to see the sandwich matrix and the estimated working correlation structure. Finally, change `type=ind` to `type=exch` and resubmit. This uses an exchangeable working correlation structure. Here is what you are to turn in (no computer output please).

a. The parameter estimates and standard errors from regular logistic regression.
b. The parameter estimates and sandwich standard errors from GEE with the independent working correlation structure and the 2 by 2 sandwich covariance matrix estimate.
c. The parameter estimates and sandwich standard errors from GEE with the exchangeable working correlation structure and the 2 by 2 sandwich covariance matrix estimate.
d. For this last situation also include the estimated working correlation structure for a litter of size 6.

Chapter 8
Hypothesis Tests under Misspecification and Relaxed Assumptions

8.1 Introduction

In Chapter 6 we gave the classical asymptotic chi-squared results for T_{W}, T_{S}, and T_{LR} under ideal conditions, that is, when the specified model density $f(y;\boldsymbol{\theta})$ is correct. In this chapter we give asymptotic results for the case when the specified model density is incorrect, and also show how to adjust the statistics to have the usual asymptotic chi-squared results. For Wald and score tests, we give these adjusted statistics in the general M-estimation context. We also discuss the quadratic inference functions of Lindsay and Qu (2003) that are related to the Generalized Method of Moments (GMM) approach found in the econometrics literature.

8.2 Likelihood-Based Tests under Misspecification

For purposes of deriving T_{W}, T_{S}, and T_{LR}, we assume that $Y_1,\ldots,Y_n$ are from a density $f(y;\boldsymbol{\theta})$ that satisfies the usual regularity conditions for asymptotic normality of $\widehat{\boldsymbol{\theta}}_{\mathrm{MLE}}$ and asymptotic chi-squared results for T_{W}, T_{S}, and T_{LR} under $H_0 : \boldsymbol{\theta}_1 = \boldsymbol{\theta}_{10}$, where $\boldsymbol{\theta}_1$ is $r \times 1$ and $\boldsymbol{\theta}$ is $b \times 1$, $r \le b$. However, suppose that the true density of Y_i is $g(y)$, where $g(y) \neq f(y;\boldsymbol{\theta})$ for any $\boldsymbol{\theta} \in \Omega$, but there is a well-defined $\boldsymbol{\theta}_g$ that satisfies

$$E_g\left\{\frac{\partial}{\partial\boldsymbol{\theta}}\log f(Y_1;\boldsymbol{\theta}_g)\right\} = \int \frac{\partial}{\partial\boldsymbol{\theta}}\log f(y;\boldsymbol{\theta}_g)g(y)\,dy = \mathbf{0}. \tag{8.1}$$

(In Chapter 7, we used the notation $\boldsymbol{\theta}_0$, but here we want to emphasize the dependence on g with $\boldsymbol{\theta}_g$.) We show that all three statistics converge in distribution under $H_0 : \boldsymbol{\theta}_{g1} = \boldsymbol{\theta}_{10}$ to $\sum_{i=1}^r c_i Z_i^2$, where $Z_1,\ldots,Z_r$ are iid standard normal random variables, and $c_1,\ldots,c_r$ are the eigenvalues of

$$\left(\boldsymbol{A}_{11} - \boldsymbol{A}_{12}\boldsymbol{A}_{22}^{-1}\boldsymbol{A}_{21}\right)\left(\boldsymbol{A}^{-1}\boldsymbol{B}\boldsymbol{A}^{-1}\right)_{11}, \tag{8.2}$$

D.D. Boos and L.A. Stefanski, *Essential Statistical Inference: Theory and Methods*, Springer Texts in Statistics, DOI 10.1007/978-1-4614-4818-1_8,

and $\boldsymbol{A}$ and $\boldsymbol{B}$ are the sandwich formula matrices

$$\boldsymbol{A}(\boldsymbol{\theta}_g) = E_g\left\{-\frac{\partial^2}{\partial\boldsymbol{\theta}\,\partial\boldsymbol{\theta}^T}\log f(Y_1;\boldsymbol{\theta}_g)\right\}$$

and

$$\boldsymbol{B}(\boldsymbol{\theta}_g) = E_g\left\{\frac{\partial}{\partial\boldsymbol{\theta}^T}\log f(Y_1;\boldsymbol{\theta}_g)\right\}\left\{\frac{\partial}{\partial\boldsymbol{\theta}}\log f(Y_1;\boldsymbol{\theta}_g)\right\}.$$

In this section we will continue the practice begun in Chapter 7 of using $\boldsymbol{A}$ instead of the information matrix $\boldsymbol{I}$ even when $\boldsymbol{\psi}$ is the likelihood score function. Note that when convenient, such as in (8.2), we suppress the dependence of $\boldsymbol{A}$ and $\boldsymbol{B}$ on the true value $\boldsymbol{\theta}_g$. Eigenvalues of a matrix $\boldsymbol{H}$ are solutions of determinant$(\boldsymbol{H}-c\boldsymbol{I}) = 0$ and of course equal to H if H is a scalar ($r = 1$ case).

If the model were correctly specified, $g(y) = f(y;\boldsymbol{\theta})$ for some $\boldsymbol{\theta} \in \Omega$, then $\boldsymbol{A}(\boldsymbol{\theta}) = \boldsymbol{B}(\boldsymbol{\theta}) = \boldsymbol{I}(\boldsymbol{\theta})$ and (8.2) is the $r\times r$ identity matrix with $c_1 = \cdots = c_r = 1$, and $\sum_{i=1}^r c_i Z_i^2 \stackrel{d}{=} \chi_r^2$. For the general misspecified case, it is easiest to look at T_{W} in order to see the origin of $\sum_{i=1}^r c_i Z_i^2$. Recall that the Wald test statistic for $H_0 : \boldsymbol{\theta}_{g1} = \boldsymbol{\theta}_{10}$ is

$$\mathrm{T_W} = \sqrt{n}(\widehat{\boldsymbol{\theta}}_1 - \boldsymbol{\theta}_{10})^T\left(\widehat{\boldsymbol{A}}_{11} - \widehat{\boldsymbol{A}}_{12}\widehat{\boldsymbol{A}}_{22}^{-1}\widehat{\boldsymbol{A}}_{21}\right)\sqrt{n}(\widehat{\boldsymbol{\theta}}_1 - \boldsymbol{\theta}_{10}), \tag{8.3}$$

where $\widehat{\boldsymbol{A}}_{ij} = \boldsymbol{A}(\widehat{\boldsymbol{\theta}})_{ij}$. Using Theorem 7.2 (p. 328) and Problem 5.10 (p. 265) under the conditions below in Theorem 8.2, we have

$$\mathrm{T_W} \xrightarrow{d} \boldsymbol{X}^T\left(\boldsymbol{A}_{11} - \boldsymbol{A}_{12}\boldsymbol{A}_{22}^{-1}\boldsymbol{A}_{21}\right)\boldsymbol{X}, \tag{8.4}$$

where $\boldsymbol{X}$ is $N_r\left(\boldsymbol{0}, \{\boldsymbol{A}^{-1}\boldsymbol{B}\boldsymbol{A}^{-1}\}_{11}\right)$. Then we apply the following well-known result (see, e.g., Graybill, 1976, p. 136) to get this limiting distribution in the form $\sum_{i=1}^k c_i Z_i^2$.

Theorem 8.1. *Suppose that $\boldsymbol{X}$ is distributed as $N_k(\boldsymbol{0}, \boldsymbol{\Sigma})$, where $\boldsymbol{\Sigma}$ has rank k. If $\boldsymbol{D}$ is a symmetric matrix of rank k, then $\boldsymbol{X}^T\boldsymbol{D}\boldsymbol{X}$ has the same distribution as $\sum_{i=1}^k c_i Z_i^2$, where $c_1, \ldots, c_k$ are the eigenvalues of $\boldsymbol{D}\boldsymbol{\Sigma}$ and $Z_1, \ldots, Z_k$ are iid standard normal random variables.*

We now give the main result for our three likelihood-based tests under possible misspecification of the density and for the partitioned null hypothesis $H_0 : \boldsymbol{\theta}_{g1} = \boldsymbol{\theta}_{10}$. Recall that $\widetilde{\boldsymbol{\theta}}$ refers to the maximum likelihood estimator under H_0, given by $\widetilde{\boldsymbol{\theta}} = (\boldsymbol{\theta}_{10}, \widetilde{\boldsymbol{\theta}}_2)^T$, where $\boldsymbol{S}_2(\widetilde{\boldsymbol{\theta}}) = \boldsymbol{0}$, $\boldsymbol{S}(\boldsymbol{\theta}) = \sum_{i=1}^n \partial\log f(Y_i;\boldsymbol{\theta})/\partial\boldsymbol{\theta}^T$. A version of Theorem 8.2 is found in Kent (1982).

Theorem 8.2. *Suppose that $Y_1, \ldots, Y_n$ are iid from density $g(y)$, but T_{W}, T_{S}, and T_{LR} are derived under the assumption that the density is $f(y;\boldsymbol{\theta})$, $\boldsymbol{\theta} \in \Theta$, $\boldsymbol{\theta}^T = (\boldsymbol{\theta}_1^T, \boldsymbol{\theta}_2^T)$, $H_0 : \boldsymbol{\theta}_1 = \boldsymbol{\theta}_{10}$, and $H_a : \boldsymbol{\theta}_1 \neq \boldsymbol{\theta}_{10}$. Suppose further that $\boldsymbol{\psi}(y, \boldsymbol{\theta}) = \partial\log f(y;\boldsymbol{\theta})/\partial\boldsymbol{\theta}^T$ and $g(y)$ satisfy the conditions of Theorem 7.2 (p. 328), and* *that $\boldsymbol{A}(\boldsymbol{\theta})$ is continuous. Then, under $\boldsymbol{\theta}_{g1} = \boldsymbol{\theta}_{10}$ each of T_W, T_S, and T_{LR} converges in distribution as $n \to \infty$ to $\sum_{i=1}^{r} c_i Z_i^2$, where $c_1, \ldots, c_r$ are the eigenvalues of (8.2, p. 339), and $Z_1, \ldots, Z_r$ are iid standard normal random variables.*

Proof. As explained above, the result for T_W is straightforward using the asymptotic normality of $\widehat{\boldsymbol{\theta}}$ from Theorem 7.2 (p. 328). For T_S we follow the proof of Theorem 6.9 (p. 288) and obtain from Theorem 7.2 (p. 328) that $\widetilde{\boldsymbol{\theta}}_2$ has the approximation-by-averages representation

$$\widetilde{\boldsymbol{\theta}}_2 - \boldsymbol{\theta}_{g2} = \frac{1}{n}\sum_{i=1}^{n} \left\{\boldsymbol{A}(\boldsymbol{\theta}_g)_{22}\right\}^{-1} \frac{\partial}{\partial \boldsymbol{\theta}_2^T} \log f(Y_i;\boldsymbol{\theta}_g) + R_n, \tag{8.5}$$

where $\sqrt{n}R_n \xrightarrow{p} \mathbf{0}$ as $n \to \infty$. Using (8.5) and Condition 2 of Theorem 7.2 (p. 328) with Theorem 5.28 (p. 249), we have that $\boldsymbol{S}_1(\widetilde{\boldsymbol{\theta}})$ is $\text{AN}\big(\mathbf{0}, n\boldsymbol{V}_{gS_1}\big)$, where

$$\boldsymbol{V}_{gS_1} = \boldsymbol{B}_{11} - \boldsymbol{A}_{12}\boldsymbol{A}_{22}^{-1}\boldsymbol{B}_{21} - \boldsymbol{B}_{12}\boldsymbol{A}_{22}^{-1}\boldsymbol{A}_{21} + \boldsymbol{A}_{12}\boldsymbol{A}_{22}^{-1}\boldsymbol{B}_{22}\boldsymbol{A}_{22}^{-1}\boldsymbol{A}_{21}. \tag{8.6}$$

Patterned matrix manipulations reveal that $\boldsymbol{V}_{gS_1}$ can be reexpressed as

$$\boldsymbol{V}_{gS_1} = \left(\boldsymbol{A}_{11} - \boldsymbol{A}_{12}\boldsymbol{A}_{22}^{-1}\boldsymbol{A}_{21}\right)\left(\boldsymbol{A}^{-1}\boldsymbol{B}\boldsymbol{A}^{-1}\right)_{11}\left(\boldsymbol{A}_{11} - \boldsymbol{A}_{12}\boldsymbol{A}_{22}^{-1}\boldsymbol{A}_{21}\right). \tag{8.7}$$

Putting (8.7) together with the form of T_S in the partitioned case,

$$T_S = n^{-1}\boldsymbol{S}_1(\widetilde{\boldsymbol{\theta}})^T\left(\widetilde{\boldsymbol{A}}_{11} - \widetilde{\boldsymbol{A}}_{12}\widetilde{\boldsymbol{A}}_{22}^{-1}\widetilde{\boldsymbol{A}}_{21}\right)^{-1}\boldsymbol{S}_1(\widetilde{\boldsymbol{\theta}}), \tag{8.8}$$

and Theorem 8.1 (p. 340), yields the result. Finally, the argument for T_{LR} follows closely the line of reasoning in the proof of Theorem 6.10 (p. 288), yielding

$$T_{LR} \xrightarrow{d} \boldsymbol{Z}^T\left(\boldsymbol{A}_{11} - \boldsymbol{A}_{12}\boldsymbol{A}_{22}^{-1}\boldsymbol{A}_{21}\right)^{-1}\boldsymbol{Z} \qquad \text{as } n \to \infty, \tag{8.9}$$

where here $\boldsymbol{Z}$ is $N_r(\mathbf{0}, \boldsymbol{V}_{gS_1})$. Then using the form (8.7) for $\boldsymbol{V}_{gS_1}$ and Theorem 8.1 (p. 340), the result follows. ■

Now we give some simple examples to illustrate Theorem 8.2 (p. 340). The first two are related to Section 3.2.4 (p. 132) and Section 7.2.2 (p. 302).

Example 8.1 (Tests of a normal mean). The assumed model is $\text{N}(\mu, \sigma^2)$, $H_0 : \mu = \mu_0$, but the true density $g(y)$ is possibly nonnormal with third and fourth central moments μ_3 and μ_4, respectively. In this case $\theta_{g1} = \mu$ and $\theta_{g2} = \sigma^2$ have the same interpretation (mean, variance) under both the assumed $\text{N}(\mu, \sigma^2)$ density and the true density g. From Section 3.2.4 (p. 132) we have $T_W = n(\overline{Y} - \mu_0)^2/s_n^2$, and T_S and T_{LR} are simple functions of T_W. Clearly, $T_W \xrightarrow{d} \chi_1^2$ under H_0 as $n \to \infty$ by the Central Limit Theorem. This also follows from Theorem 8.2 (p. 340) after noting from Example 7.2.2 (p. 302) (with the parameterization $\theta_2 = \sigma^2$) that

$$A(\boldsymbol{\theta}_g) = \begin{pmatrix} \dfrac{1}{\sigma^2} & 0 \\ 0 & \dfrac{1}{2\sigma^4} \end{pmatrix} \qquad B(\boldsymbol{\theta}_g) = \begin{pmatrix} \dfrac{1}{\sigma^2} & \dfrac{\mu_3}{2\sigma^6} \\ \dfrac{\mu_3}{2\sigma^6} & \dfrac{\mu_4 - \sigma^4}{4\sigma^8} \end{pmatrix}.$$

Then $\left(A_{11} - A_{12}A_{22}^{-1}A_{21}\right) = 1/\sigma^2$, $\left(A^{-1}BA^{-1}\right)_{11} = \sigma^2$, (8.2, p. 339) = 1, and Theorem 8.2 (p. 340) shows the limiting distribution is Z_1^2, which is a χ_1^2 random variable. Thus, in this case, the misspecification of not having a normal distribution has no asymptotic effect on test levels. In fact, virtually all tests about sample means under normal distribution assumptions, as found in ANOVA and regression situations, have this "Type I error robustness" to model misspecification. ♦

Example 8.2 (Tests of a normal variance). The assumed model is $N(\mu, \sigma^2)$, but now let $H_0 : \sigma^2 = \sigma_0^2$. Then $T_W = n(s_n^2 - \sigma_0^2)^2/(2s_n^4)$, and $T_W \xrightarrow{d} [(\text{Kurt} - 1)/2]\chi_1^2$ under H_0 as $n \to \infty$ by results from Chapter 5 about the asymptotic normality of the sample variance. Recall that Kurt $= \mu_4/\mu_2^2 = \mu_4/\sigma^4$, and Kurt $= 3$ for normal distributions. To obtain this limiting distribution from Theorem 8.2 (p. 340), we need to switch the order of the parameter to be (σ^2, μ). Then $\left(A_{11} - A_{12}A_{22}^{-1}A_{21}\right) = 1/(2\sigma^4)$, $\left(A^{-1}BA^{-1}\right)_{11} = \mu_4 - \sigma^4$, and (8.2, p. 339) = $(\mu_4 - \sigma^4)/(2\sigma^4) = (\text{Kurt} - 1)/2$. In contrast to tests about a mean, the asymptotic level of normal theory tests about a variance parameter is directly affected by nonnormality. For example, the true asymptotic level of a nominal $\alpha = .05$ one-sided test would be $P(Z > \sqrt{2/(\text{Kurt} - 1)}1.645)$. If Kurt $= 5$, then the asymptotic level is .122. In general this nonrobustness to nonnormality pervades tests about variances in ANOVA and regression situations (see, e.g., Boos and Brownie, 2004). But fortunately, these tests are not used very often in practice. ♦

Example 8.3 (Clustered binary data). Binary (0-1) data are often sampled in clusters due to aggregation of results by families, communities, etc. (see, e.g., Problem 7.14, p. 336) where there is possibly correlation among cluster members. Each data point Y_i is a sum of the binary random variables in the cluster. Ignoring the correlation, the assumed model for $Y_1, \ldots, Y_n$ is binomial$(m; p)$, where for simplicity we have set all clusters to have the same size $m > 1$. Suppose that the true density $g(y)$ has mean mp but variance $mp(1-p)d$, where $d > 1$ means Y_i is over-dispersed relative to a binomial distribution. If the Y_i have the beta-binomial distribution (see Problem 7.14, p. 336), then $d = 1 + (m-1)\rho$, where ρ is the correlation coefficient between members of the same cluster. Under the assumed binomial$(m; p)$ distribution and $H_0 : p = p_0$, we have (c.f. Example 3.2, p. 129) $T_W = nm\left(\widehat{p} - p_0\right)^2/\widehat{p}(1-\widehat{p})$ and $T_S = nm\left(\widehat{p} - p_0\right)^2/p_0(1-p_0)$, where $\widehat{p} = \sum_{i=1}^n Y_i/nm$. From the Central Limit Theorem we have $T_W \xrightarrow{d} d\chi_1^2$ under H_0 as $n \to \infty$. This result also follows from Theorem 8.2 (p. 340) by noting that $A = m/p(1-p)$ and $B = md/p(1-p)$. In the usual case with $\rho > 0$

(cluster members positively correlated), $d > 1$ and thus the limiting distribution of $\mathrm{T_W}$ is stochastically greater than when $\rho = 0$. This means that using χ_1^2 instead of $d\chi_1^2$ results in inflated Type I error rates. ♦

In the above three examples, the asymptotic random variable is fairly simple since it is just a multiple of the χ_1^2 random variable. When the tests involve more than one degree of freedom, then the asymptotic distribution is more complicated. The next example combines the first two examples above.

Example 8.4 (Tests of a normal mean and variance). The assumed model is $\mathrm{N}(\mu, \sigma^2)$ with $H_0 : \mu = \mu_0, \sigma^2 = \sigma_0^2$. For this fully specified null hypothesis, the Wald test is simply the sum of the Wald test statistics for the first two examples above, $\mathrm{T_W} = n(\overline{Y} - \mu_0)^2/s_n^2 + n(s_n^2 - \sigma_0^2)^2/(2s_n^4)$. After noting that here $A_{11} = A$ and $B_{11} = B$ and (8.2, p. 339) = BA^{-1}, the eigenvalue problem is to solve

$$\begin{vmatrix} 1 - c & \dfrac{\mu_3}{\sigma^2} \\ \dfrac{\mu_3}{2\sigma^4} & \dfrac{\mu_4 - \sigma^4}{2\sigma^4} - c \end{vmatrix} = 0. \tag{8.10}$$

When $\mu_3 = 0$, it is not difficult to show that the solutions are $c_1 = 1$ and $c_2 = (\mu_4 - \sigma^4)/(2\sigma^4) = (\mathrm{Kurt} - 1)/2$. The limiting distribution of $\mathrm{T_W}$ (and $\mathrm{T_S}$ and $\mathrm{T_{LR}}$ as well) under H_0 is that of $Z_1^2 + [(\mathrm{Kurt} - 1)/2]\, Z_2^2$. If $\mu_3 \neq 0$, then the distribution is more complicated. ♦

For problems like the first three examples above, it is fairly straightforward, at least conceptually, to estimate c_1 of the limiting random variable and then adjust the statistics, e.g., $\mathrm{T_{LR}}/\widehat{c}_1$, or the critical values, $\widehat{c}_1\chi_1^2(\alpha)$, so that the test procedure has asymptotic level α.

However, in this last example, the adjustment is not as straightforward. One approach is to estimate c_1 and c_2 defined by (8.10, p. 343), and then estimate the percentiles of $\widehat{c}_1 Z_1^2 + \widehat{c}_2 Z_2^2$. This approach works for all three test statistics but is not overly appealing because of the steps involved: solving the eigenvalue problem, estimating the c_i, and getting the percentiles of a weighted sum of chi-squared random variables (a nontrivial *analytical* exercise although straightforward to do by simulation). For $\mathrm{T_W}$, an alternative procedure is to replace $\left(\widehat{\boldsymbol{A}}_{11} - \widehat{\boldsymbol{A}}_{12}\widehat{\boldsymbol{A}}_{22}^{-1}\widehat{\boldsymbol{A}}_{21}\right)$ in (8.3, p. 340) by an estimate of $\left\{\left(\boldsymbol{A}^{-1}\boldsymbol{B}\boldsymbol{A}^{-1}\right)_{11}\right\}^{-1}$ and use critical values from χ_r^2. For $\mathrm{T_S}$, we need to replace $\left(\widetilde{\boldsymbol{A}}_{11} - \widetilde{\boldsymbol{A}}_{12}\widetilde{\boldsymbol{A}}_{22}^{-1}\widetilde{\boldsymbol{A}}_{21}\right)^{-1}$ in (8.8, p. 341) by an estimate of $\boldsymbol{V}_{gS_1}^{-1}$ (8.6, p. 341). We give these modified statistics in the next section. For the case $r > 1$, there is no obvious way to modify $\mathrm{T_{LR}}$ to have a limiting χ_r^2 distribution under H_0.

8.3 Generalized Wald, Score, and Likelihood Ratio Tests

Motivated by Theorem 8.2 (p. 340) and Chapter 7, we now give Wald, score, and likelihood ratio statistics that are related to M-estimators and their defining equations (7.1, p. 298)

$$\sum_{i=1}^{n} \boldsymbol{\psi}(Y_i, \widehat{\boldsymbol{\theta}}) = \mathbf{0}. \tag{8.11}$$

These include modifications to T_W, T_S, and T_{LR} mentioned at the end of the previous section.

Assume that $\boldsymbol{\theta}^T = (\boldsymbol{\theta}_1^T, \boldsymbol{\theta}_2^T)$ and the true density is again $g(y)$ such that $E_g\{\boldsymbol{\psi}(Y_1, \boldsymbol{\theta}_g)\} = 0$, where $\boldsymbol{\theta}_1$ is of dimension $r \times 1$ and $\boldsymbol{\theta}_2$ is of dimension $(b-r) \times 1$, and $\boldsymbol{\psi}^T = (\boldsymbol{\psi}_1^T, \boldsymbol{\psi}_2^T)$ is partitioned similarly. The null hypothesis is $H_0 : \boldsymbol{\theta}_{g1} = \boldsymbol{\theta}_{10}$. Generalized Wald test statistics are quadratic forms constructed by putting an estimate of the inverse of the asymptotic covariance matrix of $\widehat{\boldsymbol{\theta}}_1$ between $(\widehat{\boldsymbol{\theta}}_1 - \boldsymbol{\theta}_{10})^T$ and $(\widehat{\boldsymbol{\theta}}_1 - \boldsymbol{\theta}_{10})$. Generalized score test statistics are quadratic forms based on $\sum \boldsymbol{\psi}_1(Y_i, \widetilde{\boldsymbol{\theta}})$ and estimates of its covariance matrix, where $\widetilde{\boldsymbol{\theta}}$ refers to the solution of $\sum \boldsymbol{\psi}_2(Y_i, \widetilde{\boldsymbol{\theta}}) = \mathbf{0}$ with the first r components equal to $\boldsymbol{\theta}_{10}$.

When there exists a $Q(\boldsymbol{Y}, \boldsymbol{\theta})$ such that $\partial Q(\boldsymbol{Y}, \boldsymbol{\theta})/\partial \boldsymbol{\theta}^T = \sum_{i=1}^n \boldsymbol{\psi}(\boldsymbol{Y}_i, \boldsymbol{\theta})$, $Q(\boldsymbol{Y}, \boldsymbol{\theta})$ can be used as a log likelihood resulting in generalized likelihood ratio statistics of the form $T_{GLR} = 2[Q(\boldsymbol{Y}, \widehat{\boldsymbol{\theta}}) - Q(\boldsymbol{Y}, \widetilde{\boldsymbol{\theta}})]/\widehat{c}$. These statistics are not as widely available as the generalized Wald and score statistics because there may not exist such a Q, but they have been found useful in generalized linear model and robust regression contexts. Moreover, certain versions of the generalized score and likelihood ratio statistics are invariant under parameter transformations, and thus have a theoretical advantage over generalized Wald statistics.

8.3.1 *Generalized Wald Tests*

For M-estimators $\widehat{\boldsymbol{\theta}}$ that satisfy (8.11, p. 344), we know that $\widehat{\boldsymbol{\theta}}$ is $AN\left(\boldsymbol{\theta}_g, \boldsymbol{V}(\boldsymbol{\theta}_g)/n\right)$ where $\boldsymbol{V}(\boldsymbol{\theta}_g) = \boldsymbol{A}(\boldsymbol{\theta}_g)^{-1}\boldsymbol{B}(\boldsymbol{\theta}_g)\{\boldsymbol{A}(\boldsymbol{\theta}_g)^{-1}\}^T$. Thus, the obvious form for the generalized Wald statistic in the case $H_0 : \boldsymbol{\theta}_{g1} = \boldsymbol{\theta}_{10}$ is

$$T_{GW} = n(\widehat{\boldsymbol{\theta}}_1 - \boldsymbol{\theta}_{10})^T \left(\widehat{\boldsymbol{V}}_{11}\right)^{-1} (\widehat{\boldsymbol{\theta}}_1 - \boldsymbol{\theta}_{10}), \tag{8.12}$$

where $\widehat{\boldsymbol{V}}$ could be any consistent estimate of $\boldsymbol{V}(\boldsymbol{\theta}_g)$. Natural choices for $\widehat{\boldsymbol{V}}$ are the empirical sandwich estimator $\boldsymbol{V}_n(\boldsymbol{Y}, \widehat{\boldsymbol{\theta}})$ of (7.12, p. 302) or expected value versions $\boldsymbol{V}(\widehat{\boldsymbol{\theta}})$ that might depend on some modeling assumptions.

Recall that the constraint version of the null hypothesis is $H_0 : \boldsymbol{h}(\boldsymbol{\theta}_g) = \mathbf{0}$, where $\boldsymbol{h}$ is $r \times 1$ and differentiable with $\boldsymbol{H}(\boldsymbol{\theta}) = \partial \boldsymbol{h}(\boldsymbol{\theta})/\partial \boldsymbol{\theta}^T$ having dimension $r \times b$. In this case, the natural generalization of (6.14, p. 287) is

$$T_{GW} = n\boldsymbol{h}(\widehat{\boldsymbol{\theta}})^T \left\{\boldsymbol{H}(\widehat{\boldsymbol{\theta}})\widehat{\boldsymbol{V}}\boldsymbol{H}(\widehat{\boldsymbol{\theta}})^T\right\}^{-1} \boldsymbol{h}(\widehat{\boldsymbol{\theta}}). \tag{8.13}$$

The generalization of Theorem 6.8 (p. 287) is then

Theorem 8.3. *Suppose that* $Y_1, \ldots, Y_n$ *are iid with density* $g(y)$. *Assume that* $\boldsymbol{\psi}(y, \boldsymbol{\theta})$ *and* $g(y)$ *satisfy the conditions of Theorem 7.2 (p. 328) under* $H_0 : \boldsymbol{h}(\boldsymbol{\theta}) = \boldsymbol{0}$, $\boldsymbol{h}$ *satisfies the conditions of Theorem 5.20 (p. 239),* $\widehat{\boldsymbol{V}} \xrightarrow{p} \boldsymbol{V}(\boldsymbol{\theta}_g)$, *and* $\boldsymbol{H}(\boldsymbol{\theta}_g)\boldsymbol{V}(\boldsymbol{\theta}_g)\boldsymbol{H}(\boldsymbol{\theta}_g)^T$ *is nonsingular. Then* $T_{\mathrm{GW}} \xrightarrow{d} \chi^2_r$ *as* $n \to \infty$.

8.3.1a Examples

Let us first look briefly at the four examples from the previous section. For the examples assuming $Y_1, \ldots, Y_n$ are from $N(\mu, \sigma^2)$, Examples 8.1, 8.2, and 8.4, first recall that

$$\boldsymbol{V}_n(\boldsymbol{Y}, \widehat{\boldsymbol{\theta}}) = \begin{pmatrix} s_n^2 & m_3 \\ m_3 & m_4 - s_n^4 \end{pmatrix},$$

from Example 7.2.2 (p. 302). Then, for $H_0 : \mu = \mu_0$, (8.12, p. 344) gives

$$T_{\mathrm{GW}} = \mathrm{T_W} = n(\overline{Y} - \mu_0)^2/s_n^2.$$

For $H_0 : \sigma^2 = \sigma_0^2$, after remembering to switch the parameter order to (σ^2, μ), (8.12, p. 344) yields

$$T_{\mathrm{GW}} = n(s_n^2 - \sigma_0^2)^2/(m_4 - s_n^4), \tag{8.14}$$

where $m_k = \sum(Y_i - \overline{Y})^k$. At this point we might point out that generalized Wald tests inherit the non-invariance to parameter changes and null hypothesis specification of regular Wald tests. For example, in this last example, if we change $H_0 : \sigma^2 = \sigma_0^2$ to $H_0 : \sigma = \sigma_0$, then $T_{\mathrm{GW}} = 4s_n^2 n(s_n - \sigma_0)^2/(m_4 - s_n^4)$.

For the completely specified $H_0 : \mu = \mu_0, \sigma^2 = \sigma_0^2$,

$$T_{\mathrm{GW}} = n\left[(\overline{Y} - \mu_0)^2(m_4 - s_n^4) - 2(\overline{Y} - \mu_0)(s_n^2 - \sigma_0^2)m_3 + (s_n^2 - \sigma_0^2)^2 s_n^2\right]/D, \tag{8.15}$$

where $D = s_n^2(m_4 - s_n^4) - m_3^2$.

For Example 8.3 (p. 342), taking $\psi(Y_i, p) = Y_i/m - p$ results in $\widehat{p} = \sum_{i=1}^n Y_i/nm$. Thus, similar to the normal mean result, $T_{\mathrm{GW}} = n(\widehat{p} - p_0)^2/s_n^2$, where here $s_n^2 = n^{-1}\sum_{i=1}^n (Y_i/m - \widehat{p})^2$.

In all our theorems here (and in Chapters 6 and 7) we are only giving the iid results for reasons of simplicity, but note that similar results hold in nonindependent cases such as regression. In this next example we give a two-sample problem that is not covered by Theorem 8.3 (p. 345) but is covered by a regression version.

Example 8.5 (Two independent samples of clustered binary data). Similar to the setting of Example 8.3 (p. 342), suppose that we now have two independent samples of "summed" clustered binary data, $Y_1, \ldots, Y_{n_1}$, and $Y_{n_1+1}, \ldots, Y_{n_1+n_2}$. Ignoring the correlation among the correlated summands in Y_i, the natural distributional assumption would be that $Y_i|m_i$ is binomial(m_i, p_1) for the first sample and binomial(m_i, p_2) for the second sample, leading to the estimators $\widehat{p}_1 = \sum_{i=1}^{n_1} Y_i/m_{1.}$ and $\widehat{p}_2 = \sum_{i=n_1+1}^{n_1+n_2} Y_i/m_{2.}$, where $m_{1.} = \sum_{i=1}^{n_1} m_i$, $m_{2.} = \sum_{i=n_1+1}^{n} m_i$, and $n = n_1+n_2$. To use the M-estimation regression framework, these estimators are solutions of the generalized linear model equations (7.34, p. 321)

$$\sum_{i=1}^{n}(Y_i - m_i \boldsymbol{x}_i^T \boldsymbol{p})\boldsymbol{x}_i = \boldsymbol{0},$$

where $\boldsymbol{x}_i^T = (1, 0)$ for the first sample, $\boldsymbol{x}_i^T = (0, 1)$ for the second sample, and $\boldsymbol{p}^T = (p_1, p_2)$. The estimated $\boldsymbol{A}$ and $\boldsymbol{B}$ matrices are

$$\boldsymbol{A}_n(\boldsymbol{X}, \widehat{\boldsymbol{p}}) = \frac{1}{n}\begin{pmatrix} m_{1.} & 0 \\ 0 & m_{2.} \end{pmatrix}$$

and

$$\boldsymbol{B}_n(\boldsymbol{X}, \widehat{\boldsymbol{p}}) = \frac{1}{n}\begin{pmatrix} \sum_{i=1}^{n_1}(Y_i - m_i\widehat{p}_1)^2 & 0 \\ 0 & \sum_{i=n_1+1}^{n}(Y_i - m_i\widehat{p}_2)^2 \end{pmatrix},$$

where for simplicity we have used n instead of $n-2$ as was suggested in Ch. 5 for the estimated $\boldsymbol{B}$ matrix. Finally, using $h(p_1, p_2) = p_1 - p_2$ in (8.13, p. 344), we have

$$T_{\text{GW}} = \frac{(\widehat{p}_1 - \widehat{p}_2)^2}{\sum_{i=1}^{n_1}(Y_i - m_i\widehat{p}_1)^2/m_{1.}^2 + \sum_{i=n_1+1}^{n}(Y_i - m_i\widehat{p}_2)^2/m_{2.}^2}. \tag{8.16}$$

♦

Improvements to the use of T_{GW} with χ_1^2 critical values are given in Mancl and DeRouen (2001) and Fay and Graubard (2001).

8.3.2 Generalized Score Tests

Sen (1982) first derived generalized score tests based on the M-estimation formulation. White (1982) introduced the likelihood-related versions into the econometrics literature. Breslow (1989,1990) and Rotnitzky and Jewell (1990) introduced these tests in GEE contexts, and Boos (1992) gives an overview. In this section we give a formal theorem for the iid case and then discuss examples.

As mentioned previously, generalized score tests are based on $\sum \boldsymbol{\psi}_1(Y_i, \widetilde{\boldsymbol{\theta}})$, where $\widetilde{\boldsymbol{\theta}} = (\boldsymbol{\theta}_{10}, \widetilde{\boldsymbol{\theta}}_2)^T$ satisfies $\sum \boldsymbol{\psi}_2(Y_i, \widetilde{\boldsymbol{\theta}}) = \boldsymbol{0}$. The main conceptual hurdle is to find the appropriate variance matrix to invert and put between $\sum \boldsymbol{\psi}_1(Y_i, \widetilde{\boldsymbol{\theta}})^T$ and $\sum \boldsymbol{\psi}_1(Y_i, \widetilde{\boldsymbol{\theta}})$. But this follows exactly as in the proof of Theorem 8.2 (p. 340) when we replace $\boldsymbol{S}$ by $\sum \boldsymbol{\psi}$. That is, from the proof of Theorem 7.2 (p. 328) under H_0,

$$\widetilde{\boldsymbol{\theta}}_2 - \boldsymbol{\theta}_{20} = \frac{1}{n}\sum_{i=1}^{n} \{\boldsymbol{A}(\boldsymbol{\theta}_0)_{22}\}^{-1} \boldsymbol{\psi}_2(Y_i, \boldsymbol{\theta}_0) + \boldsymbol{R}_n, \tag{8.17}$$

where $\sqrt{n}\boldsymbol{R}_n \xrightarrow{p} \boldsymbol{0}$ as $n \to \infty$. Then using Theorem 5.28 (p. 249), we have

$$\frac{1}{n}\sum_{i=1}^{n} \boldsymbol{\psi}_1(Y_i, \widetilde{\boldsymbol{\theta}}) = \left(\boldsymbol{I}_r, -\boldsymbol{A}_{12}\boldsymbol{A}_{22}^{-1}\right) \begin{pmatrix} n^{-1}\sum_{i=1}^{n} \boldsymbol{\psi}_1(Y_i, \boldsymbol{\theta}_0) \\ \\ n^{-1}\sum_{i=1}^{n} \boldsymbol{\psi}_2(Y_i, \boldsymbol{\theta}_0) \end{pmatrix} + \boldsymbol{R}_{n2}, \tag{8.18}$$

where $\boldsymbol{I}_r$ is the $r \times r$ identity matrix and $\sqrt{n}\boldsymbol{R}_{n2} \xrightarrow{p} \boldsymbol{0}$ as $n \to \infty$. The variance of the linear form of this last display gives essentially $\boldsymbol{V}_{gS_1}$ of (8.6, p. 341) and (8.7, p. 341), but here we do not assume that $\boldsymbol{A}$ is symmetric (since it may not come from a likelihood). Thus, we rename it here as $\boldsymbol{V}_{\boldsymbol{\psi}_1}$,

$$\boldsymbol{V}_{\boldsymbol{\psi}_1} = \boldsymbol{B}_{11} - \boldsymbol{A}_{12}\boldsymbol{A}_{22}^{-1}\boldsymbol{B}_{21} - \boldsymbol{B}_{12}\{\boldsymbol{A}_{22}^{-1}\}^T\boldsymbol{A}_{12}^T + \boldsymbol{A}_{12}\boldsymbol{A}_{22}^{-1}\boldsymbol{B}_{22}\{\boldsymbol{A}_{22}^{-1}\}^T\boldsymbol{A}_{12}^T. \tag{8.19}$$

In the one-dimensional case, we set $\boldsymbol{A}_{12} = 0$ so that the above expression is just the scalar B. If the null hypothesis is the completely specified case $H_0 : \boldsymbol{\theta}_g = \boldsymbol{\theta}_0$, then again $\boldsymbol{A}_{12} = 0$ and the expression reduces to $\boldsymbol{B}_{11} = \boldsymbol{B}$. The generalized score statistic for $H_0 : \boldsymbol{\theta}_{g1} = \boldsymbol{\theta}_{10}$ is

$$T_{\text{GS}} = n^{-1}\left\{\sum \boldsymbol{\psi}_1(Y_i, \widetilde{\boldsymbol{\theta}})^T\right\} \widetilde{\boldsymbol{V}}_{\boldsymbol{\psi}_1}^{-1} \left\{\sum \boldsymbol{\psi}_1(Y_i, \widetilde{\boldsymbol{\theta}})\right\}, \tag{8.20}$$

where $\widetilde{\boldsymbol{V}}_{\boldsymbol{\psi}_1}$ is an estimator of $\boldsymbol{V}_{\boldsymbol{\psi}_1}$. A theorem whose proof follows closely the above development is now given.

Theorem 8.4. *Suppose that $Y_1, \ldots, Y_n$ are iid with density $g(y)$. Assume that $\boldsymbol{\psi}(y, \boldsymbol{\theta})$ and $g(y)$ satisfy the conditions of Theorem 7.2 (p. 328) and $\widetilde{\boldsymbol{V}}_{\boldsymbol{\psi}_1} \xrightarrow{p} \boldsymbol{V}_{\boldsymbol{\psi}_1}$ as $n \to \infty$. Then $T_{\text{GS}} \xrightarrow{d} \chi^2_r$ as $n \to \infty$.*

We left the specification of $\widetilde{\boldsymbol{V}}_{\boldsymbol{\psi}_1}$ very general in our definition of T_{GS}. There are a number of possibilities depending on whether we use the purely empirical versions of the $\boldsymbol{A}$ and $\boldsymbol{B}$ estimators,

$$\boldsymbol{A}_n(\boldsymbol{Y}, \widetilde{\boldsymbol{\theta}}) = \frac{1}{n}\sum_{i=1}^{n} \left\{-\boldsymbol{\psi}'(Y_i, \widetilde{\boldsymbol{\theta}})\right\}, \quad \boldsymbol{B}_n(\boldsymbol{Y}, \widetilde{\boldsymbol{\theta}}) = \frac{1}{n}\sum_{i=1}^{n} \boldsymbol{\psi}(Y_i, \widetilde{\boldsymbol{\theta}})\boldsymbol{\psi}(Y_i, \widetilde{\boldsymbol{\theta}})^T,$$

the expected value versions $\boldsymbol{A}(\widetilde{\boldsymbol{\theta}})$ and $\boldsymbol{B}(\widetilde{\boldsymbol{\theta}})$, where $\boldsymbol{A}(\boldsymbol{\theta}_g) = E_g\left\{-\boldsymbol{\psi}'(Y_1, \boldsymbol{\theta}_g)\right\}$, and $\boldsymbol{B}(\boldsymbol{\theta}_g) = E_g\left\{\boldsymbol{\psi}(Y_1, \boldsymbol{\theta}_g)\boldsymbol{\psi}(Y_1, \boldsymbol{\theta}_g)^T\right\}$. (Actually, our general approach in Chapter 7 allows even more options for the expected value versions depending on modeling assumptions.) Thus, here we can see four possible versions of T_{GS}, where $\widetilde{\boldsymbol{V}}_{\boldsymbol{\psi}_1}$ is based on

$$A_n(Y, \widetilde{\theta}), B_n(Y, \widetilde{\theta}) \tag{8.21}$$

$$A(\widetilde{\theta}), B_n(Y, \widetilde{\theta}) \tag{8.22}$$

$$A(\widetilde{\theta}), B(\widetilde{\theta}) \tag{8.23}$$

$$A_n(Y, \widetilde{\theta}), B(\widetilde{\theta}). \tag{8.24}$$

Using the purely empirical estimator based on (8.21) is the simplest and most general way to define $\widetilde{V}_{\psi_1}$ and T_{GS}. However, Rotnitzky and Jewell (1990) and Boos (1992) point out that the combination estimator based on (8.22) is invariant to reparameterization and thus has some philosophical appeal. It is the version used most often in GEE applications. If the parameter transformation is linear or the null hypothesis is the completely specified $H_0 : \theta_g = \theta_0$, then the version of T_{GS} based on (8.21) is also invariant.

8.3.2a Examples

Continuation of Example 8.1 (p. 341). The first example $H_0 : \mu = \mu_0$ is straightforward if we use $\psi_1(Y_i, \theta) = Y_i - \mu$ and $\psi_2(Y_i, \theta) = (Y_i - \mu)^2 - \sigma^2$. In that case, $A(\widetilde{\theta})$ = the identity and

$$B_n(Y, \widetilde{\theta}) = \begin{pmatrix} s_n^2(\mu_0) & m_3(\mu_0) - (\overline{Y} - \mu_0)s_n^2(\mu_0) \\ m_3(\mu_0) - (\overline{Y} - \mu_0)s_n^2(\mu_0) & m_4(\mu_0) - s_n^4(\mu_0) \end{pmatrix},$$

where $s_n^2(\mu_0) = n^{-1} \sum (Y_i - \mu_0)^2$ and $m_k(\mu_0) = n^{-1} \sum (Y_i - \mu_0)^k$. Then

$$T_{\mathrm{GS}} = \frac{n(\overline{Y} - \mu_0)^2}{s_n^2(\mu_0)} = \frac{n(\overline{Y} - \mu_0)^2}{s_n^2 + (\overline{Y} - \mu_0)^2},$$

the same as $\mathrm{T_S}$ found in Example 3.2.4 (p. 132). Note also that we get the same $\mathrm{T_S}$ if we just start with $\psi_1(Y_i, \theta) = Y_i - \mu$ and ignore ψ_2. However, suppose instead that we use the ψ functions associated with maximum likelihood for the parameterization $\theta^T = (\mu, \sigma^2)$: $\psi_1(Y_i, \theta) = (Y_i - \mu)/\sigma^2$ and $\psi_2(Y_i, \theta) = [(Y_i - \mu)^2 - \sigma^2]/(2\sigma^4)$. Then

$$A_n(Y, \widetilde{\theta}) = \begin{pmatrix} \dfrac{1}{s_n^2(\mu_0)} & \dfrac{\overline{Y} - \mu_0}{s_n^4(\mu_0)} \\[2ex] \dfrac{\overline{Y} - \mu_0}{s_n^4(\mu_0)} & \dfrac{1}{2s_n^4(\mu_0)} \end{pmatrix},$$

and

$$\boldsymbol{B}_n(\boldsymbol{Y},\widetilde{\boldsymbol{\theta}}) = \begin{pmatrix} \dfrac{1}{s_n^2(\mu_0)} & \dfrac{m_3(\mu_0)}{2s_n^6(\mu_0)} - \dfrac{\overline{Y}-\mu_0}{2s_n^4(\mu_0)} \\ \dfrac{m_3(\mu_0)}{2s_n^6(\mu_0)} - \dfrac{\overline{Y}-\mu_0}{2s_n^4(\mu_0)} & \dfrac{1}{2s_n^4(\mu_0)} \end{pmatrix}.$$

Using the result that $E_g Y_i = \mu_0$ and the convention that we use $m_3(\mu_0)$ to estimate μ_3 under $H_0 : \mu = \mu_0$, the expected-value versions $\boldsymbol{A}(\widetilde{\boldsymbol{\theta}})$ and $\boldsymbol{B}(\widetilde{\boldsymbol{\theta}})$ are the same as the above except with the terms involving $\overline{Y} - \mu_0$ set equal to zero. Thus, the version of T_{GS} based on (8.22, p. 348) is again $T_{\text{GS}} = n(\overline{Y}-\mu_0)^2/\left[s_n^2 + (\overline{Y}-\mu_0)^2\right]$ because $\boldsymbol{A}(\widetilde{\boldsymbol{\theta}})_{12} = 0$. However, using the above $\boldsymbol{A}_n(\boldsymbol{Y},\widetilde{\boldsymbol{\theta}})$ and $\boldsymbol{B}_n(\boldsymbol{Y},\widetilde{\boldsymbol{\theta}})$ leads to

$$T_{\text{GS}} = \frac{n(\overline{Y}-\mu_0)^2}{s_n^2 + (\overline{Y}-\mu_0)^2}\left[1 - 2\left(\frac{\overline{Y}-\mu_0}{s_n(\mu_0)}\right)\widetilde{\text{Skew}} + \left(\frac{\overline{Y}-\mu_0}{s_n(\mu_0)}\right)^2(\widetilde{\text{Kurt}}+1)\right]^{-1},$$

where $\widetilde{\text{Skew}} = m_3(\mu_0)/s_n(\mu_0)^3$ and $\widetilde{\text{Kurt}} = m_4(\mu_0)/s_n(\mu_0)^4$. Thus, use of the purely empirical estimators leads to the above form that is unappealing since estimates of skewness and kurtosis are not necessary here to have a valid test statistic. Of course, use of just ψ_1 by itself leads again to the simpler statistic.

Continuation of Example 8.2 (p. 342). For $H_0 : \sigma^2 = \sigma_0^2$, we switch the order of the parameters, $\boldsymbol{\theta}^T = (\sigma^2, \mu)$, and obtain for $\psi_1(Y_i, \boldsymbol{\theta}) = (Y_i - \mu)^2 - \sigma^2$ and $\psi_2(Y_i, \boldsymbol{\theta}) = Y_i - \mu$, or for the likelihood-based $\boldsymbol{\psi}$, that $\boldsymbol{A}_n(\boldsymbol{Y},\widetilde{\boldsymbol{\theta}})$ and $\boldsymbol{A}(\widetilde{\boldsymbol{\theta}})$ are diagonal matrices since $\widetilde{\boldsymbol{\theta}} = (\sigma_0^2, \overline{Y})^T$. Thus, the only relevant quantity for computing T_{GS} is the (1,1) element of the estimated $\boldsymbol{B}$. For $\boldsymbol{B}_n(\boldsymbol{Y},\widetilde{\boldsymbol{\theta}})$, this leads to

$$T_{\text{GS}} = \frac{n(s_n^2 - \sigma_0^2)^2}{m_4 - 2s_n^2\sigma_0^2 + \sigma_0^4}.$$

For $\boldsymbol{B}(\widetilde{\boldsymbol{\theta}})$ in place of $\boldsymbol{B}_n(\boldsymbol{Y},\widetilde{\boldsymbol{\theta}})$, we get $T_{\text{GS}} = n(s_n^2 - \sigma_0^2)^2/(m_4 - s_n^4)$, similar to the first of the generalized Wald statistics in (8.14, p. 345). Here, reparameterizations such as $\boldsymbol{\theta} = (\sigma^2, \mu)^T$ or $(\sigma, \mu)^T$ make no difference to T_{GS} although use of $\boldsymbol{B}(\widetilde{\boldsymbol{\theta}})$ in place of $\boldsymbol{B}_n(\boldsymbol{Y},\widetilde{\boldsymbol{\theta}})$ does make a small difference.

Continuation of Example 8.4 (p. 343). For $H_0 : \mu = \mu_0, \sigma^2 = \sigma_0^2$ and $\psi_1(Y_i, \boldsymbol{\theta}) = Y_i - \mu$ and $\psi_2(Y_i, \boldsymbol{\theta}) = (Y_i - \mu)^2 - \sigma^2$, we have

$$\boldsymbol{B}_n(\boldsymbol{Y},\boldsymbol{\theta}_0) = \begin{pmatrix} s_n^2(\mu_0) & m_3(\mu_0) - (\overline{Y}-\mu_0)\sigma_0^2 \\ m_3(\mu_0) - (\overline{Y}-\mu_0)\sigma_0^2 & m_4(\mu_0) - s_n^4(\mu_0) \end{pmatrix},$$

where $s_n^2(\mu_0) = n^{-1}\sum(Y_i - \mu_0)^2$ and $m_k(\mu_0) = n^{-1}\sum(Y_i - \mu_0)^k$. Then $T_{GS} = \{\overline{Y} - \mu_0, s_n^2(\mu_0)\}\{\boldsymbol{B}_n(\boldsymbol{Y}, \boldsymbol{\theta}_0)\}\{\overline{Y} - \mu_0, s_n^2(\mu_0)\}^T$, which is a bit long to write out but it has a similar basic form to T_{GW} in (8.15, p. 345). Use of $\boldsymbol{B}(\boldsymbol{\theta}_0)$ in place of $\boldsymbol{B}_n(\boldsymbol{Y}, \boldsymbol{\theta}_0)$ would simplify the expression by putting zero in place of $\overline{Y} - \mu_0$ in the diagonal elements of $\boldsymbol{B}_n(\boldsymbol{Y}, \boldsymbol{\theta}_0)$.

Continuation of Example 8.3 (p. 342). For $H_0 : p = p_0$, we have $T_{GS} = n(\widehat{p} - p_0)^2/s_n^2(p_0)$, where $s_n^2(p_0) = n^{-1}\sum_{i=1}^{n}(Y_i/m - p_0)^2$.

Continuation of Example 8.5 (p. 345). For $H_0 : p_1 = p_2$, we use

$$\sum_{i=1}^{n}(Y_i - m_i \boldsymbol{x}_i^T \boldsymbol{\beta})\boldsymbol{x}_i = \boldsymbol{0},$$

where $\boldsymbol{x}_i^T = (1, 0)$ for the first sample, $\boldsymbol{x}_i^T = (1, -1)$ for the second sample, and $(\beta_1, \beta_2) = (p_1, p_1 - p_2)$. After switching subscripts in the definition of T_{GS} since the null is $H_0 : \beta_2 = 0$, we obtain the same form as T_{GW} in (8.16, p. 346) except that $\widetilde{p} = \sum_{i=1}^{n} Y_i / \sum_{i=1}^{n} m_i$ replaces $\widehat{p}_1$ and $\widehat{p}_2$ in the denominator of (8.16, p. 346),

$$T_{GS} = \frac{(\widehat{p}_1 - \widehat{p}_2)^2}{\sum_{i=1}^{n_1}(Y_i - m_i\widetilde{p})^2/m_{1.}^2 + \sum_{i=n_1+1}^{n}(Y_i - m_i\widetilde{p})^2/m_{2.}^2}. \tag{8.25}$$

8.3.2b Derivation of T_{GS} by M-Estimation

We would like to point out how the M-estimation approach of Chapter 7 leads to the correct generalized score test statistic. In a sense, the $\boldsymbol{A}$ and $\boldsymbol{B}$ matrix formulation automatically does the Taylor expansion and computes the variance of the appropriate linear approximations.

To that end, let $\widehat{\boldsymbol{\theta}}_1^* = n^{-1}\sum_{i=1}^{n}\boldsymbol{\psi}_1(Y_i, \widetilde{\boldsymbol{\theta}})$ be an M-estimator that solves $\sum\left\{\boldsymbol{\psi}_1(Y_i, \widetilde{\boldsymbol{\theta}}) - \widehat{\boldsymbol{\theta}}_1^*\right\} = 0$. Then, thinking of $\boldsymbol{\theta}_1^*$ as a parameter that is the limit in probability of $\widehat{\boldsymbol{\theta}}_1^*$, the parameter for this new problem is $\boldsymbol{\theta}^*$ composed of $\boldsymbol{\theta}_1^*$ and $\boldsymbol{\theta}_2$; $\boldsymbol{\theta}_{10}$ is fixed and not a parameter in the new problem. The associated $\boldsymbol{\psi}$ functions are $\boldsymbol{\psi}_1^*(Y_i, \boldsymbol{\theta}^*) = \boldsymbol{\psi}_1(Y_i, \boldsymbol{\theta}) - \boldsymbol{\theta}_1^*$ and $\boldsymbol{\psi}_2^*(Y_i, \boldsymbol{\theta}^*) = \boldsymbol{\psi}_2(Y_i, \boldsymbol{\theta})$. Taking derivatives with respect to $\boldsymbol{\theta}^*$ and expectations, we find that

$$\boldsymbol{A}^* = \begin{pmatrix} \boldsymbol{I}_r & \boldsymbol{A}_{12} \\ \boldsymbol{0} & \boldsymbol{A}_{22} \end{pmatrix} \quad \text{and} \quad \boldsymbol{B}^* = \boldsymbol{B} = \begin{pmatrix} E\{\boldsymbol{\psi}_1\boldsymbol{\psi}_1^T\} & E\{\boldsymbol{\psi}_1\boldsymbol{\psi}_2^T\} \\ E\{\boldsymbol{\psi}_2\boldsymbol{\psi}_1^T\} & E\{\boldsymbol{\psi}_2\boldsymbol{\psi}_2^T\} \end{pmatrix},$$

where $\boldsymbol{I}_r$ is the $r \times r$ identity matrix and $\boldsymbol{A}$ and $\boldsymbol{B}$ without asterisks (*) refer to their form in the original problem. Finally, the (1, 1) element of $(\boldsymbol{A}^*)^{-1}\boldsymbol{B}^*\{(\boldsymbol{A}^*)^{-1}\}^T$ is (8.19, p. 347).

8.3.2c T_{GS} under Different Formulations of H_0

It is often more convenient to use the null hypothesis formulation $H_0 : \boldsymbol{h}(\boldsymbol{\theta}_g) = \boldsymbol{0}$ than $H_0 : \boldsymbol{\theta}_{g1} = \boldsymbol{\theta}_{10}$. However, it is not totally obvious in the M-estimation context what is meant by $\widetilde{\boldsymbol{\theta}}$ under $\boldsymbol{h}(\boldsymbol{\theta}_g) = \boldsymbol{0}$. In the case of maximum likelihood where $\boldsymbol{\psi}$ is obtained by differentiating the log likelihood, we can define $\widetilde{\boldsymbol{\theta}}$ as the solution of

$$\sum_{i=1}^{n} \boldsymbol{\psi}(Y_i, \widetilde{\boldsymbol{\theta}}) - \boldsymbol{H}(\widetilde{\boldsymbol{\theta}})^T \widetilde{\lambda} = 0, \quad \boldsymbol{h}(\widetilde{\boldsymbol{\theta}}) = 0,$$

where $\boldsymbol{H}(\boldsymbol{\theta}) = \partial \boldsymbol{h}(\boldsymbol{\theta})/\partial \boldsymbol{\theta}^T$ exists and has full row rank r, and λ is an $r \times 1$ vector of Lagrange multipliers. Of course, if $\boldsymbol{h}(\boldsymbol{\theta}_g) = \boldsymbol{0}$ has the equivalent $\boldsymbol{\theta}_{g1} = \boldsymbol{\theta}_{10}$ formulation, then $\widetilde{\boldsymbol{\theta}}$ has the usual meaning as the solution of $\sum_{i=1}^{n} \boldsymbol{\psi}_2(\widetilde{\boldsymbol{\theta}}) = 0$ with $\widetilde{\boldsymbol{\theta}}_1 = \boldsymbol{\theta}_{10}$.

Assuming that $\widetilde{\boldsymbol{\theta}}$ is well-defined, T_{GS} under $H_0 : \boldsymbol{h}(\boldsymbol{\theta}_g) = \boldsymbol{0}$ is

$$T_{\text{GS}} = n^{-1} \left\{ \sum \boldsymbol{\psi}(Y_i, \widetilde{\boldsymbol{\theta}}) \right\}^T \widetilde{\boldsymbol{V}}_h \left\{ \sum \boldsymbol{\psi}(Y_i, \widetilde{\boldsymbol{\theta}}) \right\}, \tag{8.26}$$

where

$$\widetilde{\boldsymbol{V}}_h = \left(\widetilde{\boldsymbol{A}}^{-1}\right)^T \widetilde{\boldsymbol{H}}^T \left\{ \widetilde{\boldsymbol{H}} \widetilde{\boldsymbol{A}}^{-1} \widetilde{\boldsymbol{B}} \left(\widetilde{\boldsymbol{A}}^{-1}\right)^T \widetilde{\boldsymbol{H}}^T \right\}^{-1} \widetilde{\boldsymbol{H}} \widetilde{\boldsymbol{A}}^{-1}, \tag{8.27}$$

and matrices with tildes (˜) over them refer to consistent estimates under $H_0 : \boldsymbol{h}(\boldsymbol{\theta}_g) = \boldsymbol{0}$. As before, an invariant version of T_{GS} is obtained by using (8.22, p. 348) to estimate $\boldsymbol{V}_h$. As an example, we leave for Problem 8.8 (p. 358) to show that (8.25, p. 350) follows from (8.26) above with $h(p_1, p_2) = p_1 - p_2$ when using the original parameterization of Example 8.5 (p. 345).

Some null hypotheses are best described in terms of one set of parameters, whereas the alternative is easiest to describe in terms of another set. Lack-of-fit or goodness-of-fit tests are often of this type. For example, consider binary data in b groups where $\theta_1, \ldots, \theta_b$ are the success probabilities but a proposed model is $\theta_i = w_i(\boldsymbol{\beta}), i = 1, \ldots, b$, where $\boldsymbol{\beta}$ is a $b - r$ vector of parameters. Example 3.4.3b (p. 147) about the adequacy of the logistic regression model is a good illustration of this situation.

The general description of this reparameterization approach is to let $\boldsymbol{\theta} = \boldsymbol{w}(\boldsymbol{\beta})$, where $\boldsymbol{w}$ is a b-valued vector function of the new parameter $\boldsymbol{\beta}$ that has length $b - r$. The null hypothesis is $H_0 : \boldsymbol{\theta}_g = \boldsymbol{w}(\boldsymbol{\beta}_g)$, where $\boldsymbol{W}(\boldsymbol{\beta}) = \partial \boldsymbol{w}(\boldsymbol{\beta})/\partial \boldsymbol{\beta}^T$ is assumed to have full column rank $b - r$. The generalized score statistic in this formulation is

$$T_{\text{GS}} = n^{-1} \left\{ \sum \boldsymbol{\psi}(Y_i, \widetilde{\boldsymbol{\theta}}) \right\}^T \widetilde{\boldsymbol{V}}_w \left\{ \sum \boldsymbol{\psi}(Y_i, \widetilde{\boldsymbol{\theta}}) \right\}, \tag{8.28}$$

where $\widetilde{\boldsymbol{V}}_w$ is an estimator of

$$\boldsymbol{V}_w = \widetilde{\boldsymbol{B}}^{-1} - \widetilde{\boldsymbol{B}}^{-1} \widetilde{\boldsymbol{A}} \widetilde{\boldsymbol{W}} \left(\widetilde{\boldsymbol{W}}^T \widetilde{\boldsymbol{A}}^T \widetilde{\boldsymbol{B}}^{-1} \widetilde{\boldsymbol{A}} \widetilde{\boldsymbol{W}} \right)^{-1} \widetilde{\boldsymbol{W}}^T \widetilde{\boldsymbol{A}}^T \widetilde{\boldsymbol{B}}^{-1}. \tag{8.29}$$

Note that in this case $\widetilde{\boldsymbol{B}}$ needs to be invertible (in contrast to the usual M-estimation framework). Again, one should use (8.22, p. 348) in the estimate of $\boldsymbol{V}_w$ in order to have an invariant version.

8.3.3 Adjusted Likelihood Ratio Statistics

As mentioned in the first part of this chapter, it is not so easy to adjust likelihood ratio statistics to have asymptotic chi-squared distributions. Moreover, there is no general definition of what is meant by a likelihood-ratio-type test when presented with only an estimating equation. That is, the M-estimation defining equation $\sum_{i=1}^{n} \boldsymbol{\psi}(\boldsymbol{Y}_i, \boldsymbol{\theta}) = \mathbf{0}$ need not have come from differentiating a likelihood.

However, suppose that $\sum_{i=1}^{n} \boldsymbol{\psi}(\boldsymbol{Y}_i, \boldsymbol{\theta}) = \partial Q(\boldsymbol{Y}, \boldsymbol{\theta})/\partial \boldsymbol{\theta}^T$ for some $Q(\boldsymbol{Y}, \boldsymbol{\theta})$, typically a log likelihood for some model, not necessarily the correct model. Then there are two situations where likelihood-ratio-type statistics might be used. The first is when the null hypothesis involves a single degree of freedom, and thus there is just one eigenvalue c of (8.2, p. 339) to be estimated. The second, and more important, is when all of the eigenvalues of (8.2, p. 339) are the same and can be estimated. In either case the difference of the log likelihoods can be divided by the estimated constant resulting in limiting chi-squared distributions under the null hypothesis. There seems to be no general theory to encompass all situations of interest. Therefore in this section we merely point out several well-known examples where likelihood-ratio-type statistics have been used even though the estimating equations do not come from the likelihood of the correct model.

The most well-known examples are perhaps from the generalized linear model (GLM) log likelihood (2.15, p. 54), with canonical link for simplicity, given by

$$\log L(\boldsymbol{\beta}, \phi \mid \{Y_i, \boldsymbol{x}_i\}_{i=1}^{n}) = \sum_{i=1}^{n} \left\{ \frac{Y_i \boldsymbol{x}_i^T \boldsymbol{\beta} - b(\boldsymbol{x}_i^T \boldsymbol{\beta})}{a_i(\phi)} + c(Y_i, \phi) \right\}.$$

It turns out that as long as the mean and variance are correctly specified, $E(Y_i) = b'(\boldsymbol{x}_i^T \boldsymbol{\beta})$, $\text{Var}(Y_i) = b''(\boldsymbol{x}_i^T \boldsymbol{\beta}) a_i(\phi)$, then two times a difference of this log likelihood under nested hypotheses is asymptotically χ^2 under the smaller model. In the GLM literature, the articles by Wedderburn (1974) and McCullagh (1983) and the name *quasi-likelihood* are associated with this approach. The constant $a_i(\phi)$ is usually estimated by the chi-squared goodness-of-fit statistic divided by $n - p$, where p is the dimension of $\boldsymbol{\beta}$ in the larger model. The resulting log-likelihood-ratio-type statistics are often divided by r and compared to an F distribution with r and $n - p$ degrees of freedom, and have good Type I error properties. For normal models, $a_i(\phi) = \sigma^2$ is estimated by the residual mean square. This approach arose in the bioassay literature with the name *heterogeneity factor* for ϕ in over-dispersed binomial(m_i, p_i)-type data where $a_i(\phi) = \phi/m_i$.

A second situation of interest is robust regression for the additive linear model

$$Y_i = \boldsymbol{x}_i^T \boldsymbol{\beta} + e_i, \tag{8.30}$$

where $e_1, \ldots, e_n$ are iid but there is uncertainty about the distribution, possibly normal with random outliers thrown in. Huber M-estimators use the estimating equation

$$\sum_{i=1}^{n} \psi_k \left(\frac{Y_i - \boldsymbol{x}_i^T \boldsymbol{\beta}}{\sigma} \right) = 0,$$

where ψ_k is the "Huber" ψ function defined in Example 7.4.1 (p. 311), but a variety of other ψ functions are also used in practice. The scale factor σ can be estimated in several ways, but adding the equation

$$\frac{1}{n-p}\sum_{i=1}^{n}\psi_k^2\left(\frac{Y_i - \boldsymbol{x}_i^T\boldsymbol{\beta}}{\sigma}\right) = \int_{-\infty}^{\infty}\psi_k^2(x)\phi(x)dx$$

is the most common. Here $\phi(x)$ is the standard normal density function, and this scale estimation approach is called Huber's Proposal 2. Note that the Huber ψ_k comes from differentiating a log density $-\rho_k$, but it is not assumed that e_i are drawn from this density. Thus, one might call this a situation where a misspecified likelihood is taken for granted.

Initially, generalized Wald tests were exclusively used in practice for this situation, but even with a variety of corrections they were found to be mildly liberal in small samples. Schrader and Hettmansperger (1980) suggested using the likelihood-ratio-type statistic

$$T_{\text{GLR}} = 2\widehat{c}^{-1}\left\{\sum_{i=1}^{n}\rho_k\left(\frac{Y_i - \boldsymbol{x}_i^T\widetilde{\boldsymbol{\beta}}}{\widehat{\sigma}}\right) - \sum_{i=1}^{n}\rho_k\left(\frac{Y_i - \boldsymbol{x}_i^T\widehat{\boldsymbol{\beta}}}{\widehat{\sigma}}\right)\right\}, \tag{8.31}$$

where

$$\widehat{c} = \frac{1}{n-p}\sum_{i=1}^{n}\psi_k^2\left(\frac{Y_i - \boldsymbol{x}_i^T\widehat{\boldsymbol{\beta}}}{\widehat{\sigma}}\right)\Big/\frac{1}{n}\sum_{i=1}^{n}\psi_k'\left(\frac{Y_i - \boldsymbol{x}_i^T\widehat{\boldsymbol{\beta}}}{\widehat{\sigma}}\right), \tag{8.32}$$

and $\widehat{\sigma}$ is the scale estimate for the larger model. In Problem 8.12 (p. 359) we ask for the details on how the form of c follows from results in Example 7.5.4 (p. 320) and (8.2, p. 339). Schrader and Hettmansperger (1980) suggest dividing T_{GLR} in (8.31) by r and comparing it to an F distribution with r and $n-p$ degrees of freedom.

To illustrate the Type I error performance of T_{GW}, T_{GS}, and T_{GLR} in robust regression and ANOVA situations, we report briefly some simulation results from the 1995 NC State dissertation of Lap-Cheung Lau. The three particular models used, and the corresponding hypotheses tested are:

1. Simple linear regression
   $Y_i = \beta_1 + \beta_2 x_i + e_i, i = 1,\ldots,20$ with $\beta_1 = 2.5$ and $x_i = 0.5 \times (i-1)$. The hypotheses are $H_0 : \beta_2 = 0$ vs. $H_a : \beta_2 \neq 0$.
2. One-way ANOVA
   $Y_{ij} = \mu + \tau_i + e_{ij}, i = 1,\ldots,4, j = 1,\ldots,5$ with $\mu = 2.5$ and $\tau_4 = 0$. The hypotheses are $H_0 : \tau_1 = \tau_2 = \tau_3 = 0$ vs. H_a : not all τ's are zero.
3. Randomized Complete Block Design (RCBD)
   $Y_{ij} = \mu + \alpha_i + \tau_j + e_{ij}, i = 1,\ldots,5, j = 1,\ldots,4$ with $\mu = 2.5, \alpha_1 = 0.2$, $\alpha_2 = 0.4$, $\alpha_3 = 0.6$, $\alpha_4 = 0.8$, $\alpha_5 = 0$, and $\tau_4 = 0$. The hypotheses are $H_0 : \tau_1 = \tau_2 = \tau_3 = 0$ vs. H_a : not all τ's are zero.

Table 8.1 Estimated Type I error probabilities for the classical F and Huber M-estimation robust tests in simple linear regression, one-way ANOVA, and RCBD with N(0,1) and $t_3/\sqrt{3}$ errors. Nominal level is $\alpha = .05$. Standard error of entries $\approx .002$.

		Simple Linear		One-Way		RCBD	
Tests	Cut-Off	N(0,1)	$t_3/\sqrt{3}$	N(0,1)	$t_3/\sqrt{3}$	N(0,1)	$t_3/\sqrt{3}$
classical F	F	.049	.045	.054	.042	.053	.038
T_{GS}	χ^2	.049	.050	.041	.039	.036	.033
T_{GW}	F	.061	.055	.076	.070	.073	.060
T_{GLR}	F	.062	.057	.071	.057	.065	.052

The choices of β_1 and μ have no effect on the results. The errors e_i follow either the standard normal distribution N(0,1) or the standardized t_3 distribution, i.e., $\sqrt{3}e_i \sim t_3$, with $\text{Var}(e_i) = 1$, in each case. Therefore, there is a total of six cases. The three robust test statistics are based on the Huber ψ_k function with $k = 1$ that gives asymptotic efficiency compared to least squares of approximately 90% when the errors are normal and 1.88 when the errors are t_3. T_{GLR} is as described above in (8.31, p. 353) with $k = 1$ and using F critical values. The version of T_{GW} used is (8.12, p. 344) multiplied by a correction factor $\{1 + (p/n)(1 - \widehat{\eta})/\widehat{\eta}\}^{-2}$ suggested by (Huber, 1981, p. 173) where $\widehat{\eta} = n^{-1}\sum_{i=1}^{n} \psi_k'(\widehat{e}_i/\widehat{\sigma})$. This version of T_{GW} is then divided by r so that F critical values may be used.

The version of T_{GS} used here reverses the subscripts so that the part of $\boldsymbol{\beta}$ tested is the last r components,

$$T_{\text{GS}} = n^{-1}\left\{\sum_{i=1}^{n} \psi_k\left(\frac{\widetilde{e}_i}{\widetilde{\sigma}}\right)\boldsymbol{x}_{i2}\right\}^T \widetilde{\boldsymbol{V}}_{\boldsymbol{\psi}_2}^{-1}\left\{\sum_{i=1}^{n} \psi_k\left(\frac{\widetilde{e}_i}{\widetilde{\sigma}}\right)\boldsymbol{x}_{i2}\right\}, \tag{8.33}$$

where

$$\widetilde{\boldsymbol{V}}_{\boldsymbol{\psi}_2} = \widetilde{E\psi_k'}\left\{(\boldsymbol{X}^T\boldsymbol{X})_{22} - (\boldsymbol{X}^T\boldsymbol{X})_{21}(\boldsymbol{X}^T\boldsymbol{X})_{11}^{-1}(\boldsymbol{X}^T\boldsymbol{X})_{12}\right\},$$

and $\widetilde{e}_i = Y_i - \boldsymbol{x}_{i1}^T\widetilde{\boldsymbol{\beta}}$ and $\widetilde{E\psi_k'} = (n-p+r)^{-1}\sum_{i=1}^{n}\psi_k'(\widetilde{e}_i/\widetilde{\sigma})$.

Here we present only Type I empirical rejection rates based on 10,000 simulations under the null hypotheses. These estimated Type I error probabilities in Table 8.1 (p. 354) have standard errors close to $\sqrt{(.05)(.95)/10000} = .002$.

Generally, we see that the results in Table 8.1 are similar to previous comparisons for tests of means, like in Section 3.2.4 (p. 132). That is, the score tests tend to be a bit conservative using chi-squared critical values. The Wald and likelihood ratio tests are a bit liberal even after using modifications and the F distribution that make them less liberal. T_{GLR} seems to hold its level better than T_{GW}. Of course the results in Table 8.1 are each based on only $n = 20$ data points; in larger samples, true levels are closer to .05. Note that the classical F test has quite robust Type I error properties in the face of heavy-tailed errors; the reason to use robust tests is for improved power under alternatives (up to 30% for t_3 errors, but these not displayed in Table 8.1).

8.3.3a A Note about Confidence Intervals

A key point in the discussion of tests under likelihood misspecification, or for the general estimating equation situation, is that likelihood-ratio-type statistics are not as easy to correct (or even define in general) as Wald and score tests. The GLM and robust example regression situations just discussed are exceptions, but usually there is no simple way to adjust for misspecification when using $\mathrm{T_{LR}}$ except to estimate the limiting distribution $\sum_{i=1}^{r} c_i Z_i^2$ given by Theorem 8.2 (p. 340).

Constructing confidence intervals for a single parameter, however, turns out to be a problem that allows a general solution for adjusting profile likelihood ratio statistics. So first consider constructing a profile likelihood confidence interval for a single parameter by inverting $\mathrm{T_{LR}}$. Without loss of generality, suppose that the parameter of interest is the first component of $\boldsymbol{\theta}$, say θ_1. The profile likelihood interval for θ_1 is from (3.19, p. 144)

$$\left\{\theta_1 : \mathrm{T_{LR}}(\theta_1, \widetilde{\boldsymbol{\theta}}_2(\theta_1)) \leq \chi_1^2(1-\alpha)\right\}, \tag{8.34}$$

where $\widetilde{\boldsymbol{\theta}}_2(\theta_1)$ is the maximum likelihood estimate of $\boldsymbol{\theta}_2$ with θ_1 held fixed. Of course $\mathrm{T_{LR}}(\theta_1, \widetilde{\boldsymbol{\theta}}_2(\theta_1))$ also depends on the unrestricted maximum likelihood estimator as well, but since that does not change with θ_1 we have suppressed it in the notation. To adjust this interval for misspecification, we use

$$\left\{\theta_1 : \mathrm{T_{LR}}(\theta_1, \widetilde{\boldsymbol{\theta}}_2(\theta_1)) \leq \widetilde{c}\chi_1^2(1-\alpha)\right\}, \tag{8.35}$$

where from (8.2, p. 339) we estimate the scalar

$$\widetilde{c} = \left(\widetilde{A}_{11} - \widetilde{\boldsymbol{A}}_{12}\widetilde{\boldsymbol{A}}_{22}^{-1}\widetilde{\boldsymbol{A}}_{21}\right)\left(\widetilde{\boldsymbol{A}}^{-1}\widetilde{\boldsymbol{B}}\widetilde{\boldsymbol{A}}^{-1}\right)_{11}.$$

Note that $\widetilde{\boldsymbol{\theta}}_2(\theta_1))$ and $\widetilde{c}$ must be recomputed for every trial value of θ_1. Although the computing may seem difficult, in a 2003 NC State dissertation Chris Gotwalt showed that this robustified profile likelihood approach works well in a variety of problems. Inverting T_{GS} is also hard computationally, but it did not perform quite as well as (8.35). Nevertheless, both are preferable to using the simpler T_{GW} intervals,

$$\widehat{\theta}_1 \pm \sqrt{\left(\widehat{\boldsymbol{A}}^{-1}\widehat{\boldsymbol{B}}\left\{\widehat{\boldsymbol{A}}^{-1}\right\}^T\right)_{11}} z_{1-\alpha/2}/\sqrt{n},$$

and both will be invariant to parameter transformation if the expected value estimate of $\boldsymbol{A}$ is used along with the empirical estimate of $\boldsymbol{B}$.

8.3.4 Quadratic Inference Functions and Score Tests

When presented with an estimating equation $\sum_{i=1}^{n} \boldsymbol{\psi}(\boldsymbol{Y}_i, \boldsymbol{\theta}) = \boldsymbol{0}$ that did not come from differentiating a log likelihood, it is not obvious how to define a quantity to take the place of the likelihood. This may not be a concern since it is often fine to just have the solution $\widehat{\boldsymbol{\theta}}$, some notion of its standard deviation, and perhaps a generalized Wald or score statistic for hypothesis testing.

On the other hand, there can be particular advantages to having an objective function to minimize. For example, if there are multiple solutions to $\sum_{i=1}^{n} \boldsymbol{\psi}(\boldsymbol{Y}_i, \boldsymbol{\theta}) = \boldsymbol{0}$, then an objective function could help decide which one to choose. In econometrics problems, there are often cases where the dimension of $\boldsymbol{\psi}$ is greater than that of the parameter vector. In such cases an objective function is critical for defining an appropriate solution to $\sum_{i=1}^{n} \boldsymbol{\psi}(\boldsymbol{Y}_i, \boldsymbol{\theta}) = \boldsymbol{0}$. Finally, an objective function would allow the definition of log-likelihood-ratio-type statistics.

Thus, Lindsay and Qu (2003) define the quadratic inference function (QIF)

$$Q^2(\boldsymbol{\theta}) = \overline{\boldsymbol{\psi}}^T \widehat{\boldsymbol{C}}_{\boldsymbol{\theta}}^{-1} \overline{\boldsymbol{\psi}}, \tag{8.36}$$

where $\overline{\boldsymbol{\psi}} = n^{-1} \sum_{i=1}^{n} \boldsymbol{\psi}(\boldsymbol{Y}_i, \boldsymbol{\theta})$, and $\widehat{\boldsymbol{C}}_{\boldsymbol{\theta}}$ estimates the covariance of $\overline{\boldsymbol{\psi}}$. We only consider the natural non-model-based estimator

$$\widehat{\boldsymbol{C}}_{\boldsymbol{\theta}} = \frac{1}{n} \boldsymbol{B}_n(\boldsymbol{Y}, \boldsymbol{\theta}) = \frac{1}{n^2} \sum_{i=1}^{n} \boldsymbol{\psi}(\boldsymbol{Y}_i, \boldsymbol{\theta}) \boldsymbol{\psi}(\boldsymbol{Y}_i, \boldsymbol{\theta})^T. \tag{8.37}$$

The QIF estimator $\widehat{\boldsymbol{\theta}}$ minimizes $Q^2(\boldsymbol{\theta})$. Notice that if the dimension of $\boldsymbol{\theta}$ is the same as the dimension of $\boldsymbol{\psi}$, then the QIF estimator is the same as the M-estimator that solves $\sum_{i=1}^{n} \boldsymbol{\psi}(\boldsymbol{Y}_i, \boldsymbol{\theta}) = \boldsymbol{0}$ because that estimator makes $Q^2(\boldsymbol{\theta}) = 0$, its minimum value. Hansen (1982) introduced the QIF estimating method in the econometrics literature, and a large number of publications have developed the theory in that context. Lindsay and Qu (2003) is an expository article introducing the approach to the statistics literature.

We focus here mainly on the testing situation where the dimension of $\boldsymbol{\theta}$ is the same as the dimension of $\boldsymbol{\psi}$. But first, let us give an example from Lindsay and Qu (2003) where the dimension of $\boldsymbol{\psi}$ is larger than that of $\boldsymbol{\theta}$. Suppose we are interested in the center or measure of location of a single distribution from which we have an iid sample $Y_1, \ldots, Y_n$. The estimating equation with $\psi_1(Y_i, \theta) = Y_i - \theta$ leads to the sample mean, and the estimating equation with $\psi_2(Y_i, \theta) = 1/2 - I(Y_i \le \theta)$ leads to the sample median. But what if the distribution is assumed to be symmetric so that both estimators are consistent for the center? By minimizing the QIF using these two score functions, we obtain an estimator that is the asymptotically optimal combination of the sample mean and median for estimating the center of the distribution. Thus, if the true distribution is normal, then asymptotically only the mean would be used. If the true distribution is Laplace, then asymptotically only the median would be used. In general, an optimal linear combination would result. The QIF is an extremely simple but powerful method for combining these estimators. Another motivating example may be found in Lindsay and Qu (2003), where the QIF was used in a GEE setting to optimally use a set of working correlation matrices.

Returning to the testing situation where the dimension of $\boldsymbol{\theta}$ is the same as the dimension of $\boldsymbol{\psi}$, consider first the completely specified null hypothesis $H_0 : \boldsymbol{\theta} = \boldsymbol{\theta}_0$. The QIF likelihood ratio type statistic is given by (Lindsay and Qu, 2003, p. 398) to be

$$T_{\text{QIF}} = Q^2(\boldsymbol{\theta}_0) - Q^2(\widehat{\boldsymbol{\theta}}) = Q^2(\boldsymbol{\theta}_0) \tag{8.38}$$

because $Q^2(\widehat{\boldsymbol{\theta}}) = 0$ in this case. The limiting distribution of T_{QIF} under H_0 is χ^2_r where $r = b$ in this fully specified case. Note that here T_{QIF} is the same as the generalized score test using $\boldsymbol{B}_n(\boldsymbol{Y}, \boldsymbol{\theta})$. For Example 8.1 (p. 341), use of $\psi_1(Y_i, \theta) = Y_i - \theta$ leads to

$$T_{\text{QIF}} = \frac{n(\overline{Y} - \mu_0)^2}{s_n^2 + (\overline{Y} - \mu_0)^2}.$$

Similarly in Example 8.3 (p. 342) we get $T_{\text{QIF}} = n(\widehat{p} - p_0)^2/s_n^2(p_0)$, where $s_n^2(p_0) = n^{-1}\sum_{i=1}^n (Y_i/m - p_0)^2$. And for Example 8.4 (p. 343) we also get the same result as T_{GS}.

For the composite null hypothesis $H_0 : \boldsymbol{\theta}_1 = \boldsymbol{\theta}_{10}$, (Lindsay and Qu, 2003, p. 398) give

$$T_{\text{QIF}} = Q^2(\boldsymbol{\theta}_{10}, \widetilde{\boldsymbol{\theta}}_2) - Q^2(\widehat{\boldsymbol{\theta}}) \tag{8.39}$$

$$= \min_{\boldsymbol{\theta}_2} Q^2(\boldsymbol{\theta}_{10}, \boldsymbol{\theta}_2), \tag{8.40}$$

where again we have taken advantage of $Q^2(\widehat{\boldsymbol{\theta}}) = 0$ for the case we are considering of equal dimensions for $\boldsymbol{\theta}$ and $\boldsymbol{\psi}$. The limiting distribution of T_{QIF} under H_0 is χ^2_r. Here we can see an interesting difference between T_{GS} and T_{QIF}. T_{GS} uses the estimators $\widetilde{\boldsymbol{\theta}}$ that solve $\sum_{i=1}^n \boldsymbol{\psi}_2(\boldsymbol{Y}_i, \widetilde{\boldsymbol{\theta}}) = \boldsymbol{0}$ where $\widetilde{\boldsymbol{\theta}} = (\boldsymbol{\theta}_{10}, \widetilde{\boldsymbol{\theta}}_2)^T$, whereas for T_{QIF} we minimize $Q^2(\boldsymbol{\theta}_{10}, \boldsymbol{\theta}_2)$ in $\boldsymbol{\theta}_2$. Moreover, the matrix that shows up in the middle of T_{GS} is the inverse of $\boldsymbol{V}_{\boldsymbol{\psi}_1}$ given by (8.19, p. 347), whereas the matrix that shows up in the middle of T_{QIF} is the inverse of the simpler $\widehat{C}_{\boldsymbol{\theta}}$ given by (8.37, p. 356). In the simple problem of testing a standard deviation $H_0 : \sigma = \sigma_0$, Example (8.2, p. 342), T_{GS} and T_{QIF} give almost identical results in large samples.

Thus, there are tradeoffs for computing each of T_{GS} and T_{QIF}, but more theory is needed before a definitive recommendation can be made between them. Certainly, the QIF theory gives a very important new testing method that has similarities to log likelihood ratio tests as well as to generalized score tests. Here we have only scratched the surface of potential applications.

8.4 Problems

8.1. Using (8.5, p. 341) and Theorem 5.28 (p. 249), verify that expression (8.6, p. 341) is correct.

8.2. Give arguments to support the result in (8.9, p. 341).

8.3. For Example 8.3 (p. 342), show that c_1 of Theorem 8.2 (p. 340) is equal to d.

8.4. Use the quadratic equation to write down the general solution of (8.10, p. 343). Then set $\mu_3 = 0$ and show that the solutions reduce to $c_1 = 1$ and $c_2 = (\mu_4 - \sigma^4)/(2\sigma^4)$.

8.5. Provide the details and derive (8.16, p. 346).

8.6. In the continuation of Example 8.3.2a (p. 350) for T_{GS}, obtain the estimates of $\boldsymbol{A}$ and $\boldsymbol{B}$ that lead to (8.25, p. 350).

8.7. For the setup of problem 7.15 (p. 336), show that the generalized score test for $H_0 : \beta_2 = 0$ is given by

$$T_{GS} = \frac{\left[\sum_{i=1}^{k}(Y_{i.} - n_{i.}\overline{\overline{Y}})d_i\right]^2}{\sum_{i=1}^{k}\left[(d_i - \overline{d})^2 \sum_{j=1}^{m_i}(Y_{ij} - n_{ij}\overline{\overline{Y}})^2\right]}.$$

Note that you need to reverse the subscripts of the formulas given for the generalized score test because here the null hypothesis concerns the second part of the vector.

8.8. Using the original parameterization of Example 8.5 (p. 345), show that (8.25, p. 350) follows from (8.26, p. 351) with $h(p_1, p_2) = p_1 - p_2$.

8.9. For simple linear regression with heteroscedastic errors, model (7.17, p. 315) with $\boldsymbol{x}_i^T = (1, d_i)$, shows that the generalized score test for $H_0 : \beta_2 = 0$ is given by

$$T_{GS} = \frac{\left[\sum_{i=1}^{n}(Y_i - \overline{Y})d_i\right]^2}{\frac{n}{n-1}\sum_{i=1}^{n}(Y_i - \overline{Y})^2(d_i - \overline{d})^2},$$

where we have simplified things by rewriting the model with $d_i^* = d_i - \overline{d}_i$. Then the matrix manipulations are much simpler. How does this form compare to the usual F test when homogeneity of variance is assumed?

8.10. Suppose we have data $X_1, \ldots, X_n$ that are iid. The sign test for H_0: median=0 is to count the number of X's above 0, say Y, and compare Y to a binomial $(n, p = 1/2)$ distribution. Starting with the defining M-estimator equation for the sample median (see Example 7.4.2, p. 312, with p=1/2),

a. Derive the generalized score statistic T_{GS} for H_0: median=0 and note that it is the large sample version of the two-sided sign test statistic.
b. Using the expression for the asymptotic variance of the sample median, write down the form of a generalized Wald statistic T_{GW}, and explain why it is not as attractive to use here as T_{GS}.

8.11. Using the general score equation (7.34, p. 321) for generalized linear models, but assuming only that the mean and variance are correctly specified, let $a_i(\phi) = \phi/m_i$ as used for over-dispersed binomial data and show that $c_1 = \cdots = c_r = \phi$ that is obtained from (8.2, p. 339). In other words, show that (8.2, p. 339) is ϕ times the $r \times 1$ identity matrix.

8.12. Similar to the expressions in Example 7.5.4 (p. 320) for the $\sigma = 1$ known case, if we make the homogeneity assumption that the errors $e_1, \ldots, e_n$ in (8.30, p. 352) all have the same distribution, and use $\sum_{i=1}^n \sigma^{-1}\psi_k(e_i/\sigma)\boldsymbol{x}_i = 0$ as the defining equation, then $\boldsymbol{A}_n(X, \boldsymbol{\beta}, \sigma) = \sigma^{-2}\mathrm{E}\psi_k'(e_1)\boldsymbol{X}^T\boldsymbol{X}/n$ and $\boldsymbol{B}_n(\boldsymbol{X}, \boldsymbol{\beta}, \sigma) = \sigma^{-2}\mathrm{E}\psi_k(e_1/\sigma)^2\boldsymbol{X}^T\boldsymbol{X}/n$ are the expected value versions of $\boldsymbol{A}$ and $\boldsymbol{B}$. Use them to show that (8.2, p. 339) gives $c_1 = \cdots = c_r = c$ that is estimated by (8.32, p. 353). The reason to use $\sum_{i=1}^n \sigma^{-1}\psi_k(e_i/\sigma)\boldsymbol{x}_i = 0$ instead of $\sum_{i=1}^n \psi_k(e_i/\sigma)\boldsymbol{x}_i = 0$ is because the former is the derivative of $\sum_{i=1}^n \rho_k(e_i/\sigma)$ with respect to $\boldsymbol{\beta}$ that is used to define T_{GLR} in (8.31, p. 353).

Part V
Computation-Based Methods

Chapter 9
Monte Carlo Simulation Studies

9.1 Introduction

Modern statistical problems require a mix of analytical, computational and simulation techniques for implementing and understanding statistical inferential methods. Previous chapters dealt with analytical and computational techniques, and this chapter focuses on simulation techniques. We are primarily interested in Monte Carlo simulation studies in which the sampling properties of statistical procedures are investigated. We discuss a few important principles in the design and analysis of these Monte Carlo studies. These principles are perhaps obvious after reflection, but senior, as well as junior, statistics researchers often neglect one or more of them. Secondly, we give some practical hints about choosing the Monte Carlo size N and analyzing and presenting results from Monte Carlo studies.

We start with a brief discussion of Monte Carlo estimation that is at the core of all simulation techniques. "Monte Carlo" methods essentially refer to any use of random simulation. David (1998) reports that the name was coined by famous mathematician and computer scientist John von Neumann and his Los Alamos colleague S.M. Ulam.

9.1.1 *Monte Carlo Estimation*

Monte Carlo estimation is used to estimate expected values of random variables and functions of random variables when analytic calculation is not feasible. Monte Carlo methods work because good random number generators generate surrogate random variables, or pseudo-random variables, that mimic many of the key properties of truly-random, random variables. In particular, the classical large sample theorems of probability and statistics such as the Weak and Strong Laws of Large Numbers and the Central Limit Theorems generally apply to sequences of pseudo-random

D.D. Boos and L.A. Stefanski, *Essential Statistical Inference: Theory and Methods*, Springer Texts in Statistics, DOI 10.1007/978-1-4614-4818-1_9,

variables. So if $T_1, \ldots, T_N$ is an independent and identically distributed sequence of pseudo-random variables, then

$$\overline{T} \xrightarrow{p} \mathrm{E}(T), \quad s_{N-1}^2 \xrightarrow{p} \mathrm{Var}(T) \quad \text{and} \quad \frac{\overline{T} - \mathrm{E}(T)}{s_{N-1}/N^{1/2}} \xrightarrow{d} \mathrm{N}(0,1), \qquad (9.1)$$

provided the necessary moments of T exist. Thus, for example, it is obvious that $\mathrm{E}(T)$ can be estimated arbitrarily well by $\overline{T}$ by taking N sufficiently large, and that bounds on the estimation error can be derived from the normal approximation. Because the same reasoning applies to transformed pseudo-random variables, e.g., T_i^k, $I(T_i \le t)$ and $\exp(sT_i)$, virtually any quantity related to the distribution of T can be estimated arbitrarily well, e.g., $\mu_k' = E(T^k)$, $P(T \le t) = E\{I(T \le t)\}$ and $\mathrm{mgf}_T(s) = E\{\exp(sT)\}$, provided, of course, that the necessary moments exist. Finally, we note that Monte Carlo methods apply not only to scalar-valued random variables as depicted above, but also to random vectors.

In applications to classical statistical inference, $T_1, \ldots, T_N$ are often statistics, estimators or test statistics, that are obtained by applying a statistical procedure to simulated data sets. That is, $T_i = T(X_{i,1}, \ldots, X_{i,n})$, where $T(\cdot)$ is the function that defines the statistic and $\{X_{i,1}, \ldots, X_{i,n}\}$ is the ith pseudo-random data set generated from a probability model. In such applications Monte Carlo methods are used to estimate quantities such as:

- the bias and variance of an estimator;
- the percentiles of a test statistic or pivotal quantity;
- the power function of a hypothesis test;
- the mean length and coverage probability of a confidence interval.

An important use of Monte Carlo estimation methods is to estimate the posterior distribution in a Bayes analysis by generating a sequence of pseudo-random variables that converge in distribution to the posterior distribution of interest. These Markov chain Monte Carlo (MCMC) methods are discussed in Chapter 4. Monte Carlo estimation also plays a prominent role in bootstrap resampling inferential methods discussed in Chapter 11. In both MCMC and bootstrap methods, Monte Carlo estimation is used to implement the analysis of a single data set. Thus while Monte Carlo estimation is crucial to both MCMC and bootstrap methods, neither application constitutes a Monte Carlo study as defined below.

9.1.2 Monte Carlo Studies

The focus of this chapter is on the design, implementation and analysis of a *Monte Carlo study* which we define as the application of Monte Carlo estimation methods for the purpose of studying and comparing the probabilistic and statistical properties of statistics and how these properties depend on factors such as sample size, distributional assumptions, parameter values, and so on. A Monte Carlo study is an experiment to which the fundamental concepts of experimental design — randomization, replication and blocking — apply. Many Monte Carlo studies are often factorial in nature, and the techniques and methods of analysis for multifactor experiments are relevant. In particular, Monte Carlo studies should be carefully planned with due attention paid to the analysis of the generated data.

Often a single pseudo-random data set $X_1, \ldots, X_n$ is generated for the purpose of illustrating a new method, especially with regard to calculating an estimate $T = T(X_1, \ldots, X_n)$. In complicated problems, this can be useful to explain the procedure that results in T. However, such an exercise is merely an illustration of the method of calculation, and is not a Monte Carlo study. Technically, it is a Monte Carlo study of size $N = 1$, but of no value for estimating statistical properties of a procedure such as bias, variance, power, etc., and therefore does not satisfy our definition of a Monte Carlo study. A Monte Carlo study, by definition, must involve many replications in order to estimate properties of the sampling distribution of the statistic.

We now describe a simple Monte Carlo study that is sufficient to illustrate many of the features alluded to above, and also the interplay between analytic analysis and the design of the experiment.

Example 9.1 (Comparing Location Estimators). Location estimators play an important role in statistics, especially in robust statistics, and therefore their sampling properties are often of interest. Suppose, for example, that we want to compare three location estimators, the sample mean $\overline{X}$, the sample 20% trimmed mean $\overline{X}_{.20}$ (trim 20% from each end of the ordered sample and average the remaining ones), and the sample median $\widehat{\eta}_{.5}$ for a variety of sample sizes ($n = 15, 30$ and 120) and population distributions with different tail behavior (normal, Laplace and t_5).

So far we have identified three factors in the study, distribution at three levels (normal, Laplace and t_5), sample size at three levels (15, 30 and 120), and of course, estimator also at three levels (mean, trimmed mean and median). Data sets of sizes $n = 15, 30$ and 120 will be generated from $\mathrm{N}(\mu, \sigma^2)$, $\mathrm{Laplace}(\mu, \sigma^2)$ and $t_5(\mu, \sigma^2)$ distributions where μ and σ^2 are the true mean and variance of the generated data. Should μ and σ^2 be factors in our experiment and varied at different levels? Consider that the three location estimators are all location-scale invariant, i.e., $T(a + bX_1, \ldots, a + bX_n) = a + bT(X_1, \ldots, X_n)$. Thus the sampling properties of the estimators for one set of (μ, σ^2) are completely determined by the sampling properties when the mean and variance are $(0, 1)$, e.g., $\mathrm{Var}\{T(a + bX_1, \ldots, a + bX_n)\} = b^2\mathrm{Var}\{T(X_1, \ldots, X_n)\}$. Therefore it makes no sense to study different values of (μ, σ^2), and these would typically be set to $(0, 1)$ without loss of generality.

Bias, variance and mean squared error of the estimators are typically studied. However, for our location-estimator study, bias is not relevant because all of the population distributions studied are symmetric about 0, and the estimators are odd functions of the data. So we know that the estimators are unbiased, and thus comparisons should focus on variability, and possibly kurtosis, or more generally on the entire sampling distribution of the estimators. Had we included some non-symmetric distributions in the study, then bias would be of interest, although in this case there is an attendant definitional issue as the estimators would not be estimating the same quantities.

The Monte Carlo study as we have set it up would most likely be carried out by generating N independent N(0, 1) data sets of a given size, and calculating all three estimators on each data set. This would be repeated for the different sample sizes; and then the entire previous process repeated for the two other distributions, so that all of the generated data sets are independent of one another, both within and across the nine combinations of distribution and sample size. Thus estimators calculated from different data sets are independent, whereas estimators calculated on the same data set are dependent. The latter dependence is exploited in the analysis of the output as comparisons of estimators are typically very precise because of the blocking induced by the data sets. Note that a total $9N$ data sets are generated; and $27N$ estimators are calculated and should be saved for later analysis.

The choice of N remains. As in any statistical sample-size determination both practical constraints (computing time) and statistical considerations (variance and power) should be considered. With regard to statistical considerations, Monte Carlo experiments have the advantage that preliminary results are usually available to guide the choice of sample size. For example, by running the study for a preliminary sample size N_0 for some fraction of the experimental design, one can estimate the sample size needed for adequate power in, say, the paired t-test of equality of variances (recall that bias $= 0$ by design, so that $\text{Var}(\widehat{\theta}_r)-\text{Var}(\widehat{\theta}_s) = E(\widehat{\theta}_r^2)-E(\widehat{\theta}_s^2)$ and thus equality of variances can be tested by testing the equality of means of the squared statistics). As for practical considerations, note that computing time is typically linear in N. So one can run the Monte Carlo simulation program for say $N = N_0$ where N_0 is chosen so that the program takes on the order of a few minutes. Then determine N so that the full simulation runs in an appropriate allotted time, e.g., overnight.

Below is R code for a fraction of the Monte Carlo study described above corresponding to $n = 15$ and the standard normal population distribution, with $N = 1000$.

```
sim.samp<-function(nrep,n,DIST,...){
# simulates nrep samples from DIST of size n
data <- matrix(DIST(n * nrep, ...), ncol = n, nrow = nrep)
}
set.seed(346)                     # sets the random number seed
sim.samp(1000,15,rnorm)->z        # 1000 N(0,1) samples, n=15
apply(z,1,mean)->out.m            # mean for each sample
trim20<-function(x){mean(x,.2)}   # 20% trimmed mean function
apply(z,1,trim20)->out.t20        # trim20 for each sample
apply(z,1,median)->out.med        # median for each sample
# Save all 1000 blocks of 3 estimators in a data frame
data.frame(mean=out.m,trim20=out.t20,median=out.med)->norm15
```

Table 9.1 Variance estimates (times n) for location estimators. Distributions are standardized to have variance 1

	Normal			Laplace			t_5		
	$\overline{X}$	$\overline{X}_{.20}$	Med	$\overline{X}$	$\overline{X}_{.20}$	Med	$\overline{X}$	$\overline{X}_{.20}$	Med
$n = 15$	0.99	1.11	1.47	1.00	0.70	0.71	1.02	0.85	1.06
$n = 30$	1.02	1.16	1.53	0.99	0.67	0.64	1.00	0.81	1.00
$n = 120$	1.01	1.15	1.57	0.99	0.65	0.57	1.00	0.83	1.05

Note: Entries are based on 10,000 replications and have standard errors in the range 0.01 to 0.02. The Laplace density is $(1/2)\exp(-|y|)$.

Now that we have a program to compute the basic estimates, the next step is to analyze the data, including computing:

1. the sample variances of the saved location estimators—these sample variances are the key quantities of interest in this example, and
2. the standard errors of these sample variances and possibly standard errors of differences of sample variances.

Recall that a standard error of any estimator, in this case a Monte Carlo-based estimator, is an estimate of its standard deviation. In Sections 9.3 (p. 369) and 9.4 (p. 372), we discuss the computation of standard errors for a number of basic quantities like the sample variance. Finally, we want to think about how to present the results. To complete this example, Table 9.1 illustrates one possible presentation of the results. We decided to multiply the variance estimates by sample size n so that we could compare different sample sizes. Also, after computing a few standard errors, we decided to raise the number of Monte Carlo replications to $N = 10{,}000$. ♦

9.2 Basic Principles for Monte Carlo Studies

The first principle is especially for statisticians, who as a group, are used to advising others about experiments but often ignore their own advice.

Principle 1. A Monte Carlo experiment is just like any other experiment. It requires careful planning in addition to carrying out the experiment and analyzing the results.

There is no harm in jumping in and making preliminary runs to get a feel for what the results will look like. But for a serious study to be later published or shared with others, one should carefully consider the following.

a) The choice of factors to be varied during the experiment, such as the sample size n and the distributions used to generate the data, is critical for it determines the generality of the results and hence the utility of the study. A very limited design may be unconvincing at best, and misleading at worst, if the conditions under which data are generated are not sufficiently representative of real data. However, the desire for generality must be tempered by the fact that the number of factor-level combination in a full factorial study grows rapidly with the numbers of factors and their levels.

b) The Monte Carlo sample size N must be chosen large enough to obtain satisfactory precision but also small enough to accommodate time and machine constraints. The ability to repeat a study is important if modifications to the design or to the statistics under investigation are required. Thus choosing N larger than necessary to obtain acceptable precision should be avoided. It is usually possible to add more independent replicates of the study if necessary to increase precision.
c) Some consideration about what information is to be saved from the experiment, and how it is to be analyzed to accomplish the objectives of the study is important. In some studies only certain summary statistics (means, variance, covariances) need to be saved, and these can often be aggregated on the fly using well-known updating formulas.

Provided data storage is not a limiting factor, the relevancy of the latter recommendation can be essentially eliminated by adhering to the following principle.

Principle 2. Save all of the statistics (the individual T values inside the main loop) calculated from each data set whenever possible.

The obvious advantage of Principle 2 is the ability to calculate any desired summary statistic *after* the completion of the study. Thus the post-study analysis is not limited by pre-study decisions about what information to save from individual runs. This is often useful when heavy-tailed data or outliers in the simulation output are possible, the latter possibly due to extreme data sets or to the lack of convergence of nonlinear iterative numerical methods. For then it is possible to identify problems with the output data, and adapt the method of analysis accordingly. When problems with non-convergence are likely, keeping track of algorithm "exit codes" may be helpful for identifying the nature of the numerical problems. Finally, it may be advantageous to analyze the study output using software (e.g., SAS or R) different from that used to do the simulation.

Principle 3. Keep the Monte Carlo sample size N small at first until the basic study plan is clear.

Most simulation programs and designs undergo extensive development to ensure the correctness of the program and maximize the utility of the information so obtained. A too-common problem is making runs with large N before the program is known to be bug-free or the design is finalized, thereby wasting time and effort. Thus development is usually facilitated by making initial runs with small N. After development is complete, then make the "production" runs with large N that will constitute the final study. However, keep in mind that Monte Carlo studies by statisticians are just as susceptible to bias as non-Monte Carlo studies by non-statisticians. Thus, in the course of developing a simulation program and design, care must be taken to ensure that conditions (factors, levels, seeds) are not identified (consciously or subconsciously) that are unfairly favorable or unfavorable to the any of the methods under study.

Principle 4. Organize your work and document both the simulation code and the results.

It is important to document the computer code with comments so that you or someone else (e.g., a colleague or an advisor) can check and understand what the code does, both at the time of the initial study, and a year later when you receive reviews on a paper you wrote asking for additional simulation results.

It is especially important to keep track of the seeds used in the random number generation. Set a different seed for each run and keep records of the seeds. This ensures that runs are reproducible and essentially guarantees independence from run to run.

9.3 Monte Carlo Sample Size

Determining the sample size in a Monte Carlo study is no different than for other types of studies. However, with a Monte Carlo study it is usually easier to get a good preliminary estimate of experimental error by making test runs. Basically one needs to determine the acceptable standard deviation of estimates or power of tests and invert analytical functions of sample size set equal to those values. Here are a few examples of such calculations.

9.3.1 Bias Estimation

Suppose that one computes the parameter estimate $\widehat{\theta}$ for N independent replications resulting in $\widehat{\theta}_1, \ldots, \widehat{\theta}_N$. In a table we may present entries of $N^{-1} \sum \widehat{\theta}_i$ along with the true value θ_0, or actually present the bias estimate $N^{-1} \sum \widehat{\theta}_i - \theta_0$. In either case the standard deviation is $\sigma_n / N^{1/2}$, and a standard error is $s_N / N^{1/2}$, where $\sigma_n^2 = \text{Var}(\widehat{\theta})$. So, given a guess of σ_n, say $\widetilde{\sigma}$, and an acceptable value, say d, for the standard deviation of our estimate, we solve to get $N = \widetilde{\sigma}^2 / d^2$. One way to get $\widetilde{\sigma}$ is from a preliminary run of size N_0 samples, where then $\widetilde{\sigma}$ could just be the sample standard deviation of $\widehat{\theta}_1, \ldots, \widehat{\theta}_{N_0}$.

To compare the estimates of bias of two different estimators, say $\widehat{\theta}^{(1)}$ and $\widehat{\theta}^{(2)}$, using the same Monte Carlo data, one should take into account the blocking induced and use $[\text{Var}\{\widehat{\theta}^{(1)} - \widehat{\theta}^{(2)}\}]^{1/2} / N^{1/2}$ as one would use with the paired t test.

9.3.2 *Variance Estimation*

When studying a new estimator $\widehat{\theta}$, one usually wants to estimate its variance with the sample variance,

$$s_{N-1}^2 = \frac{1}{N-1}\sum_{i=1}^{N}(\widehat{\theta}_i - \overline{\widehat{\theta}})^2,$$

where $\overline{\widehat{\theta}} = N^{-1}\sum_{i=1}^{N}\widehat{\theta}_i$. This is the Monte Carlo estimator of $\text{Var}(\widehat{\theta})$. For large N, the approximate variance of the sample variance is close to $\sigma_n^4(\text{Kurt}(\widehat{\theta}) - 1)/N$, where $\sigma_n^2 = \text{var}(\widehat{\theta})$ and $\text{Kurt}(\widehat{\theta})$ is the kurtosis of the distribution of $\widehat{\theta}$.

Many estimators are approximately normal provided the sample size is not too small. Thus, if the size n of the sample used to calculate $\widehat{\theta}$ is not small, $\widehat{\theta}$ is approximately normal, and so $\text{Kurt}(\widehat{\theta}) \approx 3$, leading to the approximation $\text{var}(s_{N-1}^2) \approx 2\sigma_n^4/N$. So the approximate standard deviation of the variance estimate is $(2/N)^{1/2}\sigma_n^2$. For acceptable d, we invert to get $N = 2\sigma_n^4/d^2$.

If one prefers to think in terms of relative error, letting d_p be the acceptable proportional error, we have that the approximate standard deviation of s_{N-1}^2/σ^2 is $(2/N)^{1/2}$, leading to $N = 2/d_p^2$. For example, if $d_p = .10$ (10 percent relative error), we get $N = 2/.01 = 200$.

If one considers the sample standard deviation s_{N-1} instead of the sample variance s_{N-1}^2, then the approximate standard deviation of s_{N-1} is $\sigma_n/(2N)^{1/2}$ (by the delta theorem). Thus, letting d_s denote an acceptable standard deviation for s_{N-1}, then $N = \sigma_n^2/(2d_s^2)$. Similarly, if d_{sp} denotes the value of an acceptable proportional error, then $N = 1/(2d_{sp}^2)$. So if $d_{sp} = .10$ (10 percent relative error), then $N = 50$. Thus for getting a 10 percent relative error for estimating the standard deviation, one needs only 1/4 of the Monte Carlo sample size as is needed to get a 10 percent relative error for the variance.

9.3.3 *Power Estimation*

For a new test procedure of the form "reject the null hypothesis if $T > c_\alpha$," we often estimate the power at a particular alternative by

$$\widehat{\text{pow}} = \frac{1}{N}\sum_{i=1}^{N} I(T_i > c_\alpha),$$

where T_i is the test statistic for the ith Monte Carlo sample, c_α is a given critical value, and I is the indicator function having value 1 if $T_i > c_\alpha$ and 0 otherwise.

This is binomial sampling, and the worst variance of our estimate (occurring at power = 1/2) is given by $1/(4N)$. Setting $d = 1/(2N^{1/2})$ yields $N = 1/(4d^2)$.

For $d = .05$ we get $N = 100$, but for $d = .01$ we get $N = 2500$. If we want to compare the power of two test statistics, $T^{(1)}$ and $T^{(2)}$, then one should take into account the blocking induced by computing both procedures on the same Monte Carlo samples. This could be accomplished by estimating the variance of

$$I(T^{(1)} > c_\alpha^{(1)}) - I(T^{(2)} > c_\alpha^{(2)})$$

in some preliminary runs.

In some cases, one may have a critical value c_α from asymptotic theory but would like to estimate the finite-sample c_α and use it in the power estimate. This can be important when comparing two test procedures that have different true significance levels. In this case, Zhang and Boos (1994) suggest using an N for estimating the critical value that is 10 times the size of the N used for estimating the power at specific alternatives.

9.3.4 Confidence Intervals

Monte Carlo studies of confidence intervals have similarities to bias estimation (Section 9.3.1, p. 369) and to power estimation (Section 9.3.3, p. 370). *Coverage probability* and *average confidence interval length* are both important quantities that should be reported whenever studying confidence intervals. Obviously we would like intervals that achieve the nominal $1 - \alpha$ coverage (like 95%) and are short on average. If one only reports estimated coverage probabilities, then we have no idea about the value of the interval because the interval $(-\infty, \infty)$ is always best in the coverage sense, achieving 100% coverage. In some situations, it is useful to report one-sided errors rather than coverage because we might consider using one-sided confidence bounds instead of intervals or we might prefer intervals that have symmetric errors. For example, we might prefer an interval that misses on the left and right of the parameter with equal probability .025 compared to one that misses on the left with probability .01 and on the right with probability .04.

For sample size considerations, note that we obtain a sample of lengths $L_1, \ldots, L_N$ and a sample $I_1, \ldots, I_N$ of binary indicators of whether the interval contains the true parameter value. For the lengths we need a preliminary estimate $\widetilde{\sigma}$ of the standard deviation of the lengths and an acceptable value d for the standard error of our estimate of average length. Then, just as in Example 9.3.1 (p. 369), we obtain $N = \widetilde{\sigma}^2/d^2$. For coverage estimation or one-sided error estimation, we would invert $d = (\alpha(1-\alpha)/N)^{1/2}$, where d is the acceptable standard deviation for our coverage estimate, to get $N = \alpha(1-\alpha)/d^2$. This is slightly different from power estimation where we often use the worst case scenario of power $= 1/2$ to get $N = 1/(4d^2)$.

9.4 Analyzing the Results

The proper analysis of Monte Carlo studies begins with correctly estimating the standard deviation of the estimates. In some cases one might want to run an analysis of variance to quantify the importance of different factors. When using a statistical procedure to compare different methods, we usually want to take into account the blocking induced by the generated data sets. That is, two estimators computed on the same data set are generally correlated. Thus, we might use paired t tests or randomized complete block analysis of variance. In the case of comparing two tests, a paired t on the binary indicators of rejection is essentially equivalent to McNemar's test (see Section 12.10.2, p. 509) for correlated proportions. Here we illustrate some possible analyses, focusing on situations where asymptotic analysis is useful. In practice, it is often simplest to use jackknife methods (see Chapter 10) for variance estimation.

9.4.1 *Comparing Two Variance Estimators*

Consider a situation like Example 9.1 (p. 365) where we have two estimators for the same quantity, and we are mainly interested in estimating the variance of these estimators. As mentioned in Section 9.3, the asymptotic variance (as $N \to \infty$ with n fixed) of $s_{N-1}^2 = (N-1)^{-1}\sum_{i=1}^{N}(\widehat{\theta}_i - \overline{\widehat{\theta}})^2$ is $\sigma_n^4(\text{Kurt}(\widehat{\theta}) - 1)/N$, where $\sigma_n^2 = \text{var}(\widehat{\theta})$ and $\text{Kurt}(\widehat{\theta})$ is the kurtosis of the distribution of $\widehat{\theta}$. Moreover, because most estimators are approximately normal, assuming that $\text{Kurt}(\widehat{\theta}) \approx 3$ leads to a simplification of the asymptotic variance of s_{N-1}^2 to $2\sigma_n^4/N$. Instead of using $\text{Kurt}(\widehat{\theta}) \approx 3$, we can simply estimate $\text{Kurt}(\widehat{\theta})$ with the sample kurtosis of the estimators $\widehat{\theta}_1, \ldots, \widehat{\theta}_N$,

$$\widehat{\text{Kurt}}(\widehat{\theta}) = N^{-1}\sum_{i=1}^{N}(\widehat{\theta}_i - \overline{\widehat{\theta}})^4 \Big/ \left\{N^{-1}\sum_{i=1}^{N}(\widehat{\theta}_i - \overline{\widehat{\theta}})^2\right\}^2.$$

Suppose, however, that we want to compare the estimates of variance for two different estimators like the mean and the trimmed mean. Because the variance estimators, say $s_{1,N-1}^2$ and $s_{2,N-1}^2$, are both computed from the same original data sets, these estimators are correlated, and the correlation must be accounted for when we estimate the standard deviation of their ratio. Using the approximation by averages results for sample variances, it follows that $(s_{1,N-1}^2, s_{2,N-1}^2)^T$ are jointly asymptotically normal as $N \to \infty$ with asymptotic mean $(\sigma_{1,n}^2, \sigma_{2,n}^2)^T$ and asymptotic covariance matrix with diagonal elements $N^{-1}\sigma_{1,n}^4(\text{Kurt}(\widehat{\theta}_1) - 1)$ and $N^{-1}\sigma_{2,n}^4(\text{Kurt}(\widehat{\theta}_2) - 1)$, respectively, and off-diagonal element

$$N^{-1}\text{Cov}_{12,n} = N^{-1}\text{Cov}\left(\left\{\widehat{\theta}_1 - \text{E}(\widehat{\theta}_1)\right\}^2, \left\{\widehat{\theta}_2 - \text{E}(\widehat{\theta}_2)\right\}^2\right).$$

Then, using the Delta method, Theorem 5.19 (p. 238), the asymptotic variance of $s^2_{1,N-1}/s^2_{2,N-1}$ is

$$\frac{1}{N}\frac{\sigma^4_{1,n}}{\sigma^4_{2,n}}\left\{\text{Kurt}(\widehat{\theta}_1)+\text{Kurt}(\widehat{\theta}_2)-2-2\frac{\text{Cov}_{12,n}}{\sigma^2_{1,n}\sigma^2_{2,n}}\right\}. \tag{9.2}$$

From the sample data at the end of Example 9.1 (p. 365), we find that the estimated variance of the sample mean divided by that of the 20% trimmed mean for normal data is $s^2_{1,N-1}/s^2_{2,N-1} = .86$. Plugging sample estimates into (9.2) and taking the square root yields a standard error of .02. Although the estimated kurtoses in this case are close to 3 (3.11 and 3.26, respectively), substituting $\text{Kurt}(\widehat{\theta}_1) = \text{Kurt}(\widehat{\theta}_2) = 3$ changes the standard error to .01 because the last piece subtracted in (9.2) due to correlation of $s^2_{1,N-1}$ and $s^2_{2,N-1}$ is quite large. When rerunning for $N = 10{,}000$, both methods gave the same standard error to three decimals, .006. Thus, this is a case where it can make a difference to estimate the kurtoses. Following is R code illustrating these calculations.

```
N<-1000; z<-norm15; var1<-var(z[,1]); var2<-var(z[,2])
sig12<-cov((z[,1]-mean(z[,1]))^2,(z[,2]-mean(z[,2]))^2)
kurt<-function(x){mean((x-mean(x))^4)/(mean((x-mean(x))^2)^2)}
K1<-kurt(z[,1]); K2<-kurt(z[,2])
cat("var1/var2=",round(var1/var2,3),fill=T)
sd1<-(var1/var2)*sqrt(K1+K2-2-2*sig12/(var1*var2))/sqrt(N)
cat("sd of var1/var2 =",round(sd1,3),fill=T)
sd2<-(var1/var2)*sqrt(4-2*sig12/(var1*var2))/sqrt(N)
cat("sd of var1/var2 assuming Kurt=3 =",round(sd2,3),fill=T)

> source("ex1.delta.r")
var1/var2 = 0.861 sd of var1/var2 = 0.019
sd of var1/var2 assuming Kurt=3 = 0.009
```

9.4.2 Comparing Estimated True Variance and the Mean of Estimated Variance Estimators

We are often interested in the quality of standard errors that we get from an estimated information matrix or from the sandwich estimator $\boldsymbol{V}_n(\boldsymbol{Y},\widehat{\boldsymbol{\theta}})$ or from some other variance estimation method. To illustrate for a simple scalar estimator $\widehat{\theta}$, we might calculate an estimator and an estimate of its variance, denoted $\widehat{\sigma}^2_n$, for each Monte Carlo sample resulting in the data $(\widehat{\theta}_i,\widehat{\sigma}^2_{i,n})$, $i = 1,\ldots,N$. We are hoping that the variance estimator $\widehat{\sigma}^2_n$ is unbiased for the true variance of $\widehat{\theta}$, say σ^2_n. Thus we form the ratio

$$R_N = \frac{\frac{1}{N}\sum_{i=1}^{N}\widehat{\sigma}^2_{i,n}}{s^2_{N-1}}. \tag{9.3}$$

where the numerator is estimating $\mathrm{E}(\widehat{\sigma}_n^2)$, and the denominator is estimating the true variance of $\widehat{\theta}$. R_N has a natural interpretation. For example, $R_N = 1.07$ means that the variance estimator $\widehat{\sigma}_n^2$ is 7% too large on average, whereas $R_N = 0.83$ means that the variance estimator is too small on average. To be confident in these judgments, the researcher needs to know the variability of R_N. Applying the Delta theorem, Theorem 5.19 (p. 238), to the joint asymptotic distribution of $\left(s_{N-1}^2, N^{-1}\sum_{i=1}^N \widehat{\sigma}_{i,n}^2\right)^T$ we obtain the asymptotic variance as $N \to \infty$ (with n fixed) of R_N,

$$\frac{1}{N}\frac{\{\mathrm{E}(\widehat{\sigma}_n^2)\}^2}{\sigma_n^4}\left\{\mathrm{Kurt}(\widehat{\theta}) - 1 - 2\frac{\mathrm{Cov}_n}{\sigma_n^2\mathrm{E}(\widehat{\sigma}_n^2)} + \frac{\mathrm{Var}(\widehat{\sigma}_n^2)}{\{\mathrm{E}(\widehat{\sigma}_n^2)\}^2}\right\}, \tag{9.4}$$

where $\mathrm{Cov}_n = \mathrm{Cov}(\left\{\widehat{\theta} - \mathrm{E}(\widehat{\theta})\right\}^2, \widehat{\sigma}_n^2)$. To get a standard error for R_N, we just substitute in (9.4) sample estimates based on $(\widehat{\theta}_i, \widehat{\sigma}_{i,n}^2)$, $i = 1, \ldots, N$ and take the square root.

A simpler expression for the standard error can be obtained from the M-estimator approach using

$$\boldsymbol{\psi}(\widehat{\theta}_i, \widehat{\sigma}_{i,n}^2, \boldsymbol{\eta}) = \begin{pmatrix} \widehat{\theta}_i - \eta_1 \\ \\ \eta_2(\widehat{\theta}_i - \eta_1)^2 - \widehat{\sigma}_{i,n}^2 \end{pmatrix}, \tag{9.5}$$

where $\boldsymbol{\eta} = (\eta_1, \eta_2)^T$ is used in place of $\boldsymbol{\theta}$ in the M-estimator notation so as not to get confused with the data in this case. Now, the standard error for R_N based on the estimated sandwich matrix (7.12, p. 302) is then

$$\frac{1}{N^{1/2}}\left[\frac{1}{N}\sum_{i=1}^{N}\left\{\frac{R_N(\widehat{\theta}_i - \overline{\widehat{\theta}})^2 - \widehat{\sigma}_{i,n}^2}{s_{N-1}^2}\right\}^2\right]^{1/2}. \tag{9.6}$$

Often the Monte Carlo variation of the numerator $N^{-1}\sum_{i=1}^N \widehat{\sigma}_{i,n}^2$ of R_N is much less than the denominator s_{N-1}^2 because sample means (even of variance estimates) tend to have smaller variances than sample variances. Thus, treating $N^{-1}\sum_{i=1}^N \widehat{\sigma}_{i,n}^2$ as constant in R_N results in dropping the last two terms of (9.4, p. 374) and gives the simpler standard error expression $R_N\{(\widehat{\mathrm{Kurt}} - 1)/N\}^{1/2}$. Furthermore, making the approximations $\mathrm{Kurt}(\widehat{\theta}) \approx 3$ and $R_N \approx 1$, a suitable standard error is simply $(2/N)^{1/2}$. Note that if you want to use $1/R_N$ instead of R_N, then just replace $\sigma_n^4/\left\{\mathrm{E}(\widehat{\sigma}_{1,n}^2)\right\}^2$ by $\left\{\mathrm{E}(\widehat{\sigma}_{1,n}^2)\right\}^2/\sigma_n^4$ in (9.4), and $(2/N)^{1/2}$ is also an approximate standard error for $1/R_N$.

Using the Example (9.1, p. 365) setting, the R program below computes R_N and the four standard errors for R_N for an estimator of the variance of the 20% trimmed mean: based on (9.4), (9.6), $R_N\{(\widehat{\mathrm{Kurt}} - 1)/N\}^{1/2}$, and $(2/N)^{1/2}$, respectively.

The first two standard errors are essentially the same (0.053 for $N = 1000$ and 0.016 for $N = 10000$) and the last two are a bit smaller as expected. The estimates $R_N = 1.05$ for $N = 1000$ and $R_N = 1.04$ for $N = 10000$ suggest that the variance estimator is slightly biased upwards.

```
set.seed(346);N<-1000;n<-15
sim.samp(N,n,rnorm)->z
trim20<-function(x){mean(x,.2)} # 20% trimmed mean function
trim.var<-function(x,trim){      # var. est. for trim20
  n<-length(x);h<-floor(trim*n)
  tm<-mean(x,trim);sort(x)->x
  ss<-h*((x[h+1]-tm)^2+(x[n-h]-tm)^2)
  ss<-ss+sum((x[(h+1):(n-h)]-tm)^2)
  t.var<-ss/((n-2*h)*(n-2*h-1))
  return(t.var)
}
trim20.var<-function(x){trim.var(x,.2)}
apply(z,1,trim20)->t20            # trim20 for each sample
apply(z,1,trim20.var)->t20v       # var. est. for each sample
RN<-mean(t20v)/var(t20)
t1<-kurt(t20)-1
t2<-cov((t20-mean(t20))^2,t20v)/(var(t20)*mean(t20v))
t3<-var(t20v)/(mean(t20v)^2) se1<-(RN/sqrt(N))*sqrt(t1-2*t2+t3)
se1m<-sqrt(mean((RN*(t20-mean(t20))^2-t20v)^2))/(var(t20)*sqrt(N))
se2<-RN*sqrt(t1/N)
se3<-sqrt(2/N)
print(round(data.frame(N,n,RN,se1,se1m,se2,se3),3))

> source("cv2.ex.r")
     N  n    RN   se1  se1m  se2   se3
1 1000 15 1.049 0.053 0.053 0.05 0.045

> source("cv2.ex.r")              # repeated for N=10000
      N  n    RN   se1  se1m   se2   se3
1 10000 15 1.041 0.016 0.016 0.015 0.014
```

9.4.3 When Should Mean Squared Error Be Reported?

Much of statistical estimation theory involves some trade-off between bias and variance. Although, we might report both bias and variance of an estimator, in some situations it is appropriate to also report mean squared error (MSE) = Variance + bias^2. Here we suggest that when variance estimation is the focus, MSE is not appropriate because it rewards underestimation too much. That is, if we are considering a variance estimate $\widehat{\sigma}^2$ for some estimator, typically we would prefer not to have $\widehat{\sigma}^2$ much smaller than the true variance. For example, approximate

Table 9.2 Mean Squared Error of Functions of the Sample Variance and Stein's Loss Function for a Normal(μ, 1) Sample of Size $n = 10$

Estimator	MSE of s_a^2	MSE of s_a	MSE of $\log(s_a^2)$	$\mathrm{E}\{L_s(s_a^2, 1)\}$
s_{n-3}^2	.45	.080	.27	.15
s_{n-2}^2	.30	.062	.25*	.12
s_{n-1}^2	.22	.055	.26	.12*
s_n^2	.19	.055*	.30	.12
s_{n+1}^2	.18*	.059	.35	.13
s_{n+2}^2	.19	.065	.41	.15

confidence intervals based on such a variance estimate could have coverage probabilities much less than the nominal value $1-\alpha$. On the other hand, the variance of variance estimators is often proportional to its expectation so that the MSE of negatively biased variance estimators is low.

To illustrate these ideas, assume that $Y_1, \ldots, Y_n$ is an iid sample from a $N(\mu, \sigma^2)$ distribution. A fact often noted when estimating σ^2 is that $s_{n+1}^2 = (n+1)^{-1} \sum (Y_i - \overline{Y})^2$ has minimum mean squared error (MSE) among estimators of the form cs_n^2, where s_n^2 is the maximum likelihood estimator of σ^2. (See Casella and Berger 2002, p. 350–351.) However, as mentioned above, MSE may not be the most appropriate loss function for deciding which estimator is best. Thus we also consider MSE on the square root scale and on the log scale, and also the expected value of a loss function due to Stein (Casella and Berger 2002, p. 351)

$$L_S(V, \sigma^2) = V/\sigma^2 - \log(V/\sigma^2) - 1.$$

In Table 9.2 we have computed MSE, MSE(s_a), MSE($\log(s_a^2)$), and $\mathrm{E}(L_S)$ for s_{n-3}^2, s_{n-2}^2, s_{n-1}^2, s_n^2, s_{n+1}^2, and s_{n+2}^2 for sample size $n = 10$. In calculating these quantities, it is helpful to know that

$$\mathrm{E}\left\{\left(\chi_\nu^2\right)^{r/2}\right\} = 2^{r/2} \frac{\Gamma((\nu + r)/2)}{\Gamma(\nu/2)}.$$

Also, the rth cumulant of $\log(Y)$, where Y is a standard gamma variable, is $\psi^{(r-1)}(\alpha)$, where ψ is the digamma function.

The asterisks in Table 9.2 mark the minimum value for each criterion. Stein's loss picks s_{n-1}^2 which corresponds to common practice when estimating variance. In simulation experiments, MSE on the log or square root scale is easy to use and much preferable to ordinary MSE. Here minimum MSE on the log scale picks s_{n-2}^2 which is slightly biased upwards in the s^2 scale. Also it might be preferable to just report the mean and standard deviation of the variance estimates or R_N of the previous section.

9.5 Presenting Results

There is a growing statistical literature on how to present results of studies (both Monte Carlo and other types) in tables and graphs. A few notable references are Ehrenberg (1977, 1978, 1981), Wainer (1993, 1997a,b), and Gelman et al. (2002). Although statisticians are in the business of informing others how to summarize information, they are sometimes cavalier about following basic common sense and statistical principles in presenting their own experimental results. Here we mention a few basic principles and then discuss presentation of testing and confidence interval studies.

9.5.1 Presentation Principles

- Use graphs whenever possible. It is fine to have both a graph and a table, subject to space restrictions.
- Always give some idea of the standard error of each estimated entry in tables, and in graphs when feasible. Generally, one can report a range or average standard error in a note at the bottom of tables and in the caption of figures.
- It is best to use at most two significant digits in table entries, and seldom are more than three required.
- It makes little sense to include digits beyond the standard error of the entry. For example, suppose the computer gives .04586 for an entry but the standard error of the entry is .002. Then there is no reason to report more than .046. (One possible exception is when the difference of entries has a much smaller standard error than the individual entry standard error.)

9.5.2 Presenting Test Power Results

We now illustrate the presentation of results for power studies using an example discussed previously. Recall the dose-response situation of Problem 7.15 (p. 336) and Problem 8.7 (p. 358): k dose levels $d_1, \ldots, d_k$, where at the ith dose we observe $\{Y_{ij}, n_{ij}, j = 1, \ldots, m_i\}$, and we assume that

$$\mathrm{E}\left(Y_{ij}/n_{ij}|n_{ij}\right) = p_i(\boldsymbol{\beta}) = [1 + \exp(-\beta_1 - \beta_2 d_i)]^{-1}.$$

In this notation, the Cochran-Armitage score test for $H_0 : \beta_2 = 0$ under the assumption that the Y_{ij} are binomially distributed is given by

$$T_S = \frac{\left[\sum_{i=1}^{k} (Y_{i.} - n_{i.}\overline{\overline{Y}})d_i\right]^2}{\sum_{i=1}^{k} (d_i - \overline{d})^2 n_{i.}\overline{\overline{Y}}(1 - \overline{\overline{Y}})}, \tag{9.7}$$

where $Y_{i.} = \sum_{j=1}^{m_i} Y_{ij}$, $n_{i.} = \sum_{j=1}^{m_i} n_{ij}$, $\overline{\overline{Y}} = \sum_{i=1}^{k} Y_{i.} / \sum_{i=1}^{k} n_{i.}$, and $\overline{d} = \sum_{i=1}^{k} n_{i.}d_i / \sum_{i=1}^{k} n_{i.}$. The generalized score test, given in Problem 8.7 (p. 358), is

$$T_{GS} = \frac{\left[\sum_{i=1}^{k} (Y_{i.} - n_{i.}\overline{\overline{Y}})d_i\right]^2}{\sum_{i=1}^{k} \left[(d_i - \overline{d})^2 \sum_{j=1}^{m_i} (Y_{ij} - n_{ij}\overline{\overline{Y}})^2\right]}. \tag{9.8}$$

The generalized score test allows for clustering or overdispersion in the Y_{ij} so that the test statistic has approximately a χ_1^2 distribution under $H_0 : \beta_2 = 0$. Some obvious questions are:

1. How do the tests perform when there is no clustering? That is, for binomial Y_{ij}, what are the true α levels under $H_0 : \beta_2 = 0$ using χ_1^2 critical values, and what are their power curves under alternatives?
2. How do the tests perform when there is clustering?

To answer these questions, we decided to simulate the simple situation given in Problem 8.7 (p. 358) with $k = 15$ and associated doses

$$(12, 12, 12, 24, 24, 24, 48, 48, 48, 96, 96, 96, 192, 192, 192) \tag{9.9}$$

and cluster sizes n_{ij}

$$(10, 9, 7, 9, 8, 6, 9, 9, 8, 3, 9, 9, 7, 9, 10). \tag{9.10}$$

Perhaps it would be more realistic to make the n_{ij} random, but we have kept them fixed. We set the true probabilities according to the logistic model on the logarithm of dose,

$$p_i = [1 - \exp\{-\beta_1 - \beta_2 \log(d_i)\}]^{-1},$$

where $\beta_1 = -2$ and β_2 varied from 0 to .9. The first two rows of Table 9.3 display the estimated power at nominal level $\alpha = .05$ (critical value $= 3.84$), and the third row gives the difference of powers. Clearly, use of T_{GS} entails some loss of power when there is no clustering and the data are binomially distributed. Figure 9.1 displays the powers graphically. It is often much easier to see differences in a graph than in a table.

Table 9.3 Estimates of Power for Cochran-Armitage Tests for Binomial and Beta-Binomial Data at $\alpha = .05$, $k = 15$, and intercept $\beta_1 = -2$. Doses and Sample Sizes are in (9.9) and (9.10) on p. 378

		$\rho = 0$ (Binomial)					
$\beta_2 =$	0	.2	.3	.4	.5	.7	.9
T_S	.042	.16	.34	.55	.75	.94	.97
T_{GS}	.047	.14	.28	.45	.62	.85	.89
$T_S - T_{GS}$		.02	.06	.09	.13	.09	.08
		$\rho = .3$ (Beta-Binomial)					
T_S^*	.051	.09	.14	.23	.34	.50	.60
T_{GS}	.029	.07	.12	.19	.26	.41	.46
$T_S - T_{GS}$		.02	.02	.04	.08	.09	.14

Note: Power estimates are based on 1000 replications and have standard deviation bounded by $(4000)^{-1/2} = .016$. The standard deviations of differences are .010 to .014. T_S refers to the usual Cochran-Armitage statistic in (9.7, p. 378). T_{GS} refers to the generalized version in (9.8, p. 378). T_S^* refers to use of T_S with critical value = 12.17.

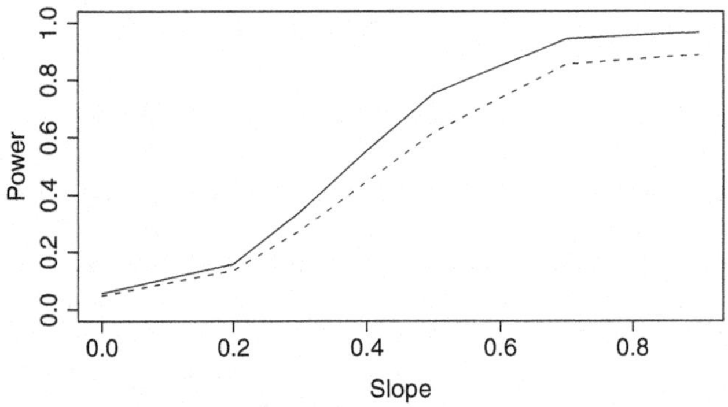

Fig. 9.1 Power Estimates for Cochran-Armitage Score Tests from Table 9.3 when Data are Binomial. Solid line, T_S; dashed line, T_{GS}

Next we generated the Y_{ij} from the beta-binomial distribution with intra-class correlation ρ, where ρ is just the regular correlation between any two binary outcomes (see, for example, Donald and Donner, 1990). To generate Y_{ij}, we first generate a beta (a, b) random variable U with $a = p_i(1-\rho)/\rho$ and $b = (1-p_i)(1-\rho)/\rho$, and then generate Y_{ij} as binomial(n_{ij}, prob. $= U$). Thus, (unconditionally) Y_{ij} has mean $n_{ij}\{a/(a+b)\} = n_{ij}p_i$ and variance

$$\text{Var}(Y_{ij}) = n_{ij}p_i(1-p_i) + n_{ij}(n_{ij}-1)\rho.$$

For $\beta_1 = -2$ and $\beta_2 = 0$, the estimated rejection probabilities for the Cochran-Armitage test (T_S) were .14, .28, and .39 for $\rho = .1, .3$, and .5, respectively (not displayed). Thus, the usual Cochran-Armitage test cannot achieve the nominal level in the face of clustering. The related estimated rejection probabilities for T_{GS} were .05, .03, and .02, suggesting a conservative trend in T_{GS} as ρ gets larger.

Just as comparing the variances of two statistics makes little sense when their biases differ, it makes little sense to compare the powers of two test statistics whose sizes differ. However, we can estimate a critical value for T_S and use that instead of 3.84. Thus, we generated 10,000 data sets with $\beta_1 = -2$, $\beta_2 = 0$, and $\rho = .3$, and estimated the 95th percentile of T_S to be 12.17. Using that critical value for T_S and the usual 3.84 for T_{GS}, we obtained the results in the lower section of Table 9.3. These results are not really fair to T_{GS} because the whole point of using T_{GS} is to adjust T_S so that it has the correct size without knowledge of the underlying distributional type (here beta-binomial), but it illustrates perhaps the cost of using T_{GS} compared to the best one could do in this clustered binomial situation. In fact, the quasi-likelihood approach with an estimated ϕ for the overdispersion would be more powerful here and perhaps close to this best power.

9.5.3 Reporting Results for Confidence Intervals

This section is intended to illustrate some of the principles useful in describing a Monte Carlo study comparing confidence intervals. Recall from Section 9.3.4 (p. 371) that *coverage probability* and *average confidence interval length* are both important quantities that should be reported whenever studying confidence intervals. Ideally, intervals have the nominal $1 - \alpha$ coverage and are short on average. Here we look at a series of three tables for the same data, each an improvement on the one before it.

Zhang and Boos (1997) discuss new methods to handle clustered binomial data appearing in a series of 2×2 tables. The details are not important here, but one of the paper's sections investigated confidence intervals for the common odds ratio obtained by inverting a new Mantel-Haenszel test statistic. In a Monte Carlo study, this new confidence interval C_U was compared to another confidence interval proposed by Liang (1985), here denoted C_L. Table 9.4 is a portion of the first attempt at presenting the results in a table.

This first attempt commits several sins; the first is reporting too many digits in the table entries. It has been argued that humans have trouble dealing visually with more than two digits in the entries of a table (Wainer 1993). Moreover, the Monte Carlo sample size was 1000, and therefore the standard deviation of the coverage estimates is approximately $\{(.05)(.95)/1000\}^{1/2} = .007$. Thus, the third decimal of the coverage estimates is, statistically, mainly random noise. Table 9.5 shows the improvement when only two digits are reported. For space considerations, we have left off notes about standard errors in these tables. Generally, it is best to mention standard errors and Monte Carlo sample size in each table.

Table 9.4 Coverage and length of 95% confidence intervals for data from the beta-binomial(ρ) distribution with odds ratio = 1.5

Number of strata		n_{ij}, m_{ij} =	$\rho = 0.0$ 5	5–10	5–15	$\rho = 0.2$ 5	5–10	5–15
$k = 5$	C_L	Coverage	.977	.977	.977	.984	.980	.975
		Mean Length	1.93	1.48	1.25	5.15	2.88	3.95
	C_U	Coverage	.969	.964	.968	.977	.968	.961
		Mean Length	0.95	0.77	0.69	1.42	1.29	1.26

Table 9.5 Coverage and length of 95% confidence intervals for data from the beta-binomial(ρ) distribution with odds ratio = 1.5

Number of strata		n_{ij}, m_{ij} =	$\rho = 0.0$ 5	5–10	5–15	$\rho = 0.2$ 5	5–10	5–15
$k = 5$	C_L	Coverage	.98	.98	.98	.98	.98	.98
		Mean Length	1.9	1.5	1.3	5.2	2.9	4.0
	C_U	Coverage	.97	.96	.97	.98	.97	.96
		Mean Length	1.0	0.8	0.7	1.4	1.3	1.3

Table 9.6 Coverage and length of 95% confidence intervals for data from the beta-binomial(ρ) distribution with odds ratio = 1.5

Number of strata		n_{ij}, m_{ij} =	$\rho = 0.0$ 5	5–10	5–15	$\rho = 0.2$ 5	5–10	5–15
$k = 5$	Coverage	C_L	.98	.98	.98	.98	.98	.98
		C_U	.97	.96	.97	.98	.97	.96
	Mean Length	C_L	1.9	1.5	1.3	5.2	2.9	4.0
		C_U	1.0	0.8	0.7	1.4	1.3	1.3

Taken from Zhang and Boos (1997, p. 1193)

Finally, we note that it helps to reorder the rows so that similar items are together, in other words to facilitate comparing "apples with apples." Table 9.6 makes this adjustment. Now it is clear that both intervals have coverage larger than 95%, but C_U is much better in terms of average length.

9.6 Problems

9.1. In the context of Example 9.1 (p. 365) the MSE could be estimated by $\widehat{\text{MSE}} = N^{-1}\sum_{i=1}^{N}(\widehat{\theta}_i - \theta_0)^2$.

a. What is the standard deviation of $\widehat{\text{MSE}}$?
b. Based on a preliminary run of size N_0, how large should N be to have the standard deviation of $\widehat{\text{MSE}} \leq d$?

9.2. Provide the details that lead to the sample size estimates $N = 2\sigma_n^4/d^2$ and $N = 2/d_p^2$ in Section 9.3.2.

9.3. Use Theorem 5.19 (p. 238) or the M-estimator method to derive the asymptotic variance of $s_{1,N-1}^2/s_{2,N-1}^2$ found in (9.2, p. 373).

9.4. Use Theorem 5.19 (p. 238) or the M-estimator method to derive the asymptotic variance of R_N found in (9.4, p. 374).

9.5. Verify that (9.6, p. 374) is the correct form of the standard error for R_N based on the M-estimator approach.

9.6. When reporting results about the variances of variance estimators, it is sometimes useful to report the standardized variance (variance divided by the square of the mean) because the variance estimators may be estimating different quantities. For example, if V is a variance estimator and $V^* = V/2$, then clearly the variance of V^* is 1/4 of the variance of V. Thus, we define the standardized variance (SV) to be $\text{Var}(V)/\{E(V)\}^2$. Given a Monte Carlo sample $V_1, \ldots, V_N$ of a variance estimator V, $\widehat{SV} = s_V^2/\overline{V}^2$, where $\overline{V}$ and s_V^2 are the sample mean and variance of the V's, respectively. Using Theorem 5.19 (p. 238) or the M-estimator method, show that the asymptotic variance of $\widehat{SV}$ is

$$\frac{\sigma_n^4}{\mu_n^4}\left(\text{Kurt}(V) - 1 - 4\frac{\sigma_n}{\mu_n}\text{Skew}(V) + 4\frac{\sigma_n^2}{\mu_n^2}\right),$$

where $\sigma_n^2 = \text{Var}(V)$ and $\mu_n = \text{E}(V)$.

9.7. Related to the SV of the last problem, one might be interested in reporting instead the coefficient of variation (CV), which is just the square root of $\widehat{SV}$, $\widehat{CV} = s_V/\overline{V}$. Thus, use the Delta theorem with the asymptotic variance expression for $\widehat{SV}$, to obtain the asymptotic variance for $\widehat{CV}$:

$$\frac{\sigma_n^2}{\mu_n^2}\left(\frac{\text{Kurt}(V) - 1}{4} - \frac{\sigma_n}{\mu_n}\text{Skew}(V) + \frac{\sigma_n^2}{\mu_n^2}\right).$$

9.8 (Monte Carlo Integration). The integral $\int_0^1 g(u)\,du < \infty$ can be estimated by $N^{-1}\sum_{i=1}^N g(U_i)$, where $U_1, \ldots, U_N$ is an iid sample from the Uniform(0,1) distribution. If $\int_0^1 g^2(u)\,du < \infty$, derive the bias and variance of this estimator, and state why consistency and asymptotic normality hold as $N \to \infty$.

9.9 (Monte Carlo Integration via Importance Sampling). The integral $\int_{-\infty}^{\infty} g(y)\,dy$ can be estimated by $N^{-1}\sum_{i=1}^N g(Y_i)/f(Y_i)$, where $Y_1, \ldots, Y_N$ is an iid sample from the density $f(y)$ whose support is the whole real line. Find conditions on g and f such that $g(Y_1)/f(Y_1)$ has finite variance. Then derive the bias and variance of this estimator, and state why consistency and asymptotic normality hold as $N \to \infty$.

9.10. Make a new version of Table 9.1 (p. 367) by transposing estimators and sample sizes. That is, let the estimators be rows and the sample sizes be columns. What advantage might this version have? Then take this new version and divide the $\overline{X}_{.20}$ and Median entries by the $\overline{X}$ entry. What advantage might this version have?

9.11. Table 9.3 (p. 379) was created using the R programs `mc.ca1.r.txt` and `mc.ca2.r.txt` on the text website

http://www4.stat.ncsu.edu/~boos/Essential.Statistical.Inference

a. Repeat the first half of Table 9.3 (p. 379) with a different intercept.
b. Repeat the first half of Table 9.3 (p. 379) with a different set of doses.
c. Repeat the first half of Table 9.3 (p. 379) with a different set of sample sizes.

9.12. Repeat the second half of Table 9.3 (p. 379) with $\rho = .1$. This requires finding the 95th percentile of T_S and substituting for 12.17 in `mc.ca2.r.txt`. To find this percentile, run `mc.ca2.r.txt` with `nmc<-10000` followed by `quantile(t1,.95)`.

9.13. In Section 9.5.3 (p. 380), the phrase "mainly random noise" was used to describe the third decimal place of estimated coverage probabilities based on $N = 1000$ Monte Carlo replications. To quantify this phrase, suppose that Y is binomial(1000,.95) and $Y/1000$ is the estimate of interest. Model the act of rounding from three decimal places to two as adding an independent uniform random variable U on $(-.005, .005)$ to $Y/1000$. Show that the increase in the standard deviation of $Y/1000 + U$ compared to the standard deviation of $Y/1000$ is 8.4%.

Chapter 10
Jackknife

The *jackknife* was developed by Quenouille (1949, 1956) as a general method to remove bias from estimators. Tukey (1958) noticed that the approach also led to a method for estimating variances. Since that time, the jackknife has been used more for variance estimation than for bias estimation. Thus our focus is mainly on jackknife variance estimation, although we begin below in the historical order with bias estimation.

We show that the jackknife variance estimation method is essentially an empirical or "algorithmic" version of the influence curve or delta method approach for estimation of asymptotic variances. Thus, the jackknife allows one to get standard errors that are theoretically grounded in the approximation by averages but without needing to know any of that theory or even how to take a derivative. In addition, Efron and Stein (1981) (see Section 10.8.2, p. 405) showed for many estimators the jackknife variance estimates tend to be slightly biased upward for any sample size, thus providing insurance against worry that the methods are biased low in small samples. The result is that jackknife variance estimators are very attractive nonparametric estimators for many situations. Monte Carlo comparisons with bootstrap variance estimators are discussed in Chapter 11.

10.1 Basics

10.1.1 Definitions

If $\widehat{\theta}$ is an estimator based on an iid sample $Y_1, \ldots, Y_n$, then let $\widehat{\theta}_{[i]}$ denote the "leave-1-out" estimator obtaining by computing $\widehat{\theta}$ with Y_i deleted from the sample. Denote the average of these "leave-1-out" estimators by $\overline{\theta}_1 = n^{-1} \sum_{i=1}^{n} \widehat{\theta}_{[i]}$ and define the pseudo-values by

$$\widehat{\theta}_{ps,i} = n\widehat{\theta} - (n-1)\widehat{\theta}_{[i]}. \tag{10.1}$$

D.D. Boos and L.A. Stefanski, *Essential Statistical Inference: Theory and Methods*, Springer Texts in Statistics, DOI 10.1007/978-1-4614-4818-1_10,

We show that these pseudo-values are related to the influence curve of $\widehat{\theta}$. For the moment, a rough heuristic is that the pseudo-value is the part of $\widehat{\theta}$ that depends on Y_i; i.e., $(n-1)\widehat{\theta}_{[i]}$, which does not depend on Y_i, is subtracted from $n\widehat{\theta}$, thus leaving only the part of $\widehat{\theta}$ that depends on Y_i. The average of these pseudo-values is the bias-adjusted jackknife estimator

$$\begin{aligned}\widehat{\theta}_{\mathrm{J}} &= \frac{1}{n}\sum_{i=1}^{n}\widehat{\theta}_{ps,i} = n\widehat{\theta} - (n-1)\overline{\theta}_1 \\ &= \widehat{\theta} - (n-1)(\overline{\theta}_1 - \widehat{\theta}).\end{aligned}$$

10.1.2 Bias Estimation

The jackknife estimator of bias is $(n-1)(\overline{\theta}_1 - \widehat{\theta})$, and the estimator $\widehat{\theta}_{\mathrm{J}}$ is the bias-corrected jackknife estimator of θ. Motivation for $\widehat{\theta}_{\mathrm{J}}$ comes from an assumed and often valid expansion of the expected value of $\widehat{\theta}$:

$$\mathrm{E}\{\widehat{\theta}\} = \theta + \frac{\beta_1}{n} + \frac{\beta_2}{n^2} + O(n^{-3}). \tag{10.2}$$

Now, write $\widehat{\theta}$ and $\overline{\theta}_1$ as response variables in a simple linear regression model after grouping the higher-order terms into error:

$$\widehat{\theta} = \theta + \frac{\beta_1}{n} + e_1, \qquad \overline{\theta}_1 = \theta + \frac{\beta_1}{n-1} + e_2.$$

In these regression models the intercept $= \theta$, the slope $= \beta_1$, and the predictor is inverse sample size, n^{-1} and $(n-1)^{-1}$. Using the two points $(x_1, y_1) = (1/n, \widehat{\theta})$ and $(x_2, y_2) = (1/(n-1), \overline{\theta}_1)$, the line segment connecting them has slope

$$\widehat{\beta}_1 = \frac{y_2 - y_1}{x_2 - x_1} = n(n-1)(\overline{\theta}_1 - \widehat{\theta}),$$

and the intercept is

$$\frac{y_1 x_2 - y_2 x_1}{x_2 - x_1} = \widehat{\theta} - (n-1)(\overline{\theta}_1 - \widehat{\theta}) = \widehat{\theta} - \widehat{\beta}_1/n = \widehat{\theta}_{\mathrm{J}},$$

which we saw above is also the average of the pseudo-values. Here, we can see that $\widehat{\theta}_{\mathrm{J}} = \widehat{\theta} - \widehat{\beta}_1/n$ is a bias-corrected estimator where $\widehat{\beta}_1/n = (n-1)(\overline{\theta}_1 - \widehat{\theta})$ is the estimated bias subtracted from $\widehat{\theta}$.

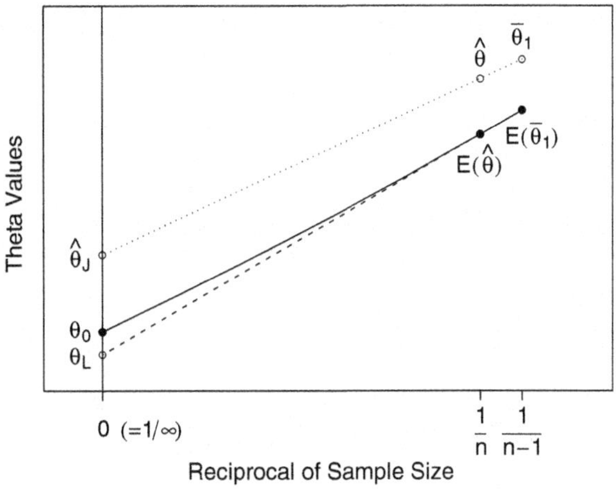

Fig. 10.1 From Example 10.1 (p. 387), the jackknife estimator $\widehat{\theta}_J$ as the intercept estimate or extrapolant at 0 from fitting a line to the two points $(1/n, \widehat{\theta})$ and $(1/(n-1), \overline{\theta}_1)$. The true value is θ_0. The solid curve is the function $E\{\exp(\overline{Y})\} = \exp\{\mu + \sigma^2/(2n)\}$ as a function of n^{-1}, and the dashed line below it is the line segment connecting the values of the function at n^{-1} and $(n-1)^{-1}$, $E\{\widehat{\theta}\}$ and $E\{\overline{\theta}_1\}$, respectively

Example 10.1 (Exponential of the sample mean). Let $Y_1, \ldots, Y_n$ be an iid sample from a normal(μ, σ^2) distribution, and consider estimation of $\theta = \exp(\mu)$ by $\widehat{\theta} = \exp(\overline{Y})$. Using the moment generating function of the normal distribution, it is easy to show that $E\{\exp(\overline{Y})\} = \exp\{\mu + \sigma^2/(2n)\}$. Viewing this latter expression as a function of $x = 1/n$, second order Taylor expansion in x around $x = 0$ yields

$$E\{\exp(\overline{Y})\} \approx e^{\mu} + \frac{\sigma^2 e^{\mu}}{2}\frac{1}{n} + \frac{\sigma^4 e^{\mu}}{4}\frac{1}{n^2},$$

so that $\beta_1 = \sigma^2 e^{\mu}/2$ and $\beta_2 = \sigma^4 e^{\mu}/4$ in (10.2). For $\mu = 0$, $\sigma^2 = 4$, $n = 10$, the solid curve in Figure 10.1 is $\exp\{\mu + \sigma^2/(2n)\}$ with points on the curve at (0,1), $(n^{-1}, E\{\widehat{\theta}\})$, and $((n-1)^{-1}, E\{\overline{\theta}_1\})$. Note that $\theta_0 = 1$ is the true value of $\theta = \exp(\mu)$ when $\mu = 0$. The dashed line below the solid curve is the straight line defined by the right-most two points and extended to $x = 0$ where the intersection is labeled θ_L to denote the *linear* extrapolation to 0. The dotted line goes through the points $(0, \widehat{\theta}_J)$, $(n^{-1}, \widehat{\theta})$, and $((n-1)^{-1}, \overline{\theta}_1)$ and was obtained as the line segment connecting the right-most two points and extended to $x = 0$ where it intersects at $\widehat{\theta}_J$. Thus, the bias-corrected estimator $\widehat{\theta}_J$ is the extrapolatant at 0 of this line. $\widehat{\theta}_J$ has two sources of error: (i) it uses a linear approximation in place of the true bias curve; and (ii) $\widehat{\theta}$ and $\overline{\theta}_1$ replace the unknown $E\{\widehat{\theta}\}$ and $E\{\overline{\theta}_1\}$. A related plot is found in Efron (1982, p. 7).

This general approach for correcting bias in estimators is based on reducing information, which here is proportional to n. To see how the estimator changes with sample size, we plot the estimators $\widehat{\theta}$ and $\overline{\theta}_1$ versus the inverse of sample size used in calculating them, n and $n-1$, respectively, and then extrapolate to 0, which is the inverse of $n = \infty$. A general version of this correction for bias in measurement error problems was introduced by Cook and Stefanski (1994) and Stefanski and Cook (1995) under the name SIMEX and further discussed in Carroll et al. (2006, Ch. 5) ♦

It is not hard to show using (10.2, p. 386) (see problem 10.2) that when (10.2) holds

$$\mathrm{E}\{\widehat{\theta}_{\mathrm{J}}\} = \theta - \frac{\beta_2}{n^2} + O(n^{-2}),$$

and thus the bias of θ has been reduced from $O(n^{-1})$ to $O(n^{-2})$. It is possible to extend this approach by leaving out two or more observations at a time (see Schucany et al., 1971), but apparently this is not often used in practice (Efron 1982 p. 8). In fact, even the first-order bias-corrected estimator $\widehat{\theta}_{\mathrm{J}}$ may not be an improvement over $\widehat{\theta}$ in terms of mean squared error (see Kim and Singh, 1998, for examples) and is not in routine use. On the other hand, the jackknife variance estimator is an important practical method used extensively in survey sampling and elsewhere. Thus we now focus our attention on it.

10.1.3 Variance Estimation

The jackknife variance estimator for $\widehat{\theta}$ can be defined in terms of the sample variances of the leave-1-out $\widehat{\theta}_{[i]}$ or of the pseudo-values (10.1, p. 385) as follows,

$$\widehat{V}_{\mathrm{J}} = \frac{(n-1)^2}{n}\frac{1}{n-1}\sum_{i=1}^{n}\left(\widehat{\theta}_{[i]} - \overline{\theta}_1\right)^2 = \frac{1}{n}\frac{1}{n-1}\sum_{i=1}^{n}(\widehat{\theta}_{ps,i} - \widehat{\theta}_{\mathrm{J}})^2. \qquad (10.3)$$

To understand $\widehat{V}_{\mathrm{J}}$, we examine the pseudo-values closely. In the case that $\widehat{\theta} = \overline{Y}$, the pseudo-values are just the sample values themselves, and the jackknife variance estimator is exactly the same as the standard variance estimate s_{n-1}^2/n. (With s_{n-1}^2 we continue our notation for referring to the sample variance with $n-1$ in the divisor. Below, $n-1$ in the subscript refers to leave-1-out estimators when the estimator based on sample size n is written with a subscript n.) More generally, we have

$$\widehat{\theta}_{ps,i} - \widehat{\theta} = (n-1)(\widehat{\theta} - \widehat{\theta}_{[i]}) = \frac{\widehat{\theta}_{[i]} - \widehat{\theta}}{-\frac{1}{n-1}}, \qquad (10.4)$$

which looks a bit like an approximate derivative. In fact, the empirical distribution function of the sample without Y_i is $F_{n-1,[i]}(y) = F_n(y) + \epsilon_n(\delta_{Y_i}(y) - F_n(y)$, where $\epsilon_n = -(n-1)^{-1}$ and $\delta_{Y_i}(y)$ is the distribution function of a constant random variable with value Y_i. Then, for functional type estimators $\widehat{\theta} = T(F_n)$, we have

$$\widehat{\theta}_{ps,i} - \widehat{\theta} = \frac{T(F_{n-1,[i]}) - T(F_n)}{-\dfrac{1}{n-1}} = \frac{T(F_n + \epsilon_n(\delta_{Y_i} - F_n)) - T(F_n)}{-\dfrac{1}{n-1}}. \qquad (10.5)$$

Looking back at the definition of the Influence Curve (5.13, p. 244), the above formula looks like an empirical form of the definition, and it turns out that $\widehat{\theta}_{ps,i} - \widehat{\theta}$ is an approximation to the influence curve evaluated at Y_i. Thus, to understand the jackknife estimator of variance, it helps to consider how the influence curve is used to get variance estimators.

10.2 Connections to the Influence Curve

Suppose that $\widehat{\theta}$ has the familiar approximation-by-averages representation

$$\widehat{\theta} - \theta_0 = \frac{1}{n}\sum_{i=1}^{n} IC(Y_i, \theta_0) + R_n, \tag{10.6}$$

where $\sqrt{n}R_n \xrightarrow{p} 0$ as $n \to \infty$. Here we have changed notation to the more common IC in place of our previous notation h in Theorem 5.23 (p. 242). If $\mathrm{E}_F\{IC(Y_1, \theta_0)\} = 0$ and $\mathrm{E}_F\{IC(Y_1, \theta_0)\}^2 = \int\{IC(y, \theta_0)\}^2 dF(y) = \sigma^2_{\mathrm{IC}}$ exists, then by the Central Limit Theorem, $\widehat{\theta}$ is $\mathrm{AN}(\theta, \sigma^2_{\mathrm{IC}}/n)$. An estimate of σ^2_{IC} is the empirical estimate

$$\widehat{\sigma}^2_{\mathrm{IC}} = \int \left\{\widehat{IC}(y, \widehat{\theta})\right\}^2 dF_n(y) = \frac{1}{n}\sum_{i=1}^{n}\left\{\widehat{IC}(Y_i, \widehat{\theta})\right\}^2, \tag{10.7}$$

where $\widehat{IC}$ means that the influence curve has been estimated in addition to replacing θ_0 by $\widehat{\theta}$. This leads to the asymptotic variance estimate $\widehat{\sigma}^2_{\mathrm{IC}}/n$. In some situations one may want to replace the divisor n in (10.7) by $n-1$ or even by $n-b$ in vector situations to correct for small sample bias. In general, though, we just use n for the divisor, and this estimator usually corresponds to plug-in estimators for σ^2_{IC}.

The Influence Curve Method is a two-stage procedure:

1. Approximate $\widehat{\theta} - \theta_0$ by the sample average of $IC(Y_i, \theta_0)$ and estimate $IC(Y_i, \theta_0)$, replacing θ_0 by $\widehat{\theta}$ and estimating other unknown quantities where necessary to obtain $\widehat{IC}(Y_i, \widehat{\theta})$.
2. Estimate σ^2_{IC} using the sample variance of the $\widehat{IC}(Y_i, \widehat{\theta})$.

In computing the sample variance in (10.7) we have not subtracted off the mean since theoretically the influence curve has expectation 0. With M-estimators it does not matter since the sample mean of the $\widehat{IC}(Y_i, \widehat{\theta})$ is usually zero by definition. Recall that M-estimators based on vector $\boldsymbol{\psi}$ have influence curves $\boldsymbol{A}(\boldsymbol{\theta}_0)^{-1}\boldsymbol{\psi}(y, \boldsymbol{\theta}_0)$ estimated by $\boldsymbol{A}_n(\boldsymbol{Y}, \widehat{\boldsymbol{\theta}})^{-1}\boldsymbol{\psi}(Y_i, \widehat{\boldsymbol{\theta}})$, and thus the vector version of (10.7, p. 389),

$$\frac{1}{n}\sum_{i=1}^{n} \widehat{\boldsymbol{IC}}(Y_i, \widehat{\boldsymbol{\theta}})\widehat{\boldsymbol{IC}}(Y_i, \widehat{\boldsymbol{\theta}})^T, \tag{10.8}$$

leads directly to the empirical sandwich estimator $\boldsymbol{V}_n(\boldsymbol{Y}, \widehat{\boldsymbol{\theta}})$ of (7.12, p. 302). Similarly, delta theorem estimators of asymptotic variance may also be viewed as Influence Curve Method variance estimators.

Now, we claim that

$$\widehat{\theta}_{ps,i} - \widehat{\theta} \approx \widehat{IC}(Y_i, \widehat{\boldsymbol{\theta}}) \tag{10.9}$$

and that the Jackknife Variance Method is just an approximation to the Influence Curve Method. That is, the pseudo-values minus $\widehat{\theta}$ play the role of the estimated influence curve, and $\widehat{V}_\mathrm{J}$ is just the sample variance of the pseudo-values divided by n. One motivation for (10.9) is the heuristic provided by (10.5, p. 388). Also, using the approximation-by-averages representation (10.6, p. 389), we have

$$\widehat{\theta}_{ps,i} - \widehat{\theta} = IC(Y_i, \theta_0) - \overline{IC} + (n-1)(R_n - R_{ni}),$$

where $\overline{IC} = n^{-1} \sum_{i=1}^{n} IC(Y_i, \theta_0)$ and R_{ni} refers to the remainder in (10.6) when $\widehat{\theta}_{[i]}$ is used rather than $\widehat{\theta}$. In Section 10.8.1 (p. 403) we discuss smallness of $R_{ni} - R_n$ as the key condition for consistency of $\widehat{V}_\mathrm{J}$ for the asymptotic variance of $\widehat{\theta}$. Here we are satisfied to illustrate with examples.

10.3 Examples

10.3.1 Sample Moments

The sample mean is an obvious first example, and the sample variance is the simplest nonlinear statistic.

Example 10.2 (Sample mean). For the sample mean $\widehat{\theta} = \widehat{\mu} = \overline{Y}$, the leave-1-out estimators are $\widehat{\theta}_{[i]} = (n\overline{Y} - Y_i)/(n-1)$, and the pseudo-values are $n\overline{Y} - (n\overline{Y} - Y_i) = Y_i$, the observations themselves. Thus $\widehat{V}_\mathrm{J} = s_{n-1}^2/n$.

The influence curve is $y - \mu_0$ with $\widehat{IC}(Y_i, \overline{Y}) = Y_i - \overline{Y}$, and the empirical estimate of σ_{IC}^2 is

$$s_n^2 = \frac{1}{n} \sum_{i=1}^{n} (Y_i - \overline{Y})^2,$$

leading to variance estimate s_n^2/n. Note that $\widehat{V}_\mathrm{J}$ and the Influence Curve Method would be the same if $n-1$ were used in the denominator of (10.7, p. 389). ♦

Example 10.3 (Sample variance). For the sample variance $\widehat{\theta} = s_n^2 = n^{-1} \sum_{i=1}^{n} (Y_i - \overline{Y})^2$, the leave-1-out estimators are

$$\begin{aligned}
\widehat{\theta}_{[i]} &= \frac{1}{n-1}\sum_{\substack{j=1\\ j\neq i}}^{n}(Y_j - \overline{Y}_{n-1,i})^2 \\
&= \frac{1}{n-1}\sum_{\substack{j=1\\ j\neq i}}^{n}\{Y_j - \overline{Y} - (\overline{Y}_{n-1,i} - \overline{Y})\}^2 \\
&= \frac{1}{n-1}\sum_{\substack{j=1\\ j\neq i}}^{n}(Y_j - \overline{Y})^2 - (\overline{Y}_{n-1,i} - \overline{Y})^2 \\
&= \frac{1}{n-1}\sum_{\substack{j=1\\ j\neq i}}^{n}(Y_j - \overline{Y})^2 - \frac{1}{(n-1)^2}(\overline{Y} - Y_i)^2,
\end{aligned}$$

where we have used $\overline{Y}_{[i]} - \overline{Y} = (n-1)^{-1}(\overline{Y} - Y_i)$ in this last step. Next, the pseudo-values are

$$\begin{aligned}
\widehat{\theta}_{ps,i} = n\widehat{\theta} - (n-1)\widehat{\theta}_{[i]} &= \sum_{j=1}^{n}(Y_j - \overline{Y})^2 - \sum_{\substack{j=1\\ j\neq i}}^{n}(Y_j - \overline{Y})^2 + \frac{1}{n-1}(\overline{Y} - Y_i)^2 \\
&= (Y_i - \overline{Y})^2 + \frac{1}{n-1}(\overline{Y} - Y_i)^2 \\
&= \left(\frac{n}{n-1}\right)(Y_i - \overline{Y})^2.
\end{aligned}$$

Taking the average of the pseudo-values, we obtain

$$\widehat{\theta}_{\mathrm{J}} = \frac{1}{n}\sum_{i=1}^{n}\widehat{\theta}_{ps,i} = \frac{1}{n-1}\sum_{i=1}^{n}(Y_i - \overline{Y})^2,$$

the usual unbiased version of the sample variance. This illustrates how $\widehat{\theta}_{\mathrm{J}}$ is less biased than $\widehat{\theta}$ where in this case $\mathrm{E}_F\{\widehat{\theta}\} = \{(n-1)/n\}\sigma^2$. In fact, Efron (1982, p. 11) asserts that the true justification of $\widehat{\theta}_{\mathrm{J}}$ is that it completely eliminates bias in quadratic estimators.

Note also that

$$\widehat{\theta}_{ps,i} - \widehat{\theta}_{\mathrm{J}} = \left(\frac{n}{n-1}\right)\left\{(Y_i - \overline{Y})^2 - s_n^2\right\},$$

which is very similar to $\widehat{IC}(Y_i, s_n^2) = (Y_i - \overline{Y})^2 - s_n^2$. Finally, the sample average of $\widehat{\theta}_{ps,i}$ is $\widehat{\theta}_J$ and taking the ($n-1$ version) sample variance of $\widehat{\theta}_{ps,i}$ divided by n, we get

$$\widehat{V}_J = \frac{1}{n(n-1)}\sum_{i=1}^{n}\left(\frac{n}{n-1}\right)^2\left\{(Y_i - \overline{Y})^2 - s_n^2\right\}^2 = \left(\frac{n}{n-1}\right)^3\frac{(m_4 - s_n^4)}{n},$$

where recall $m_k = n^{-1}\sum_{i=1}^{n}(Y_i - \overline{Y})^k$. Since s_n^2 has influence curve $(y-\mu_0)^2 - \sigma_0^2$ and $\widehat{IC}(Y_i, s_n^2) = (Y_i - \overline{Y})^2 - s_n^2$, the empirical estimate of σ_{IC}^2 is

$$\frac{1}{n}\sum_{i=1}^{n}\left\{(Y_i - \overline{Y})^2 - s_n^2\right\}^2 = m_4 - s_n^4.$$

Thus the Influence Curve Method yields $(m_4 - s_n^4)/n$ as an estimator of the asymptotic variance of s_n^2, and is smaller than $\widehat{V}_J$ by the factor $\{(n-1)/n\}^3$. ◆

The asymptotic variance of any sample moment m_k is given in (5.11, p. 243). Therefore, the Influence Curve method is easy to implement by just plugging in sample moments for any unknown population moments in that asymptotic variance expression. However, functions of sample moments such as $\widehat{\text{Skew}} = m_3/\{m_2^{3/2}\}$ have complicated asymptotic variance expressions. Fortunately, the jackknife can trivially handle such functions. Here we illustrate with a Monte Carlo study application.

Example 10.4 (Monte Carlo study example). In Section 9.4.1 (p. 372) we discussed obtaining standard errors for the ratio of two Monte Carlo sample variances. The asymptotic variance formula (9.2, p. 373) is straightforward to obtain and use, but some time is required to make sure the calculations are correct. In contrast, the jackknife is extremely easy to use in such situations. As an illustration, we take the same example as shown at the end of Section 9.4.1 (p. 372), where the delta method gives a standard error of 0.019 for the ratio of estimated variances of the sample mean and sample 20% trimmed mean. The first R function below is modified from one mentioned in the Appendix of Efron and Tibshirani (1993).

```
jack.se<-function(x, theta, ...){
      call <- match.call()
      n <- length(x)
      u <- rep(0, n)
      for(i in 1:n) {u[i] <- theta(x[ - i], ...)}
      jack.se <- sqrt(((n - 1)/n) * sum((u - mean(u))
        ^2))
      return(jack.se)
}

ratio.var<-function(index,xdata){
```

```
        # form that jackknife can use x=1:n
        var(xdata[index,1])/var(xdata[index,2])
}

# put the following code in a file named mc.ex
# data norm15 is from code in Example 1 of MC Chapter
z<-norm15
ratio.var(1:N,cbind(z[,1],z[,2]))->ratio12
jack.se(1:N,ratio.var,cbind(z[,1],z[,2]))->sd.ratio12
cat("ratio12 =",round(ratio12,3),fill=T)
cat("sd.ratio12 =",round(sd.ratio12,3),fill=T)

> source("mc.ex")
ratio12 = 0.861
sd.ratio12 = 0.019
```

We see that the jackknife reproduces the results obtained from (9.2, p. 373). The only requirement besides the basic jackknife function is to write a function such as `ratio.var` above that calculates the statistic of interest in terms of the rows of the data. Note that the call to `jack.se` uses the integers from 1 to N as the official data and passes the real data z as an additional third argument. ◆

Because moments and functions of moments play such a large role in statistics, we give here a theorem of Shao and Tu (1995, p. 25) to justify the use of $\widehat{V}_J$ with such statistics.

Theorem 10.1. *Suppose that $\boldsymbol{Y}_1, \ldots, \boldsymbol{Y}_n$ are iid vectors with finite mean $E(\boldsymbol{Y}_1) = \boldsymbol{\mu}$ and covariance $\boldsymbol{\Sigma}$. If g is a real-valued function with $g'(\boldsymbol{\mu}) \neq \boldsymbol{0}$ and $g'(\boldsymbol{y})$ is continuous at $\boldsymbol{\mu}$, then for $T = g(\overline{\boldsymbol{Y}})$, $n\widehat{V}_J \xrightarrow{wp1} \sigma^2_{IC}$ as $n \to \infty$, where $\sigma^2_{IC} = g'(\boldsymbol{\mu})\boldsymbol{\Sigma} g'(\boldsymbol{\mu})^T$.*

The theorem is deceptively simple because it is phrased in terms of a single multivariate sample mean. However, it covers the same ground as Theorem 5.26 (p. 246) where the statistic of interest is given specifically as $T = g(\overline{q}_1, \ldots, \overline{q}_k)$, where each component is a sample mean of a function of an underlying sample $X_1, \ldots, X_n$. Thus, $\boldsymbol{\mu}$ above is $[\mathrm{E}\{q_1(X_1)\}, \ldots, \mathrm{E}\{q_k(X_1)\}]^T$ in this notation, and $\boldsymbol{\Sigma}$ is the covariance matrix of $\{q_1(X_1), \ldots, q_k(X_1)\}^T$. The limiting variance form σ^2_{IC} is from the second delta method theorem, Theorem 5.19 (p. 238). The most obvious examples of components are sample raw moments, $m'_j = n^{-1}\sum_{i=1}^n X_i^j$. Of course, the sample moments $m_j = n^{-1}\sum_{i=1}^n (X_i - \overline{X})^j$ are covered because they are functions of the raw moments. Most summary statistics from Monte Carlo studies are covered by Theorem 10.1.

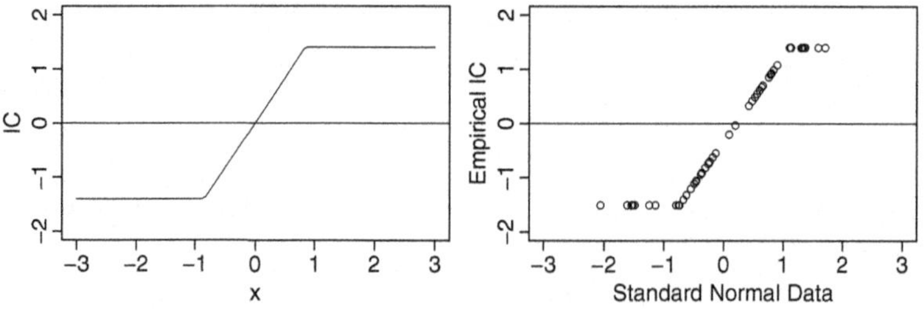

Fig. 10.2 Influence Curve of 20% trimmed mean at standard normal (Left Panel) and the jackknife empirical influence curve of a sample of size $n = 50$ from a standard normal (Right Panel)

10.3.2 Trimmed Means

This next example further illustrates the close connection between the pseudo-values and the influence curve. Although it is possible to handle the following jackknife calculations analytically, we use a purely computational approach.

Example 10.5 (20% Trimmed mean). The 20% trimmed mean is the mean of the middle 60% of the data after ordering, and we studied its variance in Example (9.1, p. 365). The influence curve is similar to the Huber ψ_k function of Example 7.4.1 (p. 311) (see Huber 1981, p. 58). In the left panel of Figure 10.2 (p. 394), we have plotted this influence curve at the standard normal. In the right panel we have plotted the values $\widehat{\theta}_{ps,i} - \widehat{\theta}$ versus the Y_i for a sample of $n = 50$ from the standard normal. We might call this plot a *jackknife empirical influence curve*. It is striking how close the plots are. We could make the plots even closer if we eliminate the randomness by using the quantiles of a standard normal, say $\Phi^{-1}(i/(n+1))$, in place of the sample. Thus, the jackknife provides a simple way to view the influence curve of any suitably smooth estimator.

To further note the similarities in the jackknife and Influence Curve Methods, the standard errors for the 20% trimmed mean for the $n = 50$ data set in Figure 10.2 (p. 394) are (.162,.155) for the jackknife and Influence Curve Method, respectively. For an $n = 20$ generated data set, they are (.158,.167), and for an $n = 100$ data set they are (.112,.112). These calculations and those in Figure 10.2 (p. 394) were made using the `trim.var` R function of Section 9.4.2 (p. 373) and the R function `jack.se` given in Example 10.4, p. 392. ♦

10.3.3 Contingency Tables

Consider a contingency table of counts where the cell counts can be viewed as arising from a single multinomial vector with sample size n. For example, a 3-by-2 table with entries N_{ij} might have arisen from a multinomial$(n; \boldsymbol{p})$ with

$k = 6$ cells. In some situations, such a table might arise from three independent binomials or from two independent multinomials with three cells each. Such "product multinomial" designs require multisample jackknife methods. Here we only discuss the situations where the single multinomial describes the whole table.

What is the correct way to calculate the leave-1-out estimators based on a multinomial vector? The standard approach for the jackknife is to leave out each of the smallest independent pieces of data. Those are obviously the individual Y_i with a simple random sample $Y_1, \ldots, Y_n$. Here, it is important to note that the multinomial vector $(N_{11}, N_{12}, \ldots)$ making up the table is actually built from summing the individual independent multinomial vectors: there are N_{11} of the form $(1, 0, \ldots, 0)$, N_{12} of the form $(0, 1, \ldots, 0)$, etc. Thus, each of these individual vectors must be dropped out in turn. Note that any statistic of interest is based on $(N_{11}, N_{12}, \ldots)$ or equivalently on $\widehat{p} = (N_{11}/n, N_{12}/n, \ldots)$. But the latter is just a sample mean of the individual independent multinomial vectors. Thus, Theorem 10.1 (p. 393) applies and $\widehat{V}_{\text{J}}$ is consistent. Here we illustrate use of $\widehat{V}_{\text{J}}$ with a simple two by two table example.

Example 10.6 (Capital punishment survey.). In a class of graduate students in 1996, the $n = 25$ students were asked to turn in a sheet of paper answering whether they were US citizens or not (US or INT) and whether they were in favor of capital punishment or not (FOR or AGAINST). The results were as follows

	Against	For	Total
US	10	4	14
International	4	7	11
Total	14	11	25

Let us consider the difference of proportions of those not in favor of capital punishment,

$$\widehat{\theta}_1 = \widehat{p}_{\text{US}} - \widehat{p}_{\text{INT}} = \frac{10}{14} - \frac{4}{11} = .35,$$

and the related odds ratio

$$\widehat{\theta}_2 = \left(\frac{\widehat{p}_{\text{US}}}{1 - \widehat{p}_{\text{US}}}\right) / \left(\frac{\widehat{p}_{\text{INT}}}{1 - \widehat{p}_{\text{INT}}}\right) = \frac{(10)(7)}{(4)(4)} = 4.375.$$

Both estimators are nonlinear since the denominators of the proportions are random here. But since the estimators are functions of multinomial proportion estimators, the delta method is straightforward (see 5.9, p. 239). On the other hand, the jackknife variance estimate can be calculated by hand very easily since there are at most four distinct leave-1-out values.

For $\widehat{\theta}_1$, there are 10 leave-1-out values equal to 9/13–4/11, 4 equal to 10/13–4/11, 4 equal to 10/14–3/10, and 7 equal to 10/14–4/10, leading to $(\widehat{V}_{\text{J}})^{1/2} = 0.202$. The usual delta method standard error is a little smaller, $\{(10/14)(4/14)/14 + (4/11)(7/11)/11\}^{1/2} = 0.189$.

There are three distinct leave-1-out values for $\widehat{\theta}_2$, (9)(7)/{(4)(4)} $= 63/16$, 70/12, and 60/16, leading to $(\widehat{V}_J)^{1/2} = 4.52$. The asymptotic variance of $\log(\widehat{\theta}_2)$ is given in Problem 5.25 (p. 267) leading to the delta method standard error for $\widehat{\theta}_2$ of 3.77. ♦

10.3.4 Sample Quantiles

The leave-1-out jackknife method is known to be inconsistent for estimating the variance of sample quantiles. To illustrate we calculate the jackknife variance estimate for the median of a data set.

Example 10.7 (Median). Consider a random sample of rainfall in Raleigh in February drawn from the years 1948 to 1992,

3.74 1.57 3.68 3.93 3.24 5.75 3.45 4.76 1.00 5.50 1.43 6.00

Ordering the 12 observations gives

1.00 1.43 1.57 3.24 3.45 3.68 3.74 3.93 4.76 5.50 5.75 6.00

The median is $(3.68 + 3.74)/2 = 3.71$. The leave-1-out estimators of the median are either equal to 3.74 (the 7th ordered value) or 3.68 (the 6th ordered value). The resulting leave-1-out jackknife standard error for the median is 0.10, much less than it ought to be (e.g., compare to the bootstrap standard error .46 of the next chapter). ♦

The jackknife works as an approximate Influence Curve method, first approximating the nonlinear statistic by a linear one, and then getting a variance estimate for the linear approximation. In the case of the median (and other quantiles), the jackknife linear approximation is not good enough to make the variance estimate consistent. Problem 10.13 (p. 410) shows what happens to $\widehat{V}_J$ for the median in large samples.

10.4 Delete-*d* Jackknife

Although the usual jackknife variance estimator for the median is inconsistent, it is possible to use instead the delete-*d* jackknife. If we compute $\widehat{\theta}$ for all $N_{n,d} = \binom{n}{d}$ possible samples of size $n - d$ and compute the $1/N_{n,d}$ version of the sample variance among these $\widehat{\theta}$ values, call it $s^2_{n,d}$, then the "delete-*d* jackknife" variance estimator is

$$\left(\frac{n-d}{d}\right) s^2_{n,d}.$$

Since this requires a huge computational burden, it is suggested that one just sample m of these possible samples and use instead

$$\left(\frac{n-d}{d}\right) s_m^2,$$

where here s_m^2 is again the $1/m$ version of the sample variance of the $\widehat{\theta}$ values. For the median with the Example 10.7 data, here are some runs with different m and d values:

```
        Leave-6-out                          Leave-3-out
m=10          s.e. = .45            m=10          s.e. = .23
m=20          s.e. = .49            m=20          s.e. = .29
m=40          s.e. = .56            m=40          s.e. = .25
m=100         s.e. = .59            m=100         s.e. = .23
m=100         s.e. = .59            m=100         s.e. = .23
```

In order for the delete-d variance estimator to be consistent, d needs to go to infinity along with n (see Shao and Tu 1995, Section 2.3). Thus it makes sense that the $d = 6$ results above look reasonable, but the $d = 3$ results are too small.

10.5 The Information Reduction Approach for Variance Estimation

Recall that for bias estimation, one way to derive the estimator $\widehat{\theta}_J$ is to use the prediction at $1/n = 1/\infty = 0$ from the line defined by the two points $(x_1, y_1) = (1/n, \widehat{\theta})$ and $(x_2, y_2) = (1/(n-1), \overline{\theta}_1)$. The related variance estimate procedure is to use the points $(x_1, y_1) = (1/n, 0)$ and $(x_2, y_2) = (1/(n-1), s_{-1}^2)$, where we define

$$s_{-1}^2 = n^{-1} \sum_{i=1}^{n} \left(\widehat{\theta}_{[i]} - \overline{\theta}_1\right)^2.$$

The intercept from the line defined by these two points is $-\widehat{V}_J$,

$$-\text{intercept} = \frac{-(y_1x_2 - y_2x_1)}{x_2 - x_1} = \frac{s_{-1}^2/n}{\dfrac{1}{n-1} - \dfrac{1}{n}} = \frac{n-1}{n} \sum_{i=1}^{n} \left(\widehat{\theta}_{[i]} - \overline{\theta}_1\right)^2 = \widehat{V}_J.$$

The idea is to treat $\widehat{\theta}$ as a "population" of size one of estimators based on size n with $y_1 = 0 =$ the population variance of the one value $\widehat{\theta}$, leading to the point $(1/n, 0)$. The second point $(1/(n-1), s_{-1}^2)$ is based on treating the $\overline{\theta}_1$ values as a population of estimators of size $n - 1$, hence $y_2 = s_{-1}^2$ is the "population" variance of those n

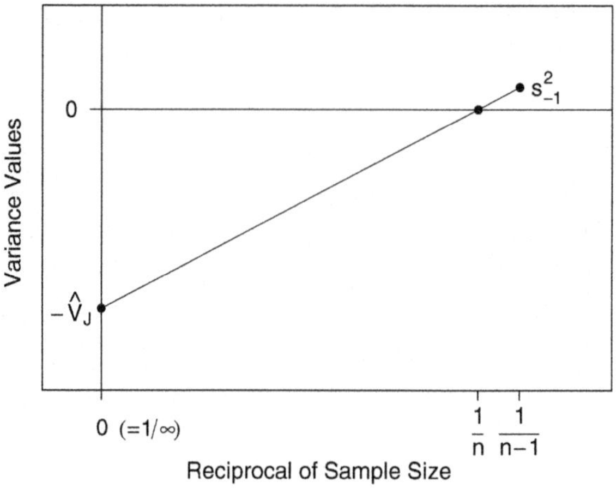

Fig. 10.3 Information reduction approach for $\widehat{V}_J$. The jackknife variance estimator $\widehat{V}_J$ is the negative of the intercept estimate or extrapolant at 0 from fitting a line to the two points $(1/n, 0)$ and $(1/(n-1), s^2_{-1})$. Data are from Example 10.1 (p. 387)

values. Figure 10.3 (p. 398) illustrates the approach with data from the situation of Example 10.1 (p. 387) (cf. Figure 10.1, p. 387). The delete-d variance estimator is derived in exactly the same way using the points $(1/n, 0)$ and $(1/(n-1), s^2_{n,d})$ or $(1/(n-1), s^2_m)$. For example,

$$-\text{intercept} = \frac{-(y_1x_2 - y_2x_1)}{x_2 - x_1} = \frac{s^2_{n,d}/n}{\dfrac{1}{n-d} - \dfrac{1}{n}} = \left(\frac{n-d}{d}\right) s^2_{n,d}.$$

10.6 Grouped Jackknife

The delete-d jackknife in Section 10.4 (p. 396) is a fairly sophisticated technique that can handle nonsmooth statistics like the median. A much older and simpler procedure, called the grouped jackknife, is based on dividing the data into m groups of size g each, $n = mg$. Then the grouped jackknife variance estimator is

$$\widehat{V}_{\text{GJ}} = \frac{m-1}{m} \sum_{i=1}^{m} \left(\widehat{\theta}_{n-g,i} - \overline{\theta}_{\text{G1}}\right)^2 = \frac{1}{m(m-1)} \sum_{i=1}^{m} \left(\widehat{\theta}_{ps,g,i} - \widehat{\theta}_{\text{GJ}}\right)^2, \quad (10.10)$$

where $\widehat{\theta}_{n-g,i}$ is the estimator with the ith group left out, $\overline{\theta}_{\text{G1}} = m^{-1} \sum_{i=1}^{m} \widehat{\theta}_{n-g,i}$, the pseudo-values are $\widehat{\theta}_{ps,g,i} = m\widehat{\theta} - (m-1)\widehat{\theta}_{n-g,i}$, and $\widehat{\theta}_{\text{GJ}}$ is the average of the pseudo-values. In fact, (10.10) maybe viewed as the general expression, and the more common delete-1 jackknife variance estimator in (10.3) is a special case with $g = 1$. In certain situations $g > 1$ may be preferred because of computational reason, but usually the standard delete-1 ($g = 1$) jackknife is preferable. This preference can easily be demonstrated when $\widehat{\theta}$ is the sample mean. In that case, it is easy to see the pseudo-values are the averages of the deleted groups, denote them by $\overline{Y}_i$, and $\overline{\theta}_{\mathrm{GI}} = \widehat{\theta}_{\mathrm{GJ}} = \overline{Y}$. Thus,

$$\widehat{V}_{\mathrm{GJ}} = \frac{1}{m(m-1)} \sum_{i=1}^{m} \left(\overline{Y}_i - \overline{Y}\right)^2 .$$

Clearly, $\mathrm{E}\{\widehat{V}_{\mathrm{GJ}}\} = (\sigma^2/g)/m = \sigma^2/n$, and $\widehat{V}_{\mathrm{GJ}}$ is thus unbiased for $\mathrm{Var}(\overline{Y})$. Moreover, using (1.23, p. 22),

$$\mathrm{Var}(\widehat{V}_{\mathrm{GJ}}) = \frac{\sigma^4}{n^2}\left\{\frac{2}{m-1} + \frac{\mathrm{Kurt}(Y_1) - 3}{n}\right\}. \tag{10.11}$$

The delete-1 usual jackknife variance estimator for $\overline{Y}$ is s_{n-1}^2/n, and its variance has exactly the same form as $\mathrm{Var}(\widehat{V}_{\mathrm{GJ}})$ (see 1.23, p. 22) except that $2/(m-1)$ is replaced by $2/(n-1)$. In the case of normal data with $\mathrm{Kurt}(Y_1) = 3$, we have $\mathrm{Var}(\widehat{V}_{\mathrm{GJ}})/\mathrm{Var}(s_{n-1}^2/n) = (n-1)/(m-1) \approx g$. Thus, the delete-1 jackknife estimator is much more efficient than the grouped jackknife variance estimator.

10.7 Non-iid Situations

Thus far, the examples we considered can be viewed in terms of an iid random sample. The key principle for extension to other situations is that one should seek to drop out each of the smallest, independent pieces that make up the data. In the case of a sample $Y_1, \ldots, Y_n$, dropping out each Y_i is the correct jackknife approach. As mentioned for the grouped jackknife, it is possible to drop out independent groups of observations like $(Y_1, Y_2), (Y_3, Y_4), \ldots (Y_{n-1}, Y_n)$, but the resulting variance estimator is not as efficient as the standard delete-1 $\widehat{V}_{\mathrm{J}}$. The delete-d estimator is different, because there we are dropping out all $\binom{n}{d}$ pieces, and they are not independent.

A nice illustration of the general principle is for two-stage survey sampling designs where it is natural to drop out the primary sampling units of the first stage. All the units selected within an individual primary sampling unit are correlated and should be kept together as a unit. A good discussion is given in Chapter 6 of Shao and Tu (1995). Here we illustrate jackknifing in three other non-iid situations: the k independent samples problem, the linear model, and times series data.

10.7.1 k Independent Samples

The data consists of k independent samples $Y_{11}, \ldots, Y_{1n_1}, Y_{21}, \ldots, Y_{2n_2}, \ldots, Y_{k1}, \ldots, Y_{kn_k}$, with associated distribution functions $F_1, \ldots, F_k$. The usual one-way ANOVA setup is a common situation with data like these. An estimator of interest could be a pooled variance estimator or the overall sample mean.

The individual data values Y_{ij} could be random vectors. In Section 10.3.3 (p. 394) we discussed r by s contingency tables when the data come from a single multinomial random vector $(N_{11}, N_{12}, \ldots, N_{rs})$, which can then be viewed as arising from n iid random vectors. However, the rows (or columns) of the contingency table are often distributed as independent multinomial vectors, $(N_{11}, N_{12}, \ldots, N_{1s})$, $(N_{21}, N_{22}, \ldots, N_{2s})$, ..., $(N_{r1}, N_{r2}, \ldots, N_{rs})$. But these rows can again be each viewed as arising from independent multinomial vectors of size 1. Thus, here we have $k = r$ samples of independent vectors. Statistics such as the odds ratio or estimates of correlation measures are commonly used with such data.

What is the appropriate way to use the jackknife to obtain variance estimators? Clearly, we see how to drop out the smallest independent pieces. But how should we define pseudo-values and put these together to get $\widehat{V}_J$? The best guide is to remember that in the iid case, the pseudo-values approximate the influence curve. Unfortunately, we have not discussed approximation by averages for the k-sample problem except in Problem 5.40 (p. 270).

Thus, we begin with the simple $k = 2$ situation and $\widehat{\theta} = g(\overline{Y}_1, \overline{Y}_2)$. By Taylor expansion, we have

$$\begin{aligned} g(\overline{Y}_1, \overline{Y}_2) - g(\mu_1, \mu_2) &\approx \frac{\partial g(\boldsymbol{\mu})}{\partial \mu_1}(\overline{Y}_1 - \mu_1) + \frac{\partial g(\boldsymbol{\mu})}{\partial \mu_2}(\overline{Y}_2 - \mu_2) \\ &= \frac{1}{n_1}\sum_{i=1}^{n_1} \frac{\partial g(\boldsymbol{\mu})}{\partial \mu_1}(Y_{1i} - \mu_1) + \frac{1}{n_2}\sum_{j=1}^{n_2} \frac{\partial g(\boldsymbol{\mu})}{\partial \mu_2}(Y_{2j} - \mu_1). \end{aligned}$$

The variance of this latter expression is then

$$\left\{\frac{\partial g(\boldsymbol{\mu})}{\partial \mu_1}\right\}^2 \frac{\sigma_1^2}{n_1} + \left\{\frac{\partial g(\boldsymbol{\mu})}{\partial \mu_2}\right\}^2 \frac{\sigma_2^2}{n_2},$$

where σ_1^2 and σ_2^2 are the variances in each sample, respectively. Plugging in sample means and variances would then yield the delta method/influence curve estimator of asymptotic variance.

Let us try defining

$$\begin{aligned} \widehat{\theta}^{(1)}_{ps,i} &= n_1\widehat{\theta} - (n_1 - 1)\widehat{\theta}^{(1)}_{[i]} \\ \widehat{\theta}^{(2)}_{ps,j} &= n_2\widehat{\theta} - (n_2 - 1)\widehat{\theta}^{(2)}_{[j]}, \end{aligned}$$

where the superscripts refer to the two different samples. For a simple linear example, $\widehat{\theta} = \overline{Y}_2 - \overline{Y}_1$, we have

$$\widehat{\theta}^{(1)}_{ps,i} = n_1(\overline{Y}_2 - \overline{Y}_1) - (n_1 - 1)\left\{\overline{Y}_2 - \left(\frac{n_1\overline{Y}_1 - Y_{1i}}{n_1 - 1}\right)\right\} = \overline{Y}_2 - Y_{1i}.$$

Similarly, $\widehat{\theta}^{(2)}_{ps,j} = -\overline{Y}_1 + Y_{2i}$. Then, define

$$\widehat{V}_{\text{J}} = \frac{s^2_{1,ps}}{n_1} + \frac{s^2_{2,ps}}{n_2},$$

where $s^2_{1,ps}$ and $s^2_{2,ps}$ are the respective sample variances of the pseudo-values. For $\widehat{\theta} = \overline{Y}_2 - \overline{Y}_1$, we then have $\widehat{V}_{\text{J}} = s_1^2/n_1 + s_2^2/n_2$ as one would get from the delta method expansion. Another example is the nonlinear estimator $\widehat{\theta} = \overline{Y}_2/\overline{Y}_1$, where in Problem 10.11 (p. 410) one is asked to show that $\widehat{V}_{\text{J}}$ is approximately equal to the delta method variance estimator.

The generalization to k samples should be clear: define pseudo-values separately in each sample and then let

$$\widehat{V}_{\text{J}} = \sum_{j=1}^{k} \frac{s^2_{j,ps}}{n_j}. \tag{10.12}$$

Arvesen (1969) first proposed the above $\widehat{V}_{\text{J}}$ for $k = 2$, but otherwise it has not been discussed much in the literature. A proper appreciation and general proof requires the k-sample approximation by averages

$$\widehat{\theta} - \theta_0 = \sum_{i=1}^{k} \frac{1}{n_i} \sum_{j=1}^{n_i} IC^{(i)}(Y_{ij}, \theta_0) + R, \tag{10.13}$$

where the ith partial influence curve $IC^{(i)}(y, \theta_0)$ is defined by

$$IC^{(i)}(y, \theta_0) = \frac{\partial}{\partial \epsilon} T(F_1, \ldots, F_{i-1}, \delta_y - F_i, F_{i+1}, \ldots, F_k)\bigg|_{\epsilon=0+}.$$

10.7.2 Linear Model

Recall the standard linear model with full rank design matrix

$$Y_i = \boldsymbol{x}_i^T \boldsymbol{\beta} + e_i,$$

where $e_1, \ldots, e_n$ are independent with mean 0 and variance σ_i^2. Here, we think of the data as pairs $(\boldsymbol{x}_1, Y_1), \ldots, (\boldsymbol{x}_n, Y_n)$ regardless of whether the $\boldsymbol{x}_i$ are random or not. The general principle above suggests again leaving out each pair $(\boldsymbol{x}_i, Y_i)$ in turn. Miller (1974) showed for the least squares estimator $\widehat{\boldsymbol{\beta}}_{\mathrm{LS}} = (\boldsymbol{X}^T\boldsymbol{X})^{-1}\boldsymbol{X}^T\boldsymbol{Y}$ that $\widehat{V}_{\mathrm{J}}$ is consistent, but $\widehat{\boldsymbol{\beta}}_{\mathrm{J}}$ is not $\widehat{\boldsymbol{\beta}}_{\mathrm{LS}}$, and Hinkley (1977) demonstrates that $\widehat{\boldsymbol{\beta}}_{\mathrm{J}}$ has higher variance than $\widehat{\boldsymbol{\beta}}_{\mathrm{LS}}$. Thus Hinkley (1977) suggested defining the weighted pseudo-values

$$\widehat{\boldsymbol{\beta}}_{wps,i} = \widehat{\boldsymbol{\beta}}_{\mathrm{LS}} + n(1-h_i)(\widehat{\boldsymbol{\beta}}_{\mathrm{LS}} - \widehat{\boldsymbol{\beta}}_{\mathrm{LS,i}}),$$

where $h_i = \boldsymbol{x}_i^T(\boldsymbol{X}^T\boldsymbol{X})^{-1}\boldsymbol{x}_i$. The average of these $\widehat{\boldsymbol{\beta}}_{wps,i}$ is $\widehat{\boldsymbol{\beta}}_{\mathrm{LS}}$ and the resulting variance estimator is

$$\widehat{V}_{\mathrm{HJ}} = (\boldsymbol{X}^T\boldsymbol{X})^{-1}\left(\frac{n}{n-p}\sum_{i=1}^{n}\widehat{e}_i^2\boldsymbol{x}_i\boldsymbol{x}_i^T\right)(\boldsymbol{X}^T\boldsymbol{X})^{-1},$$

where $\widehat{e}_i = Y_i - \boldsymbol{x}_i^T\widehat{\boldsymbol{\beta}}_{\mathrm{LS}}$ are the usual residuals and p is the dimension of $\boldsymbol{\beta}$. Notice that $\widehat{V}_{\mathrm{HJ}}$ is the same expression that occurred naturally in the M-estimation approach, (7.21, p. 317). It allows for variance heterogeneity of the errors e_i, in contrast to the usual covariance estimator $(\boldsymbol{X}^T\boldsymbol{X})^{-1}\widehat{\sigma}^2$, where $\widehat{\sigma}^2 = (n-p)^{-1}\sum_{i=1}^n \widehat{e}_i^2$. Wu (1986) suggests a variant of $\widehat{V}_{\mathrm{HJ}}$,

$$\widehat{V}_{\mathrm{WJ}} = (\boldsymbol{X}^T\boldsymbol{X})^{-1}\left(\sum_{i=1}^{n}\frac{\widehat{e}_i^2}{1-h_i}\boldsymbol{x}_i\boldsymbol{x}_i^T\right)(\boldsymbol{X}^T\boldsymbol{X})^{-1},$$

that is exactly unbiased for estimating $\mathrm{Var}(\widehat{\boldsymbol{\beta}}_{\mathrm{LS}})$ under the assumptions $\mathrm{Var}(e_i) = \sigma_i^2$ and $\boldsymbol{x}_i^T(\boldsymbol{X}^T\boldsymbol{X})^{-1}\boldsymbol{x}_j = 0$ for any pair (i, j) where $\sigma_i^2 \neq \sigma_j^2$.

The solutions suggested above apply only to least squares although one might try something similar for other types of estimators and regression situations.

10.7.3 Dependent Data

Time series and spatial data situations pose special problems for variance estimation because typically there is no independent replication. That is, all the observations in a data set are correlated. The key assumption needed to use a jackknife method is for the correlation between observations to get small as the distance between observations grows. For example, in the first-order autoregressive time series of Example 5.4 (p. 216), the correlation between X_i and X_{i+h} is ρ^h, where typically $0 < \rho < 1$. Carlstein (1986) proposed splitting the data into nonoverlapping blocks of size g, much like the grouped jackknife approach, whereas Künsch (1989) suggested sequential or moving blocks of the form $(Y_1, \ldots, Y_g)$, $(Y_2, \ldots, Y_{g+1})$, $\ldots$, $(Y_{n-g+1}, \ldots, Y_n)$. The latter approach is more similar to the delete-d jackknife, but only these $n - g$ consecutive blocks are used compared to the $\binom{n}{g}$ possible blocks of the delete-d jackknife. For consistency of the jackknife variance estimators, it is essential that the size of the blocks $\to \infty$ as well as the number of blocks $\to \infty$ as $n \to \infty$. Shao and Tu (1995, Ch. 9) survey these and other jackknife approaches for dependent data.

10.8 Several Technical Results for the Jackknife

In this section we give a general result on asymptotic consistency of $\widehat{V}_{\mathrm{J}}$ and a few details of the Efron and Stein (1981) theory concerning the upward bias of $\widehat{V}_{\mathrm{J}}$.

10.8.1 Asymptotic Consistency of $\widehat{V}_{\mathrm{J}}$

We restrict ourselves to the simple case of an iid sample $Y_1, \ldots, Y_n$ and a real estimator $\widehat{\theta}$ that has the usual approximation-by-averages representation,

$$\widehat{\theta} - \theta_0 = \frac{1}{n}\sum_{i=1}^{n} IC(Y_i, \theta_0) + R_n, \tag{10.14}$$

where $\sqrt{n}R_n \xrightarrow{p} 0$ as $n \to \infty$, $\mathrm{E}_F\{IC(Y_1, \theta_0)\} = 0$, and $\mathrm{E}_F\{IC(Y_1, \theta_0)\}^2 = \sigma^2_{\mathrm{IC}}$. Of course, for the leave-1-out estimators we also have

$$\widehat{\theta}_{[i]} - \theta_0 = \frac{1}{n-1}\sum_{\substack{j=1 \\ j \neq i}}^{n} IC(Y_i, \theta_0) + R_{ni},$$

where it is important to note that the remainder R_{ni} here is indexed with an i. From the approximation by averages, we have that $\widehat{\theta}$ is $\mathrm{AN}(\theta_0, \sigma^2_{\mathrm{IC}}/n)$ as $n \to \infty$. Thus, we would like to show that

$$\frac{\widehat{V}_{\mathrm{J}}}{\{\sigma^2_{\mathrm{IC}}/n\}} \xrightarrow{p} 1 \qquad \text{as } n \to \infty, \tag{10.15}$$

or equivalently that $n\widehat{V}_{\mathrm{J}} \xrightarrow{p} \sigma^2_{\mathrm{IC}}$. To that end, we substitute the approximation by average representations into $\widehat{V}_{\mathrm{J}}$ as follows,

$$
\begin{aligned}
n\widehat{V}_{\mathrm{J}} &= (n-1)\sum_{i=1}^{n}\left(\widehat{\theta}_{[i]} - \frac{1}{n}\sum_{j=1}^{n}\widehat{\theta}_{n-1,j}\right)^2 \\
&= (n-1)\sum_{i=1}^{n}\left(\frac{1}{n-1}\sum_{\substack{j=1\\ j\neq i}}^{n} IC(Y_j,\theta_0) + R_{ni} - \overline{IC} - \overline{R}_n\right)^2 \\
&= (n-1)\sum_{i=1}^{n}\left(\frac{\overline{IC} - IC(Y_i,\theta_0)}{n-1} + R_{ni} - \overline{R}_n\right)^2 \\
&= \frac{1}{n-1}\sum_{i=1}^{n}\left\{IC(Y_i,\theta_0) - \overline{IC}\right\}^2 + (n-1)\sum_{i=1}^{n}\left(R_{ni} - \overline{R}_n\right)^2 \\
&\quad -2\sum_{i=1}^{n}\left\{IC(Y_i,\theta_0) - \overline{IC}\right\}\left(R_{ni} - \overline{R}_n\right),
\end{aligned}
$$

where $\overline{IC} = n^{-1}\sum_{i=1}^{n} IC(Y_i,\theta_0)$ and $\overline{R}_n = n^{-1}\sum_{i=1}^{n} R_{ni}$. The first term above is a sample variance of the $IC(Y_i,\theta_0)$ and thus converges in probability to σ^2_{IC}. Using the Cauchy-Schwarz inequality, $\left\{n^{-1}\sum a_i b_i\right\}^2 \le \left\{n^{-1}\sum a_i^2\right\}\left\{n^{-1}\sum b_i^2\right\}$, we have for the cross product term

$$
\begin{aligned}
\left|\sum_{i=1}^{n}\left\{IC(Y_i,\theta_0) - \overline{IC}\right\}\left(R_{ni} - \overline{R}_n\right)\right| &\le \frac{1}{n-1}\sum_{i=1}^{n}\left\{IC(Y_i,\theta_0) - \overline{IC}\right\}^2 \\
&\quad \times (n-1)\sum_{i=1}^{n}\left(R_{ni} - \overline{R}_n\right)^2 .
\end{aligned}
$$

Putting these pieces together, we have $n\widehat{V}_{\mathrm{J}} \xrightarrow{p} \sigma^2_{\mathrm{IC}}$ if

$$
(n-1)\sum_{i=1}^{n}\left(R_{ni} - \overline{R}_n\right)^2 \xrightarrow{p} 0 \qquad \text{as } n \to \infty. \tag{10.16}
$$

Since $\sum_{i=1}^{n}\left(R_{ni} - \overline{R}_n\right)^2 \le \sum_{i=1}^{n}\left(R_{ni} - R_n\right)^2$, an alternate sufficient condition is

$$
(n-1)\sum_{i=1}^{n}\left(R_{ni} - R_n\right)^2 \xrightarrow{p} 0, \qquad \text{as } n \to \infty. \tag{10.17}
$$

The latter condition may be easier to work with. Thus, we have proved the following theorem.

Theorem 10.2. *Suppose that $Y_1, \ldots, Y_n$ are iid and (10.14, p. 403) holds for $\widehat{\theta}$, where $\sqrt{n}R_n \xrightarrow{p} 0$ as $n \to \infty$, $E_F\{IC(Y_1, \theta_0)\} = 0$, and $E_F\{IC(Y_1, \theta_0)\}^2 = \sigma^2_{IC} < \infty$. If* (10.16) *or* (10.17) *holds, then $n\widehat{V}_J \xrightarrow{p} \sigma^2_{IC}$ as $n \to \infty$.*

By the Markov inequality and the fact that $R_{n1}, \ldots, R_{nn}$ are iid, a sufficient condition for (10.17) and thus (10.16) is

$$n^2\mathrm{E}(R_{ni} - R_n)^2 \longrightarrow 0 \qquad \text{as } n \to \infty. \tag{10.18}$$

Shao and Wu (1989) refer to $n^2\mathrm{E}(R_{ni} - R_n)^2$ as a measure of smoothness of the linear approximation (10.14, p. 403), the faster it approaches 0 the more smooth is the approximation.

As a simple illustration, recall from (5.5, p. 211) for s_n^2 that $R_n = -(\overline{Y}_n - \mu)^2$. Somewhat tedious calculations reveal that $\mathrm{E}(R_{ni} - R_n)^2 = O(n^{-3})$ and $\mathrm{E}(R_{ni} - \overline{R}_n)^2 = O(n^{-3})$, although the first is easier to show than the second. Thus, Theorem 10.2 holds if Y_1 has a finite fourth moment. Of course, in Example 10.3 we derived $n\widehat{V}_\mathrm{J} = \{n/(n-1)\}^3(m_4 - s_n^4)$, which is easily shown to converge to $\mu_4 - \sigma^4$ by laws of large numbers. The development of Theorem 10.2 above was partially taken from Shao and Wu (1989) and Shao and Tu (1995, p. 51). Using similar methods, these authors also prove consistency of the delete-d jackknife and give alternative methods for proving consistency of $\widehat{V}_\mathrm{J}$ based on differentiability of statistical functionals $T(F)$.

10.8.2 Positive Bias of $\widehat{V}_\mathrm{J}$

In examples it is often seen that $\widehat{V}_\mathrm{J}$ is larger than the influence curve method of variance estimation. Also, in simulation studies $\widehat{V}_\mathrm{J}$ tends to be larger than the true variance (see for example, Table 11.3, p. 423, and Table 11.5, p. 423). Efron and Stein (1981) provided some elegant theory to support such empirical observations. Here we give a few details of their approach.

Suppose that $\widehat{\theta} = \widehat{\theta}(Y_1, \ldots, Y_n)$ is a symmetric function of the iid observations $Y_1, \ldots, Y_n$ and has finite variance. Following Hoeffding (1948), Efron and Stein (1981) define

$$\alpha(y_i) = n\left\{\mathrm{E}(\widehat{\theta}|Y_i = y_i) - \mathrm{E}(\widehat{\theta})\right\},$$

$$\beta(y_i, y_j) = n^2\left\{\mathrm{E}(\widehat{\theta}|Y_i = y_i, Y_j = y_j) - \frac{\alpha(y_i)}{n} - \frac{\alpha(y_j)}{n} - \mathrm{E}(\widehat{\theta})\right\},$$

and similarly $\gamma(y_i, y_j, y_k), \ldots, \eta(y_1, \ldots, y_n)$ (see Serfling p. 178 for related quantities) so that

$$\begin{aligned}\widehat{\theta} &= \mathrm{E}(\widehat{\theta}) + \frac{1}{n}\sum_{i=1}^{n}\alpha(Y_i) + \frac{1}{n^2}\sum_{i<j}\beta(Y_i, Y_j) \\ &+ \frac{1}{n^3}\sum_{i<j<k}\gamma(Y_i, Y_j, Y_k), + \cdots + \frac{1}{n^n}\eta Y_1, \ldots, Y_n.\end{aligned}$$

To illustrate, for $\widehat{\theta} = s_n^2$, $\alpha(Y_i) = \{(n-1)/n\}\{(Y_i - \mu)^2 - \sigma^2\}$, and $\beta(Y_i, Y_j) = -2(Y_i - \mu)(Y_j - \mu)$, leading to

$$\begin{aligned}s_n^2 &= \left(\frac{n-1}{n}\right)\sigma^2 + \frac{1}{n}\sum_{i=1}^{n}\left(\frac{n-1}{n}\right)\{(Y_i - \mu)^2 - \sigma^2\} \\ &+ \frac{1}{n^2}\sum_{i<j}\{-2(Y_i - \mu)(Y_j - \mu)\}.\end{aligned}$$

Note that this expansion is different from the usual approximation by averages form

$$s_n^2 = \sigma^2 + \frac{1}{n}\sum_{i=1}^{n}\{(Y_i - \mu)^2 - \sigma^2\} - (\overline{Y} - \mu)^2.$$

In fact, the terms $\alpha(Y_i)$, $\beta(Y_i, Y_j)$, etc., have been carefully constructed to have mean 0 and to be uncorrelated with one another so that

$$\mathrm{Var}(\widehat{\theta}) = \frac{\sigma_\alpha^2}{n} + \binom{n-1}{1}\frac{\sigma_\beta^2}{2n^2} + \binom{n-1}{2}\frac{\sigma_\gamma^2}{3n^5} + \cdots + \frac{\sigma_\eta^2}{n^{2n}}, \tag{10.19}$$

where $\mathrm{Var}\{\alpha(Y_i)\} = \sigma_\alpha^2$, $\mathrm{Var}\{\beta(Y_i, Y_j)\} = \sigma_\beta^2, \cdots$.

Next Efron and Stein (1981) show that

$$\begin{aligned}\widehat{\theta}_{[i]} - \widehat{\theta}_{[j]} &= \frac{1}{n-1}\{\alpha(Y_i) - \alpha(Y_j)\} \\ &+ \frac{1}{(n-1)^2}\sum_{\substack{k=1 \\ k\neq i,j}}^{n}\{\beta(Y_j, Y_k) - \beta(Y_i, Y_k)\} + \cdots,\end{aligned}$$

leading to

$$\mathrm{E}\left(\widehat{\theta}_{[i]} - \widehat{\theta}_{[j]}\right)^2 = \frac{2\sigma_\alpha^2}{(n-1)^2} + \binom{n-2}{1}\frac{2\sigma_\beta^2}{(n-1)^4} + \binom{n-2}{2}\frac{2\sigma_\gamma^2}{(n-1)^6} + \cdots.$$

Now, recall the U-statistic form of s^2_{n-1} in (5.25, p. 251) whereby $\sum_{i=1}^{n}(a_i - \overline{a})^2 = n^{-1}\sum_{i<j}(a_i - a_j)^2$. Thus, summing both sides of the last equation and dividing by n (for the right-hand side just multiplying by $(n-1)/2$), we have

$$E\left\{\sum_{i=1}^{n}(\widehat{\theta}_{[i]} - \overline{\theta}_1)^2\right\} = E\left\{\left(\frac{n}{n-1}\right)\widehat{V}_J\right\} = \frac{\sigma_\alpha^2}{n-1} + \binom{n-2}{1}\frac{\sigma_\beta^2}{(n-1)^3} + \binom{n-2}{2}\frac{\sigma_\gamma^2}{(n-1)^5} + \cdots .$$

Then subtracting $\text{Var}(\widehat{\theta}_{[i]})$ (use equation (10.19) with $n-1$ in place of n) from this last equation yields

Theorem 10.3 (Theorem 1 of Efron and Stein (1981)). *Suppose that* $\widehat{\theta} = \widehat{\theta}(Y_1, \ldots, Y_n)$ *is a symmetric function of the iid observations* $Y_1, \ldots, Y_n$ *and has finite variance. Then*

$$E\left\{\left(\frac{n}{n-1}\right)\widehat{V}_J\right\} - \text{Var}(\widehat{\theta}_{[i]}) = \frac{1}{2}\binom{n-2}{1}\frac{\sigma_\beta^2}{(n-1)^3} + \frac{2}{3}\binom{n-2}{2}\frac{\sigma_\gamma^2}{(n-1)^5} + \cdots . \tag{10.20}$$

This result relates $\widehat{V}_J$ based on a sample of size n times $n/(n-1)$ to the variance of $\widehat{\theta}$ based on a sample of size $n-1$ and shows the form of the difference. In particular, this difference is nonnegative since all the terms on the right-hand side of (10.20) are nonnegative. Now, if

$$\text{Var}(\widehat{\theta}) \le \left(\frac{n-1}{n}\right)\text{Var}(\widehat{\theta}_{[i]}), \tag{10.21}$$

then we can multiply the left-hand side of (10.20) by $(n-1)/n$ and obtain the main result of interest,

$$E(\widehat{V}_J) \ge \text{Var}(\widehat{\theta}). \tag{10.22}$$

Hoeffding (1948) showed that (10.21) holds for U-statistics. Efron and Stein (1981) claim that (10.22) also holds for V-statistics (see Serfling 1980, p. 174) and is asymptotically true for quadratic estimators, that is, estimators with only $\alpha(Y_i)$ and $\beta(Y_i, Y_j)$ terms in the basic expansion. Thus, we have firm results showing that (10.22) holds for certain classes of statistics and empirical evidence from examples and simulations that it often holds more generally.

10.9 Summary

The jackknife is an important tool for nonparametric bias and variance estimation. In practice the variance estimators are more important than the bias estimators. Asymptotically, $\widehat{V}_{\rm J}$ may be viewed as a computational method, without requiring calculus or analytic calculations, to implement the Influence Curve method (which includes the delta method and the sandwich method of M-estimators). We find it especially appealing for summary statistics from Monte Carlo studies as illustrated by Example 10.4 (p. 392). Even for veteran statisticians, it is much simpler to add one or two lines of code to a simulation program than to think through the appropriate delta/influence curve method.

In addition, the Efron and Stein (1981) theory suggests that $\widehat{V}_{\rm J}$ is often biased upwards relative to the true variance, which is generally an appealing property for variance estimates. For example, approximate confidence intervals based on $\widehat{V}_{\rm J}$ tend to have good coverage properties. Thus, $\widehat{V}_{\rm J}$ may be preferred on both ease of implementation as well as on statistical performance. In Chapter 11 we compare $\widehat{V}_{\rm J}$ to the bootstrap and Influence Curve method in some simple examples.

We have found that the best sources for understanding advanced features of the jackknife are Efron's 1982 monograph *The Jackknife, the Bootstrap, and Other Resampling Plans* and Shao and Tu's 1995 book *The Jackknife and Bootstrap*.

10.10 Problems

10.1. Show that the least squares estimator of θ based on $(\widehat{\theta}, 1/n)$ and $(\overline{\theta}_1, 1/(n-1))$ is $\widehat{\theta}_{\rm J}$, and the slope estimator of β_1 is $n(n-1)(\widehat{\theta} - \overline{\theta}_1)$.

10.2. If $\mathrm{E}_F\{\widehat{\theta}\} = \theta + \beta_1/n + \beta_2/n^2 + O(n^{-3})$, then show that $\mathrm{E}_F\{\widehat{\theta}_J\} = \theta - \beta_2/n^2 + O(n^{-2})$ and thus that the bias of $\widehat{\theta}$ has been reduced from $O(n^{-1})$ to $O(n^{-2})$.

10.3. For the estimator $\widehat{\theta} = \overline{Y}^2$ based on an iid sample $Y_1, \ldots, Y_n$, find the exact bias for estimating μ^2 and the bias of $\widehat{\theta}_J$.

10.4. In Example 10.1 (p. 387), the data used in Figure 10.1 (p. 387) is

```
0.67 0.66 -1.33 -0.35 -0.71 -1.08 -1.60  0.00  1.70
4.54
```

with $\widehat{\theta} = \exp(\overline{Y}) = \exp(0.25) = 1.284$ and $\overline{\theta}_1 = 1.306$. Show that $\widehat{\theta}_{\rm J} = 1.09$ by finding the intercept of the line between the points $(1/10, 1.284)$ and $(1/9, 1.306)$. The mean of $m = 10{,}000$ delete-5 estimates is 1.508. Add the point $(1/8, 1.508)$ and find the intercept for fitting all three points by least squares.

10.5. Let $\widehat{\theta} = n^{-1}\sum_{i=1}^{n} I(Y_i \le c)$, where c is a known constant. Give the influence curve estimator of the variance of $\widehat{\theta}$ (just write it down) and derive the jackknife estimator of variance, $\widehat{V}_J$.

10.6. Let $Y_1, \ldots, Y_n$ be iid random variables with mean μ and finite variance. Consider $\widehat{\theta} = g(\overline{Y})$, where g is differentiable and $0 < |g'(\mu)| < \infty$.

a. Use Taylor expansion to determine the influence curve of $\widehat{\theta}$. Then show that the influence curve method of variance estimation is exactly the same as the delta method.
b. Note that $\overline{Y}_{n-1,i} = (n\overline{Y} - Y_i)/(n-1)$. Using the Taylor series approximation $g(\overline{Y}) - g(\overline{Y}_{n-1,i}) \approx g'(\overline{Y})(\overline{Y} - \overline{Y}_{n-1,i})$, show that the jackknife variance estimate is approximately the delta method estimate.

10.7. Use the `sim.samp` R program from Chapter 9 (p. 366) to produce a matrix of 1000 rows of length 20 from the standard exponential distribution (`rexp`). Compute the 20% trimmed mean and its jackknife variance estimate with

```
apply(z,1,mean,trim=.2)->est
apply(z,1,jack.var,theta=mean,trim=.2)->v.est
```

Then compute the ratio of the mean of the variance estimates to the variance of the estimates and give the jackknife Monte Carlo standard error of this ratio. Here is sample code:

```
var.est<-function(est,v.est){
ratio1<-function(index,xdata){
      mean(xdata[index,2])/var(xdata[index,1])
}
x<-data.frame(est,v.est)
nmc<-nrow(x)
r1<-ratio1(1:nmc,x)
r1.var<-jack.var(1:nmc,theta=ratio1,xdata=x)
cat("mean of var est./var of est=",round(r1,3),
  "se=",round(sqrt(r1.var),3),fill=T)
}
var.est(est,v.est)
```

10.8. For the data in Example 10.6, p. 395, show that $\widehat{\theta}_{\mathrm{J}}$ is equal to $\widehat{\theta}_1 = .35$ for the difference in proportions and $\widehat{\theta}_{\mathrm{J}} = 1.575$ for the odds ratio. Try to explain why for the difference in proportions we get the same estimator as the original estimator. Does $\widehat{\theta}_{\mathrm{J}} = 1.575$ for the odds ratio seem unusually small compared to $\widehat{\theta}_2 = 4.375$?

10.9. In a clinical trial of the effect of eyedrops on reducing the incidence of glaucoma, 33 of 204 patients receiving no eyedrops developed glaucoma after the end of 7 years, and 17 of 204 patients taking the eyedrops developed glaucoma. Find $(\widehat{V}_{\mathrm{J}})^{1/2}$ and the delta method standard error for the natural logarithm of relative risk, $\widehat{\theta} = \log(\widehat{p}_1/\widehat{p}_2)$.

10.10. In a case-control study of the effect of taking statin drugs on glaucoma, 42 of 667 glaucoma cases had taken statin drugs for at least 24 months, and 660 of 6667 controls (no glaucoma) had been taking statin drugs. For the odds ratio, calculate the jackknife and delta method standard errors using the independent binomials assumption. (The delta method standard error is the same as in the full multinomial case of Example 10.3.3, p. 394.)

10.11. For $\widehat{\theta} = \overline{Y}_2/\overline{Y}_1$ in the case of two independent samples, show that $\widehat{V}_{\rm J}$ of (10.12, p. 401) is approximately the same as the delta method variance

$$\left(\frac{\overline{Y}_2}{\overline{Y}_1^2}\right)^2 \frac{s_1^2}{n_1} + \left(\frac{1}{\overline{Y}_1}\right)^2 \frac{s_2^2}{n_2}.$$

Hint: use Taylor expansion on the first sample pseudo-values.

10.12. Suppose we have two independent samples and put this in the linear model framework, $Y_i = \boldsymbol{x}_i^T\boldsymbol{\beta} + e_i$, by setting $x_{i1} = 1, i = 1, \ldots n$; $x_{i2} = 0, i = 1, \ldots, n_1$; $x_{i2} = 1, i = n_1 + 1, \ldots, n$; and $n = n_1 + n_2$. Show that for $\widehat{\beta}_2 = \overline{Y}_2 - \overline{Y}_1$

$$\widehat{V}_{\rm HJ} = \frac{n}{n-2}\left\{\left(\frac{n_1 - 1}{n_1^2}\right) s_1^2 + \left(\frac{n_2 - 1}{n_2^2}\right) s_2^2\right\}$$

(taken from Hinkley 1977, p. 287) and $\widehat{V}_{\rm WJ} = s_1^2/n_1 + s_2^2/n_2$.

10.13. In Pyke (1965) we find the following theorem: Let $Y_1, \ldots, Y_n$ be iid with distribution function F and positive density f at $a = F^{-1}(u)$, where $i/n \to u$ as $n \to \infty$. Then $n(Y_{(i)} - Y_{(i-1)})$ converges in distribution to an exponential random variable with mean $1/f(a)$, where $Y_{(1)}, \ldots, Y_{(n)}$ are the ordered sample values. Use that theorem to prove the following result for the jackknife estimate of the variance of the sample median:

$$n\widehat{V}_J \xrightarrow{d} \frac{1}{4f^2(\theta)}\left[\frac{\chi_2^2}{2}\right]^2,$$

where θ is the population median and we assume that $f(\theta) > 0$. For simplicity consider only even sample sizes $n = 2m$, and first verify that

$$\widehat{V}_J = \frac{n-1}{4}\left[Y_{(m+1)} - Y_{(m)}\right]^2.$$

(This result was given in Efron 1982, p. 16.)

10.14. Using (1.23, p. 22), verify that (10.11, p. 399) is correct.

10.15. Consider a randomized complete block with n blocks (rows) and k treatments. The typical data might be modeled by $X_{ij} = \mu + \alpha_i + \beta_j + e_{ij}$, $i = 1, \ldots n$, $j = 1, \ldots, k$, where e_{ij} are independent with mean 0. The estimator of interest

$\widehat{\theta}$ might be a treatment contrast or something more complex. Assume that the α_i are fixed treatment effects, and the block effects β_j may be a) fixed or b) random. Explain how to obtain a jackknife variance estimator for $\widehat{\theta}$ under both scenarios, β_j is fixed or random; in other words, what pieces of data should be dropped out for each case?

10.16. Verify that $\alpha(Y_i)$ and $\beta(Y_i, Y_j)$ defined in Section 10.8.2 have mean 0 and are uncorrelated.

10.17. For $\widehat{\theta} = \overline{Y}^2$, show that

$$\alpha(Y_i) = 2\mu(Y_i - \mu) + \frac{1}{n}\{(Y_i - \mu)^2 - \sigma^2\}$$

with variance $\sigma_\alpha^2 = 4\mu^2\sigma^2 + 4\mu\mu_3/n + (\mu_4 - \sigma^4)/n^2$, and $\beta(Y_i, Y_j) = 2(Y_i - \mu)(Y_i - \mu)$ with variance $\sigma_\beta^2 = 4\sigma^4$. Using the formulas in Section 10.8.2, find $\text{Var}(\overline{Y}^2)$ and $\text{E}(\widehat{V}_\text{J})$ and show directly that $\text{E}(\widehat{V}_\text{J}) \geq \text{Var}(\overline{Y}^2)$.

Chapter 11
Bootstrap

11.1 Introduction

The *bootstrap* is a general technique for estimating unknown quantities associated with statistical models. Often the bootstrap is used to find

1. standard errors for estimators,
2. confidence intervals for unknown parameters,
3. p-values for test statistics under a null hypothesis.

Thus the bootstrap is typically used to estimate quantities associated with the sampling distribution of estimators and test statistics.

There are several ways to view the bootstrap. The first is called the "plug-in" description. Basically we can think of what the bootstrap estimates as a functional (function of a function) of the underlying distribution function F; call the functional $Q(F)$. In a sense to be made clearer in Section 11.2.1 (p. 419), we substitute the empirical distribution function F_n in place of F to get the bootstrap estimator $Q(F_n)$.

The popularity of the bootstrap stems more from the second description of the bootstrap based on visualizing a "bootstrap world" where the data analyst knows everything. In this parallel world the true sampling design of the data is reproduced as closely as possible, and unknown aspects of the statistical model are replaced by sample estimates. In the real world we have one sample from the population of interest. In the bootstrap world, we can draw as many bootstrap samples (also called *resamples*) from the bootstrap-world population as we want. Thus, in this world, the data analyst can obtain any quantity of interest by simulation. For example, if the variance of a complicated parameter estimate in this world is desired, one just computer generates B replicate samples, computes the estimate for each sample, and then uses the sample variance of the B estimates as an approximation to the variance. As B grows large, this sample variance converges to the true variance in the bootstrap world. Of course in terms of the estimator based on the original data (the real world), this limiting sample variance is just an estimator of the true

D.D. Boos and L.A. Stefanski, *Essential Statistical Inference: Theory and Methods*, Springer Texts in Statistics, DOI 10.1007/978-1-4614-4818-1_11,

Table 11.1 Yearly Maximum Flow Rates (gallons per second) at a Gauging Station in North Carolina

5550	4380	2370	3220	8050	4560	2100	6840	5640	3500	1940	7060
7500	5370	13100	4920	6500	4790	6050	4560	3210	6450	5870	2900
5490	3490	9030	3100	4600	3410	3690	6420	10300	7240	9130	

variance of the estimate. Thus we create a bootstrap world where anything can be computed, at least up to Monte Carlo error. Those true quantities calculated in the bootstrap world are estimates of the parallel quantities in the real world. In effect this "bootstrap world" simulation approach opened up complicated statistical methods to anybody with a computer and a random number generator.

Efron and Tibshirani (1993) and Davison and Hinkley (1997) are standard texts on the bootstrap. See also Boos (2003), from which portions of this chapter are borrowed.

11.1.1 Sample Bootstrap Applications

Before getting to the details of the bootstrap in Sections 11.2–11.7, we now give several real consulting examples illustrating where the bootstrap might be used.

Example 11.1. Recall in Chapter 1 the consulting example in Section 1.4 (p. 8) where a graduate student in civil engineering wanted to model the relationship between watershed area and the maximum flow over a 100 year period at gauging stations on rivers in North Carolina. He had a model $R = kA^{\eta-1}$ relating the 100 year maximum flow rate at a station (R) to the watershed area (A) at the station; k and η are unknown parameters. Taking logarithms leads to a simple linear model. He had values of A for 140 stations, but the R measurement for each station was the maximum flow during the time the station had been keeping records. These lengths of time varied between 6 and 83 years; so they really were not comparable and also were not appropriate for the maximum over 100 years.

It was discovered that the student could get yearly maximums for a number of stations. The data in Table 11.1 are $n = 35$ yearly maximum flow rates at one particular station (also found in Problem 2.2, p. 107). Assuming year-to-year independence for the yearly maximums, the distribution function of the maximum of 100 yearly maximums is $P(R_{(100)} \le t) = [F(t)]^{100}$, where $F(t)$ is the distribution of a single yearly maximum. Thus we proposed that we estimate the median, say t_0, of the distribution function $[F(t)]^{100}$ to be used as the response variable in the regression model. Setting $[F(t_0)]^{100} = 1/2$ implies that $F(t_0) = (1/2)^{.01} = .993$. Thus, t_0 is actually the .993 quantile of the yearly maximum distribution. Because the sample sizes were too small to estimate this quantile nonparametrically, we suggested a parametric model for the yearly maximum. So, for the data of Table 11.1, we assumed a location-scale, extreme value, model (see 1.9, p. 9).

This assumption was supported with a QQ plot, and then the data were fit by maximum likelihood obtaining $\widehat{\mu} = 4395.1$ and $\widehat{\sigma} = 1882.5$. The estimate of the .993 quantile is then

$$\widehat{\sigma}\{-\log[-\log(.993)]\} + \widehat{\mu} = 13729.2.$$

Using the inverse of the estimated Fisher information and the delta method applied to the above function, we obtained a standard error of 1375.3. The idea would be for the student to do this estimation at a number of stations and possibly use the standard errors as weights in the regression fit.

The classical tools used here are quite adequate: QQ plot, maximum likelihood, and delta method. But let us see what the bootstrap can do. We may first confirm the extreme value assumption by generating $B = 100$ data sets of size $n = 35$ from the fitted distribution and computing the Anderson-Darling goodness-of-fit statistic (AD) for each sample:

$$\text{AD} = n \int_{-\infty}^{\infty} \left[F(y;\widehat{\boldsymbol{\theta}}) - F_n(y)\right]^2 \left[F(y;\widehat{\boldsymbol{\theta}})(1 - F(y;\widehat{\boldsymbol{\theta}}))\right]^{-1} f(y;\widehat{\boldsymbol{\theta}})\,dy, \quad (11.1)$$

where $F_n(y)$ is the empirical distribution function and $\widehat{\boldsymbol{\theta}} = (\widehat{\mu}, \widehat{\sigma})^T$ is the maximum likelihood estimator. It is helpful to know that there is a convenient computing formula for AD:

$$\text{AD} = -n - \frac{1}{n}\sum_{i=1}^{n}(2i - 1)\left[\log\left\{F(Y_{(i)};\widehat{\boldsymbol{\theta}})\right\} + \log\left\{1 - F(Y_{(n+1-i)};\widehat{\boldsymbol{\theta}})\right\}\right],$$

where $Y_{(1)} \leq \cdots \leq Y_{(n)}$ are the sample ordered values. Since 95 of these bootstrap AD values were larger than the value AD = .178 for the data in Table 11.1, the parametric bootstrap p-value is .95. (The term *parametric bootstrap* is used here because the bootstrap samples are generated from a fitted parametric family instead of sampled directly from the data.) If the p-value had been fairly small we would have taken B to be much larger. The null distribution of AD for this situation has been tabled (Stephens, 1977), but the bootstrap has made tabling such distributions unnecessary (except for use in computer packages). In fact the bootstrap distribution here is exact up to Monte Carlo error because the null distribution of AD does not depend on the values of the parameter. This is true for any location-scale family. We also kept track of the bootstrap parameter estimates and the estimated .993 quantile for each bootstrap resample. The mean of the quantile estimates was 13572.2, illustrating some negative bias of the estimate since 13572.2 is smaller than 13729.2, the true value in the bootstrap world. The standard deviation of the quantile estimates was 1386.0, a parametric bootstrap standard error quite close to the information-matrix delta value of 1375.3. Finally, we made a histogram and QQ plot of the 100 bootstrap .993 quantile estimates and observed that they are approximately normally distributed (suggesting that the .993 quantile estimate

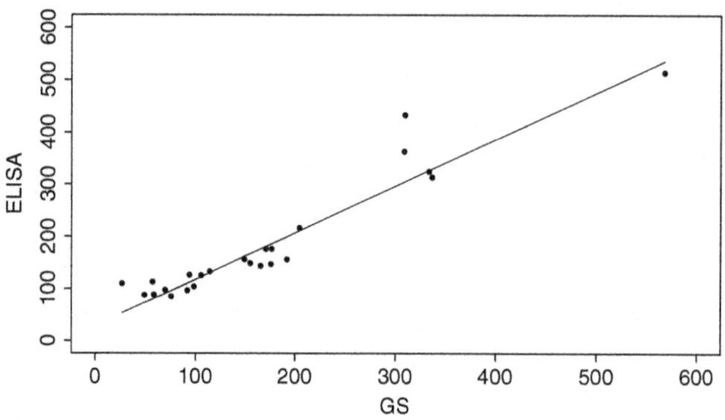

Fig. 11.1 Two methods for detecting PCB's overlaid with the least squares line

Table 11.2 GS and ELISA values for 24 Samples

GS	76	50	59	92	70	99	28	58	106	94	115	166
ELISA	81	83	84	92	93	100	106	109	121	122	129	140
GS	176	156	150	192	171	177	205	337	334	309	310	568
ELISA	143	145	152	152	172	172	212	309	320	358	429	510

from different stations is a statistically well-behaved response variable in the linear regression). These bootstrap analyses are modest additions to the original analysis, but they do add some insight. For some people, just avoiding the Fisher information and delta theorem calculus would be attractive. ♦

Example 11.2 (Detecting PCB's). A local Raleigh company came to us several years ago with data on two methods of detecting polychlorinated biphenyls (PCB's). They were developing a solid phase fluoroimmunoassay method that we call ELISA, and the other method we refer to as GS for gold standard because its results were accepted as truth. Figure 11.1 displays the data, $n = 24$ pairs of (GS,ELISA) values given in Table 11.2. The company was interested in detecting PCB levels of 200 ppb (parts per billion) or more using an "action limit" of 100 ppb for their ELISA method. That is, they planned to declare that PCB was present whenever ELISA was ≥ 100.

After some discussion, we determined that they wanted the probability of a false negative when GS = 200, $P(\text{ELISA} < 100|\text{GS} = 200)$. In screening test terminology, this probability would be one minus the *sensitivity* of the test. They also were interested in the probability of a false positive (one minus the *specificity* of the test) for different levels of GS. Since all of these calculations are similar, we focus here on the sensitivity.

To make the problem simple, we decided to model the ELISA results as a linear function of GS with normal homogeneous errors,

$$\text{ELISA} = \alpha + \beta(\text{GS}) + \sigma Z, \tag{11.2}$$

where Z is a standard normal random variable. Then

$$\begin{aligned} P(\text{ELISA} < 100|\text{GS} = 200) &= P(\alpha + \beta(\text{GS}) + Z < 100|\text{GS} = 200) \\ &= P\left(Z < \frac{100 - \alpha - \beta(200)}{\sigma}\right) \\ &= \Phi\left(\frac{100 - \alpha - \beta(200)}{\sigma}\right), \end{aligned} \tag{11.3}$$

where Φ is the standard normal distribution function.

Since (α, β, σ) were unknown, we used the least squares estimates from a fit to the data in Figure 11.1 to substitute in (11.3). The estimate of (11.3) was .002. But how should the variation due to the parameter estimates be accounted for? One could use the joint asymptotic normal distribution of the least squares estimates and the delta method to get a standard error for the estimate and an approximate confidence interval. But it was much simpler to just generate 10,000 samples from the model (11.2) with the least squares estimates in place of (α, β, σ). As in the previous example, this is a parametric bootstrap. The bootstrap standard error was .005 and the upper 95% probability bound was .013 using the percentile method (see Section 9.4). We worried a bit about the normality assumption, and decided to try a t distribution with five degrees of freedom as an alternative to the normal distribution. The probability estimate was then .007 with bootstrap standard error .009 and upper 95% probability bound of .023. So there is clearly some sensitivity to the normal assumption. (Actually, it appears that a Weibull distribution might be more appropriate for the errors, but we did not pursue distributional alternatives further because of time and cost.) One could also worry about variance heterogeneity and other model inadequacies or use a better bootstrap confidence interval method. ♦

11.2 Bootstrap Standard Errors

The influence curve and jackknife variance estimation methods are focused on estimating the asymptotic variance of an estimator. In contrast, the bootstrap method attempts to directly estimate the variance of an estimator. The basic idea is to

1. write the variance of the estimator in terms of the unknown distribution function F of the data: $\text{Var}_F\{\widehat{\theta}\}$ (see equation 11.4, p. 420);
2. and then conceptually substitute an estimate $\widehat{F}$ for F into the variance expression, resulting in $\text{Var}_{\widehat{F}}\{\widehat{\theta}\}$.

In this respect the bootstrap estimator is called a "plug-in" estimator as mentioned in the introduction to this chapter because $\widehat{F}$ is plugged into the variance expression wherever needed. In fact the influence curve method is really a "plug-in" method for the asymptotic variance.

An alternative view of the bootstrap method is based on sampling from a fictional pseudo-population that we called the "bootstrap world." As mentioned in the chapter introduction, it is this view that makes the bootstrap easy to understand and implement. The basic idea is to

1. First create a pseudo-population from the sample values. In a simple situation, we just consider the set of sample values $\{Y_1, \ldots, Y_n\}$ to be a population. (Usually, we are not thinking of a finite population situation; so if necessary, think of this population as consisting of an infinite number of Y_1 values, an infinite number of Y_2 values, ..., an infinite number of Y_n values, each one occurring $1/n$ proportion of the time.)
2. Then one conceives of drawing a random sample (often called a *resample*) from this pseudo-population, mimicking the true sampling process as closely as possible. This is random sampling in the bootstrap world. For the simple population above and iid sampling, this resampling means drawing samples **with replacement** from $\{Y_1, \ldots, Y_n\}$.
3. Because we know everything about the pseudo-population, we can theoretically calculate the variance of our estimator when the sample is drawn from the pseudo-population. This variance is the bootstrap variance estimator.

In practice the calculation of the last step is too difficult except for very simple cases, and we approximate the variance calculation by repeatedly sampling from the pseudo-population, computing the estimator for each bootstrap resample, and then computing a sample variance of these estimators. In other words, we estimate the theoretical variance in the bootstrap world by Monte Carlo methods. Remember that the theoretical variance of our estimator in the bootstrap world is actually an estimate of the variance when viewed from the real world. Thus, a Monte Carlo approach gives an estimate of the bootstrap estimate of variance.

Example 11.3 (Food consumption of female rats). Data from an experiment on food consumption of female rats treated with zinc-calcium EDTA are taken from Brownie and Brownie (1986). The ordered sample values are

```
> sort(rats)
 [1] 5.35 5.37  5.53  5.95  6.20  7.12 7.22
 [8] 7.62 7.63  7.63  7.67  7.97  8.43 8.68
[15] 9.20 9.63 11.32 11.52 15.27 15.90
```

The 10th and 11th ordered values are 7.63 and 7.67, respectively, and thus the sample median is 7.65. Now let us consider this sample as the pseudo-population and draw $B = 10$ random samples from it. To do this, we merely draw values independently with replacement from the set of sample values. For illustration, the *subscripts* of the first sorted resample are

```
> set.seed(200)
> sort(sample(20,replace=T))
 [1]  3  4  4  7  8  8  9  9 10 11
[11] 11 11 11 12 14 14 15 15 16 19
```

Notice a number of repeats. Now the actual resample values are

```
> set.seed(200)
> sort(rats[sample(20,replace=T)])
 [1] 5.53 5.95 5.95  7.22 7.62 7.62 7.63 7.63
 [9] 7.63 7.67 7.67  7.67 7.67 7.97 8.68 8.68
[17] 9.20 9.20 9.63 15.27
```

Here, the third value 5.53 appears once, the fourth value 5.95 appears twice, etc. The median of this resample is $\widehat{\theta}_1^* = 7.67$. Repeating this nine more times gives an iid set of $B = 10$ sample medians (iid in the bootstrap world):

```
7.67  8.68  7.80  7.62  8.325  8.43  7.67  8.68  8.03
7.63
```

The sample standard deviation of these ten values is 0.44. This is a Monte Carlo (random sampling) approximation to the true bootstrap standard deviation estimate which itself is a theoretical calculation based on the above pseudo-population (see Problem 11.1). To show how variable our approximation is, three more sets of $B = 10$ bootstrap samples led to estimated bootstrap standard errors of 0.37, .07, and 0.40. A typical recommendation is to use at least $B = 100$ bootstrap samples. Repeating $B = 100$ four times gave estimated bootstrap standard errors of 0.43, 0.42, 0.53, and 0.32. Repeating four times at $B = 1000$ gave 0.42, 0.53, 0.49, and 0.44. Keep in mind of course that the true bootstrap standard error (actually 0.46 if $B \to \infty$) has quite a large sampling variance itself, much larger than the approximation error due to using $B = 100$ instead of $B = \infty$ resamples. Thus, if we obtained a new sample of 20 rat results, the bootstrap standard error would likely be quite different from 0.46. ♦

In Efron and Tibshirani (1993, Ch. 19), details are given about the decomposition of variability due to the original sample and to the bootstrap resampling. The basic result is that the variance of the estimated bootstrap standard error is approximately $c_1/n^2 + c_2/nB$, where c_1 and c_2 depend on the underlying population and type of estimator. The sampling error due to resampling B times can be made arbitrarily small by making B large, but it rarely is worth the effort to go beyond $B = 100$ or $B = 1000$.

11.2.1 Plug-In Interpretation of Bootstrap Variance Estimates

Now we explain the plug-in approach that gives a more theoretical understanding of the bootstrap method. For simplicity we start with an iid sample $Y_1, \ldots, Y_n$ with each Y_i having distribution function F, and a real parameter θ is estimated by $\widehat{\theta}$. When necessary, we think of $\widehat{\theta}$ as a function of the sample, $\widehat{\theta}(Y_1, \ldots, Y_n)$. The variance of $\widehat{\theta}$ is then

$$\mathrm{Var}_F(\widehat{\theta}) = \int \left\{\widehat{\theta}(y_1,\ldots,y_n) - \mathrm{E}_F(\widehat{\theta})\right\}^2 dF(y_1)\cdots dF(y_n), \tag{11.4}$$

where

$$\mathrm{E}_F(\widehat{\theta}) = \int \widehat{\theta}(y_1,\ldots,y_n)\, dF(y_1)\cdots dF(y_n).$$

The nonparametric bootstrap estimate of $\mathrm{Var}(\widehat{\theta})$ is just to replace F by the empirical distribution function $F_n(y) = n^{-1}\sum_{i=1}^n I(Y_i \le y)$:

$$\mathrm{Var}_{F_n}(\widehat{\theta}) = \int \left\{\widehat{\theta}(y_1,\ldots,y_n) - \mathrm{E}_{F_n}(\widehat{\theta})\right\}^2 dF_n(y_1)\cdots dF_n(y_n).$$

Now the general expression for these expectations is ugly. For example,

$$\mathrm{E}_{F_n}(\widehat{\theta}) = \int \widehat{\theta}(y_1,\ldots,y_n)\, dF_n(y_1)\cdots dF_n(y_n) = \frac{1}{n^n}\sum_{i_1=1}^{n}\cdots\sum_{i_n=1}^{n}\widehat{\theta}(Y_{i_1},\ldots,Y_{i_n}).$$

Recall that integrating a function $g(y)$ in one dimension with respect to dF is $\int g(y)dF(y) = \int g(y)f(y)dy$ for continuous Y_1 where f is the derivative of F, and $\int g(y)dF(y) = \sum g(y_i)P(Y_1 = y_i)$ for discrete Y_1 with possible values $y_1, y_2, \ldots$ Thus, replacing $F(y)$ by the empirical distribution function $F_n(y) = n^{-1}\sum_{i=1}^n I(Y_i \le y)$ gives $\int g(y)dF_n(y) = n^{-1}\sum_{i=1}^n g(Y_i)$ because the empirical distribution function corresponds to a discrete random variable with probability $1/n$ at the observed data values. Of course an n-fold integral leads to an n-fold sum.

For very simple estimators, we can make the above calculations exactly. For example, if $\widehat{\theta} = \overline{Y}$, we know that $\mathrm{E}_F(\overline{Y}) = \mathrm{E}_F(Y_1)$ and therefore

$$\mathrm{E}_{F_n}(\overline{Y}) = \mathrm{E}_{F_n}(Y_1) = \int y\, dF_n(y) = \overline{Y}.$$

Similarly, we know that $\mathrm{Var}_F(\overline{Y}) = \mathrm{Var}_F(Y_1)/n = \left[\mathrm{E}_F(Y_1^2) - \{\mathrm{E}_F(Y_1)\}^2\right]/n$ and then

$$\mathrm{Var}_{F_n}(\overline{Y}) = \frac{\mathrm{E}_{F_n}(Y_1^2) - \{\mathrm{E}_{F_n}(Y_1)\}^2}{n} = \frac{1}{n}\left(\frac{1}{n}\sum_{i=1}^n Y_i^2 - \overline{Y}^2\right) = \frac{s_n^2}{n}.$$

Thus s_n^2/n is the nonparametric bootstrap estimator of the variance of the sample mean.

Suppose that we make the parametric assumption that $F(y) = F(y;\sigma) = 1 - \exp(-y/\sigma)$, an exponential distribution with mean σ. Then the maximum likelihood estimator of F is $\widehat{F} = F(y;\overline{Y}) = 1 - \exp(-y/\overline{Y})$, and the parametric bootstrap estimate of the variance of the sample mean is just $\mathrm{Var}_{\widehat{F}}\overline{Y} = \mathrm{Var}_{\widehat{F}}Y_1/n = \overline{Y}^2/n$ since the square of the mean is the variance for an exponential distribution. The parametric bootstrap is not really in the spirit of Chapters 7 (M-estimation) and 10 (jackknife) that emphasized standard errors obtained under minimal assumptions. However, it is an important technique in cases where parametric assumptions are reasonable, but analytic variance calculations are difficult. When using maximum likelihood estimation, the parametric bootstrap typically yields the maximum likelihood estimator of the quantity being estimated. Note that for nonlinear estimators, the parametric bootstrap estimate of variance is different from those obtained from information matrices which are in fact asymptotic variance estimators.

It might be good at this time to note more clearly that the influence curve method of variance estimation is really a bootstrap method for estimating asymptotic variance. Recall that given an estimator $\widehat{\boldsymbol{\theta}}$ with influence curve $IC(y, \boldsymbol{\theta}_0)$, the asymptotic variance of $\widehat{\boldsymbol{\theta}}$ is just

$$\Sigma_F/n = \mathrm{E}_F \{IC(Y_1, \boldsymbol{\theta}_0) IC(Y_1, \boldsymbol{\theta}_0)\}^T /n.$$

Suppose that we write $\boldsymbol{\theta}_0$ as $\boldsymbol{\theta}(F)$ and assume that $\widehat{\boldsymbol{\theta}}$ may be thought of as a function $\boldsymbol{\theta}(F_n)$ of the empirical distribution function F_n. Then for a real-valued estimator, we may write Σ_F as

$$\Sigma_F = \mathrm{E}_F [IC\{Y_1, \theta(F)\}]^2 = \int [IC\{y, \theta(F)\}]^2 \, dF(y). \tag{11.5}$$

Replacing F by F_n (which also turns IC into $\widehat{IC}$) in (11.5) yields

$$\int \left[\widehat{IC}\{y, \theta(F_n)\}\right]^2 dF_n(y) = \frac{1}{n}\sum_{i=1}^{n} \left[\widehat{IC}\{Y_i, \theta(F_n)\}\right]^2 = \frac{1}{n}\sum_{i=1}^{n} \left\{\widehat{IC}(Y_i, \widehat{\theta})\right\}^2, \tag{11.6}$$

the expression that was given previously in vector form in Chapter 10 (10.8, p. 389). Thus we may think of the Influence Curve variance estimator as a bootstrap plug-in estimator of asymptotic variance.

11.3 Comparison of Variance Estimation Methods

The three main nonparametric methods of variance estimation are 1) the influence curve method (direct estimation of asymptotic variance), 2) the jackknife method, and the bootstrap method. As mentioned previously, Efron and Stein (1981) showed theoretically that the jackknife variance estimate is typically biased upwards, that is, usually $\mathrm{E}(\widehat{V}_J) \geq \mathrm{Var}(\widehat{\theta})$. There are no similar general results for the influence curve method or the bootstrap. Moreover, first-order asymptotic analysis of these three variance estimation methods has not provided generally useful insights for recommending one method over another, except to identify those cases where the jackknife does not work, such as the median. Thus, Monte Carlo studies are used for insight.

Efron (1982) reports on a number of simulation comparisons. His Tables 3.1–3.3 suggest:

1. that the influence curve variance estimate can be badly biased downwards in some cases;
2. that the jackknife variance estimate tends to be more variable than either the bootstrap or the influence curve estimates.

Here, for an iid sample $Y_1, \ldots, Y_n$, we give simulation results for variance estimators of (i) the usual sample variance, $\widehat{\theta} = s_n^2$, (ii) the 20% trimmed mean, and (iii) the L-moment coefficient of skewness.

Recall that the asymptotic variance of $\text{Var}(s_n^2)$ is $(\mu_4 - \sigma^4)/n$, leading to the influence curve estimator $\widehat{V}_{\text{IC}} = (m_4 - s_n^4)/n$. The jackknife variance estimator in this case was derived in Chapter 10 to be $[n/(n-1)]^3(m_4 - s_n^4)/n$. The exact variance is

$$\text{Var}(s_n^2) = \left(\frac{n-1}{n}\right)^2 \left(\frac{2\sigma^4}{n-1} + \frac{\mu_4 - 3\sigma^4}{n}\right),$$

leading to the bootstrap variance estimator

$$\widehat{V}_{\text{Boot}} = \left(\frac{n-1}{n}\right)^2 \left(\frac{2s_n^4}{n-1} + \frac{m_4 - 3s_n^4}{n}\right).$$

Table 11.3 displays the variance estimator results for $n = 10$. For each distribution, there are two columns labeled $\widehat{V}/V$ and CV, respectively. The column labeled $\widehat{V}/V$ refers to the average of the 10,000 variance estimates for a method divided by the true variance V. (In Tables 11.4 and 11.5, V is estimated by the sample variance of the 1,000 $\widehat{\theta}$ values.) The CV column is the sample coefficient of variation of the 10,000 variance estimates for a method, that is, the Monte Carlo sample standard deviation of the 10,000 variance estimates divided by their sample mean. Although all three variance estimators have a similar form for this simple situation in Table 11.3 with $\widehat{\theta} = s_n^2$, the constants make quite a difference in small samples, and the influence curve and bootstrap variance estimators are too small on average, seriously too small for the normal and exponential cases. In contrast, the jackknife variance estimator is too large for the uniform, but fairly unbiased for the normal and exponential. Moving to $n = 30$ (not displayed), all three estimators improve considerably: 10,000 reps yielded $(\widehat{V}_{\text{IC}}/V, \widehat{V}_{\text{J}}/V, \widehat{V}_{\text{Boot}}/V) = (0.98, 1.08, 0.99)$ for the uniform, (0.93, 1.03, 0.90) for the normal, and (0.91, 1.01, 0.87) for the exponential. So the same basic pattern emerges that the influence curve and bootstrap variance estimators tend to be a bit small for the normal and exponential, whereas the jackknife estimator is a bit too large for the uniform and generally the best of the methods.

Table 11.3 Monte Carlo average of estimators of $\mathrm{Var}(s_n^2)$ for $n = 10$ divided by true variance V ($\widehat{V}/V$). CV is the estimated coefficient of variation for the variance estimator. Based on 10,000 Monte Carlo replications

	Uniform		Normal		Exponential	
Distribution	$\widehat{V}/V$	CV	$\widehat{V}/V$	CV	$\widehat{V}/V$	CV
Influence Curve	0.91	0.55	0.80	1.20	0.70	4.1
Jackknife	1.24	0.55	1.09	1.20	0.96	4.1
Bootstrap	0.93	0.51	0.74	1.15	0.60	4.0
Average s.e.'s	0.01	0.01	0.01	0.02	0.03	0.3

Table 11.4 Monte Carlo average of estimators of the variance of the 20% trimmed mean for $n = 20$ divided by true variance V ($\widehat{V}/V$). CV is the estimated coefficient of variation for the variance estimator. Based on 1,000 Monte Carlo replications

	Uniform		Normal		Exponential	
Distribution	$\widehat{V}/V$	CV	$\widehat{V}/V$	CV	$\widehat{V}/V$	CV
Influence Curve	0.96	0.41	0.95	0.47	1.02	0.69
Jackknife	0.87	0.34	0.93	0.42	1.04	0.70
Bootstrap ($B = 100$)	0.91	0.38	0.99	0.43	1.12	0.67
Average s.e.'s	0.04	0.01	0.04	0.01	0.05	0.04

Table 11.5 Monte Carlo average of estimators of the variance of the L-moment skewness estimator for $n = 20$ divided by true variance V ($\widehat{V}/V$). CV is the estimated coefficient of variation for the variance estimator. Based on 1,000 Monte Carlo replications

	Uniform		Normal		Exponential	
Distribution	$\widehat{V}/V$	CV	$\widehat{V}/V$	CV	$\widehat{V}/V$	CV
Jackknife	1.12	0.39	1.31	0.68	1.32	0.97
Bootstrap ($B = 100$)	1.17	0.34	1.07	0.37	1.03	0.44
Average s.e.'s	0.05	0.01	0.06	0.04	0.06	0.03

Note that we have reported average values of estimated variances. Efron (1982) reports average values of the estimated square root of variances, that is, averages of standard deviation estimators. Averages of standard deviation estimators naturally appear less biased than averages of variance estimators.

Perhaps we have been a bit unfair to the influence curve method in Table 11.3 because we used n^{-1} in the definition of $\widehat{V}_{\mathrm{IC}}$ instead of $(n-1)^{-1}$ as in the jackknife. To be more fair, we now present results for the 20% trimmed mean where the influence curve variance estimator has been tuned a bit (see `trim20.var` in Section 9.4.2, p. 373). In Table 11.4, we use the same distributions as in the other tables but change to $n = 20$ and 1,000 replications because we use algorithm versions of the jackknife and bootstrap ($B = 100$). Here we see that all three variance estimators perform fairly similarly.

Finally, in Table 11.5 we compare the jackknife and the bootstrap variance estimators for the L-moment coefficient of skewness estimator that is a ratio similar to $\widehat{\text{Skew}}$ but more robust (see Hosking, 1990). Here we see a clear superiority for the bootstrap over the jackkknife in terms of both bias ($\widehat{V}/V$) and variability (CV). In fact, the coefficient of variation results for the bootstrap can even be reduced some by taking $B = 1000$ instead of $B = 100$.

We have studied variance estimators for the simple nonlinear estimator s_n^2 and for the 20% trimmed mean and L-moment skewness estimator. In the first example, Table 11.3 (p. 423), the jackknife was the best performer. In the second example, Table 11.4 (p. 423), no clear winner emerges. The bootstrap is a clear winner in Table 11.5, and we note that several of the examples in Efron (1982) where the bootstrap is preferred are for estimators involving ratios. We did not include the influence curve method in Table 11.5 because of the additional work involved in deriving it. In general, the jackknife and bootstrap variance estimators are much simpler to use because they require no analytic derivation. We should also point out that when sample sizes are large, such as when analyzing Monte Carlo simulation output, all three methods give similar answers, and the easiest method to implement should be chosen, usually the jackknife.

We tend to agree with conclusions given by Shao and Tu (1995, end of Ch. 3) that for obtaining standard errors, the jackknife is often the preferred method because of its simplicity and the fact that it is rarely too small on average. The bootstrap may be reserved for use in more complicated settings or for problems like confidence intervals or hypothesis tests.

11.4 Bootstrap Asymptotics

Singh (1981) and Bickel and Freedman (1981) initiated study of the asymptotic properties of bootstrap procedures. The basic result is that under fairly weak conditions the bootstrap estimate of the standardized distribution of an estimator converges almost surely to the same asymptotic distribution as for the estimator. The most usual case is when the estimator or statistic is asymptotically normal, and then a typical proof method is to use approximation by averages. To show that the bootstrap estimator of variance is consistent requires more conditions. See Shao and Tu (1995, Ch. 3) for details.

To illustrate the almost sure convergence in distribution of the bootstrap distribution estimate, we state Theorem 3.1 of Bickel and Freedman (1981). Their result is about the V-statistic of degree $m = 2$

$$V_n = \frac{1}{n^2}\sum_{i=1}^{n}\sum_{j=1}^{n} v(Y_i, Y_j),$$

where ν is a symmetric kernel with $\nu(x, y) = \nu(y, x)$. Previously, we discussed in Chapter 5 the related U-statistic

$$U_n = \frac{2}{n(n-1)} \sum_{i=1}^{n} \sum_{j=1}^{n} \nu(Y_i, Y_j).$$

An example is $\nu(x, y) = (x - y)^2/2$ for which $V_n = s_n^2$ and $U_n = s_{n-1}^2$. The asymptotic variance times n for both V_n and U_n is

$$\sigma_\nu^2 = 4 \left[\int \left\{ \int \nu(x, y)\, dF(y) \right\}^2 dF(y) - \theta_\nu^2 \right],$$

where $\theta_\nu = \mathrm{E}\{\nu(Y_1, Y_2)\}$ is their limit in probability. In the following result we conceive of bootstrap samples $Y_1^*, \ldots, Y_n^*$ that are iid with distribution function equal to F_n, the empirical distribution function of the given sample $Y_1, \ldots, Y_n$, and V_n^* is the statistic based on a bootstrap sample. In this bootstrap world, the true parameter is $\theta_{\nu,n} = \mathrm{E}_{F_n}\{\nu(Y_1^*, Y_2^*)\} = V_n$.

Theorem 11.1 (Bickel and Freedman 1981, Theorem 3.1). *Suppose that $Y_1, \ldots, Y_n$ are iid with distribution function F, $E\{\nu(Y_1, Y_2)\}^2 < \infty$, $E\{\nu(Y_1, Y_1)\}^2 < \infty$, and $\sigma_\nu^2 > 0$. Then for almost all $Y_1, Y_2, \ldots$, given $(Y_1, \ldots, Y_n)$, as $n \to \infty$*

$$\sqrt{n}(V_n^* - \theta_{\nu,n}) \quad \textit{converges weakly to} \quad N(0, \sigma_\nu^2).$$

Remembering that convergence in distribution or *weak convergence* is about convergence of distribution functions, this result means that

$$P^*\left\{\sqrt{n}(V_n^* - \theta_{\nu,n}) \le x\right\} \stackrel{wp1}{\longrightarrow} \Phi(x/\sigma_A) \qquad \text{as } n \to \infty,$$

where we write the above probability as P^* to emphasize that the probability is conditional on the given sample, or a probability in the bootstrap world. Thus, first order bootstrap asymptotics involves verifying that the statistic in the bootstrap world (in the above case V_n^*) has the same asymptotic normality as the parallel statistic V_n in the real world. The proof usually involves approximation by averages and then a Central Limit Theorem that takes into account the fact that the underlying true distribution F_n in the bootstrap world is changing with n.

These first-order asymptotic results such as Theorem 11.1 suggest that since the bootstrap distribution converges, then related quantities used in inference like bootstrap distribution quantiles or variances also converge. But often, such results are not sufficient for analyzing variations in bootstrap methods or comparison with other types of methods. Thus, Edgeworth expansions of bootstrap distribution functions such as $P^*\{\sqrt{n}(V_n^* - \theta_{\nu,n}) \le x\}$ are used for such analyses. To illustrate, suppose that $\widehat{\theta}$ is AN$(\theta, \sigma/n)$. A one-term Edgeworth expansion for $\widehat{\theta}$ would give

$$P\{\sqrt{n}(\widehat{\theta}-\theta) \le x\} = \Phi(x/\sigma) + \frac{c}{\sqrt{n}} + o(n^{-1/2}), \tag{11.7}$$

as $n \to \infty$ for each x, where Φ is the standard normal distribution function. Analogously, in the bootstrap world, we should have the related result

$$P^*\{\sqrt{n}(\widehat{\theta}^* - \widehat{\theta}) \le x\} = \Phi(x/\sigma_n^*) + \frac{c_n}{\sqrt{n}} + o(n^{-1/2}), \tag{11.8}$$

almost surely as $n \to \infty$ for each x, where $\sigma_n^* \xrightarrow{p} \sigma$ and $c_n \xrightarrow{p} c$. Thus, subtracting (11.8) from (11.7), we have

$$\begin{aligned} P\{\sqrt{n}(\widehat{\theta}-\theta) \le x\} - P^*\{\sqrt{n}(\widehat{\theta}^* - \widehat{\theta}) \le x\} &= \Phi(x/\sigma) - \Phi(x/\sigma_n^*) + o_p(n^{-1/2}) \\ &= O_p(n^{-1/2}), \end{aligned}$$

where in this last step we have used Taylor expansion of Φ and assumed that $\sigma_n^* - \sigma = O_p(n^{-1/2})$. Thus the bootstrap distribution of $\sqrt{n}(\widehat{\theta}^* - \widehat{\theta})$ is within $O_p(n^{-1/2})$ of the distribution of $\sqrt{n}(\widehat{\theta}-\theta)$.

For comparison, let us now consider the analogous expansions for the t-like quantities $t = (\widehat{\theta}-\theta)/\widehat{\sigma}$ and $t^* = (\widehat{\theta}^* - \widehat{\theta})/\widehat{\sigma}^*$, where $\widehat{\sigma}$ is an estimator of σ and we assume convergence to the standard normal:

$$P(t \le x) = \Phi(x) + \frac{d}{\sqrt{n}} + o(n^{-1/2})$$

$$P^*(t^* \le x) = \Phi(x) + \frac{d_n}{\sqrt{n}} + o_p(n^{-1/2}).$$

Now, assuming that $d_n - d = O_p(n^{-1/2})$, we have $P(t \le x) - P(t^* \le x) = O_p(n^{-1})$, a faster convergence rate than for the unstudentized quantities. These type expansions have provided an important way to compare confidence intervals that we study in the next section. A good source for understanding the expansions is Hall (1992).

11.5 Bootstrap Confidence Intervals

Bootstrap confidence intervals have been the focus of a large part of the bootstrap research literature. Basically there have been three main lines of development: Efron's (1979) original percentile method and its improvements resulting in the bias-corrected, accelerated interval (BC_a), the bootstrap t interval introduced in Efron (1982) and analyzed in Hall (1988), and the double bootstrap interval introduced in Hall (1986). We briefly review these methods in the context of a simple example, the odds ratio in a 2×2 table. Note that we work with the

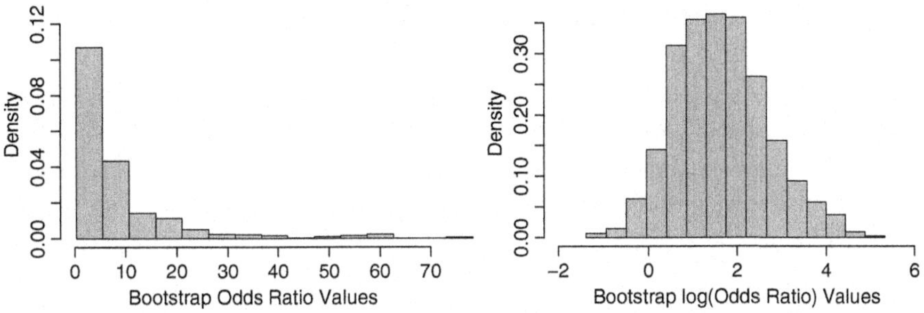

Fig. 11.2 Histograms of Bootstrap Odds Ratios and log Odds Ratios for Death Penalty Data

nonparametric bootstrap although the methods to be discussed also apply to the parametric bootstrap.

Example 10.6 (p. 395) had data on death penalty views of American and foreign students: 10 of 14 American students were against the death penalty, whereas only 4 of 11 foreign students were against the death penalty. The estimated odds ratio is $\widehat{\theta} = (10)(7)/(4)(4) = 70/16 = 4.375$. We seek a 95% confidence interval for the true odds ratio. Figure 11.2 gives a histogram of 2000 bootstrap $\widehat{\theta}^*$ values and also of $\log(\widehat{\theta}^*)$. It is well known that working on the log scale with the odds ratio is better for standard methods; one reason to use this example is to see what difference the transformation makes to bootstrap procedures. Figure 11.2 is valuable first to illustrate that the log transformation is approximately a normalizing transformation; $(\widehat{\text{Skew}}, \widehat{\text{Kurt}}) = (5.77, 55.2)$ for the first histogram and $(\widehat{\text{Skew}}, \widehat{\text{Kurt}}) = (0.38, 3.17)$ for the log transformed values. Also, we use $B = 2000$ bootstrap resamples as typically recommended by Efron for confidence intervals (although $B = 1999$ is more appropriate based on Section 11.6.2, p. 440).

Note that in order to use standard R code for the nonparametric bootstrap (R package `bootstrap`), one can write out the data as 25 rows with two variables, one indicating group membership and for the other, 1 = "against the death penalty" and 2 = "in favor of the death penalty." Also note that resampling with replacement from these 25 rows of data is equivalent to drawing samples from a multinomial with the estimated probabilities as true values; thus, one could call this a parametric bootstrap.

The delta method estimated asymptotic variance ($\widehat{V}_{\text{IC}}$) of $\widehat{\theta}$ is 14.219 leading to the Wald interval $4.375 \pm (1.96)\sqrt{14.219}$, $(-3.0, 11.8)$. A much better Wald interval is to take the interval on the log scale, $\log(4.375) \pm (1.96)\sqrt{.743}$, and exponentiate it to get $(0.81, 24.7)$ (Agresti 2002, p. 71). The exact interval based on the conditional noncentral hypergeometric (Agresti 2002, p. 99) is (0.62,32.9). It is known to be conservative due to the discreteness of the conditional distribution. Nevertheless, good intervals ought to have left endpoint around 0.6 to 0.8 and right endpoint between 25 and 33.

11.5.1 The Percentile Interval

Efron's (1979) original $100(1-2\alpha)\%$ bootstrap *percentile* interval is to just take the empirical 100α and $100(1-\alpha)$ percentiles from the bootstrap values $\theta_1^*, \ldots, \theta_B^*$ as the left and right endpoints, respectively. If $\widehat{K}_B$ is the empirical distribution function of the bootstrap values, then the $100(1-2\alpha)\%$ percentile interval is

$$\left(\widehat{K}_B^{-1}(\alpha), \widehat{K}_B^{-1}(1-\alpha)\right). \tag{11.9}$$

For the odds ratio example, the (.025)(2000) = 50th ordered value from the left panel of Figure 11.2 is 0.86, and the (.975)(2000) = 1950th ordered value is 52.5 leading to (0.86,52.5) as the 95% bootstrap percentile interval. Note that the percentile interval from the log transformed values (right panel of Figure 11.2), and then exponentiated, is exactly the same (0.86,52.5). Thus, the percentile interval is transformation invariant; transforming the data does not change the interval. This invariance is in dramatic contrast to the standard Wald intervals $(-3.0, 11.8)$ and $(0.81, 24.7)$ based on the original and then on the log transformed data, respectively. In effect, the bootstrap is automatically finding a transformation, and using it; on the other hand, knowing a good transformation does not help the percentile interval. Justification for the percentile interval follows in the next section.

We should mention that there are a wide range of definitions of sample pth quantiles. The bootstrap programs in the R package `bootstrap` use the kth order statistic, where $k = [B * p]$ and $[\cdot]$ is the greatest integer function. From here on we use Definition 6 from Hyndman and Fan (1996) that is implemented in the R version 2 quantile function, a linear interpolation of the kth and $(k+1)$th order statistics, where $k = [(B+1)p]$. Using this definition, the percentile interval is (0.86,52.8).

11.5.2 Heuristic Justification of the Percentile Interval

Efron's motivation for the percentile interval is based on assuming the existence of an increasing transformation g such that

$$P\left\{g(\widehat{\theta}) - g(\theta) \le x\right\} = H(x), \tag{11.10}$$

where H is the distribution function of a random variable symmetric about 0, i.e., $H^{-1}(\alpha) = -H^{-1}(1-\alpha)$ for $0 < \alpha < 1$. (Typically, Efron describes $H(x)$ as a mean 0 normal distribution function, $\Phi(x/\sigma)$, but only symmetry is used in the derivation.) Similarly, in the bootstrap world, assume that (11.10) holds approximately,

$$P^*\left\{g(\widehat{\theta}^*) - g(\widehat{\theta}) \le x\right\} \approx H(x), \tag{11.11}$$

where recall that the subscript * is used to denote calculations in the bootstrap world where $\widehat{\theta}$ is the true value. Suppose for a moment that g is known. Then (11.10) leads to $P(g^{-1}\{g(\widehat{\theta}) - x\} \leq \theta) = H(x)$. Setting this probability equal to $1-\alpha$ gives $x = H^{-1}(1-\alpha)$ and substituting for x gives

$$P\left(g^{-1}\{g(\widehat{\theta}) - H^{-1}(1-\alpha)\} \leq \theta\right) = 1-\alpha.$$

So under (11.10) and knowledge of g, $(g^{-1}\{g(\widehat{\theta}) - H^{-1}(1-\alpha)\}, \infty)$ is an exact one-sided interval with coverage probability $1-\alpha$. Because the upper $1-\alpha$ confidence bound is derived similarly, we concentrate on this lower bound. Now, the goal is to show that $g^{-1}\{g(\widehat{\theta}) - H^{-1}(1-\alpha)\}$ is actually estimated by the left endpoint of the percentile interval. To do that, recall that we denote the empirical distribution function of the bootstrap values $\theta_1^*, \ldots, \theta_B^*$ by $\widehat{K}_B$ so that the left endpoint of the percentile interval is simply $L_\alpha = \widehat{K}_B^{-1}(\alpha)$. Then

$$\begin{aligned}
\alpha &= P^*(\widehat{\theta}^* \leq L_\alpha) \\
&= P^*\{g(\widehat{\theta}^*) \leq g(L_\alpha)\} \\
&= P^*\{g(\widehat{\theta}^*) - g(\widehat{\theta}) \leq g(L_\alpha) - g(\widehat{\theta})\} \\
&\approx H\{g(L_\alpha) - g(\widehat{\theta})\},
\end{aligned}$$

where the approximation in the last step is from (11.11, p. 428). Finally, solving $\alpha \approx H\{g(L_\alpha) - g(\widehat{\theta})\}$ for L_α yields

$$L_\alpha \approx g^{-1}\{H^{-1}(\alpha) + g(\widehat{\theta})\} = g^{-1}\{g(\widehat{\theta}) - H^{-1}(1-\alpha))\}$$

because $H^{-1}(\alpha) = -H^{-1}(1-\alpha)$. This displayed expression is the same as the exact lower bound given above. Although existence of g satisfying (11.10, p. 428) and (11.11, p. 428) was assumed in this derivation, g is not used in the definition of the percentile interval. Rigorous theoretical justification of the interval follows from appropriate asymptotic convergence of $\widehat{K}_B$.

11.5.3 Asymptotic Accuracy of Confidence Intervals

Consider a nominal $1-\alpha$ one-sided confidence interval $(L_n(\alpha), \infty)$ for θ, i.e., $L_n(\alpha)$ is a $1-\alpha$ lower bound. If the non-coverage probability

$$P\{\theta < L_n(\alpha)\} = \alpha + O(n^{-k/2}),$$

the interval is said to be kth-order accurate. Thus $P\{\theta < L_n(\alpha)\} = \alpha + c/\sqrt{n}$ is called first order accurate, and $P\{\theta < L_n(\alpha\} = \alpha + d/n$ is called second order accurate. A similar definition holds for upper bounds and two-sided intervals. It is not hard to show that the one-sided standard Wald intervals and the percentile interval are both first order accurate. The two-sided versions are second order accurate for statistics that are asymptotically normal. In future sections we see bootstrap intervals that have better asymptotic accuracy than the percentile interval. These higher-order results require inversion of the Edgeworth expansions discussed briefly at the end of Section (11.4, p. 424). Hall (1986, 1988) initiated the study of the higher-order accuracy of bootstrap intervals.

11.5.4 The BC Interval

Efron's (1982, Section 10.7) first improvement of the percentile interval is called the bias-corrected (BC) percentile interval. It is based on observing that the estimator under consideration may not be median-unbiased. That is, the median of the distribution of $\widehat{\theta}$ is θ if $\widehat{\theta}$ is median-unbiased. (Note that mean-unbiasedness, $\mathrm{E}\{\widehat{\theta}\} = \theta$, is the more common notion of unbiasedness.) Deviations from median-unbiasedness can be estimated in the bootstrap world since there $\widehat{\theta}$ is the true value, and the true distribution function is just the limit of the empirical distribution function of $\theta_1^*, \ldots, \theta_B^*$, $\widehat{K}_B$, as $B \to \infty$. That is, if $\widehat{\theta}$ is median-unbiased, then $\widehat{K}_B(\widehat{\theta}) \approx 1/2$. To correct for median-bias, Efron (1982) assumes a more complicated model than (11.10) that allows for a shift in the distribution of $g(\widehat{\theta})$ by an unknown amount z_0,

$$P\left\{g(\widehat{\theta}) - g(\theta) + z_0 \le x\right\} = \Phi(x). \tag{11.12}$$

Similar to (11.11, p. 428) we also assume in the bootstrap world

$$P^*\left\{g(\widehat{\theta}^*) - g(\widehat{\theta}) + z_0 \le x\right\} \approx \Phi(x). \tag{11.13}$$

Notice that in contrast to (11.10, p. 428) and (11.11, p. 428), here we specify $\Phi(x)$ instead of a general distribution function $H(x)$. This difference is important because the definition of z_0 involves Φ. Actually, the derivation below goes through if we use $\Phi(x/\sigma)$ in place of $\Phi(x)$, but for simplicity we just assume σ is absorbed in g.

To understand z_0 better, we have $P(\widehat{\theta} \le \theta) = P\{g(\widehat{\theta}) \le g(\theta)\} = P\{g(\widehat{\theta}) - g(\theta) + z_0 \le z_0\} = \Phi(z_0)$. So if $\widehat{\theta}$ is median-unbiased, then $P(\widehat{\theta} \le \theta) = 1/2 = \Phi(z_0)$ and $z_0 = 0$. Otherwise, we estimate z_0 from the bootstrap distribution function estimator $\widehat{K}_B$ as follows:

$$\widehat{K}_B(\widehat{\theta}) = P^*(\widehat{\theta}^* \le \widehat{\theta})$$

$$= P^* \left\{g(\widehat{\theta}^*) \le g(\widehat{\theta})\right\}$$

$$= P^* \left\{g(\widehat{\theta}^*) - g(\widehat{\theta}) + z_0 \le z_0\right\}$$

$$\approx \Phi(z_0),$$

where this last step is from (11.11). Thus $\widehat{K}_B(\widehat{\theta}) \approx \Phi(z_0)$, and $\widehat{z}_0 = \Phi^{-1}(\widehat{K}_B(\widehat{\theta}))$. The $1 - 2\alpha$ BC percentile interval is

$$\left(\widehat{K}_B^{-1}(\alpha_1), \widehat{K}_B^{-1}(1-\alpha_2)\right), \tag{11.14}$$

where $\alpha_1 = \Phi\{2\widehat{z}_0 + \Phi^{-1}(\alpha)\}$, $\alpha_2 = 1 - \Phi\{2\widehat{z}_0 + \Phi^{-1}(1-\alpha)\}$.

Justification of (11.14) is as follows. Similar to the manipulations of the last section, the exact lower endpoint of an exact interval based on (11.12) and knowledge of g and z_0 is

$$g^{-1}\left\{g(\widehat{\theta}) + z_0 - \Phi^{-1}(1-\alpha)\right\}. \tag{11.15}$$

We shall show that the left endpoint of (11.14) approximates this exact endpoint. Using (11.13), for any $0 < \beta < 1$

$$\beta \approx P^* \left\{g(\widehat{\theta}^*) - g(\widehat{\theta}) + z_0 \le \Phi^{-1}(\beta)\right\}$$

$$= P^* \left\{g(\widehat{\theta}^*) \le g(\widehat{\theta}) - z_0 + \Phi^{-1}(\beta)\right)$$

$$= P^* \left\{\widehat{\theta}^* \le g^{-1}\left(g(\widehat{\theta}) - z_0 + \Phi^{-1}(\beta)\right)\right\}$$

$$= \widehat{K}_B \left\{g^{-1}\left(g(\widehat{\theta}) - z_0 + \Phi^{-1}(\beta)\right)\right\}.$$

Now substitute $\beta = \widetilde{\alpha}_1 = \Phi\{2z_0 + \Phi^{-1}(\alpha)\}$ and apply the inverse of $\widehat{K}_B$ to both sides of the last equation to get

$$\widehat{K}_B^{-1}(\widetilde{\alpha}_1) \approx g^{-1}\left\{g(\widehat{\theta}) - z_0 + 2z_0 + \Phi^{-1}(\alpha)\right\}$$

$$= g^{-1}\left\{g(\widehat{\theta}) + z_0 - \Phi^{-1}(1-\alpha)\right\}.$$

This last expression is the same as the exact endpoint (11.15), and $\widehat{K}_B^{-1}(\widetilde{\alpha}_1)$ is the same as the left endpoint of the BC interval except that $\widetilde{\alpha}_1$ uses z_0 instead of $\widehat{z}_0$.

For calculating $\widehat{z}_0$, Efron and Tibshirani (1993, p. 186) uses the left-continuous version of $\widehat{K}_B$ so that $\widehat{z}_0 = \Phi^{-1}\{\text{proportion of } \widehat{\theta}* \text{ values} < \widehat{\theta}\}$. In the example, 883 of the 2000 values are less than $\widehat{\theta} = 4.375$ leading to $\widehat{z}_0 = \Phi^{-1}(883/2000) =$

$-.147$. Since 26 of the 2000 $\widehat{\theta}*$ values are exactly 4.375, using the more standard right continuous version of $\widehat{K}_B$ leads to $\widehat{z}_0 = \Phi^{-1}(909/2000) = -.114$. We continue, however, to use Efron's left continuous version here. Then $\alpha_1 = \Phi\{2(-.147) - 1.96\} = 0.012$ and $\alpha_2 = 1 - \{\Phi(2(-.147) + 1.96\} = 0.048$, leading to $\left(\widehat{K}_B^{-1}(.012), \widehat{K}_B^{-1}(.952)\right) = (0.63, 32.7)$. This simple correction of adjusting 0.025 to 0.048 has brought the right endpoint of the percentile interval into the more reasonable value 32.7, and the adjustment of 0.025 to 0.012 has moved the left endpoint from 0.86 to 0.63. Similar to the percentile interval, the BC interval is also transformation invariant. Its one-sided versions are only first order accurate, but Liu and Singh (1987) show that the BC interval has some asymptotic superiority to the standard Wald interval (see also, Shao and Tu, p. 153).

11.5.5 The BC$_a$ Interval

Efron's (1987) second improvement to the percentile interval is called the bias-corrected, accelerated interval, BC$_a$ for short. It is based on assuming that

$$P\left\{\frac{g(\widehat{\theta}) - g(\theta)}{1 + ag(\theta)} + z_0 \le x\right\} = \Phi(x), \tag{11.16}$$

where z_0 is as before and the "acceleration" constant a is related to the 3rd moment skewness coefficient of the influence curve of $\widehat{\theta}$,

$$a = \frac{1}{6}\text{Skew}\left\{\sum_{i=1}^{n} IC(Y_i, \theta)\right\} = \frac{\text{Skew}\{IC(Y_1, \theta)\}}{6\sqrt{n}}.$$

From (11.16) we see that a relates to how quickly the standard deviation of $g(\widehat{\theta})$ is changing as a linear function of $g(\theta)$. In practice one common way to estimate a is with the jackknife approximations to the influence curve,

$$\widehat{a} = \frac{\dfrac{1}{n}\displaystyle\sum_{i=1}^{n}(\overline{\theta}_1 - \widehat{\theta}_{[i]})^3}{6\sqrt{n}\left\{\dfrac{1}{n}\displaystyle\sum_{i=1}^{n}(\overline{\theta}_1 - \widehat{\theta}_{[i]})^2\right\}^{3/2}}, \tag{11.17}$$

where $\widehat{\theta}_{[i]}$ are the leave-one-out estimators with $\overline{\theta}_1 = n^{-1}\sum_{i=1}^{n}\widehat{\theta}_{[i]}$. Using (11.16) and manipulations similar to those for deriving the percentile interval and BC interval, the $1 - 2\alpha$ BC$_a$ interval is

$$\left(\widehat{K}_B^{-1}(\alpha_1), \widehat{K}_B^{-1}(1 - \alpha_2)\right), \tag{11.18}$$

where

$$\alpha_1 = \Phi\left[\widehat{z}_0 + \frac{\widehat{z}_0 + \Phi^{-1}(\alpha)}{1 - \widehat{a}\{\widehat{z}_0 + \Phi^{-1}(\alpha)\}}\right],$$

$$\alpha_2 = 1 - \Phi\left[\widehat{z}_0 + \frac{\widehat{z}_0 + \Phi^{-1}(1-\alpha)}{1 - \widehat{a}\{\widehat{z}_0 + \Phi^{-1}(1-\alpha)\}}\right],$$

and $\widehat{z}_0 = \Phi^{-1}[\widehat{K}_B(\widehat{\theta})]$ as for the BC interval, and $\widehat{a}$ is given by (11.17). Notice that if $\widehat{a} = 0$, then the BC_a interval reduces to the BC interval, and if both $\widehat{a} = 0$ and $z_0 = 0$, then it reduces to the percentile interval. For the odds ratio example, $\widehat{a} = -0.025$, $\widehat{z}_0 = -.147$ as before, $\alpha_1 = .0089$, $\alpha_2 = .056$, and the BC_a interval is (.60,27.5).

The BC_a interval is transformation invariant like the percentile and BC intervals and is generally second order accurate. Efron (1987) and DiCiccio and Efron (1996) give a number of examples where the performance of the BC_a is very good. On the other hand, discussants for those papers give examples where the coverage probability of BC_a intervals is too small. It may be that the acceleration constant estimate $\widehat{a}$ is biased or too variable in small samples. For example, for the odds ratio data we generated 10 bootstrap samples and computed $\widehat{a}$ for each one. The mean and standard deviation of the 10 $\widehat{a}$ values was $-.046$ and .03, respectively.

11.5.6 The Double Bootstrap (Calibrated Percentile) Interval

The improvements to the percentile interval given in the last two sections are based on simple to compute adjustments to the α and $1-\alpha$ used in the percentile interval. Another approach introduced by Hall (1986), and Loh (1987) is to adjust α and $1 - \alpha$ by a second level of bootstrap resampling. That is, suppose that for each of the B bootstrap resamples, an additional C resamples are drawn, the percentile endpoints are calculated resulting in B confidence intervals, and the proportions of those B intervals that are completely to the left and right of $\widehat{\theta}$ (the true value in the bootstrap world) are recorded. Call those estimates ML(α) and MU(α), respectively, for "miss left" and "miss right." If the percentile interval is working correctly, then we would expect ML(α) $\approx \alpha$ and MR(α) $\approx \alpha$. Now suppose that we compute the B second stage intervals and the associated miss proportions for a fine grid of α values. We then find values α_1 and α_2 such that MR(α_1) $\approx \alpha$ and ML(α_2) $\approx \alpha$. The calibrated percentile interval is

$$\left(\widehat{K}_B^{-1}(\alpha_1), \widehat{K}_B^{-1}(1-\alpha_2)\right). \tag{11.19}$$

This interval is transformation invariant and second-order accurate. The main drawback is the large number BC of resamples required. Booth and Hall (1994) show that C ought to be chosen much smaller than B, on the order of $\sqrt{B}$. Booth and Presnell (1998, Table 1) give some approximately optimal combinations: if $BC = 10{,}000$, use $(B = 217, C = 46)$; if $BC = 100{,}000$, use $(B = 1020, C = 98)$; if $BC = 1{,}000{,}000$, use $(B = 4717, C = 212)$. Although this method holds a lot of promise, it does not seem to be used much in practice because of the computations.

11.5.7 Reflected Percentile and Bootstrap-t Intervals

One of the earliest examples in Efron (1979) was about approximating the distribution of $\widehat{\theta}-\theta$ by the bootstrap distribution of $\widehat{\theta}^*-\widehat{\theta}$. Since the α and $1-\alpha$ estimated quantiles of $\widehat{\theta}^* - \widehat{\theta}$ based on $\theta_1^*,\ldots,\theta_B^*$ are $\widehat{K}_B^{-1}(\alpha) - \widehat{\theta}$ and $\widehat{K}_B^{-1}(1-\alpha) - \widehat{\theta}$, respectively, an approximate $1-2\alpha$ confidence interval would manipulate the set

$$\left\{\theta : \widehat{K}_B^{-1}(\alpha) - \widehat{\theta} \le \widehat{\theta} - \theta \le \widehat{K}_B^{-1}(1-\alpha) - \widehat{\theta}\right\}$$

to get the set

$$\left\{\theta : 2\widehat{\theta} - \widehat{K}_B^{-1}(1-\alpha) \le \theta \le 2\widehat{\theta} - \widehat{K}_B^{-1}(\alpha)\right\}.$$

The resulting interval,

$$\left(2\widehat{\theta} - \widehat{K}_B^{-1}(1-\alpha), 2\widehat{\theta} - \widehat{K}_B^{-1}(\alpha)\right), \tag{11.20}$$

is called the *reflected percentile* interval, the *hybrid* interval (Shao and Tu, 1995, p. 141) and the *basic* interval (Davison and Hinkley, 1997, p. 29). We find the derivation of this interval much easier to understand than the percentile interval, but it often seems to have the wrong shape. For our example, recall the percentile interval is (0.86,52.8), and thus the reflected interval is $(2(4.375)-52.8, 2(4.375)-0.86) = (-44.1, 7.89)$, clearly a poor interval in this situation. This interval is not transformation invariant, and is vastly better in this case if computed on the log(odds ratio) scale and exponentiated, leading to (0.36,22.3).

A generally much better interval, the bootstrap-t interval (or percentile-t interval), is based on using the bootstrap distribution of $t^* = (\widehat{\theta}^* - \widehat{\theta})/\widehat{\sigma}^*$ as an estimate of the distribution of $t = (\widehat{\theta} - \theta)/\widehat{\sigma}$, where $\widehat{\sigma}$ is some estimate of the standard deviation of $\widehat{\theta}$. Then, manipulation of the set

$$\left\{\theta : \widehat{F}_{t^*}^{-1}(\alpha) \le \frac{\widehat{\theta} - \theta}{\widehat{\sigma}} \le \widehat{F}_{t^*}^{-1}(1-\alpha),\right\}$$

where $\widehat{F}_{t^*}^{-1}(p)$ refers to the bootstrap pth quantile estimates of t^*, leads to the bootstrap-t interval

$$\left(\widehat{\theta} - \widehat{\sigma}\widehat{F}_{t^*}^{-1}(1-\alpha), \widehat{\theta} - \widehat{\sigma}\widehat{F}_{t^*}^{-1}(\alpha)\right). \qquad (11.21)$$

Note that this derivation is exactly the same as used in deriving the usual t interval for the mean, but the t interval looks a bit different because we take advantage of the symmetry of the t distribution to just use the $1-\alpha$ quantile in the representation $\overline{Y} \pm s_{n-1}t(n-1, 1-\alpha)$. Here a major advantage of this bootstrap-t interval is that it allows for asymmetry. Hall (1988) shows that the one-sided bootstrap-t intervals are second-order accurate, but in small samples they require a good estimate of standard deviation and perform best when $t = (\widehat{\theta} - \theta)/\widehat{\sigma}$ is approximately pivotal, i.e., when the distribution of t is not dependent on unknown parameters. In some situations it performs very well, but in other situations it tends to have very long length. The R package `bootstrap` program `boott` uses a bootstrap standard deviation estimate if no external function is provided. Thus, for the odds ratio example, for each of the $B = 2000$ bootstrap samples, another 25 resamples were used to get the bootstrap $\widehat{\sigma}^*$ resulting in $2000(25) = 50,000$ calculations of the statistic. The program `boott` gave $(-17.2, 80.8)$ for the 95% bootstrap-t interval, clearly too long. If we first transform to log(odds ratio), get the bootstrap-t interval, and then exponentiate, the interval is (0.30,23.4). Thus the bootstrap-t interval is not transformation invariant, and performs much better on some scales than on others.

11.5.8 Summary of Bootstrap Confidence Intervals

Efron and coworkers developed the bias-corrected (BC) and bias-corrected accelerated (BC_a) intervals as improvements to the original bootstrap percentile interval. All three are motivated by transformations to normality but do not require knowledge of the transformation. In fact, all three intervals are transformation invariant, that is, transforming first, forming the interval, and then backtransforming has no effect on these intervals. Examples show that all three can work well in many situations, but they all fail to give adequate coverage in a number of well-known examples such as for $\widehat{\theta} = s_n^2$, the sample variance. In DiCiccio and Tibshirani (1987), for samples of size $n = 20$ drawn from a normal distribution, the estimated coverage probabilities for nominal 90% intervals was .76, .81, and .81 for the percentile, BC, and BC_a intervals, respectively. Note, however, the parametric bootstrap versions, resamples drawn from a $\mathrm{N}(\overline{Y}, s_{n-1}^2)$ distribution, resulted in coverages of .89, .89, and .90, respectively. Recall that in this section we have been exclusively talking about the nonparametric bootstrap, although all the methods introduced are valid when drawing parametric bootstrap resamples. Parametric bootstrap confidence intervals typically perform better than the nonparametric bootstrap intervals.

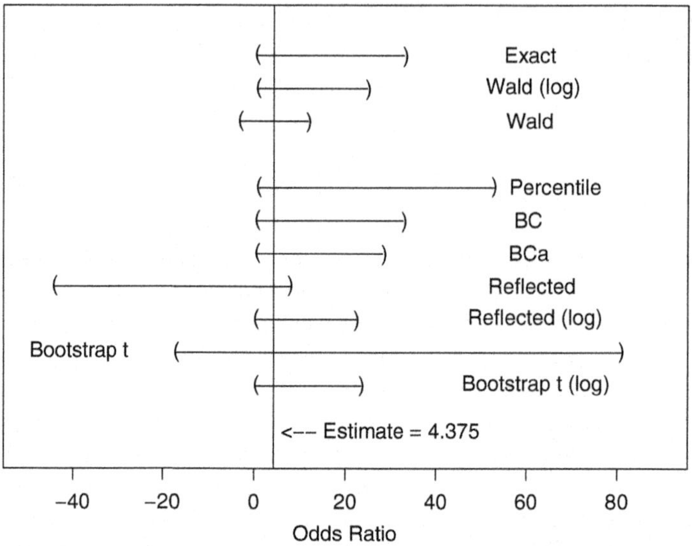

Fig. 11.3 Confidence Intervals for Odds ratio Example Data

Figure 11.3 shows all the confidence intervals discussed so far for the odds ratio example. We consider the top two as "gold standards," that is, approximating where good intervals should be. The Wald interval on the original scale is too short and inappropriately symmetric. The right tail of the percentile interval is too far to the right, but the BC and BC_a nicely adjust it. The motivation and second order asymptotic accuracy of the bootstrap-t interval are very appealing. Figure 11.3 shows, though, that the right (log) scale is very important. In fact, even the reflected method is pretty good when based on the log scale.

A lot of work has gone into developing bootstrap confidence intervals. The BC_a, the bootstrap-t, and calibrated percentile (double bootstrap) methods are asymptotically second-order accurate and work well in many situations. However, counter-examples show that an automatic bootstrap confidence interval method has yet to be discovered.

11.6 Bootstrap Resampling for Hypothesis Tests

11.6.1 The Basics

An understanding of bootstrap resampling for obtaining a standard error or confidence interval does not necessarily provide intuition concerning how to resample in a hypothesis testing situation. The key point is that for getting a p-value, resampling must be performed under an appropriate null hypothesis, whereas for standard errors and confidence intervals, resampling is unrestricted.

To make this clear, consider the case of two independent samples $X_1, \ldots, X_m$ and $Y_1, \ldots, Y_n$, and suppose that we are interested in the difference in population means, say $\mu_X - \mu_Y$. For a nonparametric bootstrap confidence interval, we would merely draw independent samples from the empirical cdf's or equivalently with replacement from the sets $\{X_1, \ldots, X_m\}$ and $\{Y_1, \ldots, Y_n\}$, respectively. Instead, consider testing $\mu_X - \mu_Y = 0$ with a t statistic, say the pooled t statistic

$$t_p = \frac{\overline{X} - \overline{Y}}{\sqrt{s_p^2\left(\frac{1}{m} + \frac{1}{n}\right)}},$$

where

$$s_p^2 = \frac{(m-1)s_X^2 + (n-1)s_Y^2}{m+n-2},$$

and

$$s_X^2 = \frac{1}{m-1}\sum_{i=1}^{m}(X_i - \overline{X})^2, \quad s_Y^2 = \frac{1}{n-1}\sum_{i=1}^{n}(Y_i - \overline{Y})^2.$$

The above method of resampling from each sample separately would lead to a test with power approximately equal to the nominal level of the test regardless of the magnitude of $|\mu_X - \mu_Y|$.

Why is that resampling method wrong? Resampling in the above fashion puts no restriction on the data and thus does not generate an approximation to the *null* distribution of the t_p statistic. For confidence intervals for $\mu_X - \mu_Y$, we do not want any restriction in the bootstrap world. But for the null distribution of t_p, we need to force the means to be equal when drawing bootstrap samples.

One way to make bootstrap world true means equal is to draw both samples with replacement from the pooled set $\{X_1, \ldots, X_m, Y_1, \ldots, Y_n\}$. By doing this, in the bootstrap world we have created the null hypothesis

$$H_0 : P(X^* \le t) = P(Y^* \le t) = H_N(t), \tag{11.22}$$

where $P(X^* \le t)$ is the distribution function of an X in the bootstrap world, $P(Y^* \le t)$ is the distribution function of a Y in the bootstrap world, and $H_N(t)$ is the empirical distribution function of the pooled set with $N = m + n$. In effect we are trying to test the real world hypothesis

$$H_0 : F(t) = G(t), \tag{11.23}$$

where F and G are the distribution functions of the X and Y samples, respectively.

It might be worth pointing out that there is an exact permutation test available for (11.23), obtained by constructing all $N!/m!n!$ partitions of $\{X_1,\ldots,X_m,Y_1,\ldots,Y_n\}$ into two samples of size m and n, respectively. Then t_p is computed for each partition, and the empirical distribution of these $N!/m!n!$ values is called the permutation distribution. The statistic t_p for the original sample is then compared to this distribution to get exact tests and p-values. This elegant approach was introduced by R. A. Fisher (1934a) and is discussed fully in Chapter 12.

How do permutation tests and bootstrap tests compare? Permutation tests are limited to a relatively small number of testing situations where permutations under the null hypothesis have the same distribution. For those situations, the permutation method gives exact results for any statistic. In contrast, the scope of application for bootstrap tests is huge. The resulting tests, though, are only approximately valid and depend on asymptotics for justification (except for the parametric bootstrap in special situations as mentioned in Examples 11.1 and 11.2). Bootstrapping studentized statistics like t_p is usually much preferable to bootstrapping statistics like $\overline{X}-\overline{Y}$ due to faster convergence of the bootstrap distribution (see Hall, 1986 or 1992 and the Edgeworth expansion argument at the end of Section 11.4, p. 424). For permutation tests, however, using different statistics often leads to the same result. For example in the above problem, the permutation method applied to t_p and to $\overline{X}-\overline{Y}$ yields the same test.

To further illustrate these ideas, consider a larger null hypothesis than (11.23):

$$H_0 : \mu_{\mathrm{X}} - \mu_{\mathrm{Y}} = 0, \tag{11.24}$$

but with no other restrictions on the distributions except for finite second moments. This allows the distributions to have different variances and even totally different shapes. A suitable statistic might be Welch's t

$$t_w = \frac{\overline{X}-\overline{Y}}{\sqrt{\dfrac{s_{\mathrm{X}}^2}{m}+\dfrac{s_{\mathrm{Y}}^2}{n}}}.$$

One way to create a bootstrap world with an appropriate null hypothesis would be to draw the X resamples with replacement from $\{X_1-\overline{X},\ldots,X_m-\overline{X}\}$, and the Y resamples from $\{Y_1-\overline{Y},\ldots,Y_n-\overline{Y}\}$. This forces the X and Y distributions in the bootstrap world to have mean 0 but leaves their shapes different. Of course, we could add the same constant to both sets of resamples and not change the results (since t_w is invariant to such additions). It is easy to show that the bootstrap distribution of t_w converges with probability one to a standard normal distribution, the same limiting distribution as t_w in the real world under the null hypothesis (11.24). Thus under (11.24), the bootstrap p-value converges with probability one to a uniform random variable. The permutation method cannot handle (11.24).

Example 11.4 (Comparing two variances). We again consider the two independent samples situation: $X_1, \ldots, X_m$ and $Y_1, \ldots, Y_n$ with respective means μ_X and μ_Y and variances σ_X^2 and σ_Y^2. The null hypothesis of interest is $H_0 : \sigma_X = \sigma_Y$. A semiparametric assumption is made that both samples are from the same location-scale family, but the family is unknown. Thus, the distribution function of each X_i is $F_1(x) = F_0((x - \mu_X)/\sigma_X)$ and the distribution function of each Y_i is $F_2(x) = F_0((x - \mu_Y)/\sigma_Y)$, where F_0 is the cdf of an unknown distribution with mean 0 and variance 1. The statistic used to test H_0 is

$$T_{mn} = \left(\frac{mn}{m+n}\right)^{1/2} \left\{\log(s_X^2) - \log(s_Y^2)\right\}.$$

It is fairly straightforward to show that under H_0, the limiting distribution of T_{mn} as $m, n \to \infty$ with $m/(m+n) \to \lambda \in (0, 1)$, is normal with mean 0 and variance $\text{Kurt}(F_0) - 1$. Because we do not know F_0, we need to estimate a critical value to be used with T_{mn}. We could estimate $\text{Kurt}(F_0)$ directly, but it may be better to use the bootstrap. In the introduction to Boos et al. (1989), four bootstrap resampling plans are discussed:

I. Draw both bootstrap samples independently and with replacement from the pooled set $\{X_1, \ldots, X_m, Y_1, \ldots, Y_n\}$.
II. Draw $X_1^*, \ldots, X_m^*$ with replacement from $\{X_1, \ldots, X_m\}$ and independently draw $Y_1^*, \ldots, Y_n^*$ with replacement from $\{Y_1, \ldots, Y_n\}$.
III. Draw both bootstrap samples independently and with replacement from the pooled set of residuals $\{X_1 - \overline{X}, \ldots, X_m - \overline{X}, Y_1 - \overline{Y}, \ldots, Y_n - \overline{Y}\}$.
IV. Draw $X_1^*, \ldots, X_m^*$ with replacement from $\{X_1/s_X, \ldots, X_m/s_X\}$ and independently draw $Y_1^*, \ldots, Y_n^*$ with replacement from $\{Y_1/s_Y, \ldots, Y_n/s_Y\}$.

Plan I is not appropriate unless the means are equal, $\mu_X = \mu_Y$, an assumption that is usually not warranted. The limiting distribution of T_{mn}^* is normal with mean 0 and variance $\text{Kurt}(G) - 1$, where $G(x) = \lambda F_1(x) + (1 - \lambda)F_2(x)$, and this $\text{Kurt}(G) - 1$ can be quite different from $\text{Kurt}(F_0) - 1$ when $\mu_X \neq \mu_Y$.

For Plan II, the limiting distribution of T_{mn}^* is exactly the same as that of T_{mn} under both H_0 and any alternative. Thus, the test has approximately the correct level α under H_0, but the power is also approximately α under any alternative.

For Plan III, the limiting distribution of T_{mn}^* is normal with mean 0 and variance $\text{Kurt}(H) - 1$, where

$$\begin{aligned} H(x) &= \lambda F_1(x + \mu_X) + (1 - \lambda)F_2(x + \mu_Y) \\ &= \lambda F_0(x/\sigma_X) + (1 - \lambda)F_0(x/\sigma_Y). \end{aligned} \tag{11.25}$$

Under $H_0 : \sigma_X = \sigma_Y = \sigma$, $H(x) = F_0(x/\sigma)$, and $\text{Kurt}(H) = \text{Kurt}(F_0)$ as needed for the correct asymptotic level. Under an alternative, the kurtosis of T_{mn}^* is not the same as $\text{Kurt}(F_0)$, but because the asymptotic mean of T_{mn}^* is 0, the resulting test under an alternative has power converging to 1.

For Plan IV, the asymptotic results are similar to those for Plan III except that equal kurtoses for the two samples are not needed. Thus, Plan IV is more robust asymptotically, but in small samples it does not perform as well as Plan III because the pooling in Plan III produces faster convergence of critical values. ♦

Other examples of creating null hypotheses in the bootstrap world may be found in Beran and Srivastava (1985) and Davison and Hinkley (1997, Ch. 4).

11.6.2 The Definition of Bootstrap *P*-Values and the "99 Rule"

Suppose that T_0 is the value of a test statistic T computed for a particular sample. Then $P(T \geq T_0|H_0)$ is the definition of the p-value in situations where large values of T support the alternative hypothesis. If the null distribution of T is a discrete uniform distribution on some values $t_1, \ldots, t_k$ (each value has probability $1/k$), then the p-value is just the proportion of t_i's greater than or equal to T_0. In analogous fashion, when B resamples are made in the bootstrap world under an induced null hypothesis, define the bootstrap p-value

$$p_{\mathrm{B}} = \frac{\{\#\text{ of } T_i^* \geq T_0\}}{B},$$

where $T_1^*, \ldots, T_B^*$ are the values of T computed from the resamples. This is the definition we prefer and the one given by Efron and Tibshirani (1993, p. 221). We should note, however, that Davison and Hinkley (1997, p. 148, 161) and others prefer $(Bp_{\mathrm{B}} + 1)/(B + 1)$.

Consider a situation where the statistic T is continuous, and a parametric bootstrap gives the exact sampling distribution as B grows large (such as in Example 11.1, p. 414). In this case, $T_0, T_1^*, \ldots, T_B^*$, are iid, all $(B + 1)!$ orderings are equally likely, and p_{B} has a discrete uniform distribution,

$$P(p_{\mathrm{B}} = 0) = P(p_{\mathrm{B}} = 1/B) = (p_{\mathrm{B}} = 2/B) == \cdots = P(p_{\mathrm{B}} = 1) = \frac{1}{B+1}.$$

Thus, the test defined by the rejection region $p_{\mathrm{B}} \leq \alpha$ has exact level α if $(B + 1)\alpha$ is an integer. For example, if $\alpha = .05$, then $P(p_{\mathrm{B}} \leq .05) = 5/(99 + 1) = .05$ if $B = 99$, but $P(p_{\mathrm{B}} \leq .05) = 6/(100 + 1) = .0594$ if $B = 100$. So, for small B one should use values like $B = 19$ or 39 or 99 to get standard α levels. We call this the "99 rule," and note that simulation-based tests such as this are often called "Monte Carlo" tests (first suggested by Barnard 1963). Hall (1986) gives an approximate version of this result for the nonparametric bootstrap, and thus the "99 rule" should be followed generally in bootstrap testing situations.

When analyzing a single data set, it is often possible to use a large B where there is very little difference between using B and $B + 1$. $B = 1000$ gives a rejection rate of $51/1001 = .051$. But for studying the power function of a bootstrap test, two Monte Carlo loops are required (the outer one for replicate samples of the true data situation, the inner one for the bootstrap procedure), and computations can be time consuming. Thus $B = 59$ or 99 or 199 might be used to save time. Since the resulting power estimates are typically monotone increasing in B, one can adjust the estimates if B is taken to be small (see Boos and Zhang 2000). Davison and Hinkley (1997, Sec. 4.5) use related arguments to justify the use of 99 resamples in the inner loop of a double bootstrap procedure to get adjusted bootstrap p-values for a single data set.

11.6.3 Convergence of Parametric Bootstrap P-Values

There have been many papers on the convergence properties of nonparametric bootstrap distributions, but not as many on convergence of parametric bootstrap distributions (see, for example, Beran 1986, 1988). Robins et al. (2000) give interesting results for the parametric bootstrap. One conclusion for bootstrap p-values from their Theorem 1 is as follows. If the test statistic T is asymptotically normal $(a(\boldsymbol{\theta}), b^2(\boldsymbol{\theta})/n)$ under a null hypothesis, then under some fairly strong but general conditions, the parametric bootstrap p-value is asymptotically uniform if $a(\cdot)$ does not depend on $\boldsymbol{\theta}$, and is asymptotically conservative otherwise. The context of the result is a paper on p-values for model adequacy, and the authors suggest that the conservative property is not appealing in that context. They may have a point, but the result is comforting in terms of general usage of the bootstrap in hypothesis testing situations.

11.7 Regression Settings

We consider typical regression settings based on iid random pairs $(Y_1, \boldsymbol{X}_1), \ldots, (Y_n, \boldsymbol{X}_n)$, or $(Y_1, \boldsymbol{x}_1), \ldots, (Y_n, \boldsymbol{x}_n)$, where the explanatory vectors $\boldsymbol{x}_i$ are viewed as fixed constants.

In the random pairs case, it is natural to draw with replacement from the set of pairs resulting in a bootstrap resample $(Y_1^*, \boldsymbol{X}_1^*), \ldots, (Y_n^*, \boldsymbol{X}_n^*)$. We repeat this B times and proceed as usual to calculate coefficients $\widehat{\boldsymbol{\beta}}_1^*, \ldots, \widehat{\boldsymbol{\beta}}_B^*$ and whatever bootstrap quantities are of interest such as the estimated covariance matrix of $\widehat{\boldsymbol{\beta}}$,

$$(B-1)^{-1} \sum_{i=1}^{B} (\widehat{\boldsymbol{\beta}}_i^* - \overline{\widehat{\boldsymbol{\beta}}}^*)(\widehat{\boldsymbol{\beta}}_i^* - \overline{\widehat{\boldsymbol{\beta}}}^*)^T. \tag{11.26}$$

This bootstrap method is very general and applies to almost any regression method; for example, logistic regression, Poisson regression, and linear and nonlinear least squares. Moreover, the assumed model used to derive estimators does not need to be true in order for bootstrap estimates to be consistent. We call this method the *random pairs* bootstrap although it is really just the standard nonparametric bootstrap method.

There are a few reasons, however, to consider other bootstrap approaches in regression settings:

1. Inference in regression setting is usually carried out conditional on the explanatory vectors $\boldsymbol{X}^T = (\boldsymbol{X}_1, \ldots, \boldsymbol{X}_n)$ regardless of whether they are considered fixed or random. The random pairs bootstrap, however, gives unconditional estimates such as the covariance matrix estimate (11.26).
2. The random pairs bootstrap does not take advantage of any model assumptions such as an additive error structure with homogeneous errors. This nonparametric aspect of the random pairs bootstrap gives it strong robustness to model assumptions, but also can result in much less efficient procedures.

For these reasons, let us consider the *residual-based* bootstrap that is appropriate for additive errors models of the form $Y_i = g(\boldsymbol{x}_i, \boldsymbol{\beta}) + e_i$, where g is a known function and $e_1, \ldots, e_n$ are iid random errors. Defining the residuals as $\widehat{e}_i = Y_i - g(\boldsymbol{X}_i, \widehat{\boldsymbol{\beta}})$, draw bootstrap errors $e_i^*, \ldots, e_n^*$ with replacement from the set

$$\left\{(\widehat{e}_i - \overline{\widehat{e}})/\sqrt{1 - p/n}, i = 1, \ldots, n\right\}.$$

Then form the bootstrap responses $Y_i^* = g(\boldsymbol{X}_i, \widehat{\boldsymbol{\beta}}) + e_i^*, i = 1, \ldots, n$. If the model is linear, $g(\boldsymbol{X}_i, \boldsymbol{\beta}) = \boldsymbol{X}_i^T \boldsymbol{\theta}$, then the least squares estimator in the bootstrap world is $\widehat{\boldsymbol{\beta}}^* = (\boldsymbol{X}^T \boldsymbol{X})^{-1} \boldsymbol{X}^T \boldsymbol{Y}^*$ with variance $\text{Var}^*(\widehat{\boldsymbol{\beta}}^*) = \widehat{\sigma}^2 (\boldsymbol{X}^T \boldsymbol{X})^{-1}$, where

$$\widehat{\sigma}^2 = \frac{1}{n} \sum_{i=1}^{n} \left[(\widehat{e}_i - \overline{\widehat{e}})/\sqrt{1 - p/n}\right]^2 = \frac{1}{n - p} \sum_{i=1}^{n} (\widehat{e}_i - \overline{\widehat{e}})^2. \tag{11.27}$$

Further, if the first column of $\boldsymbol{X}$ is a column of ones, then $\overline{\widehat{e}} = 0$, and we recognize the bootstrap estimate of $\text{Var}(\widehat{\boldsymbol{\beta}})$ is the same as the usual unbiased one.

Thus, in a standard linear model setting with homogeneous errors and using least squares, there is no need to simulate at all. The residual-based bootstrap using the above adjusted residuals gives the same standard error as standard theory. In large samples, the random-pairs bootstrap gives essentially the same result. However, in a variety of settings where at least one of the usual assumptions is false or least squares is not used, then the bootstrap might be a reasonable option because standard methods may not exist or may not perform adequately. For example, if heterogeneity of errors is suspected, then the random-pairs bootstrap produces a covariance matrix estimator that is close to the weighted jackknife estimator and to the sandwich estimator (7.21, p. 317) (except for the factor $n/(n - p)$). In nonlinear regression settings, the bootstrap may be more trustworthy for certain problems as the following example illustrates.

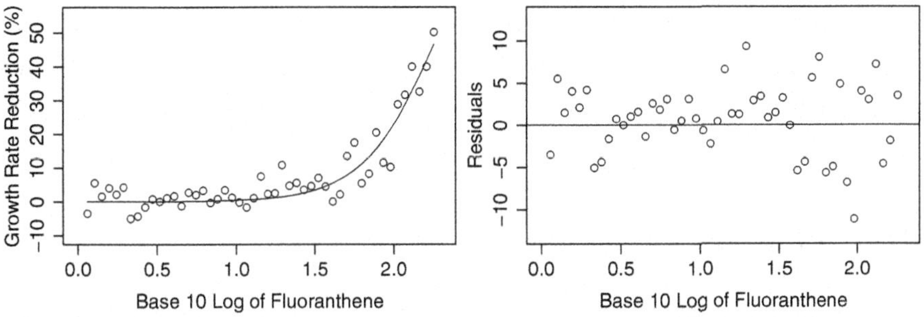

Fig. 11.4 Least squares fit of fluoranthene data and residuals

Example 11.5 (Confidence interval in nonlinear regression). The following problem was presented by Allen Olmstead, a post-doctoral fellow in the Toxicology Department at NC State University. He had obtained data on the percent growth rate reduction of young daphnids as a function of the chemical fluoranthene in micrograms per liter. The model is

$$Y_i = \frac{100}{1 + (x_m/x_i)^{\rho}} + e_i, \tag{11.28}$$

where x_m and ρ are unknown parameters to be fit to the data. Allen had fit 49 pairs of points to this model using nonlinear least squares and obtained $\widehat{x}_m = 193.7$ and $\widehat{\rho} = 1.85$. Figure 11.4 shows the fit overlaying the points with $\log_{10}$(fluoranthene) on the x axis, and also the residual plot. The fit seems pretty good although there is a hint of increasing variability in the residual plot. Allen had solved the fitted equation to find the value of fluoranthene that produced a 5% reduction in growth, 39.3 micrograms per liter. But he wanted a 95% confidence interval for the underlying true value, and the bootstrap had been suggested to him.

There were two NC State graduate students working on the project. Xianzheng Huang implemented the residual-based bootstrap with a SAS macro and obtained (32.0,48.7) for a 95% BC_a confidence interval ($B = 2000$). Liqiu Jiang implemented the random pairs bootstrap using the R program `bcanon` referred to in the back of Efron and Tibshirani (1993) and obtained (27.8,55.0) for a 95% BC_a confidence interval ($B = 2000$). Xianzheng also ran a small simulation and obtained true coverages of around 93% to 94% coverage for the residual-based method, suggesting that it is a bit liberal here. The longer length for the random-pairs method is perhaps due to the heterogeneity of the errors, but it is also likely due to the difference between conditional and unconditional inference. Reflecting these two differences, the unweighted jackknife gave 7.3 as a standard error for the estimate (an unconditional standard error), whereas the delta method gave the much smaller standard error 4.3 based on the conditional covariance estimate from the least squares output. Recall that the unconditional must be larger because

$\mathrm{Var}(\widehat{\theta}) = \mathrm{E}\left\{\mathrm{Var}(\widehat{\theta} \mid \boldsymbol{X})\right\} + \mathrm{Var}\left\{\mathrm{E}(\widehat{\theta} \mid \boldsymbol{X})\right\}$. Bootstrap confidence intervals have been studied in contexts similar to this example in Rosen and Cohen (1995) and Zeng and Davidian (1997). ♦

11.8 Problems

11.1. Consider an iid sample Y_1, Y_2, Y_3 of size $n = 3$. For simplicity in thinking about the problem, suppose that Y_i are continuous so that all three are distinct with probability 1.

a. Consider drawing samples of size $n = 3$ with replacement from the set $\{Y_{(1)}, Y_{(2)}, Y_{(3)}\}$, where $Y_{(1)} < Y_{(2)} < Y_{(3)}$. Write down the $3^3 = 27$ equally likely resamples. Actually there are only $\binom{2n-1}{n} = 10$ distinct resamples (but you will find that out empirically.)
b. Using a) write down the exact nonparametric bootstrap distribution for the sample median in terms of the ordered values $Y_{(1)} < Y_{(2)} < Y_{(3)}$.
c. From b) write down the expectation and variance of the sample median from a bootstrap resample.

11.2. For an iid sample $Y_1, \ldots, Y_n$, consider the estimator

$$\widehat{\theta} = \frac{1}{n}\sum_{i=1}^{n} q(Y_i),$$

where q is a known function and $\mathrm{var}(q(Y_1)) < \infty$. Find variance estimates for $\widehat{\theta}$ using the M-estimator method, the IC method, the jackknife method, and the bootstrap method.

11.3. Consider the estimator $\widehat{\theta} = g(\overline{Y})$, where g is a known function. The exact general expression for the variance is the n-fold integral given in (11.4). However, suppose that the data are from a gamma(α, β) distribution. Write down the exact expression for the variance of $\widehat{\theta}$ using the fact that a sum of independent gamma random variables has a gamma distribution. The result involves single integrals. Now suppose that you have found the maximum likelihood estimators of α and β. Give an expression for a parametric bootstrap estimate of the variance of $\widehat{\theta}$ without resampling.

11.4. Consider estimation of the distribution function $K_n(y) = P(\widehat{\theta} \le y)$ by the Monte Carlo estimator $\widehat{K}_{n,B}(y) = B^{-1}\sum_{i=1}^{B} I(\widehat{\theta}_i^* \le y)$ where $\widehat{\theta}_1^*, \ldots, \widehat{\theta}_B^*$ are iid and computed from resamples from $\widehat{F}$, an estimate of the distribution function F of the original iid data $Y_1, \ldots, Y_n$. Note that $\widehat{K}_{n,B}(y)$ is just an approximation to the true bootstrap estimator $K_{n,B=\infty}(y) = P^*(\widehat{\theta}^* \le y)$. Find a variance decomposition of $\text{Var}(\widehat{K}_{n,B}(y))$ in terms of means and variances of functions of $K_{n,\infty}(y)$. (Hint: use the iterated variance formula $\text{Var}(A) = \text{E}[\text{Var}(A|B)] + \text{Var}[\text{E}(A|B)]$.) Use the decomposition to suggest good values of B.

11.5. Find bootstrap standard errors for the mean, median, and 20% trimmed mean of the food consumption values in Example 11.3 (p. 418). An R function modified from the Appendix of Efron and Tibshirani (1993) is

```
boot.se <- function(x,nboot,theta, ...){
call <- match.call()
n <- length(x)
bootsam<- matrix(sample(x,size=n*nboot,replace=T),
                          nrow=nboot)
thetastar<-apply(bootsam,1,theta,...)
se<-sd(thetastar)
return(se) }
```

11.6. For the data and odds ratio estimator of Example 10.6 (p. 395), find the bootstrap standard error and compare with the influence curve (delta) method and the jackknife method given there. Below is an appropriate R function for the odds ratio when the data are given as

```
1   US  1
2   US  1
.    .  .
.    .  .
24 INT  0
25 INT  0
```

```
or<-function(index,data){
   # computes odds ratio based on expanded data
   x <- data[index,1]
   y <- data[index,2]
   n1 <- length(x[x=="US"])
   n2 <- length(x[x=="INT"])
   n11<-sum(y[x=="US"])
   n12<-n1-n11
   n21<-sum(y[x=="INT"])
   n22<-n2-n21
   if (min(n11,n12,n21,n22) > 0) or<-n11*n22/
     (n12*n21)
   else  or<-(n11+.5)*(n22+.5)/((n12+.5)*(n21+.5))
  return(or)
}
```

Remember, when the data consist of independent vectors, the official "data" are the row numbers, in this case 1 to 25, and `data=z` is used as a 3rd argument in boot.se if the data are in z. The bootstrap standard error you get should be much larger than those from the influence curve and jackknife methods. Try to explain the reason.

11.7. Construct a table similar to Table 11.3 (p. 423) for the sample coefficient of variation, $s_n/\overline{Y}$. Use sim.samp and var.est of Problem 10.7 (p. 409). For the normal distribution let the mean be 5 and variance be 1 so that the sample mean is positive with very high probability. You may choose n. An R program for $\widehat{V}_{\mathrm{IC}}$ is

```
cv.var<-function(x){
# asy. var of cv=s/xbar, from Serfling, 1980 p. 137
n<-length(x)
xbar<- mean(x)
diff<- x-xbar
mu2<- mean(diff^2)
mu3<- mean(diff^3)
mu4<- mean(diff^4)
var<- mu2^2/(xbar^4)-mu3/(xbar^3)
          +(mu4-mu2^2)/(4*xbar*xbar*mu2)
return(var/n)
}
```

11.8. Find 95% confidence intervals for the 20% trimmed mean of the population from which the food consumption values in Example 11.3 (p. 418) were obtained. Intervals to find:

(a) $\widehat{\theta} \pm 2.093(\widehat{V}_{\mathrm{Boot}})^{1/2}$, where 2.093 is from the t with 19 degrees of freedom.
(b) The nonparametric percentile interval.
(c) The nonparametric BC_a interval.
(d) The nonparametric bootstrap-t interval.

You might want to use either of the R packages bootstrap or boot. If you use boott in package bootstrap for the bootstrap-t interval, you have a choice of supplying a standard deviation estimate or letting the program get a bootstrap estimate (thus a type of double bootstrap). The latter takes longer but is simpler to use.

11.9. Simulate samples from a $N(\mu, \sigma^2)$ with $n = 20$ and estimate the non-coverage probabilities of 90% intervals for σ^2. Also estimate the proportion of times the associated one-sided 95% bounds do not contain the true σ^2. Intervals to use:

a. $\widehat{\theta} \pm 1.729(\widehat{V}_{\mathrm{Boot}})^{1/2}$, where 1.729 is from the t with 19 degrees of freedom.
b. The nonparametric percentile interval.
c. The nonparametric BC_a interval.

11.10. Derive the Wald intervals (−3.0, 11.8) and (0.81,24.7) for the odds ratio given in Section 11.5 (p. 426).

11.11. Using (11.10, p. 428) and (11.11, p. 428), show that the *right* endpoint of the percentile interval is approximately equal to the right endpoint of the corresponding exact interval.

11.12. Using (11.12, p. 430) and (11.13, p. 430), show that the *right* endpoint of the BC interval is approximately equal to the right endpoint of the corresponding exact interval.

11.13. To justify the BC$_a$ interval, first use (11.16, p. 432) and assume z_0 and a are known to show that the *left* endpoint of the exact BC$_a$ interval is

$$g^{-1}\left(\frac{g(\widehat{\theta}) + z_0 + \Phi^{-1}(\alpha)}{1 - a\{z_0 + \Phi^{-1}(\alpha)\}}\right) = g^{-1}\left(g(\widehat{\theta}) + \frac{\{1 + ag(\widehat{\theta})\}\{z_0 + \Phi^{-1}(\alpha)\}}{1 - a\{z_0 + \Phi^{-1}(\alpha)\}}\right).$$

Then, assume that (11.16, p. 432) holds approximately in the bootstrap world, and similar to the argument for the BC interval, show that $\widehat{K}_B^{-1}(\widetilde{\alpha}_1)$ is approximately equal to the above exact endpoint, where $\widetilde{\alpha}_1$ is the same as α_1 for the BC$_a$ interval except that z_0 and a are used in place of their estimates.

11.14. For Example 11.4 (p. 439), suppose that the populations are normal with means $\mu_X = 0$ and $\mu_Y = 3$, $\sigma_X^2 = \sigma_Y^2 = 1$, $m = n$, and that we use resample Plan I with T_{mn}. First verify that $G(x) = \lambda F_1(x) + (1 - \lambda)F_2(x)$ for this case has $\text{Kurt}(G) = 2.04$. Then find the asymptotic level of a bootstrap test of $H_0 : \sigma_X = \sigma_Y$ versus $H_a : \sigma_X > \sigma_Y$ based on T_{mn} using the fact that under H_0, T_{mn} has asymptotic variance $3 - 1 = 2$, and T_{mn}^* has asymptotic variance $2.04 - 1 = 1.04$.

11.15. Under the semiparametric assumptions of Example 11.4 (p. 439), T_{mn} is asymptotically normal with asymptotic mean

$$\left(\frac{mn}{m+n}\right)^{1/2} \{\log(\sigma_X^2) - \log(\sigma_Y^2)\}$$

and asymptotic variance $\text{Kurt}(F_0) - 1$. The bootstrap 0.05 level critical value from Plan III converges to $1.645\{\text{Kurt}(H) - 1\}^{1/2}$, where H is given in (11.25, p. 439). These results are true for both H_0 and an alternative. Suppose that the data are from normal distributions with $\sigma_X = 2$, $\sigma_Y = 1$, and $m = n = 10$. Based on these asymptotic approximations, show that the approximate power is 0.56.

11.16. Use the function `power.var` from the file `power.boot.variances.txt` on the website to estimate the power for the situation in the previous problem. Then set `ratio` equal to 1 in `power.var` to see the empirical level for this situation. Raw empirical power estimates have binomial variance $\widehat{\text{pow}}(1 - \widehat{\text{pow}})/N$. Repeat the `ratio=2` situation a few times with different seeds and estimate the variance of the extrapolated estimates (both the linear and quadratic estimates). Are they close to $\widehat{\text{pow}}(1 - \widehat{\text{pow}})/N$?

11.17. Verify that the residual-based bootstrap covariance estimator of the least squares estimator is $\widehat{\sigma}^2(\boldsymbol{X}^T\boldsymbol{X})^{-1}$, where $\widehat{\sigma}^2$ is given by (11.27, p. 442).

11.18. For Example 11.5 (p. 443), note that the solution of the estimated equation to give percentage growth reduction $100p$ is $\widehat{x}_m[p/(1-p)]^{1/\widehat{\rho}}$. Using Maple, show that the delta method variance for this estimate is 18.3 when $p = .05$. The covariance estimate for $(\widehat{x}_m, \widehat{\rho})$ from the least squares output is

$$\begin{pmatrix} 129.105 & -1.560 \\ -1.560 & 0.030 \end{pmatrix}.$$

Chapter 12
Permutation and Rank Tests

12.1 Introduction

In the early 1930s R. A. Fisher discovered a very general exact method of testing hypotheses based on permuting the data in ways that do not change its distribution under the null hypothesis. This *permutation* method does not require standard parametric assumptions such as normality of the data. It does require, however, certain invariance properties under the null hypothesis that restricts application to fairly simple designs. But in such situations, the method results in exact tests with level α under very weak distributional assumptions. Moreover, the method is *statistic-inclusive* in the sense that any test statistic can be used and inherits the level-α property, although some statistics are much more powerful than others.

Tests based on this method are called *permutation tests* or *randomization tests* depending on whether the data can be viewed as samples from populations or not. That is, when sampling from populations, "permutation tests" refer to use of the permutation method to obtain level α tests under weak distributional assumptions. In Fisher's words (1935, Sec. 21), these are tests of a "wider" null hypothesis (as compared to assuming normal distributions, for example).

However, experiments may be performed on units that cannot be viewed as arising from random sampling of any population. In such situations "randomization inference" refers to inference drawn based only on the physical randomization of the units to different treatments, and on the test statistic calculated at all possible randomizations of the data. The same test that we called a permutation test in random sampling contexts is now called a randomization test. Of course one needs to qualify all statements of significance about such experiments with the disclaimer that randomization inference only applies to the units used in the experiment.

Permutation tests are the foundation of classical nonparametric statistics (also called *distribution-free* statistics), which itself is often identified with rank tests. Rank tests are actually a special subclass of permutation tests with three distinct advantages:

D.D. Boos and L.A. Stefanski, *Essential Statistical Inference: Theory and Methods*, Springer Texts in Statistics, DOI 10.1007/978-1-4614-4818-1_12,

1. For data without ties, the conditional permutation distribution of a rank test is actually unconditional (does not change from sample to sample) because the ranks of a continuous data set are the same for every sample. Thus, the distribution of an important rank statistic like the Wilcoxon Rank Sum statistic can be tabulated or programmed. However, this computing advantage is less important today, and when there are ties in the data (a very common occurrence), the tabulated values are not appropriate, and the conditional permutation distribution is required for exact inference.
2. The key philosophical foundation of rank tests arises from the theory of invariant tests as described in Lehmann (1986, Ch. 5). The idea with invariant tests is to reduce the class of tests considered to those that are naturally invariant with respect to a group of transformations G on the sample space of the data. Given G, a maximal invariant is a statistic $M(\boldsymbol{x})$ with the property that any invariant test with respect to G must be a function of $\boldsymbol{x}$ only through $M(\boldsymbol{x})$. Now consider the two-sample problem with $H_0 : F_X(x) = F_Y(x)$ versus the alternative "F_Y is stochastically larger than F_X," that is, $H_a : 1 - F_Y(x) \geq 1 - F_X(x)$ for all x with strict inequality for at least one x. This alternative is more general than the usual shift alternative, $F_Y(x) = F_X(x - \Delta)$, but it certainly includes the shift alternative as a special case. Let G be the group of transformations such that each $g \in G$ is continuous and strictly increasing. For this testing problem and group G, the set of ranks of the combined X and Y samples is the maximal invariant statistic. Thus, any invariant test must be a function of the ranks. Does it make sense to require tests to be invariant with respect to monotone transformations? Whenever data are ordinal or we do not trust the measurement scale, then invariance certainly makes sense, and rank tests are the obvious choice.
3. Rank tests may be preferred in many situations because of their Type II error robustness. That is, for an appropriate data generation model, the permutation method can make any statistic Type I error robust (level α), but because rank tests are a function of the data only through the ranks, the influence of outliers is automatically limited. Thus, rank tests are power robust in outlier-prone situation. The key example is the Wilcoxon Rank Sum test that is powerful in the face of a wide variety of distributional shapes. In fact, Hodges and Lehmann (1956) showed that the asymptotic relative efficiency (ARE) of the Wilcoxon Rank Sum test to the t test satisfies the following:

   a) ARE= .955 for normal shift alternatives, and thus the Wilcoxon Rank Sum test loses little in comparison to the t where the t is best;
   b) and ARE $\geq$ .864 for any continuous unimodal shift alternative with finite variance, and thus the Wilcoxon Rank Sum test can never be much worse than the t-test but possibly much better.

Optimality for permutation and rank procedures is discussed in more detail later.

Although the term "nonparametric" was classically associated with permutation and rank procedures, in recent times it is more commonly used for nonparametric density and regression estimation methods based on smoothing. Thus, when describing rank or permutation procedures, it is best to use the specific names "rank" or "permutation" rather than "nonparametric." Although permutation tests are inherently defined in terms of randomization, they overlap with a variety of conditional procedures and uniformly most powerful unbiased (UMPU) "Neyman structure similar" tests based on exponential family theory (the most well known is Fisher's Exact Test).

Permutation procedures are very computationally intensive. These extensive computations prevented widespread use of the method until the 1990's. Thus, asymptotic approximations were dominant until the 1990's, although exact small-sample distributions were tabled for a number of important rank test statistics.

The asymptotic approximations are basically of three kinds: normal approximations based on the Central Limit Theorem, F or *beta* approximations based on matching permutation moments with normal theory moments, and Edgeworth expansions that improve on the normal approximations. The normal approximations have been used the most due to their simplicity. However, the F approximations initiated by Pitman (1937a,b) and Welch (1937) in the 1930s and updated by Box and Andersen (1955) are generally better for situations where they apply. The Edgeworth approximations are very good for the Wilcoxon Rank Sum and Wilcoxon Signed Rank statistics, but are somewhat more complicated for other statistics and seem not to be in general usage. Thus, we emphasize the F approximations rather than the normal or Edgeworth approximations. In fact these F approximations appear to be underused in general, but the work of Conover and Iman (1981) may have rekindled their use. Asymptotic normal theory remains important for comparing different methods according to asymptotic power, rather than for finding critical values. We give an overview of these results and then a few technical details in an appendix. There are excellent texts such as Hajek and Sidak (1967) and Randles and Wolfe (1979) that carefully explain asymptotic normality proof techniques for rank statistics. We add that most nonparametric texts of the last forty years are mainly about rank statistics, although Lehmann (1975) and Pratt and Gibbons (1981) have portions devoted to permutation tests. Puri and Sen (1971) emphasize the theory of permutation tests in multivariate settings.

In our current situation of extensive computing power, Monte Carlo approximations are the most important alternative to exact calculations. By Monte Carlo approximation we mean random sampling from the set of all permutations. This method can be used for any statistic in a situation where permutation methods are appropriate. Moreover, the error of approximation can be reduced by just adding more replications. This sampling (or resampling) in the "permutation world" is very similar to sampling in the bootstrap world; the main difference is that bootstrap p-values are typically approximate, even using the limit as the number of resamples B goes to ∞. In contrast, the limiting p-value in the permutation world is exact, and even the finite B estimated p-value has an exact interpretation.

Thus, our treatment of nonparametric methods is quite a bit different from most texts written in the last half of the twentieth century, which have emphasized rank tests and asymptotic normal approximations. We believe the basic permutation approach is the most important idea because it provides Type I error robustness for any statistic. Monte Carlo approximations can handle any problem for which the exact permutation distribution is too difficult to compute. Rank methods are still very important, but now because they provide Type II error robustness (good power in the face of outliers), not because they are easy to use or their distributions are tabled.

We start first with the two-sample problem to illustrate the basic permutation test approach. We then give some general theory for permutation tests along with approximations and discuss optimality results. Then we review results for the most important designs admitting permutation tests, their use in contingency tables, and estimators and confidence procedures derived from inverting permutation and rank tests.

12.2 A Simple Example: The Two-Sample Location Problem

We illustrate here the basic permutation approach with a simple two treatment experiment.

A clever middle school student believes that she has discovered a new method for teaching fractions to third graders. To test her hypothesis, she selects six students from her father's third grade class and randomly assigns four to learn the new method and two to use the standard method. After training both groups, they are given twenty test problems. The scores for the standard method group are $x_1 = 6$, $x_2 = 8$ and for the new method group are $y_1 = 7$, $y_2 = 18$, $y_3 = 11$, $y_4 = 9$. The results look promising for the new method, but how shall we assess statistical significance?

One possible test statistic is the standard two-sample t,

$$t(\boldsymbol{X}, \boldsymbol{Y}) = \frac{\overline{Y} - \overline{X}}{\sqrt{s_p^2\left(\frac{1}{m} + \frac{1}{n}\right)}}, \tag{12.1}$$

where $s_p^2 = \{\sum(X_i - \overline{X})^2 + \sum(Y_j - \overline{Y})^2\}/(m + n - 2)$. If t is large, then one might be convinced that the new method is better than the standard one.

Another commonly used statistic is W = the sum of the ranks of the Y values when both X and Y samples are thrown together and ranked from smallest to largest. Let $\boldsymbol{Z}$ denote the joint sample of both $\boldsymbol{X}$ and $\boldsymbol{Y}$ together: $\boldsymbol{Z} = (\boldsymbol{X}, \boldsymbol{Y})$ with observed values here $(6, 8, 7, 18, 11, 9)$. The ranks of these observed values are then $(1, 3, 2, 6, 5, 4)$ and $W = 2 + 6 + 5 + 4 = 17$, the sum of the Y ranks. If the new teaching method is better, then on average we would expect W to be large. Assuming that either t or W are reasonable statistics for our testing problem, we still need to agree on what is a proper reference distribution for each. A simple but very general approach is to recognize that there were actually $\binom{6}{2} = 15$ different ways that two students could have been selected from the original six to go in the X sample (with the remaining four assigned to the Y sample). Table 12.1 is a listing of the possible samples and the values of t and W for both.

Table 12.1 All Possible Permutations for Example Data

	X Sample		Y Sample				$\sum Y_i$	t	W
1.	6	8	7	18	11	9	45	1.17	17
2.	7	8	6	18	11	9	44	0.91	16
3.	18	8	7	6	11	9	33	−1.36	12
4.	11	8	7	18	6	9	40	0.12	13
5.	9	8	7	18	11	6	42	0.49	14
6.	6	7	8	18	11	9	46	1.47	18
7.	6	18	7	8	11	9	35	−0.84	14
8.	6	11	7	18	8	9	42	0.49	15
9.	6	9	7	18	11	8	44	0.91	16
10.	7	18	6	8	11	9	34	−1.08	13
11.	18	11	7	6	8	9	30	−2.98	10
12.	11	9	7	18	6	8	39	−0.06	12
13.	7	11	6	18	8	9	41	0.30	14
14.	7	9	6	18	11	8	43	0.69	15
15.	18	9	7	6	11	8	32	−1.72	11

Table 12.2 Permutation Distribution of t

t	−2.98	−1.72	−1.36	−1.08	−0.84	−0.06	0.12
$P(t)$	$\frac{1}{15}$	$\frac{1}{15}$	$\frac{1}{15}$	$\frac{1}{15}$	$\frac{1}{15}$	$\frac{1}{15}$	$\frac{1}{15}$
t	0.30	0.49	0.69	0.91	1.17	1.47	
$P(t)$	$\frac{1}{15}$	$\frac{2}{15}$	$\frac{1}{15}$	$\frac{2}{15}$	$\frac{1}{15}$	$\frac{1}{15}$	

If the treatments produce identical results, then the outcomes for each student would have been exactly the same for any of the 15 possible randomizations. Thus, a suitable reference distribution for t or W is just the possible 15 values of t or W along with the probability 1/15 of each. This reference distribution for t, called the permutation distribution, is in Table 12.2.

Note that the permutation distribution of t is discrete even when sampling from a continuous distribution. (Here the distribution of the data is also discrete because the possible test scores are 0, 1, …, 20).

Using the distribution in Table 12.2, a conditional test for this experiment with $\alpha = 1/15$ would be to reject if $t \geq 1.47$. A one-sided p-value for the observed value of $t = 1.17$ is 2/15. Similarly a conditional $\alpha = 1/15$ level test based on the rank sum W would reject if $W \geq 18$, and the one-sided p-value is 2/15.

In general, the tests based on t and W would not give exactly the same results. For example, suppose the original data had been the 14th permutation, (7,9,6,18,11,8). Then the permutation p-value for t would be $5/15 = .33$, whereas the permutation

p-value for W would be $6/15 = .40$. Note, however, the column in Table 12.1 (p. 453) for the sum of the Y values. Comparing the $\sum Y_i$ and t values, one can see that the permutation p-values from $\sum Y_i$ and t are identical if the original data had been any of the 15 permutations. In such a case, we say that the two statistics are permutationally equivalent because they give exactly the same testing results.

In Problem 12.1 (p. 523) we ask for the permutation distribution of W from Table 12.1 (p. 453). A unique feature of rank statistics when there are no ties in the data is that the permutation distribution is the same for every such data set. That is, although the data values would change for every data set, as long as there are no ties in the 6 data points, the ranks would always be (1,2,3,4,5,6). Thus, the results for W in Table 12.1 (p. 453) would be exactly the same except in a different order, and therefore the distribution would be the same. This is one reason that rank statistics gained popularity: without ties, the exact distribution does not change and can then be tabled for easy lookup.

For simplicity we purposely started with a data set having no ties. However, ties occur frequently in real data even in continuous data settings due to rounding or inaccurate measurement. The standard way to rank data with ties is to assign the average rank to each of a set of tied values. For example, suppose our second X data point had been 7 instead of 8. Then the Z vector would have been (6,7,7,18,11,9), and instead of (1,3,2,6,5,4) for the ranks we would have (1,2.5,2.5,6,5,4). These are now called the *midranks*. We have taken the values 7 and 7 that would have occupied ranks 2 and 3 and replaced them by $(2+3)/2 = 2.5$. If the first X data point had also been a 7, then the midrank vector would have been (2,2,2,6,5,4), where we have used $(1+2+3)/3 = 2$ for the first three midranks. The use of midranks has no effect on the general permutation approach, but tabling distributions as mentioned in the previous paragraph is no longer possible since every configuration of tied values has a different permutation distribution.

12.3 The General Two-Sample Setting

The two-sample problem assumes that N experimental units (rats, for example) are available to compare two treatments A and B. First, m units are randomly assigned to receive treatment A, and the $n = N - m$ remaining units are assigned to receive treatment B. After the experiment is run, we obtain realizations of some measurement $X_1, \ldots, X_m$ for treatment A and $Y_1, \ldots, Y_n$ for treatment B. The null hypothesis H_0 is that both treatments are the same or have identical effects on the rats. In other words, if the third rat in group A whose measurement is X_3 had been assigned to group B instead, the X_3 would still have been the result under H_0 for that rat, but now it would have a Y label. In fact, we can think of all possible $\binom{N}{m}$ random assignments of m rats to group A and n rats to group B, and assume that under H_0 the individual results would be the same regardless of group assignment.

We might then formulate a test procedure as follows.

1. Randomly assign m units to A and n units to B.
2. Run the experiment to obtain $X_1, \ldots, X_m$ and $Y_1, \ldots, Y_n$.
3. Think of the collection $\boldsymbol{Z} = (X_1, \ldots, X_m, Y_1, \ldots, Y_n)$ as fixed and order the $M_N = \binom{N}{m}$ values of some statistic T calculated for each $\boldsymbol{Z}^*$ obtained by permuting $\boldsymbol{Z}$ to have different sets of m first coordinates. Call these ordered values $T_{(1)} \le T_{(2)} \le \ldots \le T_{(M_N)}$, and let $T_0 = T(\boldsymbol{X}, \boldsymbol{Y})$ be the statistic calculated for the original data.
4. Reject H_0 if $T_0 > T_{(k)}$.

This test, conditional on $\boldsymbol{Z}$, has conditional α-level

$$1 - \frac{k}{M_N}$$

if $T_{(k)} < T_{(k+1)}$ (not tied) since $M_N - k$ values of T are larger than $T_{(k)}$. The exact conditional p-value is the proportion of values greater than or equal to T_0,

$$\frac{[\#T_{(i)} \ge T_0]}{M_N}. \tag{12.2}$$

When T is the t statistic in (12.1, p. 452), the above two-sample permutation procedure was proposed by Pitman (1937a). The credit for the permutation approach, however, goes to R. A. Fisher who had earlier introduced the permutation approach in the fifth edition of *Statistical methods for Research Workers* (2×2 table example) published in 1934 and in the first edition of *The Design of Experiments* (one-sample t example) in 1935.

Besides computational problems, the main drawback of the procedure described in points 1.–4. outlined above is that:

a) the results pertain to the N units obtained and not to a larger population;
b) computations of test power are difficult.

Thus, it is often useful to assume a population sampling model of the usual form

$$X_1, \ldots, X_m \quad \text{iid} \quad F_X(x) = P(X_1 \le x),$$

$$Y_1, \ldots, Y_n \quad \text{iid} \quad F_Y(x) = P(Y_1 \le x),$$

with $H_0 : F_X(x) = F_Y(x)$. Under this model we can show that the conditional permutation test actually has exact size α unconditionally, i.e.,

$$P(\text{rejection} \mid H_0) = \alpha.$$

The permutation approach has the advantage that no assumption regarding distributions of random variables is required. Moreover, one can often show using permutational Central Limit Theorems (e.g., Theorem 12.2, p. 465) that the conditional distribution of $T(\boldsymbol{X}, \boldsymbol{Y})$ properly standardized converges to a standard normal as $\min(m,n) \to \infty$. Thus, in large samples one can use normal critical values rather than list all M_N possible values of T. Alternatively, one can randomly sample B of the possible permutations and base a test on the ordered values of $T_1, \ldots, T_B$. First we give the general theory of permutation tests and then discuss these approximations as well as the Box-Andersen F approximation.

12.4 Theory of Permutation Tests

12.4.1 Size α Property of Permutation Tests

In this subsection we show that permutation tests used in random sampling contexts can have exact size α when randomizing on rejection region boundaries, and otherwise has level α when the test is carried out without such randomization. Recall that a size α test is one for which $\sup_{H_0} P(\text{reject}H_0) = \alpha$ and level α means $\sup_{H_0} P(\text{reject}H_0) \leq \alpha$. The reference to *randomization* merely refers to flipping a biased coin for sample points on the boundary between the rejection and acceptance region in order to obtain size α and has nothing to do with the randomization used in the definition of a permutation test.

To prove size-α results rigorously, we need some additional notation. Two useful sources are Hoeffding (1952) and Puri and Sen (1971). Let $\boldsymbol{Z} = (Z_1, \ldots, Z_N)^T$ have joint distribution function $F_{\boldsymbol{Z}}(\boldsymbol{z})$ and sample space S. Let G be a group of M_N transformations of S onto S such that under H_0 the distribution of each $g_i(\boldsymbol{Z})$, $g_i \in G, i = 1, \ldots, M_N$, is exactly the same as the distribution of $\boldsymbol{Z}$. Two examples of such groups are as follows.

Permutations: G consists of all $N!$ permutations of $\boldsymbol{Z}$. If $\boldsymbol{Z}$ is exchangeable or iid, then $g_i(\boldsymbol{Z}) \stackrel{d}{=} \boldsymbol{Z}$. Although, in the two-sample problem (two independent samples), we usually consider only the $\binom{N}{m}$ partitions into two groups since the statistics used do not change by permuting elements within each sample. In the k-sample problem (k independent samples), we consider only the

$$\binom{N}{n_1 n_2 \ldots n_k} = \frac{N!}{n_1! \cdots n_k!}$$

partitions into k groups, where $n_1 + n_2 + \cdots + n_k = N$. The group of $N!$ permutations is relevant for the two-sample, k-sample, and correlation problems.

Sign Changes: G consists of all 2^N sign change transformations, $g_1(\boldsymbol{Z}) = (Z_1, Z_2, \ldots, Z_N), g_2(\boldsymbol{Z}) = (-Z_1, Z_2, \ldots, Z_N), g_3(\boldsymbol{Z}) = (Z_1, -Z_2, Z_3, \ldots, Z_N)$, etc. If the Z_i's are independently (but not necessarily identically) distributed, where each Z_i is symmetrically distributed about 0, then $g_i(\boldsymbol{Z}) \stackrel{d}{=} \boldsymbol{Z}$. The sign change group is relevant for the paired two-sample problem and the one-sample symmetry problem.

The following development is due to Hoeffding (1952). Because the permutation distribution is discrete, it is not possible to achieve arbitrarily chosen α-levels like $\alpha = .05$ without using a randomized testing procedure. This makes the details seem harder than they really are.

Let $T(z)$ be a real-valued function on S such that for each $z \in S$

$$T_{(1)}(z) \le T_{(2)}(z) \le \cdots T_{(M_N)}(z)$$

are the ordered values of $T(g_i(z)), i = 1, \dots, M_N$. Given α, $0 < \alpha < 1$, let k be defined by

$$k = M_N - [M_N \alpha],$$

where $[\cdot]$ is the greatest integer function. Let $M_N^+(z)$ and $M_N^0(z)$ be the numbers of $T_{(j)}(z), j = 1, \dots, M_N$, which are greater than $T_{(k)}(z)$ and equal to $T_{(k)}(z)$, respectively. Define

$$a(z) = \frac{M_N \alpha - M_N^+(z)}{M_N^0(z)}.$$

Then define the test function $\phi(z)$ by

$$\phi(z) = \begin{cases} 1, & \text{if } T(z) > T_{(k)}(z); \\ a(z), & \text{if } T(z) = T_{(k)}(z); \\ 0, & \text{if } T(z) < T_{(k)}(z). \end{cases}$$

Note that for a test function, $\phi(z) = 1$ means rejection of H_0, $\phi(z) = 0$ means acceptance of H_0, and $\phi(z) = \pi$ means to randomly reject H_0 with probability π. The test defined by ϕ is an exact conditional level α test by construction. The following theorem tells us that under $g_i(\boldsymbol{Z}) \overset{d}{=} \boldsymbol{Z}$ for each $g_i \in G$, the test is unconditionally a size-α test.

Theorem 12.1. *(Hoeffding). Let the data* $\boldsymbol{Z} = (Z_1, \dots, Z_N)$ *and the group* G *of transformations be such that* $g_i(\boldsymbol{Z}) \overset{d}{=} \boldsymbol{Z}$ *for each* $g_i \in G$ *under* H_0. *Then the test defined above by* $\phi(\boldsymbol{Z})$ *has size* α.

Proof. First note that by the definition of $a(z)$ and ϕ, we have for each $z \in S$

$$\frac{1}{M_N} \sum_{i=1}^{M_N} \phi(g_i(z)) = \frac{M_N^+ + a(z) M_N^0(z)}{M_N} = \alpha.$$

Now since $g_i(\boldsymbol{Z}) \overset{d}{=} \boldsymbol{Z}$ and G is a group, $\mathrm{E}_{H_0}\phi(\boldsymbol{Z}) = \mathrm{E}_{H_0}\phi(g_i(\boldsymbol{Z}))$ for each i, and

$$P_{H_0}(\text{rejection}) = E_{H_0}\phi(\boldsymbol{Z}) = \frac{1}{M_N} \sum_{i=1}^{M_N} \mathrm{E}_{H_0}\phi(g_i(\boldsymbol{Z}))$$

$$= \mathrm{E}_{H_0}\left[\frac{1}{M_N} \sum_{i=1}^{M_N} \phi(g_i(\boldsymbol{Z}))\right] = \alpha.$$

■

The above proof is deceptively simple. The key fact that makes it work is that $\mathrm{E}_{H_0}\phi(g_i(\boldsymbol{Z}))$ is the same for each g_i including $g(\boldsymbol{Z}) = \boldsymbol{Z}$. This fact rests on the identical distribution of $g_i(\boldsymbol{Z})$ for each i and on the group nature of G. The identical distribution requirement is intuitive, but why do we need G to be a group? Recall that the test procedure consists of computing T for each member of G and then rejecting if $T(\boldsymbol{Z})$ is larger than an order statistic of the $T(g_i(\boldsymbol{Z}))$ values. Now $\phi(g_i(\boldsymbol{Z}))$ is the test that computes $T(g_j(g_i(\boldsymbol{Z})))$, $j = 1, \ldots, M_N$, orders all of them, and rejects if $T(g_i(\boldsymbol{Z}))$ is larger than one of the ordered values. If G is not a group, then the set of ordered values will not be the same for each test $\phi(g_i(\boldsymbol{Z}))$ because $g_j(g_i)$ will not be in G for some i and j. Since the sets of ordered values could be different, there would be no basis for believing that a test based on $g_i(\boldsymbol{Z})$ would have the same expectation as that based on $\boldsymbol{Z}$.

Note also that the use of $a(z)$ in $\phi(z)$ is a way of randomizing to get an exact size-α test. In practice we might just define $\phi(z)$ to be one if $t(z) > t_{(k)}(z)$ and zero otherwise. The resulting unconditional level is a weighted average of the discrete levels less than or equal to α and will usually be less than α.

The conditional test procedure described in 1) – 4) may be used for any test statistic, but the rejection region in Step 4) should be modified to correspond to the situation. For example, the alternative hypothesis might be that the mean of A is less than that of B. We would then look for small values of t. Or the test could be two-sided and we would reject if $t < t_{(k)}$ or if $t > t_{(m)}$.

12.4.2 Permutation Moments of Linear Statistics

The exact permutation distribution may be difficult to compute. For certain linear statistics, though, we can calculate the moments of the permutation distribution quite easily. These moments are then used in the various normal and F approximations found in later sections.

We consider general results for situations associated with the group of transformations consisting of all permutations. These situations include the two-sample and k-sample situations, and bivariate data $(X_1, Y_1), \ldots, (X_N, Y_N)$ where correlation and regression of Y on X are of interest. Let $\boldsymbol{a} = (a_1, \ldots, a_N)$ and $\boldsymbol{c} = (c_1, \ldots, c_N)$ be two vectors of real constants. We select a random permutation of the a values, call them $A_1, \ldots, A_N$, and form the statistic

$$T = \sum_{i=1}^{N} c_i A_i. \tag{12.3}$$

In applications $\boldsymbol{a}$ is actually the observed vector $\boldsymbol{Z}$ (or a function of $\boldsymbol{Z}$ such as the rank vector), and $\boldsymbol{c}$ is chosen for the particular problem at hand. For example, in the two-sample problem, with $\boldsymbol{a} = \boldsymbol{Z}$ and $c_i = 0$ for $i = 1, \ldots, m$ and 1 otherwise, the observed value of T for the original data is $\sum_{i=1}^{n} Y_i$, and here $T = \sum_{i=m+1}^{N} A_i$ is a sum of the last n elements of a random permutation of $\boldsymbol{Z}$. A very important subclass of (12.3) are the linear rank statistics given in the next section.

Assuming that each permutation of $\boldsymbol{A}$ is equally likely and thus has probability $1/N!$, it is easy to see that

$$P(A_i = a_s) = \frac{1}{N} \quad \text{for } s = 1, \ldots, N,$$

and

$$P(A_i = a_s, A_j = a_t) = \frac{1}{N(N-1)} \quad \text{for } s \neq t = 1, \ldots, N.$$

Then, using those two results, we get

$$\mathrm{E}(A_i) = \frac{1}{N}\sum_{i=1}^{N} a_i \equiv \overline{a}, \quad \text{for } i = 1, \ldots, N,$$

$$\mathrm{Var}(A_i) = \frac{1}{N}\sum_{i=1}^{N}(a_i - \overline{a})^2, \quad \text{for } i = 1, \ldots, N,$$

and

$$\mathrm{Cov}(A_i, A_j) = \frac{-1}{N(N-1)}\sum_{i=1}^{N}(a_i - \overline{a})^2, \quad \text{for } i \neq j = 1, \ldots, N.$$

Finally, putting these last three results together, we get

$$\mathrm{E}(T) = N\overline{c}\,\overline{a},$$

and

$$\mathrm{Var}(T) = \frac{1}{N-1}\sum_{i=1}^{N}(c_i - \overline{c})^2 \sum_{j=1}^{N}(a_j - \overline{a})^2, \tag{12.4}$$

where $\overline{a}$ and $\overline{c}$ are the averages of the a's and c's, respectively. These first two moments of T are sufficient for normal approximations based on the asymptotic normality of T as $N \to \infty$. In some cases it may be of value to use more complex approximations involving the third and fourth moments of T. Thus, the central third moment is

$$\mathrm{E}\{T - \mathrm{E}(T)\}^3 = \frac{N}{(N-1)(N-2)}\sum_{i=1}^{N}(c_i - \overline{c})^3 \sum_{j=1}^{N}(a_j - \overline{a})^3,$$

and the standardized third moment (skewness coefficient) is

$$\mathrm{Skew}(T) = \frac{\mathrm{E}\{T - \mathrm{E}(T)\}^3}{\{\mathrm{Var}(T)\}^{3/2}} = \frac{(N-1)^{1/2}}{(N-2)} \frac{\mu_3(\boldsymbol{c})\mu_3(\boldsymbol{a})}{\{\mu_2(\boldsymbol{c})\mu_2(\boldsymbol{a})\}^{3/2}},$$

where we have introduced the notation $\mu_q(\boldsymbol{c}) = N^{-1}\sum_{i=1}^{N}(c_i - \overline{c})^q$ for $q \geq 2$. Similarly the standardized central fourth moment (kurtosis coefficient) is

$$\begin{aligned}\text{Kurt}(T) = \frac{\text{E}\{T - \text{E}(T)\}^4}{\{\text{Var}(T)\}^2} &= \frac{(N+1)(N-1)}{N(N-2)(N-3)}\frac{\mu_4(\boldsymbol{c})\mu_4(\boldsymbol{a})}{\{\mu_2(\boldsymbol{c})\mu_2(\boldsymbol{a})\}^2} \\ &- \frac{3(N-1)^2}{N(N-2)(N-3)}\left[\frac{\mu_4(\boldsymbol{c})}{\{\mu_2(\boldsymbol{c})\}^2} + \frac{\mu_4(\boldsymbol{a})}{\{\mu_2(\boldsymbol{a})\}^2}\right] \\ &+ \frac{3(N^2-3N+3)(N-1)}{N(N-2)(N-3)}.\end{aligned}$$

12.4.3 Linear Rank Tests

Many popular rank tests have the general form

$$T = \sum_{i=1}^{N} c(i)a(R_i) \tag{12.5}$$

of a *linear rank statistic*, where $c(1), \ldots, c(N)$ are called the *regression constants* and $a(1), \ldots, a(N)$ are called the *scores*, and $\boldsymbol{R}$ is the vector of ranks (possibly midranks due to ties) of some data vector $\boldsymbol{Z}$. There is a room for confusion here in the use of the notation for $\boldsymbol{c}$ and $\boldsymbol{a}$, because in the general notation of the last section, $(c_1, \ldots, c_N)$ and $(a_1, \ldots, a_N)$ are vectors of real numbers, but here $c(\cdot)$ and $a(\cdot)$ are functions so that $c_1 = c(1), \ldots, c_N = c(N)$ and $a_1 = a(1), \ldots, a_N = a(N)$. This function notation just makes it easier to work with rank statistics. In particular, the score functions $a(\cdot)$ are typically derived from *scores generating functions* ϕ via $a(i) = \phi(i/(N+1))$. In tied rank situations, $a(\cdot)$ needs to be defined for non-integer values.

The simplest setting is the two-sample problem where $\boldsymbol{Z}^T = (X_1, \ldots, X_m, Y_1, \ldots, Y_n)$ and the c values are all zeroes for the Xs and ones for the Ys or vice-versa. A different situation covered by T, though, is for trend alternatives, where $c(1), \ldots, c(N)$ are the integers $1, \ldots, N$ and $T = \sum_{i=1}^{N} iR_i$ will tend to be large when Z_{i+1} tends to be larger than Z_i. A related problem is for N independent pairs $(X_1, Y_1), \ldots, (X_N, Y_N)$. Here, tests based on Spearman's Correlation (Section 12.7, p. 487) are equivalent to ones having the same null distribution as $T = \sum_{i=1}^{N} iR_i$.

Clearly T in (12.5) is a subclass of the linear permutation statistics given in (12.3, p. 458). Thus results for that class are inherited by T. For example, if $\boldsymbol{R}$ is uniformly distributed on the permutations of $1, \ldots, N$ (no tied ranks), then

$$\text{E}(T) = N\overline{c}\,\overline{a},$$

and

$$\mathrm{Var}(T) = \frac{1}{N-1}\sum_{i=1}^{N}(c(i)-\overline{c})^2\sum_{j=1}^{N}(a(j)-\overline{a})^2,$$

where of course $\overline{c}$ and $\overline{a}$ are the means of the c and a values, respectively. For a tied rank situation with observed vector of midranks $\boldsymbol{R}$, the expressions above still hold but with $a(j)$ replaced by $a(R_j)$.

For deciding on a score function in a given problem, we first select a parametric family and then derive an optimal score function for that family. An overview of how to do this is given in Section 12.5 (p. 473). The most important linear rank statistic is the Wilcoxon Rank Sum. So we give a few more details about it in the next section.

12.4.4 Wilcoxon-Mann-Whitney Two-Sample Statistic

For two independent samples $X_1,\ldots,X_m$ and $Y_1,\ldots,Y_n$, Wilcoxon (1945) introduced the linear rank statistic

$$W = \sum_{i=m+1}^{N} R_i, \tag{12.6}$$

where $R_1,\ldots,R_N$ are the joint rankings of $\boldsymbol{Z} = (X_1,\ldots,X_m,Y_1,\ldots,Y_n)^T$, $N = m+n$. The Wilcoxon Rank Sum test has a number of optimal properties that are mentioned in Section 12.5 (p. 473). Along with the Wilcoxon Signed Rank test for paired data (Section 12.8.3, 494), it is the simplest and most important rank test.

Independently, Mann and Whitney (1947) proposed the equivalent statistic

$$W_{\mathrm{YX}} = \sum_{i=1}^{m}\sum_{j=1}^{n} I(Y_j < X_i), \tag{12.7}$$

where $I(\cdot)$ is the indicator function. In the absence of ties $W_{\mathrm{YX}} = mn + n(n+1)/2 - W$. Another equivalent version is

$$W_{\mathrm{XY}} = \sum_{i=1}^{m}\sum_{j=1}^{n} I(Y_j > X_i), \tag{12.8}$$

with $W_{\mathrm{XY}} = W - n(n+1)/2$. We prefer this latter version and define the U-statistic estimator of $\theta_{\mathrm{XY}} = P(Y_1 > X_1)$

$$\widehat{\theta}_{\mathrm{XY}} = \frac{W_{\mathrm{XY}}}{mn} = \frac{1}{mn}\sum_{i=1}^{m}\sum_{j=1}^{n} I(Y_j > X_i). \tag{12.9}$$

In a clinical trial, θ_{XY} can be viewed as the probability of a more favorable response for a randomly selected patient getting Treatment 2 compared to another patient getting Treatment 1. For screening tests where a "positive" is declared if $Y > c$ for a diseased subject or if $X > c$ for a non-diseased subject, then θ_{XY} is the area under the receiver operating characteristic (ROC) curve. This interpretation is developed in Problem 12.8 (p. 525).

For hand computations, W is much easier to handle than these U-statistic versions. The null moments follow easily from Section 12.4.2 (p. 458) after noting that $c(1) = \cdots = c(m) = 0$ and $c(m+1) = \cdots = c(N) = 1$ lead to $\overline{c} = n/N$ and $\sum_{i=1}^{N}(c(i) - \overline{c})^2 = mn/N$. The null mean is $n(N+1)/2$ whether there are ties or not. The variance follows from (12.4, p. 459). With no ties, we have

$$\text{Var}(W) = \frac{mn(N+1)}{12}. \tag{12.10}$$

With ties so that $(R_1, \ldots, R_N)$ are the tied ranks, we have

$$\text{Var}(W) = \frac{mn}{N(N-1)}\left\{\sum_{i=1}^{N} R_i^2 - \frac{N(N+1)^2}{4}\right\}. \tag{12.11}$$

Lehmann (1975, p. 20) gives a different expression for the variance of W in the face of ties,

$$\text{Var}(W) = \frac{mn(N+1)}{12} - \frac{mn\sum_{i=1}^{e}(d_i^3 - d_i)}{12N(N-1)}, \tag{12.12}$$

where e are the number of tied groups, and d_i is the number of tied observations in each group. For example, with the simple example data modified to $(\{6, 7\}, \{7, 18, 11, 9\})$, the midranks are $(1, 2.5, 2.5, 6, 5, 4)$ and $e = 1$, $d_1 = 2$; so $\text{Var}(W) = (2)(4)(6+1)/12 - (2)(4)[2^3 - 2]/[12(6)(5)] = 4.53$. Expression (12.12) may be easier to use by hand than (12.11), but its main value may be to show that the variance of W for tied data is always smaller than (12.10) for untied data.

The U-statistic versions in (12.7)–(12.9) are useful for easy calculation of moments and derivation of asymptotic normality under non-null distributions. For example, using equation (3.4.7, p. 91) of Randles and Wolfe (1979) for the variance of a two-sample U-statistic from independent iid samples, we have that

$$\text{Var}(\widehat{\theta}_{XY}) = \frac{1}{mn}\left\{(m-1)(\gamma_{0,1} - \theta_{XY}^2) + (n-1)(\gamma_{1,0} - \theta_{XY}^2) + \gamma_{1,1} - \theta_{XY}^2\right\}, \tag{12.13}$$

where in the absence of ties $\gamma_{0,1} = P(Y_1 > X_1, Y_1 > X_2)$, $\gamma_{1,0} = P(Y_1 > X_1, Y_2 > X_1)$, and $\gamma_{1,1} = \theta_{XY} = P(Y_1 > X_1)$. If the X and Y have identical continuous distributions, then it is easy to show that $\gamma_{0,1} = \gamma_{1,0} = 1/3$ and $\gamma_{1,1} = \theta_{XY} = 1/2$ and (12.13) reduces to (12.10).

In the presence of ties, the U-statistic quantities need to be modified by adding $I(Y_j = X_i)/2$ to the indicators in the sums. For example,

$$\widehat{\theta}_{\mathrm{XY}} = \frac{W_{\mathrm{XY}}}{mn} = \frac{1}{mn}\sum_{i=1}^{m}\sum_{j=1}^{n}\{I(Y_j > X_i) + I(Y_j = X_i)/2\}. \qquad (12.14)$$

The relationships $W_{\mathrm{YX}} = mn + n(n+1)/2 - W$ and $W_{\mathrm{XY}} = W - n(n+1)/2$ then continue to hold. The definitions of $\gamma_{0,1}$, $\gamma_{1,0}$, and $\gamma_{1,1}$ for use in (12.13) have to be modified in the face of ties; see, for example, Boos and Brownie (1992, p. 72). In the next section we give the basic asymptotic normal results for linear statistics under the null hypothesis of identical populations. Those general results are useful for approximate critical regions for permutation and rank statistics. However, the Wilcoxon statistics are special because they are related to the U-statistic $\widehat{\theta}_{\mathrm{XY}}$ for which a large body of theory exists. In particular, $\widehat{\theta}_{\mathrm{XY}}$ is $\mathrm{AN}\left\{\theta_{\mathrm{XY}}, \mathrm{Var}(\widehat{\theta}_{\mathrm{XY}})\right\}$, and this follows from basic U-statistic theory with no assumptions except that $X_1, \ldots, X_m$ are iid with any distribution function $F(x)$, and $Y_1, \ldots, Y_n$ are iid with any distribution function $G(x)$. Because this asymptotic result is not just for null situations, it helps us think about i) the form of the alternative hypothesis, ii) the classes of distribution functions for which the Wilcoxon Rank Sum is consistent, in other words, rejects with probability converging to 1, and iii) asymptotic power and sample size determination. We now discuss these ideas.

In general, the null hypothesis of interest is

$$H_0 : F(x) = G(x), \text{ each } x \in (-\infty, \infty).$$

However, the alternative hypothesis can be formulated in several ways. The most common way is to assume the shift model $G(x) = F(x - \Delta)$, and then the alternative hypothesis is purely in terms of Δ, for example

$$H_1 : \Delta > 0.$$

Another popular, more nonparametric, way to phrase the alternative is

$$H_2 : F(x) \geq G(x), \text{ each } x \in (-\infty, \infty),$$

and with strict inequality for at least one x. Here, G is said to be *stochastically larger* than F. Clearly, H_2 is a larger class of alternatives since $(F, G) \in H_1$ implies $(F, G) \in H_2$. Lastly, the natural alternative when thinking in terms of $\widehat{\theta}_{\mathrm{XY}}$ is

$$H_3 : \theta_{\mathrm{XY}} > \frac{1}{2}.$$

Now if F and G are continuous distribution functions and $(F, G) \in H_2$, then $(F, G) \in H_3$. This follows from

$$\theta_{XY} = P(Y_1 > X_1) = \int\int I(y > x)\, dF(x)\, dG(y) = \int \{1 - G(x)\}\, dF(x),$$

after noting that if continuous distribution functions satisfy $F(x) > G(x)$ for at least one x, then this strict inequality must hold for an interval of x values, and $\int F(x)\, dF(x) = 1/2$. Assuming that H_3 holds, then the Wilcoxon Rank Sum test is consistent because of the general asymptotic normality result mentioned above. This also means that it is also consistent under alternatives H_1 and H_2.

Lastly, following Noether (1987), the approximate power of a one-sided α level test when $\theta_{XY} > \frac{1}{2}$ is given by

$$1 - \Phi\left\{\frac{1/2 - \theta_{XY}}{\rho\sigma_0} + \frac{\Phi^{-1}(1-\alpha)}{\rho}\right\}, \tag{12.15}$$

where σ_0 is the square root of the null variance of W (12.10, p. 462), ρ is the ratio of the square root of the non-null variance of W (m^2n^2 times eq. 12.13, p. 462) to σ_0, and Φ is the standard normal distribution function. Typically, ρ is close to 1. Letting $\rho = 1$ and $m = \lambda N$, the total sample size N required to have power $1 - \beta$ for alternative θ_{XY} is given by Noether (1987) to be

$$N = \frac{\{\Phi^{-1}(1-\alpha) + \Phi^{-1}(1-\beta)\}^2}{12\lambda(1-\lambda)(\theta_{XY} - 1/2)^2}. \tag{12.16}$$

This is a fairly simple formula, but it might be preferable to state power and sample size in terms of the shift model. Plugging in $G(x) = F(x - \Delta)$, we have

$$\theta_{XY} = P(Y_1 > X_1) = \int \{1 - F(x - \Delta)\}\, dF(x).$$

For example, if we wanted shifts of size Δ/σ in a normal(μ, σ^2) population, then a simple R program to get θ_{XY} using the midpoint rule is

```
theta.xy<-function(delta,n=10000){
# u-stat parameter for normal shift delta/sigma
# for sigma=1
# n is the number of points for midpoint rule
    points<-(2*(1:n)-1)/(2*n)
    mean(1-pnorm(qnorm(points)-delta))
}
```

If $\Delta/\sigma = .5$, then

```
> theta.xy(.5,10000)
[1] 0.6381632
```

so that $\theta_{XY} = .638$. Choosing $\alpha = .05$, $\beta = .80$, and $\lambda = 1/2$, we find $N = 108$ or $m = n = 54$.

12.4.5 Asymptotic Normal Approximation

Approximate normal distributions for linear statistics have been the most popular approximation to permutation distributions, especially for rank statistics. Here we use the following permutation Central Limit Theorem for $T = \sum_{i=1}^{N} c_i A_i$, introduced in (12.3, p. 458), directly from Puri and Sen (1971, p. 73) who give credit to Wald and Wolfowitz (1944), Noether (1949), and Hoeffding (1951). The notation $\mu_q(\boldsymbol{c})$ is for the qth central moment $N^{-1}\sum_{i=1}^{N}(c_i - \overline{c})^q$.

Theorem 12.2 (Wald-Wolfowitz-Noether-Hoeffding). *If for $N \to \infty$*

(i)

$$\frac{\mu_q(\boldsymbol{c})}{\mu_2(\boldsymbol{c})^{q/2}} = O(1) \quad \textit{for all } q = 3, 4, \ldots$$

(ii)

$$\frac{\mu_q(\boldsymbol{a})}{\mu_2(\boldsymbol{a})^{q/2}} = o(N^{r/2-1}) \quad \textit{for all } q = 3, 4, \ldots,$$

then

$$\frac{T - E(T)}{\sqrt{Var(T)}} \xrightarrow{d} N(0,1).$$

In a particular problem either or both of the vectors $\boldsymbol{c}$ and $\boldsymbol{a}$ may be random, that is, calculated from the data $\boldsymbol{Z}$. In such cases we would need to show that the appropriate conditions (i) and/or (ii) hold $wp1$ with respect to the random vector $\boldsymbol{Z}$. Moreover, the conclusion of Theorem 12.2 is that the permutation distribution of the standardized T converges to a standard normal distribution with probability one with respect to $\boldsymbol{Z}$.

In the case of linear rank statistics without ties, we can give a much simpler theorem due to Hajek (1961). We follow the exposition given in Randles and Wolfe (1979, Ch. 8) and state their version of Hajek's theorem.

Theorem 12.3 (Hajek). *Let $T = \sum_{i=1}^{N} c(i)a(R_i)$ be the linear rank statistic, where the rank vector $\boldsymbol{R}$ comes from data vector $\boldsymbol{Z}$ that is continuous (no ties with probability one) and exchangeable, the constants $c(1), \ldots, c(N)$ satisfy the Noether condition*

$$\frac{\sum_{i=1}^{N}(c(i) - \overline{c})^2}{\max_{1 \le i \le N}(c(i) - \overline{c})^2} \to \infty \quad \textit{as } N \to \infty, \tag{12.17}$$

and the scores have the form $a(i) = \phi(i/(N+1))$, where ϕ can be written as the difference of two nondecreasing functions and $0 < \int_0^1 \phi(t)^2 dt < \infty$ and $\int_0^1 |\phi(t)| dt < \infty$. Then T is $AN\{N\overline{c}\,\overline{a}, Var(T)\}$ as $N \to \infty$.

It has been customary to use the normal approximation with rank statistics, often with a continuity correction. For example, in the two-sample problem, consider the Wilcoxon Rank Sum W of (12.6, p. 461). Note that for application of Theorem 12.3 above, $\phi(u) = u$, and the theorem actually applies directly to $W/(N+1)$. For the simple example of Section 1.2 where $\boldsymbol{z} = (\boldsymbol{x}, \boldsymbol{y}) = (6, 8, 7, 18, 11, 9)$ with ranks $\boldsymbol{R} = (1, 3, 2, 6, 5, 4)$, we find $W = 17$, $\mathrm{E}(W) = 4(6+1)/2 = 14$, $\mathrm{Var}(W) = (2)(4)(6+1)/12 = 14/3$ (from 12.10, p. 462), and the normal approximation p-value is

$$p \approx P\left(N(0,1) \geq \frac{17-14}{\sqrt{14/3}}\right) = P(N(0,1) \geq 1.39) = 0.08.$$

With continuity correction the normal approximation p-value is

$$p \approx P\left(N(0,1) \geq \frac{17-14-1/2}{\sqrt{14/3}}\right) = P(N(0,1) \geq 1.16) = 0.12.$$

Lehmann (1975, p. 16) cites Kruskal and Wallis (1952, p. 591) with the recommendation that the continuity correction be used when the probability is above 0.02. Recall that the exact null distribution of W can be obtained from Table 12.1 leading to the usual p-value $P(W \geq 17) = 2/15 = 0.13$ which is closer to the continuity corrected value.

When there are tied values, we can still use the normal approximation with W, but we must be sure to use the null variance from (12.11, p. 462) or (12.12, p. 462) and not from (12.10, p. 462). Lehmann (1975, p. 20) does not use the continuity correction in the presence of ties.

We can also look at approximations to the permutation p-value of $T = \sum_{i=1}^{n} Y_i$ which is permutationally equivalent to the two-sample t statistic. For the simple example $\boldsymbol{c} = (0, 0, 1, 1, 1, 1)$ and $\boldsymbol{a} = \boldsymbol{z} = (6, 8, 7, 18, 11, 9)$. Thus, $\mathrm{E}(T) = (6)(4/6)(59/6) = 39.33$, $\mathrm{Var}(T) = 25.23$, and the normal approximation p-value is

$$p \approx P\left(N(0,1) \geq \frac{45-39.33}{\sqrt{25.23}}\right) = P(N(0,1) \geq 1.13) = 0.13.$$

This seems almost too good an approximation to the true permutation p-value of $2/15 = 0.13$. Usually the t approximation p-value is more accurate, but here it is $P(t_4 \geq 1.17) = 0.15$.

12.4.6 Edgeworth Approximation

Edgeworth approximations were mentioned briefly in Ch. 3 (5.6, p. 219) and Ch. 9 (11.7, p. 426). Basically, an Edgeworth expansion is an approximation to the distribution function of an asymptotically normal statistic. It is based on

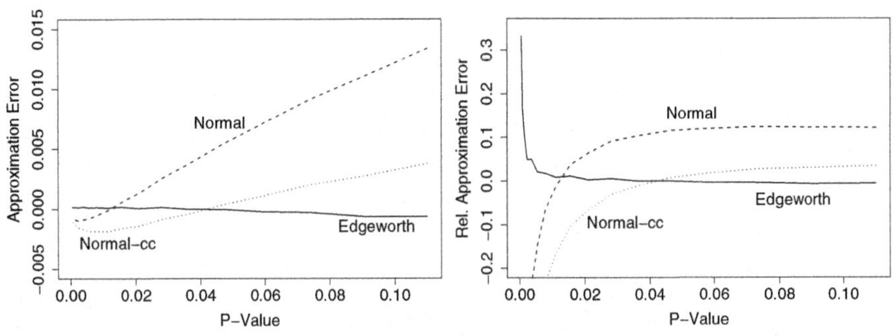

Fig. 12.1 Error (Left Panel) and relative error (Right Panel) of approximations to Wilcoxon Rank Sum p-values for $m = 10, n = 6$: normal approximation, normal approximation with continuity correction, and the Edgeworth approximation in (12.18, p. 467)

estimation of Skew and/or Kurt and other higher moments of the statistic. Rigorous development of Edgeworth expansions for general permutation statistics under the null hypothesis may be found in Bickel (1974), Bickel and van Zwet (1978), and Robinson (1980). However, it has not proved of much practical use for obtaining critical values or p-values of permutation statistics except in the special case of the Wilcoxon Rank Sum W and of the one-sample Wilcoxon signed rank statistic.

Here we give the approximation for W originally due to Fix and Hodges (1955). For $W = \sum_{i=1}^{n} R_i$,

$$P(W \geq w) \approx 1 - \Phi(t) - \left\{ \frac{m^2 + n^2 + mn + m + n}{20mn(m + n + 1)} \right\} (t^3 - 3t)\phi(t), \quad (12.18)$$

where ϕ and Φ are the standard normal density and distribution function, respectively, and $t = \{w - \mathrm{E}(W) - 1/2\}/\sqrt{\mathrm{Var}(W)}$, $\mathrm{E}(W) = n(N + 1)/2$, $\mathrm{Var}(W) = mn(N + 1)/12$.

Figure 12.1 gives the error = true p-value − (12.18) and the relative error = [true p-value − (12.18)]/(true p-value) of (12.18) compared to the true p-value and similar quantities for the normal approximations. The range of the p-values is most of the right tail of the distribution function of W plotted in reverse order, that is, 0.0005 to 0.11. The Edgeworth approximation is excellent for p-values larger than 0.0024, but then deteriorates as the p-value gets very small. For example, when the true p-value is 0.00087, the Edgeworth approximation is 0.00073, and at 0.00025 it is 0.00009. The right panel of Figure 12.1 is especially helpful for illuminating what happens at small p-values. The normal approximation is much cruder, and below 0.02 we can see that the continuity correction is no longer useful.

Figure 12.1 suggests that (12.18) can be used for most values of W, thus essentially replacing tabled values of the distribution of W. However, when there are ties in the data, (12.18) as well as tabled values are no longer correct, and the exact permutation distribution (or a Monte Carlo approximation) is required.

12.4.7 Box-Andersen Approximation

Pitman (1937a,b) and Welch (1937) pioneered an approximation to permutation distributions that was modernized by Box and Andersen (1955) and Box and Watson (1962). These later authors mainly used the approach to show the Type I error robustness of F statistics for tests comparing means and the nonrobustness of tests comparing variances. However, we follow the Box and Andersen (1955) formulation since it is the most straightforward.

The basic idea of the approximation is to get F statistics into their equivalent "beta" version, then match the first two permutation moments of this beta version to what one gets from the first two moments of a beta distribution with degrees of freedom multiplied by a constant d. Solving for d leads to the approximation of the permutation distribution of the F statistics by an F distribution with usual degrees of freedom multiplied by d. We develop the approximation here for the two-sample problem and later give it for one-way and two-way ANOVA situations.

The square of the t statistic in (12.1, p. 452) may be written in the one-way ANOVA F form

$$t^2 = \frac{m(\overline{X}-\overline{Z})^2 + n(\overline{Y}-\overline{Z})^2}{s_p^2} = \frac{\text{SSTR}}{\text{SSE}/(N-2)}, \tag{12.19}$$

where recall we use the Z's to denote all the X and Y values thrown together, and SSTR and SSE are sums of squares for treatments and error, respectively. Using the fact that $\sum_{i=1}^{N}(Z_i - \overline{Z})^2 = \text{SSTR} + \text{SSE}$, we have for the beta version of the F statistic

$$b(t^2) = \frac{t^2}{t^2 + N - 2} = \frac{\text{SSTR}}{\sum_{i=1}^{N}(Z_i - \overline{Z})^2}.$$

Note that for normal data under the null hypothesis, $b(t^2)$ has a beta$(1/2, (N-2)/2)$ distribution. Originally $b(t^2)$ was used with the beta critical values rather than t^2 with $F(1, N-2)$ critical values. Although, t^2 and $b(t^2)$ are equivalent test statistics, for permutation analysis $b(t^2)$ is much simpler because the denominator is constant over permutations. Thus, the first permutation moment is

$$E_P\{b(t^2)\} = \frac{m\text{Var}_P(\overline{X}) + n\text{Var}_P(\overline{Y})}{\sum_{i=1}^{N}(Z_i - \overline{Z})^2} = \frac{1}{N-1},$$

where we have used (12.4, p. 459) to get

$$\text{Var}_P(\overline{X}) = \frac{n\sum_{i=1}^{N}(Z_i - \overline{Z})^2}{mN(N-1)} \qquad \text{Var}_P(\overline{Y}) = \frac{m\sum_{i=1}^{N}(Z_i - \overline{Z})^2}{nN(N-1)}.$$

Note also that under normal theory $\mathrm{E}\{b(t^2)\} = 1/2/(1/2+(N-2)/2) = 1/(N-1)$ from the beta distribution. Thus, the normal theory and permutation first moments of $b(t^2)$ are both $1/(N-1)$. The next step is to calculate the permutation variance of $b(t^2)$ (involving fourth moments), equate it to the variance of a beta$(d/2, d(N-2)/2)$ distribution, $2(N-2)/[d(N-1)(N+3)]$, and solve for d. Box and Andersen (1955, p. 13) give d for the general one-way ANOVA situation with k groups and sample sizes $n_1, n_2, \ldots, n_k$:

$$d = 1 + \left(\frac{N+1}{N-1}\right)\frac{c_2}{(N^{-1}+A)^{-1} - c_2}, \tag{12.20}$$

where

$$A = \frac{N+1}{2(k-1)(N-k)}\left(\frac{k^2}{N} - \sum_{i=1}^{k}\frac{1}{n_i}\right),$$

$c_2 = k_4/k_2^2$,

$$k_2 = \frac{1}{N-1}\sum_{i=1}^{N}(Z_i - \overline{Z})^2, \tag{12.21}$$

$$k_4 = \frac{N(N+1)\sum_{i=1}^{N}(Z_i - \overline{Z})^4 - 3(N-1)\left\{\sum_{i=1}^{N}(Z_i - \overline{Z})^2\right\}^2}{(N-1)(N-2)(N-3)}. \tag{12.22}$$

The statistics k_2 and k_4 are unbiased estimators of the population cumulants introduced in Chapter 1.

For our two-sample t^2, $k = 2$, $n_1 = m$, $n_2 = n$, $m + n = N$, and the Pitman-Welch-Box-Andersen approximation is to compare t^2 to an $F(d, d(m+n-2))$ distribution. Box and Andersen (1955) show that $\mathrm{E}(d) \approx 1 + (\text{Kurt} - 3)/N$ under the null hypothesis of sampling from equal populations with kurtosis Kurt. Thus, t^2 with the usual $F(1, (m+n-2))$ is quite Type I error robust to nonnormality since the correction d is relatively small for moderate size N. Also, for long-tailed distributions with thicker tails than the normal distribution, Kurt >3 and thus $d > 1$, so that using the $F(1, (m+n-2))$ critical values results in conservative tests, that is, true test levels less than the nominal α values. For example, with Laplace data, Kurt $= 6$ and $d \approx 1+3/N$; at $m = n = 10$ $d \approx 1.15$, and a nominal $\alpha = .05$ level test would actually have true level approximately .043. For continuous uniform data, Kurt $= 1.8$; at $m = n = 10$ $d \approx .94$ and a nominal $\alpha = .05$ level test would have true level approximately .053. Since these deviations from α are small, common practice is to just use the standard $F(1, (m+n-2))$ reference distribution with the t^2 statistic rather than the permutation distribution or an approximation to it.

Although t^2 is Type I error robust in the face of outliers, it loses power because outliers inflate the variance estimate in the denominator of t^2. Thus t^2 is not Type II error robust when sampling from distributions heavier-tailed than the normal. In contrast, as we mentioned in the Chapter introduction, the Wilcoxon Rank Sum statistic W is Type II error robust, and later we use asymptotic power calculations to verify its superiority to t^2. But for the moment, we note that W is related to t^2 applied to the ranks of the data, and therefore inherits robustness to outliers because the ranks themselves are resistant to the effects of outliers. This relationship also allows us to use the above approximation for the permutation distribution of W.

Define the standardized Wilcoxon Rank Sum statistic by

$$W_S = \frac{W - E(W)}{\{\text{Var}(W)\}^{1/2}}.$$

Then, t^2 applied to the ranks of the observations, that is, the X ranks $R_1, \ldots, R_m$ replacing $X_1, \ldots, X_m$, and the Y ranks $R_{m+1}, \ldots, R_N$ replacing $Y_1, \ldots, Y_n$, results in

$$t_R^2 = \frac{(N-2)W_S^2}{N - 1 - W_S^2}.$$

Thus t_R^2 and W are equivalent test statistics and we can apply the Box-Andersen approximation to t_R^2 using $d \approx 1 + (1.8 - 3)/N$ because the ranks are a uniform distribution on the integers 1 to N and thus have Kurt $\approx$ 1.8, the kurtosis of a continuous uniform distribution. For example, in the case of $m = 10$ and $n = 6$ given in Figure 12.1 (p. 467), the Box-Andersen approximation along with the continuity correction gives results that are considerably better than the normal approximation with continuity correction but not quite as good as the Edgeworth approximation. In later sections we see that the Box-Andersen approximation is very good in one-way and two-way ANOVA situations when the number of treatments is greater than two.

12.4.8 Monte Carlo Approximation

In the previous sections, approximations to permutation distributions were given for statistics based on linear forms, and essentially rely on the Central Limit Theorem and its extensions. However, the simplest and most important approximation to a permutation distribution is to randomly sample from the set of all possible permutations, and directly estimate the permutation distribution. This approach can be used for any statistic T, and its accuracy is determined simply by the number B of random permutations used. This resampling of permutations is very similar to resampling in the bootstrap world, and we suggest sampling with replacement because of simplicity although sampling without replacement could be used.

Suppose that T calculated on all permutations has distinct values $t_1, \ldots, t_k$. For example, in Table 12.1 (p. 453) the t statistic has $k = 13$ distinct values $-2.98, -1.72, -1.36, -1.08, -0.84, -0.06, 0.12, 0.30, 0.49, 0.69, 0.91, 1.17, 1.47$, corresponding to the 15 permutations (0.49 and 0.91 appeared twice). The Monte Carlo approach is to randomly select B times from the 15 possible permutations, calculate the statistic for each random selection, say $T_1^*, \ldots T_B^*$, and let the number of T^*s equal to t_i be denoted N_i, $i = 1, \ldots, k$. If we select permutations with replacement, then $(N_1, \ldots, N_k)$ is multinomial$(B; p_1, \ldots, p_k)$, where p_i is the permutation distribution probability of obtaining t_i. The estimates N_i/B have binomial variances $p_i(1 - p_i)/B$. Thus, if we were trying to estimate the probabilities in Table 12.2 (p. 453), most of the estimates would have variance $(1/15)(14/15)/B$ although two of them would have variance $(2/15)(13/15)/B$ because of the duplication of values 0.49 and 0.91.

In typical applications, we are not interested in the whole permutation distribution, but merely want to estimate the p-value given in (12.2, p. 455) using

$$\widehat{p} = \frac{\{\#T_i^* \geq T_0\}}{B},$$

where T_0 is the value of the statistic for the original data. In the simple example, $T_0 = 1.17$. Recall that in this case the true permutation p-value is $2/15 = .13$. Thus, $B = 1000$ would yield an estimate with standard deviation $\{(.13)(.87)/1000\}^{1/2} = .01$ that would be adequate for most purposes. However, if the p-value were smaller, say .005, then we would want to take B larger so that the standard deviation of the estimate would be a small fraction of the p-value, say not more than 10–20%. For example, setting $.001 = \{(.005)(.995)/B\}^{1/2}$ would suggest $B = 4975$. When the estimated p-value is to be used with rejection rules like "reject H_0 if $\widehat{p} \leq \alpha$," then it is wise to choose B so that $(B + 1)\alpha$ is an integer as was discussed in the bootstrap Section 11.6.2 (p. 440) as the"99 rule". Mainly this would be used in Monte Carlo simulation studies where $B = 99$ or $B = 199$ might be used to save computing time. However, in situations where computations of the test statistic are extremely expensive, one may view the random partitions as part of the test itself, and the procedure "reject H_0 if $\widehat{p} \leq \alpha$" is called a Monte Carlo test, not just an approximation to the permutation test. This approach was first introduced by Barnard (1963) and later studied by Hope (1968), Jöckel and Jockel (1986), and Hall and Titterington (1989).

12.4.9 Comparing the Approximations in a Study of Two Drugs

A new drug regimen (B) was given to 16 subjects, and one week later each subject's status was assessed. A second independent group of 13 subjects received the standard drug regimen (A). Both sets of measurements were compared to baseline measurements taken before the treatment period began. The difference from baseline data is given in Figure 12.2. This is real data but the actual details are confidential. The drug company wanted to prove that regimen B involving their new drug had larger differences from baseline than the standard. In terms of means of the differences, the testing situation is $H_0 : \mu_B = \mu_A$ versus $H_a : \mu_B > \mu_A$.

Fig. 12.2 Change from Baseline for Drugs A and B

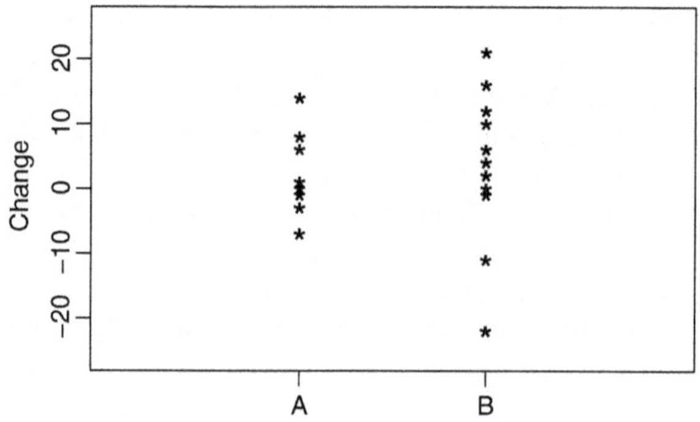

The sample means and standard deviations are $\overline{X} = .92, \overline{Y} = 3.19, s_X = 5.45, s_Y = 10.21$. The standard pooled t from (12.1, p. 452) is .72 with one-sided p-value .24 from the t distribution. The exact permutation t p-value is 0.249, but with a large p-value like this, the t distribution approximation is adequate and agrees with the Type I error robustness mentioned previously. The Box-Andersen $d = 1.074$ leading to an adjusted t p-value of .245.

However, Figure 12.2 reveals that most of the Drug B subjects have positive changes from baseline whereas the Drug A changes are more centered around 0. The two large negative values −22 and −11 have a strong effect on the t statistic. The Wilcoxon Rank Sum statistic W is less affected by outliers, and might paint a different picture. First we compute the midranks and list them with the data ordered within samples.

A:	−7	−3	−3	−1	−1	−1	−1	0	0	1	6	8	14
Rank:	3	4.5	4.5	9	9	9	9	14	14	16	21.5	23	27

B:	−22	−11	−1	−1	−1	0	2	2	4	4	6	10	10
Rank:	1	2	9	9	9	14.0	17.5	17.5	19.5	19.5	21.5	24.5	24.5

B:	12	16	21
Rank:	26	28	29

Then $W = 1 + 2 + \ldots + 28 + 29 = 271.5$. The null mean of W is $(16)(16 + 13 + 1)/2 = 240$. To compute the null variance using the formula for ties, (12.12, p. 462), note that there are $e = 16$ distinct values and 2 values tied at −3, 7 tied at −1, 3 tied at 0, 2 tied at 2, 2 tied at 4, 2 tied at 6, and 2 tied at 10. Thus the null variance is

$$\frac{(16)(13)(16+13+1)}{12} - \frac{(16)(13)}{(12)(29)(29-1)}\left[(7^3-7)+(3^3-3)+5(2^3-2)\right]$$
$$= 520 - 8.325 = 511.675.$$

The approximate normal statistic is $(271.5 - 240)/\sqrt{511.675} = 1.39$ with p-value .082. The t statistic on the ranks is 1.42 with p-value .084. The Box and Andersen (1955) degrees of freedom approximation with $d = (1 - 1.2/29) = 0.96$ does not change that latter p-value until the fourth decimal. The Edgeworth approximation p-value is .084 without continuity correction and .087 with continuity correction.

Unfortunately, because of the ties we cannot trust the exact tables or a continuity correction or the Edgeworth approximation. Thus, it seems wise to either calculate the exact permutation p-value or estimate it by Monte Carlo methods. With $B = 10,000$ we got $\widehat{p} = .085$ with 95% confidence interval (.080,.090). Rather than make B larger, in this case it is fairly easy to get the exact p-value $= .0849$ with existing software. Summarizing the one-sided p-values, we have

Statistic	Method	P-value
t	Exact Permutation	0.2490
	$t(m+n-2)$	0.239
	Box-Andersen	0.245
W	Exact Permutation	0.0849
	Normal	0.082
	$t(m+n-2)$	0.084
	Box-Andersen	0.084
	Edgeworth	0.084
	Edgeworth (with cc)	0.087
	Monte Carlo (B=10,000)	0.085

So this is a situation where the Wilcoxon Rank Sum statistic might be preferred to the t because of its robustness to outliers. Here it apparently downweighted the outliers -22 and -11 enough to have a much lower p-value than the t statistic. The normal and t approximations to the W p-value are quite reasonable here, but we would not know that without getting the exact p-value $= .0849$ or by estimating it fairly accurately.

12.5 Optimality Properties of Rank and Permutation Tests

There are actually very few results available on the optimality properties of permutation tests. The main source is Lehmann and Stein (1949), see also Lehmann (1986, Ch. 5), who give the form of the most powerful permutation test for shift alternatives and note that it depends on a variety of unknown quantities including the form of the distribution. In the particular case of normal data with common unknown variance, they show that the most powerful permutation statistic is $\overline{Y}$ or equivalently $\overline{Y} - \overline{X}$ or the pooled two sample t statistic. Thus general optimality results are not available, but a general approach is clear: derive an (asymptotically) optimal parametric test statistic under a specific parametric family assumption (your best guess), and use the permutation approach for critical values. The resulting permutation test is valid under the null hypothesis for any distribution as long as the conditions of Theorem 12.1 (p. 457) hold, and is close to optimal if the distribution of the data is close to the one used to derive the test statistic.

For rank statistics there are two main bodies of results: locally most powerful rank tests and asymptotically most powerful rank tests based on Pitman Asymptotic Relative Efficiency (ARE). Here we briefly give the flavor of these approaches and main results leaving technical details for the Appendix.

12.5.1 Locally Most Powerful Rank Tests

For simplicity we focus on the two-sample shift model where $X_1, \ldots, X_m$ are iid with distribution function F, and $Y_1, \ldots, Y_n$ are iid with distribution $G(y) = F(y - \Delta)$. We assume that F is continuous with density f. Consider

$$H_0 : \Delta = 0 \quad \text{versus} \quad H_a : \Delta > 0.$$

If there exists a rank test that is uniformly most powerful of level α for some $\epsilon > 0$ in the restricted testing problem

$$H_0 : \Delta = 0 \quad \text{versus} \quad H_{a,\epsilon} : 0 < \Delta < \epsilon,$$

then we say that the test is the *locally most powerful rank test* for the original testing problem.

The basic approach to finding a locally most powerful rank test is to take a Taylor expansion of the probability of the rank vector as a function of Δ and maximize its derivative at $\Delta = 0$. For sufficiently small Δ, the values of the rank vector that are ordered by its probability under the alternative Δ are the same as those ordered by its derivative at $\Delta = 0$. Thus, we need only obtain an expression for the derivative and maximize it. These details are left for the Appendix.

For the two-sample shift problem, the locally most powerful rank test rejects for large values of

$$T = \sum_{i=m+1}^{N} a(R_i),$$

where $a(i) = \mathrm{E}\{\phi(U_{(i)}, f)\}$,

$$\phi(u, f) = -\frac{f'(F^{-1}(u))}{f(F^{-1}(u))} \tag{12.23}$$

is called the optimal score function, and $U_{(1)} \leq U_{(2)} \leq \cdots \leq U_{(N)}$ are the order statistics from a uniform (0,1) distribution. Recall that $R_{m+1}, \ldots, R_N$ are the ranks of the Y values in the joint ranking of all the X's and Y's together. We see in the next section that a closely related statistic, $\sum_{i=m+1}^{N} \phi(R_i/(N+1), f)$, is asymptotically equivalent and comes naturally from asymptotic relative efficiency considerations.

If F is the logistic distribution, then we are led to the Wilcoxon Rank Sum as the locally most powerful rank test for shift alternatives because $-f'(x)/f(x) = 2F(x) - 1$ and $\mathrm{E}\{U_{(i)}\} = i/(N+1)$. When F is a normal distribution, then the optimal score function is $\phi(u, f) = \Phi^{-1}(u)$, and the locally most powerful test is based on the *normal scores*

$$a(i) = \mathrm{E}\{\Phi^{-1}(U_{(i)})\} = \mathrm{E}\{Z_{(i)}\},$$

where $Z_{(i)}$ is a standard normal order statistic. For shifts in the scale of an exponential distribution, $F(x; \sigma) = 1 - \exp(-x/\sigma)$, we can turn it into a shift in location of the negative of an extreme value distribution, $F(x) = 1 - \exp\{-\exp(x)\}$, by taking the natural logarithm of the exponential data. The resulting optimal test has score

$$a(i) + 1 = \sum_{j=N+1-i}^{N} \frac{1}{j},$$

where the latter sum is the expected value of the ith order statistic from a standard exponential distribution. These are called *Savage* scores from Savage (1956). In censored data situations, the analogous test is called the logrank test.

Lehmann (1953) studied alternatives of the form

$$F_\Delta(x) = (1 - \Delta)F(x) + \Delta F^2(x),$$

and showed that the Wilcoxon Rank Sum is the locally most powerful rank test for these alternatives. In general, alternatives of the form $F_\Delta(x) = h_\Delta(F(x))$ for some function $h_\Delta(u)$, are called *Lehmann alternatives*. They have the property that two-sample rank tests have the same distribution under an alternative Δ for all continuous F.

Johnson et al. (1987) consider locally most powerful rank tests using Lehmann alternatives for the nonresponder problem where only a fraction of subjects respond to treatment. Conover and Salsburg (1988) consider other locally most powerful rank tests for the nonresponder problem. Additional situations where locally most powerful rank tests are considered include Doksum and Bickel (1969) and Bhattacharyya and Johnson (1973).

The optimal score functions (12.23, p. 475) appear in the k-sample problem, Section 12.6 (p. 480), and in the correlation problem, Section 12.7 (p. 487).

Analogous results are also available in the one-sample location or matched pairs problem, Section 12.7 (p. 487), and are mentioned there.

Theoretical development and rigorous theorems on locally most powerful rank tests may be found in Hajek and Sidak (1967, Ch. 2), Conover (1973), and Randles and Wolfe (1979, Chs. 4 and 9).

12.5.2 Pitman Asymptotic Relative Efficiency

Perhaps the most useful way to evaluate and compare rank tests is due to Pitman (1948) and further developed by Noether (1955) and others. The basic idea is that Pitman Asymptotic Relative Efficiency (ARE) is the ratio of sample sizes for two different tests to have the same power at a sequence of alternatives converging to the null hypothesis.

Let S and T be two test statistics for $H : \theta = \theta_0$ where θ_k is a sequence of alternatives converging to θ_0 as $k \to \infty$. If we can choose sample sizes N_{S_k} and N_{T_k} and critical values c_{S_k} and c_{T_k} for S and T, respectively, such that $S > c_{S_k}$ and $T > c_{T_k}$ have levels that converge to α and their powers under θ_k converge to β, $\alpha < \beta < 1$, then the Pitman asymptotic relative efficiency of S to T is given by

$$\mathrm{ARE}(S,T) = \lim_{k\to\infty} \frac{N_{T_k}}{N_{S_k}}.$$

Note that if $\mathrm{ARE}(S,T) > 1$, then S is preferred to T because it takes fewer observations (N_{S_k} is less than N_{T_k}) to achieve the same power. Technical conditions in the Appendix and $P(S_k > c_{S_k}) \to \beta < 1$ require that the alternatives have a specific form: for some $\delta > 0$

$$\theta_k = \theta_0 + \frac{\delta}{\sqrt{N_{S_k}}} + o\left(\frac{1}{\sqrt{N_{S_k}}}\right) \quad \text{as } k \to \infty. \tag{12.24}$$

Such sequences of alternatives are called *Pitman alternatives*. Another important quantity arising from the technical details is the *efficacy* of a test statistic S,

$$\mathrm{eff}(S) = \lim_{k\to\infty} \frac{\mu'_{S_k}(\theta_0)}{\sqrt{N_{S_k}\sigma^2_{S_k}(\theta_0)}},$$

where $\mu_{S_k}(\theta_0)$ and $\sigma_{S_k}(\theta_0)$ are the asymptotic mean of S and standard deviation of S. Thus, the efficacy of a test is the rate of change of its asymptotic mean at the null hypothesis relative to its asymptotic standard deviation (the factor $1/\sqrt{N_{S_k}}$ is introduced in the derivative because of 12.24). A powerful test in the Pitman sense is one that is able to detect changes in the parameter value near the null hypothesis. The ARE of S to T turns out to be

$$\mathrm{ARE}(S,T) = \left\{\frac{\mathrm{eff}(S)}{\mathrm{eff}(T)}\right\}^2.$$

Table 12.3 ARE(W, t) for the Two-Sample Shift Model

Distribution	ARE(W, t)
Lower Bound	0.864
Normal	0.955
Uniform	1.00
Logistic	1.10
Laplace	1.50
t_6	1.16
t_3	1.90
t_1 (Cauchy)	∞
Exponential	3.00

The Pitman ARE is both a limiting ratio of sample sizes required to give the same power and the square of the ratio of the test efficacies. High efficacies lead to high ARE's.

In the Appendix we give details for finding efficacies in the one-sample problem, but here we use similar standard results on efficacies for the two-sample problem from Randles and Wolfe (1979, Chs. 5 and 9). The most important comparison is between the two-sample t test and the Wilcoxon Rank Sum test. The efficacy of the t test is

$$\text{eff}(t) = \frac{\sqrt{\lambda(1-\lambda)}}{\sigma},$$

where σ is the standard deviation of the X distribution function $F(x)$ and of the Y distribution function $G(y) = F(x - \Delta)$, and $\lambda = \lim_{\min(m,n)\to\infty} m/(m+n)$. For the Wilcoxon Rank Sum statistic W we have

$$\text{eff}(W) = \sqrt{12\lambda(1-\lambda)} \int_{-\infty}^{\infty} f^2(x)\, dx,$$

where f is the density of $F(x)$, and the integral is assumed to exist. Putting these efficacies together, we have that the Pitman ARE of W to t is

$$\text{ARE}(W, t) = 12\sigma^2 \left\{ \int_{-\infty}^{\infty} f^2(x)\, dx \right\}^2. \tag{12.25}$$

We put ARE(W, t) into Table 12.3 for a number of distributions. Remember that ARE(W, t) > 1 means that the Wilcoxon Rank Sum test is preferred to the t test. The first number is the lower bound 0.864 derived by Hodges and Lehmann (1956) which shows that the Wilcoxon Rank Sum cannot do much worse than the t test for any continuous unimodal distribution. The second number 0.955 is for the normal distribution and shows that the Wilcoxon loses very little efficiency at the normal distribution where the t test is optimal. At the uniform distribution, the tests perform equivalently, and at the remaining examples in Table 12.3, the Wilcoxon is preferred.

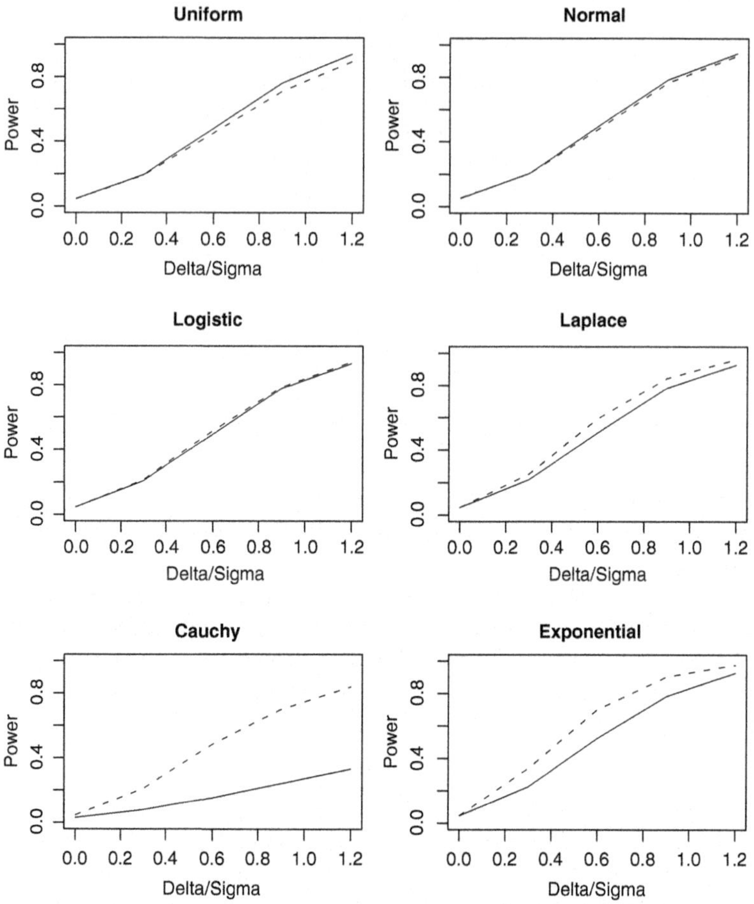

Fig. 12.3 *Power of Wilcoxon Rank Sum* ($\cdots$) *and t* (——) *for* $m = n = 15$ *from Table 4.1.10 of Randles and Wolfe (1979)*

One might think that these ARE results are just asymptotic and may not relate to small sample results. To supplement the ARE results, in Figure 12.3 we plot power results for $m = n = 15$ taken from Table 4.1.10 of Randles and Wolfe (1979, p. 118–119). They simulated the power of the t and Wilcoxon using 1000 replications. Here we see good correspondence between small sample power and the ARE results of Table 12.3. For the normal, uniform, and logistic distributions, there is little power difference as one might expect from ARE values of .955, 1.00, and 1.10, respectively. For the Laplace, the Wilcoxon has a significant power advantage, perhaps not quite as large at the ARE$(W, t) = 1.5$ would imply. The t_1 (Cauchy) and exponential power results strongly favor the Wilcoxon and are consistent with the large ARE values.

We should mention that the Laplace distribution with density $f(x) = (1/2)\exp(-|x|)$ has been used quite a bit in the rank literature as a model for data, especially for ARE comparisons and simulation studies. But it may not be very useful as a model for real data, and ARE results for it are not as consistent with simulation results in small samples as with other densities. The optimal rank test for the Laplace uses scores $a(i) = 1$ for $i > (N+1)/2$ and 0 otherwise, and is called the two-sample median test. However, its power performance in small samples, even when simulating from the Laplace distribution, is poor. Freidlin and Gastwirth (2000) show by simulation that the Wilcoxon Rank Sum test outperforms the median test at the Laplace distribution for samples sizes $m = n$ less than or equal to 25. They recommend that the median test "be retired" from general usage, and we agree.

It turns out that in the scale problem mentioned briefly in Section 12.6.6 (p. 486), ARE values are overly optimistic when compared to small sample power results. This may reflect the fact that measuring scale (standard deviation) is an inherently harder problem that is not as well suited to rank statistics. Klotz (1962) pointed out this discrepancy between small sample power and ARE results. Fortunately, ARE results have been used mainly in location comparisons where they yield good intuition about the qualitative behavior of tests.

Another result from Randles and Wolfe (1979, p. 307) is that under suitable regularity results on the score functions, the efficacy of any linear rank test $S = \sum_{i=m+1}^{N} \phi(R_i/(N+1))$ in the two-sample shift model is given by

$$\text{eff}(S) = \sqrt{\lambda(1-\lambda)} \frac{\displaystyle\int_0^1 \phi(u)\phi(u,f)\,du}{\left[\displaystyle\int_0^1 \{\phi(u) - \overline{\phi}\}^2\,du\right]^{1/2}}, \tag{12.26}$$

where $\phi(u,f)$ is given in (12.23, p. 475). Expression (12.26) now justifies the name *optimal score function* since the efficacy in (12.26) is optimized by choosing $\phi(u) = \phi(u,f)$. This can be seen by noting that

$$\int_0^1 \phi^2(u,f)\,du = \int_{-\infty}^{\infty} \left\{\frac{f'(x)}{f(x)}\right\}^2 f(x)\,dx = I(f),$$

where $I(f)$ is the Fisher information for the model $f(x;\theta) = f(x-\theta)$. Now, noting that $\int_0^1 \phi(u,f)\,du = 0$, (12.26) can be reexpressed as

$$\text{eff}(S) = \sqrt{\lambda(1-\lambda)I(f)}\,\text{Corr}(\phi(U), \phi(U,f)), \tag{12.27}$$

where U is a uniform random variable and Corr is the correlation. Clearly, the correlation is maximized by choosing $\phi(u) = \phi(u,f)$. Moreover, it can also be shown that $\sqrt{\lambda(1-\lambda)I(f)}$ is not only the largest possible efficacy among linear rank tests but also among all α-level tests. Thus, optimal linear rank tests are asymptotically equivalent in terms of Pitman ARE to the best possible tests, say likelihood ratio or score or Wald tests for the shift model in a parametric framework. Of course, this optimality in either the rank test or the parametric test requires that the assumed family is correct.

In the next sections we consider i) the k-sample problem that is a generalization of the two-sample problem to $k > 2$ samples; ii) the correlation or regression problem; and then iii) the matched pairs or one-sample symmetry problem. The Pitman ARE analysis has to be adjusted to handle each situation, but the numbers found in Table 12.3 (p. 477) continue to hold for these situations as well. Thus Wilcoxon procedures, in other words rank methods using scores $a(i) = i$, tend to give very good results across a wide range of distributions in each of these situations.

12.6 The k-sample Problem, One-way ANOVA

The extension of the two-sample case to k samples or treatments is straightforward. Suppose that we have available k independent random samples $\{Y_{i1}, \ldots, Y_{in_i};\ i = 1, \ldots, k\}$, where in each sample the Y_{ij} $(j = 1, \ldots, n_i)$ are iid with distribution function $F_i(x)$, and $N = n_1 + \cdots + n_k$. The linear model representation is

$$Y_{ij} = \mu + \alpha_i + e_{ij}. \tag{12.28}$$

If the errors e_{ij} all come from the same distribution, then (12.28) is an extension of the shift model for two-sample data.

For example, the following are data on the ratio of Assessed Value to Sale Price for single family dwellings ($n_1 = 27$), two-family dwellings ($n_2 = 22$), three-family dwellings ($n_3 = 17$), and four or more family dwellings ($n_4 = 14$) in Fitchburg, Massachusetts, in 1979.

1 Family			2 Family			3 Family		4 or More	
46	74	87	55	85	129	51	100	22	119
60	75	87	60	86	150	64	107	44	120
65	75	87	67	90	203	73	111	71	129
67	77	89	73	94	730	82	112	85	143
68	78	92	76	96		83	126	89	487
69	81	95	77	97		85	134	90	
70	82	95	80	98		89	140	98	
71	84	100	80	100		95	195	102	
73	85	121	82	113		100		113	

The null hypothesis of interest is of identical distribution functions,

$$H_0 : F_1(y) = F_2(y) = \cdots = F_k(y), \tag{12.29}$$

which arises most naturally if we randomly assigned N experimental units to k treatment groups with sample sizes $n_1, n_2, \ldots, n_k$. (The above data are not of this type.) There are

$$M_N = \binom{N}{n_1 n_2 \cdots n_k} = \frac{N!}{n_1! n_2! \cdots n_k!}$$

possible assignments, which of course is the relevant number of permutations even if the data do not come from a randomized experiment. Pitman (1938) proposed the permutation approach for the ANOVA F statistic

$$F = \frac{\dfrac{1}{k-1}\displaystyle\sum_{i=1}^{k} n_i(\overline{Y}_{i.} - \overline{Y}_{..})^2}{\dfrac{1}{N-k}\displaystyle\sum_{i=1}^{k}\sum_{j=1}^{n_i}(Y_{ij} - \overline{Y}_{i.})^2}, \tag{12.30}$$

where $\overline{Y}_{i.} = n_i^{-1}\sum_{j=1}^{n_i} Y_{ij}$, and $\overline{Y}_{..} = N^{-1}\sum_{i=1}^{k} n_i \overline{Y}_i$. The number of permutations M_N gets large very fast. For example, with $k = 3, N = 15, n_1 = n_2 = n_3 = 5$, we get $M_N = \binom{15}{5\,5\,5} = 756,756$. Thus Monte Carlo or asymptotic approximations are more important than in the two-sample case. For the above housing data, the ANOVA F in (12.30) is $F = 1.24$ with p-value = .30 from the $F(3, 75)$ distribution. The exact permutation p-value is obtained by computing F for each of the 1.9×10^{44} distinct allocations of $\{Y_{i1}, \ldots, Y_{in_i}; i = 1, \ldots, 4\}$ to samples of size $n_1 = 27$, $n_2 = 22$, $n_3 = 17$, and $n_4 = 14$, and finding the proportion of these greater to or equal to $F = 1.24$. A Monte Carlo estimate of the exact permutation p-value is .267 based on 100,000 resamples with standard error = .0014. Because the housing ratios are quite skewed with a number of large observations, it is not surprising that F is small. Now we turn to rank methods that naturally limit the effect of outliers.

12.6.1 Rank Methods for the k-Sample Location Problem

Kruskal and Wallis (1952) proposed the rank extension of the Wilcoxon Rank Sum statistic to the k-sample situation. The rank approach is to put all N observations together and rank them; let R_{ij} be the rank of Y_{ij} in the combined sample. Further define the sample sums

$$S_i = \sum_{j=1}^{n_i} a(R_{ij}),$$

where the scores $a(i)$ could be of any form for permutational analysis, but for asymptotic results we assume $a(i) = \phi(i/(N+1))$ and ϕ is a scores generating function as in Theorem 12.3 (p. 465). The Kruskal-Wallis statistic uses $a(i) = i$ or equivalently $a(i) = i/(N+1)$. Note that S_i is just a two-sample linear rank statistic for comparing the ith population to all the others combined. The general linear rank statistic form for comparing the k populations is then

$$Q = \sum_{i=1}^{k} \frac{1}{s_a^2 n_i}(S_i - n_i\overline{a})^2 = \sum_{i=1}^{k} \left(\frac{N-n_i}{N}\right) \frac{(S_i - \mathrm{E}S_i)^2}{\mathrm{Var}(S_i)}, \qquad (12.31)$$

where $s_a^2 = (N-1)^{-1}\sum_{i=1}^{N}\{a(i) - \overline{a}\}^2$, $\overline{a} = \sum_{i=1}^{N} a(i)$, and $\mathrm{Var}(S_i)$ is given by (12.4, p. 459) with the constants c_i in that expression equal to 1 for n_i of them and 0 otherwise. The reason for giving the second form in (12.31) is that it is then clear that $\mathrm{E}(Q) = k - 1$ under the null hypothesis of equal populations. The Kruskal-Wallis statistic that allows for ties is explicitly given by

$$H = \frac{(N-1)\left\{\sum_{i=1}^{k} n_i \left(\overline{R}_{i.} - \frac{N+1}{2}\right)^2\right\}}{\left(\sum_{i=1}^{k}\sum_{j=1}^{n_i} R_{ij}^2\right) - N(N+1)^2/4},$$

where $\overline{R}_{i.} = n_i^{-1}\sum_{j=1}^{n_i} R_{ij}$. If there are no ties in the data, then

$$\sum_{i=1}^{k}\sum_{j=1}^{n_i} R_{ij}^2 = N(N+1)(2N+1)/6,$$

and H reduces to the more familiar form

$$H = \frac{12}{N(N+1)} \sum_{i=1}^{k} n_i \left(\overline{R}_{i.} - \frac{N+1}{2}\right)^2.$$

Under the null hypothesis (12.29, p. 480), standard asymptotic theory similar to Theorem 12.3 (p. 465) yields that $Q \xrightarrow{d} \chi^2_{k-1}$ as $\min\{n_1, \ldots, n_k\} \to \infty$. The χ^2_{k-1} approximation is not very good in small samples, but fortunately the F statistic on the scores $a(R_{ij})$ is a monotone function of Q,

$$F_{\mathrm{R}} = \left(\frac{N-k}{k-1}\right)\left(\frac{Q}{N-1-Q}\right),$$

and using $F(k-1, N-k)$ as a reference distribution or the Box-Andersen adjusted $F(d(k-1), d(N-k))$ distribution yields excellent results. For the housing data above, $H = 9.8856$ with p-value = 0.020 from the χ^2_3 distribution. $F_R = 3.6283$

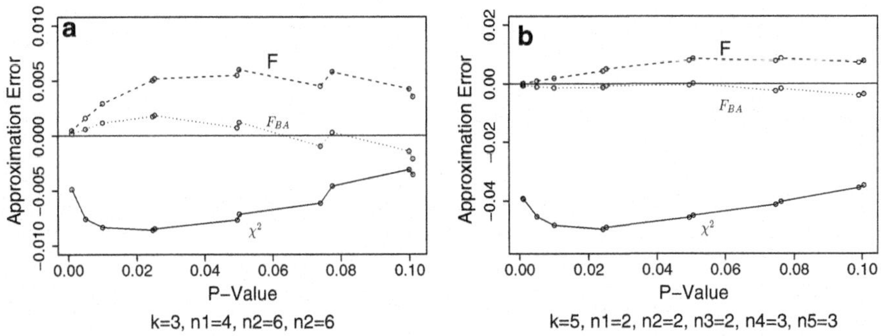

Fig. 12.4 (Exact P-Values − Approximate P-Values) versus Exact P-Values for Kruskal-Wallis Statistic. $F = F(k-1, N-k)$, $F_{BA} = F(d(k-1), d(N-k))$, and $\chi^2 = \chi^2_{k-1}$

with p-value 0.017 from the $F(3, 75)$ distribution. The Box-Andersen d=0.9876, and so the adjustment is very minor, only in the fourth decimal place. A Monte Carlo approximation to the exact p-value is .017 based on 100,000 samples with standard error .0004. So here the F distribution approximation is right on target to 3 decimals, but the χ^2 approximation is not bad due to the fairly large samples.

In Figure 12.4 we look at much smaller sample sizes for $k = 3$ and $k = 5$. Figure 12.4 shows the difference between the exact permutation p-value and each approximation versus the exact p-value for the Kruskal-Wallis statistic. Note that the left panel is more expanded in the vertical scale than the right panel and actually has less error. Nevertheless, the Box-Andersen approximation is the best in both plots and is generally very good for $k > 2$. The χ^2_{k-1} approximation gets more conservative as k gets larger. This can be explained by the following large-k asymptotic results.

12.6.2 Large-k Asymptotics for the ANOVA F Statistic

Brownie and Boos (1994) show under the null hypothesis of equal populations that

$$\sqrt{k}(F_{\mathrm{R}} - 1) \xrightarrow{d} \mathrm{N}\left(0, \frac{2n}{n-1}\right), \tag{12.32}$$

for equal sample sizes $n_1 = n_2 = \cdots = n_k = n$ and $k \to \infty$ with n fixed. Note that the usual result with $n \to \infty$ and k fixed is $(k-1)F_{\mathrm{R}} \xrightarrow{d} \chi^2_{k-1}$, similar to the result for Q. The "large k" asymptotic result (12.32) implies that

$$\sqrt{k}\left(\frac{Q}{k-1} - 1\right) \xrightarrow{d} \mathrm{N}\left(0, \frac{2(n-1)}{n}\right), \tag{12.33}$$

as $k \to \infty$ with n fixed, using

$$Q = \frac{(N-1)F_{\mathrm{R}}}{(N-k)/(k-1) + F_{\mathrm{R}}} \tag{12.34}$$

(see Problem 12.17, p. 527). Note that comparing Q to a χ^2_{k-1} is asymptotically ($k \to \infty$) like comparing $Q/(k-1)$ to a N$\{1, 2/(k-1)\}$ because a χ^2_{k-1} random variable obeys the Central Limit Theorem (it is a sum of χ^2_1 random variables). However, (12.33) says that $Q/(k-1)$ should be compared to a N$\{1, 2(n-1)/(kn)\}$ distribution. Because $2(n-1)/(kn) < 2/(k-1)$, using the χ^2_{k-1} distribution with Q results in conservative true levels. For example, if $k = 5$ and $n = 5$, then the large sample 95th percentile from N$\{1, 2/(k-1)\}$ is $1 + (2/4)^{1/2}1.645 = 2.16$, and the approximate true level of a nominal $\alpha = .05$ test is

$$P(Q \geq \chi^2_4(.95)) \approx P(1 + (8/25)^{1/2}Z \geq 2.16) = P(Z \geq 2.05) = .02.$$

In contrast, use of F_{R} with an $F(k-1, N-k)$ reference distribution is supported by (12.32) under $k \to \infty$ and by the usual asymptotics $(k-1)F_{\mathrm{R}} \xrightarrow{d} \chi^2_{k-1}$ when $n \to \infty$ with k fixed. We leave those details for Problem 12.18 (p. 527). Thus, it is not surprising that the F approximations in Figure 12.4 are much better than the χ^2_{k-1} ones.

12.6.3 Comparison of Approximate P-Values – Data on Cadmium in Rat Diet

Nation et al. (1984) studied the effect of diets containing cadmium (Cd) on the neurobehavior of adult rats. The data consists of the number of platform descents during a passive-avoidance training scheme for 27 rats randomly assigned to three groups:

										$\overline{Y}$	s_{n-1}
Control:	82	80	77	75	72	68	59	47	42	67	14
Cd1:	86	66	60	51	44	41	38	29	10	47	22
Cd5:	81	67	38	36	32	29	20	17	14	37	23

The control group had no Cd in the diet, and Cd1 and Cd5 refer to daily diets containing 1 milligram and 5 milligrams, respectively, of Cd per kilogram of body weight. The usual one-way ANOVA $F = 5.10$, and the permutation p-value F statistic is $\widehat{p} = 0.016$ based on 100,000 random permutations. The $F(2, 24)$ distribution gives p-value = .014, and the Box-Andersen correction factor is $d = .954$ leading to p-value = .016. The Kruskal-Wallis rank statistic is $Q = 8.18$ with permutation p-value $\widehat{p} = .012$ based on 100,000 random permutations. The χ^2_2

approximation gives p-value = .017. The associated F statistic is $F_R = 5.51$ with p-value = .011. The Box-Andersen correction factor is $d = 1 - 1.2/24 = .95$ leading to p-value = .012. A summary is as follows:

Statistic	Method	P-value
F	Monte Carlo (B=100,000)	0.016
	$F(2, 24)$	0.014
	Box-Andersen	0.016
KW	Monte Carlo (B=100,000)	0.012
	χ_2^2	0.017
	$F(2, 24)$	0.011
	Box-Andersen	0.012

As expected the F approximations give excellent p-values.

12.6.4 Other Types of Alternative Hypotheses

The k-sample F statistic and Kruskal-Wallis statistic are used to compare the centers or locations of the k populations. Other statistics could be used for that purpose, perhaps ones more suited to long-tailed or skewed populations. The logrank or Savage scores, for example, are asymptotically optimal for detecting shifts in the scale parameter of exponential populations (or the shift parameter of extreme value distributions).

Other types of alternatives may also be of interest. For example, there may be an implied order in the populations, say increasing doses, and there may be interest in trends in location. There might also be interest in comparing the spread of the populations or even the skewness.

These latter alternatives present a problem to permutation and rank methods because the null hypothesis of interest may not be the one of identical populations. For comparing spread, the usual null hypothesis of interest would be equal spread rather than identical populations. In such a situation, use of the permutation approach would require subtraction of unknown location parameters. We first discuss ordered alternatives in location.

12.6.5 Ordered Means or Location Parameters

Recall Section 3.6.1a (p. 151) where we discussed likelihood-based methods for ordered alternatives. Here we discuss permutation methods with simple statistics in the context of a Phase I toxicology study where there seems to be trends in both the means and variances with dose:

Dose					$\overline{Y}$	s_{n-1}
0	1.44	1.63	1.40	1.59	1.52	0.11
1	1.27	1.50	1.45	1.57	1.45	0.13
2	1.26	1.07	1.38	1.75	1.37	0.29
3	1.04	1.14	1.46	1.06	1.18	0.19
4	1.37	0.79	1.32	1.42	1.23	0.29

The F statistic for comparing means is $F = 1.77$, and the usual $F(4, 16)$ distribution and the Box-Andersen approximation give p-value = 0.19. Similarly, a Monte Carlo estimated p-value based on 10,000 random permutations gives $\widehat{p} = 0.19$. The Kruskal-Wallis statistic is $H = 6.73$ with χ^2_4 p-value = 0.15. The F approximation from $F_R = 2.06$ and the Box-Andersen approximation both give p-value = 0.14. A Monte Carlo estimated p-value based on 10,000 random permutations gives $\widehat{p} = 0.14$. So the global comparison of location is not significant at usual levels.

Suppose that we consider H_0 : identical populations versus H_a : means are decreasing. The permutation approach with $M_N = \binom{20}{44444}$ permutations may be used with the t statistic from a regression of the observations on dose or equivalently Pearson's correlation coefficient (see also the next section). Pearson's correlation coefficient is $r = -0.53$ with Monte Carlo estimated p-value $\widehat{p} = 0.007$ based on 10,000 random permutations. Spearman's correlation coefficient is –0.56 with $\widehat{p} =$ 0.005. Another statistic that could have been used is the likelihood ratio statistic for decreasing means assuming the data are normally distributed (see Section 3.6.1a, p. 151). In addition to Spearman's correlation coefficient, the standard rank-based statistic is the Jonckheere-Terpstra statistic based on summing pairwise Wilcoxon Rank Sum statistics in increasing order, $\sum_{i<j} W_{ij}$, where W_{ij} is the Wilcoxon Rank Sum for comparing dose group i with dose group j (see Lehmann 1975, p. 233). Its value here is -2.458 with exact permutation p-value = 0.0069. So we can be pretty confident that there is a downward trend in means or other location measures.

12.6.6 Scale or Variance Comparisons

Motivated by the apparent increase in variances for the dose-response data above, we now discuss hypotheses about variances or scale parameters. Unfortunately, there is a philosophical dilemma for using permutation procedures here. Usually, the typical set of hypotheses when testing for unequal variances is for a semiparametric model, $P(Y_{ij} \leq y) = F_0((y - \mu_i)/\sigma_i)$, $j = 1, \ldots, n_i$; $i = 1, \ldots, k$, where F_0 is an unknown distribution function. Note that if $F_0(x)$ has mean 0 and variance 1, then μ_i is the ith population mean, and σ_i^2 is the ith population variance. In any case, under this semiparametric model, the ith standard deviation is $c\sigma_i$ for some constant c, and we can always refer to σ_i as a scale parameter. The hypotheses for increasing scale are then $H_0 : \sigma_1 = \cdots = \sigma_k$ versus $H_a : \sigma_1 \leq \cdots \leq \sigma_k$ with at least one inequality. The reason for this hypothesis formulation is that we often know that the means are different; therefore it makes little sense to assume identical populations when testing for variance differences. Basically, we usually want to test for variance differences in the presence of location differences.

Unfortunately, the permutation argument requires that the null hypothesis be one of identical populations. It makes intuitive sense to center the data first by subtracting means, but these residuals $Y_{ij} - \overline{Y}_i$ no longer satisfy exchangeability required for using Theorem 12.1 (p. 457). The permutation distribution is correct asymptotically, but the exact level-α property no longer holds. An overview of the scale testing problem is given in Boos and Brownie (2004). The best method that has emerged for comparing scales is to use t or F statistics on the data Y_{ij} replaced by $|Y_{ij} - M_i|$, where M_i is the ith sample median.

One way to avoid the centering problem for the dose-response data is to reduce the data to the sample standard deviations (or some other scale estimator) and then calculate an appropriate statistic for the $5! = 120$ permutations possible. For the correlation between dose and standard deviation we get $r = 0.79$ and p-value $= 7/120 = .058$. If we use the likelihood ratio test for increasing variances for normal distributions, we get p-value = 5/120=.042. There is a loss of information when the number of permutations get reduced so much, from $M_N = \binom{20}{44444}$ to $M_N = 120$; perhaps the loss of information is just a discreteness problem caused by having too few permutations. This can be seen more clearly by calculating the exact permutation test on the data reduced to the five means; the correlation is higher than when using all the data, but the p-value = 2/120 = .017 is much larger than the .007 value we obtained previously with the whole data set.

We note that the use of rank statistics for scale comparisons has not been very successful. The subtraction of means or medians ruins the permutation argument as mentioned above. However, rank statistics for scale based on centered data are asymptotically distribution free if the samples are symmetrically distributed. The larger problem is that rank tests for scale tend to have low power in small samples. Although rank tests for location perform well in small samples and are consistent with asymptotic relative efficiency comparisons, the opposite is true for rank tests for scale. The latter statistics are not as powerful in small samples as would be expected from asymptotic relative efficiency calculations.

12.7 Testing Independence and Regression Relationships

Regression methods are among the most important tools of statistics. Unfortunately, permutation methods can really be applied in only the simplest setting of (X, Y) pairs; that is, correlation or simple regression (not necessarily linear). Here we discuss that simple situation and mention at the end of the section why permutation methods cannot handle the more interesting case of multiple explanatory variables.

Suppose that we have iid random pairs $(X_1, Y_1), \ldots, (X_n, Y_n)$ and permute each coordinate independently to get $n!$ different pairings. In reality, we need only permute one of the coordinates to obtain all the different pairings. For example, suppose that $n = 3$ with pairs $(1, 2.5), (2, 3.7), (3, 6.4)$. Then the 6 possible permutations are

1	2	3	4	5	6
(1,2.5)	(1,3.7)	(1,6.4)	(1,2.5)	(1,3.7)	(1,6.4)
(2,3.7)	(2,2.5)	(2,3.7)	(2,6.4)	(2,6.4)	(2,2.5)
(3,6.4)	(3,6.4)	(3,2.5)	(3,3.7)	(3,2.5)	(3,3.7)

Pitman (1937b) suggested that a test for independence of X and Y based on the sample correlation

$$r = \frac{\sum_{i=1}^{n}(X_i - \overline{X})(Y_i - \overline{Y})}{\left[\sum_{i=1}^{n}(X_i - \overline{X})^2 \sum_{i=1}^{n}(Y_i - \overline{Y})^2\right]^{1/2}}$$

use this permutation distribution for critical values. A permutationally equivalent statistic is the least squares slope estimate $\widehat{\beta} = \sum_{i=1}^{n}(X_i - \overline{X})(Y_i - \overline{Y}) / \sum_{i=1}^{n}(X_i - \overline{X})^2$. Other popular measures that could be used to test independence are Kendall's rank correlation and Spearman's rank correlation. Spearman's estimated correlation coefficient r_S is simply to replace X_i by its rank among $X_1, \ldots, X_n$ and Y_i by its rank among $Y_1, \ldots, Y_n$, and compute the Pearson correlation r between these pairs of ranks. It is important to keep in mind that the null hypothesis is independence of X and Y and not zero correlation. Independence is needed for the $n!$ different pairings to have the same distribution and thus for Theorem 12.1 (p. 457) to apply.

Typical approximations to the permutation distribution of r (and similarly of r_S) are to compare $(n-1)^{1/2} r$ to a standard normal distribution or $(n-2)^{1/2} r/(1-r^2)^{1/2}$ to a $t(n-2)$ distribution. Pitman (1937b) gave the first two permutation moments of r^2, $E_P(r^2) = 1/(n-1)$, and

$$E_P(r^4) = \frac{3}{(n-1)(n+1)} + \frac{(n-2)(n-3)}{n(n+1)(n-1)^3} \left\{\frac{k_4(X)}{k_2(X)^2}\right\} \left\{\frac{k_4(Y)}{k_2(Y)^2}\right\},$$

where the sample cumulants k_2 and k_4 were given in (12.21, p. 469) and (12.22, p. 469), respectively. Note that these moments are straightforward from the results in Section 12.4.2 (p. 458) since the numerator of r has the form (12.3, p. 458) of a linear statistic, and the denominator is constant over permutations. If the pairs are iid with a bivariate normal distribution, then r^2 has a beta$(1/2, n/2 - 1)$ distribution with $E(r^2) = 1/(n-1)$ and $E(r^4) = 3/(n-1)(n+1)$. Because the permutation moments and normal theory moments are so close, Pitman (1937b) suggested using the beta approximation, which is equivalent to comparing $(n-2)r^2/(1-r^2)$ to an $F(1, n-2)$ distribution. Box and Watson (1962) generalized these results to the full p regressor case for the test that all regressors are independent of Y. They derived the adjusted F approximation (see Box and Watson 1962, p. 100), which for the $p = 1$ case here is to compare $(n-2)r^2/(1-r^2)$ to an $F(d, d(n-2))$ distribution, where

$$\frac{1}{d} = 1 + \frac{(n+1)\alpha_1}{n-1-2\alpha_1}, \quad \alpha_1 = \frac{n-3}{2n(n-1)} \left\{ \frac{k_4(X)}{k_2(X)^2} \right\} \left\{ \frac{k_4(Y)}{k_2(Y)^2} \right\}.$$

In large samples, $d \approx 1 + \{\text{Kurt}(X) - 3\}\{\text{Kurt}(Y) - 3\}/2n$, revealing a double Type I error robustness to nonnormality: if either X or Y is approximately normally distributed, then the usual F approximation is very good. To numerically illustrate, recall $r = -.53$ from the dose-response data (p. 486) where the Monte Carlo estimated one-sided p-value was $\widehat{p} = .007$. Taking half of the $F(1, 18)$ p-value approximation for $18r^2/(1-r^2) = 7.03$, we get p-value = .008. Similarly, for Spearman's $r_S = -.56$ we obtained previously $\widehat{p} = .005$. Using one half of the $F(1, 18)$ p-value for $18r_S^2/(1-r_S^2) = 8.22$ yields p-value = .005.

Now let us move to the more complicated situation of the linear model,

$$Y_i = \beta_0 + \beta_1 X_{1i} + \beta_2 X_{2i} + e_i, \quad i + 1, \ldots, n,$$

where we assume $e_1, \ldots, e_n$ are iid from some distribution and independent of all the X_{ij}. As mentioned above, permuting the Y's under the assumption $H_0 : \beta_1 = \beta_2 = 0$ yields a suitable permutation distribution for testing independence of Y and (X_1, X_2). Unfortunately, we are usually much more interested in testing $H_0 : \beta_2 = 0$ with β_0 and β_1 unrestricted. Without knowledge of β_1, however, an exact permutation procedure for $H_0 : \beta_2 = 0$ is not possible. (Actually, it is possible to take the maximum over permutation p-values for each value of β_1 in a confidence interval under H_0 as described in Berger and Boos (1994), but the loss in power is typically not worth the gain in exactness.) Anderson and Robinson (2001) review a number of different proposals that use residuals from first fitting the reduced model, and show that they are asymptotically correct but do not satisfy the assumptions of Theorem 12.1 (p. 457). Fortunately, standard linear model and rank-based linear model testing procedures have good Type I error robustness properties in general. The rank-based linear model methods given in Ch. 5 of Hettmansperger (1984) have good Type II error robustness properties as well. Similarly, the M-estimation regression methods discussed in Ch. 5 also have good robustness properties.

We conclude this section with an example that illustrates how easy it is to use Monte Carlo approximation in an autocorrelation setting.

Example 12.1 (Raleigh snowfall). Is the total snowfall in one year independent of the total snowfall in other years? The left panel of Figure 12.5 plots Raleigh, NC, annual snowfall for 1962–1991 versus year. The right panel plots each year's

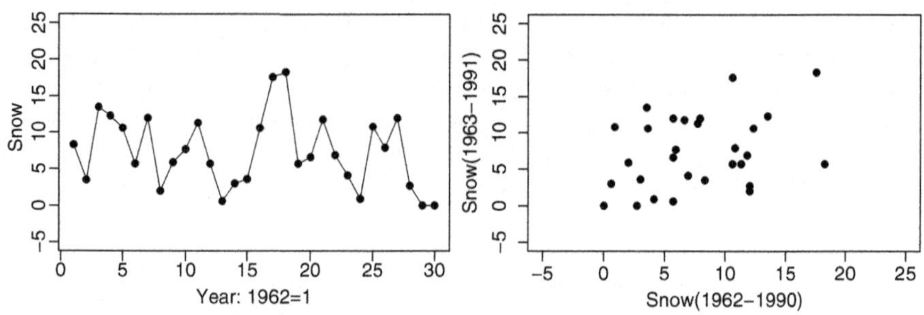

Fig. 12.5 Annual snowfall in Raleigh, NC, 1962–1991 (left panel) and annual snowfall versus annual snowfall of previous year (right panel)

snowfall versus the previous year's snowfall. The sample correlation from the right panel is $r = .32$. Does that suggest nonzero autocorrelation? The null hypothesis for a permutation approach is that the sequence of yearly snowfalls is iid or at least exchangeable. Below we give R code for sampling B permutations from the set of 30! possible permutations, computing the lag-1 sample correlation for each, and estimating the one-sided p-value for a positive autocorrelation. Using $B = 10,000$, we get $\widehat{p} = .027$ with standard error .0016. Thus there is good evidence of a positive autocorrelation. The main point here is to illustrate how easy it is to carry out the permutation test.

```
r.auto<-function(x){
     n<-length(x)
     cor(x[1:(n-1)],x[2:n])
}
perm1<-function(b, x, stat, ...){
  # Gives est. permutation $p$-value for vector x.
  # Assumes test rejects for large values of stat.
        call <- match.call()
        n <- length(x)
        t0 <- stat(x)
        res <- numeric(b)
        for(i in 1:b) {
                perm.xx <- sample(x)
                res[i] <- stat(perm.xx)
        }
        pvalue <- sum(res >= t0)/b
        se<-sqrt(pvalue*(1-pvalue)/b)
        return(list(call=call,results=data.frame
          (nperm=b, stat0=round(t0,4),pvalue=pvalue,
          se=round(se,5))))
}
> set.seed(2458)
```

```
> perm1(10000,raleigh.snow$snow,r.auto)
  nperm  stat0 pvalue      se
1 10000 0.3245 0.0269 0.00162
```

♦

12.8 One-Sample Test for Symmetry about θ_0 or Matched Pairs Problem

Fisher (1935) introduced the permutation approach for the matched-pairs problem in a discussion of Darwin's data on self-fertilized and cross-fertilized plants. There were 15 pairs of plants, and the differences

$$49, -67, 8, 16, 6, 23, 28, 41, 14, 29, 56, 24, 75, 60, -48$$

have mean $\overline{D} = 20.933$, $s = 37.744$, and $t = 2.148$ for testing $H_0 : \mu_D = 0$ versus $H_a : \mu_D \neq 0$, where μ_D is the population mean difference. The two-sided p-value is .0497 from the t table with 14 degrees of freedom. Alternatively, consider Fisher's permutation argument. There were 2^{15} possible random assignments of types of seeds to the 15 blocks of size 2. Thus, Fisher considered all 2^{15} sums $\sum_{i=1}^{15} D_i$, where D_i is the ith difference, and found only 835+28 = 863 which are greater than or equal to the observed sum = 314. The two-sided p-value is (2)(863)/32,768 = .0527 (by symmetry there are 863 sums ≤ -314). Note that $t = \sqrt{n}\overline{D}/s$ is permutationally equivalent to $\sum_{i=1}^{15} D_i$ because t is a monotonic function of $\sum_{i=1}^{15} D_i$ that depends on $\sum_{i=1}^{15} D_i^2$, which is constant over all 2^{15} permutations.

Let us consider the theory behind Fisher's approach. The population null model is that the differences $D_1, \ldots, D_n$ are independent, each with a symmetric distribution about some θ_0; often $\theta_0 = 0$. The distributions do not need to be the same, merely symmetric about θ_0. Thus

$$H_0 : D_i - \theta_0 \stackrel{d}{=} \theta_0 - D_i, \quad i = 1, \ldots, n. \tag{12.35}$$

The group of transformations to be used with Theorem 12.1 (p. 457) is the set of 2^n sign changes applied to the data with θ_0 subtracted. For notational simplicity, let $D_{i0} = D_i - \theta_0$, $i = 1, \ldots, n$. Then, for example, if $n = 4$, one such transformation is $(-, +, +, -)$. It would transform

$$(D_{10}, D_{20}, D_{30}, D_{40}) \tag{12.36}$$

into

$$(-D_{10}, D_{20}, D_{30}, -D_{40}). \tag{12.37}$$

Because of (12.35) and independence, all 2^n transformations of the original data have the same distribution. That is, under (12.35) and independence, the joint distribution of (12.36) is the same as (12.37), etc. Thus, the conditions of Theorem 12.1 (p. 457) apply with the group of sign changes, and Fisher's original method is a valid permutation approach.

12.8.1 Moments and Normal Approximation

Now let us abstract the above situation slightly in order to compute moments and approximations. Suppose that $d_1, \ldots, d_n$ is a sequence of real constants, playing the role of the observed $D_i - \theta_0$ above. Let $c_1, \ldots, c_n$ be iid random variables with $P(c_i = 1) = P(c_i = -1) = 1/2$; these play the role of making the sign changes. Now consider the linear statistic $T = \sum_{i=1}^n c_i d_i$. Note that the c_i are symmetrically distributed around 0 so that all odd moments of c_i are 0 and all even moments equal to 1. Then T is also symmetrically distributed about 0 with odd moments 0 and $\mathrm{E}(T^2) = \mathrm{Var}(T) = \sum_{i=1}^n d_i^2$ and $\mathrm{E}(T^4) = 3(\sum_{i=1}^n d_i^2)^2 - 2\sum_{i=1}^n d_i^4$. Now we give a Central Limit Theorem for T. A more general version and proof are given in Hettmansperger (1984, p. 302–303).

Theorem 12.4. . *Suppose that $d_1, \ldots, d_n$ and $c_1, \ldots, c_n$ are defined as above and*

$$\frac{1}{n}\sum_{i=1}^n d_i^2 \longrightarrow \sigma^2 < \infty \qquad \text{as } n \to \infty.$$

Then

$$\frac{T}{\sqrt{Var(T)}} = \frac{\sum_{i=1}^n c_i d_i}{\left(\sum_{i=1}^n d_i^2\right)^{1/2}} \stackrel{d}{\longrightarrow} N(0,1) \qquad \text{as } n \to \infty.$$

Now we apply this theorem to the permutation distribution of $\sum_{i=1}^n D_i$ when sampling from a population.

Theorem 12.5. *Suppose that $D_1, \ldots, D_n$ are iid random variables satisfying (12.35) and with variance $\sigma^2 < \infty$. Then the permutation distribution function of $\sum_{i=1}^n (D_i - \theta_0)$ under the group of sign changes satisfies*

$$P^* \left\{ \sum_{i=1}^n (D_i - \theta_0)/\sqrt{n}\sigma \right\} \stackrel{wp1}{\longrightarrow} N(0,1) \qquad \text{as } n \to \infty.$$

We have used the notation P^* to emphasize that the probability is taken with respect to the permutation distribution holding $D_1, \ldots, D_n$ fixed. An alternative statement of the result is that the permutation distribution of $\sum_{i=1}^{n}(D_i - \theta_0)/\sqrt{n}\sigma$ converges in distribution to a standard normal distribution with probability 1. Note also that we could just as well have put $\{\sum_{i=1}^{n}(D_i - \theta_0)^2\}^{1/2}$ in place of $\sqrt{n}\sigma$ in the conclusion, giving

$$\frac{\sum_{i=1}^{n}(D_i - \theta_0)}{\left\{\sum_{i=1}^{n}(D_i - \theta_0)^2\right\}^{1/2}} \xrightarrow{d^*} \mathrm{N}(0,1) \qquad \text{as } n \to \infty \quad wp1. \tag{12.38}$$

The result follows from Theorem 12.4 because for each infinite sequence $D_1(\omega)$, $D_2(\omega), \ldots$ where $\omega \in \Omega$ with $P(\Omega) = 1$,

$$\frac{1}{n}\sum_{i=1}^{n}(D_i(\omega) - \theta_0)^2 \longrightarrow \sigma^2 \qquad \text{as } n \to \infty$$

by the Strong Law of Large Numbers. For each of these sequences, Theorem 12.4 holds, and thus the convergence in distribution holds with probability 1.

12.8.2 Box-Andersen Approximation

The Box-Andersen adjusted F approximation to the permutation distribution of $\sum_{i=1}^{n}(D_i - \theta_0)$ uses the *beta* version of $t^2 = n(\overline{D} - \theta_0)^2/s^2$,

$$b(t^2) = \frac{t^2}{n-1+t^2} = \frac{n(\overline{D} - \theta_0)^2}{\sum_{i=1}^{n}(D_i - \theta_0)^2}.$$

Under an iid normal distribution assumption for $D_1, \ldots, D_n$, $b(t^2)$ has a $beta(1/2, (n-1)/2)$ distribution with mean $1/n$ and variance $2(n-1)/\{n^2(n+2)\}$. Using the results in the previous section for $T = \sum_{i=1}^{n} c_i d_i$, where $d_i = (D_i - \theta_0)/n$, the permutation moments of $b(t^2)$ are $\mathrm{E_P}\{b(t^2)\} = 1/n$ and

$$\mathrm{Var_P}\{b(t^2)\} = \frac{2(n-1)}{n^2(n+2)}\left(1 - \frac{f_2 - 3}{n-1}\right), \tag{12.39}$$

where $f_2 = (n+2)\sum_{i=1}^{n}(D_i - \theta_0)^4/\{\sum_{i=1}^{n}(D_i - \theta_0)^2\}^2$. Equating the permutation moments to those of a $beta(d/2, d(n-1)/2)$ distribution leads to

$$d = 1 + \frac{f_2 - 3}{n\{1 - f_2/(n+2)\}}. \tag{12.40}$$

In the above derivation we have followed the notation in Box and Andersen (1955, p. 9), but their W is $1 - b(t^2)$, and we relabeled their b_2 as f_2. Note that f_2 is close to the sample kurtosis of the $D_i - \theta_0$, and thus $d \approx 1 + \{\text{Kurt}(D) - 3\}/n$.

For the Darwin data, $d = .94$ and the F adjusted two-sided p-value is .053. Recall from previous analysis that the exact two-sided permutation p-value is .0527. The normal approximation here is $Z = 1.9282$ with two-sided p-value$= .054$. Thus, the normal approximation is surprisingly good here, better than the $F = t^2$ approximation that Fisher gave (.0497), but the Box-Andersen adjustment has made the F approximation slightly better than the normal approximation.

12.8.3 Signed Rank Methods

Now we turn to signed rank methods. Here again for simplicity we use the notation D_{i0} for $D_i - \theta_0$. Let R_i be the rank of $|D_{i0}|$ among $|D_{10}|, \ldots, |D_{n0}|$. Let the sign function be defined by $\text{sign}(x) = I(x > 0) - I(x < 0)$ if x is nonzero and $\text{sign}(0) = 0$. Then the signed rank of D_{i0} is $\text{sign}(D_{i0})R_i$ although some authors use $I(D_{i0} > 0)R_i$ as the definition of the signed rank. We illustrate with a simple data set from Wilcoxon (1945) on the difference between wheat yields in two treatments in 8 blocks:

D_{i0}	58	32	30	5	−7	6	11	10
R_i	8	7	6	1	3	2	5	4
$sign(D_{i0})R_i$	8	7	6	1	−3	2	5	4
$I(D_{i0} > 0)R_i$	8	7	6	1	0	2	5	4

Then define $W^+ = \sum_{i=1}^n I(D_{i0} > 0)R_i$, $W^- = \sum_{i=1}^n I(D_{i0} < 0)R_i$ and $W = \sum_{i=1}^n \text{sign}(D_{i0})R_i$. As long as there are no ties in the data, then all three of these are equivalent and $W = W^+ - W^-$. For the above sample we have $W^+ = 33$, $W^- = 3$, and $W = 30$. It is perhaps more standard to call W^+ the Wilcoxon Signed Rank statistic. Under (12.35, p. 491) and continuity of the data (implying no ties with probability 1), the basic facts are that:

1. $\text{sign}(D_{10}), \ldots, \text{sign}(D_{n0})$ and $I(D_{10} > 0), \ldots, I(D_{n0} > 0)$ are independent of $|D_{10}|, \ldots, |D_{n0}|$ and thus also independent of $R_1, \ldots, R_n$;
2. $W^+ \stackrel{d}{=} W^- \stackrel{d}{=} \sum_{i=1}^n I(D_{i0} > 0)i$, and $I(D_{10} > 0), \ldots, I(D_{n0} > 0)$ are independent Bernoulli(1/2) random variables;
3. $W \stackrel{d}{=} \sum_{i=1}^n \text{sign}(D_{i0})i$, and $\text{sign}(D_{10}), \ldots \text{sign}(D_{n0})$ are iid with $P(\text{sign}(D_{i0}) = 1) = 1/2$;
4.

$$\text{E}(W^+) = \frac{1}{2}\sum_{i=1}^{n} i = \frac{n(n+1)}{4}, \quad \text{Var}(W^+) = \frac{1}{4}\sum_{i=1}^{n} i^2 = \frac{n(n+1)(2n+1)}{24};$$

5.

$$\mathrm{E}(W) = 0, \quad \mathrm{Var}(W) = \sum_{i=1}^{n} i^2 = \frac{n(n+1)(2n+1)}{6}.$$

For the simple example above with $n = 8$, we have $\mathrm{E}(W^+) = (8)(9)/4 = 18$ and $\mathrm{Var}(W^+) = (8)(9)(17)/24 = 51$ leading to the standardized value $(33 - 18)/\sqrt{51} = 2.1$, which is clearly the same for W^- and W as well. From a normal table, we get the right-tailed p-value .018, whereas the exact permutation p-value for the signed rank statistics is $5/256 = .01953$.

Although the Wilcoxon Signed Rank is by far the most important of the signed rank procedures, the general signed rank procedures are $T^+ = \sum_{i=1}^n I(D_{i0} > 0)a(R_i)$, $T^- = \sum_{i=1}^n I(D_{i0} < 0)a(R_i)$, and

$$T = \sum_{i=1}^{n} \mathrm{sign}(D_{i0})a(R_i), \tag{12.41}$$

where the scores $a(i)$ could be of any form. The analogues of the above properties for W hold for the general signed rank statistics. In particular $T \stackrel{d}{=} \sum_{i=1}^n \mathrm{sign}(D_{i0})a(i)$ simplifies the distribution and moment calculations in the case of no ties. In the case of ties, the permutation variance of T, given the midranks $R_1, \ldots, R_n$, is $\sum_{i=1}^n \{a(R_i)\}^2$. Thus, for the normal approximation, it is simplest to use the form

$$Z = \sum_{i=1}^{n} \mathrm{sign}(D_{i0})a(R_i) \Big/ \left[\sum_{i=1}^{n} \{a(R_i)\}^2\right]^{1/2}, \tag{12.42}$$

that automatically adjusts for ties (see Section 12.8.6, p. 497, for a discussion of ties).

The most well-known score functions are $a(i) = i$ for the Wilcoxon, the quantile normal scores $a(i) = \Phi^{-1}(1/2 + i/[2(n+1)])$, and the sign test $a(i) = 1$. These are asymptotically optimal for shifts in the center of symmetry D_0 of the logistic distribution, the normal distribution, and the Laplace distribution, respectively. For asymptotic analysis we assume $a(i) = \phi^+(i/(n+1))$, where $\phi^+(u)$ is nonnegative and nonincreasing and $\int_0^1 [\phi^+(u)]^2 du < \infty$. The asymptotically optimal general form for data with density $f(x - \theta_0)$ and $f(x) = f(-x)$ is

$$\phi^+(u) = -\frac{f'\left\{F^{-1}\left(\frac{1}{2} + \frac{u}{2}\right)\right\}}{f\left\{F^{-1}\left(\frac{1}{2} + \frac{u}{2}\right)\right\}}.$$

Asymptotic normality is similar to Theorem 12.5 (p. 492) (see for example, Theorem 10.2.5, p. 333 of Randles and Wolfe, 1979). The Edgeworth expansion for W^+ and T^+ may be found on p. 37 and p. 89, respectively, of Hettmansperger (1984).

Table 12.4 Pitman ARE's for the One-Sample Symmetry Problem

Distribution	ARE(S, t)	ARE(S, W^+)	ARE(W^+, t)
Normal	0.64	0.67	0.955
Uniform	0.33	0.33	1.00
Logistic	0.82	0.75	1.10
Laplace	2.00	1.33	1.50
t_6	0.88	0.76	1.16
t_3	1.62	0.85	1.90
t_1 (Cauchy)	∞	1.33	∞

12.8.4 Sign Test

The sign test mentioned in the last section as (12.41) with $a(i) = 1$ is usually given in the form $T^+ = \sum_{i=1}^{n} I(D_{i0} > 0)$, the number of positive differences. Under the null hypothesis (12.35, p. 491), T^+ has a binomial$(n, 1/2)$ distribution and is extremely easy to use. Because of this simple distribution, T^+ is often given early in a nonparametric course to illustrate exact null distributions.

The sign test does not require symmetry of the distributions to be valid. It can be used as a test of H_0 : median of $D_i - \theta_0 = 0$, where it is assumed only that $D_1, \ldots, D_n$ are independent, each with the same median. Thus, the test is often used in skewed distributions to test that the median has value θ_0. This generality, though, comes with a price because typically the sign test is not as powerful as the signed rank or t test in situations where all three are valid. If there are zeroes in $D_1, \ldots, D_n$, the standard approach is remove them before applying the sign test.

12.8.5 Pitman ARE for the One-Sample Symmetry Problem

In the Appendix, we give some details for finding expressions for the efficacy and Pitman efficiency of tests for the one-sample symmetry problem. Here we just report some Pitman ARE's in Table 12.4 for the sign test, the t test, and the Wilcoxon signed rank. The comparison of the signed rank and the t are very similar to those given in Table 12.3 (p. 477) for the two-sample problem. The only difference is that skewed distributions are allowed in the shift problem but not here.

The general message from Table 12.4 is that the tails of the distribution must be very heavy compared to the normal distribution in order for the sign test to be preferred. This is a little unfair to the sign test because symmetry of f is not required for the sign test to be valid, whereas symmetry is required for the Wilcoxon signed rank test. In fact Hettmansperger (1984, p. 10–12) shows that the sign test is uniformly most powerful among size-α tests if no shape assumptions are made about the density of f. Moreover, in the matched pairs situation where symmetry is justified by differencing, the uniform distribution is not possible, and that is where the sign test performs so poorly.

Monte Carlo power estimates in Randles and Wolfe (1979, p. 116) show that generally the ARE results in Table 12.4 correspond qualitatively to power comparisons. For example, at $n = 10$ and normal alternative $(\theta_0 + .4)/\sigma$, the Wilcoxon signed rank has power .330 compared to .263 for the sign test. The ratio $.263/.330 = .80$ is not too far from ARE$= .64$. The estimated power ratio at $n = 20$ is $.417/.546 = .76$. The Laplace distribution AREs in Table 12.4 are not as consistent. For example, at $n = 20$ for a similar alternative, the ratio is $.644/.571 = 1.13$, not all that close to ARE$= 2.00$.

The Wilcoxon signed rank test is seen to have good power relative to the sign test and to the t test. The Hodges and Lehmann (1956) result that ARE$(W^+, t) \geq .864$ also holds here for all symmetric unimodal densities. Coupled with the fact that there is little loss of power relative to the t test at the normal distribution (ARE$(W^+, t) = 0.955$), W^+ should be the statistic of choice in many situations.

12.8.6 *Treatment of Ties*

The general permutation approach is not usually bothered by ties in the data, although rank methods typically require some thought about how to handle the definition of ranks in the case of ties. For the original situation of n pairs of data and a well-defined statistic like the paired t statistic, the 2^n permutations of the data merely yield redundance if members of a pair are equal. For example, consider $n = 3$ and the following data with all 8 permutations (1 is the original data pairing):

1	2	3	4	5	6	7	8
3,5	5,3	3,5	5,3	3,5	5,3	3,5	5,3
2,2	2,2	2,2	2,2	2,2	2,2	2,2	2,2
7,4	7,4	4,7	4,7	7,4	7,4	4,7	4,7

Permutations 1–4 are exactly the same as permutations 5–8 because permuting the 2nd pair has no effect. Thus, a permutation p-value defined from just permutations 1–4 is exactly the same as for using the full set 1–8. After taking differences between members of each pair, the 2^n sign changes work in the same way by using sign$(0) = 0$; that is, there is the same kind of redundancy in that there are really just 2^{n-n_0} unique permutations, where n_0 is the number of zero differences.

For signed rank statistics, there are two kinds of ties to consider after converting to differences, multiple zeros and multiple non-zero values. For the non-zero multiple values, we just use mid-ranks (average ranks) as before. For the multiple zeros, there are basically two recommended approaches:

Method 1: Remove the differences that are zero and proceed with the reduced sample in the usual fashion. This is the simplest approach and the most powerful for the sign statistic (see Lehmann 1975, p. 144). Pratt and Gibbons (1981, p. 169) discuss anomalies when using this procedure with W^+.

Method 2: First rank all $|D_{10}|, \ldots, |D_{n0}|$. Then remove the ranks associated with the zero values before getting the permutation distribution of the rank statistic, *but do not change the ranks associated with the non-zero values.* However, as above, since the permutation distribution is the same with and without the redundancy, it really just makes the computing easier to remove the ranks associated with the zero values. The normal approximation in (12.42, p. 495) automatically eliminates the ranks associated with the zero values because sign(0) $= 0$. For the Box-Andersen approximation, the degrees of freedom are different depending on whether the reduced set is used or not. It appears best to use the reduced set for the Box-Andersen approximation although a few zero values make little difference.

Example 12.2 (Fault rates of telephone lines). Welch (1987) gives the difference (times 10^5) of a transformation of telephone line fault rates for 14 matched areas. We modify the data by dividing by 10 and rounding to 2 digits leading to

D_{i0}	−99	31	27	23	20	20	19	−14	11	9	8	−8	6	0
$\text{sign}(D_{i0})R_i$	−14	13	12	11	9.5	9.5	8	−7	6	5	3.5	−3.5	2	0

Notice that there two ties in the absolute values 20 and 8 for which the midranks are given. The exact right-tailed permutation p-value based on the t statistic is .38, whereas the t tables gives .33 and the Box-Andersen approximation is .40. The large outlier −99 essentially kills the power of the t statistic. The sign test first removes the 0 value and then the binomial probability of getting 10 or more positives out of 13 is .046. Welch (1987) used the sample median as a statistic and for these data we get exact p-value .062. Note that the mean and sum and t statistic are all permutationally equivalent, but the median is not permutationally equivalent to using a Wald statistic based on the median. So, the properties of using the median as a test statistic are not totally clear.

For the Wilcoxon Signed Rank, no tables can be used because of the ties and the 0. However, it is straightforward to get the permutation after choosing one of the methods above for dealing with the 0 difference.

Method 1: First remove the 0, then rank. The remaining data are

D_{i0}	−99	31	27	23	20	20	19	−14	11	9	8	−8	6
$\text{sign}(D_{i0})R_i$	−14	13	12	11	9.5	9.5	8	−7	6	5	3.5	−3.5	2

The exact p-value based on the $\text{sign}(D_{i0})R_i$ values above (for example, just insert the signed ranks into the R program below) is.048, the normal approximation is .047, and the Box-Andersen approximation is .049.

Method 2: Rank the data first, then throw away the signed rank associated with the 0. The exact p-value is .044 Recall, for the permutation p-value, it does not matter whether we drop the 0 or not after ranking. Similarly, the normal approximation p-value .042 based on (12.42, p. 495) automatically handles the 0 value. For the Box-Andersen approximation, we get .0437 based on all 14 signed ranks and .0441 after throwing out the 0; so it matters very little whether we include the 0 or not. ◆

For problems with $n \leq 20$, the following R code modified from Venables and Ripley (1997, p. 189-190) gives the exact permutation p-value for signed statistics:

```
perm.sign<-function(d,stat,pr=FALSE, ...){
 # Exact perm. $p$-value for one-sample problem.
 # Assumes test rejects for large values of stat.
 # Looks at all 2^n sign change samples.
 # Use only for small n.
 # Need the following obscure function
bi<-function(x,digits=if(x>0)1+
             floor(log(x,base=2)) else 1){
  ans<-0:(digits-1)
  (x %/% 2^ans) %% 2
  }        # note %/% and %% are different
# The main program
  t0<-stat(d, ...)
  digits<-length(d)
  b <- 2^digits
  res <- numeric(b)
  for(i in 1:b){
    x <- d*2*(bi(i,digits=digits) - 0.5)
    res[i] <- stat(x, ...)
    if(pr)cat(i,x,res[i],fill=T) # prints
  }
  pvalue <- sum(res >= t0)/b
  sum(res==t0)->co
  return(data.frame(b=b,stat0=round(t0,4),
    eq.t0=co,rt.pvalue=pvalue,pv2=2*pvalue))
}
```

12.9 Randomized Complete Block Data—the Two-Way Design

Blocking is one of the most important techniques for reducing variation in experimental designs. The usual Randomized Complete Block design may be viewed as a generalization of the matched pairs to situations with more than two treatments. To use the permutation argument with blocked data, we do not need for the treatments to be assigned randomly, but it is most natural to discuss blocked data in that context. The key assumption required under H_0 is that the data are exchangeable within blocks.

Suppose that k treatments are to be assigned at random within each block of size k. For n blocks, there are $(k)^n$ possible permutations of the data corresponding to permuting independently among treatments within each block. In the following table there are $k = 4$ blocks with $n = 10$ treatments, thus $M_N = 24^{10} = 6.34\times 10^{13}$ possible permutations. These data are actually treatments 6–15 from an example of aphid infestation of crepe myrtle cultivars given in Table 1 of Brownie and Boos (1994). The response variable is the number of aphids on the three most heavily infested leaves plus the percent of foliage covered with sooty mold.

	Treatments									
Block	1	2	3	4	5	6	7	8	9	10
1	0	0	93	78	5	1	0	21	1	1
2	0	24	0	3	2	180	0	0	3	9
3	0	2	10	0	0	3	2	3	3	140
4	0	4	2	2	0	0	1	47	1	52

The linear model representation is

$$Y_{ij} = \mu + \beta_i + \alpha_j + e_{ij}, \tag{12.43}$$

where $\alpha_1, \ldots, \alpha_k$ are the treatment effects, and $\beta_1, \ldots \beta_n$ are the block effects. Note that we have switched subscripts on Y_{ij} compared to the one-way model (12.28, p. 480) so that the blocks can be the rows. Often the block effects are assumed random, but the nonparametric literature typically considers them fixed effects.

The usual ANOVA F statistic could be used with these data:

$$F = \frac{\dfrac{1}{k-1}\displaystyle\sum_{j=1}^{k} n(\overline{Y}_{.j} - \overline{Y}_{..})^2}{\dfrac{1}{(k-1)(n-1)}\displaystyle\sum_{i=1}^{n}\sum_{j=1}^{k}(Y_{ij} - \overline{Y}_{i.} - \overline{Y}_{.j} + \overline{Y}_{..})^2}, \tag{12.44}$$

where $\overline{Y}_{i.} = k^{-1}\sum_{j=1}^{k} Y_{ij}$, $\overline{Y}_{.j} = n^{-1}\sum_{i=1}^{n} Y_{ij}$, and $\overline{Y}_{..} = n^{-1}\sum_{i=1}^{n} \overline{Y}_{i.}$. For the above data $F = 0.80$ with p-value $= 0.62$ from an F distribution with 9 and 27 degrees of freedom. Since the F distribution approximates the permutation distribution, the value 0.62 should be satisfactory. A Monte Carlo approximation to the exact permutation p-value based on 10,000 samples gave .60 with standard error .005, thus confirming the Type I error robustness of the usual F procedure. However, the nonnormality of the response variable is cause for concern because the F statistic is not Type II error robust in the face of outliers. Transformations are an obvious approach, and F on $\log(Y_{ij} + 1)$ resulted in p-value = .29. Fortunately, with rank procedures we do not have to guess the correct transformation.

12.9.1 Friedman's Rank Test

The standard rank procedure was introduced by Friedman (1937). For the untied case, it has the form

$$T = \frac{12n}{k(k+1)} \sum_{j=1}^{k} \left(\overline{R}_{.j} - \frac{k+1}{2} \right)^2, \tag{12.45}$$

where R_{ij} is the rank of Y_{ij} within the ith row, and $\overline{R}_{.j} = n^{-1} \sum_{i=1}^{n} R_{ij}$ is the jth treatment mean rank. Note that $(k+1)/2$ is $\overline{R}_{..}$ since the average of the integers 1 to k is $(k+1)/2$. The within-row ranks R_{ij} for the above table are

	Treatments									
Block	1	2	3	4	5	6	7	8	9	10
1	2	2	10	9	7	5	2	8	5	5
2	2.5	9	2.5	6.5	5	10	2.5	2.5	6.5	8
3	2	4.5	9	2	2	7	4.5	7	7	10
4	2	8	6.5	6.5	2	2	4.5	9	4.5	10

We see immediately that there are numerous ties in the data. The form of the Friedman statistic that accommodates ties is (see, for example, Conover and Iman, 1981, p. 126)

$$T = \frac{(k-1)n^2 \sum_{j=1}^{k} \left(\overline{R}_{.j} - \frac{k+1}{2} \right)^2}{\left(\sum_{i=1}^{n} \sum_{j=1}^{k} R_{ij}^2 \right) - \frac{nk(k+1)^2}{4}}. \tag{12.46}$$

Under the null hypothesis of identical treatments, T converges to a χ^2_{k-1} distribution as $n \to \infty$ and k remains fixed. For the above data, $T = 13.7732$, and comparing to a χ^2_9 distribution gives p-value = .13. However, as in the one-way design, the χ^2 approximation becomes increasingly conservative as the number of treatments gets large relative to the number of blocks. F distribution p-values provide much better approximations and can be justified by either asymptotic theory or the Box-Andersen permutation moment approximations.

12.9.2 *F* Approximations

Friedman (1937, pp. 694–695) conjectured that the Friedman statistic is asymptotically normal as $k \to \infty$ with mean $k-1$ and variance $2(n-1)(k-1)/n$ (a proof may be found in Lemma 4 of Brownie and Boos, 1994). Similar to the one-way design, this asymptotic normal result is consistent with applying the F statistic (12.44, p. 500) to the within-row Friedman ranks and then using the $F(k-1,(k-1)(n-1))$ distribution for p-values. This argument is to be fleshed out in Problem 12.22 (p. 528). Of course, the F distribution should be used in practice; the asymptotic normal result just supports use of the F distribution.

From Box and Andersen (1955, p. 14-15), we may approximate the permutation distribution of F of (12.44, p. 500) or of the same F applied to the within-row Friedman ranks by a $F(d(k-1), d(k-1)(n-1))$ distribution, where

$$d = 1 + \frac{(nk - n + 2)V_2 - 2n}{n(k-1)(n - V_2)},$$

$$V_2 = \frac{1}{n-1}\sum_{i=1}^{n}(s_i^2 - \overline{s}^2)^2/(\overline{s}^2)^2,$$

and the s_i^2 are the within-row variances, and $\overline{s}^2 = n^{-1}\sum_{i=1}^{n} s_i^2$. In the case of the Friedman ranks with no ties in the data, $d = 1 - 2/\{n(k-1)\}$. For the Crepe Myrtle data this latter expression is $d = .944$, the same (to three decimals) as the actual d value from the tied ranks. We summarize the various approximations in the following table:

Approximate P-Values
for the Crepe Myrtle Data

	Monte Carlo	$F(9,27)$	Box-And. $F(9d,27d)$	χ^2_9
Friedman	.10			.13
F_R	.10	.10	.11	
F on Y	.60	.62	.63	
F on $\log(Y+1)$	.29	.29	.30	

The Monte Carlo estimates are based on 10,000 random permutations and have standard error bounded by .005. The F approximations are good, but the Box-Andersen adjustments do not help here. Interestingly, $d = 1.08$ for the usual F (row 3), but the p-value is adjusted upwards because the $F = .80$ is so small. Typically, a d value greater than 1 lowers the p-value from the F approximation.

Table 12.5 Pitman ARE of the Friedman Test to the F Test

	k = Number of Treatments					
Distribution	2	3	4	5	10	∞
Normal	0.64	0.72	0.76	0.80	0.87	0.955
Uniform	0.67	0.75	0.80	0.83	0.91	1.000
t_3	1.27	1.42	1.52	1.58	1.73	1.900

12.9.3 Pitman ARE for Blocked Data

From van Elteren and Noether (1959) we find the surprising result that the Pitman asymptotic relative efficiency of the Friedman test to the ANOVA F depends on the number of treatments k,

$$\text{ARE(Friedman}, F) = \left\{\frac{k}{k+1}\right\} 12\sigma^2 \left\{\int_{-\infty}^{\infty} f^2(x)\,dx\right\}^2, \qquad (12.47)$$

where σ^2 is the variance of the observations. Expression (12.47) is just $k/(k+1)$ times the ARE(W, t) in (12.25, p. 477). Table 12.5 gives a few values of (12.47) for several distributions.

The value .64 at $k = 2$ for the normal distribution is the same as the ARE of the sign test to the t in Table 12.4 (p. 496). That is no accident. It turns out that for $k = 2$, the Friedman test is equivalent to the sign test. (The other values in Table 12.4, p. 496, do not correspond to the $k = 2$ values in Table 12.5 because Table 12.4 refers to the distribution after taking differences, whereas Table 12.5 is for the distribution of the individual treatment results, not the difference of treatment results. For the normal distribution, the difference of normal random variables is also normally distributed; so for the normal the results are the same in both tables.)

The reason for the low efficiency in Table 12.5 is that ranking within rows (intrablock ranking) takes no advantage of between block (interblock) information. For the $k = 2$ case, the Wilcoxon signed rank statistic uses interblock information by ranking the absolute differences (note the improved efficiencies in Table 12.4, p. 496, for the signed rank test compared to the sign test). In the next section we discuss some rank approaches that use interblock information.

12.9.4 Aligned Ranks and the Rank Transform

Many approaches have been used to remedy the low efficiency in Table 12.5 for small values of k. Perhaps the earliest approach (and still one of the best) is the aligned rank method due to Hodges and Lehmann (1962). The aligned rank approach is to first subtract the block mean (or any other location measure such as the median) from each observation Y_{ij}, then rank all the resulting nk residuals together. These latter ranks on the residuals, denoted $\widehat{R}_{ij}$, are called *aligned ranks*. We suggest using F of (12.44, p. 500) on these aligned ranks.

Actually, Sen (1968) and Lehmann (1975, p. 272) use

$$\widehat{Q} = \frac{n^2(k-1)\sum_{j=1}^{k}\left(\overline{\widehat{R}}_{.j} - \frac{nk+1}{2}\right)^2}{\sum_{i=1}^{n}\sum_{j=1}^{k}\left(\widehat{R}_{ij} - \overline{\widehat{R}}_{i.}\right)^2}, \tag{12.48}$$

a statistic that is asymptotically χ^2_{k-1} under H_0. The justification for the form (12.48) comes from noting that the permutation mean of $\overline{\widehat{R}}_{.j}$ is $(nk+1)/2$, and the permutation covariance matrix of $(\overline{\widehat{R}}_{.1}, \ldots, \overline{\widehat{R}}_{.k})$ is

$$\frac{\sigma^2 k}{k-1}\text{diag}\left(\boldsymbol{I}_k - \frac{\mathbf{1}_k\mathbf{1}_k^T}{k}\right), \tag{12.49}$$

where $\boldsymbol{I}_k$ is the k-dimensional identity matrix, $\mathbf{1}_k$ is a vector of ones, and

$$\sigma^2 = \frac{1}{n^2k}\sum_{i=1}^{n}\sum_{j=1}^{k}(\widehat{R}_{ij} - \overline{\widehat{R}}_{i.})^2 \tag{12.50}$$

is the permutation variance of $\overline{\widehat{R}}_{.j}$. $\widehat{Q}$ in (12.48) is the appropriate quadratic form in $(\overline{\widehat{R}}_{.1}, \ldots, \overline{\widehat{R}}_{.k})$ upon noting that $(k-1)\boldsymbol{I}_k/(k\sigma^2)$ is a generalized inverse of the covariance matrix (12.49).

Other authors (Fawcett and Salter, 1984, and O'Gorman, 2001) use a one-way ANOVA F on the aligned ranks, but we prefer the two-way F of (12.44, p. 500) because the Box-Andersen adjustment is readily available. All three statistics, $\widehat{Q}$ and the two F statistics on the aligned ranks, are permutationally equivalent to the numerator of $\widehat{Q}$; so if exact or Monte Carlo approximations are used, it does not matter which of the three statistics is chosen. Clearly, either of the two Fs gives better approximate p-values than $\widehat{Q}$ with χ^2_{k-1} p-values.

Mehra and Sarangi (1967) give somewhat complicated formulas for the Pitman ARE of the aligned rank approach to the usual F and to Friedman's statistic, but the bottom line is that the AREs of the aligned rank procedure to the usual F are close to the last column of Table 12.5 (p. 503). Thus, the aligned rank approach is able to recover most of the interblock information.

Another approach to recovering the interblock information is to just rank all the observations together and apply F of (12.44, p. 500) on the resulting ranks. This *rank transform* approach, due to Conover and Iman (1981) works well as long as the block effects are not strong. When the block effects are strong, then this approach is similar to Friedman's test. Hora and Iman (1988) give Pitman ARE results for this approach.

There is an extensive literature on rank methods in block models. Mahfoud and Randles (2005) and Kepner and Wackerly (1996) are several places that briefly review many of the approaches. The latter also gives extensions to incomplete blocks.

12.9.5 Replications within Blocks

In the preceding discussion we have been talking about cases where there is just one observation per cell, nk total observations for n blocks and k treatments, and no block by treatment interaction. Consider the $k = 2$ case and n blocks where there are m_i Xs for the first treatment in block i and n_i Ys for the second treatment, $i = 1, \ldots, n$. These type data arise naturally in clinical trials at n centers or sites. The sites might be hospitals or clinics or individual doctors. The usual rank approach is the van Elteren statistic (van Elteren, 1960, or Lehmann 1975, p. 145), a weighted sum of individual Wilcoxon rank sum statistics W_i within each block,

$$W_{\mathrm{VE}} = \sum_{i=1}^{n} \frac{W_i}{m_i + n_i + 1}.$$

van Elteren (1960) showed that the weights $1/(m_i + n_i + 1)$ are asymptotically optimal among all linear combinations of the W_i. This optimality makes sense if we write the standardized version of W_{VE} as

$$\sum_{i=1}^{n} \frac{1}{\sigma_0^2(\widehat{\theta}_i)} \left(\widehat{\theta}_i - \frac{1}{2}\right) \Bigg/ \left\{\sum_{i=1}^{n} \frac{1}{\sigma_0^2(\widehat{\theta}_i)}\right\}^{1/2}, \qquad (12.51)$$

where $\widehat{\theta}_i$ is the Mann-Whitney estimator of $\theta_i = P(Y_{i1} > X_{i1}) + (1/2)P(Y_{i1} = X_{i1})$ given in (12.14, p. 463) (here we have dropped the XY subscript for simplicity), and $\sigma_0^2(\widehat{\theta}_i)$ is the variance of $\widehat{\theta}_i$ under the null hypothesis of identical X and Y populations. In the completely nonparametric case (in the absence of the shift model), θ_i is the underlying parameter of interest for Wilcoxon statistics. For continuous data (no ties), $\sigma_0^2(\widehat{\theta}_i) = (m_i + n_i + 1)/(12m_i n_i)$. Thus, the numerator of the standardized version of W_{VE} is a weighted average of $\widehat{\theta}_i - 1/2$, where the weights are inversely proportional to null variances.

The analogous t procedure is based on standardizing

$$\sum_{i=1}^{n} \frac{m_i n_i}{m_i + n_i} (\overline{Y}_i - \overline{X}_i). \qquad (12.52)$$

Thus, the t procedure uses a weighted linear combination of the difference of sample means, where the weights are inversely proportional to $\text{Var}\left(\overline{Y}_i - \overline{X}_i\right) = \sigma^2(1/m_i + 1/n_i)$.

The standard permutation approach is to consider all possible

$$M_N = \prod_{i=1}^{n} \binom{m_i + n_i}{n_i}$$

independent permutations within sites. The normal approximation for W_{VE} should be very good if $\sum_{i=1}^{n} m_i$ and $\sum_{i=1}^{n} n_i$ are reasonably large and therefore is widely used in practice. In the case that $\sum_{i=1}^{n} m_i$ and $\sum_{i=1}^{n} n_i$ converge to ∞, Hodges and Lehmann (1962) give the Pitman ARE of (12.51) to (12.52) for normal data as

$$.955 \sum_{i=1}^{n} \frac{m_i n_i}{m_i + n_i + 1} \Big/ \sum_{i=1}^{n} \frac{m_i n_i}{m_i + n_i}.$$

Thus, if $m_i + n_i$ is reasonably large, then the ARE is close to the best value .955. For example, if $m_i + n_i = 10$ for each site, then the ARE is .955(10/11).

For the case that there are small numbers of replications per block (site), we are led back to the procedures of the previous section, aligned ranks and possibly the rank transform. With replications within blocks, however, we now have the ability to test for block by treatment interactions. Unfortunately, standard permutation procedures are not available for testing the no interaction hypothesis in the face of main effects. A large literature exists evaluating and criticizing the rank transform approach for testing interactions. See, for example, Akritas (1990, 1991) and Thompson (1991). In general, for more complicated fixed effects models with interaction, to achieve robustness via rank methods, we feel it is better to use the general R-estimation linear model approach mentioned at the end of Section 12.7 (p. 487).

Boos and Brownie (1992) argue that a mixed model approach is usually more appropriate, allowing inferences to be made to a larger population, but the mixed model leads away from van Eltern's statistic (12.51, p. 505) and permutation inference.

12.10 Contingency Tables

12.10.1 2 x 2 Table – Fisher's Exact Test

The first use of the permutation method was given by Fisher (1934a, *Statistical Methods for Research Workers*, fifth edition) in an analysis of 2×2 tables. Fisher's example was of 13 identical twins and 17 fraternal twins (of the same sex) who had at least one of the pair convicted of a crime. Of the 13 identical twins only 3 had a twin free of conviction. Of the 17 fraternal twins 15 had a twin free of conviction. Thus the table is as follows,

	Both Convicted	One Convicted	Total
Identical	10	3	13
Fraternal	2	15	17
Total	12	18	30

To fix notation, a general 2×2 table is,

	Category 1	Category 2	Total
Group 1	N_{11}	N_{12}	$N_{1.}$
Group 2	N_{21}	N_{22}	$N_{2.}$
Total	$N_{.1}$	$N_{.2}$	N

A standard analysis of these data assumes that N_{11} is binomial $(N_{1.}, p_1)$ and independent of N_{21} assumed to be binomial $(N_{2.}, p_2)$. The usual statistic for testing $H_0 : p_1 = p_2$ is the pooled Z, the square root of the score statistic found in Section 3.2.9 (p. 142),

$$Z = \frac{\widehat{p}_1 - \widehat{p}_2}{\left\{ \frac{\widetilde{p}(1-\widetilde{p})}{N_{1.}} + \frac{\widetilde{p}(1-\widetilde{p})}{N_{2.}} \right\}^{1/2}},$$

where $\widehat{p}_1 = N_{11}/N_{1.}$, $\widehat{p}_2 = N_{21}/N_{2.}$, and $\widetilde{p} = N_{.1}/N$. To test $H_a : p_1 > p_2$, the standard approach would be to compare Z to z_α, the $1-\alpha$ quantile of the standard normal.

Instead of this approximate procedure, Fisher noted that conditional on the margins $N_{.1}$ and $N_{.2}$ held fixed in addition to $N_{1.}$ and $N_{2.}$, that a given table has hypergeometric probability of $(n_{11}, n_{12}, n_{21}, n_{22})$ given by

$$\frac{\binom{N_{1.}}{n_{11}}\binom{N_{2.}}{n_{21}}}{\binom{N}{N_{.1}}} = \frac{N_{1.}!N_{2.}!N_{.1}!N_{.2}!}{N!n_{11}!n_{12}!n_{21}!n_{22}!}.$$

This hypergeometric probability is easily obtained if one thinks about an urn with $N_{.1}$ balls of type 1 and $N_{.2}$ of type 2. If we draw out $N_{1.}$ balls without replacement, then the above probability is the probability of getting n_{11} of type 1 and n_{21} of type 2.

One can also think of the above table arising in the two-sample problem where the data consists of just 1's and 0's. Although there are $\binom{N}{N_{1.}}$ permutations of interest, many of them yield the same table. The numerator of the above hypergeometric probability just gives the number of permutations which lead a given table.

Now a variety of statistics can be used to order the possible tables from supporting H_0 to strongly rejecting H_0 and to calculate a p-value. Or one can just use intuition for the ordering: most people would agree that for testing $H_a : p_1 > p_2$, the table below is more extreme than the original.

	Category 1	Category 2	Total
Group 1	$N_{11} + 1$	$N_{12} - 1$	$N_{1.}$
Group 2	$N_{21} - 1$	$N_{22} + 1$	$N_{2.}$
Total	$N_{.1}$	$N_{.2}$	N

Thus, a one-tailed p-value would be obtained by summing up the hypergeometric probabilities of those tables as extreme or more extreme than the original table $(N_{11}, N_{12}, N_{21}, N_{22})$. A number of seemingly different ways of ordering the tables lead to the same definition of "more extreme" and are called Fisher's Exact Test. The simplest way to order is either the intuitive notion above or to order via the pooled Z statistic.

For the twins data, Fisher noted that the two more extreme tables have $N_{11} = 11$, $N_{12} = 2$, $N_{21} = 1$, $N_{22} = 16$ and $N_{11} = 12$, $N_{12} = 1$, $N_{21} = 0$, $N_{22} = 17$. Thus the p-value is the probability of the original table plus the probability of these two more extreme tables:

$$\frac{13!17!12!18!}{30!}\left\{\frac{1}{10!3!2!15!} + \frac{1}{11!2!1!16!} + \frac{1}{12!1!0!17!}\right\} = \frac{619}{1330665} = .000465.$$

The definition of a two-sided p-value is not so clear, but the usual practice is to add in the probabilities of tables as extreme or more extreme in the other direction (having probabilities less than or equal to the probability of the observed table). In the above example we would need to add the probabilities of tables with $N_{11} = 0$, $N_{12} = 13$, $N_{21} = 12$, $b_{22} = 5$ and $N_{11} = 1$, $N_{12} = 12$, $N_{21} = 11$, $N_{22} = 6$ but not $N_{11} = 2$, $N_{12} = 11$, $N_{21} = 10$, $N_{22} = 7$ since it has higher probability than the original table.

When accompanied by a randomization rule to yield exact α levels, Fisher's Exact Test is uniformly most powerful unbiased as discussed in Lehmann (1986, Ch. 4). But many people have noted how conservative it is when p-values are used with the rule: reject H_0 when p-value $\leq \alpha$. In this case the discreteness of the permutation distribution does prove costly in terms of power.

Barnard (1945, 1947), Boschloo (1970), and Suissa and Shuster (1985) proposed unconditional tests in the 2 x 2 table that are typically more powerful than the

Fisher Exact Test without randomization. See Berger (1996) for details and power comparisons.

We have given Fisher's Exact Test in the context of two independent binomials and $H_0 : p_1 = p_2$. It also applies in the context of multinomial data where the data consists of a pair of binary variables (X, Y) with values x_1 and x_2 and y_1 and y_2, respectively:

		Y		
		y_1	y_2	Total
X	x_1	N_{11}	N_{12}	$N_{1.}$
	x_2	N_{21}	N_{22}	$N_{2.}$
	Total	$N_{.1}$	$N_{.2}$	N

The entries $(N_{11}, N_{12}, N_{21}, N_{22})$ are multinomial$(N; p_{11}, p_{12}, p_{21}, p_{22})$ with associated parameters

		Y		
		y_1	y_2	Total
X	x_1	p_{11}	p_{12}	$p_{1.}$
	x_2	p_{21}	p_{22}	$p_{2.}$
	Total	$p_{.1}$	$p_{.2}$	1

In this paired variable context, the null hypothesis for Fisher's Exact Test is independence of X and Y,

$$H_0 : p_{ij} = p_{i.}p_{.j}, \quad i = 1, 2; j = 1, 2. \tag{12.53}$$

Of course, if $p_{11} = p_{1.}p_{.1}$, then all the other equalities such as $p_{12} = p_{1.2}p_{.2}$ hold as well.

12.10.2 Paired Binary Data – McNemar's Test

In the context of paired binary data introduced in the last section, we might expect association between X and Y, but our main interest could be in their marginal probabilities. In particular, the null hypothesis is often

$$H_0 : p_{1.} = p_{.1}. \tag{12.54}$$

A typical application is in matched pair studies such as the following well-known case-control data from Miller (1980),

		Sibling (Control)		
		Tons.	No Tons.	Total
Hodgkin's	Tons.	26	15	41
Patient	No Tons.	7	37	44
	Total	33	52	85

where Hodgkin's patients were paired with a sibling and it was determined whether they each had a tonsillectomy or not. If the marginal estimates $\widehat{p}_{1.} = N_{1.}/N = 41/85$ and $\widehat{p}_{.1} = N_{.1}/N = 33/85$ differ significantly, then incidence of tonsillectomies may be associated with contracting Hodgkin's disease. Noting that $\widehat{p}_{1.} - \widehat{p}_{.1} = N_{12}/N - N_{21}/N$ has multinomial variance $\{p_{12} + p_{21} - (p_{12} - p_{21})^2\}/N = (p_{12} + p_{21})/N$ under H_0, the score statistic is

$$Z = \frac{N_{12} - N_{21}}{(N_{12} + N_{21})^{1/2}}.$$

Exact inference follows by noting that under (12.54, p. 509), $N_{12}|N_{12} + N_{21}$ has a binomial$(N_{12} + N_{21}, 1/2)$ distribution. Thus, $Z = 1.71$ has approximate normal one-sided p-value $= .044$, but $P(\text{binomial}(22, 1/2) \geq 15) = .067$. These procedures are generally referred to as McNemar's test.

What do these tests have to do with permutation and rank statistics? Let $X = 1$ denote that a Hodgkin's patient had a tonsillectomy, and $X = 0$ denote that he/she did not, and similarly $Y = 1$ and $Y = 0$ for the sibling control. Then the paired data and their differences are

Pair	Hodgkin's Patient	Sibling (Control)	Diff.
1	1	1	0
.	.	.	.
.	.	.	.
26	1	1	0
27	1	0	1
.	.	.	.
.	.	.	.
41	1	0	1
42	0	1	−1
.	.	.	.
.	.	.	.
48	0	0	0
49	0	0	0
.	.	.	.
.	.	.	.
85	0	0	0

Note that there are $N_{12} = 15$ positive differences out of $N_{12} + N_{21} = 22$ nonzero differences. Thus, the exact binomial procedure above is just the sign test for the differences, and Z is exactly (12.42, p. 495) for $a(i) = 1$. In fact, since all the nonzero absolute differences are identically 1, the exact signed rank test (assuming zeroes are deleted) yields the same binomial procedure, and Z is also (12.42, p. 495) with $a(i) = i$.

12.10.3 *I* by *J* Tables

We now consider the general I by J contingency table

		Y					
		y_1	.	.	.	y_J	Total
	x_1	N_{11}	.	.	.	N_{1J}	$N_{1.}$
	.	.	.	.	.	.	.
X	.	.	.	.	.	.	.
	.	.	.	.	.	.	.
	x_I	N_{I1}	.	.	.	N_{IJ}	$N_{J.}$
	Total	$N_{.1}$	.	.	.	$N_{.J}$	N

The distribution of these data could be a full multinomial with IJ cells or I independent rows of multinomial data. In either case, exact permutation analysis is achieved by conditioning on the marginal totals resulting in a multiple hypergeometric for the joint distribution of the entries N_{ij} having probability $P(N_{ij} = n_{ij}, i = 1, \ldots, I; j = 1, \ldots, J \mid N_{1.}, \ldots, N_{I.}, N_{.1}, \ldots, N_{.J})$ given by

$$\frac{\left(\prod_{i=1}^{I} N_{i.}!\right)\left(\prod_{j=1}^{J} N_{.j}!\right)}{N! \prod_{i=1}^{I} \prod_{j=1}^{J} n_{ij}!}.$$

The question remains as to what statistic should be used. If both X and Y have nominal categories, then the chi-squared goodness-of-fit statistic is natural, but not very interesting. If X and Y have numerical scores or are at least ordered, then some type of association or correlation statistic should be used. For example, one might use Pearson's r or Spearman's rank correlation. If X has nominal categories and Y has numerical categories, then ANOVA type comparisons among the row means makes sense. If X has nominal categories and Y has ordered categories, then the Kruskal-Wallis test might be a good choice of statistic. Moreover, all these situations can be generalized to multi-way tables, say I by J by K tables, usually viewed as stratified comparisons of X and Y.

All these options for statistics in two-way and multiway tables come under the general purview of *Generalized Cochran-Mantel-Haenszel statistics*. Expositions of these statistics may be found in Landis et al. (1978) and Agresti (2002, Section 7.5.3) and implementation is found in `SAS PROC FREQ`.

12.11 Confidence Intervals and R-Estimators

Confidence intervals can be obtained from permutation and rank test statistics in the same way as for other types of statistics: choose values of θ appearing in a null hypothesis such that the statistic $T(\theta)$ viewed as a function of θ does not reject the null hypothesis (see 3.19, p. 144). We often refer to this approach as "inverting a test statistic." For example, in the one-sample problem with data $D_1, \ldots, D_n$ assumed to be symmetrically distributed about θ_0, a two-sided permutation t test could just as well be based on $T(\theta_0) = |\sum_{i=1}^{n}(D_i - \theta_0)|$. The permutation distribution depends on the 2^n sign change configurations of $D_i - \theta_0, \ldots, D_n - \theta_0$; we reject if $T(\theta_0)$ is larger than the largest α of the 2^n values of $T(\theta_0)$ computed on those permutations. So the $1-\alpha$ confidence interval can be found by trial and error, but it would seem to be a pretty laborious task because the permutation distribution changes with each θ_0. A somewhat easier computing method is suggested in Lehmann (1986, p. 263), but in general, the usual t interval is close enough to the permutation interval that it is mostly used in practice.

Inverting the signed rank statistic W^+ leads to an interval $[W_{(k_1)}, W_{(k_2)}]$, where $W_{(1)} \le W_{(2)} \cdots \le W_{(n(n+1)/2)}$ are the ordered values of the *Walsh averages*

$$W_{ij} = \frac{D_i + D_j}{2}, \qquad 1 \le i \le j \le n. \tag{12.55}$$

The order number k_2 is such that $P(W^+ \ge k_2) \le \alpha/2$, and $k_1 = n(n+1)/2+1-k_2$. We have specified a closed interval so that the probability of coverage is at least $1-\alpha$ for tied data situations (see Randles and Wolfe, 1979, p. 181-183). For example, at $n = 7$ with continuous data and $\alpha = .05$, $P(W^+ \ge 26) = P(W^+ \le 2) = .0234$, and thus the interval $[W_{(3)}, W_{(26)}]$ has exact confidence level $1-.0468 = .9532$. Often k_1 and k_2 are taken from the normal approximation to the permutation distribution of W^+. For example, $k_1 = q + 1$ and $k_2 = n(n+1)/2 - q$, where q is the closest integer to

$$\frac{n(n+1)}{4} - z_{\alpha/2}\left\{\frac{1}{4}\sum_{i=1}^{n} R_i^2\right\}^{1/2}.$$

In the $n = 7$ example above, this latter calculation gives 2.4, and thus $q = 2$, $k_1 = 3$, and $k_2 = 28 - 2 = 26$ as before. For the sample $-1.11, 2.23, 3.35, 4.67, 5.34, 6.17, 7.44$, the interval is $[W_{(3)}, W_{(26)}] = [1.12, 6.39]$.

Inverting the sign test leads to an interval of order statistics

$$(D_{(k)}, D_{(n-k+1)}), \quad 1 \le k \le n - k + 1.$$

This interval has exact coverage probability $C_n(k) = 1 - (1/2)^{n-1} \sum_{i=0}^{k-1} \binom{n}{i}$ for the population median from any continuous, not necessarily symmetric distribution. To obtain at least the same coverage for any discrete distribution, we need to again change to the closed interval $[D_{(k)}, D_{(n-k+1)}]$. An interesting addendum to these intervals due to Guilbaud (1979) is that the average of two such intervals,

$$\left[\frac{D_{(k)} + D_{(k+t)}}{2}, \frac{D_{(n-k-t+1)} + D_{(n-k+1)}}{2}\right], \quad k + t \le n - k - t + 1,$$

has guaranteed coverage $\{C_n(k) + C_n(k + t)\}/2$ for any distribution. This latter interval is useful for small n because it give more options for the confidence level than given by $C_n(k)$ alone. A more practical solution is given byHettmansperger and Sheather (1986), who interpolate between adjacent order statistics to get an interval with approximately the specified confidence, say 95%. The intervals are no longer distribution-free, but the confidence is close to the specified value.

Moving to the two-sample problem, the permutation interval based on the two-sample t is hard to compute, similar to the one-sample interval, and the usual t interval is mostly used in practice. Inversion of the Wilcoxon Rank Sum statistic for the shift model $G(x) = F(x - \Delta)$ leads to a confidence interval for Δ of the form $[U_{(k_1)}, U_{(k_2)}]$, where $U_{(1)} \le U_{(2)} \cdots \le U_{(mn)}$ are the ordered values of the pairwise differences

$$U_{ij} = Y_j - X_i, \quad i = 1, \dots, m; j = 1, \dots, n. \qquad (12.56)$$

Similar to the one-sample case, k_2 is chosen so that $P(W \ge k_2 + n(n+1)/2) = \alpha/2$ and $k_1 = mn + 1 - k_2$. In practice, one often uses the normal approximation interval with $k_1 = q + 1$ and $k_2 = mn - q$, where q is the integer closest to

$$\frac{mm}{2} - z_{\alpha/2} \{\text{Var}(W)\}^{1/2},$$

where $\text{Var}(W)$ is given by (12.10, p. 462) or (12.11, p. 462).

Point estimators obtained from rank test statistics were introduced by Hodges and Lehmann (1963). These *R-estimators* inherit some of the natural robustness properties of rank methods; see, for example Huber (1981) and Serfling (1980, Ch. 9), Randles and Wolfe (1979, Ch. 7), and Hettmansperger (1984, Ch. 5). The most well known are: i) the one-sample center of symmetry estimator $\widehat{\theta} = \text{median}\{W_{ij}\}$, where the W_{ij} are in (12.55, p. 512); and ii) the two-sample shift estimator $\widehat{\Delta} = \text{median}\{U_{ij}\}$, where the U_{ij} are in (12.56, p. 513). Asymptotic relative efficiency comparisons for confidence intervals and estimators derived from rank tests are exactly the same as for the associated rank tests.

12.12 Appendix – Technical Topics for Rank Tests

12.12.1 Locally Most Powerful Rank Tests

Recall from Section 12.5.1 (p. 474) that for $H_0 : \Delta = 0$ versus $H_a : \Delta > 0$, if there exists a rank test that is uniformly most powerful of level α for some $\epsilon > 0$ in the restricted testing problem $H_0 : \Delta = 0$ versus $H_{a,\epsilon} : 0 < \Delta < \epsilon$, we say that the test is the *locally most powerful rank test* for the original testing problem. By using a Taylor expansion of the probability of the rank vector $\boldsymbol{R}$ as a function of Δ, $L_{\boldsymbol{r}}(\Delta) \equiv P_\Delta(\boldsymbol{R} = \boldsymbol{r})$, we need only obtain an expression for the derivative of $L_{\boldsymbol{r}}(\Delta)$ and maximize it.

To see this consider the Taylor expansion

$$L_{\boldsymbol{r}}(\Delta) = L_{\boldsymbol{r}}(0) + L'_{\boldsymbol{r}}(0)\Delta + o(|\Delta|),$$

and a rank test with $\alpha = k/N!$ based on maximizing $L'_{\boldsymbol{r}}(0)$. Let $\boldsymbol{r}^{(1)}$ be the rank configuration that makes $L'_{\boldsymbol{r}}(0)$ largest among all $N!$ rank configurations, $\boldsymbol{r}^{(2)}$ makes $L'_{\boldsymbol{r}}(0)$ second largest among all $N!$ rank configurations, etc. Such a rank test has power

$$\beta(\Delta) = \sum_{j=1}^{k} L_{\boldsymbol{r}^{(j)}}(\Delta) = \sum_{j=1}^{k} \left[\frac{1}{N!} + L'_{\boldsymbol{r}^{(j)}}(0)\Delta + o(|\Delta|) \right].$$

For each rank configuration $\boldsymbol{r}^{(j)}$, we can choose Δ_j small enough so that $L_{\boldsymbol{r}^{(j)}}(\Delta)$ is also the jth largest among $L_{\boldsymbol{r}^{(1)}}(\Delta), \ldots, L_{\boldsymbol{r}^{(N!)}}(\Delta)$ for all $0 < \Delta < \Delta_j$. Now take ϵ to be smaller than all of the Δ_j. This shows that for $0 < \Delta < \epsilon$, the power of the test that places points in the rejection region as ordered by $L'_{\boldsymbol{r}}(0)$ also puts points in the rejection as ordered by $P_\Delta(\boldsymbol{R} = \boldsymbol{r}) = L_{\boldsymbol{r}}(\Delta)$; in other words, it is the locally most powerful rank test.

Let us now consider the two-sample problem where $X_1, \ldots, X_m$ are iid with distribution function $F(x)$, and $Y_1, \ldots, Y_n$ are iid with distribution function $G(x)$. Suppose that F and G have densities $f(x)$ and $g(x)$, respectively, whose support is contained in that of a density $h(x)$. This means that $h(x)$ is positive whenever $f(x)$ and $g(x)$ are positive; for example, when all three densities have support on $(-\infty, \infty)$. From Theorem 12.6, (p. 515), we have

$$P(\boldsymbol{R} = \boldsymbol{r}) = \frac{1}{N!} \mathrm{E} \left[\frac{\prod_{i=1}^{m} f(V_{(r_i)}) \prod_{i=m+1}^{N} g(V_{(r_i)})}{\prod_{i=1}^{m} h(V_{(r_i)}) \prod_{i=m+1}^{N} h(V_{(r_i)})} \right],$$

where $V_{(1)} < \cdots < V_{(N)}$ are the order statistics of an iid sample of size N from $h(x)$.

Shift alternatives have the form $g(x) = f(x - \Delta)$ so that the X distribution has the same shape as the Y distribution but shifted Δ to the right of it. If $f(x)$ has support on $(-\infty, \infty)$, then we may take $h(x) = f(x)$ and obtain

$$P_\Delta(\boldsymbol{R} = \boldsymbol{r}) = \frac{1}{N!}\mathrm{E}\left[\frac{\prod_{i=m+1}^{N} f(V_{(r_i)} - \Delta)}{\prod_{i=m+1}^{N} f(V_{(r_i)})}\right], \tag{12.57}$$

where now $V_{(1)} < \cdots < V_{(N)}$ are order statistics for a random sample from f. Now suppose that $f(x)$ is differentiable and that we can take the derivative inside the expectation in (12.57) . Then,

$$L'_{\boldsymbol{r}}(0) = \frac{\partial}{\partial \Delta} P_\Delta(\boldsymbol{R} = \boldsymbol{r})\bigg|_{\Delta=0} = \frac{1}{N!}\sum_{i=m+1}^{N} \mathrm{E}\left[\frac{-f'(V_{(r_i)})}{f(V_{(r_i)})}\right]. \tag{12.58}$$

The locally most powerful rank test places points in the rejection region according to large values of this latter expression.

If we let $V_{(1)} < \cdots < V_{(N)}$ be replaced by $F^{-1}(U_{(1)}) < \cdots < F^{-1}(U_{(N)})$ where the $U_{(i)}$ are uniform order statistics from an iid sample $U_1, \ldots, U_N$, then the locally most powerful rank test rejects for large values of

$$T = \sum_{i=m+1}^{N} a(R_i),$$

where $a(i) = \mathrm{E}\phi(U_{(i)}, f)$, and $\phi(u, f) = -f'(F^{-1}(u))/f(F^{-1}(u))$ is given in (12.23, p. 475) and called the optimal score function.

12.12.2 Distribution of the Rank Vector under Alternatives

A version of the following result first appeared in Hoeffding (1951).

Theorem 12.6. *Suppose that $Z_1, \ldots Z_N$ are independent continuous random variables with respective densities $f_1, \ldots, f_N$. Let $\boldsymbol{R} = (R_1, \ldots, R_N)^T$ be the corresponding rank vector. If h is the density of a continuous random variable whose support contains the support of each of $f_1, \ldots, f_N$, then*

$$P(\boldsymbol{R} = \boldsymbol{r}) = \frac{1}{N!}E\left[\frac{\prod_{i=1}^{N} f_i(V_{(r_i)})}{\prod_{i=1}^{N} h(V_{(r_i)})}\right],$$

where $V_{(1)} < \cdots < V_{(N)}$ are the order statistics of an iid sample from h.

Proof. Let $C = \{\boldsymbol{t} : t_i \text{ has rank } r_i\}$. Then by definition

$$P(\boldsymbol{R} = \boldsymbol{r}) = \int \cdots \int I(\boldsymbol{t} \in C)\left\{\prod_{i=1}^{N} f_i(t_i)\right\} dt_1 dt_2 \cdots dt_N.$$

Now let $v_{(r_i)} = t_i$ so that $v_{(1)} < \cdots < v_{(N)}$. On the set C this is just a 1-to-1 change of variable, but its implications are important. For a given vector $\boldsymbol{t}$ suppose that t_1 has rank $r_1 = 3$; that is, t_1 is third from the bottom when the components of $\boldsymbol{t}$ are ranked. Then $v_{(r_1)} = v_{(3)} = t_1$. If t_2 has rank $r_2 = 9$, then $v_{(r_2)} = v_{(9)} = t_2$. Now we make the change of variable, and multiply and divide by $N! \prod_{i=1}^{N} h(v_{(r_i)})$ to get

$$P(\boldsymbol{R} = \boldsymbol{r}) = \frac{1}{N!} \int \cdots \int \left[\frac{\prod_{i=1}^{N} f_i(v_{(r_i)})}{\prod_{i=1}^{N} h(v_{(r_i)})} \right] I(v_{(1)} < \cdots < v_{(N)}) N!$$
$$\times \left\{ \prod_{i=1}^{N} h(v_{(i)}) \right\} dv_{(1)} dv_{(2)} \cdots dv_{(N)}.$$

The result follows by noticing that $I(v_{(1)} < \cdots < v_{(N)}) N! \prod_{i=1}^{N} h(v_{(i)})$ is the density of the order statistic vector from h. ■

12.12.3 Pitman Efficiency

Recall from Section (12.5.2, p. 476) that the Pitman asymptotic relative efficiency of test S to test T is given by

$$\text{ARE}(S, T) = \lim_{k \to \infty} \frac{N'_k}{N_k},$$

where N_k and N'_k are the sample sizes required for the two tests to have the same limiting level α and power β under the sequence of alternatives

$$\theta_k = \theta_0 + \frac{\delta}{\sqrt{N_k}} + o\left(\frac{1}{\sqrt{N_k}}\right) \quad \text{as } k \to \infty. \tag{12.59}$$

These sequences of alternatives are called *Pitman alternatives*, and the basic approach is due to Pitman (1948) and Noether (1955). In the following we have drawn heavily from the accounts in Lehmann (1975) and Randles and Wolfe (1979).

We assume in Theorem 12.7 below that both test statistics satisfy 1–7 below. For simplicity we state the conditions for just S and then give a result on asymptotic power before giving the main theorem.

In the following $\mu_{S_k}(\theta)$ and $\sigma_{S_k}(\theta)$ refer to sequences of constants associated with S_k under θ. They might be the means and standard deviations, but need not be.

1.

$$\theta_k \to \theta_0 \text{ as } k \to \infty.$$

2.

$$N_k \to \infty \text{ as } k \to \infty.$$

3. Under $\theta = \theta_0$

$$\frac{S_k - \mu_{S_k}(\theta_0)}{\sigma_{S_k}(\theta_0)} \xrightarrow{d} \mathrm{N}(0,1) \text{ as } k \to \infty.$$

4. Under $\theta = \theta_k$

$$\frac{S_k - \mu_{S_k}(\theta_k)}{\sigma_{S_k}(\theta_k)} \xrightarrow{d} \mathrm{N}(0,1) \text{ as } k \to \infty.$$

5. The derivative $\mu'_{S_k}(\theta)$ exists in a neighborhood of $\theta = \theta_0$ with $\mu'_{S_k}(\theta_0) > 0$ and

$$\frac{\mu'_{S_k}(\theta_k^*)}{\mu'_{S_k}(\theta_0)} \to 1 \text{ for all } \theta_k^* \to \theta_0 \text{ as } k \to \infty.$$

6.

$$\frac{\sigma_{S_k}(\theta_k)}{\sigma_{S_k}(\theta_0)} \to 1 \text{ as } k \to \infty.$$

7. There exists a positive constant c such that

$$c = \lim_{k\to\infty} \frac{\mu'_{S_k}(\theta_0)}{\sqrt{N_k \sigma^2_{S_k}(\theta_0)}}.$$

This constant c is called the efficacy of S and denoted eff(S). Based on these conditions we first give a result on asymptotic power. The result shows that the higher the efficacy of a test, the more power it has. The result also gives a way to approximate the power of a test based on S. Let Z be a standard normal random variable, and let z_α be its upper $1-\alpha$ quantile.

Theorem 12.7. *Suppose that the test that rejects for $S_k > c_k$ has level $\alpha_k \to \alpha$ as $k \to \infty$ under $H_0 : \theta = \theta_0$.*

a) If Conditions 1–7 and (12.59, p. 516) hold, then

$$\beta_k = P(S_k > c_k) \to P(Z > z_\alpha - c\delta) \text{ as } k \to \infty, \tag{12.60}$$

where δ is given in (12.59, p. 516).

b) If Conditions 1–7 and (12.60) *hold, then (12.24, p. 476) holds.*

Proof. Note first that if Condition 3. holds, then since $\alpha_k \to \alpha$

$$\frac{c_k - \mu_{S_k}(\theta_0)}{\sigma_{S_k}(\theta_0)} \to z_\alpha \text{ as } k \to \infty.$$

Now $P(S_k > c_k)$ is given by

$$P\left(\frac{S_k - \mu_{S_k}(\theta_k)}{\sigma_{S_k}(\theta_k)} > \left[\frac{c_k - \mu_{S_k}(\theta_0)}{\sigma_{S_k}(\theta_0)} - \frac{\mu_{S_k}(\theta_k) - \mu_{S_k}(\theta_0)}{\sigma_{S_k}(\theta_0)}\right]\frac{\sigma_{S_k}(\theta_0)}{\sigma_{S_k}(\theta_k)}\right)$$

$$\rightarrow P(Z > z_\alpha - c\delta) \text{ as } k \rightarrow \infty.$$

To see this last step, note that by the mean value theorem there exists a θ_k^* such that

$$\frac{\mu_{S_k}(\theta_k) - \mu_{S_k}(\theta_0)}{\sigma_{S_k}(\theta_0)} = \frac{\mu'_{S_k}(\theta_k^*)(\theta_k - \theta_0)}{\sigma_{S_k}(\theta_0)}$$

$$= \frac{\mu'_{S_k}(\theta_k^*)}{\mu'_{S_k}(\theta_0)} \frac{\mu'_{S_k}(\theta_0)}{\sqrt{N_k \sigma^2_{S_k}(\theta_0)}} \sqrt{N_k}(\theta_k - \theta_0) \rightarrow c\delta.$$

For part b) we just work backwards and note that (12.60) and Conditions 1–7 force the convergence to $c\delta$ which means that $\sqrt{N_k}(\theta_k - \theta_0) \rightarrow \delta$ which is equivalent to (12.59, p. 516). ■

Now we give the main Pitman ARE theorem.

Theorem 12.8. *Suppose that the tests that reject for $S_k > c_k$ and $T_k > c'_k$ based on sample sizes N_k and N'_k, respectively, have levels α_k and α'_k that converge to α under $H : \theta = \theta_0$ and their powers under θ_k both converge to β, $\alpha < \beta < 1$. If conditions 1–7 hold and their efficacies are c =eff(S) and c'=eff(T), respectively, then the Pitman asymptotic relative efficiency of S to T is given by*

$$\text{ARE} = \left\{\frac{\text{eff}(S)}{\text{eff}(T)}\right\}^2.$$

Proof. By Theorem 12.7 (p. 517) b), $\beta = P(Z > z_\alpha - c\delta) = P(Z > z_\alpha - c'\delta')$. Thus $c\delta = c'\delta'$ and

$$\text{ARE}(S,T) = \lim_{k\rightarrow\infty} \frac{N'_k}{N_k}$$

$$= \lim_{k\rightarrow\infty} \left(\frac{\sqrt{N'_k}(\theta_k - \theta_0)}{\sqrt{N_k}(\theta_k - \theta_0)}\right)^2$$

$$= \left(\frac{\delta'}{\delta}\right)^2 = \left(\frac{c}{c'}\right)^2.$$

■

To apply Theorem 12.8 it would appear that we have to verify Conditions 3–6 above for arbitrary subsequences θ_k converging to θ_0 and then compute the efficacy in 7 for such sequences. However, if Conditions 1–7 and (12.60, p. 517) hold, we know by Theorem 12.7 (p. 517) that (12.24, p. 476) holds. Thus, we really only need to assume Condition 2 and verify Conditions 3–6 for alternatives of the form (12.59, p. 516). Moreover, the efficacy need only be computed for a simple sequence N converging to ∞ since the numerator and denominator in Condition 7 only involve θ_0.

12.12.4 *Pitman ARE for the One-Sample Location Problem*

Using the notation of Section 12.8 (p. 491) let $D_1, \ldots, D_N$ be iid from $F(x-\theta)$, where $F(x)$ has density $f(x)$ that is symmetric about 0, $f(x) = f(-x)$. Thus D_i has density $f(x-\theta)$ that is symmetric about θ. The testing problem is $H_0 : \theta = \theta_0$ versus $H_a : \theta = \theta_k$, where θ_k is given by (12.59).

12.12.4a Efficacy for the One-Sample *t*

The one-sample t statistic is

$$t = \frac{\sqrt{N}(\overline{D} - \theta_0)}{s},$$

where s is the $n-1$ version of the sample standard deviation. The simplest choice of standardizing constants are

$$\mu_{t_k}(\theta_k) = \frac{\sqrt{N_k}(\theta_k - \theta_0)}{\sigma}$$

and $\sigma_{t_k}(\theta_k) = 1$, where σ is the standard deviation of D_1 (under both $\theta = \theta_0$ and $\theta = \theta_k$). To verify Conditions 3 and 4 (p. 517), we have

$$\begin{aligned}\frac{t_k - \mu_{t_k}(\theta_0)}{\sigma_{t_k}(\theta_0)} &= \frac{\sqrt{N_k}(\overline{D} - \theta_0)}{s} - \frac{\sqrt{N_k}(\theta_k - \theta_0)}{\sigma} \\ &= \frac{\sqrt{N_k}(\overline{D} - \theta_k)}{\sigma}\left(\frac{s}{\sigma}\right) + \sqrt{N_k}(\theta_k - \theta_0)\left(\frac{1}{s} - \frac{1}{\sigma}\right).\end{aligned}$$

Under both $\theta = \theta_0$ and $\theta = \theta_k$, s has the same distribution and converges in probability to σ if D has a finite variance. Thus, under $\theta = \theta_k$ the last term in the latter display converges to 0 in probability since (12.59) forces $\sqrt{N_k}(\theta_k - \theta_0)$ to converge to δ. Of course under $\theta = \theta_0$ this last term is identically 0. The standardized means converge to standard normals under both $\theta = \theta_0$ and $\theta = \theta_k$ by Theorem 5.33 (p. 262). Two applications of Slutsky's Theorem then gives

Conditions 3 and 4 (p. 517). Since the derivative of $\mu_{t_k}(\theta)$ is $\mu'_{t_k}(\theta) = \sqrt{N_k}/\sigma$ for all θ, Condition 5 (p. 517) is satisfied. Since $\sigma_{t_k}(\theta_k) = 1$, Condition 6 (p. 517) is satisfied. Finally, dividing $\mu'_{t_k}(\theta_0) = \sqrt{N_k}/\sigma$ by $\sqrt{N_k}$ yields

$$\text{eff}(t) = \frac{1}{\sigma}.$$

It should be pointed out that this efficacy expression also holds true for the permutation version of the t test because the permutation distribution of the t statistic also converges to a standard normal under $\theta = \theta_0$.

12.12.4b Efficacy for the Sign Test

The sign test statistic is the number of observations above θ_0,

$$S = \sum_{i=1}^{N} I(D_i > \theta_0).$$

S has a binomial$(N, 1/2)$ distribution under $\theta = \theta_0$ and a binomial$(N, 1-F(\theta_0-\theta))$ distribution under general θ. Let $\mu_{S_k}(\theta) = N[1 - F(\theta_0 - \theta)]$ and $\sigma^2_{S_k}(\theta) = N[1 - F(\theta_0 - \theta)]F(\theta_0 - \theta)$. Conditions 3. and 4. (p. 517) follow again by Theorem 5.33 (p. 262), and $\mu'_{S_k}(\theta) = Nf(\theta_0 - \theta)$. Since F is continuous, Condition 6 (p. 517)is satisfied, and if f is continuous, then Condition 5 (p. 517) is satisfied, and the efficacy is

$$\text{eff}(S) = \lim_{N\to\infty} \frac{Nf(0)}{\sqrt{N^2/4}} = 2f(0).$$

Now we are able to compute the Pitman ARE of the sign test to the t test:

$$\text{ARE}(S,t) = 4\sigma^2 f^2(0).$$

Table 12.4 (p. 496) gives values of ARE(S,t) for some standard distributions.

12.12.4c Efficacy for the Wilcoxon Signed Rank Test

Recall that the signed rank statistic is

$$W^+ = \sum_{i=1}^{N} I(D_i > \theta_0)R_i^+,$$

where R_i^+ is the rank of $|D_i - \theta_0|$ among $|D_1 - \theta_0|, \ldots, |D_N - \theta_0|$. The asymptotic distribution of W^+ under θ_k requires more theory than we have developed so far, but Olshen (1967) showed that the efficacy of W^+ is

$$\sqrt{12}\int_{-\infty}^{\infty} f^2(x)dx$$

under the condition that $\int_{-\infty}^{\infty} f^2(x)dx < \infty$. Thus the Pitman asymptotic relative efficiency of the sign test to the Wilcoxon Signed Rank test is

$$\text{ARE}(S, W^+) = \frac{f^2(0)}{3\left(\int_{-\infty}^{\infty} f^2(x)dx\right)^2}.$$

Similarly, the Pitman asymptotic relative efficiency of the Wilcoxon Signed Rank test to the t test is

$$\text{ARE}(W^+, t) = 12\sigma^2\left(\int_{-\infty}^{\infty} f^2(x)dx\right)^2.$$

Table 12.4 (p. 496) displays these AREs for a number of distributions.

12.12.4d Power approximations for the One-Sample Location problem

Theorem 12.7 (p. 517) gives the asymptotic power approximation

$$P(Z > z_\alpha - c\delta) = 1 - \Phi\left(z_\alpha - c\sqrt{N}(\theta - \theta_0)\right)$$

based on setting $\delta = \sqrt{N}(\theta - \theta_0)$ in (12.60, p. 517), where θ is the alternative of interest at sample size N.

For example, let us first consider the t statistic with $c = 1/\sigma$ and $\theta_0 = 0$. The power approximation is then

$$1 - \Phi\left(z_\alpha - \sqrt{N}\theta/\sigma\right).$$

This is the exact power we get for the Z statistic $\sqrt{N}(\overline{X} - \theta_0)/\sigma$ when we know σ instead of estimating it. At $\theta/\sigma = .2$ and $N = 10$, we get power 0.16, which may be compared with the estimated exact power taken from the first four distributions in Randles and Wolfe (1979, p. 116): .14, .15, .16, .17. These latter estimates were based on 5000 simulations and have standard deviation around .005. At $\theta/\sigma = .4$ and $N = 10$, the approximate power is 0.35, and the estimated exact powers for those first four distributions in Randles and Wolfe (1979, p. 116) are .29, .33, .35, and .37, respectively. So here our asymptotic approximation may be viewed as substituting a Z for the t, and the approximation is quite good. Of course, for the normal distribution we could easily have used the noncentral t distribution to get the exact power.

For the sign test, the approximation is

$$1 - \Phi\left(z_\alpha - \sqrt{N} 2 f(0)\theta\right) = 1 - \Phi\left(z_\alpha - \sqrt{N} 2 f_0(0)\theta/\sigma\right),$$

where we have put f in the form of a location-scale model $f(x) = f_0((x - \theta)/\sigma)/\sigma$, where $f_0(x)$ has standard deviation 1, and thus σ is the standard deviation. For the uniform distribution, $f_0(x) = I(-\sqrt{3} < x < \sqrt{3})/\sqrt{12}$, so that $2f_0(0) = 2/\sqrt{12}$. The approximate power at $\theta/\sigma = .2, .4, .6, .8$ and $N = 10$ is then .10, .18, .29, .43, respectively. The corresponding Randles and Wolfe (1979, p. 116) estimates are .10, .19, .30, and .45, respectively. Here of course we could calculate the power exactly using the binomial. The approximate power we have used is similar to the normal approximation to the binomial but not the same because our approximation has replaced the difference of $p = F(0) = 1/2$ and $p = F(\theta)$ by a derivative times θ (Taylor expansion) and also used the null variance. It is perhaps surprising how good the approximation is.

The most interesting case is the signed rank statistic because we do not have any standard way of calculating the power. The approximate power for an alternative θ when $\theta_0 = 0$ is

$$\begin{aligned} P(Z > z_\alpha - c\delta) &= 1 - \Phi\left(z_\alpha - \theta\sqrt{12N}\int_{-\infty}^{\infty} f^2(x)dx\right) \\ &= 1 - \Phi\left(z_\alpha - \frac{\theta}{\sigma}\sqrt{12N}\int_{-\infty}^{\infty} f_0^2(x)dx\right). \end{aligned}$$

Here again in the second part we have substituted so that σ is the standard deviation of $f(x)$. For example, at the standard normal $\int_{-\infty}^{\infty} f_0^2(x)dx = 1/\sqrt{4\pi}$, and the approximate power is

$$1 - \Phi\left(z_\alpha - \sqrt{\frac{3N}{\pi}}\frac{\theta}{\sigma}\right).$$

Plugging in $\theta/\sigma = .2, .4, .6$, and .8 at $N = 10$, we obtain .15, .34, .58, and .80, respectively. The estimates of the exact powers from Randles and Wolfe (1979, p. 116) are .14, .32, .53, and .74. Thus the asymptotic approximation is a bit too high, especially at the larger θ/σ values.

Although the approximation is a little high, it could easily be used for planning purposes. For example, suppose that a clinical trial is to be run with power = .80 at the $\alpha = .05$ level against alternatives expected to be around $\theta/\sigma = .5$. Since the FDA requires two-sided procedures, we use $z_{.025} = 1.96$ and solve $\Phi^{-1}(1 - .8) = 1.96 - \sqrt{3N/\pi}(.5)$ to get

$$N = \left[\frac{1.96 - \Phi^{-1}(.2)}{.5}\right]^2 \frac{\pi}{3} = 32.9.$$

Notice that if we invert the Z statistic power formula used above for approximating the power of the t statistic, the only difference from the last display is that the factor $\pi/3$ does not appear. Thus for the t the calculations result in 31.4 observations. Of course this ratio $3/\pi = 31.4/32.9$ is just the ARE efficiency of the signed rank test to the t test at the normal distribution.

12.13 Problems

12.1. For the permutations in Table 12.1 (p. 453), give the permutation distribution of the Wilcoxon Rank Sum statistic W.

12.2. For the two-sample problem with samples $X_1, \ldots, X_m$ and $Y_1, \ldots, Y_n$, show that the permutation test based on $\sum_{i=1}^{n} Y_i$ is equivalent to the permutation tests based on $\sum_{i=1}^{m} X_i$, $\sum_{i=1}^{n} Y_i - \sum_{i=1}^{m} X_i$, and $\overline{Y} - \overline{X}$.

12.3. A one-way ANOVA situation with $k = 3$ groups and two observations within each group ($n_1 = n_2 = n_3 = 2$) results in the following data. Group 1: 37, 24; Group 2: 12, 15; Group 3: 9, 16. The ANOVA $F = 5.41$ results in a p-value of .101 from the F table. If we exchange the 15 in Group 2 for the 9 in Group 3, then $F = 7.26$.

a. What are the total number of ways of grouping the data that are relevant to testing that the means are equal?
b. Without resorting to the computer, give reasons why the permutation p-value using the F statistic is 2/15.

12.4. In a one-sided testing problem with continuous test statistic T, the p-value is either $F_H(T_{\text{obs.}})$ or $1 - F_H(T_{\text{obs.}})$ depending on the direction of the hypotheses, where F_H is the distribution function of T under the null hypothesis H, and $T_{\text{obs.}}$ is the observed value of the test statistic. In either case, under the null hypothesis the p-value is a uniform random variable as seen from the probability integral transformation. Now consider the case where T has a discrete distribution with values $t_1, \ldots, t_k$ and probabilities $P(T = t_i) = p_i, i = 1, \ldots, k$ under the null hypothesis H_0. If we are rejecting H_0 for small values of T, then the p-value is $p = P(T \le T_{\text{obs.}}) = p_1 + \cdots + P(T = T_{\text{obs.}})$, and the mid-$p$ value is $p - (1/2)P(T = T_{\text{obs.}})$. Under the null hypothesis H_0, show that E(mid-p)=1/2 and thus that the expected value of the usual p-value must be greater than 1/2 (and thus greater than the expected value of the p-value in continuous cases).

12.5. Consider a finite population of values $a_1, \ldots, a_N$ and a set of constants $c_1, \ldots, c_N$. We select a random permutation of the a values, call them $A_1, \ldots, A_N$, and form the statistic

$$T = \sum_{i=1}^{N} c_i A_i .$$

The purpose of this problem is to derive the first two permutation moments T given in Section 12.4.2 (p. 458).

a. First show that

$$P(A_i = a_s) = \frac{1}{N} \quad \text{for } s = 1, \ldots, N,$$

and

$$P(A_i = a_s, A_j = a_t) = \frac{1}{N(N-1)} \quad \text{for } s \neq t = 1, \ldots, N.$$

(Hint: for the first result there are $(N-1)!$ permutations with a_s in the ith slot out of a total of $N!$ equally likely permutations.)

b. Using a. show that

$$\mathrm{E}(A_i) = \frac{1}{N}\sum_{i=1}^{N} a_i \equiv \overline{a}, \quad \mathrm{Var}(A_i) = \frac{1}{N}\sum_{i=1}^{N}(a_i - \overline{a})^2, \quad \text{for } i = 1, \ldots, N,$$

and

$$\mathrm{Cov}(A_i, A_j) = \frac{-1}{N(N-1)}\sum_{i=1}^{N}(a_i - \overline{a})^2, \quad \text{for } i \neq j = 1, \ldots, N.$$

c. Now use b. to show that

$$\mathrm{E}(T) = N\overline{c}\,\overline{a} \quad \text{and} \quad \mathrm{Var}(T) = \frac{1}{N-1}\sum_{i=1}^{N}(c_i - \overline{c})^2 \sum_{j=1}^{N}(a_j - \overline{a})^2,$$

where $\overline{a}$ and $\overline{c}$ are the averages of the a's and c's, respectively.

12.6. As an application of the previous problem, consider the Wilcoxon Rank Sum statistic W = sum of the ranks of the Y's in a two-sample problem where we assume continuous distributions so that there are no ties. The c values are 1 for $i = m + 1, \ldots, N = m+n$ and 0 otherwise. With no ties the a's are just the integers $1, \ldots, N$ corresponding to the ranks. Show that

$$\mathrm{E}(W) = \frac{n(m+n+1)}{2}$$

and

$$\mathrm{Var}(W) = \frac{mn(m+n+1)}{12}.$$

12.7. In Section 12.4.4 (p. 461), the integral

$$P(X_1 < X_2) = \mathrm{E}\{I(X_1 < X_2)\} = \int\int I(x_1 < x_2)\, dF(x_1)\, dF(x_2)$$
$$= \int F(x)\, dF(x)$$

arises, where X_1 and X_2 are independent with distribution function F. If F is continuous, argue that $P(X_1 < X_2) = 1/2$ since $X_1 < X_2$ and $X_1 > X_2$ are equally likely. Also use iterated expectations and the probability integral transformations to get the same result. Finally, let $u = F(x)$ in the final integral to get the result.

12.8. Suppose that X and Y represent some measurement that signals the presence of disease via a threshold to be used in screening for the disease. Assume that Y has distribution function $G(y)$ and represents a diseased population, and X has distribution function $F(x)$ and represents a disease-free population. A "positive" for a disease-free subject is declared if $X > c$ and has probability $1 - F(c)$, where $F(c)$ is called the *specificity* of the screening test. A "positive" for a diseased subject is declared if $Y > c$ and has probability $1 - G(c)$, called the *sensitivity* of the test. The receiver operating characteristic (ROC) curve is a plot of $1 - G(c_i)$ versus $1 - F(c_i)$ for a sequence of thresholds $c_1, \ldots, c_k$. Instead of a discrete set of points, we may let $t = 1 - F(c)$, solve to get $c = F^{-1}(1 - t)$, and plug into $1 - G(c)$ to get the ROC curve $R(t) = 1 - G(F^{-1}(1 - t))$. Show that

$$\int_0^1 R(t)\, dt = \int \{1 - G(u)\}\, dF(u) = \theta_{\mathrm{XY}}$$

for continuous F and G.

12.9. Use the asymptotic normality result for $\widehat{\theta}_{\mathrm{XY}}$ to derive (12.15, p. 464).

12.10. Use (12.15, p. 464) to prove that the power of the Wilcoxon Rank Sum Test goes to 1 as m and n go to ∞ and m/N converges to a number λ between 0 and 1. You may assume that the F and G are continuous.

12.11. Use (12.15, p. 464) to derive (12.16, p. 464).

12.12. Suppose that $\widehat{\theta}_{\mathrm{XY}}$ is .7 and $m = n$. How large should $m = n$ be in order to have approximately 80% power at $\alpha = .05$ with the Wilcoxon Rank Sum Test?

12.13. Suppose that two normal populations with the same standard deviation σ differ in means by $\Delta/\sigma = .7$. How large should $m = n$ be in order to have approximately 80% power at $\alpha = .05$ with the Wilcoxon Rank Sum Test?

12.14. The number of permutations needed to carry out a permutation test can be computationally overwhelming. Thus the typical use of a permutation test involves estimating the true permutation p-value by randomly selecting $B = 1,000$, $B = 10,000$, or even more of the possible permutations. If we use sampling with replacement, then $B\widehat{p}$ has a binomial distribution with the true p-value p being the probability in the binomial. Consider the following situation where an approach of questionable ethics is under consideration. A company has just run a clinical trial comparing a placebo to a new drug that they want to market, but unfortunately the estimated p-value based on $B = 1000$ shows a p-value of around $\widehat{p} = .10$. Everybody is upset because they "know" the drug is good. One clever doctor suggests that they run the simulation of $B = 1000$ over and over again until they get a $\widehat{p}$ less than .05. Are they likely to find a run for which $\widehat{p}$ is less than .05 if the true p-value is $p = .10$? Use the following calculation based on k separate (independent) runs resulting in $\widehat{p}_1, \ldots, \widehat{p}_k$:

$$\begin{aligned} P(\min_{1 \le i \le k} \widehat{p}_i \le .05) &= 1 - P(\min_{1 \le i \le k} \widehat{p}_i > .05) \\ &= 1 - [1 - P(\widehat{p}_1 \le .05)]^k \\ &= 1 - [1 - P(\text{Bin}(1000,.1) \le 50)]^k. \end{aligned}$$

Plug in some values of k to find out how large k would need to be to get a $\widehat{p}$ under .05 with reasonably high probability.

12.15. The above problem is for given data, and we were trying to estimate the true permutation p-value conditional on the data set and therefore conditional on the set of test statistics computed for every possible permutation. In the present problem we want to think in terms of the overall unconditional probability distribution of $B\widehat{p}$ where we have two stages: first the data is generated and then we randomly select $T_1^*, \ldots, T_B^*$ from the set of permutations. The calculation of importance for justifying Monte Carlo tests is the unconditional probability $P(\widehat{p} \le \alpha) = P(B\widehat{p} \le B\alpha)$ that takes both stages into account.

a. First we consider a simpler problem. Suppose that we get some data that seems to be normally distributed and decide to compute a t statistic, call it T_0. Then we discover that we have lost our t tables, but fortunately we have a computer. Thus we can generate normal data and compute $T_1^*, \ldots, T_B^*$ for each of B independent data sets. In this case $T_0, T_1^*, \ldots, T_B^*$ are iid from a continuous distribution so that there are no ties among them with probability one. Let $\widehat{p} = \sum_{i=1}^{B} I(T_i^* \ge T_0)/B$ and prove that $B\widehat{p}$ has a discrete uniform distribution on the integers $(0, 1, \ldots, B + 1)$. (Hint: just use the argument that each ordering has equal probability $1/((B + 1)!)$. For example, $B\widehat{p} = 0$ occurs when T_0 is the largest value. How many orderings have T_0 as the largest value?)
b. The above result also holds if $T_0, T_1^*, \ldots, T_B^*$ have no ties and are merely exchangeable. However, if we are sampling $T_1^*, \ldots, T_B^*$ with replacement from a finite set of permutations, then ties occur with probability greater than one. Think of a way to randomly break ties so that we can get the same discrete uniform distribution.
c. Assuming that $B\widehat{p}$ has a discrete uniform distribution on the integers $(0, 1, \ldots, B)$, show that $P(\widehat{p} \le \alpha) = \alpha$ as long as $(B + 1)\alpha$ is an integer.

12.16. From (12.20, p. 469), $d = .933$ for the Wilcoxon Rank Sum statistic for $m = 10$ and $n = 6$ and assuming no ties. This corresponds to $\boldsymbol{Z}$ being the integers 1 to 16. For no ties and $W = 67$, the exact p-value for a one-sided test is .0467. Show that the normal approximation p-value is .0413 and the Box-Andersen p-value is .0426. Also find the Box-Andersen p-values using the approximations $d = 1 + (1.8 - 3)/(m + n)$ and $d = 1$.

12.17. Show that the result "$Q/(k-1)$ of (12.31, p. 482) is AN$\{1, 2(n-1)/(kn)\}$ as $k \to \infty$ with n fixed" follows from (12.32, p. 483) and writing

$$\sqrt{k}\left(\frac{Q}{k-1} - \frac{nF_{\mathrm{R}}}{n-1+F_{\mathrm{R}}}\right) = \frac{\sqrt{k}\{(N-1)/(k-1) - n\}F_{\mathrm{R}}}{(n-1)\left(\dfrac{k}{k-1}\right) + F_{\mathrm{R}}}$$

$$+ \sqrt{k}(nF_{\mathrm{R}})\left(\frac{1}{(n-1)\left(\dfrac{k}{k-1}\right) + F_{\mathrm{R}}} - \frac{1}{n-1+F_{\mathrm{R}}}\right).$$

Then show that each of the above two pieces converges to 0 in probability and use the delta theorem on $nF_{\mathrm{R}}/(n-1+F_{\mathrm{R}})$. (Keep in mind that n is a fixed constant.)

12.18. Justify the statement: "use of F_{R} with an $F(k-1, N-k)$ reference distribution is supported by (12.32, p. 483) under $k \to \infty$ and by the usual asymptotics $(k-1)F_{\mathrm{R}} \xrightarrow{d} \chi^2_{k-1}$ when $n \to \infty$ with k fixed." Hint: for the $k \to \infty$ asymptotics, write an $F(k-1, N-k)$ random variable as an average of $k-1$ χ^2_1 random variables divided by an independent average of $k(n-1)$ χ^2_1 random variables. Then subtract 1, multiply by $\sqrt{k}$ and use the Central Limit Theorem and Slutsky's Theorem.

12.19. From Section 12.8.1 (p. 492), show that for $T = \sum_{i=1}^n c_i d_i$, $\mathrm{E}(T^4) = 3(\sum_{i=1}^n d_i^2)^2 - 2\sum_{i=1}^n d_i^4$. (Hint: first show that

$$\left(\sum c_i d_i\right)^4 = \sum c_i^4 d_i^4 + 6\sum_{i<j} c_i^2 d_i^2 c_j^2 d_j^2$$

plus sums of odd moments.)

12.20. Verify (12.39, p. 493) and (12.40, p. 493) for the Box-Andersen approximation in the matched pairs problem.

12.21. Using results in Section 12.4.2 (p. 458), show that $\mathrm{E}\{\overline{R}_{.j}\} = (k+1)/2$, $\mathrm{Var}\{\overline{R}_{.j}\} = (k^2-1)/(12n)$, and $\mathrm{Cov}\{\overline{R}_{.j}, \overline{R}_{.m}\} = -(k^2-1)/\{12n(k-1)\}$, where $R_{i1}, \ldots R_{ik}$ are Friedman ranks in the ith block randomly assigned to the integers 1 to k and independent of the ranks in the other blocks. Putting these results together, the covariance matrix of $\overline{\boldsymbol{R}} = (\overline{R}_{.1}, \ldots, \overline{R}_{.k})^T$ is $\{k(k+1)/(12n)\}C_k$, where $C_k =$

$\mathrm{diag}\left(\boldsymbol{I}_k - \frac{\boldsymbol{1}_k \boldsymbol{1}_k^T}{k}\right)$. Using the fact that C_k is idempotent, find a generalized inverse of the covariance matrix of $\overline{\boldsymbol{R}}$, call it G, and show that (12.45, p. 501) is given by $\overline{\boldsymbol{R}}^T G \overline{\boldsymbol{R}}$.

12.22. Similar to Problem 12.18, explain why asymptotic normality of the Friedman statistic (12.45, p. 501) supports use of the F in (12.44, p. 500) on the within row Friedman ranks with an $F(k-1,(k-1)(n-1))$ reference distribution.

12.23. From Section 12.9.4 (p. 503) verify the permutation moments in (12.49, p. 504) and (12.50, p. 504). Use results from Section 12.4.2 (p. 458) under the assumption that permutations are independently carried out within rows.

12.24. From Section 12.10.1 (p. 506) consider the two independent binomial testing problem where $m = 12\ (N_{11} + N_{12})$ for Group 1 and $n = 4\ (N_{21} + N_{22})$ for Group 2, and we want to test $H_0 : p_1 = p_2$ versus $H_a : p_1 < p_2$, where p_1 and p_2 are the respective probabilities of falling in Category 1. Suppose that $T = 4\ (N_{11} + N_{21})$ is observed. Write down the conditional probability distribution of $N_{11}|T = 4$ (just the hypergeometric probabilities for $n_{11} = 0, 1, 2, 3, 4$). Also, letting each of $0, 1, 2, 3, 4$ be considered observed values for N_{11}, list:

a. the Fisher Exact p-values
b. the Fisher Exact mid-p values.

12.25. For a multinomial vector $(N_{11}, N_{12}, N_{21}, N_{22})$, $N_{11}+N_{12}+N_{21}+N_{22} = N$, with associated probabilities $(p_{11}, p_{12}, p_{21}, p_{22})$, show that the variance of $N_{12}-N_{21}$ is $N\{p_{12} + p_{21} - (p_{12} - p_{21})^2\}$.

12.26. Show that (12.58, p. 515) follows from (12.57, p. 515) if the derivative can be taken inside the expectation.

12.27. Show why $\alpha_k \to \alpha$ and Condition 3. (p. 517) imply that

$$\frac{c_k - \mu_{S_k}(\theta_0)}{\sigma_{S_k}(\theta_0)} \to z_\alpha \text{ as } k \to \infty.$$

(Hint: it helps to use Pólya's result on uniform convergence, Theorem 5.6, p. 222.)

12.28. Verify that Theorem 5.33 (p. 262) applies to $\overline{X}$ when $X_1^*, \ldots, X_{N_k}^*$ are iid from $F(x)$ having mean 0 and finite variance σ^2, and $X_i = X_i^* + \delta/\sqrt{N_k}, i = 1, \ldots, N_k$.

12.29. Verify that Theorem 5.33 (p. 262) applies to $S = \sum_{i=1}^N I(X_i > 0$ when $X_1^*, \ldots, X_{N_k}^*$ are iid from $F(x)$ having median 0 and $X_i = X_i^* + \delta/\sqrt{N_k}, i = 1, \ldots, N_k$.

12.30. The data are $Y_1, \ldots, Y_n$ iid with median θ. For $H_0 : \theta = 0$ versus $H_a : \theta > 0$, use the normal approximation to the binomial distribution to find a power approximation for the sign test and compare to the expression

$1 - \Phi\left(z_\alpha - \sqrt{N}2f(0)\theta_a\right)$ derived from Theorem 12.7 (p. 517), where θ_a is an alternative. Where are the differences?

12.31. For the Wilcoxon Signed Rank statistic, calculate an approximation to the power of a .05 level test for a sample of size $N = 20$ from the Laplace distribution with a shift of .6 in standard deviation units. Compare with the simulation estimate .63 from Randles and Wolfe (1979, p.116).

12.32. Consider the two-sample problem where $X_1, \ldots, X_m$ and $Y_1, \ldots, Y_n$ are iid from $F(x)$ under H_0, but the Y's are shifted to the right by $\Delta_k = \delta/\sqrt{N_k}$ under a sequence of the Pitman alternatives. Verify Conditions 3.-6 (p. 517), making any assumptions necessary and show that the efficacy of the two-sample t test is given by $\text{eff}(t) = \sqrt{\lambda(1-\lambda)}/\sigma$, where σ is the standard deviation of F.

12.33. Consider a variable having a Likert scale with possible answers 1,2,3,4,5. Suppose that we are thinking of a situation where the treatment group has answers that tend to be spread toward 1 or 5 and away from the middle. Can we design a rank test to handle this? Here is one formulation. For the two-sample problem suppose that the base density is a beta density of the following form:

$$\frac{\Gamma(2(1-\theta))}{\Gamma(1-\theta)\Gamma(1-\theta)} x^{-\theta}(1-x)^{-\theta}, \quad 0 < x < 1, \quad \theta < 1.$$

A sketch of this density shows that it spreads towards the ends as θ gets large. Using the LMPRT theory, find the optimal score function for $H_0 : \theta = \theta_0$ versus $H_a : \theta > \theta_0$, where $0 \le \theta_0 < 1$. At $\theta_0 = 0$, the score function simplifies to $\phi(u) = -2 - \log[u(1-u)]$. Sketch this score function and comment on whether a linear rank statistic of the form $S = \sum_{i=1}^{m} \phi(R_i/(N+1))$ makes sense here.

12.34. For the two-sample problem with $G(x) = (1-\Delta)F(x) + \Delta F^2(x)$ and $H_0 : \Delta = 0$ versus $H_a : \Delta > 0$, show that the Wilcoxon Rank Sum test is the locally most powerful rank test. (You may take $h(x) = f(x)$ in the expression for $P(\boldsymbol{R} = \boldsymbol{r})$.)

12.35. In some two-sample situations (treatment and control), only a small proportion of the treatment group responds to the treatment. Johnson et al. (1987) were motivated by data on sister chromatid exchanges in the chromosomes of smokers where only a small number of units are affected by a treatment, that is, where the treatment group seemed to have a small but higher proportion of large values than the control group. For this two-sample problem, they proposed a mixture alternative,

$$G(x) = (1-\Delta)F(x) + \Delta K(x),$$

where $K(x)$ is stochastically larger than $F(x)$, i.e., $K(x) \le F(x)$ for all x, and Δ refers to the proportion of responders. For $H_0 : \Delta = 0$ versus $H_a : \Delta > 0$, verify that the locally most powerful rank test has optimal score function

$k(F^{-1}(u))/f(F^{-1}(u)) - 1$. Let $F(x)$ and $K(x)$ be normal distribution functions with means μ_1 and μ_2, respectively, $\mu_2 > \mu_1$, and variance σ^2. Show that the optimal score function is

$$\phi(u) = \exp(-\delta^2/2)\exp(\delta\Phi^{-1}(u)) - 1, \tag{12.61}$$

where $\delta = (\mu_2 - \mu_1)/\sigma$.

12.36. Related to the previous problem, Johnson et al. (1987) give the following example data:

```
X: 9  9 10 10 14 14 14 15 16 20
Y: 6 10 13 15 18 21 22 23 30 37
```

By sampling from the permutation distribution of the linear rank statistic $\sum_{i=m+1}^{m+n} \phi(R_i/(m+n+1))$ with score function in (12.61), estimate the one-sided permutation p-values with $\delta = 1$ and $\delta = 2$. For comparison, also give one-sided p-values for the Wilcoxon rank sum (exact) and pooled t-tests (from t table).

12.37. Similar in motivation to problem 12.35 (p. 529), Conover and Salsburg (1988) proposed the mixture alternative

$$G(x) = (1 - \Delta)F(x) + \Delta\{F(x)\}^a .$$

Note that $\{F(x)\}^a$ is the distribution function of the maximum of a random variables with distribution function $F(x)$. For $H_0 : \Delta = 0$ versus $H_a : \Delta > 0$, verify that the locally most powerful rank test has optimal score function u^{a-1}.

12.38. For the data in Problem 12.36 (p. 530), by sampling from the permutation distribution of the linear rank statistic $\sum_{i=m+1}^{m+n} \phi(R_i/(m+n+1))$ with score function $\phi(u) = u^{a-1}$, estimate the one-sided permutation p-value with $a = 5$. For comparison, also give one-sided p-values for the Wilcoxon rank sum (exact) and pooled t-tests (from t table).

12.39. Conover and Salsburg (1988) gave the following example data set on changes from baseline of serum glutamic oxaloacetic transaminase (SGOT):

```
X:   -50  -17  -10   -3    4    7    8   12   26   37
Y:  -116  -56   20   24   29   29   35   35   37   41
```

Plot the data and decide what type of test should be used to detect larger values in some or all of the Y's. Then, give the one-sided p-value for that test and for one other possible test.

12.40. Use `perm.sign` to get the exact one-sided p-value 0.044 for the data give in Example 12.2 (p. 498). Then by trial and error get an exact confidence interval for the center of the distribution with coverage at least 90%. Also give the exact confidence interval for the median based on the order statistics with coverage at least 90%.

Appendix A
Derivative Notation and Formulas

A.1 Notation

Suppose that $\boldsymbol{x} = (x_1, \ldots, x_k)^T$ is a $k \times 1$ column vector and g is a scalar-valued function of $\boldsymbol{x}$. That is, $g : R^k \to R^1$. Then the derivative of g with respect to $\boldsymbol{x}$ is the $1 \times k$ row vector of partial derivatives denoted by any of the following equivalent expressions:

$$g'(\boldsymbol{x}) = \frac{\partial g}{\partial \boldsymbol{x}} = \left(\frac{\partial g}{\partial x_1}, \ldots, \frac{\partial g}{\partial x_k} \right). \tag{A.1}$$

Accordingly, the transpose of this vector is denoted

$$g'(\boldsymbol{x})^T = \left(\frac{\partial g}{\partial \boldsymbol{x}} \right)^T = \frac{\partial g}{\partial \boldsymbol{x}^T} = \begin{pmatrix} \frac{\partial g}{\partial x_1} \\ \vdots \\ \frac{\partial g}{\partial x_k} \end{pmatrix}. \tag{A.2}$$

We often have need for the matrix of mixed, partial second derivatives denoted as follows:

$$g''(\boldsymbol{x}) = \frac{\partial}{\partial \boldsymbol{x}} \left(\frac{\partial g}{\partial \boldsymbol{x}^T} \right) = \frac{\partial^2 g}{\partial \boldsymbol{x} \partial \boldsymbol{x}^T} = \begin{pmatrix} \frac{\partial^2 g}{\partial^2 x_1} & \frac{\partial^2 g}{\partial x_2 \partial x_1} & \cdots & \frac{\partial^2 g}{\partial x_k \partial x_1} \\ \vdots & & & \\ \frac{\partial^2 g}{\partial^2 x_1 \partial x_k} & \frac{\partial^2 g}{\partial^2 x_2 \partial x_k} & \cdots & \frac{\partial^2 g}{\partial^2 x_k} \end{pmatrix}$$

$$= \left(\frac{\partial^2 g}{\partial x_j \partial x_i} \right)_{i=1,\ldots,k,\ j=1,\ldots,k}. \tag{A.3}$$

D.D. Boos and L.A. Stefanski, *Essential Statistical Inference: Theory and Methods*, Springer Texts in Statistics, DOI 10.1007/978-1-4614-4818-1,

In this book we encounter only functions g for which

$$\frac{\partial^2 g}{\partial x_j \partial x_i} = \frac{\partial^2 g}{\partial x_i \partial x_j}, \qquad \text{for all } i \text{ and } j.$$

In this case the matrices in (A.3) are symmetric, and thus, it is also true that

$$g''(\boldsymbol{x}) = \left(\frac{\partial^2 g}{\partial x_i \partial x_j}\right)_{i=1,\ldots,k,\ j=1,\ldots,k}.$$

The two rightmost expressions in (A.3) are special cases of the notation for the derivative of a vector-valued function of a vector argument. Suppose now that $\boldsymbol{h} = (h_1, \ldots, h_s)^T$ is an $s \times 1$ column vector-valued function of $\boldsymbol{x}$. That is, $h : R^k \to R^s$. Then the derivative of $\boldsymbol{h}$ with respect to $\boldsymbol{x}$ is the $s \times k$ matrix of partial derivatives denoted as follows:

$$\boldsymbol{h}'(\boldsymbol{x}) = \frac{\partial \boldsymbol{h}}{\partial \boldsymbol{x}} = \begin{pmatrix} \frac{\partial h_1}{\partial x_1} & \frac{\partial h_1}{\partial x_2} & \cdots & \frac{\partial h_1}{\partial x_k} \\ \vdots & \vdots & & \vdots \\ \frac{\partial h_s}{\partial x_1} & \frac{\partial h_s}{\partial x_2} & \cdots & \frac{\partial h_s}{\partial x_k} \end{pmatrix}$$

$$= \left(\frac{\partial h_i}{\partial x_j}\right)_{i=1,\ldots,s,\ j=1,\ldots,k}. \tag{A.4}$$

A.2 Definition and Taylor Approximations

Suppose now that $\boldsymbol{h} : R^k \to R^s$. Regardless of the dimensions k and s, the derivative of $\boldsymbol{h}$ at a particular point $\boldsymbol{x}_0$ provides a linear approximation to $\boldsymbol{h}(x)$ for $\boldsymbol{x}$ close to $\boldsymbol{x}_0$. The linear approximation has the form

$$\underbrace{\boldsymbol{h}(\boldsymbol{x})}_{s\times 1} \approx \underbrace{\boldsymbol{h}(\boldsymbol{x}_0)}_{s\times 1} + \underbrace{\boldsymbol{h}'(\boldsymbol{x}_0)}_{s\times k}\underbrace{(\boldsymbol{x}-\boldsymbol{x}_0)}_{k\times 1} \tag{A.5}$$

and is best in the sense that the derivative $\boldsymbol{h}'$ has the defining property

$$\lim_{\boldsymbol{\delta}\to\boldsymbol{0}} \frac{||\boldsymbol{h}(\boldsymbol{x}+\boldsymbol{\delta}) - \{\boldsymbol{h}(\boldsymbol{x}) + \boldsymbol{h}'(\boldsymbol{x})\boldsymbol{\delta}\}\,||}{||\boldsymbol{\delta}||} = 0, \tag{A.6}$$

where $||\cdot||$ is the usual Euclidean norm, $||\boldsymbol{v}_{n\times 1}|| = \left(\boldsymbol{v}^T\boldsymbol{v}\right)^{1/2}$. The relevant vector and matrix dimensions in (A.5) are included for clarity and to point out that remembering the form of the linear approximation in (A.5) makes it easy to remember the dimensions of $\boldsymbol{h}'$ for arbitrary $\boldsymbol{h}$ in (A.4).

Occasionally, a second-order Taylor approximation is required, usually only when $s = 1$, that is, $h : R^k \to R^1$. In this case

$$\underbrace{h(\boldsymbol{x})}_{1\times 1} \approx \underbrace{h(\boldsymbol{x}_0)}_{1\times 1} + \underbrace{h'(\boldsymbol{x}_0)}_{1\times k}\underbrace{(\boldsymbol{x}-\boldsymbol{x}_0)}_{k\times 1} + \frac{1}{2}\underbrace{(\boldsymbol{x}-\boldsymbol{x}_0)^T}_{1\times k}\underbrace{h''(\boldsymbol{x}_0)}_{k\times k}\underbrace{(\boldsymbol{x}-\boldsymbol{x}_0)}_{k\times 1}. \tag{A.7}$$

A.3 Working with Derivatives

We now describe certain vector-derivative formulas and expressions commonly encountered in statistical applications. In the following $\boldsymbol{x} = (x_1, \ldots, x_k)^T$ is always a $k \times 1$ column vector, but the dimensions of the domains and ranges of the function $\boldsymbol{g}$ and $\boldsymbol{h}$ vary as stated.

Product Rule

Suppose that $\boldsymbol{g} : R^k \to R^s$ and $\boldsymbol{h} : R^k \to R^s$. Thus, we know that $\boldsymbol{g}'$ and $\boldsymbol{h}'$ are $s\times k$ matrices. Define the scalar-valued function r as the product $r(\boldsymbol{x}) = \boldsymbol{g}(\boldsymbol{x})^T\boldsymbol{h}(\boldsymbol{x})$. Then $r'(\boldsymbol{x})$ is the $1 \times k$ row vector function

$$\underbrace{r'(\boldsymbol{x})}_{1\times k} = \underbrace{\boldsymbol{h}^T(\boldsymbol{x})}_{1\times s}\underbrace{\boldsymbol{g}'(\boldsymbol{x})}_{s\times k} + \underbrace{\boldsymbol{g}^T(\boldsymbol{x})}_{1\times s}\underbrace{\boldsymbol{h}'(\boldsymbol{x})}_{s\times k} \tag{A.8}$$

Chain Rule

Suppose that $\boldsymbol{g} : R^s \to R^p$ and $\boldsymbol{h} : R^k \to R^s$. Define the function $\boldsymbol{r}$ as the composition $\boldsymbol{r}(\boldsymbol{x}) = \boldsymbol{g}(\boldsymbol{h}(\boldsymbol{x}))$. Thus, we know that $\boldsymbol{g}'$ is $p \times s$, $\boldsymbol{h}'$ is $s \times k$, and $\boldsymbol{r}'$ is $p \times k$. Furthermore,

$$\underbrace{\boldsymbol{r}'(\boldsymbol{x})}_{p\times k} = \underbrace{\boldsymbol{g}'(\boldsymbol{h}(\boldsymbol{x}))}_{p\times s}\underbrace{\boldsymbol{h}'(\boldsymbol{x})}_{s\times k}. \tag{A.9}$$

Make-Your-Own Rule

There are other versions of these two rules that are sometimes needed. However, in such cases, it is often easier to derive the relevant expression than it is to identify and apply the relevant rule. We illustrate with a simple example. Suppose that $\boldsymbol{r}(\boldsymbol{x}) = g(\boldsymbol{x})\boldsymbol{h}(\boldsymbol{x})$ where $g : R^k \to R^1$ and $\boldsymbol{h} : R^k \to R^s$. Thus, $\boldsymbol{r} : R^k \to R^s$, and we know that $\boldsymbol{r}'$ is $s \times k$. To obtain the form of $\boldsymbol{r}'$, consider that for small $\boldsymbol{\delta}$,

$$
\begin{aligned}
\boldsymbol{r}(\boldsymbol{x}+\boldsymbol{\delta})-\boldsymbol{r}(\boldsymbol{x}) &= g(\boldsymbol{x}+\boldsymbol{\delta})\boldsymbol{h}(\boldsymbol{x}+\boldsymbol{\delta})-g(\boldsymbol{x})\boldsymbol{h}(\boldsymbol{x}) \\
&\approx \left\{g(\boldsymbol{x})+g'(\boldsymbol{x})\boldsymbol{\delta}\right\}\left\{\boldsymbol{h}(\boldsymbol{x})+\boldsymbol{h}'(\boldsymbol{x})\boldsymbol{\delta}\right\}-g(\boldsymbol{x})\boldsymbol{h}(\boldsymbol{x}) \\
&= g(\boldsymbol{x})\boldsymbol{h}(\boldsymbol{x})+g(\boldsymbol{x})\boldsymbol{h}'(\boldsymbol{x})\boldsymbol{\delta}+g'(\boldsymbol{x})\boldsymbol{\delta}\boldsymbol{h}(\boldsymbol{x}) \\
&\quad + g'(\boldsymbol{x})\boldsymbol{\delta}\boldsymbol{h}'(\boldsymbol{x})\boldsymbol{\delta}-g(\boldsymbol{x})\boldsymbol{h}(\boldsymbol{x}) \\
&= \left\{g(\boldsymbol{x})\boldsymbol{h}'(\boldsymbol{x})+\boldsymbol{h}(\boldsymbol{x})g'(\boldsymbol{x})\right\}\boldsymbol{\delta}+g'(\boldsymbol{x})\boldsymbol{\delta}\boldsymbol{h}'(\boldsymbol{x})\boldsymbol{\delta} \\
&\approx \left\{g(\boldsymbol{x})\boldsymbol{h}'(\boldsymbol{x})+\boldsymbol{h}(\boldsymbol{x})g'(\boldsymbol{x})\right\}\boldsymbol{\delta}. \qquad \text{(A.10)}
\end{aligned}
$$

The first approximation results from applying (A.5) to both g and $\boldsymbol{h}$. Then after two steps of algebra, the second approximation arises by dropping term(s) in which $\boldsymbol{\delta}$ appears more than once, as these are negligible compared to terms in which $\boldsymbol{\delta}$ appears only once (for small $\boldsymbol{\delta}$). In this case, there is only one such term $g'(\boldsymbol{x})\boldsymbol{\delta}\boldsymbol{h}'(\boldsymbol{x})\boldsymbol{\delta}$. Comparing the two end expressions in (A.10) to the form of the first-order Taylor approximation $\boldsymbol{r}(\boldsymbol{x}+\boldsymbol{\delta}) \approx \boldsymbol{r}(\boldsymbol{x})+\boldsymbol{r}'(\boldsymbol{x})\boldsymbol{\delta}$, derived from (A.5), shows that $\boldsymbol{r}'(\boldsymbol{x}) = g(\boldsymbol{x})\boldsymbol{h}'(\boldsymbol{x})+\boldsymbol{h}(\boldsymbol{x})g'(\boldsymbol{x})$.

A.4 Problems

A.1. Suppose that $\boldsymbol{r}(\boldsymbol{x}) = g(\boldsymbol{v}(\boldsymbol{x}))\boldsymbol{w}(\boldsymbol{x})^T\boldsymbol{h}(\boldsymbol{x})$ where $g : R^s \to R^1$, $\boldsymbol{v} : R^k \to R^s$, $\boldsymbol{w} : R^k \to R^p$, and $\boldsymbol{h} : R^k \to R^p$. Apply the make-your-own rule to derive an expression for $\boldsymbol{r}'$ in terms of g, $\boldsymbol{v}$, $\boldsymbol{w}$, $\boldsymbol{h}$, and their derivatives.

A.2. Suppose that $\boldsymbol{h} : R^k \to R^s$ has component functions $h_j : R^k \to R^1$, $j = 1,\ldots,s$. That is, $\boldsymbol{h} = (h_1,\ldots,h_s)^T$. Use the approximation in (A.7) componentwise to derive the approximation

$$
\boldsymbol{h}(\boldsymbol{x}) \approx \boldsymbol{h}(\boldsymbol{x}_0)+\boldsymbol{h}'(\boldsymbol{x}_0)(\boldsymbol{x}-\boldsymbol{x}_0)+\frac{1}{2}\begin{pmatrix}(\boldsymbol{x}-\boldsymbol{x}_0)^T\boldsymbol{h}_1''(\boldsymbol{x}_0)(\boldsymbol{x}-\boldsymbol{x}_0)\\ \vdots \\ (\boldsymbol{x}-\boldsymbol{x}_0)^T\boldsymbol{h}_s''(\boldsymbol{x}_0)(\boldsymbol{x}-\boldsymbol{x}_0)\end{pmatrix}. \qquad \text{(A.11)}
$$

A.3. Suppose that $\mu : R^p \to R^1$ is a mean function of the $p \times 1$ parameter $\boldsymbol{\theta}$. That is, $E(Y) = \mu(\boldsymbol{\theta})$. For a fixed y, define $r(\boldsymbol{\theta}) = \{y-\mu(\boldsymbol{\theta})\}^2$. Derive $r'(\boldsymbol{\theta})$ and $r''(\boldsymbol{\theta})$.

References

Agresti, A. (2002). *Categorical Data Analysis*. John Wiley & Sons, New Jersey.

Agresti, A. and Coull, B. A. (1998). Approximate is better than "exact" for interval estimation of binomial proportions. *The American Statistician*, 52:119–126.

Agresti, A. and Min, Y. (2005). Frequentist performance of Bayesian confidence intervals for comparing proportions in 2 × 2 contingency tables. *Biometrics*, 61(2):515–523.

Aitchison, J. and Silvey, S. D. (1958). Maximum-likelihood estimation of parameters subject to restraints. *The Annals of Mathematical Statistics*, 29:813–828.

Akaike, H. (1973). Information theory and an extension of the maximum likelihood principle. In Petrov, B. N. e. and Czaki, F. e., editors, *2nd International Symposium on Information Theory*, pages 267–281. Akademiai Kiado.

Akritas, M. G. (1990). The rank transform method in some two-factor designs. *Journal of the American Statistical Association*, 85:73–78.

Akritas, M. G. (1991). Limitations of the rank transform procedure: A study of repeated measures designs. Part I. *Journal of the American Statistical Association*, 86:457–460.

Anderson, M. J. and Robinson, J. (2001). Permutation tests for linear models. *Australian & New Zealand Journal of Statistics*, 43(1):75–88.

Andrews, D. W. K. (1987). Asymptotic results for generalized Wald tests. *Econometric Theory*, 3:348–358.

Arvesen, J. N. (1969). Jackknifing u-statistics. *The Annals of Mathematical Statistics*, 40:2076–2100.

Bahadur, R. R. (1964). On Fisher's bound for asymptotic variances. *The Annals of Mathematical Statistics*, 35:1545–1552.

Bahadur, R. R. (1966). A note on quantiles in large samples. *The Annals of Mathematical Statistics*, 37:577–580.

Barlow, R. E., Bartholomew, D. J., B. J. M., and Brunk, H. D. (1972). *Statistical Inference under order restrictions: the theory and application of isotonic regression*. John Wiley & Sons.

Barnard, G. A. (1945). A new test for 2 × 2 tables. *Nature*, 156:177.

Barnard, G. A. (1947). Significance tests for 2 × 2 tables. *Biometrika*, 34:123–138.

Barnard, G. A. (1963). Discussion on "the spectral analysis of point processes". *Journal of the Royal Statistical Society, Series B: Statistical Methodology*, 25:294–294.

Barndorff-Nielsen, O. (1978). *Information and Exponential Families in Statistical Theory*. John Wiley & Sons.

Barndorff-Nielsen, O. (1982). Exponential families. In Banks, D. L., Read, C. B., and Kotz, S., editors, *Encyclopedia of Statistical Sciences (9 vols. plus Supplement), Volume 2*, pages 587–596. John Wiley & Sons.

D.D. Boos and L.A. Stefanski, *Essential Statistical Inference: Theory and Methods*, Springer Texts in Statistics, DOI 10.1007/978-1-4614-4818-1,

Barndorff-Nielsen, O. and Cox, D. R. (1979). Edgeworth and saddle-point approximations with statistical applications (C/R p299-312). *Journal of the Royal Statistical Society, Series B: Methodological*, 41:279–299.

Bartholomew, D. J. (1957). A problem in life testing. *Journal of the American Statistical Association*, 52:350–355.

Bartholomew, D. J. (1959). A test of homogeneity for ordered alternatives. *Biometrika*, 46:36–48.

Benichou, J. and Gail, M. H. (1989). A delta method for implicitly defined random variables (C/R: 90V44 p58). *The American Statistician*, 43:41–44.

Beran, R. (1986). Simulated power functions. *The Annals of Statistics*, 14:151–173.

Beran, R. (1988). Prepivoting test statistics: A bootstrap view of asymptotic refinements. *Journal of the American Statistical Association*, 83:687–697.

Beran, R. and Srivastava, M. S. (1985). Bootstrap tests and confidence regions for functions of a covariance matrix (Corr: V15 p470-471). *The Annals of Statistics*, 13:95–115.

Berger, J. O. and Wolpert, R. L. (1984). *The Likelihood Principle*. Institute of Mathematical Statistics.

Berger, R. L. (1996). More powerful tests from confidence interval p values. *The American Statistician*, 50:314–318.

Berger, R. L. and Boos, D. D. (1994). P values maximized over a confidence set for the nuisance parameter. *Journal of the American Statistical Association*, 89:1012–1016.

Berndt, E. R. and Savin, N. E. (1977). Conflict among criteria for testing hypotheses in the multivariate linear regression model. *Econometrica*, 45:1263–1277.

Best, D. J. and Rayner, J. C. W. (1987). Welch's approximate solution for the Behrens-Fisher problem. *Technometrics*, 29:205–210.

Bhattacharyya, G. K. and Johnson, R. A. (1973). On a test of independence in a bivariate exponential distribution. *Journal of the American Statistical Association*, 68:704–706.

Bickel, P. J. (1974). Edgeworth expansions in nonparametric statistics. *The Annals of Statistics*, 2:1–20.

Bickel, P. J. and Doksum, K. A. (1981). An analysis of transformations revisited. *Journal of the American Statistical Association*, 76:296–311.

Bickel, P. J. and Freedman, D. A. (1981). Some asymptotic theory for the bootstrap. *The Annals of Statistics*, 9:1196–1217.

Bickel, P. J. and van Zwet, W. R. (1978). Asymptotic expansions for the power of distribution free tests in the two-sample problem (Corr: V6 p1170-1171). *The Annals of Statistics*, 6:937–1004.

Billingsley, P. (1999). *Convergence of Probability Measures*. John Wiley & Sons.

Birmbaum, L. S., Morrissey, R. E., and Harris, M. W. (1991). Teratogenic effects of 2,3,7,8-tetrabromodibenzo-p-dioxin and three polybrominated dibenzofurans in c57bl/6n mice. *Toxicology and Applied Pharmacology*, 107:141–152.

Boos, D., Janssen, P., and Veraverbeke, N. (1989). Resampling from centered data in the two-sample problem. *Journal of Statistical Planning and Inference*, 21:327–345.

Boos, D. D. (1992). On generalized score tests (Com: 93V47 p311-312). *The American Statistician*, 46:327–333.

Boos, D. D. (2003). Introduction to the bootstrap world. *Statistical Science*, 18(2):168–174.

Boos, D. D. and Brownie, C. (1992). A rank-based mixed model approach to multisite clinical trials (Corr: 94V50 p322). *Biometrics*, 48:61–72.

Boos, D. D. and Brownie, C. (2004). Comparing variances and other measures of dispersion. *Statistical Science*, 19(4):571–578.

Boos, D. D. and Hughes-Oliver, J. M. (2000). How large does n have to be for Z and t intervals? *The American Statistician*, 54(2):121–128.

Boos, D. D. and Monahan, J. F. (1986). Bootstrap methods using prior information. *Biometrika*, 73:77–83.

Boos, D. D. and Zhang, J. (2000). Monte Carlo evaluation of resampling-based hypothesis tests. *Journal of the American Statistical Association*, 95(450):486–492.

Booth, J. and Presnell, B. (1998). Allocation of Monte Carlo resources for the iterated bootstrap. *Journal of Computational and Graphical Statistics*, 7:92–112.

Booth, J. G. and Hall, P. (1994). Monte Carlo approximation and the iterated bootstrap. *Biometrika*, 81:331–340.

Boschloo, R. D. (1970). Raised conditional level of significance for the 2 × 2-table when testing the equality of two probabiities. *Statistica Neerlandica*, 24:1–35.

Box, G. E. P. and Andersen, S. L. (1955). Permutation theory in the derivation of robust criteria and the study of departures from assumption. *Journal of the Royal Statistical Society, Series B: Statistical Methodology*, 17:1–16.

Box, G. E. P. and Cox, D. R. (1964). An analysis of transformations. *Journal of the Royal Statistical Society Series B-Statistical Methodology*, 26:211–252.

Box, G. E. P. and Watson, G. S. (1962). Robustness to non-normality of regression tests (Corr: V52 p669). *Biometrika*, 49:93–106.

Breslow, N. (1989). Score tests in overdispersed GLM's. In Decarli, A., Francis, B. J., Gilchrist, R., and Seeber, G. U. H., editors, *Statistical Modelling*, pages 64–74. Springer-Verlag Inc.

Breslow, N. (1990). Tests of hypotheses in overdispersed Poisson regression and other quasi-likelihood models. *Journal of the American Statistical Association*, 85:565–571.

Breslow, N. E. and Clayton, D. G. (1993). Approximate inference in generalized linear mixed models. *Journal of the American Statistical Association*, 88:9–25.

Brockwell, S. E. and Gordon, I. R. (2007). A simple method for inference on an overall effect in meta-analysis. *Statistics in Medicine*, 26(25):4531–4543.

Brown, L., Cai, T., DasGupta, A., Agresti, A., Coull, B., Casella, G., Corcoran, C., Mehta, C., Ghosh, M., Santner, T., Brown, L., Cai, T., and DasGupta, A. (2001). Interval estimation for a binomial proportion - comment - rejoinder. *Statistical Science*, 16(2):101–133.

Brown, L. D. (1986). *Fundamentals of Statistical Exponential Families: with Applications in Statistical Decision Theory*. Institute of Mathematical Statistics.

Brownie, C., Anderson, D. R., Burnham, K. P., and Robson, D. S. (1985). *Statistical Inference from Band Recovery Data: A Handbook (Second Edition)*. U.S. Fish and Wildlife Service [U.S. Department of Interior].

Brownie, C. and Boos, D. D. (1994). Type I error robustness of ANOVA and ANOVA on ranks when the number of treatments is large. *Biometrics*, 50:542–549.

Brownie, C. F. and Brownie, C. (1986). Teratogenic effect of calcium edetate (caedta) in rats and the protective effect of zinc). *Toxicology and Applied Pharmacology*, 82(3):426–443.

Carlstein, E. (1986). The use of subseries values for estimating the variance of a general statistic from a stationary sequence. *The Annals of Statistics*, 14:1171–1179.

Carroll, R. J. and Ruppert, D. (1984). Power transformations when fitting theoretical models to data. *Journal of the American Statistical Association*, 79:321–328.

Carroll, R. J. and Ruppert, D. (1988). *Transformation and weighting in regression*. Chapman & Hall Ltd.

Carroll, R. J., Ruppert, D., Stefanski, L. A., and Crainiceanu, C. M. (2006). *Measurement error in nonlinear models: a modern perspective*. Chapman & Hall.

Casella, G. and Berger, R. L. (2002). *Statistical Inference*. Duxbury Press.

Chernoff, H. (1954). On the distribution of the likelihood ratio. *Annals of Mathematical Statistics*, 25(3):573–578.

Chernoff, H. and Lehmann, E. L. (1954). The use of maximum likelihood estimates in χ^2 tests for goodness of fit. *Annals of Mathematical Statistics*, 25:579–586.

Coles, S. G. and Dixon, M. J. (1999). Likelihood-based inference for extreme value models. *Extremes*, 2:5–23.

Conover, W. J. (1973). Rank tests for one sample, two samples, and k samples without the assumption of a continuous distribution function. *The Annals of Statistics*, 1:1105–1125.

Conover, W. J. and Iman, R. L. (1981). Rank transformations as a bridge between parametric and nonparametric statistics (C/R: P129-133). *The American Statistician*, 35:124–129.

Conover, W. J. and Salsburg, D. S. (1988). Locally most powerful tests for detecting treatment effects when only a subset of patients can be expected to "respond" to treatment. *Biometrics*, 44:189–196.

Cook, J. R. and Stefanski, L. A. (1994). Simulation-extrapolation estimation in parametric measurement error models. *Journal of the American Statistical Association*, 89:1314–1328.

Cook, S. R., Gelman, A., and Rubin, D. B. (2006). Validation of software for Bayesian models using posterior quantiles. *Journal of Computational and Graphical Statistics*, 15(3):675–692.

Cox, D. R. (1970). *Analysis of binary Data*. Chapman & Hall.

Cox, D. R. and Snell, E. J. (1981). *Applied statistics: principles and examples*. Chapman & Hall Ltd.

Cox, D. R. and Snell, E. J. (1989). *Analysis of Binary Data*. Chapman & Hall Ltd.

Cramér, H. (1946). *Mathematical Methods of Statistics*. Princeton University Press.

Cressie, N. and Read, T. R. C. (1984). Multinomial goodness-of-fit tests. *Journal of the Royal Statistical Society, Series B: Methodological*, 46:440–464.

Dacunha-Castelle, D. and Gassiat, E. (1999). Testing the order of a model using locally conic parametrization: Population mixtures and stationary ARMA processes. *The Annals of Statistics*, 27(4):1178–1209.

Darmois, G. (1935). The laws of probability to exhaustive estimation. *Comptes rendus hebdomadaires des seances de l academie des sciences*, 200:1265–1266.

David, H. A. (1998). First (?) occurrence of common terms in probability and statistics — A second list, with corrections (Corr: 1998V52 p371). *The American Statistician*, 52:36–40.

Davies, R. B. (1977). Hypothesis testing when a nuisance parameter is present only under the alternative. *Biometrika*, 64:247–254.

Davies, R. B. (1987). Hypothesis testing when a nuisance parameter is present only under the alternative. *Biometrika*, 74:33–43.

Davison, A. C. and Hinkley, D. V. (1997). *Bootstrap Methods and Their Application*. Cambridge University Press.

DeGroot, M. H. (1970). *Optimal Statistical Decisions*. New York, McGraw-Hill.

Dempster, A. P., Laird, N. M., and Rubin, D. B. (1977). Maximum likelihood from incomplete data via em algorithm. *Journal of the Royal Statistical Society Series B-Methodological*, 39:1–38.

DerSimonian, R. and Laird, N. (1986). Meta-analysis in clinical trials. *Controlled Clinical Trials*, 7:177–188.

DiCiccio, T. and Tibshirani, R. (1987). Bootstrap confidence intervals and bootstrap approximations. *Journal of the American Statistical Association*, 82:163–170.

DiCiccio, T. J. and Efron, B. (1996). Bootstrap confidence intervals (Disc: P212-228). *Statistical Science*, 11:189–212.

Diggle, P. J., Heagerty, P., Liang, K.-Y., and Zeger, S. L. (2002). *Analysis of Longitudinal Data*. Oxford University Press.

Doksum, K. and Bickel, P. (1969). Test for monotone failure rate based on normalized spacing. *The Annals of Mathematical Statistics*, 40:1216–1235.

Donald, A. and Donner, A. (1990). A simulation study of the analysis of sets of 2×2 contingency tables under cluster sampling: Estimation of a common odds ratio. *Journal of the American Statistical Association*, 85:537–543.

Dubey, S. D. (1967). Some percentile estimators for Weibull parameters. *Technometrics*, 9:119–129.

Dubinin, T. M. and Vardeman, S. B. (2003). Likelihood-based inference in some continuous exponential families with unknown threshold parameters. *Journal of the American Statistical Association*, 98(463):741–749.

Dunlop, D. D. (1994). Regression for longitudinal data: A bridge from least squares regression. *The American Statistician*, 48:299–303.

Efron, B. (1979). Bootstrap methods: Another look at the jackknife. *The Annals of Statistics*, 7:1–26.

Efron, B. (1982). *The Jackknife, the Bootstrap and Other Resampling Plans*. SIAM [Society for Industrial and Applied Mathematics].

Efron, B. (1987). Better bootstrap confidence intervals (C/R: P186-200). *Journal of the American Statistical Association*, 82:171–185.

Efron, B. and Hinkley, D. V. (1978). Assessing the accuracy of the maximum likelihood estimator: Observed versus expected Fisher information (C/R: P482-487). *Biometrika*, 65:457–481.

Efron, B. and Morris, C. (1972). Empirical Bayes on vector observations: An extension of Stein's method. *Biometrika*, 59:335–347.

Efron, B. and Morris, C. (1973). Stein's estimation rule and its competitors – An empirical Bayes approach. *Journal of the American Statistical Association*, 68:117–130.

Efron, B. and Stein, C. (1981). The jackknife estimate of variance. *The Annals of Statistics*, 9:586–596.

Efron, B. and Tibshirani, R. (1993). *An Introduction to the Bootstrap*. Chapman & Hall Ltd.

Ehrenberg, A. S. C. (1977). Rudiments of numeracy (Pkg: P277-323). *Journal of the Royal Statistical Society, Series A: General*, 140:277–297.

Ehrenberg, A. S. C. (1978). *Data Reduction: Analysing and Interpreting Statistical Data (Revised Reprint)*. John Wiley & Sons.

Ehrenberg, A. S. C. (1981). The problem of numeracy. *The American Statistician*, 35:67–71.

El-Shaarawi, A. H. (1985). Some goodness-of-fit methods for the Poisson plus added zeros distribution. *Applied and environmental microbiology*, 49:1304–1306.

Embrechts, P. A. L., Pugh, D., and Smith, R. L. (1985). *Statistical Extremes and Risks. Course Notes*. Imperial College, London, Dept. of Mathematics.

Fawcett, R. F. and Salter, K. C. (1984). A Monte Carlo study of the F test and three tests based on ranks of treatment effects in randomized block designs. *Communications in Statistics: Simulation and Computation*, 13:213–225.

Fay, M. P. and Graubard, B. I. (2001). Small-sample adjustments for Wald-type tests using sandwich estimators. *Biometrics*, 57(4):1198–1206.

Feller, W. (1966). *An Introduction to Probability Theory and Its Applications, Vol. II*. John Wiley & Sons.

Fisher, R. A. (1912). On an absolute criterion for fitting frequency curves. *Messenger of Mathematics*, 41:155–160.

Fisher, R. A. (1922). On the mathematical foundations of theoretical statistics. *Philos. Trans. Roy. Soc. London Ser. A*, 222:309–368.

Fisher, R. A. (1934a). *Statistical Methods for Research Workers, fifth edition*. Oliver & Boyd.

Fisher, R. A. (1934b). Two new properties of mathematical likelihood. *Proceedings of the Royal Society of London. Series A*, 144:285–307.

Fisher, R. A. (1935). *The Design of Experiments (eighth edition, 1966)*. Hafner Press.

Fix, E. and Hodges, J. L. (1955). Significance probabilities of the wilcoxon test. *The Annals of Mathematical Statistics*, 26:301–312.

Fouskakis, D. and Draper, D. (2002). Stochastic optimization: A review. *International Statistical Review*, 70(3):315–349.

Freidlin, B. and Gastwirth, J. L. (2000). Should the median test be retired from general use? *The American Statistician*, 54(3):161–164.

Friedman, M. (1937). The use of ranks to avoid the assumption of normality implicit in the analysis of variance. *Journal of the American Statistical Association*, 32:675–701.

Fuller, W. A. (1987). *Measurement Error Models*. John Wiley & Sons.

Gallant, A. R. (1987). *Nonlinear Statistical Models*. John Wiley & Sons.

Gelfand, A. E. and Smith, A. F. M. (1990). Sampling-based approaches to calculating marginal densities. *Journal of the American Statistical Association*, 85:398–409.

Gelman, A., Pasarica, C., and Dodhia, R. (2002). Let's practice what we preach: Turning tables into graphs. *The American Statistician*, 56(2):121–130.

Geman, S. and Geman, D. (1984). Stochastic relaxation, Gibbs distributions, and the Bayesian restoration of images. *IEEE Transactions on Pattern Analysis and Machine Intelligence*, 6:721–741.

Ghosh, J. K. (1971). A new proof of the Bahadur representation of quantiles and an application. *The Annals of Mathematical Statistics*, 42:1957–1961.

Glasser, M. (1965). Regression analysis with dependent variable censored. *Biometrics*, 21:300–306.

Gnedenko, B. V. (1943). Sur la distribution limite du terme maximum d'une série aléatoire. *Annals of Mathematics*, 44:423–453.
Godambe, V. P. (1960). An optimum property of regular maximum likelihood estimation (Ack: V32 p1343). *The Annals of Mathematical Statistics*, 31:1208–1212.
Goffinet, B., Loisel, P., and Laurent, B. (1992). Testing in normal mixture models when the proportions are known. *Biometrika*, 79:842–846.
Graybill, F. A. (1976). *Theory and Application of the Linear Model.* Duxbury Press.
Graybill, F. A. (1988). *Matrices with Applications in Statistics.* Wadsworth.
Guilbaud, O. (1979). Interval estimation of the median of a general distribution. *Scandinavian Journal of Statistics*, 46:29–36.
Haberman, S. J. (1989). Concavity and estimation. *The Annals of Statistics*, 17:1631–1661.
Hadi, A. S. and Wells, M. T. (1990). A note on generalized Wald's method. *Metrika*, 37:309–315.
Hajek, J. and Sidak, Z. (1967). *Theory of Rank Tests.* Academic Press.
Hald, A. (1998). *A History of Mathematical Statistics from 1750 to 1930.* Wiley, New York.
Hall, P. (1986). On the bootstrap and confidence intervals. *The Annals of Statistics*, 14:1431–1452.
Hall, P. (1987). Edgeworth expansion for Student's t statistic under minimal moment conditions. *The Annals of Probability*, 15:920–931.
Hall, P. (1988). Theoretical comparison of bootstrap confidence intervals (C/R: P953-985). *The Annals of Statistics*, 16:927–953.
Hall, P. (1992). *The Bootstrap and Edgeworth Expansion.* Springer-Verlag Inc.
Hall, P. and Titterington, D. M. (1989). The effect of simulation order on level accuracy and power of Monte Carlo tests. *Journal of the Royal Statistical Society, Series B: Methodological*, 51:459–467.
Hampel, F. R. (1974). The influence curve and its role in robust estimation. *Journal of the American Statistical Association*, 69:383–393.
Hansen, L. P. (1982). Large sample properties of generalized method of moments estimators. *Econometrica*, 50:1029–1054.
Harville, D. (1976). Extension of the Gauss-Markov theorem to include the estimation of random effects. *The Annals of Statistics*, 4:384–395.
Harville, D. A. (1977). Maximum likelihood approaches to variance component estimation and to related problems (C/R: P338-340). *Journal of the American Statistical Association*, 72:320–338.
Hastie, T., Tibshirani, R., and Friedman, J. H. (2001). *The Elements of Statistical Learning: Data Mining, Inference, and Prediction: with 200 Full-color Illustrations.* Springer-Verlag Inc.
Hauck, W. W. and Donner, A. (1977). Wald's test as applied to hypotheses in logit analysis (Corr: V75 p482). *Journal of the American Statistical Association*, 72:851–853.
Heagerty, P. J. and Lumley, T. (2000). Window subsampling of estimating functions with application to regression models. *Journal of the American Statistical Association*, 95(449):197–211.
Hernandez, F. and Johnson, R. A. (1980). The large-sample behavior of transformations to normality. *Journal of the American Statistical Association*, 75:855–861.
Hettmansperger, T. P. (1984). *Statistical Inference Based on Ranks.* John Wiley & Sons.
Hettmansperger, T. P. and Sheather, S. J. (1986). Confidence intervals based on interpolated order statistics (Corr: V4 p217). *Statistics & Probability Letters*, 4:75–79.
Higgins, J. P. T., Thompson, S. G., and Spiegelhalter, D. J. (2009). A re-evaluation of random-effects meta-analysis. *Journal of the Royal Statistical Society, Series A: Statistics in Society*, 172(1):137–159.
Hinkley, D. V. (1977). Jackknifing in unbalanced situations. *Technometrics*, 19:285–292.
Hinkley, D. V. and Runger, G. (1984). The analysis of transformed data (C/R: P309-320). *Journal of the American Statistical Association*, 79:302–309.
Hirji, K. F., Mehta, C. R., and Patel, N. R. (1987). Computing distributions for exact logistic regression. *Journal of the American Statistical Association*, 82:1110–1117.
Hobert, J. P. and Casella, G. (1996). The effect of improper priors on Gibbs sampling in hierarchical linear mixed models. *Journal of the American Statistical Association*, 91:1461–1473.

Hodges, J. L., J. and Lehmann, E. L. (1962). Rank methods for combination of independent experiments in the analysis of variance. *The Annals of Mathematical Statistics*, 33:482–497.

Hodges, J. L. and Lehmann, E. L. (1956). The efficiency of some nonparametric competitors of the t-test. *The Annals of Mathematical Statistics*, 27:324–335.

Hodges, J. L. and Lehmann, E. L. (1963). Estimates of location based on rank tests (Ref: V42 p1450-1451). *The Annals of Mathematical Statistics*, 34:598–611.

Hoeffding, W. (1948). A class of statistics with asymptotically normal distribution. *The Annals of Mathematical Statistics*, 19:293–325.

Hoeffding, W. (1951). A combinatorial central limit theorem. *The Annals of Mathematical Statistics*, 22:558–566.

Hoeffding, W. (1952). The large-sample power of tests based on permutations of observations. *The Annals of Mathematical Statistics*, 23:169–192.

Hoerl, A. E. and Kennard, R. W. (1970). Ridge regression: Biased estimation for nonorthogonal problems. *Technometrics*, 12:55–67.

Hope, A. C. A. (1968). A simplified Monte Carlo significance test procedure. *Journal of the Royal Statistical Society, Series B: Methodological*, 30:582–598.

Hora, S. C. and Iman, R. L. (1988). Asymptotic relative efficiencies of the rank-transformation procedure in randomized complete block designs. *Journal of the American Statistical Association*, 83:462–470.

Hosking, J. R. M. (1990). L-moments: Analysis and estimation of distributions using linear combinations of order statistics. *Journal of the Royal Statistical Society, Series B: Methodological*, 52:105–124.

Huber, P. J. (1964). Robust estimation of a location parameter. *The Annals of Mathematical Statistics*, 35:73–101.

Huber, P. J. (1967). The behavior of maximum likelihood estimates under nonstandard conditions. In Neyman, J., editor, *Proceedings of the Fifth Berkeley Symposium on Mathematical Statistics and Probability, Volume 1*, pages 221–233. University of California Press.

Huber, P. J. (1973). Robust regression: Asymptotics, conjectures and Monte Carlo. *The Annals of Statistics*, 1:799–821.

Huber, P. J. (1981). *Robust Statistics*. John Wiley & Sons.

Hyndman, R. J. and Fan, Y. (1996). Sample quantiles in statistical packages. *The American Statistician*, 50:361–365.

Iverson, H. K. and Randles, R. H. (1989). The effects on convergence of substituting parameter estimates into U-statistics and other families of statistics. *Probability Theory and Related Fields*, 81:453–471.

James, W. and Stein, C. (1961). Estimation with quadratic loss. In Neyman, J., editor, *Proceedings of the Fourth Berkeley Symposium on Mathematical Statistics and Probability, Volume 1*, pages 361–379. University of California Press.

Jeffreys, H. (1961). *Theory of Probability*. Oxford University Press.

Jöckel, K.-H. and Jockel, K.-H. (1986). Finite sample properties and asymptotic efficiency of Monte Carlo tests. *The Annals of Statistics*, 14:336–347.

Johansen, S. (1979). *Introduction to the Theory of Regular Exponential Families*. University of Copenhagen.

Johnson, R. A., Verrill, S., and Moore, D. H., I. (1987). Two-sample rank tests for detecting changes that occur in a small proportion of the treated population. *Biometrics*, 43:641–655.

Jones, G. L. and Hobert, J. P. (2004). Sufficient burn-in for Gibbs samplers for a hierarchical random effects model. *The Annals of Statistics*, 32(2):784–817.

Kackar, R. N. and Harville, D. A. (1984). Approximations for standard errors of estimators of fixed and random effects in mixed linear models. *Journal of the American Statistical Association*, 79:853–862.

Kass, R. E. and Steffey, D. (1989). Approximate Bayesian inference in conditionally independent hierarchical models (parametric empirical Bayes models). *Journal of the American Statistical Association*, 84:717–726.

Kass, R. E. and Wasserman, L. (1996). The selection of prior distributions by formal rules (Corr: 1998V93 p412). *Journal of the American Statistical Association*, 91:1343–1370.

Kent, J. T. (1982). Robust properties of likelihood ratio tests (Corr: V69 p492). *Biometrika*, 69:19–27.

Kenward, M. G. and Roger, J. H. (1997). Small sample inference for fixed effects from restricted maximum likelihood. *Biometrics*, 53:983–997.

Kepner, J. L. and Wackerly, D. D. (1996). On rank transformation techniques for balanced incomplete repeated-measures designs. *Journal of the American Statistical Association*, 91:1619–1625.

Khatri, C. G. (1963). Some results for the singular normal multivariate regression models. *Sankhyā, Series A*, 30:267–280.

Kilgore, D. L. (1970). The effects of northward dispersal on growth rate of young at birth and litter size in sigmodon hispidus. *American Midland Naturalist*, 84:510–520.

Kim, H.-J. and Boos, D. D. (2004). Variance estimation in spatial regression using a non-parametric semivariogram based on residuals. *Scandinavian Journal of Statistics*, 31(3):387–401.

Kim, Y. and Singh, K. (1998). Sharpening estimators using resampling. *Journal of Statistical Planning and Inference*, 66:121–146.

Klotz, J. (1962). Non-parametric tests for scale. *The Annals of Mathematical Statistics*, 33:498–512.

Koopman, B. O. (1936). On distributions admitting a sufficient statistic. *Transactions of the American Mathematical Society*, 39(3):399–409.

Kruskal, W. H. and Wallis, W. A. (1952). Use of ranks in one-criterion variance analysis. *Journal of the American Statistical Association*, 47:583–621.

Künsch, H. R. (1989). The jackknife and the bootstrap for general stationary observations. *The Annals of Statistics*, 17:1217–1241.

Laird, N. M. and Ware, J. H. (1982). Random-effects models for longitudinal data. *Biometrics*, 38:963–974.

Lambert, D. (1992). Zero-inflated Poisson regression, with an application to defects in manufacturing. *Technometrics*, 34:1–14.

Landis, J. R., Heyman, E. R., and Koch, G. G. (1978). Average partial association in three-way contingency tables: A review and and discussion of alternative tests. *International Statistical Review*, 46:237–254.

Larsen, R. J. and Marx, M. L. (2001). *An Introduction to Mathematical Statistics and Its Applications*. Prentice-Hall Inc.

Lawless, J. F. (1982). *Statistical Models and Methods for Lifetime Data*. John Wiley & Sons.

Lawley, D. N. (1956). A general method for approximating to the distribution of likelihood ratio criteria. *Biometrika*, 43:295–303.

Leadbetter, M. R., Lindgren, G., and Rootzén, H. (1983). *Extremes and Related Properties of Random Sequences and Processes*. Springer-Verlag Inc.

Lehmann, E. L. (1953). The power of rank tests. *The Annals of Mathematical Statistics*, 24:23–43.

Lehmann, E. L. (1975). *Nonparametrics: Statistical Methods Based on Ranks*. Holden-Day Inc.

Lehmann, E. L. (1983). *Theory of Point Estimation*. John Wiley & Sons.

Lehmann, E. L. (1986). *Testing Statistical Hypotheses*. John Wiley & Sons.

Lehmann, E. L. and Casella, G. (1998). *Theory of Point Estimation*. Springer-Verlag Inc.

Lehmann, E. L. and Stein, C. (1949). On the theory of some non-parametric hypotheses. *The Annals of Mathematical Statistics*, 20:28–45.

Leroux, B. G. and Puterman, M. L. (1992). Maximum-penalized-likelihood estimation for independent and Markov-dependent mixture models. *Biometrics*, 48:545–558.

Liang, K.-Y. (1985). Odds ratio inference with dependent data. *Biometrika*, 72:678–682.

Liang, K.-Y. and Zeger, S. L. (1986). Longitudinal data analysis using generalized linear models. *Biometrika*, 73:13–22.

Lindley, D. V. and Phillips, L. D. (1976). Inference for a Bernoulli process (A Bayesian view). *The American Statistician*, 30:112–119.

Lindsay, B. G. (1994). Efficiency versus robustness: The case for minimum Hellinger distance and related methods. *The Annals of Statistics*, 22:1081–1114.

Lindsay, B. G. and Qu, A. (2003). Inference functions and quadratic score tests. *Statistical Science*, 18(3):394–410.

Little, R. J. A. and Rubin, D. B. (1987). *Statistical Analysis with Missing Data*. J. Wiley & Sons.

Liu, R. Y. and Singh, K. (1987). On a partial correction by the bootstrap. *The Annals of Statistics*, 15:1713–1718.

Liu, X. and Shao, Y. (2003). Asymptotics for likelihood ratio tests under loss of identifiability. *The Annals of Statistics*, 31(3):807–832.

Loh, W.-Y. (1987). Calibrating confidence coefficients. *Journal of the American Statistical Association*, 82:155–162.

Magee, L. (1990). R^2 measures based on Wald and likelihood ratio joint significance tests. *The American Statistician*, 44:250–253.

Mahfoud, Z. R. and Randles, R. H. (2005). Practical tests for randomized complete block designs. *Journal of Multivariate Analysis*, 96(1):73–92.

Makelainen, T., Schmidt, K., and Styan, G. P. H. (1981). On the existence and uniqueness of the maximum-likelihood estimate of a vector-valued parameter in fixed-size samples. *Annals of Statistics*, 9:758–767.

Mancl, L. A. and DeRouen, T. A. (2001). A covariance estimator for GEE with improved small-sample properties. *Biometrics*, 57(1):126–134.

Mann, H. B. and Whitney, D. R. (1947). On a test of whether one of two random variables is stochastically larger than the other. *The Annals of Mathematical Statistics*, 18:50–60.

McCullagh, P. (1983). Quasi-likelihood functions. *The Annals of Statistics*, 11:59–67.

McCullagh, P. (1997). Linear models, vector spaces, and residual likelihood. In Gregoire, T. G., Brillinger, D. R., Diggle, P. J., Russek-Cohen, E., Warren, W. G., and Wolfinger, R. D., editors, *Modelling Longitudinal and Spatially Correlated Data: Methods, Applications, and Future Directions. Lecture Notes in Statistics, Vol. 122*, pages 1–10. Springer-Verlag Inc.

McLachlan, G.J. & Krishnan, T. (1997). *The EM Algorithm and Extensions*. J. Wiley & Sons.

Mehra, K. L. and Sarangi, J. (1967). Asymptotic efficiency of certain rank tests for comparative experiments. *The Annals of Mathematical Statistics*, 38:90–107.

Memon, M. A., Cooper, N. J., Memon, B., Memon, M. I., and Abrams, K. R. (2003). Meta-analysis of randomized clinical trials comparing open and laparoscopic inguinal hernia repair. *British Journal of Surgery*, 90:1479–1492.

Messig, M. A. and Strawderman, W. E. (1993). Minimal sufficiency and completeness for dichotomous quantal response models. *The Annals of Statistics*, 21:2149–2157.

Miller, R. G., J. (1974). An unbalanced jackknife. *The Annals of Statistics*, 2:880–891.

Miller, J. J. (1977). Asymptotic properties of maximum likelihood estimates in the mixed model of the analysis of variance. *The Annals of Statistics*, 5:746–762.

Miller, R. G. (1980). Combining 2×2 contingency tables. In Miller, R. G. e., Efron, B. e., Brown, B. W., J. e., and Moses, L. E. e., editors, *Biostatistics Casebook*, pages 73–83. John Wiley & Sons.

Monahan, J. F. (2001). *Numerical Methods of Statistics*. Cambridge University Press.

Monahan, J. F. and Boos, D. D. (1992). Proper likelihoods for Bayesian analysis. *Biometrika*, 79:271–278.

Moore, D. S. (1977). Generalized inverses, Wald's method, and the construction of chi-squared tests of fit. *Journal of the American Statistical Association*, 72:131–137.

Moore, D. S. (1986). Tests of chi-squared type. In D'Agostino, R. B. and Stephens, M. A., editors, *Goodness-of-fit techniques*, pages 63–95. Marcel Dekker Inc.

Nation, J. R., Bourgeois, A. E., Clark, D. E., Baker, D. M., and Hare, M. F. (1984). The effects of oral cadmium exposure on passive avoidance performance in the adult rat. *Toxicology Letters*, 20:41–47.

Nelder, J. A. and Wedderburn, R. W. M. (1972). Generalized linear models. *Journal of the Royal Statistical Society, Series A: General*, 135:370–384.

Neyman, J. and Pearson, E. S. (1928). On the use and interpretation of certain test criteria for purposes of statistical inference. *Biometrika*, 20A:175–240, 263–294.

Neyman, J. and Pearson, E. S. (1933). On the problem of the most efficient tests of statistical hypotheses. *Philosophical Transactions of the Royal Society of London, Ser. A*, 231:289–337.

Neyman, J. and Scott, E. L. (1948). Consistent estimates based on partially consistent observations. *Econometrica*, 16:1–32.

Noether, G. E. (1949). On a theorem by wald and wolfowitz. *The Annals of Mathematical Statistics,*, 20:455–458.

Noether, G. E. (1955). On a theorem of pitman. *The Annals of Mathematical Statistics,*, 26:64–68.

Noether, G. E. (1987). Sample size determination for some common nonparametric tests. *Journal of the American Statistical Association*, 82:645–647.

O'Gorman, T. W. (2001). A comparison of the F-test, Friedman's test, and several aligned rank tests for the analysis of randomized complete blocks. *Journal of Agricultural, Biological, and Environmental Statistics*, 6(3):367–378.

Olshen, R. A. (1967). Sign and Wilcoxon tests for linearity. *The Annals of Mathematical Statistics*, 38:1759–1769.

Pace, L. and Salvan, A. (1997). *Principles of Statistical Inference: from a Neo-Fisherian Perspective*. World Scientific Publishing Company.

Pirazzoli, P. A. (1982). Maree estreme a venezia (periodo 1872-1981). *Acqua Aria*, 10:1023–1029.

Pitman, E. J. G. (1936). Sufficient statistics and intrinsic accuracy. *Proc. Camb. Phil. Soc.*, 32:567–579.

Pitman, E. J. G. (1937a). Significance tests which may be applied to samples from any populations. *Supplement to the Journal of the Royal Statistical Society*, 4:119–130.

Pitman, E. J. G. (1937b). Significance tests which may be applied to samples from any populations. ii. the correlation coefficient test. *Supplement to the Journal of the Royal Statistical Society*, 4:225–232.

Pitman, E. J. G. (1938). Significance tests which may be applied to samples from any populations. iii. the analysis of variance test. *Biometrika*, 29:322–335.

Pitman, E. J. G. (1948). *Notes on Non-Parametric Statitistical Inference*. Columbia University (duplicated).

Portnoy, S. (1988). Asymptotic behavior of likelihood methods for exponential families when the number of parameters tends to infinity. *The Annals of Statistics*, 16:356–366.

Pratt, J. W. and Gibbons, J. D. (1981). *Concepts of Nonparametric Theory*. Springer-Verlag Inc.

Prentice, R. L. (1988). Correlated binary regression with covariates specific to each binary observation. *Biometrics*, 44:1033–1048.

Presnell, B. and Boos, D. D. (2004). The ios test for model misspecification. *Journal of the American Statistical Association*, 99(465):216–227.

Puri, M. L. and Sen, P. K. (1971). *Nonparametric Methods in Multivariate Analysis*. John Wiley & Sons.

Pyke, R. (1965). Spacings (with discussion). *Journal of the Royal Statistical Society, Series B: Statistical Methodology*, 27:395–449.

Qu, A., Lindsay, B. G., and Li, B. (2000). Improving generalised estimating equations using quadratic inference functions. *Biometrika*, 87(4):823–836.

Quenouille, M. H. (1949). Approximate use of correlation in time series. *Journal of the Royal Statistical Society, Series B*, 11:18–84.

Quenouille, M. H. (1956). Notes on bias in estimation. *Biometrika*, 43:353–360.

Quesenberry, C. P. (1975). Transforming samples from truncation parameter distributions to uniformity. *Communications in Statistics*, 4:1149–1156.

Radelet, M. L. and Pierce, G. L. (1991). Choosing those who will die: race and the death penalty in florida. *Florida Law Review*, 43:1–34.

Randles, R. H. (1982). On the asymptotic normality of statistics with estimated parameters. *Annals of Statistics*, 10:462–474.

Randles, R. H. and Wolfe, D. A. (1979). *Introduction to the Theory of Nonparametric Statistics*. John Wiley & Sons.

Randolph, P. A., Randolph, J. C., Mattingly, K., and Foster, M. M. (1977). Energy costs of reproduction in the cotton rat sigmodon hispidus. *Ecology*, 58:31–45.

Rao, C. R. (1948). Large sample tests of statistical hypotheses concerning several parameters with application to problems of estimation. *Proceedings of the Cambridge Philosophical Society*, 44:50–57.

Rao, C. R. (1973). *Linear Statistical Inference and Its Applications*. John Wiley & Sons.

Rao, C. R. and Wu, Y. (2001). On model selection (Pkg: P1-64). In *Model selection [Institute of Mathematical Statistics lecture notes-monograph series 38]*, pages 1–57. IMS Press.

Read, T. R. C. and Cressie, N. A. C. (1988). *Goodness-of-fit Statistics for Discrete Multivariate Data*. Springer-Verlag Inc.

Reid, N. (1988). Saddlepoint methods and statistical inference (C/R: P228-238). *Statistical Science*, 3:213–227.

Ridout, M., Hinde, J., and Demétrio, C. G. B. (2001). A score test for testing a zero-inflated Poisson regression model against zero-inflated negative binomial alternatives. *Biometrics*, 57(1):219–223.

Robert, C. P. (2001). *The Bayesian Choice: from Decision-theoretic Foundations to Computational Implementation*. Springer-Verlag Inc.

Roberts, M. E., Tchanturia, K., Stahl, D., Southgate, L., and Treasure, J. (2007). A systematic review and metaanalysis of set-shifting ability in eating disorders. *Psychol. Med.*, 37:1075–1084.

Robertson, T., Wright, F. T., and Dykstra, R. (1988). *Order Restricted Statistical Inference*. John Wiley & Sons.

Robins, J. M., van der Vaart, A., and Ventura, V.

Robinson, J. (1980). An asymptotic expansion for permutation tests with several samples. *The Annals of Statistics*, 8:851–864.

Rosen, O. and Cohen, A. (1995). Constructing a bootstrap confidence interval for the unknown concentration in radioimmunoassay (Disc: P953-953). *Statistics in Medicine*, 14:935–952.

Rotnitzky, A. and Jewell, N. P. (1990). Hypothesis testing of regression parameters in semiparametric generalized linear models for cluster correlated data. *Biometrika*, 77:485–497.

Ruppert, D. (1987). What is kurtosis - an influence function-approach. *American Statisitican*, 41:1–5.

Sampson, A. R. (1974). A tale of two regressions. *Journal of the American Statistical Association*, 69:682–689.

Santner, T. J. (1998). Teaching large-sample binomial confidence intervals. *Teaching Statistics*, 20:20–23.

Savage, I. R. (1956). Contributions to the theory of rank order statistics-the two-sample case. *The Annals of Mathematical Statistics*, 27:590–615.

Schrader, R. M. and Hettmansperger, T. P. (1980). Robust analysis of variance based upon a likelihood ratio criterion. *Biometrika*, 67:93–101.

Schucany, W. R., Gray, H. L., and Owen, D. B. (1971). On bias reduction in estimation. *Journal of the American Statistical Association*, 66:524–533.

Schwarz, G. (1978). Estimating the dimension of a model. *The Annals of Statistics*, 6:461–464.

Searle, S. R. (1971). *Linear models*. John Wiley & Sons.

Seber, G. A. F. and Wild, C. J. (1989). *Nonlinear Regression*. John Wiley & Sons.

Self, S. G. and Liang, K.-Y. (1987). Asymptotic properties of maximum likelihood estimators and likelihood ratio tests under nonstandard conditions. *Journal of the American Statistical Association*, 82:605–610.

Sen, P. K. (1968). On a class of aligned rank order tests in two-way layouts. *The Annals of Mathematical Statistics*, 39:1115–1124.

Sen, P. K. (1982). On M tests in linear models. *Biometrika*, 69:245–248.

Serfling, R. J. (1980). *Approximation Theorems of Mathematical Statistics*. Wiley, New York.

Shanbhag, D. N. (1968). Some remarks concerning Khatri's result on quadratic forms. *Biometrika*, 55:593–595.

Shao, J. and Tu, D. (1995). *The Jackknife and Bootstrap*. Springer-Verlag Inc.

Shao, J. and Wu, C. F. J. (1989). A general theory for jackknife variance estimation. *The Annals of Statistics*, 17:1176–1197.

Sidik, K. and Jonkman, J. N. (2007). A comparison of heterogeneity variance estimators in combining results of studies. *Statistics in Medicine*, 26(9):1964–1981.

Siegel, A. F. (1985). Modelling data containing exact zeroes using zero degrees of freedom. *Journal of the Royal Statistical Society, Series B: Methodological*, 47:267–271.

Silvapulle, M. J. and Sen, P. K. (2005). *Constrained Statistical Inference: Inequality, Order, and Shape Restrictions*. Wiley-Interscience.

Simpson, D. G. (1987). Minimum Hellinger distance estimation for the analysis of count data. *Journal of the American Statistical Association*, 82:802–807.

Singh, K. (1981). On the asymptotic accuracy of Efron's bootstrap. *The Annals of Statistics*, 9:1187–1195.

Smith, R. L. (1985). Maximum likelihood estimation in a class of nonregular cases. *Biometrika*, 72:67–90.

Smith, T. C., Spiegelhalter, D. J., and Thomas, A. (1995). Bayesian approaches to random-effects meta-analysis: A comparative study. *Statistics in Medicine*, 14:2685–2699.

Stacy, E. W. (1962). A generalization of the gamma distribution. *The Annals of Mathematical Statistics*, 33:1187–1191.

Stefanski, L. A. and Boos, D. D. (2002). The calculus of M-estimation. *The American Statistician*, 56(1):29–38.

Stefanski, L. A. and Carroll, R. J. (1987). Conditional scores and optimal scores for generalized linear measurement-error models. *Biometrika*, 74:703–716.

Stefanski, L. A. and Cook, J. R. (1995). Simulation-extrapolation: The measurement error jackknife. *Journal of the American Statistical Association*, 90:1247–1256.

Stephens, M. A. (1977). Goodness of fit for the extreme value distribution. *Biometrika*, 64:583–588.

Student (Gosset, W. S. (1908). The probable error of a mean. *Biometrika*, 6(1):1–25.

Styan, G. P. H. (1970). Notes on the distribution of quadratic forms in singular normal variables. *Biometrika*, 57:567–572.

Suissa, S. and Shuster, J. J. (1985). Exact unconditional sample sizes for the 2 by 2 binomial trial. *Journal of the Royal Statistical Society, Series A: General*, 148:317–327.

Tanner, M. A. and Wong, W. H. (1987). The calculation of posterior distributions by data augmentation (C/R: P541-550). *Journal of the American Statistical Association*, 82:528–540.

Tarone, R. E. (1979). Testing the goodness of fit of the binomial distribution (Corr: V77 p668). *Biometrika*, 66:585–590.

Tarone, R. E. and Gart, J. J. (1980). On the robustness of combined tests for trends in proportions. *Journal of the American Statistical Association*, 75:110–116.

Teo, K. K., Yusuf, S., Collins, R., Held, P. H., and Peto, R. (1991). Effects of intravenous magnesium in suspected acute myocardial infarction: Overview of randomised trials. *British Medical Journal*, 303:1499–1503.

Thompson, G. L. (1991). A note on the rank transform for interactions (Corr: 93V80 p711). *Biometrika*, 78:697–701.

Tobin, J. (1958). Estimation of relationships for limited dependent-variables. *Econometrica*, 26:24–36.

Tsiatis, A. A. and Davidian, M. (2001). A semiparametric estimator for the proportional hazards model with longitudinal covariates measured with error. *Biometrika*, 88:447–458.

Tukey, J. (1958). Bias and confidence in not quite large samples (abstract). *The Annals of Mathematical Statistics*, 29:614–614.

van den Broek, J. (1995). A score test for zero inflation in a Poisson distribution. *Biometrics*, 51:738–743.

van der Vaart, A. W. (1998). *Asymptotic Statistics*. Cambridge University Press.

van Elteren, P. H. (1960). On the combination of independent two-sample tests of wilcoxon. *Bulletin of the International Statistical Institute*, 37:351–361.

van Elteren, P. H. and Noether, G. E. (1959). The asymptotic efficiency of the χ_r^2-test for a balanced incomplete block design. *Biometrika*, 46:475–477.

Venables, W. N. and Ripley, B. D. (1997). *Modern Applied Statistics with S-Plus*. Springer-Verlag Inc.

Verbeke, G. and Molenberghs, G. (2003). The use of score tests for inference on variance components. *Biometrics*, 59(2):254–262.

von Mises, R. (1947). On the asymptotic distribution of differentiable statistical functions. *The Annals of Mathematical Statistics*, 18:309–348.

Wainer, H. (1993). Tabular presentation. *Chance*, 6(3):52–56.

Wainer, H. (1997a). Improving tabular displays, with NAEP tables as examples and inspirations. *Journal of Educational and Behavioral Statistics*, 22:1–30.

Wainer, H. (1997b). *Visual Revelations: Graphical Tales of Fate and Deception from Napoleon Bonaparte to Ross Perot*. Springer-Verlag Inc.

Wald, A. (1943). Tests of statistical hypotheses concerning several parameters when the number of observations is large. *Transactions of the American Mathematical Society*, 54(3):426–482.

Wald, A. (1949). Note on the consistency of maximum likelihood estimate. *The Annals of Mathematical Statistics*, 20:595–601.

Wald, A. and Wolfowitz, J. (1944). Statistical tests based on permutations of the observations. *The Annals of Mathematical Statistics,*, 15:358–372.

Warn, D. E., Thompson, S. G., and Spiegelhalter, D. J. (2002). Bayesian random effects meta-analysis of trials with binary outcomes: Methods for the absolute risk difference and relative risk scales. *Statistics in Medicine*, 21(11):1601–1623.

Wedderburn, R. W. M. (1974). Quasi-likelihood functions, generalized linear models, and the Gauss-Newton method. *Biometrika*, 61:439–447.

Weir, B. S. (1996). *Genetic Data Analysis 2: Methods for Discrete Population Genetic Data*. Sunderland: Sinauer Associates.

Welch, B. L. (1937). On the z-test in randomized blocks and latin squares. *Biometrika*, 29:21–52.

Welch, W. J. (1987). Rerandomizing the median in matched-pairs designs. *Biometrika*, 74:609–614.

White, H. (1981). Consequences and detection of misspecified nonlinear regression models. *Journal of the American Statistical Association*, 76:419–433.

White, H. (1982). Maximum likelihood estimation of misspecified models. *Econometrica*, 50:1–26.

Wilcoxon, F. (1945). Individual comparisons by ranking methods. *Biometrics Bulletin*, 6:80–83.

Wilks, S. S. (1938). The large-sample distribution of the likelihood ratio for testing composite hypotheses. *Annals of Mathematical Statistics*, 9:60–62.

Wilson, D. H. (2002). *Signed Scale Measures: An Introduction and Application*. Ph.D Thesis, NC State University.

Wu, C. F. J. (1983). On the convergence properties of the EM algorithm. *The Annals of Statistics*, 11:95–103.

Wu, C. F. J. (1986). Jackknife, bootstrap and other resampling methods in regression analysis (C/R: P1295-1350; Ref: V16 p479). *The Annals of Statistics*, 14:1261–1295.

Wu, C.-t., Gumpertz, M. L., and Boos, D. D. (2001). Comparison of GEE, MINQUE, ML, and REML estimating equations for normally distributed data. *The American Statistician*, 55(2):125–130.

Zehna, P. W. (1966). Invariance of maximum likelihood estimations. *The Annals of Mathematical Statistics*, 37:744–744.

Zeng, Q. and Davidian, M. (1997). Bootstrap-adjusted calibration confidence intervals for immunoassay. *Journal of the American Statistical Association*, 92:278–290.

Zhang, J. and Boos, D. D. (1994). Adjusted power estimates in Monte Carlo experiments. *Communications in Statistics: Simulation and Computation*, 23:165–173.

Zhang, J. and Boos, D. D. (1997). Mantel-Haenszel test statistics for correlated binary data. *Biometrics*, 53:1185–1198.

Zhao, L. P. and Prentice, R. L. (1990). Correlated binary regression using a quadratic exponential model. *Biometrika*, 77:642–648.

Author Index

D.D. Boos and L.A. Stefanski, *Essential Statistical Inference: Theory and Methods*,
Springer Texts in Statistics, DOI 10.1007/978-1-4614-4818-1,

Example Index

D.D. Boos and L.A. Stefanski, *Essential Statistical Inference: Theory and Methods*, Springer Texts in Statistics, DOI 10.1007/978-1-4614-4818-1,

R-code Index

D.D. Boos and L.A. Stefanski, *Essential Statistical Inference: Theory and Methods*, Springer Texts in Statistics, DOI 10.1007/978-1-4614-4818-1,

Subject Index

D.D. Boos and L.A. Stefanski, *Essential Statistical Inference: Theory and Methods*, Springer Texts in Statistics, DOI 10.1007/978-1-4614-4818-1,

W

Z

CPSIA information can be obtained
at www.ICGtesting.com
Printed in the USA
LVOW05*1626060817
544027LV00008B/429/P

9 781461 448174